T0224362

BASIC STRUCTURES OF MATTER

SUPERGRAVITATION UNIFIED THEORY

STOYAN SARG

 www.trafford.com

North America & international
toll-free: 1 888 232 4444 (USA & Canada)
fax: 812 355 4082

FOREWORD

The book "Basic Structures of Matter – Supergravitation Unified Theory" is one original attempt to provide a view on some relations among the fundamental laws in Nature. It is different from the other unified theories and hypothesis by using an approach in which the formulated initial framework and all theoretical analysis of the models are in real tree-dimensional space with unidirectional time. In such approach, the principle of causality (the cause precedes the effect) is preserved, which facilitates the logical understanding. This leads to an interesting interpretation of some Quantum Mechanical phenomena for which the human logic appeared inadequate. At first glance, the presented theory supports the existence of material Ether, but the proposed original model is not in conflict with the relativistic phenomena. The initial framework of the theory is based on the assumption that at the Plank's scale of the microcosmos there are two fundamental particles of indestructible super-dense matter, which in a pure empty space environment (a space without any physical properties) interact by forces inverse proportional to the cube of the distance. The analysis of the developed models and their comparison to experimental data leads to interesting results in the field of microcosmos - the elementary particles might contain a geometrical structure of super-dense matter. The author extends the analysis through formation of atoms and molecules and he proposes an interesting hypothetical scenario for cosmological processes in the Universe. Using the suggested fundamental particles, he provides also a model of a hypothetical underlying structure of the physical vacuum, which has features of quantum space and space-time properties. The proposed structure allows the existence and propagation of fields, which the author identifies as gravitational, electrical and magnetic. From the point of view of the derived physical models, the author provides reinterpretation of experiments and observations from different fields of physics and he makes conclusions and predictions, the further examination of which is quite reasonable.

Prof. Asparuh Petrakiev
Dr. of Science in Physics
PhD in Chemistry
Academician of International Academy of Ecology and Life Protection Sciences,
St. Petersburg, Russia

PREFACE

The search of alternative concept about space (physical vacuum) has been abandoned for 100 years, considering that this issue is solved once for ever. At the same time Nature surprises the physicists with unexplainable phenomena. My 30 years experience in different fields of physics and technology gave me a confidence to try one original approach. For many years, my mind was bothered by the curiosity of Nature. Every atom, either here or in a distant galaxy has exactly the same complex spectral signature. Even for the simplest atom, the hydrogen, we need a complex mathematics to describe its spectrum. Then the logical question is: How the laws of Physics are written in the language of Nature? This issue obviously has physical and philosophical aspect. My endeavour to get a logical answer evolved into a deep rational vision about some fundamental principles in Nature. I arrived to ideas that may seem speculative from a first gland, but if the curious reader examines the analysis and arguments, he will find that the conclusions and the predictions are reasonable. Some of the most intriguing questions are: In what kind of space we live? What is behind the inertia and gravitation? For a first time, the reader will find such issue discussed in understandable way, free of abstractness. The analysis of some enigmatic phenomena from a new point of view leads to conclusions and predictions, which have not been discussed so far. They might invoke a new technological advance. This book provides the physical base for supporting of such statement. The presented theoretical work falls into the category of Unified Theories.

The most common definition of Unified Theory is a theory, which describes the gravitational, electromagnetic, strong and weak forces within a single, all-encompassing framework. Attempts for building of such theory are made for over 100 years. Let define first what should be the criteria for a successful Unified Theory:

- It must suggest a physical model of fundamental particles in a real 3-dimensional space, working in all fields of Physics without contradictions and able to explain the relations between the gravitational, electrical and magnetic fields
- The number of initially adopted postulates and fundamental particles must be as smaller as possible
- The derived models must be able to explain all kind of experiments and observations in all fields of Physics and in the range from micro-cosmos to Cosmology
- The energy conservation principle must be strictly observed

One additional criterion of philosophical aspect is that the theory must show how the laws of Physics are written in the language of Nature.

Presently, the existing theoretical models pretending to have features of Unified theory, could not suggest a model satisfying the above-defined criteria. What is proposed is an enormous number of particles, postulates and rules, often contradictable between themselves or using an unrealistic speculative multidimensional space. Since the suggested models suffer from logical understanding, it has been claimed that the human logic fails, so a mathematical one must replace it. Such statement, criticized in the past (1900-1930), now is silently accepted. It became convenient for development of an enormous number of abstractive theories, while the main issue is unsolved.

The treatise Basic Structures of Matter, a Super Gravitation Unified Theory unveils the relation between the forces in Nature by adopting of the following framework:

- Empty Euclidian space without any physical properties and restrictions
- Two super dens fundamental particles, able to vibrate and congregate
- A fundamental law of Super Gravitation (SG) – an inverse cubic law valid in empty space.

These two fundamental particles driven by the fundamental SG law are able to congregate to geometrical formations. The enormous abundance of such mixture with energy above some critical level form self-organised hierarchical levels of geometrical formations, which leads deterministically to creation of space with quantum properties - physical vacuum and a galaxy as observable matter. All known laws of Physics are embedded in the underlying structure of the physical vacuum and the structure of the elementary particles. The fundamental SG law is behind the gravitational, electric and magnetic fields and governs all kind of interactions between the elementary particles in the space of physical vacuum.

The treatise Basic Structures of Matter (BSM) is based on an original idea about physical vacuum, which has never been investigated so far. It follows the recommendation of James Clerk Maxwell, the father of Modern Physics, for search of the structure of the space. In his "Treatise on Electricity and Magnetism" vol. II he writes:

… If something is transmitted from one particle to another at a distance, what is its condition after it has left the one particle and before it has reached to the other?... Hence, all these theories lead to the conception of a medium in which the propagation takes place, and if we admit this medium as an hypotreatise, I think it ought to occupy a prominent place in our investigations, and that we ought to endeavor to construct a mental representation of all the details of its action, and this has been my constant aim in this treatise

In his book "Sidelights on Relativity" (1922) Albert Einstein" states in favor of Ether quoting *"According to General Theory of Relativity, space without ether is unthinkable"*. In the same book he states that all attempts to explain the Maxwell's electrodynamics by a physical model of material ether has failed, but he doesn't suggest a proof that this is impossible. In fact the Casimir forces have been discovered 35 years later, although nobody so far has guessed that they might be the detectable signature of the inverse cubic SG law, so such a law could be used for building a physical model of the material Ether.

The space concept suggested in BSM allows explanation of all enigmatic phenomena in Particle Physics, Quantum mechanics, Relativity and Cosmology using reliable classical methods. The derived theoretical models show excellent agreement with the experiments and observations. This opens a new opportunity for investigating the micro and macro Cosmos. One of the practically useful results is the unveiled structure of the elementary particles and atomic nuclei, graphically presented in the Appendix *Atlas of Atomic Nuclear Structures*. They could be used in the fields of structural chemistry, nanotechnology, biomolecules and for further understanding of the nuclear reactions and radioactivity. Some conclusions have important ecological aspect. One important issue, for example, is the envisioned new kind of danger from a nuclear explosion (discussed in chapter 12), which has not been recognized so far. Another example is the opportunity to investigate the tornado phenomenon from a new point of view.

The BSM is particularly useful for researchers in the fields of the Zero Point Energy, gravitation and inertia. Among the most important results, summarized in the last Chapter 13, are the following: Theoretical proof for existence of the most universal law in Nature – the Law of Supergravitation. It is behind all physical phenomena and directly related to a hidden space energy, which appears to be a primary source of the nuclear energy; Predicting the possibility and the potential physical mechanisms for control of the gravitation and inertia of a material object; Predicting of hidden non-EM waves and a possibility for Supercommunication by them.

I acknowledge the moral support of my family, my wife Denka, and my son Ivor, without whose tolerance and encouragement this work could not have been finished. I would like to express special thanks for my colleagues, scientists, engineers, relatives and friends, which have encouraged me to complete this monograph.

Stoyan Sarg, PhD in Physics
Toronto, Canada

CONTENTS

CONTENTS

CONTENTS

CONTENTS

CONTENTS

CONTENTS

Chapter 13. Potential and special applications of the BSM theory. New opportunities for space travels. 13-1

List of Abbreviations

BSM	Basic Structures of Matter theory
CL	Cosmic Lattice
CL space	Cosmic Lattice space
(CP)	Central Part (of twisted prism's model)
EB	Electronic Bond
EQ	Electrical Quasisphere
FOHS	First Order Helical Structure
FQHE	Fractional Quantum Hall Effect
GB	Gravitational Bond
G. S.	Ground State
SG	Super Gravitation
MQ	Magnetic Quasisphere
NRM	Node Resonance Momentum (vector)
PE	Photo Electron
RL	Rectangular Lattice
RL(R)	Rectangular Lattice (Radial)
RL(T)	Rectangular Lattice (Twisted)
SC state	Super Conductivity state
SOHS	Second Order Helical Structure
SPM	Spatially Precession Momentum (vector)
PP SPM	Phase Propagation of SPM
QM	Quantum Mechanics
(TP)	Twisted Part (of twisted prism's model)
ZPE	Zero Point Energy
ZPE-D	Dynamic ZPE
ZPE-S	Static ZPE

DISCLAIMER

The information provided in this book is for scientific advancement. The author is not responsible for any incidental, damage or injury, arising from the use or misuse of the information contained in this monograph. The special applications predicted by BSM theory are related to new fields, which have not been enough investigated, so far. Experiments in these fields should be provided by qualified researchers on their own risk and responsibility. Precautions for unexpected effects must be considered and monitoring equipment for eventual hazardous radiation must be used.

1. Chapter 1   Introduction

   Our vision about the micro-cosmos and Universe is formed through the prism of the space-time concept, which is defined by the concept of the physical vacuum. The latter has been changed four times during the history of Physics. Is the currently adopted concept a final truth? The time according to the Theory of Relativity is not absolute. Then what defines the space-time properties of the physical vacuum and why the velocity of light propagation is postulated? These and other similar questions could not be found in the textbooks. Presently our understanding of micro-cosmos and Universe is full of enigmatic problems. In the cosmology, the Big-Bang "theory" relies on the presumption that the observed red shift is of Doppler origin and the space in the Universe is homogeneous. The concepts of the Big Bang and the black wholes contain unexplainable problem of singularity. In few words, this means that the enormous matter of the Universe originate from (or end-up into) a mathematical point containing an unimaginable enormous energy. If so, one may accept an independent existence of other such points whose explosion will lead to a Universe disaster, but such phenomenon is not observed. The Big Bang concept is plagued by many observational puzzles, such as: existence of stars (located usually in the Globular clusters) older than the "age of the Universe", a signature of hidden matter, called "a dark matter", existence of a supermassive black hole in the center of every well developed galaxy and so on. Recent observations of how the Hubble "constant" changes with the z-shift led to a new enigmatic problem for the cosmologists: an "open" (disintegrating) Universe. To save the ill Big Bang concept now a distributed dark-energy component with a "negative pressure" is invented. It seems that any possible attempts are made to avoid the admission that some underlying material structure of the space (physical vacuum) may exist. At the present time, such option is avoided from discussion because this means a death of the Big Bang theory.

   In the microscale range, the Particle Physics proposes an enormous number of sub-elementary particles and contradicting rules. The "quark", for example, is a substructure of the proton but its mass is estimated as about 185 times larger than the proton's mass. Many interactions provide infinities (in an energy aspect) that are unexplainable. All these facts point to some misunderstanding of the Universe and microcosmios. Apart of this, it is not possible to connect the Quantum Mechanics with the theory of Special and General Relativity. Despite a century long efforts a satisfactory unified theory is not available. In the experimental fields the discrepancies are even more obvious. In the article "Those Scandalous Clocks", R. R. Hatch [1], a distinguished pioneer in the GPS system provides evidence of discrepancies between the observational facts (from the GPS system, the Very Long Based Interferometry and the pulsar detections) and some formulations in the Special Relativity. He provides analysis showing that such discrepancies do not exist if using the Lorenz Ether Theory (LET). His analysis led to revealing one very important effect: the speed of light is velocity dependent, but in the experiments it appears independent, because the Doppler shift and the relativistic effect of clock rate change cancel each other. Such effect means a reconsidering of light velocity experiments. This includes also the Michelson-Morley experiment, which is cited in text books as a basic proof of not existing Ether. Now not only new experiment confirm the detection of our motion through a space medium, but the original data from Michelson-Morley experiment has been reanalyzed by M. Consoli and E. Constanzo [4] using a correct method and two velocities are clearly identified: the Earth orbital velocity around the Sun and the solar system velocity around the centre of Milky Way. Among the early modern experiments are those suggested and performed by Prof. Stefan Marinov [33,34,35]. In the period of 1976 to 1986 he successfully detected and measured the orbital motion of the Earth around the Sun and the Solar system motion around the Milky Way center, using pure laboratory experiments. One of his experiments is repeated by E. W. Silvertooth (1986) and the results are confirmed. These and number of other experiments prove the existence of absolute frame, which for us is the center of the Milky Way galaxy. This requires a redefinition of the inertial frame postulated by Special Relativity (SR). At the same time, the redefinition of some postulates does not contradict to the existence of the SR effects, such as the time dilation and the relativistic mass increase.

The accumulated problems and discrepancies could not be resolved without revision of some adopted fundamental rules and postulates in Physics. All of them are dependable on the adopted concept of the physical vacuum.

The revision of the vacuum concept requires some acquaintance with the developments in pre-modern Physics. Until the 17th century, the vacuum concept was influenced by the ancient Greek philosophers Aristotle, Leucippus and Democritus [1,2]. It has been redefined after the invention of the barometer by the Evangelista Torrichelli in 1644 and the vacuum became regarded as a pure empty space for a while. After the discovering of the electromagnetic radiation, however, the Ether concept became dominant in the 19th century Physics. The 17, 18 and 19th centuries gave great physicists whose contribution to the science and the vacuum concept is enormous. In this aspect it is worth mentioning Isaac Newton, Andre-Marie Ampere, Michael Faraday, William Thomson, James Clerk Maxwell. In the beginning of 20 century, the Ether concept has been abandoned and in 1925 a space-time concept was accepted as a result of adopted Quantum mechanical postulates, known as Copenhagen formalism. The Theory of Relativity played an important role for the introduced space-time concept but Einstein did not agree with some formulations of the Copenhagen interpretation in 1925. This is evident from his article "Can Quantum-mechanical Description of Physical Reality be Considered Complete" [3], with co-authorship with B. Podolsky and N. Rosen. In the beginning of 20th century, Einstein initially denied the existence of the Ether, but later changed his opinion. In the article of Galina Granec, Haifa University, Israel an authentic material from Einstein about the physical vacuum is collected: In a letter to Lorenz dated 17 June 1916, Einstein wrote (quoted in Miller, 1986, p.55 [38]; see also Kostro, 1988, p. 238 [39]):

I agree with you that the general relativity theory admits of an ether hypothesis as does the special relativity theory. But this new ether theory would not violate the principle of relativity.[40].

In 1920 at lecture in Leiden, Einstein says [40,41]

... there is a weightly argument to be adduced in favour of the eher hypothesis. To deny the ether is ultimately to assume that the empty space has no physical qualities whatever. The fundamental facts of mechanics do not harmonized with this view.

While Einstein introduced a cosmological constant in his earlier equations, later he removed it claiming that this has been the "greatest blunder" of his career. Despite of this, the Big Bang supporters today use namely this constant as a last life belt for saving the sinking Big Bang concept.

In fact, even the enigmatic space-time concept and all relativistic effects could be better explained by an alternative but correct vacuum concept, which is closer to the Ether one than to the void space. In the last decade, the interest to the vacuum properties is significantly increased due to the unexpected results from non-conventional experiments, which are unexplainable from the point of view of contemporary Physics.

When beginning a revision of the adopted concept of the physical vacuum we must take into account the rational achievements in Classical Physics. The father of Modern Electrodynamics, James Clerk Maxwell built his famous theory with the presumption for the existence of Ether. In his "Treatise on Electricity and Magnetism" vol. II [5] he writes:

The theory I propose may therefore be called a theory of Electromagnetic Field, because it has to do with the space in the neighbourhood of the electric or magnetic bodies, and it may be called a Dynamical theory , because it assumes that in that space there is a matter in motion , by which the observed electromagnetic phenomena are produced.

... If something is transmitted from one particle to another at a distance, what is its condition after it has left the one particle and before it has reached to the other?... Hence, all these theories lead to the conception of a medium in which the propagation takes place, and if we admit this medium as an hypotreatise, <u>I think it ought to occupy a prominent place in our investigations, and that we ought to endeavor to construct a mental representation of all the details of its action, and this has been my constant aim in this treatise</u>

The original Maxwell's equations, defined for 20 field variables, are formulated in a quaternion form [6,7]. Later other physicists (Oliver Heaviside and William Gibbs, Lawrence etc.) transformed them into the known today vectors form. While the vector equations are very compact, Maxwell has not recommended them, because they are not able to describe completely physical phenomena. These tailored Maxwell equation are in all textbooks today. They are convenient but do not describe the whole truth. Recently K. J. van Vlaenderen and A. Waser in the article "Electrodynamics with scalar field" [8] shows that the electrodynamics can be efficiently formulated in biquaternion form in which the original Maxwell's concept is preserved. The major profit from this is the prediction of existence of longitudinal electroscalar waves in vacuum. A similar result is obtained independently by K. P. Butusov [9]. Longitudinal waves firstly introduced and observed by Nikola Tesla and recently confirmed by many experiments are not apparent when using the vector form of the Maxwell's equations.

The acceptance of Ether existence automatically leads to the conclusion that it should possess **two kinds of states: a steady state and a transient one.** One of the consequence from tailoring the original Maxwell's equations is the exclusion of the transient state properties of the physical vacuum. That's why some physical phenomena may look like paradoxes and some experiments are regarded as a contradiction to the "laws of Physics". In other words, the transient state of the vacuum is outside of the filed of view of Modern Physics today. One of the features of this state is the possibility for transmission of energy in a way unexplainable by the currently adopted space concept. The radiant energy discovered by Nikola Tesla 100 years is one proof for the transient state of the space. The Tesla's famous experiments about wireless and single-wire power transmission were regarded for many years as "exotic", but now they are confirmed [10,11]. Presently, the search related to the hidden vacuum energy and the effects related to Ether disturbances reached unprecedented level [12,13].

The existence of hidden energy in space, for which experimental proofs exist, could not be explained by the presently adopted concept of the physical vacuum, so a new model must be found. Such task should go parallel with a more universal one – a building of unified theory, because the new physical model about space and matter must fit to the unified vision about the forces in Nature. In order to avoid any departure from reality, some of the adopted so far rules of the Copenhagen formalism must be ignored and the principles of causality, objective reality and logical understanding must be accepted as rules. In such aspect, instead of the traditional way of studying the vacuum properties as energetic interactions, a new approach was introduced – a building of detailed physical model of the underlying structure of the physical vacuum whose elements must be also involved in the possible structures of the elementary particles. Such physical model with unveiled fundamental interactions must provide explanation about the known physical fields, forces and interactions in a real 3+1 space-time. It also must provide a vision about the connections between the different fields – gravitational, electrical and magnetic. It must find also the relation between Classical Mechanics, Quantum Mechanics and Einstein's Theory of Relativity.

In Cosmology, it was found that a supermassive black whole (with size of billion solar masses exists in the center of every galaxy and any such "hole" is in a balance with the total mass of the visible matter of the host galaxy [14]. Additionally, an existence of a hidden "dark" matter whose signature is apparent from the galactic rotational curve is rather a rule than an occasional fact [15]. These new discoveries together with many others lead to the idea that the indirectly detectable "dark" matter is in fact a signature of an underlying material structure of the physical vacuum. Such structure should exist around us and within us.

The search for the correct space-time concept required extensive study on some features of the physical vacuum such as the Zero Point Energy, the quantum fluctuations, the vacuum polarization, the Plank's length and frequency and so on. In such aspect, the theoretical articles provided by T. H. Boyer [16], H. E. Puthoff [17,18,19] H. E. Puthoff et al [20], B. Haisch et al. [21] and F. M. Meno [2] were quite useful. The articles "Experimental evidence that the gravitational constant varies with the orientation" by M. M. Gershteyn et al [22] and the "Speed of gravity revisited" by M. Ibison et al. [23] lead to the idea that the <u>Newton's law of gravitation might be derivable instead of postulated.</u> This idea obtained some theoretical treatment by H. E. Puthoff [17] (1989) who show that the Newton's law of gravitation is related to Planck's frequency, $\omega = [2\pi c^5 /(hG)]^{1/2}$. Using one hypothesis of Sakharov he shows that the Newton's law is derivable. In the development of BSM theory, an idea was conceived that the Planck's frequency could be an intrinsic parameter of some fundamental particle (or pair particles) interacting in a pure empty space by a fundamental law of Super Gravitation.

Concept of the Basic Structures of Matter (BSM) treatise

The treatise Basic Structures of Matter, a Super Gravitation Unified Theory unveils the relation between the forces in Nature by adopting of the following framework:
  - Empty Euclidian space without any physical properties and restrictions
  - Two super dens fundamental particles, able to vibrate and congregate
  - A fundamental law of Super Gravitation (SG) – an inverse cubic law valid in empty space.

Enormous abundance of these particles, driven by the fundamental SG law into self-organised hierarchical levels of geometrical formations, leads deterministically to creation of space with quantum properties - physical vacuum and a galaxy as observable matter.

The underlying structure of created space (physical vacuum) is called Cosmic Lattice (CL). It is built of two super dens sub-elementary particles, which are geometrical formations from the fundamental particles hold by SG law. The two sub-elementary particles with a shape of elongated prism are arranged in flexible nodes, each one formed by 4 prisms. Additionally, the SG field of the prisms exhibits an axial anisotropy with a right or left-hand twisting component, respectively for both types of prisms, due to their lower level structure. The observable space is filled by CL grid of alternatively arranged nodes, forming a lattice similar as the diamond atomic lattice. The estimated node distance is in order of $(1\sim2)\times10^{-20}$ (m), while the intrinsic matter density of the prisms is about 1×10^{13} time higher than the average density of the atomic matter. The individual node of the CL structure possesses a flexible geometry, a freedom to oscillate in a complex spatial mode and energy well. The complex dynamics of CL node oscillations is related to the parameters permeability and permitivity of free space. The common mode oscillations with a running phase synchronization are related to the parameters Compton wavelength and define the light velocity. The CL space exhibits quantum features and provides conditions for existence of fields: gravitational, electrical and magnetic. These fields are defined by the static and dynamic parameters of the CL nodes. The elementary particles also possess a structure in which helical formations of prisms are identifiable and called helical structures. The lowest order helical structures have denser (than CL) internal lattices with a twisted component. The SG field of this internal lattice is able to modulate in a specific way some of the oscillating parameters of the surrounding CL nodes, creating an electrical field. The same denser internal lattice causes a partial folding and displacement of the CL nodes when the particle moves in CL space. This process, in which proximity fields are also involved, defines the intrinsic inertial properties of the elementary particle. Some of the main CL space parameters are the following: a Static CL pressure, a Dynamical CL pressure and a Partial CL pressure. They are related respectively to: the mass of the elementary particles, the CL space background temperature (2.72K), and the inertia of the atomic matter. Theoretical equations of these parameters are found and expressed by the known physical constants. A mass equation applicable for the elementary particles is derived by analysis of the dynamical interactions of the electron in CL space environment. The Planck's constant and the rules of the Quantum mechanics obtain physical explanations. The velocity of light is a derivable parameter. The relativistic concept of an inertial frame obtains a logical physical explanation and the effects of the General and Special relativity are understandable. The CL space allows also a creation and propagation of virtual particles, corresponding to the Dirac's idea of virtual particles. Possessing only a charge without intrinsic matter, they could be easily confused with the real particles, which possess matter. The results from the developed models are in excellent agreement with the experiments and observations in the range from a micro to macro Cosmos.

The proposed BSM models are verified by cross validation of their output results with experimental data from different fields of physics. The unveiled low-level structures, involved in the physical vacuum and the elementary particles, further allow deciphering the structures of the atomic nuclei. It appears that the rows and column pattern of the Periodic table is a signature of the atomic nuclear configuration – the spatial arrangement of the protons and neutrons in the nucleus. The electronic orbital shapes are strictly defined by the nuclear configuration. In this aspect, one of the most useful results from the BSM theory is the illustrative appendix titled **Atlas of Atomic Nuclear Structures (ANS).** It shows the unveiled configurations of the atomic nuclei for the elements from Hydrogen to Lawrencium (Z = 103) [27].

The alternative vacuum concept changes our vision about the micro Cosmos. It appears to be well organized due to well-defined and logically consistent natural rules. The new concept leads also to alternative cosmology: The Universe must be stationary. This is in good agreement with the accumulated cosmological data. For example: The red shift periodicity in the observation of Q-stellar objects [28,29,30] that is explainable only if the galactic red shift is not of Doppler type; the observed dipole shape of the cosmic microwave background; the Lyman alpha forests [31]; the observable deviation of the Hubble law from the expected one for redshifts above 0.8; the galactic rotational curves; many observations indicating an enormous percentage (over 90%) of hidden "dark matter" and so on. A careful analysis of the observational data, based on the new space concept, leads to the conclusion that the galactic red shift is a result of small energy losses that photons exhibit when passing from one galactic space to another. According to the new space concept, this is a result from the small differences between the underlying vacuum space structures of the different galaxies, because they are from different evolutional formations. This particular result, is in a good agreement with the alpha forest observations [31]. While this phenomenon is unexplainable enigma for the Big Bang model, it excellently fit to the BSM concept (see section

12.B.12 Chapter 12 of BSM). All these cross-related phenomena change significantly our vision about the Universe. Instead of searching for a hypothetical Big Bang, the focus is moved on the individual galaxy as a family member of the Universe. In such different scenario, every galaxy should have a cycle comprised of phases, such as an active life, a collapse and a rebirth. **While the active life is only visible, the other phases are completely invisible, but they are indirectly identifiable by some cosmological phenomena.** In the hidden phases, a complete recycling of the old matter and a crystallization of a new one takes place, but some low level structures of the intrinsic matter could not be destroyed and they preserve the information defining the existence of "matter instead of antimatter". The hypothetical concept of this cycle and its phases are presented in Chapter 12 of BSM, where some insight about the fundamental SG law is also discussed. The underlying vacuum structure from the consecutive cycle periods may obtain some small differences, because the total matter of the galaxy plays a role in the formation of the new sub-elementary particles – the prisms. It is known that our galaxy contains older stars than the "age of the Universe" and they are located in the globular clusters. These formations appear to be remnants from the previous galactic life. They have escaped the galactic collapse due to the CL space break-up during this cosmological event. The Cepheids from these clusters exhibit different features. Even the motion of the stars in these clusters exhibits a strange behavior, which is explainable if they have a "lower Maxwellian energy", according to I. R. King [32]. (see section 12B.7.2.1, Chapter 12 of BSM).

It is well known that some physical phenomena in Quantum Mechanics, Particle Physics and Relativity could not be logically explained. As a result, it is accepted that the human logic fails. This is a big obstacle for building of successful unified theory. BSM theory is free of such problem. This permits to build a comprehensive logical scenario about the evolution of the matter covering the range from micro cosmos to the Universe, while using a new interpretation of the observed phenomena. Such scenario is free from the paradoxes and problems, which currently plague the Big Bang theory. Figure 1 provides a comparison between the currently existed concept about the Universe and the concept envisioned by BSM theory. In the latter case, the principles of causality and unperturbed human logic are strongly observed.

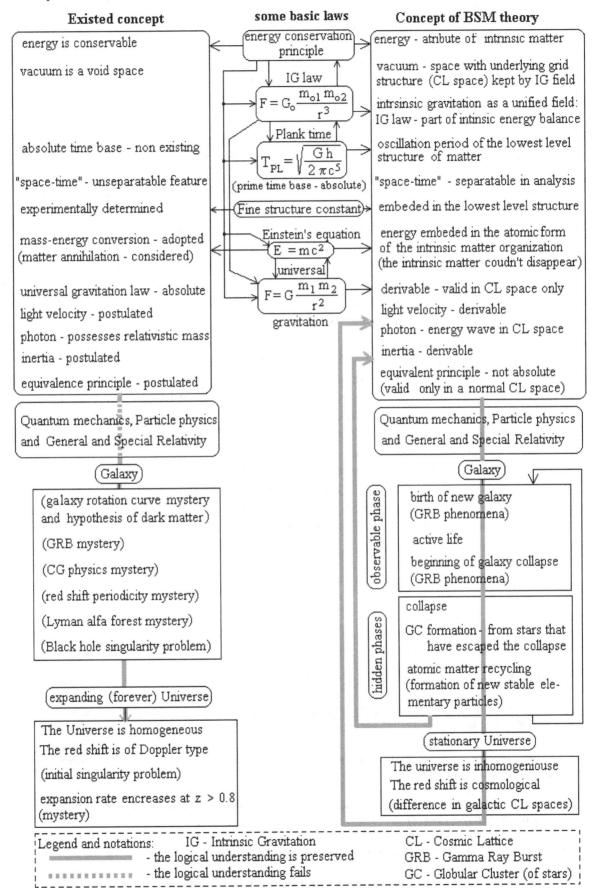

Fig. 1. Difference between the present and the new vision about the Universe according to **BSM**

The unveiled structural formations and the derived physical models are illustrated by a large number of drawings. In their analysis a straightforward mathematical methods are used. A number of new features, parameters and interactions mechanisms are discovered. They are denoted and explained in the first time they appear, while later are only referenced. For this reason the reader is advised to follow the Chapters order.

The Basic Structures of Matter monograph contains 13 Chapters and a few Appendices. The most important appendix is **The Atlas of Atomic Nuclear Structures (ANS)**

Part I of ANS provides the geometry and internal structure of the basic elementary particles, consisting of helical structures. Part II of ANS provides the atomic nuclear structure of the elements up to $Z = 103$. In order to simplify the complex three-dimensional configuration of the atom, symbolic notations are used for the proton, neutron, deuteron and helium. The electronic orbits are not shown, but their positions are well defined by the position and the proximity electrical field of the protons (or deuterons). The raw and column signature of the periodic table is well matched; the Hund's rules and Pauli exclusion principle are identifiable. Due to a drawing complexity, the twisting feature of the proton, neutron and atomic nuclei are not directly shown.

In the last Chapter 13 a summary of the potential applications is presented, while some of them are apparent from the analysis in previous chapters. In this chapter three Special applications are also discussed, which are results of the very original predictions of BSM theory: a hidden space energy of EM type – a primary source of the nuclear energy, a possibility for control the gravitation and inertia of material object, and a supercommunication by using a new type of waves.

The full version of BSM and some related articles are published initially in www.helical-structures.org (S. Sarg, 2001) and archived to the National Library of Canada [24] (first edition, 2002, second edition, 2005). A brief introduction to BSM and other related articles are published in Journal of Theoretics [26] (S. Sarg, 2003), in http://lanl.arxiv.org [25] and in Physics Essays [37].

References (for Chapter 1):

1. Ronald R. Hatch, Those Scandalous Clocks, Springer-Verlag, DOI 10.107, 30 Apr 2004 http://springerlink.com

2. F. M. Meno, A Planck-length atomistic kinetic model of physical reality, Physic Essays, 4, No. 1, 94-104, (1991)

3. A. Einstein, B. Podolsky and N. Rosen, Can Quantum-mechanical Description of Physical Reality Be Considered Complete?, Physical Review, v. 47, 777-780 (1935)

4. M. Consoli and E. Constanzo, The motion of the Solar System and the Michelson-Morley experiment, arXiv:astro0ph/0311576v1 (2003)

5. J. C. Maxwell. *A Treatise on Electricity & Magnetism*, (1893) Dover Publications, New York ISBN 0-486-60636-8 (Vol. 1) & 0-486-60637-6 (Vol. 2)

6. A. Waser, On the notation of Maxwell's field equations, www.aw-verlag.ch/EssaysE.htm (2000)

7. D. Sweetser and G. Sandri, Maxwell's vision: Electromagnetism with Hamilton's Quuaternions, Second Meeting on Quaternionic Structures, Roma, 6-10 Sep 1999

8. K. J. van Vlaenderen and A. Waser, "Electrodynamics with the scalar field, www.aw-verlag.ch/EssaysE.htm also with slight adaptations: van Vlaenderen Koen and A. Waser, "generalisation of classical electrodynamics to admit a scalar field and longitudinal waves", Hadronic Journal **24**, 609-628 (2001)

9. K. P. Butusov, Longitudinal Waves in Vacuum: Creation and Research, New Energy Technologies, Sep-Oct 2001, pp. 46-47.

10. D. S. Strebkov, S. V. Avaramenko, A. I Nekrasov, O. A. Roschin, Investigation of 20 kW, 6.8 kV, 80 mkm Single-Wire Electrical Power System, New Eenergy Technologies, Nov-Dec, 2002, pp. 52-54.

11. S. K. Avramenko, Method& Appratus for Sinle Line Electrical Transmission, US Patent 6,104,107. (Aug. 2000)

12. N. Kosinov, Power Phenomenon in Vacuum, SciTechLibrary www.sciteclibrary.com/eng/catalog/pages/2646.html

13. A. V. Frolov, Some Experimental News, New Energy Technologies, Nov-Dec 2002, pp. 1-5.

14. L. Ferrarese, D. Merrit, A fundamental relation between supermassive black holes and their host galaxies, http://arxiv.org/abs/astro-ph No. 0006053 v. **2** 9 Aug 2000

15. D. F. Roscoe, An analysis of 900 optical rotation curves: Dark matter in a corner?, Phahama - Journal of Physics, Indian Academy of Sciences, Vol. **53**, No 6, Dec 1999, p. 1033-1037

16. T. H. Boyer, The Classical Vacuum, Scientific American, Aug. 1985, p.70-78.

17. H. E. Puthoff, Gravity as a zero-point-fluctuation force, Phys. Rev. A, vol. 39, no 5, 2333-2342, (1989)

18. H. E. Puthoff, Polarizable-Vacuum (PV) Approach to General Relativity, Foundations of Physics, V. 32, No. 6, 927-943 (2002)

19. H. E. Puthoff, Can the Vacuum be Engineered for Spaceflight applications, NASA Breakthrough Propulsion Physics, conference at Lewis Res. Center, (1977)

20. H. E. Puthoff, S. Tittle, M. Ibison, Engineering the Zero-Point Field and Polarizable Vacuum for Interstellar Flight, First International Workshop in Field Propulsion, Univ. of Sussex, Brighton, UK, Jan 2001, http://www.nidsci.org/article3.html

21. B. Haisch, A. Rueda and H. E. Puthoff, Inertia as a Zero-point filed lorenz force, Phys. Rev. A, **49**, 678 (1994). See also Science 263, 612 (1994).

22. M. L. Gershteyn, L. Gershteyn, A. Gershteyn, O. Karagioz, Experimental evidence that the gravitational constant varies with orientation, (2002), http://arxiv.org/abs/physics/0202058

23. M. Ibison, H. E. Puthoff and S. R. Little. The Speed of Gravity Revisited, posted to LANL archives, http://xxx.lanl.gov/abs/physics/9910050

24. S. Sarg, "Basic Structures of Matter", monograph, (2001), http://www.helical-structures.org
also in National Library of Canada, (2002) http://www.nlc-bnc.ca/amicus/index-e.html (AMICUS No. 27105955) (first edition (2002) ISBN 0973051517; second edition (2005) ISBN 0-9730515-5-8)

25. S. Sarg, New approach for building of unified theory about the Universe and some results, http://lanl.arxiv.org/abs/physics/0205052

26. S. Sarg, Brief introduction to the Basic Structures of Matter Theory and derived atomic models, Journal of Theoretics, (2003), www.journaloftheoretics.com/Links/Papers/Sarg.pdf

27. S. Sarg, Atlas of Atomic Nuclear Structures According to the Basic Structures of Matter Theory, Journal of Theoretics (2003) www.journaloftheoretics.com/Links/Papers/Sarg2.pdf

28. G. Burbidge, Astrophysical Journal,, 147, 851 (1967)

29. G. Burbidge, Astrophysical Journal, 155, L41 (1968)

30. B. N. G. Guthrie and w. M. Napier, Astronomy and Astrophysics, 310, 353-370 (1996)

31. A. Songallia, E. M. Hu and L. L. Cowie, Nature v. 375, 124-126 (1955)

32. I. R. King, Astronomical Journal, v. 71, No 1, 64-75 (1996)

33. S. Marinov, Measurement of the Laboratory's Absolute Velocity, General Relativity and Gravitation, vol. 12, No 1, 57-65, (1980)

34. S. Marinov, The interrupted 'rotating disc' experiment, J. Phys. A: Math. Gen. **16**, 1885-1888, (1983)

35. E. W. Silvertooth, Experimental detection of the ether, Speculations in Science and Technology, Vol 10 No 1, 3-7, (1986)

36. G. F. Smoot et al. Detection of Anisotropy in the Cosmic Blackbody Radiation, Physical Review Letters, v. 39, No. 14, 898-901, (1977)

37. S. Sarg, A Physical Model of the Electron According to the Basic Structures of Matter Hypothesis, Physics Esays, **16**, No 2, 180-195, (2003)

38. A. I. Miller, *Imagery in Scientific Thought: Creating twentieth-Century Physics* (Cambridge: MIT Press, (1986)

39. L. Kostro, Einstein and the ether, Electronics & Wireless World 94, 238-239, (1988)

40. G. Granec, Einstein's Ether: F. Why did Einstein Come Back to the Ether?, Apeiron, v. 8, No 3, July (2001)

41. Albert Einstein, (documented movie footage, 1920) as a video "Free Energy the Race to Zero Point, *Sidelights on Relativity* available by Lightworks Audio & Video www.lightworksav.com (available also by amazon.com)

Chapter 2. Matter, space and fields

2.0. Criteria for successful unified theory

The most common definition of Unified Theory is a theory, which describes the gravitational, electromagnetic, strong and weak forces within a single, all-encompassing framework. Attempts for building of such theory are made for over 100 years. Let us define first what should be the criteria for a successful Unified Theory:

- It must suggest a physical model of fundamental particles in a real 3-dimensional space, working in all fields of Physics without contradictions and able to explain the relations between the gravitational, electrical and magnetic fields

- The number of initially adopted postulates and fundamental particles must be as smaller as possible

- The derived models must be able to explain all kind of experiments and observations in all fields of Physics and in the range from micro-cosmos to Cosmology

- The energy conservation principle must be strictly observed

One additional criterion of philosophical aspect is that the theory must show how the laws of Physics are written in the language of Nature.

Presently, the existing theoretical models pretending to have the properties of Unified theory, such as the Standard Model in Particle Physics and the String theories, could not suggest a model satisfying the above-defined criteria. What is proposed is an enormous number of particles, postulates and rules, often contradictable between themselves or using an unrealistic speculative multidimensional space. Since the suggested models suffer from logical understanding, it has been claimed that the human logic fails, so a mathematical one must replace it. Such statement, criticized in the past (1900-1930), now is silently accepted. It became convenient for development of an enormous number of abstractive theories, while the main issue is unsolved.

The main objective of the Basic Structures of Matter (BSM) treatise is to build a functional base for a Unified theory operating in a real three-dimensional space and unidirectional time. The successful solution of such task requires using approach in which the principles of real objectivity, causality and logical understanding are strongly observed for any kind of physical phenomena in the range from micro-cosmos to Universe. These restrictions makes the BSM approach distinguishable from other existing theories, for which they are too narrow. The major benefits, however, are enormous:

- using understandable classical methods

- using the power of unperturbed logic and understanding

- avoiding a fallacy into unrealistic imaginary concepts

The treatise Basic Structures of Matter, a Super Gravitation Unified Theory unveils the relation between the forces in Nature by adopting of the following framework:

- Empty Euclidian space without any physical properties and restrictions

- Two super dens fundamental particles, able to vibrate and congregate

- A fundamental law of Super Gravitation (SG) - an inverse cubic law valid in empty space.

It will be shown in these treatise that enormous abundance of these particles, driven by the fundamental SG law into self-organised hierarchical levels of geometrical formations, leads deterministically to creation of space with quantum properties - physical vacuum and a galaxy as observable matter.

All known laws of Physics are embedded in the underlying structure of the physical vacuum and the structure of the elementary particles. The fundamental SG law is behind the gravitational, electric and magnetic fields and governs all kind of interactions between the elementary particles in the space of physical vacuum.

The argument supporting the above underlined statement will be presented in this treatise by showing how the derived models agree with the observational and experimental data.

2.1. Alternative concept about the physical vacuum and the structure of the elementary particles.

One of the main differences of BSM from the existing so far theories is the concept about space in which we live and observe, known as a physical vacuum. BSM is based on an original alternative concept of the physical vacuum that has not been investigated so far. Such concept requires reformulation of some of the postulates and laws adopted about 100 years ago. From a methodological point of view, one of the major benefits is the use of classical methods for analysis in which the logical understanding is extremely useful. This means that tough physical phenomena from different fields could be successfully analysed by physical models at lower level.

The alternative vacuum concept admits that the space is not void but containing underlying structure of distributed hidden matter of some more primary form than the known elementary particles. These structure is quite distinctive from the concept of ideal gas or the Ether models, which dominated the physics before the 20th century.

The new approach regards the matter existence as a primary fundamental rule, while the energy is always its attribute. The matter may exist in hierarchical levels of formations. Then it is reasonable to accept that some lower levels of matter organization may exist in microscale range, which is beyond the technological limit of detection. In such case, a particular destruction of some structural formations to a level below the detection limit may seam for us as a matter annihilation.

The above considerations require more universal formulation of the fundamental concept about matter, space and energy, in accordance of the adopted principles of real objectivity, causality and understanding:

- definition of absolute space in a classical way, free of relativistic considerations
- definition of structured space as a three dimensional grid defined by a hypothetical underlying material structure - a space lattice, whose features are defined by the spatial arrangement and properties of some basic sub-elementary particles.
- a logically accepted option that the elementary particles possess a structure build by the same sub-elementary particles.

The formulated definitions must provide a bases for the following:

- understandable logical explanations of the phenomena in Quantum Mechanics, Special and General Relativity, the physics behind the space-time and quantum features of the space and the interactions between the elementary particles and the quantum space
- understandable logical explanations of the basic physical parameters as mass, inertia, and field properties, as interactions between the structured space and the elementary particles.

At some level of matter organization lying hierarchy below the level of stable elementary particles (proton, neutron, electron, positron), we may expect existence of common sub-elementary particles, which are building blocks for both - the structured space and the elementary particles. In search for such particles it is reasonable to expect that they should posses features for explanation of the field property of the matter and the quantum features of the physical vacuum.

Extensive analysis of phenomena from different fields of physics allowed to formulate the search criteria for the possible physical model of this structure. The search for the correct model took also into account a large number of published theoretical articles about the properties of the physical vacuum. They are related to some features, known as: a Zero Point Energy, quantum fluctuations and polarizability of the physical vacuum. Among some recent publications in this field are article published by T. H. Boyer, Frank Meno, H. E. Puthoff, A. Rueda, M. Ibison, B. Haisch and others.

In the article "Gravity as a zero-point fluctuation force", H. E. Puthoff (1989) begins from the equation of the Planck's frequency.

$$\omega_{PL} = \sqrt{\frac{2\pi c^5}{hG}} \qquad (2.0)$$

Using one hypothesis of Sakharov, Puthoff successfully derives the Newton's law of gravitation. This and other results of such kind served as a valuable orientation in the BSM search for correct physical model. Accepting the Planck's frequency as a real physical parameter was a step in a right direction for building of the BSM concept. The confidence about this was increased by the consistency between the derived models and the known physical constants, from one hand, and the agreement with the accumulated experimental and observational data.

The defined criteria permitted to narrow the range of search, so one of the most promising model is suggested. According to this model, **the vacuum space possesses a underlying grid structure of sub-elementary particles arranged in nodes.**

The most promising candidates for such sub-elementary particles appeared to be a **pair of profiled rods with shape close to a hexagonal prism, made respectively by two different substances of intrinsic matter. These two rods possess also gravitational anisotropy and embedded helicity in their lower level structure.** The possible lower level structure of the suggested profiled rods and how they are formed is discussed in Chapter 12. **For simplicity of the visualization and analysis, the suggested profiled rods are replaced by three-dimensional models of right handed and left handed hexagonal twisted prisms.** These models are convenient for visual presentation of these sub-elementary particles which are involved in both: the underlying material structure of the physical vacuum and the material structure of the elementary particles. For this reason, we will often use the concept of the twisted prisms during the course of BSM theory. The chosen model of twisted prism is suitable for understanding some important basic matter properties in a structured space without initial knowledge of the internal structures of these particles. While the guessed basic properties of this basic particles - prisms has been obtained iteratively, now we will formulate them. They will be confirmed later during the course of this treatise.

The shape and the relative dimensions of the two twisted prisms are shown in Fig. 2-1.

Some of their basic properties are the following:

- **The models of two basic sub-elementary particles in the observable Universe are left handed and right handed twisted prisms, made of two different substances of intrinsic matter.**
- **The right-handed prism has inverse conformation similarity with the left-handed prisms, while the dimensional scale factor between the right and left-handed prisms is 3/2.**
- **The attraction forces in a classical void space between two prisms of same handedness (a same intrinsic matter substance), is not the same as between two prisms of different**

handedness (different intrinsic matter substances)
- **The length to diameter ratio of both prisms has a minimum acceptable value dependable on the twisting angle.**
- **The two ends of the prisms are rounded**
- **There are bumps on any surface of the hexagonal rods (prisms) as shown in the Fig. 2.1, obtained during the phase of their formation.**

Fig. 2.1.a. shows sketches of left and right handed twisted prisms, which serve as models of the real prisms. Fig. 2.1.b. shows the external shape of the real prisms.

The approximate length of the longer prism (estimated in the following later analysis) is in the range of $(1 \text{ to } 10) \times 10^{-21}$ (m)

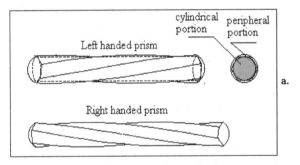

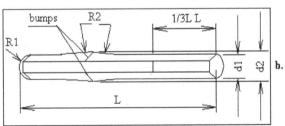

Fig. 2.1

a. Model of twisted prisms (with an external twisted shape)
b. Real shape of the real prism (with an internal twisting)

The logical question after the definition of both prisms as sub-elementary particles is how such complex shape could be obtained. The reply of this question is left in Chapter 12, where a detailed scenario is presented, showing that they can be built by the two fundamental particles in a self organised process. The reason for presenting this in Chapter 12 is that the prisms, regarded as a level of matter organisation, are closer to the elementary particles and the properties of the physical vacuum, for which we have enough experimental data. The fundamental particles lie in the bottom level of matter organisation, which can be logically inferred only after broad analysis ranging from particle physics to Cosmology.

In order to disperse some doubt in the beginning of this treatise we will show here only a summarised view of the scenario for formation of the prisms (referred also as basic particles) from the fundamental particles.

Summarised scenario of prism formation

- Two indestructible fundamental particles (FPs) of two substances of intrinsic matter with a spherical shape, a radius ratio of 2:3 and different super-high density. They have a Bell shape curve dependence of the their density on the radius, which permits a spherical type of vibrations in 3D formations. They also have extremely small but different time constants, which defines a proper resonance fraudulency. Its average value is associated with the Planck frequency.

- A fundamental law of Intrinsic Gravitation (IG), governing the interactions between FPs in a classical void space, according to which FPs from a same substance are attracted by forces inverse proportional to a cube of the distance:

$$F_{SG} = (G_0 m_{01} m_{02})/r^3$$

where: GO - intrinsic gravitational constant, m01, m02 intrinsic matter mass, r - distance

- Vibrational energy: FPs preserve a limited freedom of vibrations in formations from the same type of substance

In pure empty space, FPs may congregate in geometrical formations, held by SG forces. Such formations preserve the primary vibrational mode of the FPs, while additional modes appear with a divided frequency and spatial features defined by the 3D geometry of the formation. The vibrational modes are part of SG energy (discussed in Chapter 12) and define concentration of SG forces at the edges (similar as in the permanent magnet). The consecutive formations in one congregational order are: Tetrahedron (TH), Quasipentagon (QP) and Quasiball (QB).

60TH -> 12QB ->1Q

They are illustrated by the panels 1 and 2.

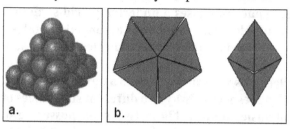

Panel 1: a. Tetrahedron (TH)
b. Quasipentagon (QP)

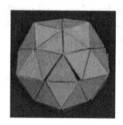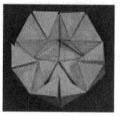

external view   internal sectional view

Panel 2: External view and internal sectional view of a Quasiball (QB)

TH has gaps, which is combined in a common gap of 7.355 deg. The common vibrational modes are oriented along the symmetrical axis. The QB has internal empty space, as shown in Panel 2. It also can be twisted in left or right-handed direction due to the gaps in the embedded QPs. This is the lowest level memory. The consecutive formations in one congregational are repeated in any upper congregational order. The Planck's frequency is divided with any consecutive formation and order by accurate division mechanism, based on the vibrational modes.

The lowest order TH is formed of one of the basic fundamental particle - a primary ball (PB). The number of PB in lowest order TH is defined by the first most stable vibrational mode, which depends on the intrinsic properties of the fundamental particles. Its signature is the fine structure constant (this is discussed in Chapter 12).

If a large number of QB of higher order are collected in a spherical shell, the SG forces coppers them. A few highest orders QBs crash consequtively, which lead to formation of prisms. Their internal structure is comprised of oriented QPs of some lower order, with distributed gaps carrying the twisting information of the crashed QBs. as illustrated in Panel 3. As a result the prism has a SG anisotropy along its his and handedness. Prisms from one type of FP carry a right-handed anisotropy and from the other - left-handed.

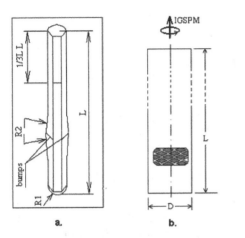

Panel 3. Mould prism

It is evident that prisms of same type and order will be comprised of exactly the same number of fundamental particles. This feature qualify them to serve as building elements for both: the structure of space and (physical vacuum) and the structure of the elementary particles. Detailed scenario for their formation is presented in Chapter 12.

2.2 Basic definitions and physical laws.

2.2.1. Adopted terms and definitions

The lattice space and the atomic matter (the elementary particles) are both built by the two basic particles - the prisms.

BSM relies strongly on the principles of real objectivity and causality. This leads to the following considerations: **All physical processes are in three dimensional space and the time is unidirectional.** This allows using a classical physical approach in the analysis of any kind of physical phenomenon. For this reason two types of spaces are defined:

- **Classical void (empty) space** (or not structured space) - the space properties are not influenced by the matter
- **Lattice space** (structured space) - the space properties are influenced by the matter

The lattice space contains gravitational lattice structure. The space properties are defined by the parameters of this structure.

The adopted above definition of two types of spaces provides a possibility to study the properties of the structured space (the vacuum), and the elementary particles. For this purpose two types of 3-dimensional frames are used:

- **Lattice frame**
- **Absolute frame**

The lattice frame corresponds to the space known in our world with its quantum properties and relativistic features.

The absolute frame is introduced for convenience in order to use classical methods for analysis. Physically it can be regarded is a classical three-dimensional space, but filled with a lattice built of prisms. This lattice, however, has unique properties, since the interactions between the prisms are govern by more fundamental law called a Law of Super Gravitation, which will be discussed later.

The existence of material lattice defines particularly the physical properties of the space. Theoretically, we may investigate any kind of physical phenomena in such space with underlying structure, if having two absolute etalons: one - for length and a second one - for time. Let us initially define the following etalons for length and time:

- **absolute length unit** - the stable length of one of the prisms
- **absolute time base** - related with some intrinsic time constant of the prisms interactions governed by the Super Gravitational law, which is related to the most fundamental time constant related to the Planck's frequency.

The meaning of the absolute time base and the fundamental time constant will be understood in the final chapter, where the Law of Super Gravitation will be discussed. During the analysis in this and following chapters, however, we will find many signatures of these time bases. One suitable **secondary time base appears to be the Compton frequency, which is found to be a signature of both: a specific oscillation frequency of the lattice node and a proper oscillation frequency of the electron.** The electron structure is unveiled and discussed in Chapter 3, where it is shown that it is a 3 body structure with two proper frequencies. The first proper frequency appears to be the well known Compton frequency.

According to a BSM approach, the structure of the vacuum and the matter are always regarded as placed in an absolute frame. The definition of a lattice frame and an absolute frame provides the opportunity to separate the space-time parameters in the analysis. This is impossible if the very basic structure of the physical vacuum is unknown as in the present state of contemporary physics. Once this structure and its properties are unveiled, the Quantum mechanical and relativistic phenomena could be successfully analysed and logically understood.

2.2.1.A Level of matter organization

It will become apparent in the following chapters that the intrinsic matter possesses a broad range of organizational complexity with clearly distinctive hierarchical levels. In such space, it is useful to introduce the term **level of matter organization**. The suggested model of twisted prisms does not belong to the lowest level of matter organization. However, it is convenient in the analysis of the structure of the space and the elementary particles, since we may use directly the present knowledge about their properties. The possible lower level structure of the prisms is discussed in Chapter 12, since some important features in the upper level structures will become apparent from the analysis in the previous chapters. It is useful at this stage to list some features about the matter organization in different levels, while they will become apparent from the later analysis.

- **The formations belonging to any lower level of matter organization are implemented in the upper higher level, while their vibrational modes and energy exchange interactions are preserved.**
- **Any lower level formation possesses higher frequency vibrational modes in respect to the upper level.**

2.2.2 Adopted postulates and basic physical laws

One of the main goals of BSM theory is to reduce the number of postulates to a minimum. The structural pyramid of matter organization, however, so big that it is not convenient to refer always the complex processes to the very basic level or postulates. One good feature of the BSM approach is that a number of postulates accepted in contemporary physics (for example the constant light velocity, Newtonian mass and inertia) appear derivable. In order to simplify the analysis while keeping a close link to the present knowledge, **two types of postulates are adopted in BSM: basic and derivable.**

The derivable postulates, rules and laws are result of analysis based on the adopted basic postulates. They are not so apparent at the beginning but are derivable from the analysis and comply to the experiments and observations.

2.2.2.1 Basic postulates and parameters

A. The energy conservation principle: the energy could not be created and it cannot disappear. It only converts from one form to another.

B. The energy could not be separated from the matter

C. The Super Gravitation is a form of energy exchange between intrinsic matter objects involved in the total energy balance of the system.

D. The fine structure constant α is one of the most fundamental physical parameter.

In Chapter 12 it will be shown that the fine structure constant is embedded in the fundamental level of matter organization, which from its hand is embedded in all upper levels.

2.2.2.2 Derivable postulates, rules and laws

A. Valid for both types of space (later defined as a classical void and a lattice space):

- The most efficient and fast energy exchange between separate systems is by multiple oscillations in a frequency range closer to the system resonance frequencies
- The oscillation frequency assuring the energy exchange is different for the different levels of matter organization. Lower level formations exchange energy at higher frequencies

B. Valid in lattice space only

- The lattice space contains distributed matter connected by a static and kinetic energy (two types of the Zero Point Energy)

- The validity of number of postulates adopted in contemporary physics is preserved (the constant light velocity, the equivalence principle, the relativistic phenomena and so on)
- The validity of the basic laws adopted in contemporary physics is preserved: Newton's gravitational law, the laws of inertia, the rules in Quantum mechanics and so on.
- The validity of the basic physical constants is preserved: μ_0, ε_0, h, q, λ_c, G, m_e, m_p
- The difference between the parameters of both types of intrinsic matter substances leads to a complex energy exchange between the formations in the higher level of matter organization

C. Laws, valid in a classical void space only

Some of the physical laws in a classical void space are different from the laws in the lattice space.

- **The most fundamental law governing the interaction between formations of intrinsic matter is the Law of Super Gravitational (SG)**
- **The inertial interactions between objects of intrinsic matter in a classical void space are different than in the lattice space of the physical vacuum**

The main difference between the Newton's law of gravitation and the SG law is in the inverse power degree of the distance. The attraction forces according to the Newton's gravitational law (known as universal) are inverse proportional to the square of the distance. The Super Gravitational forces in a pure empty space (a classical void space without quantum mechanical properties) are inverse proportional to the cube of the distance, according to Eqs. (2.1)

$$F_{SG} = G_o \frac{m_{o1} m_{o2}}{r^3} \qquad (2.1)$$

where: m_o is the Super Gravitational mass of the particle made of intrinsic matter; F_{SG} - is the Super Gravitational force in a classical void space; G_o - is the Super Gravitational constant in a classical void space;

r - is the distance in an absolute frame

It is assumed that the SG law has some relation to the theoretically known physical parameter called Planck's frequency, ω_{PL}

$$\omega_{PL} = \sqrt{\frac{2\pi c^5}{hG}} \qquad (2.0)$$

The Super Gravitational constant G_o could be different from the known gravitational constant G. The latter is measured and valid for a lattice space.

The Super Gravitational constant may also have two different values for the two substances of the matter:

G_{os} - for SG force between objects made of same intrinsic matter substance

G_{od} - for SG force between objects made of different intrinsic matter substances

The following relation between the two gravitational constants is accepted initially, but its validity becomes apparent later.

$$G_{os} > G_{od} \qquad (2.1.a)$$

The relation (2.1.a) becomes apparent in the property analysis of the two types of prisms in Chapter 6 and the discussion about the lower level structures involved in the prisms in Chapter 12.

In the SG field analysis in Chapter 12, one additional important feature of this field is envisioned. When formations of material structures are involved in a mixed lattice of both intrinsic matter substances, the SG parameter G_{OD} may change its sign (+ or -) as a result of a phase change of the oscillating SG modes. This is very important for existing of mixed type lattices, which will be presented later in this chapter.

The application of the law of inertia for the objects made of intrinsic matter like prisms is different from the objects in the macroworld. The objects in the macroworld are complex formation of prisms, so they could be considered as ordered structures. When the ordered structures are immersed in a lattice space (the space in which we live and observe), the gravitation and the inertia formulated by the Newton's laws are valid. In the lattice space the inertia is a kind of interaction between the elementary particle and the lattice structure. The empty space does not contain any lattice and the interaction between sub-elementary particles is defined only by the SG law. Then the inertial interactions are different but they can be easily defined.

Additionally, the gravitational and inertial properties of material objects are different for:

- formations from different levels of matter organizations

- formations of same level of matter organization but put in different space environments: (a classical void (empty) space or a lattice space).

In order to distinguish the above mentioned differences, the following terminology is introduced:

- **Super Gravitational mass** - involved in the law of Super Gravitation given by Eq. (2.1)

- **apparent gravitational mass (or Newtonian mass)** - the mass of ordered structures in a lattice space (the mass we are familiar with)

- **intrinsic inertial mass** - defines the inertial property of the intrinsic matter in a classical void space

- **apparent inertial mass (or inertial mass we are familiar with)** - for the ordered structures in lattice space

In order to give some preliminary vision about the mentioned differences, Fig. 2.2 illustrates two objects from one and a same level of matter organization but in two cases: a. - placed in a classical void space and b. - placed in a lattice space.

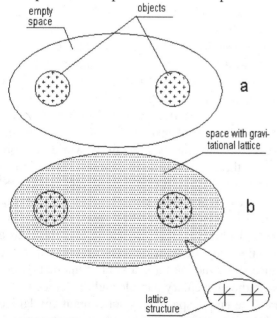

Fig. 2.2, Objects of rarefied matter in a classical void space (case a) and immersed in a lattice space (case b)

In case **a.** the two objects of rarefied matter are placed in a classical void space. In case b. the same objects are placed in a lattice space (immersed in a gravitational lattice). In the first case we have only Super Gravitational interaction between two intrinsic masses. In the second case the gravitational lattice will change significantly the Super Gravitational attraction between the immersed two objects. The inertial interactions will be also different in the two cases. The inertia in the case **b.** will be much larger, than in the case **a.** due to the increased interaction between the rarefied system from one side and the lattice from the other. The two objects may have own lattice configurations. So they may be regarded as body systems with own properties. The two body systems may affect the proximity range of the surrounding lattice, which can be regarded as a local field. In such case, the interaction between the body system with its local field and the global system **may involve folding and unfolding of the lattice nodes** (these unique features of the lattice node will be discussed later). The folded lattice nodes are able to pass through the normal lattice space, so their interactions with the rarefied objects may define the inertial properties of that object in a lattice space. The inertial properties of the particles and macrobodies in the lattice space are analysed in Chapter 10 of BSM.

2.3 Guessed property of the intrinsic matter

The guessed property of the two intrinsic matter substances are envisioned by the analysis of selected geometrical structures in a lattice space. They lead to logical explanation of fundamental physical processes.

Apriory adopted properties of the intrinsic matter:

- **Two substances of intrinsic matter exist as most fundamental indivisible super-small and dens material particles. They are not mixable due to their different parameters, such as specific intrinsic density, different intrinsic time constant and a different SG constants**

- **Both fundamental particles are able to build stable formations 3D geometrical formations** held by the SG forces (low level objects of intrinsic matter).

- **The SG attraction law for objects of intrinsic matter in a classical void space is inverse proportional to the cube of the distance between them. The SG constant for objects of different substances is different and could even reverse the sign of the SG forces**
- **Structural formations from fundamental particles, which appear stable in both, a void and a lattice space, can be denoted as basic particles. In Chapter 12 a scenario is presented, how stable basic particles of both substances can be formed by enormous pressure in a self-governed process, which may take place in some not observable phase of the galactic formation.**
- **The two basic particles and their low level structures have identical geometry, which could not be changed even at the high temperature processes in the stars.**
- **The two basic particles have external shape of hexagonal prisms. They are often referred as twisted prisms, which reflects some features of their internal structure. Prisms from the two different substances have an opposite twisting in their internal structure: left and right handed. This feature is related to the cirality, which is known as an important property of the elementary particles and their interactions.**

During the course of the BSM theory, an evidence will be provided, that the length ratio, between the two prisms is 2/3. This is the ratio between the left and right handed prisms (if the assignment of the handedness is correct).

$$L_R/L_L = 3/2 \qquad (2.2)$$

The same ratio between their circumscribed radii is accepted. The further presented concept about the structure of the elementary particles requires that the length of one of the prisms must be larger than the other. Such consideration is necessary for the existence of the gravitational lattice structures and the explanation of the crystallization process of the elementary particles. The corre-

sponding volume ratio between both types of prisms is:

$$V_R/V_L = 27/8 \qquad (2.2.a)$$

From these considerations, an initial guess about the specific weight ratio of the two substances could be obtained.

The intrinsic forces between a pair of left handed and pair of right handed prisms in a classical void space are given respectively by Eqs. (2.3) and (2.4).

$$F_L = G_{os} \frac{m_L^2}{r^3} = G_{os} \frac{V_L^2 \rho_L^2}{r^3} \qquad (2.3)$$

$$F_R = G_{os} \frac{m_R^2}{r^3} = G_{os} \frac{V_R^2 \rho_R^2}{r^3} \qquad (2.4)$$

where: m_L and m_R are the SG masses of the left and right handed prisms; v_L and v_R are their volumes, ρ_1 and ρ_2 are the specific SG weights of the two intrinsic substances.

Making a ratio between both forces and having in mind Eq. (2.2) and dimensional ratio $R_L/R_R = h_L/h_R$, one obtains the ratio between the intrinsic matter densities of the two types of prisms.

$$\frac{\rho_L}{\rho_R} \geq \frac{8}{27} ; \qquad (2.5)$$

The SG force between two prisms of different matter substances is:

$$F_{R\_L} = G_{od} \frac{V_R \rho_R (V_L \rho_L)}{r^3} \qquad (2.5.a)$$

$$F_L/F_R = 1/4 . \qquad (2.5.b)$$

This difference allows selectiveness in the attractions between prisms. **Prisms of same substance (and handedness) are attracted stronger.** This is a very feature permitting: (a) a self creation of lattice space from free prisms, (b) important condition in the phase of particle crystallization. (Additional parameters and differences are discussed in Chapter 12).

In Eq. (2.5.b) the distance parameter r is eliminated. But in the further analysis we will need this parameter. We must keep in mind also that the two types of prisms have a length ratio of 2:3. According to the definition of **absolute frame** in §2.2.1 we need a reference length, so we must select one of the prism as an absolute etalon length.

In §2.6 the underlying structure of the physical vacuum is presented. We live and observe phenomena only in a lattice space, using units valid for this space. According to contemporary physics, the unit of length is related to the light velocity. From a point of view of BSM, however, this unit it is not a primary but a secondary etalon for length, despite the fact that it influences our physical measurements. In order to analyse and determine the vacuum structure parameters, we must use a more fundamental etalon for length, which must be valid in both - a pure void and a lattice space. The **common length unit** in a prism's level is discussed in Chapter 6. It gives a bridge to the length unit used in Modern Physics today. Following the adopted concept that the same intrinsic matter substances are attracted much stronger means that for a common length unit the following relation is valid:

$$G_{os} > G_{od} \tag{2.6}$$

The above relation may obtain a logical sense if accepting that SG force between two objects of intrinsic matter are related to some **super high intrinsic frequencies** of the low level structures contained in the prisms. These frequencies could come from some intrinsic features of the two intrinsic matter substances related also to their specific intrinsic density. The two different substances should have different intrinsic frequencies. In this case the SG attraction between the same substances could be regarded as an interaction based on one and a same intrinsic frequency. In the case of different substances, the SG interactions are characterised with different intrinsic frequencies. In Chapter 12 some insight about the possible explanation of SG forces is discussed. A model of SG interaction is presented (related to a resonance type of energy interactions at superhigh intrinsic frequency).

The assumption of super high intrinsic frequency automatically leads to existence of **intrinsic time constant**. This constant is intrinsically small as we will find in Chapter 12, but it is a necessary factor for explanation of the inertial interaction processes between objects of intrinsic matter in a classical void space.

The existence of super high intrinsic frequency and intrinsic time constant of the intrinsic matter will become more apparent through the course of BSM.

The SG (Super Gravitational) masses obtain mathematical meaning according to SG law in a classical void space in a similar way as the masses in the Newton's gravitational low (valid only for a lattice space). They could be considered as attribute of SG matter. The ratio between the intrinsic masses of the prisms will be derived in Chapter 6 and additionally verified in later chapters.

For a better visual perception about the possible interactions and geometrical formations from both prisms, they are shown as externally twisted (while this is an internal structural property). From the energetic point of view of the involved interactions, the model of twisted prisms contain some features of the real prisms. The application of the twisted prism as a simplified model of the real prisms provides a possibility to denote one parameter based on a simple geometrical feature, which is common for both types of prisms. This is the ratio between the volumes of the twisted peripheral part and the internal cylindrical part of the hexagonal prism. For prisms with above mentioned length to diameter ratio, this parameter is 0.09239. This ratio appears close to some intrinsic ratio of the real prisms (discussed in Chapter 12). In such aspect, the following abbreviations are introduced.

CP - the **central part** of the prism: the cylindrical part inscribed inside of the hexagonal section of the prism plus two semispherical ends

TP - the peripheral **twisted part** of the prism

2.4 Inertia of objects made of intrinsic matter in a classical void space.

Our every day perceptions does not give us enough indications that we may live in a world of structured space. We accept the inertia as something natural without a need for deeper physical understanding. Let us consider a simple example of pair flywheels possessing the same external dimensions, but the first one is a solid body, while the second one is hollow. The solid one exhibits a larger inertia. When calculating the moment of inertia of a flywheel we reference velocity of the every mass point of the wheel to a fixed not rotational frame. Despite the fact that only the points lying on the surface are in touch with the external rest space, all the mass points in the solid flywheel contribute to the moment of inertia. Logically thinking it seams that these mass points move through some media.

According to BSM concept, we are immersed in fine structured space with a node distance much smaller than the physical dimensions of the elementary particles. Our experimental environments could never provide conditions for a classical void space. Therefore, our natural perceptions are based on observations provided and valid only for a structured space environment, which is currently known as a physical vacuum. Consequently, many adopted postulates and laws can not be automatically transferred for a classical void space conditions. For this reason we must define a more general criterion for inertia to be valid for low level structures in a classical void space. Later this criterion will be applied for the elements from which the structured space is built. The inertial criterion for material object in a classical void space could be defined by introduction of a new term called **inertial factor.**

The inertial factor between simple objects of intrinsic matter, involved in a repeatable motion, is equal to the ratio between their interaction energies and the mean gravitational energy averaged per one cycle.

$$I_F = \frac{E_I}{E_g} \qquad (2.6.a)$$

where:

I_F - is **the inertial factor**, and E_I is the interaction energy normalized to one cycle of the repeatable motion.

$E_g = F_{SG}/d$ is the average SG energy (or potential), calculated for the distance d between their centres of repeatable motion and normalized also per one cycle.

The interaction energy can be expressed as a work. The gravitational energy also could be expressed as a work for moving of one particle from its position, corresponding to a distance d, to infinity.

The inertial factor in a classical void space does not appear so simple as the inertia of a Newton's mass in a structured space. Firstly, it could not be defined for a single object. Secondly, even for simple objects the inertial factor depends on the shape of the objects, their mutual space positions, the type of motion and the relative velocity between them.

The analysis of possible simple cases of interacting intrinsic matter objects in a classical void space provides the following results for the inertial factor:

(a) a single object with any shape: the inertial factor could not be defined. There is not intrinsic interaction.

(b) two spheres: When rotating around their centres or around a common center, while keeping a constant distance between them, their inertial factor is zero. If the distance between them changes, the inertial factor obtains some finite value.

(c) two cylinders at fixed distances. (1) If they perform a rotation about a common axis which preserved symmetry, their inertial factor is zero. (2) If they perform rotation or motion as a result of which their common geometrical position changes, their inertial factor will have some finite value.

(d) two axially aligned twisted prisms, with fixed axial distance but with oscillating motion along their axes: The internal part of the prisms, regarded as inscribed cylinder (CP), will exhibit zero inertial factor for a case of rotation but some finite inertial factor for a case of axial vibration. The peripheral twisted part (TP) will exhibit a finite value (non zero) inertial factors for both : a rotational motion and an axial motion. The total inertial factor I_F could be expressed as a sum of two factors, one for the axial vibrations E_{vib} and a second one for the rotational motion E_{rot}.

$$I_F = \frac{E_{vib}}{E_g} + \frac{E_{rot}}{E_g} \qquad (2.7)$$

(e) a set of left and right handed twisted prisms arranged alternatively, but with a constrain that the distance between the centers of vibration of the same type of prisms is kept constant.

The prisms will exhibit axial and rotational (spin) interactions. The axial interaction is contributed mainly by the inscribed cylindrical part (CP). For case **e.** the following conclusion can be made:

Conclusion 2.4.1: The momentum interaction from axial vibrations, contributed by the inscribed cylindrical part (CP), has one and a same sign for the right and left handed prisms. The net algebraic sum of interaction energy for axial vibrations for a volume containing a large

number of ordered prism pairs could become equal to zero.

The momentum interaction from the axial and rotational motion, due to the twisted peripheral part (TP) of the prisms has a different sign for the right and left handed prisms. The net algebraic sum from the (TP) interactions will have a finite value different than zero.

The energy conservation principle requires the motion invoked by intrinsic interactions to be performed for a finite time. This is valid also for two intrinsic matter bodies left under SG forces. Evidently, some finite time constant of the intrinsic matter should exist. We can call it an **intrinsic time constant**. It will determine how fast the interactions between simple shape objects of intrinsic matter in a classical void space will be performed if they are suddenly left in a condition of free motion under SG forces between them. This parameter could not be zero, because in such case the interacting objects must get infinite velocities that is in a conflict with the energy conservation principle. The intrinsic time constant is related to the intrinsic super high frequency of the intrinsic matter, discussed in the previous paragraph. In the course of BSM, it will be shown that the structured space also possesses a characteristic time constant, but the intrinsic time constant of the intrinsic matter is much smaller.

Finally, we may define some of the basic properties of the accepted basic particle and their representative model: the twisted prisms.

• **Any prism looks one and a same, when viewing from both ends, although, both type prisms are always distinguishable by the direction of their twisting.**

• **Viewed from both ends, the twisted prism looks one and a same, so if rotating it at 180 deg around an axis perpendicular to the prism axis, it obtains the same indistinctive position (the same twisting direction). Then mathematically, the spatial orientation of the prisms could be described as opposite but not vanishing vectors, possessing a common point of origin.**

• **The SG field of the twisted prism is anisotropic and characterized by a vector possessing two components: axial and rotational**

The last of the above mentioned features will become apparent in the later analysis and especially in the analysis of the lowest level structures in Chapter 12.

2.5 Dynamical interactions between spatially ordered prisms

Let us analyse some simple interactions between aligned prisms in a classical void space. In Fig. 2.3.a the left and right-handed prisms are aligned in their common cylindrical axis. If the prisms L1 and L3 rotate synchronously with variable speed, the middle prism L2 will tend to rotate with the same speed and direction, without getting any translational momentum. From the other hand, the prisms R1 and R2 will get a translational momentum due to the peripheral SG interaction, caused by the different helicity (handedness). The distances d will be not affected, but the distance x will be affected by the peripheral SG interactions. We can call this type of interaction an **axial interaction**. The interaction effect is the same if the prisms L1 and L3 are fixed and R1, R2 are free for rotational and translational motions.

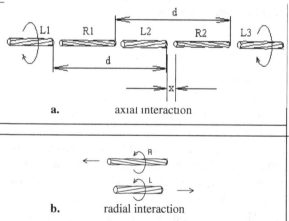

Fig. 2.3. Interaction between spatially ordered prisms in a classical void space

In Fig. 2.3.b. a pair of right and left handed prisms are parallel each other. If both prisms get angular momentums in a same direction, the interaction will cause momentums with axial components in opposite direction, so they will tend to separate. If the angular momentums are in opposite direction, they will tend to stay together. If instead of right and left handed, we have pair prisms of same handedness, the interaction will keep them together. We will call this type of interaction a **radial interaction.**

We could refer to the mentioned both effects of interaction, simply as, a **pair prism interaction.**

From the analysis of the dynamics of interactions it becomes evident that a group of moving and spinning prisms will have enhanced common inertial properties. This will lead to some preferable spatial distribution for the prisms of such group. For example: a single left handed prisms in a bundle of right handed prisms will get relative momentum with direction in order to escape the bundle. At the same time, the spinning prisms with a same handedness, in a bundle, will exhibit common inertial interactions that will keep them together.

We can summarize the following basic inertial properties of a group of spatially ordered prisms.

• **Group of spatially separated prisms of same handedeness with common axial alignment, moving and rotated in a same direction, will have common interactions tending to equalize their velocities. The intrinsic inertia of such bundle is larger than the product of the inertial interaction between pair prisms and the number of pairs in the bundle.**

• **An uniform spatial lattice structure formed of prisms will have a constant inertial factor. (this conclusion becomes apparent later)**

2.6. Gravitational lattices in a classical void space.

2.6.1. Types and general properties of the gravitational lattices

The gravitational lattice is an ordered 3 dimensional spatial structure, composed of prisms, held only by Super Gravitational forces. Ignoring firstly the boundary conditions, the basic requirements to the lattice is to be a stable. This means that the return forces acting on a unit cell should be conservative. In other words, a displaced single node should have a tendency for returning to the range of stable positions. Some types of lattices are stable if proper boundary conditions exist.

Let us consider the following three types of gravitational lattices whose existence can be possible in a classical void space: a **mono rectangular**

lattice (**RL**), a **mixed rectangular lattice** and a **cosmic lattice (CL)**. Anyone of these structures contains a unit cell that we may call a **lattice node**. It is formed of exactly defined number of prisms of a same type hold together only by SG forces. The inverse cubic law (about distance dependence of these forces) becomes apparent from the later following analysis.

Rectangular lattice composed of same type nodes is illustrated in Fig. 2.4. It could be referenced also as a monorectangular lattice.

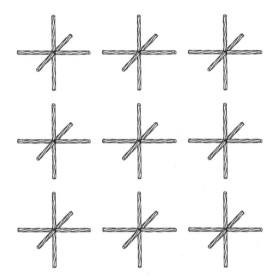

Fig. 2.4
Mono rectangular lattice
(only one layer is shown)

The mono rectangular lattice is dependable on the boundary conditions. They may cause a deviation of the lattice cell from a cubic shape. The mono rectangular lattice is able to exist inside of long cylindrical space. In such case, the unit cells of the rectangular lattice can get a shape closer to trapezoid or twisted trapezoid. Despite of this, we may still refer all type of such modification to a rectangular lattice (RL). The distortion of RL could be considered as an additional feature.

In the mixed rectangular lattice, the nodes formed of the shorter prisms are in the center of the rectangular cell, formed by the nodes of longer prisms. This type of lattice also requires boundary conditions. Fig. 2.5 shows a configuration layer of mixed rectangular lattice. Both type of nodes does not lie on one plane.

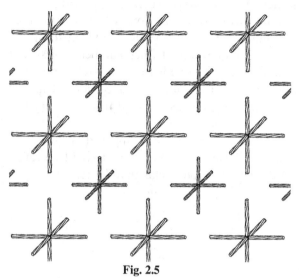

Fig. 2.5
Mixed rectangular lattice
(both type of layers are in different planes)

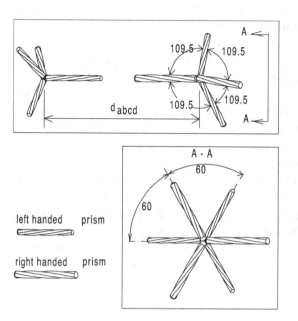

Fig. 2.6
Common position of pair nodes of CL structure

One specific feature of the rectangular lattices is that the gaps between neighbouring prisms may become zero. In case of mono RL in a cylindrical space (space defined by the cylindrical boundary conditions of a helical structure, which will be presented later) this option is possible only for the prisms with a radial alignment, but not for tangentially aligned prisms. In case of mixed RL in a spherical space (with spherical boundary conditions) the gaps between the longer size prisms are almost zero, while the gaps between the smaller size prisms are not. Consequently, the smaller size prisms have much larger freedom in this case

The least dense lattice is the **Cosmic Lattice (CL)**. It has arrangement similar to the crystal structure of the diamond, but with alternatively arranged right and left handed nodes. Each node is formed of 4 prisms of same type held by SG forces. The CL structure is more difficult to be visualize by a perspective view. Figure 2.6 illustrate only the common positions of two neighbouring left-handed and right-handed nodes.

Figure 2.7 illustrate a two dimensional projection view of thin lattice layer of CL structure. The right and left-handed prisms connected in CL nodes are shown as black and white rectangles.

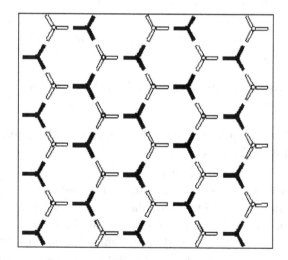

Fig. 2.7
Plane projection of Cosmic lattice layer

The hextograms in Fig. 2.7 in fact do not lie in a single plane The prisms of right and left-handed nodes lie on different tilted planes in respect to the plane of drawing. If not considering a vibrational motion of the CL nodes, the apex angles between the prisms axes are 109.5 deg but they are projected on the drawing plane as 120 deg angles.

One very important feature of the CL type of lattice is the finite gap between the prisms of the neighbouring nodes, as shown in Fig. 2.6. These feature is defined by the unique feature of the SG law, which may convert from attraction to repulsion. This feature is discussed in Chapter 12 and it is shown that it depends on the total energy of the system.

For now it is enough to say that the Cosmic Lattice is hold from a heavy material object (objects) and it may exist without external boundary conditions and at very low temperature. Its own stiffness has a finite value that is defined by the SG forces between the CL nodes. The Cosmic Lattice is spread everywhere throughout the visible Universe and penetrates even inside of the atomic particles (with exception of the contained denser lattice of the elementary particles, which will be presented later in this chapter), such as the proton and the neutron. Put in a volume of a classical void space, the Cosmic Lattice provides a structured space referred as a **CL space**.

The existence of mono rectangular lattice requires external boundary conditions. Such type of lattice exists inside the structures of the electron, positron and some subatomic particles, as pions and kaons.

Criterion for stability of the gravitational lattices

The gravitational lattice is not a momentary formation but a stable one. Based on the discussed so far conditions for stability, a **stability rule** can be formulated for the presented three types of gravitational lattices.

- **The return forces for any axis passing through the neighbouring node prisms should be conservative.**

The stable existence of the gravitational lattice and especially the CL type requires additional self repairing mechanism from disturbances caused by some external conditions. For CL type of lattice the self repairing mechanism could be of same origin as the lattice creation mechanism. So it could be the following:

Due to the difference of SG constants of the two substances, the different volumes of both type of prisms, and the SG anisotropy (including twisting) the prisms of same type are selectively attracted. This leads to formation of two types of nodes: right handed and left handed. The cosmic lattice (CL) nodes are formed of four prisms, whose axes in a geometrical equilibrium position are at 109.5 deg, each other. The rectangular lattice nodes are formed of six prisms, whose axes in a geometrical equilibrium position are at 90 deg each other (not considering a possible distortion from the boundary conditions). After first parent domain of the lattice is formed, the additional growing is facilitated by the spatially ordered SG forces, the SG anisotropy and their dynamical interaction.

The classical void space provides conditions for formation of CL type lattice only. The two rectangular lattices require a different environment condition that will be discussed later.

***Summary* about the general features** of the gravitational lattices:

(1) The cosmic lattice contains nodes of four prisms

(2) The both types of rectangular lattices contain nodes of 6 prisms

(3) Every CL node of right handed prisms have symmetrical neighbouring CL nodes of left handed prisms and vs versa.

(4) The prisms, while attached to its own node, their free ends are orientated to their neighbouring nodes.

(5) The formation of CL node and lattice building is a self sustaining process, govern by the SG energy balance between the two types of intrinsic matter from which the prisms and CL nodes are built.

(6) SG interactions between the prisms are characterised by an axial gravitational anisotropy and handedness. When using the twisted prism model, the first feature is associated with the inscribed cylindrical core, while the second one - with the twisted peripheral part. The Super Gravitation is undetectable in CL space, but it is behind the Newtonian gravitation that may exists only in CL space between ordered structures composed of prisms.

(7) The SG forces may leak through CL space in two cases: (a) close atoms or molecules (some of the Wan der Walls forces); (b) highly polished solid objects in a close proximity (Casimir forces)

(8) The rectangular gravitational lattices have much larger stiffness than the cosmic lattice.

(9) The interconnection between the CL and the RL formations is very weak.

(10) In CL space the motion of left (or right) handed prism from one CL node of the lattice invokes an opposite motion of right (or left) handed prism of neighbouring CL node due to the twisted attribute (handedness) of their SG fields.

(11) The lattice nodes could oscillate around their central points (of geometrical equilibrium), but the prism's axes of the neighbouring nodes are always aligned due to SG forces. This alignments are accurately kept even during the node resonance oscillation, due to the intrinsically small time constants of the intrinsic matter substances.

(12) The kinetic energy distributed between the oscillating nodes of CL structure provides the dynamic type of Zero Point Energy (ZPE) of the physical vacuum.

(13) Both basic types of gravitational lattices (CL and RL) have different oscillation properties, stiffness and resistence to destruction.

(14) A Massive object, containing a large quantity of prisms, causes geometrical deformation of the surrounding CL space, known as a space curvature. In this aspect, every massive object is immersed and surrounded of partially distorted CL space. The gravitational field is propagated by the cylindrical part of the twisted prisms from which the CL nodes are formed. In ensemble of few massive bodies, every one has its own local gravitational field and locally distorted CL space.

(15) The atomic matter (the matter we know) is built of elementary particles (arranged in atoms) containing fine structure which contains RL type of lattice. Assuming a constant motion in the galactic CL space the fine structure of RL type is obstacle for the CL lattice, so

the CL nodes are separated, partially folded, deviated and then returned and reconnected to the CL space. This processes is behind the inertia of the atomic matter, whose inertial factor is much larger than the inertial factor of the prisms interactions.

(16) The mass density in any macro object is not uniform. It is distributed in more or less dens zones which modulate differently the CL space in which they are immersed.

(17) Dense formations (structures) of prisms of same handedness placed in CL space affects both types of CL space nodes in a different way. The non symmetrical behaviour in this type of interaction leads to appearance of field effect. (The electric and magnetic field are discussed later in this chapter.)

(18) A Cosmic Lattice disturbance caused by the motion of dense formations of highly ordered structures built of prisms of same type appears as an energy flow, exhibiting a complex spatial and time variable pattern (the magnetic energy is discussed later in this chapter).

(19) The CL space allows propagation of diversity of waves: zero point (or order) waves, EM waves, virtual particle waves, shock type of waves (containing a longitudinal component).

(20) The Cosmic Lattice exhibits two types of energy wells related to a Zero Point Energy of the vacuum - a static type and dynamic type. Presently only the dynamic type of ZPE is recognized in Modern physics, due to the adopted concept of the physical vacuum. The static type of ZPE, envisioned by BSM, is the energy holding the integrity of the CL space. Some of its major physical parameters (derived in this chapter) are the Static and Dynamic Lattice Pressures. The Static pressure of CL space is exercised only on structures containing RL structures, because it is not penetrative even for folded CL nodes.

Explanation:
Features from 1) to 4) are geometrical properties of the suggested physical model.
Features (5): The Super Gravitational force between prisms in a classical void space is given by Eq. (2.1) discussed in §2.2.2. The validity

of the inverse cubic law responsible for this field will become apparent through the course of BSM theory.

Feature (6): The gravitation we know as a universal (Newtonian) gravitation is a far field propagation of the Super Gravitation between ordered systems in a lattice space. It is propagated by the cylindrical part of the prisms (twisted prisms model). Since in the lattice arrangements they are closer and the space between them is constant within a limited range, the Super Gravitation is propagated by them. In this case the effective range is extended. The gravitational force in such conditions appears inverse proportional to a square of the distance between the objects. This is the Newtonian gravitation. The prove of this will become apparent in the following next chapters and especially by the equation of the mass of the elementary particles, derived in Chapter 3 and used in later analysis. The propagating feature of a lattice from extended objects, for example, can be demonstrated by using ordered iron rods and permanent magnet. If we make consecutively serial connections of one, two and 3 iron rods to a magnet (even with small gaps between them), the field range is extended along the direction defined by the rods. If we make consecutively parallel connections, with the rods between a magnet and iron, the effective force becomes proportional to their number. If the rods are arranged in equal distances, the attractive force appears inverse proportional to the square of the rod distance.

In the provided example the set of rods could be spatially arranged in order to simulate the prisms arrangement in CL nodes. The guided magnetic field through the rods simulated partly the anisotropic propagation of SG forces by the prisms. In the case of CL space, however, the gaps between the prisms of the neighbouring nodes are also important.

Feature (7): SG forces between objects of intrinsic matter in a void space are proportional to inverse cubic power of the distance. In CL space, when two particles or massive objects with smooth surfaces quite close, the low of leaking SG forces is dependable on distance. In this case, the resultant force of attraction is contributed by the leaking SG forces, the forces of Newtonian gravitation and EM field forces. As a result, the attrac-

tive force, becomes inverse proportional to the distance at higher power. Figure 2.8 a. and b. illustrates how the proximity SG filed depends on a distance, for a classical void space and for CL space, respectively. While SG forces in a classical void space are proportional to inverse cubic low of the distance (case a.), in CL space an inverse power law of higher degree may take place. Practically, such effect could be detected only below some critical distance (case b.), which will depend also on the shape of the objects put in proximity.

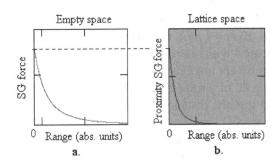

Fig. 2.8

Proximity SG forces between two objects of intrinsic matter: a. - in classical void space; b. - in lattice space

It is well known that some attractive forces between atoms and molecules exists when they are in close proximity. They are known as retarded and not retarded forces, a definition according to the London's theory (F. London, 1930), (D. Langbein, 1974). The effect of attractive forces between well polished objects at close distance is known as a Casimir effect. Casimir itself regarded the attractive forces as retarded Wan Der Wall forces. The theory of Lifshitz additionally treats the problem. The Casimir forces are experimentally confirmed, and even a change of the power law of the attractive forces at closer distance is observed. The explanation of this change as a transition between retarded and not retarded Wan Der Walls forces is not very convincing. The change of the power law is observed by J. Israelashvili and D. Tabor (1972), by experiment with two crossed mica cylinders. The transition between these objects occurs between 12 nm and 50 nm. R. Forward (1984) cites the observed effect and writes: "The data show good agreement with the $1/a^4$ Lifshitz law from 30 to 20 nm with a break in slope at 15 nm changing to a $1/a^3$ London - Van der Waals law from 10 down to

1.4 nm" (*a* is the separation distance). The $1/a^3$ law from the provided experiment becomes pure apparent below 10 nm.

According to BSM, some of the forces defined as Waals attractions between atoms or molecules are SG type of forces propagated in CL space. While at larger distance the SG forces are dimmed by the EM effects, at smaller one they become predominant. SG forces apear as Casimir forces but with inverse cubic power low. The proton, according to the BSM analysis, has a shape of Hippoped curve with dimensions: 0.667 A - length; 0.193 A width, and thickness of 8.85E-15 m. (A - angstrom unite, 10 A = 1 nm). It contains a number of helical structures with internal rectangular lattices. This is a significant amount of intrinsic matter. In the experiment of crossed mica cylinders, (having in mind that the distance between oxygen and Si atoms in SiO_2 molecule is roughly about 3 proton lengths (see the Atomic nuclear atlas)), the distance of 10 nm at which the inverse cubic law is detected corresponds to about 50 nuclear diameters. So we may expect, that the measured force is a leakage of SG force in CL space at very close distance.

The explanation of features from (12) to (20) requires detailed analysis of the dynamical properties of the gravitational lattice, which is provided later.

2.6.4 Boundary and interface layers

2.6.4.1 Boundary layers

The boundary layer is the last layer between the lattice and the void space. The boundary layer is formed by the same nodes, but some of the prisms are deviated from their usual positions.

In the case of cosmic lattice facing a void space, no any prism should be pointed out to the void space. If we look at the drawing shown in Fig. 2.7, the boundary layer nodes will be in the same positions, but the upper level nodes (white) will be rotated 180 deg to the normal of the drawing Then the free end of the prisms of this nodes will be directed toward the internal part of lattice space and away from other node prisms. The lower level nodes may go lower and its prisms may even touch the prisms of the neighbouring nodes. The node angles of black and white nodes of the layer will also be deviated from 109.5 deg. The neighbouring layers below the boundary layer also might be distorted. This will cause severe deviation from the normal oscillation properties of the nodes. For this reason we may consider that CL structure facing a void space is terminated by a **boundary zone** with a finite thickness. The oscillation property of the CL node belonging to the boundary zone are disturbed. Such zone may not reflect properly the radiated energy it gets. The lattice stiffness of the boundary zone might be much larger than the stiffness of the normal CL space. The importance of the boundary zone and its features are additionally discussed in Chapter 12 Cosmology, §12.B.5.5., where the conditions of stable existence of a CL boundary layer will become more understandable.

The mono rectangular lattice also may contain a boundary layer if the conditions for a stable lattice existence are met. The boundary layer of first order rectangular lattice is shown in Fig. 2.8.a. The left and right handed nodes and prisms are shown respectively as black and white rectangles.

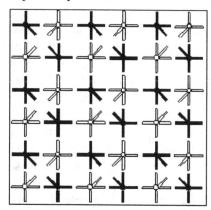

Baundary layer of first order cubic lattice

Fig. 2.8.a

Every node of the rectangular lattice contains 6 prisms. Every four nodes of the boundary layer have one angular deviated prism. The free ends of these prisms from four neighbouring nodes form one interconnection node. Two rectangular layers with slightly different node distances could be interfaced by their interconnection nodes.

2.6.4.2 Interfacing layers

The interfacing layers connect two lattice domains with different parameters: We may distinguish four types of interfacing layers:

(1) between mono rectangular lattices in a cylindrical space

(2) between CL and rectangular latices

(3) between two CL spaces formed of prisms with slightly different dimensions

The first type of interfacing layer appears in the internal lattices of the first order helical structure, which is discussed later.

The second type of interface appears in hypothetical conditions of particle crystallization discussed in chapter 12.

The third type of interface appears between CL spaces of the neighbouring galaxies (Chapter 12).

The configuration of the rectangular interface layer is similar as the boundary layer but with a difference that the deviated prisms become points of connections between the two RL structures.

2.7. Helical structures and ordered systems.

The atoms are composed of protons, neutrons and electrons. These are the basic stable elementary particles. All stable elementary particles are composed of helical structures. The helical structures itself are built by prisms of both type. Their configuration, however is not simple. They are distinguished by the helical order, by the type of internal rectangular lattices they posses and by their common spatial arrangement.

The elementary particles and their helical structures are ordered compositions of intrinsic matter, formed in a process similar to the crystallization. Such conditions exist only during the incubation process of the protogalaxy. Two basic physical laws are essentially important during this process: the low of Super Gravitation and the energy - matter balance. The process, whose possible scenario is presented in Chapter 12, is self-governing and deterministic. It involves a number of phases and mechanisms, the natural conditions for which exist only during a particular phase of matter evolution. The process of particle crystalization is self regulated and leads to final formations of stable elementary particles with a very high probability. The crystallization process and the phases of matter evolution take place in particular phases of individual galactic cycle. These phases are com-

pletely hidden for observation, since the EM radiation is blocked due to a temporal separation from the CL space of the observable Universe. They could be inferred only if the form of matter organization in the visible Universe are well understood. For this reason the analysis of the hidden evolutional processes of the matter is left for the last Chapter 12 of BSM, while the configuration of the **helical structures** with some inferred process of their crystalization are presented in the next section.

While the helical structures appear as building blocks of the elementary particles, the term **"ordered system"** is used for a more general category. The ordered system may include stable and not stable particles built of helical structures. While any static configuration of helical structures possesses a charge, the charge of composite ordered system could be dynamically compensated in order to apear neutral in a far field. In such aspect, any single atom or molecule could be considered as an ordered system.

2.7.1. Elementary helical structure

The helical structure contains a boundary helical core and internal lattice. The radial section of the helical core is comprised by 7 prisms, held by Super Gravitational (SG) forces. One may identify a single element, as a bunch 7 stacked prisms - **helical core element** (or core node element). The helical core element shown in Fig. 2.8 is built of right handed prisms. The real prisms are not externally twisted. The twisting is in their internal structure. Externally twisted prisms are only models showing some of the features of the real prisms. It is obvious that a core node of 7 prisms is the most compact option than any other option with a larger number of prisms.

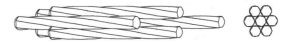

Fig. 2.8.B. Helical core element formed of 7 prisms of same type, held by SG forces. The external twisting is shown only for illustrating the prism's handedness (the twisting is embedded in the internal prism's structure)

The basic geometrical features of the core node element and the assembly of such nodes are the following:

a. The prisms of the node are stacked together along their length touching themselves by the hexagonal corner edges.

b. Each node prism has a longitudinal displacement to its adjacent prisms at 1/3 and 2/3 of the prism's length.

c. The small bumps in the peripheral part of the prism (shown in Fig. 2.1 of BSM) facilitate the correct attachment of the prisms during the process of crystalization.

d. In order to form a stable node, it is evident that common length between two attached prisms (2/3 of the prism length) should be larger than the prism diameter.

e. The neighbouring nodes are stacked together along their axis forming a node assembly without axial gaps between the prism's ends.

f. A stable node assembly can be formed only by prisms of same handedness (substance)

g. A long node assembly can be twisted in a clockwise or counter clockwise direction forming a helix. The twisting is possible due to a slight axial disalignment between prisms, kept by SG forces. The obtain formation formes the boundary of the helical structure, which has also internal lattice structure (shown later in this chapter).

All above-mentioned features are obvious from a pure geometrical considerations.

2.7.2 Types of helical structures

One may distinguish a few types of helical structures by the following attributes:

- type of prisms they are made of: left or right-handed
  - spatial positions: internal or external
  - helical order: first; second; third
  - overall shape: straight, twisted, toroidal
  - number of repeatable turns: single turn (or coil) and multiturn.

Structures of the above-mentioned types are embedded in the elementary particles: proton, neutron, electron. (positron). They are built as a result of unique crystallisation process.

Figure 2.9 shows a straight helical structures, denoted as First Order Helical Structure (FOHS).

The radius of right and left handed FOHS is determined by the size of the prisms and the condition expressed by Eq. (2.8), discussed later in §2.8.1. A left-handed FOHS structure can be freely inserted inside the a right-handed FOHS as shown in Fig. 2.9.c. The helical cores of both structures, however, do not touch each other, because every FOHS has own internal rectangular lattice. Inside of the left-handed FOHS (smaller envelope diameter) only a straight core of right handed prisms could be inserted.

If the FOHSs shown in Fig. 2.9 are long enough, they will be bent by the SG forces. While the prisms are rigid, the holding SG forces allow obtaining a small bending curvature. The bending of enough long FOHS converts it to a SOHS. The radius and the step of the SOHS is more dependent on the external lattice parameters than the radius and step of the FOHS. It is equivalent to consider that the stiffness of the higher order helical structure is lower. The stiffness, however, is not defined by the helical core stiffness, but by the internal modified rectangular lattice, described elsewhere. The structure core provides also a boundary conditions for the internal rectangular lattice.

Figure 2.10.a shows a SOHS of left-handed prisms, containing inside a core of a first order helical structure from right-handed prisms. The stiffness of this structure is defined by the stiffness of its internal rectangular lattice.

Figure 2.10.b shows a compound structure formed by two SOHSs of different type, one inside the other with a central core of the internal SOHS. The external one is right-handed, while the internal one is left-handed and the central core is right-handed. In this case, the right-handed stiffness dominate and it is kept stable by the internal lattices (shown elsewhere in BSM).

Figure 2.13.a shows a compound single coil helical structure made of one turn of right-handed second order helical structure, in which one single turn of right-handed structure is inserted.

FOHS - stands for: First Order helical Structure
SOHS - stands for: Second Order Helical Structure

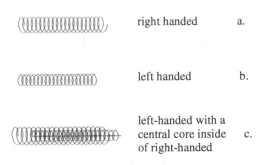

right handed　　　　a.

left handed　　　　b.

left-handed with a
central core inside　　c.
of right-handed

Fig. 2.9 First order helical structures

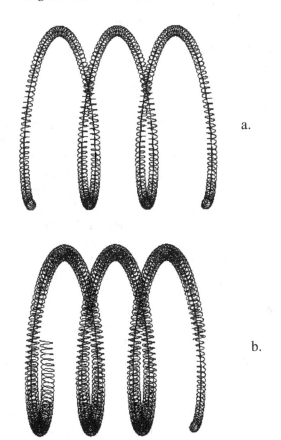

a.

b.

Fig. 2.10 Second order helical structures:
a. - single structure with a central core;
b. - compound structure

In some conditions the single turn structure may loose its internal structure as in the cases **b.** and **c.** The core stiffness of a single turn structure becomes a dominant factor, and the handedness of the structure, shown in Fig. 2.13.c, is defined by the core handedness. The handedness of the single turn structures in case **b.** is defined by the internal rectangular lattice that has a memory from the case **a.**

Consequently, the structures, shown in case **a, b, c** have one and a same second order handedness. This means, that the structures **b.** and **c.** are able to recombine again into a structure **a.**

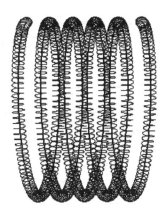

Fig. 2.12 Second order helical structure with a
central core and smaller second order helical step

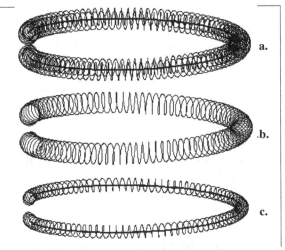

a.

.b.

c.

Fig. 2.13. Single turns of second order helical structure

The structures shown in Fig. 2.13. could be obtained from multiturn second order structure in a process of disintegration of the helical structure until only a single turn is left, which is trimmed to exact structural parameters of the electron or positron. In the case when the whole structure is destroyed, the charge only is left, which imitate a real electron or positron. This happens in some particle collision experiments discussed in Chapter 6.

Figure 2.14 shows the possible overall shapes of high order toroidal structure. A second order structure with a large number of turns will bend and may form a third order helical structure. One turn of the third order helical structure with connected ends forms a toroidal structure, illustrated in Fig. 2.14.c.

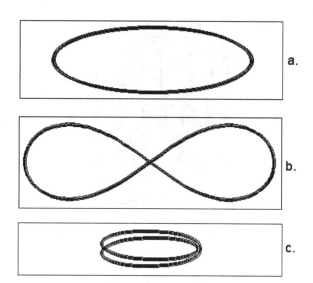

Fig. 2.14 Three conformations of one and a same higher order helical structure: a. - torus; b. - twisted torus (proton); c. folded torus (neutron)

Based on simple mechanical properties we can formulate the following static features of the helical structures.

- **Helical structures of first order without internal core posses the handedness of the prisms they are built of.**
- **The stiffness of the helical structure is defined by the stiffness of its internal rectangular lattice.**
- **The lower order helical structures have greater helical stiffness than the higher order structures.**
- **The handedness of a compound higher order helical structure containing structures of both type prisms is determined by the handedness of the lowest order structure (because it possesses a greater helical stiffness).**
- **Structures composed of a few helical substructures one inside another are built by prisms whose handedness follows the rule: right, left, right (in a direction of radius increase originated at the central core). The central core is always from right-hand**

prisms (if the right-hand prism is the larger one).
- **The smaller radius and step of helical structures of different types have different dependence on the lattice parameters.**
- **Different types of external lattices should influence the helical step and radius of the helical structure to a different degree, due to the different coupling between the external and internal lattice parameters. For this reason, a long second order helical structure may have a different step and radius in a RL and CL type of lattice space**

2.7.3.2 Symbolic notation of the helical structures.

Simple symbolic notations are introduced for identification of the different formations of helical structures, as illustrated above. **Later we will see that the prism handedness is related to the charge polarity.** This does not mean, however, that the handedness and the charge is one and a same thing. The charge is a kind of modulation of the CL space parameters from the helical structure. It is discussed later in this chapter. In order to annotate the handedness, however, we need to associate one type of prism with the positive and the other with the negative charge. In this approach the probability of correct assignment is 50% (if additional considerations are not a priory presented). Therefore, we may provide two systems of symbolic notation for formations of helical structures: **by handedness** and **by charge**. The first one is more descriptive, but the second one is more convenient if the assignment of the handedness is not correct (in fact some considerations about the assignment exists, but they are discussed later in BSM).

Let us make a selection:

right handed prisms - related to the negative charge

left handed prism - related to the positive charge

Note: **The selected notations by charge do not mean that the prisms itself posses a charge. However, they become associated to the selected charge when the helical structure is put in a CL space environment.**

Both systems of notation (by handedness, and by charge notation) are presented in Fig. 2.15. The charge notation will be not affected from a possible wrong assignment of the handedness, while the handedness notation will be affected.

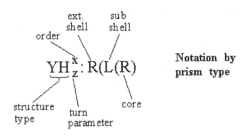

Notation by prism type

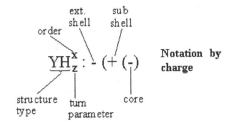

Notation by charge

Fig. 2.15. Notation systems for complex formations of helical structures

Notation parameters:

H - helical structure of twisted prisms
x - order of the structure (0 to 3):
Y - type of the structure:
    S - straight type
    T - torus type (Fig. 2.11)
        FT - folded torus (Fig. 2.14.a)
        TT - twisted torus Fig. 2.14.b)
        CT - curled torus
z - turn parameter: s - single coil
                      m - multiturn
                      # - number of turns
R - right handed prisms
L - left handed prisms

Table 2.1 provides examples of complex formations of helical structures by using the two types of notations.

Notations of formations of helical structures   Table 2.1

Notation by prism type	Notation by charge	Description
H_m^0:R	H_m^0:-	Zero order structure of right prisms (straight core)
SH_m^1:R( )	H_m^1:-( )	First order structure without core (Fig. 2.13.a)
SH_m^1:R(L(R)	H_m^1:-(+(-)	First order structure with core (Fig. 2.9.c)
SH_m^2:L(R)	H_m^{2-}:+(-)	Second order structure with external shell of L prisms (positive charge) with core of R prisms (Fig. 2.10.a)
SH_m^2:R(L(R)	H_m^{2-}:-(+(-)	Second order structure with ext. shell of R prisms (negative charge) with core of R prisms (Fig. 2.10.b)
SH_1^2:R(L(R)	SH_1^2:-(+(-)	One turn of second order structure (Fig. 2.13.a)
SH_m^3:L(R)	H_m^3:+(-)	Third order structure with positive external shell and negative core
TH_1^3:L(R)	TH_1^3:+(-)	Torus structure made of one turn of third order helix with core
TTH_1^3:L(R)	TTH_1^3:+(-)	Twisted torus from third order structure (shape as Fig. 2.14.b)
FTH_1^3:L(R)	FTH_1^3:+(-)	Folded torus from third order structure (shape as Fig. 2.14.a)

2.7.A Identification of the atomic and subatomic particles

A detailed scenario for particle crystallization is provided in Chapter 12, while many feature about such crystallization becomes apparent in the analysis of physical phenomena in the previous chapters.

2.7.A.2 Primary particles. Identification of atomic and subatomic particles.

Ones the galaxy is born, the crystallisation process of the atomic particles is over. Two major particles are product of the crystalization: a torus shaped particle with a positive external shell: TH_1^3:+(-), named by BSM as a protoneutron and a negative particle H_1^1:-(+(-) that is the electron system (known as an electron).

The primary atomic and subatomic particles in the end of crystallization process with their structures are given in Table 2.1.A

Table 2.1.A

Name	Notation	External shell	Internal structures
electron system	e^-	H_1^1:-(+(-)	e^-
positron	e^+	H_1^1:+(-)	H_1^0:(-)
degenerated electron		H_1^1:-( )	missing
pion (+)	π^+	CH_m^2:+(-)	
pion(-)	π^-	CH_m^2:-(+(-)	
Kaon	K_L^-	SH_m^1:-(+(-)	
protoneutron		TH_1^3:+(-)	# pair pions, kaon
proton	p	TTH_1^3:+(-)	# pair pions, kaon
neutron	n	FTH_1^3:+(-)	# pair pions, kaon

2.7.A.2.1 Electron system (electron structure)

The electron system H_1^1:-(+(-) is composed of external shell, made of negative prisms, containing inside an internal shell made of positive prisms. The latter one is the external shell of the positron. So the positron structure is H_1^1:+(-). Both, the electron and positron shells have internal rectangular lattices (RL), which in CL space gets twisting. Any one of this lattice contains much larger number of prisms than the helical shell. Therefore, the structure of internal lattice keeps the stiffness of the structure, while the external helical core provides the boundary conditions, which are necessary for the stable existence of the internal lattice. At the same time, the internal lattice is able to modulate the external CL space due to the highly ordered radially aligned prisms, possessing anisotropic axial SG field with a left or right handed twisting.

The internal negative RL structure of the electron allows the positron to vibrate. In some extreme conditions the positron may even come out as a free positron. In such case, the electron system

converts to a degenerated electron. It still possesses a negative charge, but does not have the oscillation properties of the normal electron. The positron is more difficult to lose its core, but if it is lost, it could be regenerated by a trapping hole effect (described later) if suitable external conditions exist.

When the composition of the helical structure is not needed to be mentioned, we may use the following simple notations:

FOHS - stands for: First Order helical Structure

SOHS - stands for: Second Order Helical Structure

2.7.A2.2 Protoneutron and its internal structure. Conversion to proton or neutron.

The protoneutron has a shape of torus. It's external shell is a TH_1^3:+(-) type structure. The overall shape of the protoneutron was shown in Fig. 2.14 c. The third order torus structure is formed of second order structure (shown in the Fig. 2.12) containing a lot of number of turns. In such aspect, the external shell of the protoneutron contains a positive FOHS with an internal mono rectangular lattice and one central core of negative prisms (see Fig. 2.15.A and 2.15.B). The external shell forms the envelop of the protoneutron. Inside of this envelope, there are pions and one central kaon. The analysis, provided by BSM and the calculations in Chapter 6 show that the most probable number of pions are two: one negative and one positive. The internal pions are also closed third order helical structures but with a larger second order step. (They could be regarded as closed loops of curled FOHSs). The negative pion, however, contains inside a positive FOHS. The central kaon is a negative FOHS, but contains inside a positive FOHS. Every FOHS contains own mono rectangular lattice. The radial section of the protoneutron core with two pions and one central kaon is show in Fig. 2.15.A., while the axial core section is shown in Fig. 2.15.B.

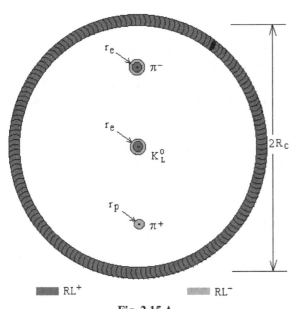

Fig. 2.15.A
Radial section of a protoneutron core

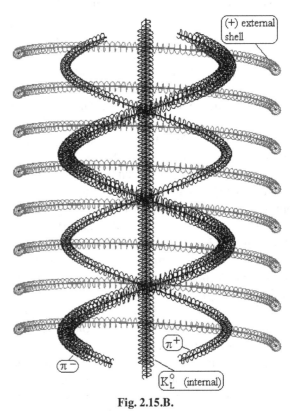

Fig. 2.15.B.
3D axial sectional view of a protoneutron structure showing the two internal pions and the central kaon (The proton and neutron, both have the same internal structure)

The internal pions, having a second order helicity, are centred around the central kaon. The dimensions in the core section of the protoneutron in Fig. 2.15.A are approximately in scale. The proton and neutron, both have the same internal structures and consequently the same core section. Two question may arise from a first gland:

- how the internal pions and the kaon are kept in their positions?

- why the FOHS diameter is so small in comparison to the core envelope?

This questions do not have a simple reply at this stage. A complete understanding comes after the reader is acquainted with the analysis in the following chapters: For now we may say that the internal pions and kaon structures are held by the proximity electrical fields.

The FOHS diameter seams small, but it is kept by the internal rectangular lattice, which is much more dense, than the CL structure.

The gaps between the FOHS turns of the external positive shell are large enough in order to allow a free penetration of CL structure inside the protoneutron (proton, neutron) envelope. So the internal space between the FOHS's is also a CL space environment.

The protoneutron shape of torus is not stable in CL space environment. Once the protoneutron is in such environment, twisting forces appear. They arise mainly from the internal pions and kaon, whose mono rectangular lattices get a helical modification in CL space. The balance of the interaction forces between the FOHS with its rectangular layer, from one side, and the CL space from the other (see Eq. (2.8) from the next paragraph §2.8) leads to appearance of twisting forces. Then the whole torus structure get twisting in one preferable direction, obtaining a shape, shown in Fig. 2.14.b. **This is the proton. The preferable twisting direction defines the proton handedness.** The proton is stable particle in CL space. The process of

2-25

conversion, however, is accompanied by the following effects:

- a small volume shrinkage of the FOHS of the pions and kaon, due to a small cell shape modification of the internal rectangular lattices

- an appearance of a far field electrical charge

In Chapter 3 we will see that the volume of the FOHS defines the Newtonian mass of the elementary particle. Therefore, the volume shrinkage leads directly to a small mass change.

The second effect is related to appearance of a detectable (far field) charge and generation of an opposite negative charge, which propagates as a quasiparticle wave - a so called Beta particle. **The Beta particle is a reaction of the CL space to the new born positive charge. For such reason, the Beta particle possesses an opposite handedness of this of the proton**.

The proton additionally could fold and get a shape shown in Fig. 2.14.a. This double folded protoneutron is a neutron. The neutron alone has a limited time of stability in CL space, unless it is not combined with a proton. When combined with the proton, it takes a symmetrical position over the proton saddle. This is the deuteron. Its shape is shown in Fig. 2.15.C)

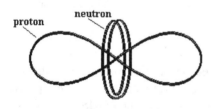

Fig. 2.15.C
Shape of Deuteron

The neutron in the deuteron is stable due to the interactions between the SG fields of the two particles. The neutron is kept centred around the proton due to the interactions between their proximity fields (in CL space environments only), whose energy is also supplied by the SG energy of the system.

The conversion of the proton to neutron is accompanied by similar effects of small mass change and far field charge disappearance with generating of Beta particle. The Beta particle according BSM is a reaction of the CL space to the birth or death of the electrical charge and always have an opposite

charge value. It is a quasiparticle wave (it will be discussed in details later). The opposite handedness of the Beta particle in respect to the proton twisting is a normal reaction of the CL space. **This effect, known as a parity violation so far, is a normal physical interaction of the helical structures of the proton (or neutron) in CL space environment, according to BSM theory.**

2.8 Modified rectangular lattice in the internal space of the helical structures

A modified version of the rectangular lattice is possible in a cylindrical space if the axial length is much larger than its radius. The internal space of a helical structure, built of same type of prisms could be considered as a curved cylindrical space.

2.8.1 General features

The process of internal lattice formation follows the process of helical core formation by crystallization. When the helical core becomes quite long, it begins to bend and continues to grow as a helix. After this helix gets a length beyond some critical value, conditions for building of internal mono rectangular lattice are created. Rectangular nodes, formed of 6 prisms of same type are attracted to the core in the internal side. At the same time, the created internal lattice provides pull-in forces, which tend to shrink the helical structure radius. The bending of the helical core, although, meets the resistance of the SG forces between the prisms of the helical core. It is evident that the radius of the structure with a completed internal lattice will be determined by the balance between all forces according to Eq. (2.8):

$$F_{int} = F_{hst} - F_{ext} \qquad (2.8)$$

where: F_{int} is the equivalent internal force, trying to shrink the helical structure diameter;

F_{hst} is the reaction force of the helical structure, caused by the bending of the helical core

F_{ext} - is the opposing external force from the SG interaction between the internal rectangular lattice and the external lattice space (comparatively small between different types of lattices)

The configuration of the internal rectangular lattice (RL) could be inferred by analysis of the possible axial and radial configuration, using pure geometrical considerations and the axial anisotropy of the prism's SG field. This is illustrated by Figure 2.16, where a. - is the axial section of the RL, b. - shows the axial section of the impenetrable volume and c. shows the radial section with the impenetrable volume. The sectional area with a light grey colour shows the volume, which is impenetrable for CL nodes, a feature explained below and confirmed by the BSM analysis in other chapters.

In the radial section (shown in a.) the RL nodes are aligned with the helical core, which serve as a boundary condition for the stable existence of the RL structure.

One conclusion is clearly evident: the boundary helical core provides simultaneously the boundary conditions for the internal RL structure and the helicity, as well. The RL structure is built of same type of prisms as the boundary helical core. The FOHS of any type, discussed in the previous paragraph, has internal rectangular lattice. It is also evident from the shown configuration that the unit cell of the internal RL structure is not exactly rectangular, but distorted.

We may identify concentric layers. The number of the radial stripes in any one of the concentric layer is a constant but different for the different layers. For the smallest radius of any layer, the gaps between tangentially aligned prisms approach zero, while for the larger radius they approach the sum of the two prisms length. The radial stripes of one layer do not posses gaps. Any neighbouring layers are connected by a thin interfacing layer, discussed in the previous paragraph. The radial thickness of any internal layer is half of the thickness of the surrounding external layer. If scanning from the boundary radius to the center, this condition is valid for all layers, until reaching another boundary condition of the cylindrical space. If a first order structure with a smaller radius is inserted, the boundary condition is defined by its external radius. In the region closer to the centre of the radial section the layer terminates due to the finite prism's length. Therefore, the RL configuration will be terminated with a central hole, as shown in Fig. 2.16.c. This is important feature of the FOHS, discussed later in this chapter.

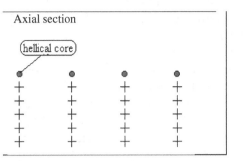

Axial section

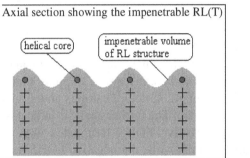

Axial section showing the impenetrable RL(T)

a.

b.

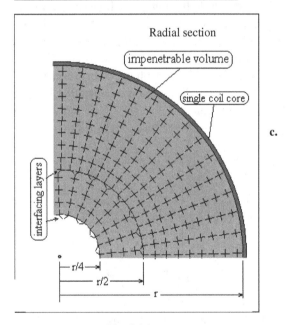

Radial section

c.

Fig. 2.16
Axial and radial sections of internal rectangular lattice

The rectangular lattice from prisms of same type possesses very unique features:
- **The SG forces from the central part of the prisms are symmetrical, so their external influence is eliminated**
- **The SG forces from the twisted part of the prisms are highly spatially ordered but their external influence is not self compensated.**

This provides a significant modulation effect on the dynamical parameters of the external CL space.

Long formation of second order helical structures also form a cylindrical space, but the boundary conditions in this case are not so uniform. As a result of this, a mixed type of rectangular lattice may be formed only in the phase of particle crystallization (a detailed scenario is presented in Chapter 12). Its formation can be terminated by a central hole.

2.8.2 Trapping hole.

When the rectangular lattice is terminated with the smallest radius, it forms a trapping hole. Such formation of RL structure exhibits an unique trapping mechanism. The further analysis indicates that only the positive FOHS terminates with a small trapping hole, while the negative FOHS terminates with a hole having a much larger diameter in which a positive FOHS is inserted. This difference is likely due to the different intrinsic parameters of both types of prisms and the force balance, according to Eq. 2.8, in a phase of crystalization. Fig. 2.17 shows a radial section of a positive FOHS with a trapping hole.

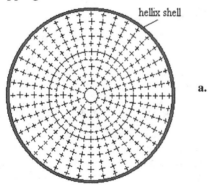

hellix shell

a.

Fig. 2.17. A radial section of a positive FOHS with a trapping hole

The trapping mechanisms in the FOHS with a same type rectangular lattice is very strong. It plays an essential role in a process of formation of smaller FOHS inside of a larger FOHS or SOHS. The highly ordered internal lattice of FOHS provides a focusing of the SG field of the twisted part of the prisms into the hole. In a such case, the strength of SG field in the hole volume is signifi-

cantly increased. Such hole obtains the ability to attract selectively prisms or node of opposite handedness. The trapped prisms or nodes are able to stack and form long core inside the trapping hole. When the formed core becomes long enough it begins to bent and get helicity. Simultaneously, the internal RL structure gets some twisting. This feature is propagated to the host FOHS, so the whole structure begins to get a helicity. Then at proper external condition, the FOHS creates a SOHS. If the turns of SOHS are closer enough, it creates a new cylindrical space in which a new environment for RL might be created.

We may even expect that the obtained new cylindrical space may create conditions for a mixed rectangular lattice. Such formation should also ends with a trapping hole in the central zone, but its properties are distinguished from the trapping holes of the same type RL formation. However it may play a termination role in the process of the particle crystalization. In the analysis provided in Chapter 12 it becomes evident that mixed lattice may exist only in a phase of crystalization (but not in a later phase when the crystalized helical structures are immersed in CL space

The trapping hole mechanism is further discussed in this chapter and in Chapter 12.

2.8.2.A. Stiffness of internal RL structure and its influence on the external structure radius

The radial section of RL structure may contain number of sublayers of same type prisms but with a different thickness. The radial thickness of any internal sublayer is a half of the thickness of the external one. Therefore, the number of sublayers depends on the ratio between prism length and the external radius. Smaller ratio means a large number of sublayers. The ratio used for the drawing in Fig. 12.16.c is quite larger than the real one.

Let us analyse the stiffness in the radial section of RL formed of prisms of same type. In order to simplify the analysis we may assume that all RL nodes from the radial section lie in one plane. Then for such section, we may define two types of linear stiffness:

- a radial one
- a tangential one

The radial stiffness is defined for any radial direction passing through the centre of the section.

The tangential stiffness is defined for any tangent at fixed radius in the radial section.

Geometrical considerations leads to the assumption that the radially aligned prisms touches each other, so the radial stiffness should be not dependent on the layer radius. The tangential stiffness however is defined by the node distance in the tangential direction. Therefore, its density and stiffness will vary between min and max value for each layer. This is illustrated in Fig. 12.17.A.

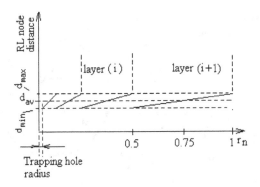

Fig. 2.17.A

Tangential node distance as a function of the normalized layer radius

The tangential node distance in the radial section of any FOHS varies between its maximal and minimal value, respectively d_{max} and d_{min}. The boundary node distances, expressed by the length L_R, of the longer (right handed) prism, are respectively:

$$d_{min} = \frac{5}{3}L_R \qquad d_{max} = 2d_{min} = \frac{10}{3}L_R \qquad d_{av} = \frac{5}{2}L_R \quad (2.9)$$

According to the SG law the tangential stiffness is inverse proportional to the cube of the node distance, so it has non linear dependence of the radius. Consequently, the tangential stiffness of the most external layer is strongly dependent on the external radius, which is the radius of the helical structure. Then the condition (2.8) could become fulfilled at proper radius of the helical structure.

- **The steeper radial dependence of the tangential stiffness in the most external layer in combination with Eq. (2.8) provides boundary value of the external radius of the new formed helical structure, grown in conditions of external rectangular lattice.**

Let us assume that a new structure of left handed prism is formed inside of the structure of right handed prisms. According to above condition, if the radius of the internal structure appears to satisfy the condition (2.8) in the range of 0.5 to 1 of the external structure radius, then it should be between 0.5 and 0.75 (see Fig. 2.17.A). From the BSM analysis, such value for the case of electron positron system, is found to be 2/3.

After the new structure with its internal lattice is completed, the obtained radius is kept stable by its internal rectangular lattice. The radial alignment of the stripes between different layers of same type of RL is provided by the interfacing nodes of the intermediate boundary layers. In such aspect the interfacing nodes are supporting elements for keeping the integrity between the layers with different radii. This kind of interconnection is possible only between layers of same type (left or right handed).

2.8.3 Helical structures in different space environments

The first order helical structures (FOHS) may apear in two forms:
- open structure: both ends are free
- closed structure: both ends are connected

In Chapter 12 it will be shown that many helical structures are built by crystalization process in an environment of external cylindrical space, where RL type of space initially exists. If an open structure appears in an external CL space, its FOHS, should get modification, according to Eq. (2.8), because the external force F_{ext} is changed. The modification involves a slight twisting of the FOHS, which involves a slight distortion of the RL cell and decrease of the small radius of the FOHS. Figure 2.18 shows a part of a radial section of FOHS in two space environments: **a.** in RL space and **b**. in CL space. Figure 2.19 shows that axial section of the FOHS in two different space environment and the distortion of the single RL cell, denoted as an unite cell. As a result of the FOHS twisting, the radial stripes are not any more normal to the helical core. We may call the modified RL as a **rectangular lattice - twisted and denote it as a RL(T)**.

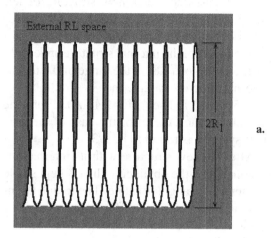

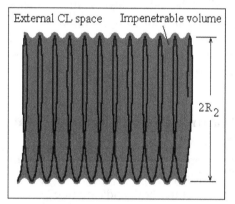

Fig. 2.18
Part of FOHS in different external space environments

In Fig. 2.19.c the SG forces in a distorted cell are shown. The difference between F_1 and F_2 forces is responsible for the additional twisting. They are balanced with the core force, opposing the bending of the helical core beyond some point. Since the radially aligned RL nodes are without gaps, they will provide a strong SG field in the external CL space with a filed lines following their twisted shape. Consequently:

The twisted rectangular lattice [RL(T)] of FOHS in CL space, provides enhanced external SG field, with a handedness defined by the type of the prisms, from which the helical structure is built.

The process of twisting of the FOHS is characterized by a slight change of the 3D geometry of the RL nodes. This means a slight squeezing of the

internal RL structure, leading to twisting and a slight decrease of the small radius of the FOHS. Consequently, once the external environment conditions are changed from RL to CL type of space, **the structure obtains a new but stable helical radius R_2** as shown in Fig. 2.18. and 2.19. The radial integrity in this modification is preserved.

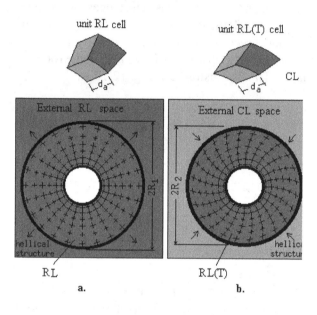

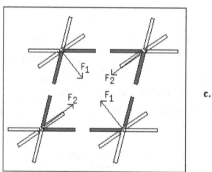

Fig. 2.19
Effect of helical structure radius dependence from the external lattice environment

Let us focus now on one important feature of the FOHS in external CL space environment: the possession of impenetrable volume for the CL space. This volume is marked by a dark grey colour in the FOHS sections in Fig. 2.18 and 2.19. This important feature becomes more understandable after the analysis of the static and dynamic behaviour of the CL nodes, provided later in this chapter.

It becomes apparent in the analysis in Chapter 3, where the equation for the Newtonian mass of the elementary particles is derived and is also confirmed by a number of other analytical results of BSM. At this moment only the following explanation will be given:

The SG field of the prisms involved in the RL(T) structure possesses vibrational modes, the frequency of which is higher than the proper resonance frequency of the CL node (discussed later in this chapter). For any FOHS, the quantity of the prisms embedded in the internal RL structure is much larger than the prisms embedded in the boundary helical core. This means that the twisting of the FOHS assures a strong SG mode rotating along the twisted RL(T) structure. This mode is an obstacle for the CL nodes (even separated and partly twisted) to penetrate the inside volume of the FOHS. This is very important feature of the twisted FOHS.

Let us analyse a FOHS with an internal lattice terminated by a trapping hole. Analysing the radial dependence of the tangential node displacement as a result of the twisting, we see, that it is much larger for the external negative layers, than for the internal positive one. For the layer near to the trapping hole, the tangential change is insignificant. The unit cell in this region could be considered not distorted. Then the axial length is also unchanged. This means, that the twisting will cause additional wounding of the FOHS, but its total length will be the same. Consequently, we arrive to the following important conclusion:

The twisting modification of FOHS in external CL space environment causes a radius and helical step change of this FOHS, while the overall length is unchanged. As a result of this, the volume of the open FOHS is slightly reduced.

The volume change is important feature, because it is directly related to the apparent mass change (see mass equation derivation in Chapter 3).

The described above effect of radius and volume change is possible only for open FOHS. If the structure is closed (both ends connected), the number of turns is fixed. Consequently, such structure could not change its helical parameters and volume in CL space. It will get only internal ten-sion. If, however, the structure is cut, it undergoes this volume change, right away. This is just what happens, in the pion to muon decay, after braking the proton (see Chapter 6). A closed loop structure, however, may also undergo a relatively small degree of twisting (or additional twisting) if its overall shape is folded or twisted (This is the case of neutron - proton or proton - neutron conversion discussed in Chapter 6).

Every higher order helical structure is built of lower order structures. Then, the following important conclusion can be made:

- **If a closed helical structure get broken in CL space, it undergoes a volume, and consequently a mass change (according to mass equation derived later in this chapter).**

The direction of RL twisting is in agreement with the handedness of the helical structure, which is the same as the handedness of FOHS core.

If one helical structure with own RL(T) is inside of opposite handedness structure with own RL(T), every one of both lattices is stronger connected to the helical structure that defines its upper boundary radius. The both lattices, in this case, cannot have interface connection between themselves, due to their opposite handedness. **Then the internal structure may oscillate in the lattice hole of the external one. Obviously, such system will have a resonance frequency.** The lattice of the external structure in this case has the following important features:

- it serves as an ideal not frictional bearing for the axial motion of the internal structure

- the twisted SG field of the internal RL(T) structure is not able to propagate through the RL(T) of external structure due to their different handedness

Consequently:

- **A system of two helical structures with internal twisted rectangular lattices is able to oscillate with its resonance frequency and accumulate a kinetic energy**

- **When one FOHS is inside another FOHS, the twisted SG filed of the internal RL(T) structure is shielded by the external one, and could not apear in the external space. Only the SG field of RL(T) of the external FOHS is able to appear in the external space and to modulate its parameters.**

The above made conclusions help to explain the electron, as a system of FOHSs of different types, and its properties in CL space. Detailed analysis of electron is provided in Chapter 3.

2.8.4 Helical structure with internal RL in a classical void space

Despite the fact that such situation is unusual, it is temporally possible in some conditions of temporally lattice destruction. Such temporally conditions, according to BSM, are created during the nuclear weapon explosion.

In a classical void space the external force F_{ext} is zero. The closed and open helical structures will be affected in a different way.

The closed helical structures (neutron, proton), have a large internal stiffness that is able to keep them stable. When appearing in a classical void space such type of structure could lose the electrical field only, but could not be destroyed. So the neutron and proton will be not affected significantly if appearing temporally in a classical void space (they are in a similar conditions in the phase after crystallization).

The situation for a open helical system, as the electron is little bit different, when it appears in empty space. The electron is a compound single coil structure shown in Fig. 2.13.a. In empty space it loses the electrical field due to absence of CL nodes. So it loses the interaction forces with external space. The single coils structures, however, have larger stiffness than multiturn structures. So we may expect, that the electron system (electron) may undergo a small modification only of its helical step. The electron posses significant kinetic energy as an oscillating system. In a lack of external EM filed we may consider, that this energy will be preserved. However, during the transition process, the whole or fraction of its energy might be dumped. It seams, that all features of the electron will be restored, when appearing again in CL space, but it may need some finite adaptation time, if its kinetic energy is dumped. There is another possibility also that the internal positive structure is biased during the transition time. Then large oscillations could be invoked, leading to CL space pumping and emission of X-ray (the emission of X-rays from the electron is discussed in Chapter 3).

2.8.5 Intrinsic mass contained in the first order helical structure (FOHS) with internal lattice

Knowing the node density distribution, and having the dimensions of the helical structure, approximative calculations for the amount of prisms can be made. Calculations made for positron show, that the number of prisms, from which the helical core of FOHS is built is insignificant in comparison to the total number of prisms contained in its internal rectangular lattice. At the same time, we will see that the apparent inertial and gravitational mass depends only on the FOHS volume (mass equation). In this case the intrinsic mass of the internal lattice appears undetected. Consequently a **vast amount of the intrinsic matter carried by the helical structures in CL space appears hidden**.

The average node density of the internal rectangular lattice is larger than the node density of the external CL space. This means, that a large energy should be applied in order to crush such system. This conclusion helps to explain why so large energies are necessary in the particle coliders in order to obtain Rege resonances. (Detailed discussion about this issue are presented in Chapter 6).

2.8.6. Mixed rectangular lattice in a spherical space

In Chapter 12, we will see, that such space may exist during the phase of particle crystallisation. More accurately this could be a part of spherical space enclosed between two concentric shperical shells. In such space, for a section passing through the geometrical center, the mixed RL will have a similar layer configuration as the radial section of a mono rectangular layer in a cylindrical space. The mixed RL, however, will have one distinctive feature: If the radial stripes are without gaps, then the nodes formed of the longer prisms (right-handed, for example) will be connected without gaps, while the lefthanded nodes will be separated by gaps (due to the shorter lefthanded prisms). Then only the lefthanded nodes will have freedom to vibrate and to posses a kinetic energy. This feature plays an important role in the process of the helical structure crystallization, which is possible and likely takes place in one of the hidden phases of the galactic evolution. This is discussed in Chapter 12.

2.9 Dynamical property of the Cosmic Lattice

Let us consider domain of uniform Cosmic Lattice (CL) space, not disturbed by mass particles and propagating waves (not considering the permanently existed Zero Point Waves discussed elsewhere). The average node distance for such CL domain is constant. The factors that will define the dynamical property of this CL domain are only two: the Super Gravitation and the internal energy. The latter will cause some oscillations of the lattice nodes, so they obviously will have a proper resonance frequency. This frequency will be determined by the mean distance between the neighbouring nodes and the average value of the intrinsic gravitational forces between them. We may consider that in a steady state the nodes will oscillate with their proper resonance frequency, v_R, which for uniform lattice will be a constant:
$v_R = const$.

Let us find out, how the motion of the node prism contributes to the inertial factor of the lattice. For this purpose we will use the model of twisted prisms instead of real prisms (in the real prisms the twisted component of SG field is inside the whole volume of the prisms, due to its lower level structure; this is discussed in Chapter 12). The average velocity contribution from the axial oscillations from the central part of the twisted prism (inscribed cylindrical part of the prism) will be eliminated according to the conclusion 2.4.1.in §2.4. This will cause elimination of the first term in Eq. (2.7). The interaction between SG field from the twisting part of the prisms however is not compensated and must be taken into account. Let us consider this type of interaction as a virtual rotation of the neighbouring prisms from the opposite CL nodes, so it can be considered as an intrinsic inertial interaction between them. Such interaction will involve SG energy. The angular momentum from this virtual rotation is:

$$L = m_{per} v_{per} r_{per} \qquad (2.11)$$

where: m_{per} is the intrinsic mass of peripheral part of the prisms (averaged value for left and right handed prisms)

v_{per} is a peripheral velocity

r_{per} is a peripheral equivalent radius

If t_p is the period of this virtual rotation, then dividing the angular momentum on t_p one obtains the inertial factor I_f:

$$I_F = \frac{L/t_p}{E_g} = \left[\frac{m_{per} v_{per} r_{per}}{E_g}\right] f_p \qquad (2.12)$$

where: E_g - is the SG energy, which is constant for a steady state CL space

Let us accept that the virtual velocity of the peripheral part of the prism has an upper limit value: $v_{per} = v_{lim}$. But this velocity and the prism rotational frequency f_p are connected. Then f_p will have upper limit constant value f_{lim}. **It is equivalent to say that the prisms in the lattice have upper limit of their virtual angular frequency.** The latter could be defined by the intrinsic time constant of the intrinsic matter, (mentioned in §2.3), the prism shape factor and the distances between neighbouring prisms. So if the virtual peripheral velocity has a limit, the inertial factor also will have an upper constant value, according to Eq. (2.12). The inertial factor of the twisted part is a predominant. Consequently it will define the interaction energy. **Then the limited value of the angular frequency will determine the finite value of the node resonance frequency.**

Conditions for stable lattice in empty space without boundary:

According to the stability rule, formulated in §2.6, the return forces should be conservative for any axes passing through the equivalent geometrical point. They should return the displaced node in a range of stable positions. The stable positions may not coincide with the geometrical equilibrium point. From the energetic point of view, the stable lattice should have a point with lowest potential. Only stable gravitational lattices could really exist.

In the simplified model of twisted prism we admitted that they have a freedom of axial rotation within the CL node assembly influencing in this way each other. This is convenient for simplifying of the analysis. In reality, a rotation of the real prisms combined in CL node or any other gravitational lattice is not necessary. In the real prisms the internal modes of SG field in fact possess rotational interactions. They are pure energy rotational modes, discussed in Chapter 12 and they namely provide the twisted SG component of the SG field of the prisms.

2.9.1 Node configuration of CL structure

Figure 2.20 illustrates a geometry of a single node in a position of geometrical equilibrium with its axes of symmetry.

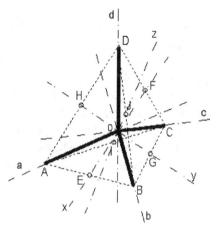

Fig. 2.20

CL node in geometrical equilibrium position
The two sets of axes of symmetry are: *abcd* and *xyz*

The thicker lines designate the four prisms of the node, each one at angle of 109.5 deg from the others. The prism ends ABCD form a tetrahedron ABCD. The four axes, at which the prisms are aligned are **a, b, c, d,**. The node has also another three axes of symmetry **x, y, z**, which passes through the middle of tetrahedron edges. These axes **x, y, z** are orthogonal each other. The **x,y,z** axes intercept the axes of **a,b,c,d** at **54.75 deg.**

If not taking into account the twisted part of the prisms, the x, y, z components do not have + and - direction. If taking into account only the twisted parts, the same axes may get + or - depending of the direction of their spin vector.

The derivation of the expressions in this section is shown in "Appendix to Chapter 2".

2.9.1.1 Conditions for a stable lattice existence.

Analysing the return forces of cosmic lattice, according to the stability rule, formulated in §2.6, it is found that:

- applying gravitational law proportional to square inverse power dependence on distance leads to unstable lattice along the axes a,b,c,d.

- **applying gravitational law, proportional to inverse cubic power of distance leads to a stable lattice along a,b,c,d.**

- the return forces along x,y,z axes have a valley symmetrical in respect to the geometrical equilibrium point.

The return force of a single node of CL along the axes x,y,z is derived, by considering that the neighbouring nodes are in fixed position. For a node distance of few prism lengths, we may accept, that the centre of mass is always in the node centre. Then the return force, normalized to the product $G_o m_n^2$ is expressed by Eq. (2.14).

$$(2.14)$$

$$F = 2\left[\frac{x + d\cos\left(\frac{\theta}{2}\right)}{\left[x^2 + d^2 + 2xd\cos\left(\frac{\theta}{2}\right)\right]^2} - \frac{d\sqrt{0.5(1 + \cos(\theta))} - x}{\left[x^2 + d^2 - 2xd\cos\left(\frac{\theta}{2}\right)\right]^2}\right]$$

where: G_o - is the intrinsic gravitational constant; m_n - is the node intrinsic mass (equivalent mass for left and right handed node); d - is the distance between neighbouring nodes of CL, θ - is the angle between prisms axes in geometrical equilibrium.

The plot of equation (2.14) is shown in Fig. 2.21, where the displacement is normalized to the node distance d.

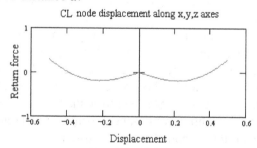

CL node displacement along x,y,z axes

Fig. 2.21

Return force for displacement along x, y, z axes

The return forces plot shows, that two stable points exist along x,y,z axes, at both sides of the geometrical equilibrium point 0.

Let us investigate now the return forces along a,b,c,d axes. The node geometry for displacement along a,b,c,d axes is shown in Fig. 2.22. The prisms in the geometrical equilibrium are shown as thick black lines, while for a displaced node they are shown as thick grey lines.

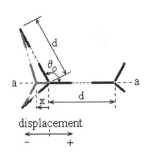

Fig. 2.22
CL node displacement along one of *abcd* axes
Note: The displacement of both nodes is symmetrical (not shown in the figure)

If not taking into account the node geometry and centre of mass change for small displacements, the return force for a left displacement according to Fig. 2.22 is given by Eq. (2.15), while for a right displacement - by Eq. (2.16).

$$F = \frac{1}{(d+x)^3} + \frac{3(x + d\cos(\theta_0))}{(d^2 + x^2 + 2dx\cos(\theta_0))^2} \qquad (2.15)$$

$$F = \frac{3(x - d\cos(\theta_0))}{(d^2 + x^2 - 2dx\cos(\theta_0))^2} - \frac{1}{(d-x)^3} \qquad (2.16)$$

The plot of the return forces for left and right displacements is shown in Fig. 2.23. The displacement is normalized to the node distance *d*.

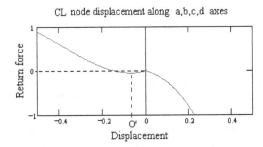

Fig. 2.23
Return force for displacement
along a, b, c, d axes

The return forces along the axes a,b,c,d, do not appear symmetrical in respect to the geometrical equilibrium. It is evident from the plot, that the reaction forces for left and right displacement are

quite different. Although, a valley exists in the left side of the geometrical centre 0, corresponding to point O'. This valley denotes a stable point for motions along a,b,c,d axes.

If investigating the motion in the intermediate axes we will see, that it is also stable. Then the CL lattice can be considered stable without boundary holding conditions.

The valley shown in Fig. 2.23 was obtained, when applying SG force proportional to inverse power of 3. If applying inverse power of 2, the valley along a,b,c,d axes is missing. Consequently a Cosmic type of lattice is not possible if SG force is proportional to inverse power of 2. This result is in agreement with the accepted inverse cubic dependence of SG force on distance in empty space.

Despite disregarding the slight node geometry and centre of mass change, the simple equations (2.15) and (2.16) demonstrate the existence of valley along a,b,c,d axes. In order to investigate more accurately the return forces and stiffness along the different axes, the above mentioned features has been also considered . In this case the equations are pretty long and will be not given here, but their plots are similar. For derivation of this equations the following considerations are made:

- the movable node is regarded as a group of four prisms, attached in their common end and always aligned to the neighbouring nodes, which are considered as fixed;

- the forces are applied between any one of neighbouring nodes and the centre of mass of anyone of the node prisms;

- the centre of mass correction is applied

The centre of mass of the prism, presented as a mass bar depends of both - the distance and the angle. This dependence for inverse cubic power is larger than for the inverse square power. Besides the long equations, some small corrections require graphical solving, fitting and iterations. The problem solving was focused on this displacement along a,b,c,d axes, for which a stable zone exists.

Fig. 2.24 shows a combined plot of return forces along the stable zone in a,b,c,d and x,y,z axes when the above mentioned factors are considered.

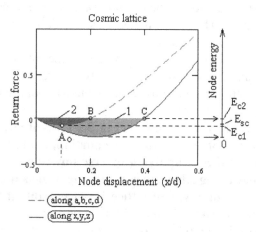

Fig. 2.24

Node energy diagram for cosmic lattice

The left scale shows return forces referenced to the geometrical equilibrium point of the node. However, this is not a point of gravitational equilibrium. The gravitational equilibrium in fact is a three dimensional surface, centred around the geometrical equilibrium point. For displacements along intermediate axes, not coinciding with the two axes set, the valley will have different height. In the right side of the plot an energy scale for estimation of the node energy is shown. The energy scale is for reference only and is not proportional. The following energy levels are defined:

0 - is a zero level corresponding to the bottom of the valley, but from displacement in all possible directions, multiplied by the probability to vibrate in these directions.

E_{1c} - is the first critical level corresponding to the zero level for motion along and around the a,b,c,d axes.

E_{2c} is the second critical level. Energy above this level will lead to a node destruction, because this is upper point for displacement along a,b,c,d axes (see Fig. 2.23). Practically this will not happen, because the individual CL node position is self adjusted. E_{2c} determines the full energy well of the CL node (maximum energy). It is equal to the sum of two energy wells, artificially separated and annotated in the figure, by 1 and 2. The energy well 1 is contributed by x,y,z axes and is much larger than the energy well 2.

E_{sc} is the superconductivity critical level. It will be discussed later.

The energy levels are very important features of the CL node. While the importance of E_{2c} is evident, the role of E_{c1} and E_{sc} are not so evident at first glance, but it is very important for the EM field propagation and play important role for understanding the superconductivity state of the matter.

The full energy well is contributed mainly by the light gray area, but taken from a solid angle around x,y,z axes. The dark grey area 2 contributes to the node energy only above the level E_{c1}. The energy quantity $(E_{c2} - E_{c1})$ is a small fraction of the node energy well, because it is contributed only of small solid angle around 4 axes (instead of 6) and with smaller deepness.

It is evident also, that the stiffness for displacements along x, y, z axes is much lower that the stiffness along a, b, c, d axes.

Fig. 2.25. shows the return forces in a,b,c,d displacement, normalized to forces in x,y,z for two node distances.

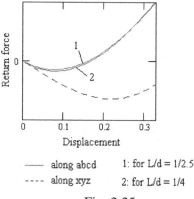

—— along abcd	1: for $L/d = 1/2.5$
---- along xyz	2: for $L/d = 1/4$

Fig. 2.25.

Return forces in case of node distance change. L - is a prism's length, d - is a node distance (considering the linear dimensions L and d as an average value from both prisms).

From the plot of the return forces in Fig. 2.25 one can make conclusion, that the energy ratio of {(node well)/$(E_{c2}-E_{c1})$} is slightly affected when the node spacing is changed. This is due to the finite dimensions of the prisms.

Analysing the return forces and stiffness for displacement along both set of axes and all possible axes between them, it is obvious that the node will be able to vibrate in a complex way. It will be able also to store a kinetic energy due to its energy well.

This is the dynamical type of the zero point energy (ZPE) of the node. The ZPE of the CL node is mostly contributed by the valley deepness for x,y,z displacement, because it is symmetrical and deeper than the valley in a,b,c,d displacement.

Let us analyse the node oscillations for two cases:

A case: The node energy is between zero and E_{c1}.

2.9.2 Node oscillations described by vectors

The dynamical properties of the lattice could be studied by the analysis of the single node dynamics and the dynamical interactions between the nodes. This process is complicated, but it can be simplified, if proper vectors are introduced.

The stiffness along the x,y,z axes is smaller than along a,b,c,d axes. The nodes tend to oscillate between the opposite valleys of x,y,z, but the trace curve could not be flat. The trace of oscillating node will not pass through the geometrical point O. It will bypass it, because the stiffness in the vicinity of this point is higher. The trace is an open three dimensional curve. Its projection on a plane *xy* is shown in Fig. 2.26.

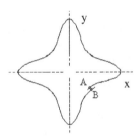

Fig. 2.26
Trace curve of the resonance cycle
of the oscillating CL node

The trace curve of a single cycle is open, so it will not pass through the same initial point. This is shown in the projection curve. Despite of this, we can formulate a period between points A and B. **The trace period is a time between two adjacent traces, at arbitrary chosen closest points.** This period determines the CL node **resonance frequency**. The node inertial factor is very small, so the **resonance frequency is to be very high.** The

nodes of lattice domain with a constant node distance will have a constant resonance frequency.

If considering consecutive periods, the node trace curve will not pass through the same trace point. For multiple periods, the trace will circumscribe a three dimensional surface. This surface will have a shape of deformed sphere with six bumps along x,y,z axes, and four deeps along a,b,c,d axes. We can call this surface a node trace quasisphere.

In any moments, the node have intrinsic inertial momentum with direction coinciding with the instant velocity vector. If integrating this momentum per one resonance period, we get node average momentum, which could be expressed also by a vector, but passing through the geometrical equilibrium point of the node. We can call this vector a **node resonance momentum vector**, or abbreviated: NRM. It is more convenient to operate with such vector, normalized to its maximum value.

The node resonance momentum is a three dimensional vector, expressing the integrated intrinsic inertial momentum of the node for one period of the resonance oscillations. The origin of the vector is at the geometrical equilibrium point of the node.

Analysing the node resonance momentum behaviour for a large number of consecutive cycles, we will find that it has a larger density in the conical spacial angles centred along x,y,z axes, than along a,b,c,d axes. Then we can introduce the node density momentum vector.

The node density momentum vector is a product of the node momentum vector for many cycles multiplied by its differential angular cross section and averaged to its maximums at x,y,z axes.

The behaviour of the density momentum vector is described by a **node density momentum quasisphere.** It also have six bumps and four deeps, but their shape is not the same as the node trace quasisphere.

The shape of the density momentum quasisphere is shown in Fig. 2.27.

In order to investigate the features of the node quasispheres and introduce other vectors, a few energy dependent cases of the node oscillations will be analysed.

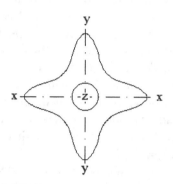

Fig. 2.27. Density momentum quasisphere of the oscillating CL node

A. Case: The node energy is between zero and E_{c1}.

If considering a single node dynamics, a node momentum along a,b,c,d axes should not exist. However the influence of the neighbouring nodes will cause some small fluctuations along this axes. In a longer period of time, the node momentum vector will oscillate in a spherical coordinates, but spending more time around the x,y,z axes. The node quasispheres can be defined, for some large number of oscillations, but circumscribed in a random way. It will be still symmetrical, but the trace density of circumscribed quasisphere will not be uniform.

B. Case: The node energy is between E_{c1} and E_{c2}. SPM mode of operation.

In comparison to the case A, the node stiffness near the axes a,b,c,d will be affected. As a result of this, the node momentum vector will be affected by small force, tending to change the vector direction continuously. This will cause a precession momentum, that will control the direction of node momentum vector in a definite way. At the same time the node momentum will flip from cone to cone around x,y,z, but more uniformly, than in case A. The consecutive cycles of the node momentums will define again quasisphere, but more systematically. The momentum vector for every consecutive cycle will be aligned in different directions, but with tendency of preserving the momentum of precession. In this case a **vector of precession momentum** could be defined, whose period will contain a large number of node momentum periods. For one period of this vector, the node

density momentum quasisphere will be completely defined. We can call this vector a **Spatial Precession Momentum vector** or **SPM vector**, and the operational mode, respectively, **SPM mode.**

The SPM vector is a spatial precession momentum of the node momentum oscillations.

In SPM mode of operation, the node density momentum quasisphere is completely defined for one period of SPM vector.

If a gradient of gravitational or other field exists across the CL space domain, the node equilibrium may be biased, so for correct SPM mode of operation the node energy should be little bit above E_{c1}.

We could accept now, and prove later, that the CL node in normal conditions (superconductivity is excluded) has a capability to accumulate and preserve the full well node energy. This is the nominal dynamical ZPE of the vacuum. It will be shown in Chapter 5, that the relict radiation is a signature of this type of CL space ZPE (BSM interpretation).

The node resonance and SPM frequency are related by the expression:

$$\nu_R = N_{RQ}\nu_{spm} \qquad (2.17)$$

where: ν_R - is the CL node resonance frequency

ν_{spm} - is the CL node SPM frequency

N_{RQ} - is the number of resonance cycle for one period of SPM

The coefficient N_R is a constant of the CL space and may have a quite large value. Estimation of its value will be made in the following sections.

The SPM effect is essential feature of the lattice space. It helps to understand some basic properties of the space in which we live. These properties include: the quantum features of the space, the electrical and magnetic fields, the light velocity, the conductivity and superconductivity states of the matter.

2.9.3 Resonance frequency stabilization effect

Although the trace of the oscillating node is a complicated curve, some quantitative analysis could be done, in order to show that some stabilization effect of the resonance frequency exists. We may distinguish two stabilization mechanisms: weak and strong stabilization.

2.9.3.1 Weak stabilization mechanism

Let us consider two cases of CL node energies E_1 and E_2 included in the range between the E_{sc} and the nominal ZPE level, as shown in Fig. 2.28.a.

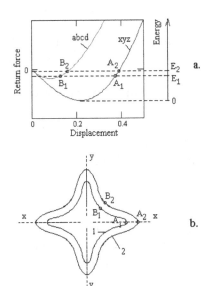

Fig. 2.28
NRM frequency stabilizing effect

Fig. 2.28.b. shows the projection of the trace of the node for one period of the resonance frequency. The projection is not of the classical type, but made in a way to preserve the path length and distance from the centre. Due to the centripetal acceleration the trace will tend to pass through the more distant points from the geometrical centre. For energy E_1 it will pass through the points A_1 and B_1, while for energy E_2 respectively through A_2 and B_2. Then the internal trace projection 1 will correspond to the lower energy E_1 and the external trace 2 to the higher energy E_2. The ratio between real traces lengths will be approximately preserved. The slop of return force in any point of the displacement curve gives the stiffness. From Fig. 2.28.b we see, that the return forces and stiffness for trace 2 are larger than for trace 1, and the stiffness of x,y,z curve is larger. Then the larger stiffness for trace curve 2 will tend to reduce the period. Consequently the change of the node stiffness of the two displacement curves exercises some kind

of stabilization effect on the node resonance frequency. Therefore we may conclude that:

For CL node energies in the range between E_{sc} and nominal ZPE, a weak stabilization effect of the node resonance frequency exists.

2.9.3.2 Strong stabilization mechanism.

In the paragraph 2.11.3 the derivation of the light velocity equation is provided. It will become evident, that the constancy of the light velocity is directly related to the resonance frequency. Then the resonance frequency should be stable in the same order that the constancy of the light velocity. Evidently a well defined and self regulated stabilization mechanism should exist.

One of the necessary condition for the strong stabilization mechanism is to work in conditions, where, the resonance frequency dependence on the ZPE is a continuous function. In this aspect the weak stabilization mechanism fulfils this conditions. In the following analysis we will see, that the change of the ZPE within the allowable limit is a continuous function of the resonance frequency.

Let us analyse the frequency of not disturbed NRM quasisphere. In §2.10.3 we will see, that such nodes are associated with continuous fluctuations, forming a so called magnetic protodomains. In such aspect the nodes with not disturbed NRM quasispheres are simply referred as magnetic quasispheres (MQ). Let us initially suppose that the ZPE of the MQ is below the normal level. If it gets some additional energy, the oscillation amplitude and its frequency will increase. But the amplitude increase may lead to overpassing the limit, corresponding to the E_{c2} level of ZPE (see Fig. 2.24). This could lead to a node geometry destruction, but the CL structure has ability to transfer the excess momentum to the neighbouring node. Due to the very small inertial factor of the node, the transfer of the excess momentum occurs for a time nor larger than one resonance cycle. Consequently the average oscillating amplitude could not exceed the value corresponding of E_{c2}. At the same time, due to the zero point waves, (discussed later), every node has a possibility to support its nominal ZPE. This is confirmed by the background temperature of 2.72 K of the CL space. (A detailed discussion, derivation and calculation this CL space parameter is presented in Chapter 5). Consequently the accuracy of the nominal ZPE determines the accuracy of the resonance frequency. There is a second factor, that also contributes to the stabilization. This is the mutual synchronisation of the neighbouring MQs as a result of their participation in magnetic protodomains. Consequently we may summarise, that:

- **The strong stabilizing mechanism of the CL resonance frequency is controlled by two factors: the nominal ZPE of the node and the CL node participation in magnetic proto domains.**

   Note: The magnetic protodomains will be discussed later. They are directly related to the permeability of the physical vacuum.

2.9.3.3 Analytical presentation of the frequency stabilization effect.

Here only a simple and aproximative analytical model will be presented. However it will give a basics for correct interpretation of the data from the Quantum Hall experiments in Chapter 4, from where the frequency stabilization effect becomes evident.

We will use two oscillating models, valid only for MQ node:

- node oscillations as a spring system
- node oscillations as an conical pendulum

A. Node oscillations model as a spring system

This option is only for a simple illustration without showing the strong stabilization mechanism. The resonance frequency is given by:

$$\nu_R = \frac{1}{2\pi}\sqrt{\frac{k}{m_n}} \qquad (2.17a)$$

where: k - is a stiffness, m_n is the node inertial mass.

In the analysis of the Quantum Hall Experiments (QHE) in Chapter 4, we will see that the node distance is unchanged, and we may consider with a first approximation, that m_n in such condition is a constant. We may determine the stiffness, by differentiating the return force along x,y,z axes (equation (2.14)) on the displacement, for a node distance normalised to unity. Then the stiffness is given by:

$$k = \frac{2}{A^2} - \frac{8\left(x + \cos^2\left(\frac{\theta}{2}\right)\right)}{A^3} + \frac{2}{B^2} + \frac{8C\left(x - \cos\left(\frac{\theta}{2}\right)\right)}{B^3} \qquad (2.17.b)$$

$$A = x^2 + 1 + 2x\cos\left(\frac{\theta}{2}\right) \qquad B = x^2 + 1 - 2x\cos\left(\frac{\theta}{2}\right)$$

$$C = 8(\sqrt{0.5 + 0.5\cos(\theta)}) - x$$

Due to the reversible direction of the return force, the stiffness curve obtains a negative part, but this is an artifact, due to the reference to the geometrical equilibrium point. To correct this we shift the curve up by the maximum of the negative value. Then substituting in (2.17a) and normalising to $(\sqrt{m_n})/2\pi$, we get the resonance frequency dependence on stiffness change.

$$\nu_R = \sqrt{k(x) + 1.33} \qquad (2.17.c)$$

The plot of Eq. (2.17.c) is shown in Fig. 2.28.A. The node displacement x could be considered as a radius, corresponding to a lattice with defined stiffness.

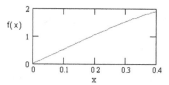

Fig. 2.28.A

B. Node oscillations model as a conical pendulum

In this case, the node oscillation properties, are simulated by a circular conical pendulum. The simulation is approximate and valid only for a MQ type of node. In this type of simulation the effect of the strong frequency stabilization will be shown.

The parameters of the circular conical pendulum (the trajectory is a circle) are shown in Fig. 2.28.B.

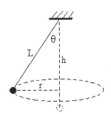

Fig. 2.28.B
Conical pendulum

We will associate the node displacement with the parameter r of the conical pendulum, the node distance - with L and the node energy - with the potential energy of the pendulum. Then for a uniform CL domain the oscillation period will depend only on the parameter r. The oscillation frequency of the conical pendulum is given by the equation

$$f = \frac{1}{2\pi}\sqrt{\frac{g}{h}} = [g[L^2 - r^2]^{-0.5}]^{0.5} \qquad (2.17.d)$$

We are interested of the node operation in the zone from the right side of point A_o, where the smallest trace curve radius is determined by the curve 2 and the larger radius by the curve 1. We may approximate the node trace curve with an equivalent circle, whose radius vary in the range between 0.23 to 0.29 of the node distance (see Fig. 2.24). Then the parameter r in eq. (2.17.d) will be substituted by $(x + 0.23)$, where x is the new argument that we may relate to the ZPE.

In order to simulate the loss of energy, when the node overpasses the ZPE level for a large displacement, we will introduce a change of the L parameter of the pendulum. This change will begin when the argument x overpasses some threshold value. So we will define a continuous function with a kink. We may call this function an **energy dumping function**. For this purpose the following exponential function is used.

$$L(x) = \left[1 + 0.0106\exp\left[-\left[\frac{(x-0.03)}{2.48\times10^{-4}}\right]^2\right]\right] \qquad (2.17.e)$$

The parameter L from Eq. (2.17.d) is substituted by L(x) of Eq. (2.17.e). In order to adjust the frequency span and frequency normalization to unity, we must select also a value for g, and offset

the frequency f with some constant value. The final frequency simulation equation takes a form:

$$f = [22000[(L(x))^2 - (x - 0.23)^2]^{-0.5}]^{0.5} - 150.4 \qquad (2.17.f)$$

The energy dumping and frequency simulation curves are plotted respectively in Fig. 2.28.C a, b. The frequency stabilization is at $x = 0.06$. For the real conical pendulum model the dumping energy will stop to grow at the point of the stabilization, and **the frequency will get a fixed value**. For this reason the frequency down slop is shown as a dashed line. We see, that the stabilization effect is very sensitive to the energy dump. From the micro-scale point of view, the frequency uniformity between the CL nodes depends only on two factors:
- prisms uniformity
- node vibrational energy (dynamic ZPE)

The high degree of prisms uniformity is accepted a priory, but its physical concept and analysis are presented in Chapter 12.

The second factor is self regulated by the zero point waves. The prove of this is the Cosmic background temperature uniformity discussed in Chapter 5.

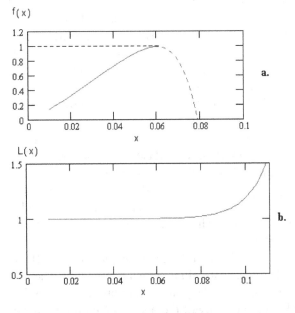

Fig. 2.28.C
Conical pendulum model for demonstration of the resonance frequency stabilization

The demonstration of the stabilization effect here, does not take into account the appearance of the quantum stabilization process in CL space, and a possible change of the node inertial momentum for the range of *x* below the maximum. In the real case, the frequency dependence on the node displacement may not be so linear. The presented simple model, however, is helpful for interpretation of some results obtained by the Quantum Hall experiments (see Chapter 4.) In such aspect it helps to understand the dynamics of the CL node oscillations.

2.9.4 SPM vector affected by external electrical field

A. Case: The node energy is between E_{sc} and nominal ZPE

In a presence of external electrical field, the node resonance momentum vector will be affected by the vector of electrical field. The NRM will get preferential distribution along the filed. On the one hand, this will cause deformation of the node quasisphere, transforming it to prolate spheroid with bumps, aligned with the vector of electrical field. On the other hand, the SPM frequency will be changed but the amount of this change will depend on the degree of the quasisphere deformation. In one of the next paragraphs, where the energy transfer is discussed, we will see that the change is in direction of decreasing the SPM frequency. While the NRM frequency is still stable, this means that the SPM quasisphere is circumscribed by a smaller number of resonance cycles.

The decrease of SPM frequency of node quasisphere in external electrical field is one important feature of the CL space.

The above mentioned feature helps to explain the electrical interactions between charge particles in CL space. This will be discussed later in this chapter.

The shape of the CL node quasisphere, affected by electrical field is shown in Fig. 2.29. The distance between the points could provide impression about the density of the node momentum vector in the particular direction

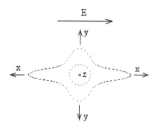

Fig. 2.29
Node density momentum
quasisphere affected by electrical field
E - direction of the field

B. Case: The node energy is below E_{c1}.

In this case the asymmetrical stiffness along a,b,c,d axes will be eliminated and will not control the node momentum vector. The external electrical field above some threshold level could be able to affect the vector, but random or controlled precession conditions do not exists. So a normal SPM effect here does not exist. The node momentum will define a node quasisphere, but severely deformed. The cross sectional momentum quasisphere will approach the shape of prolate spheroid with not so sharp bumps. Its shape is shown in Fig. 2.29.a.

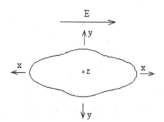

Fig. 2.29.a
Node density momentum quasisphere
in external electrical field
for node energy below E_{c1}

The deformed quasisphere will lead to increased propagation of the external electrical field in comparison to the case A. This is valid for subcritical node energy but not much below E_{c1}. For energy lower than some level, the propagation of the electrical field will start to decrease as a result of a lack of a necessary momentum. In normal conditions, however, the CL lattice tends to keep its nominal dynamic ZPE.

From the analysis of this case, it is apparent that:

- **In a case of subcritical node energy a SPM effect does not exits**
- **The propagation of the electrical field in subcritical node energy is enhanced**
- **The magnetic field is not able to propagate in a lattice domain with subcritical node energy.**

The last statement will be proved in Chapter 4.

The three features cited above are important factors in the superconductivity state of the matter. They are discussed in Chapter 4.

2.9.5 SPM frequency for node energy between E_{c1} and E_{sc}

When the node energy is below E_{sc}, the slop of the return force along a,b,c,d is gradually, but significantly reduced in comparison to the slop of x,y,z force (see Fig. 2.24). This causes diminution of the SPM frequency stabilization effect, discussed in the §2.9.3. The change of stiffness along a,b,c,d below E_{sc} leads to decrease of SPM frequency even for the normal quasisphere (i. e. at absence of external electrical field). This effect plays important role in the superconductivity state of the matter.

Summary of introduced vectors and their presentation

- Node trace
- Node trace projection
- Node trace quasisphere
- **NRM** (node resonance momentum) vector
- **NDM** (node density momentum) vector
- SP (spatial precession) effect
- **SPM** (spatial precession momentum) vector

The **SPM** vector is obtained by the spatial precession effect from the trace rotation of the **NRM** vector. This could be expressed by the operator **SP{ }**.

SP{NRM} => SPM

The control functions of the SP operator depend of the node energy conditions, discussed in the previous paragraphs.

The SPM vector can be presented by the following components and subcomponents as shown in Fig. 2.29.B.

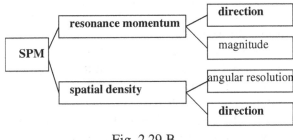

Fig. 2.29.B

The bold text is used for vector components, while the plane text is for scalars. The node quasisphere is a convenient way for graphical presentation of the NDM or SPM behaviour. The SPM quasisphere is distinguished from the NDM only by its surface uniformity, due to the SPM effect. In the future analysis, where the conditions are implicitly known, the name node quasisphere will be used.

The quasisphere is a surface in spherical coordinates, circumscribed by a radius vector with components shown in the Fig. 2.29.C

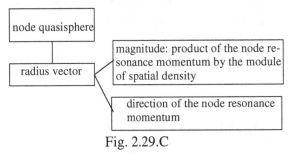

Fig. 2.29.C

2.9.6 Method of separation of intrinsic interaction processes contributed by the central or peripheral twisted part of the prisms.

When the interaction process in a lattice is analysed in a lowest level (the level of the twisted prisms), any interaction could be regarded as a contribution from two prism's interactions: interaction due to the cylindrical core of the prism, and interaction due to the twisted peripheral part. The SG although could not distinguish the interaction between the cylindrical parts of right handed from left handed prisms. Then in a stable lattice space, we could separate the interactions involving the cylindrical part of the prism from those involving the twisted part. A stable lattice space is this one, whose lattice, is in a steady state. There are processes of CL lattice destruction and rebuilding, where transition time phase exists (nuclear test explosion, for example). During such a process, the CL space is not in a steady state phase.

For SG interactions in which the cylindrical part of the prisms is involved, the following simplification could be applied.

- **From the point of view of the inertial factor, the SG interactions of the central part of the prisms of the opposite nodes are equivalent.**

- **For a domain of stable lattice space containing large number of nodes, the integrated result from the interactions between the cylindrical part of the prisms for one SPM cycle is equal to zero.**

The integrated result from the peripheral twisted part for prisms for one cycle of NRM or SPM, however, is not equal to zero.

- **SG interactions from the right handed and left handed prisms have different spatial momentum.**

- **The integrated results from interactions between the twisted part of the prisms in the neighbouring nodes of CL space are not equal to zero.**

We could apply the method of interaction separation in most of the cases related with the analysis of the electrical and magnetic fields. But when analysing the mass deficiency effect of the charged particles, for example, the interaction separation could not be applied.

In some other cases, like estimation of charges from extended helical structures, the symmetry of the lattice field beyond some critical radius should be checked.

We may apply the method of interaction separation for the NRM and SPM vectors. For this purpose we attach the following attributes to the vectors:

CP - standing for: **cylindrical part**
TP - standing for: **twisted part.**

Then the node vectors will have the following two components:

NRM(CP);   NRM(TP);
SPM(CP);   SPM(TP)

Some parameters of the CP and TP components of the vectors are equivalent, for example the period (frequency), the normalized quasispheres and so on. In the analysis, where this is implicitly understood, we can dismiss the attribute TP in the vector notation.

The method of interaction separation facilitates significantly the analysis and helps to unveil the physics behind the gravitational, electrical and magnetic field including the physics behind the unit electrical charge.

2.9.6.A Oscillating velocity distribution of the CL node

Let us consider a CL node in a not disturbed CL space (a space domain without electrical and magnetic field). The CL node will have an MQ type of SPM quasisphere, but not synchronized with its neighbours. We will consider also that it has a normal ZPE of dynamical type, i. e. its quantum wells are filled. We may assume that the acceleration is a linear. Then we can determine approximately the oscillating velocity distribution in the space volume occupied by the single CL node. In order to simplify the model, we may consider that the volume is equivalent sphere. Dividing this sphere in small unit volumes, we may estimate the instant velocity in each volume and build a histogram of the velocity distribution. The derivation of the expression giving the envelope of the histogram is quite long to be shown here, so only its plot is shown in Fig. 2.29.D

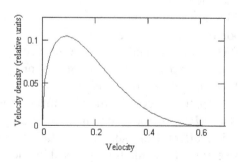

Fig. 2.29.D

Approximate shape of CL node velocity distribution in the oscillating volume

The velocity histogram in Fig. 2.29.D is shown as a velocity density distribution in relative units. Despite the accepted simplification, the shape of obtained histogram appears similar as the shape of distribution of population among the rotational states of the molecules. This feature will be additionally discussed in Chapter 9.

2.9.6.B Fine structure constant as embedded feature of the twisted prisms

The fine structure constant, denoted as α is one of very basic physical parameters. It appears not only in atomic and molecular spectrum but in many phenomena in Quantum mechanics and particle physics. Its value is experimentally estimated with a very high accuracy. It appears very often also in the derived equation of BSM theory. In Chapter 10 and 12 theoretical equations for α are presented, providing extremely a value of α extremely close to the experimentally measured one. One important conclusion in the BSM is that α is embedded in the structure of both types of prisms. This feature will become apparent in Chapter 9, where the molecular oscillations are analysed. The final understanding of the fine structure constant will become apparent in Chapter 12, where the internal structure of the prisms will be discussed. In this chapter we will show only some relations between CL space parameters in which α appears embedded.

The fine structure constant is dimensionless. Therefore it may be regarded as a ratio of parameters with a same dimension.

Let us determine the amount of the gravitational energy, E, in empty space between two equal prisms of a same type at distance r_0. It is equal to the work that is required to separate them to infinity, which is given by the integral:

$$E = -\int_{r_o}^{\infty} G_{os} \frac{m_o^2}{r^3} = G_{os} \frac{m_o^2}{2r_o^2} = G_{os} \frac{\rho_o^2 V^2}{2r_o^2} \qquad (2.A.17.A)$$

where : m_o is the intrinsic mass, ρ_0 - is the intrinsic density and V - is the prism's volume.

Using (2.A.17.A) We may estimate the ratio of E for the central and for the twisted part of the prism, as a ratio of squares of their volumes

$$E_{TP}/E_{CP} = V_{TP}^2/V_{CP}^2 \qquad (2.A17.B)$$

The twisted part is this portion of the prism, which is left when removing the inscribed cylinder. If considering twisted prisms without spherical edges, the above shown ratio is a constant that does not depend on the prism's radius or length. It's value is 0.0105386. If considering, however, the two edges having a shape of a half sphere, the ratio E_{TP}/E_{CP}, becomes dependable on the length to radius ratio L/r (r - is the radius of the inscribed cylinder). For L/r ratio in the vicinity of 6.484, the energy ratio E_{TP}/E_{CP} becomes exactly equal to the fine structure constant $\alpha = 7.29735 \times 10^{-3}$. The corresponding L/r ratio, could not be accepted as real and seams to be a small. But we have to keep in mind, that the twisted prism is distinguished by the real twisted rods, mentioned at the beginning of Chapter 2. In the real twisted rods, the twisted effect is not so much external but internal feature of the intrinsic matter structure from which the prism is built. This will be discussed in the Chapter 12.

When analysing the ratio of E_{TP}/E_{CP} in CL space, however, we have to take into account that the vectors SG(TP) and SG(CP) do not have one and a same type of propagation. Since SG(CP) vectors do not have a handedness their propagation is additive, while the propagation of SG(TP) depends on handedness of interacting prisms or CL nodes. For CL nodes of opposite handedness SG(TP) propagation is additive. Then, if the E_{TP} energy, propagated in CL space from the right and left handed prisms is denoted respectively as $E^R_{SG(TP)}$ and $E^L_{IG(TP)}$, then the proper SG energy balance will be achieved at ratio:

$$\frac{E^R_{SG}(TP) + E^L_{SG}(TP)}{E_{SG}(CP)} = \frac{2E_{SG}(TP)}{2E_{SG}(CP)} = 2\alpha \qquad (2.A.17.C)$$

where: the energy $E_{SG}(TP)$ is just the energy related to the twisted part (without referring to the right and left handed fractions), and $E_{SG}(CP) = E_{CP}$ is the energy related to the central part of the twisted prism model.

The ratio given by Eq. (2.A.17.C), matches quite well a number of calculations, related to the electrical field and charge unity (presented elsewhere), and the expressions about the atomic vibrations in the molecules, presented in Chapter 9.

The fine structure constant shows its signature in many equations derived in BSM. It is related also to one basic parameter of the FOHS - the ratio between the confine radius and the second order helical step (this to be seen in Chapter 3).

At this stage we may conclude:

- **The fine structure constant is embedded in both types of prisms**
- **The fine structure constant determines the ratio of the intrinsic energy transmitted by the twisted and the central part of the prisms. In CL space environment, the right and left handed fractions contribute additively.**
- **The fine structure constant carry a signature of the internal structure of the prisms (this issue is presented in Chapter 12).**

2.9.7 Summarized features of the Cosmic Lattice space

We can summarize the following features related to the dynamical property of the Cosmic Lattice.

- **The CL Space is able to handle intrinsic kinetic energy, which can be regarded as a dynamical type of Zero Point Energy of the physical vacuum. Its measurable parameter is the absolute temperature of the CL space (to be confirmed in Chapter 5 and 6).**
- **The CL node oscillating with its proper resonance frequency exhibits a spatial precession momentum (SPM) effect**
- **The rotation of the node resonance momentum is caused by the different stiffness (return forces) for node displacements along the two sets of axes of symmetry (x,y,z and a,b,c,d axes).**
- **The dynamical properties of the SPM vector are well characterised by the shape of the CL node quasisphere**
- **The vector SPM is directly involved in the unidirectional propagation of the photons (to be shown later in this chapter)**
- **The CL space can propagate two types of waves: quantum waves - possessing energy equally mixed between both types of involved nodes; and quasiparticle waves (or virtual particles) - possessing energy not equally mixed between both types of nodes**
- **The dynamical property of the CL structure could be investigated more effectively if**

using the CL node angular velocity as an argument.

- **The SPM effect helps to explain the quantum features of the physical vacuum - the CL space in which we are immersed.**

2.9.8 Static and dynamical features of the rectangular lattices

The rectangular lattices (in fact with slightly distorted right angles) are inside of any FOHSs. In comparison to the CL structure, they have much larger stiffness due to the larger prism density.

The other type of rectangular lattice - a mixed RL (which exists only in a hidden phase of galaxy evolution, as will see in Chapter 12) RL nodes from shorter prisms have a freedom to oscillate. The return force of a deviated RL node of such lattice has a symmetrical return forces. They have valleys along orthogonal axes at 54.7 deg relative to the prisms axes. Consequently, an oscillating RL node in such lattice will exhibit also a quasisphere behaviour, but with smaller bumps. The lack of asymmetrical return forces as for CL space, means that such nodes will not have stable precession momentum. Consequently:

(a) The RL inside of FOHS does not have node oscillating feature, but embeds strong SG energy due to higher RL node density and radially touching RL nodes.

(b) In RL of mixed type only the nodes of shorter prisms have oscillating freedom but not exhibiting synchronised SPM effect as the CL nodes.

Feature (a) is important for the trapping effect of the rectangular lattice in cylindrical space, described later. Feature (b) is important for the particle crystalization mechanism which takes place in a hidden phase of every galaxy evolution (discussed in Chapter 12).

2.10 Disturbance of the CL space surrounding the helical structures.

2.10.1 Electrical charge and electrical field

The prism is extended solid object with anisotropic SG field aligned with the prism's axis. So the vector of aligned SG field according to the inverse cubic law is.

$$F_{SG} = G_o \frac{m_o^2}{|r|^4} \dot{r}$$

where: $\hat{r}$ is the directional unit vector.

If a huge number of prisms are ordered in some spatial configuration, their effective SG component in CL space is increased. If the prisms are of same type (right or left handed) the effective component has also the same handedness. The twisted rectangular lattice [RL(T)] inside the first FOHS has an ordered configuration. So the RL(T) is able to provide a strong modulation of the the external CL space. How this modulation affects the dynamical behaviour of the oscillating CL nodes?

CL nodes (with a normal ZPE) oscillate with their resonance frequency that is much higher than their SPM frequency. When a FOHS is in such environment, a strong interaction appears between its RL(T) from one side and the oscillating external CL nodes. While the ordered structure of RL(T) has a fixed spatial position, determined of the helical structure, the NRM and SPM phases of the CL nodes are free to be adjusted. In fact the phase of NRM vectors are adjusted to the RL(T) field due to a prism - to prism interaction. On the other hand the CL nodes having a common SPM frequency tends to be synchronized by a phase propagated with the speed of light (discussed later in this Chapter). In this conditions the SG field of RL(T) interact as an AC type of prisms interaction which defines the direction of SPM phase propagation.

As a result of the above described interactions the CL node quasispheres obtain simultaneously elongation and spatial orientation. Such quasisphere is shown in Fig. 2.29. The elongated shape means that a larger number of resonance cycles are involved in the cones of elongated size than in the cone of the shrunk size of the quasisphere. The resonance cycle although is still stabilized. The kinetic energy of such quasisphere is larger, but SPM frequency is lower. There are two important features in this interaction process:

(a) If one FOHS is inside of another external FOHS, the RL(T) of the external structure only are able to modulate the external CL space

(b) quasispheres of the same type can be only affected

The feature (a) is explained by the lack of interfacing between the SG fields of the two internal RL(T)s due to their different handedness. This feature defines a rule **any FOHS can modulate the external CL space, by creating elongated**

quasispheres only by its RL(T) that is located most externally in their internal (curved) cylindrical space. In other words, the interaction is determined by the external helical shell of the RL(T).**

The feature (b) is a result of prisms to prisms interaction of parallel type and from the feature that the opposite type node can get partially complimentary motion due to the induced interaction between the neighbouring nodes having different handedness.

According to the above made considerations, we are able to provide the following definition of electrical charge and field:

(A) The electrical charge is a spatial region of CL space where the CL nodes obtain EQ type of node quasispheres arranged in particular order

(B) The electrical field is a CL space disturbance, which affects the spatial parameters of the NRM vector of CL nodes.

(C) The electrical charge invokes EQs which form electrical field lines.

(D) The electrical charge could be of static or dynamical type:

- (a) The static charge is invoked by one or group of helical structures whose charge is not compensated in proximity.

- (b) The dynamical charge can be compensated or neutral (formed of equal amount of opposite EQs) or not compensated. The latter will appear as a propagating charge.

It is evident, that in a normal CL space only a particle containing a matter may have a static electrical charge. It is also evident, that in the higher order structures, containing FOHS, **only the external helical shell may exhibit electrical field.** This is the case of the structures shown in Fig. 2.10.b and Fig. 2.13.a.

Figure 2.29.E shows the near filed electrical lines for the radial section of FOHS, whose internal RL(T) terminates by a hole where another FOHS of opposite prism type has been crystalized

(a process taking place in a hidden phase of galaxy evolution discussed in Chapter 12).

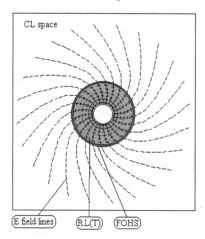

Fig. 2.29.E
Radial section of near E field of FOHS with internal RL(T)

The external filed lines follow the trend of the internal lines in the near field. In a far field of a static charge, however, they become radial due to the Zero Point Waves that tend to equalise the ZPE of the CL nodes.

Some helical structures could exhibit near electrical field, but locked in a proximity space around the particle (case of neutron discussed below). The SG potential propagated the handedness (interaction due to twisted part of prism) must overcome some critical value that depends on the neutral SG field of the helical structure (due to the cylindrical part of the prisms). If this condition is not fulfilled, the handedness field will be locked by the neutral SG field. In this case, we can say that the structure has a near field, but lacks a far field. The condition for the "far" field may depend, also, on the overall geometrical shape of the helical structure, because the overall shape influences the E-field lines orientation in the "near" field.

Stated in different words, a far electrical field is obtained, when the SG(TP) field, propagating the handedness, is able to escape from the local neutral SG(CP) field.

In this context, the structures shown in *a.* and *c.* in Fig. 2.14 will have a near, but not a far electrical field. Although, when in motion, the structure of Fig. 2.14.a (neutron shape) can gener-

ate a weak magnetic field, and consequently, the near electrical field in this case becomes unlocked. When the same structure is arranged in a configuration, shown in Fig. 14.b, it obtains a far electrical field even in a static position (proton shape), due to a spatial realignment of the lines of the near E-field, generated by the external shell of RL(T).

2.10.1.A. Graphical rule for a particle charge

In order to determine if the structure may have a far field for a lattice disturbance, we could apply the following graphical rule:

Draw tangent lines to every equidistant point of external FOHS of which the complex structure is made. Use equal line lengths larger than the size of the structure. If the endpoints of the lines are distributed not uniformly in a surface of sphere outside the structure, this helical structure will have far electrical field. If they are uniformly distributed, the structure will not have such field. Perform the checking procedure for a set of line with different lengths.

2.10.2 Interaction between particles, possessing a charge.

According to the definition of the previous paragraph we may use the term "charge particle" assuming it is comprised of helical structures possessing a FOHS with RL(T). A single charge particle elongates the quasispheres of the CL nodes of same handedness and align them to the vector of electrical field, due to its SG field propagating the handedness. The shape of distorted quasisphere was shown in Fig. 2.29.a. At the same time the electrical field decreases the SPM frequency of affected nodes.

Due to the handedness attribute of the particle's FOHSs and the CL nodes the right handed type of interactions are distinguished from the left handed one. We may consider that the interaction between FOHS SG(TP) field and the CL nodes of the same type (and handedness) is based on a prism to prism interactions (see §2.5). These type of interaction between the enhanced field from RL(TP) of the FOHS and the CL space, however, is much stronger in comparison to prism to prism interactions in empty space.

The SG field of the charge particle could be regarded as a global synchronization field for every node. **In this case every CL node could be considered as an oscillator, whose frequency is adjusted to the global SG(TP) field of the same handedness, provided by the charge particle.** For simplicity, however, we will consider that the charge synchronizes the nodes. Or in other words, the charge synchronizes the node momentum vectors of the same type nodes. This case is schematically shown in Fig. 2.30.a., where a rectangular lattice is shown for simplicity.

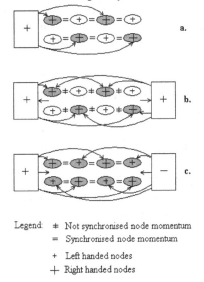

Legend: ≠ Not synchronised node momentum
         = Synchronised node momentum

         ⊕ Left handed nodes
         ⊕ Right handed nodes

Fig. 2.30 Interactions between charge particles in lattice space

Fig. 2.30.b illustrates the nodes behaviour between two charges with a same handiness (polarity). In this case, the nodes are influenced by two global fields, generated by the two charges and they are confused from which field to get synchronization at the resonance frequency. In such way the node momentum vectors are almost aligned in the direction of the external field but their phases are not synchronized. Considering only the TP component of the node momentum vectors, the random phases will bump and provide forces of repulsion. This forces will push the charges away. The pushing force is active even if the charges are in motion in an opposite direction. This is possible, because, the passed nodes get readiness quite fast, due to their small inertial factor.

Fig. 2.30.c illustrates the node behaviour between two charges of opposite polarity. In this case every node get synchronization from the global field corresponding to its polarity (handedness). There is not conflict of synchronization. The positive nodes gets node momentum synchronization provided by the positive charge, and the negative nodes - by the negative charge. The both charges although are not synchronized between themselves. How the node momentum phase can be adjusted in order all the node to be in phase and to get attraction force? The explanation is the following:

The propagated TP NRM reaches the more distant nodes of same type with some delay (determined by the light velocity), but the phase differences between the neighbouring opposite nodes will have some stable value. This constant phase difference could be eliminated, by suitable displacement of the nodes from their geometrical equilibrium points. As a result of this, the node momentums of all nodes appear synchronized and the charges get an attractive force due to the SG(TP) attraction.

Let us analyse the dynamics of the attraction forces, when the opposite charge particles are in motion. The neighbouring quasispheres (of both types) are synchronized by the external field. They eliminate the constant phase difference between themselves by a proper geometrical displacement. Consequently, the neighbouring nodes obtain a complimentary motion. As a result of this, they tend to become closer, creating forces of attraction. The mean node distance although could not be changed, as will be seen from the photon propagation analysis. The attraction forces become applied to the external charges, which are responsible for their generation. If the opposite charges are not fixed they will move under the attraction forces. The motion, although could not diminish the attraction force effect, as the phase propagation of NRM(TP) is usually much faster, than the velocity of the moving charge particle.

In a similar way, it can be considered that the CL space between same types of charges creates repulsion forces, applied to the charges.

In a case of single charge, the opposing interactions of the opposite nodes are missing, because they do not have spatial reference point in order to resist.

Summary:

- **The sign of the electrical charge is defined by the handedness of the prism, from which the external shell facing the lattice space is built.**
- **Internal FOHS (shell) can be considered to possess a hidden electrical charge, when it is completely inside of another FOHS, which is always from an opposite handedness. Once it appears in open CL space, its charge becomes active.**
- **The charge particles are able to interact between themselves due to the ZPE of the CL space.**
- **The attractive and repulsive electrical forces are reactions of the CL space, applied to the charge particles**
- **Electric and magnetic interactions are impossible at lattice temperature of absolute zero.**

If a finite amount of atomic matter in some particular conditions is at temperature of absolute zero, it does not mean, that the surrounding CL space is at the same temperature. The ZPE in open space is equalized very rapidly due to the Zero order waves (considered also as ZPE waves).

2.10.3. Node quasisphere behaviour in a permanent magnetic field

Let us consider CL domain away from any static or dynamic electrical charge. The common synchronization of the CL nodes in such domain requires a finite time. We may accept apriory that this time is larger than the resonance period of CL node. This will become apparent later. Having in mind also, that the EQ CL has a larger energy than MQ node we may conclude:

In not disturbed CL space CL nodes can not become spontaneously EQ types of nodes since this will means a creation of electrical charge.

Note; Not disturbed CL space exclude generation of opposite EQ nodes. They are feature corresponding to the Dirac see idea, but we must consider that there appearance is always invoked by some CL space disturbance.

Now let investigate the possibility for spontaneously synchronized normal quasispheres (MQ) nodes in not disturbed CL space domain but at

SPM frequency, that is much lower than the resonance one. In order to propagate an external magnetic field, the CL node energy should be above the E_{sc} level. In this case, every node exhibits stabilized resonance and SPM frequencies. Analysing the node fluctuations in this energy conditions, we find two antagonistic processes:

1) The system try to keep uniformity energy level

2) The system try to keep the lowest state by minimizing the interaction energy between the neighbouring CL nodes.

From the point of view of the neighbouring CL node interaction, the lower interaction energy corresponds to commonly synchronised nodes by their SPM frequency. As a result of this effect, it is very probable, that CL may contain commonly synchronized domains comprising of large number of nodes. Any such domain will exhibit a common SPM field that is a features of magnetic field, if they were connected in closed loops. So the field of this domains has a feature of local magnetic field, but with open lines. If their lines recombine in closed loops, an energy will start to circulate and this will change the CL energy uniformity. This means a creation of magnetic line, but this process will not be self-created without external cause (external magnetic field or moving charge).

- **The SPM synchronised CL nodes, not involved in closed loops can be regarded as magnetic protodomains.**

Without external cause, the mentioned magnetic protodomains (open loops), continuously fluctuate and recombine, i. e. their space-time parameters fluctuate, but for a large number of CL nodes they will exhibit a constant average value (this will be discussed later).

When an external disturbance with the same SPM frequency is applied, the magnetic protodomains readily accept this synchronization and organize themselves in loops, that become closed in the CL space - magnetic lines. The number of created closed loops (magnetic lines) depends on the strength of the external field, which causes the SPM synchronization.

The mutual synchronization of the CL nodes involved in magnetic protodomains, which become part of magnetic field lines if external synchronization factor exist is illustrated in Fig. 2.31. In this figure two dimensional projections of SPM synchronised CL nodes are shown. The arrows show the confinement between the direction of the SPM vectors of the neighbouring nodes. Their directions are confined in a three dimensional space, but with one very important detail.

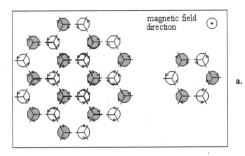

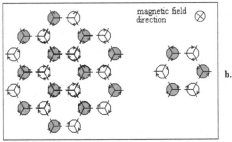

Fig. 2.31. Magnetic protodomains in CL space and their synchronization at SPM frequency

The phase between the nodes of same type is zero, but between the nodes of different type is equal to $\pi/2$. As a result of this, a wave motion synchronised with the CL node resonance frequency is possible in two opposite directions, as shown in Fig. 2.31.A. These two directions in fact define the direction of the vector magnetic field

If analysing the interaction energy transfer between the nodes, we could find that some portion of energy really circulate in the closed magnetic line. If this line is intercepted by a conductor, the circulated energy dissipates in it in a form of induced electrical field, pumping the CL nodes of the conductor space and converting them to EQ nodes. If the conductor is a closed loop an electrical current is generated in order to compensate the generated charge potential.

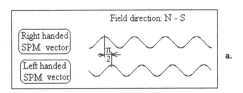

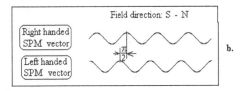

Fig. 2.31.A
Direction of the magnetic field vector, defined by the phase difference between the SPM vectors of both type of nodes

If the node quasispheres are distorted by electrical field, their directional property and the spatial synchronization between the neighbouring nodes will be disturbed. If the condition for the phase difference of $\pi/2$ has some tolerance, the distorted quasisphere also will have some critical value, beyond which the magnetic filed propagation will not be possible.

Creation of magnetic line can be invoked not only by external magnetic field (permanent magnet) but by moving electrical charge. In fact the case of permanent magnet is also invoked by properly oriented electrical charges inside the permanent.

The existence of magnetic protodomains is behind one of the important features of the CL space, corresponding to the physical parameter known as permeability of the free space μ_o.

The rectangular lattices does not have a frequency stabilised SPM effect, and consequently, could not propagate magnetic field.

Knowing the property of not distorted (symmetrical) and distorted (elongated) quasispheres of the SPM vector, in further analysis we may refer to them simply as Electric and Magnetic quasispheres. From this point of view it is convenient to introduce the following abbreviation:

EQ standing for **Electric Quasisphere**

MQ standing for **Magnetic (not disturbed) Quasisphere**

Summarizing the analysis, we can provide a definition of the magnetic field in CL space:

- **The permanent magnetic field in CL space is composed of closed loops formed of aligned MQs and synchronised with a phase propagating NRM frequency. In these loops excess energy is circulated.**
- **The magnetic field line is formed of spatially aligned MQs, connected in a closed loop. The SPM vectors of left and right handed CL nodes, along the magnetic line, are spatially synchronized, while the relative phase difference between them is equal to $\pi/2$. The sign of this phase difference defines the direction of magnetic line (for example, $\pi/2$ - for N-S direction and $-\pi/2$ - for S-N direction).**
- **A CL node quasisphere distorted beyond some point could not be included in a magnetic line. Such distortion is an attribute of electrical line.**

2.10.4 Quantum electromagnetic wave

2.10.4.1 Energy propagation between neighbouring nodes

In the return force analysis, the neighbouring nodes were considered fixed only for simplicity. In fact the oscillation of any one single node could not be isolated from its neighbour.

The dynamical model for this exchange is complicated, although an aproximative analogy could be made with a system of elliptical conical pendulums, attached to a common spring. A configuration of such system is illustrated in Fig. 2.32.

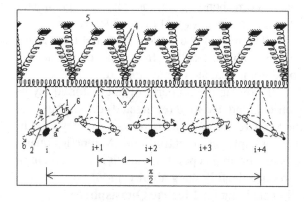

Fig. 2.32

System of conical pendulums in a common spring

The common spring is attached to the fixed pads 5 by set of other strings 4 with equal lengths. The pendulums, having equal length and mass, are mounted equidistantly. All pendulums have one and a same oscillating frequency. Let us induce, in first, a conical oscillation of the left most, pendulum, but without precession motion of the long axes 2. For enough long time interval, a steady process will occur and the neighbouring pendulums will get complimentary motion, also without precession. Let us then invoke a continuous precession in the left most pendulum i, with direction shown by the arrows b-b. In a proper experimental arrangement, a continuous energy for the precessional motion of this pendulum could be supplied by a magnetic field below the pendulum mass. Due to the spring stiffness, the neighbouring node will also obtain such precession, but in an opposite direction and with some delay. So the invoked precession will be propagated with some delay to the next pendulums. The delay will depend on the spring stiffness and lengths, on the one hand and the pendulum mass and length on the other. After many periods, we will find that the precession axes of (i+n) pendulum is in a same position as the pendulum i. Following the black reference point 6, we see that in our case the pendulum (i+4) gets delay of $\pi/4$. Then for this example we have n=16, and the precessional phase of the long axes will be repeated in distance nd, where d - is the distance between neighbouring pendulums. At the same time, all pendulums will have one and a same precessional period, but their phases will differ by a constant value. So a wave like motion of energy occurs with a spatial wavelength $\lambda = nd$.

The provided model could be used for better understanding the property of the CL space in a wave-like energy propagation. The following analogy can be used:

Pendulum model	CL node dynamics
oscillating frequency <=>	node resonance frequency
long axes momentum <=>	NRM vector
long axes precession <=>	SPM vector

When the lattice is in a steady state, it has only magnetic protodomain fluctuations, whose energy is a part of its ZPE. In this case we may consider that the energy is equally distributed between the both types of nodes. The amount of ZPE, although, could not pass the upper critical level E_{cr2}, because this will lead to a lattice break, according to the single CL node dynamics. The CL structure, however, has ability to react fast enough, so the excess energy of the node is propagated to the neighbouring node.

Let us suppose that one node gets excess energy above E_{cr2}. We have to analyse the node motion during one resonance cycle. When moving in a destruction direction along one of abcd axes, the node will affect the four neighbouring nodes whose prisms are of opposite handedness. While the intrinsic interaction from the NRM(CP) vector leads to node deviation toward the destruction a,b,c,d zones, the NRM(TP) vector try to keep the node motion away from these zones. As a result of this, the neighbouring nodes will take part of the excess energy. This energy is then propagated to their neighbours in a same way. There are two node distances in CL lattice, considered as a neighbouring node distance: distance along abcd axes, and distance along xyz axes. The second one is about twice larger. Let us analyse the momentum transfer along xyz axes between two neighbouring CL nodes, as illustrated in Fig. 2.33.

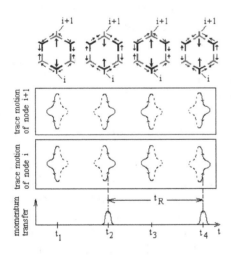

Fig. 2.33

Two neighbouring (by xyz axes) nodes (i) and (i+1) are shown in the upper part of Fig. 2.23 at four consecutive time moments. The motions as a part of resonance cycle are shown by arrows: the

big arrows are for motions along x,y,z axes and small arrow - along a,b,c,d axes. The energy between (i) and (i+1) nodes is transferred through the two arms, formed of node prisms. **While the arrow directions indicating the motion of the two arms are symmetrical, the prisms spin rotation (virtual) due to prisms interaction are not symmetrical. This will affect the balance of the momentum.** The trace of motion and the momentum transfer are shown below. We see, that the momentum from node (i) transfers to node (i+1) in the position, when they are closer. The most important fact is that the condition for momentum transfer occurs ones per resonance cycle. Consequently, we may conclude that:

If the CL node possesses an excess energy, it will be propagated with a velocity of one node distance along x,y,z axes per one cycle of NRM vector (CL node resonance cycle). The resonance cycle for the electrical quasisphere, however may be different, than the magnetic one.

The above rule is simply expressed by the equation:

$$\upsilon = d_{xyz}/t_R = d_{xyz}\nu_R \quad \text{(m/s)} \qquad (2.17.A)$$

where: d_{xyz} - is a node distance in x,y,z direction, t_r and ν_R - are respectively the resonance time and frequency for EQ.

According to the analysis of the previous paragraph, the CL nodes in a steady state space domain are connected in temporally magnetic protodomains, phase synchronized by a propagating (with a light velocity) SPM vectors. This condition favours the start of wave energy propagation. In the steady state of wave propagation, the phase difference of the SPM vectors of neighbouring nodes is very small, and their quasispheres are aligned. Therefore, the propagation of the excess energy between the nodes is in a moment when their bumps are aligned with a very small phase difference. In such conditions, the efficiency of the momentum transfer is maximal. Then we may conclude, that the propagated excess energy will influence the phase of the SPM vector.

Vector of Running SPM quasisphere

From the above analysis we see that the excess energy is carried by the EQ nodes (electrical quasispheres) and is propagated with a speed of one node distance per resonance cycle. **It is more convenient to analyse the process of the energy propagation by considering a running EQ SPM and MQ SPM vector instead of the stationary one**. This will facilitate the analysis of the quantum wave propitiations in CL space for unveiling the structure of the photon wavetrain.

Definition of Running EQ and MQ SPM vectors:

(a) A Running EQ SPM vector is a EQ SPM vector of a stationary CL node, in a time interval equal to one SPM cycle.

(b) A running MQ SPM vector is a MQ SPM vector of a stationary CL node, in a time interval equal to one SPM cycle.

Notation:

REQ SPM - running EQ SPM vector

RMQ SPM - running MQ SPM vector

(c) Both REQ SPM and MQ SPM vectors preserve the shape of the stationary node EQ or MQ SPM vectors.

(d) Both REQ and MEQ SPM vectors have apparent frequency referenced to the photon frequency according to expression

$$E = h\nu \qquad (2.18)$$

where E -s the photon energy, h - Plankc's constant and ν - a quantum frequency.

(e) RMQ SPM vectors may have one cycle for one cycle of a stationary MQ or $1/n$ cycles, where n - is a subharmonic number (integer 1, 2, 3, ...) to be discussed later, but not a phase difference. REQ SPM will have the same number of cycles as REQ SPM, but with a phase difference.

Photons are propagated without energy loss. Consequently, the photon wavetrain should have a boundary condition. Since MQs not involved in a magnetic line do not carry excess energy and RMQs have the same parameters, we may conclude that ERQs should be enveloped by RMQs.

The propagation of excess CL node energy may involve a large, but finite number of REQs. In §2.10.4.1 it was mentioned that the resonance frequency (NRM frequency) of the EQ nodes may be different than the MQ nodes and could depend on the EQ elongation. A different NRM frequency will mean a different SPM frequency. In proper spatial configuration of the REQs, the frequency or phase difference between the EQs and MQs may be

spatially distributed along a large number of EQs with gradually reduced elongation (eccentricity). This means that the major axis of the eccentricity between the neighbouring REQ's will have small spatial deviation. Since RMQs do not carry excess momentum, we may conclude:

- **the spatial energy distribution of aligned REQs should terminate with RMQs.**
- **RMQs having the same SPM frequency like stationary MQs will provide isolation of the energy carried by the REQs. In other words RMQs will provide the boundary conditions of the propagating quantum wave - photon.**

The above conclusions are very important for conservation of the photon's energy in a finite volume of its wavetrain, while propagating with the velocity of light. The boundary conditions are very important isolating factor. Since the photon energies are different but quantized by the Planck's constant, their wavetrain should have a common spatial configuration with some difference reflecting their quantized energy. This problem can be solved if the the boundary conditions of the wavetrain are formed of RMQs which have cycles 1/n cycles (n = 1, 2, 3 ...) per one cycle of a stationary MQ SPM frequency. In such conditions the separation between the energy propagating REQs from the MQs of the surrounding space is optimal, or the photon energy will be preserved.

Note: The cycle of the RQM and REQs is an apparent cycle, referenced to the photons frequency according to Eq. (2.18). The stationary CL node involved in any particular moment in the photon wavetrain possess its own SPM cycle.

The above consideration becomes apparent, if analysing the SPM vector behaviour of the boundary RMQs. Since RMQs is phase synchronized with the stationary MQs, we may analyse the SPM vector behaviour of the stationary CL nodes at the boundary of the wavetrain. Let us define a reference point of the wavetrain, for example, when the SPM vector passes through its maximum in one fixed spatial point. We have to estimate the SPM vector of a stationary node residing in a plane passing through the above selected point and normal to the axis of the wavetrain but in a distance corresponding to the boundary conditions. We have to estimate how many apparent SPM cycles of the boundary RMQs will correspond for one SPM cycle of a stationary node for photons with different quantum frequencies. The results are given in the Table 2.2.

Table 2.2

Subharmonic (frequency) number	Wavelength	Number of the apparent cycles of the boundary RMQs per one SPM cycle of MQs
1	λ_{SPM}	1
2	$2\lambda_{SPM}$	1/2
3	$3\lambda_{SPM}$	1/3
n	$n\lambda_{SPM}$	1/n

From the view point of the quantum wave, the boundary conditions are provided by the RMQs that are at the boundary of the wavetrain volume. Their apparent SPM vector appears as a subharmonic of the SPM frequency of the external CL space. Another requirement is the integrity of the wavetrain volume. Then we may conclude:

The wavetrain of the quantum wave is characterised by two basic conditions: boundary conditions and wavetrain integrity.

The two basic conditions can be formulated as:

• **Boundary conditions are provided by the RMQs whose apparent SPM frequency is a subharmonic of the SPM frequency of the external CL space**
• **The integrity of the quantum wave could be provided by synchronization between the REQs and RMQs.**

From the second conclusion it follows that the stationary EQ nodes may differ from the stationary MQ nodes only by a phase, while having a same frequency. If taking into account that the NRM and SPM vectors spend a longer time in a narrow spatial angles entered along xyz axes the possibility for a phase difference is quite reasonable.

According to the accepted notations in Table 1A, the shortest quantum wave will have a frequen-

cy equal to the SPM frequency of the CL node. From the point of view of terminology consistency with the quantum waves, that are subharmonics, this quantum wave is often referenced as a first harmonic of SPM frequency. Its energy according to Eq. (2.18) is 511 KeV. For the Earth local field its wavelength is equal to the Compton wavelength. This is the shortest wavelength generated by the electron system (discussed in Chapter 3) and defines an important limiting condition in the wavetrain configuration (will be shown later). A periodical motion of a helical structures in steady state CL space generates a quantum wave, whose frequency is a subharmonic of the SPM frequency. (The subharmonic rule is not valid for the gamma rays, which posses larger energies. This is discussed in §2.10.7.).

The SPM vectors behaviour of the nodes, involved in the energy propagation for first and second subharmonics are shown respectively in Fig. 2.34.a and b. The RMQs and REQs are shown by their envelopes. The circular shape corresponds to a RMQ, whose apparent SPM frequency can be only v_c/n (integer by n = 1, 2, 3..), but its symmetrical shape is preserved. The ellipse corresponds to a REQ for which both parameters: the apparent SPM frequency and the quasisphere shape are affected. The first one favours the magnetic-like field propagation, while the second one - the electrical one.

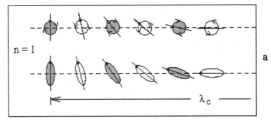

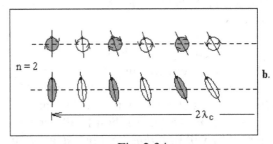

Fig. 2.34

The case **a.** shows a fragment of a wavetrain in which the excess energy propagates in one direction at a distance equal to the SPM wavelength for one period of SPM vector. From the BSM analysis it becomes evident that the SPM frequency is equal to the well known Compton frequency. (In fact this is accurately valid for the Earth local gravitational field). Then the SPM wavelength is equal to the Compton's wavelength experimentally estimated as $\lambda_c = 2.426 \times 10^{-12}$ m). In other words, the velocity of the wave propagation is $\lambda_c / \nu_c = c$, where ν_c - is the Compton (SPM) frequency and c - is the velocity of light. In case **b.** the excess energy propagates at a distance twice the SPM length (a distance equal to twice the Compton wavelength), but for two SPM cycles of MQ. So the propagating velocity is again equal to the velocity of light.

The above example is very important for understanding that wavetrain containing REQ and RMQ with subharmonic number of n = 1, 2, 3... propagate with one and a same velocity equal to the velocity of light.

The REQs, involved in the wave motion, may have a range of distortion between a prolate spheroid and a symmetrical shape of RMQs.

One particular feature of the neighbouring nodes that are of opposite handedness is that, the coordinate system of their x,y,z axes is rotated at 90 degrees each other, but the xyz axes assignment is relative (and dynamics on + and - of each axis is symmetrical), so the rectangular set of the neighbouring CL nodes coincide. Therefore, **the x,y,z coordinates of all CL nodes are aligned and their axes pass trough the bumps of the node quasispheres**. This fact facilitates the analysis of the propagated EM wave. We may consider that the EM wave propagates in a common orthogonal system XYZ, in which CL nodes are aligned, while keeping in mind the actual propagation of the CL node momentum is through abcd axes, defined by prism axes.

The excess energy of a CL node is obviously synchronized with the NRM and SPM vectors. The EM waves are wavetrains containing many repeatable cycles of the involved REQs and RMQs. Then we arrive to the following conclusion:

(A). If REQs have a slightly different frequency, they will be aligned not in a straight line but in a helix. Then the wavetrain of the quantum wave will have a helical momentum.

The conclusion (A) is very important for unveiling some features of the photons, such as the different types of polarization.

Let us consider two options for the possible alignment of RMQs in a propagated wavetrain of the photon. (a) the RMQsa are aligned in a straight line along the axes z; (b) the RMQs are aligned in a helical curve centred along the z axes. The both cases are shown in Fig. 2.37.a and b. respectively, where 1 is a trace of the electrical quasispheres and 2- a trace of the magnetic quasispheres.

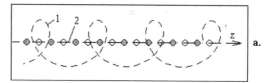

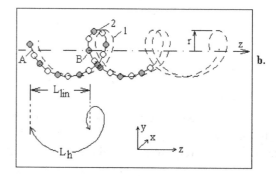

Fig. 2.34.A

From the analysis of RMQs and REQs alignment in Fig. 2.34 we found that the axial propagation for subharmonics n = 1, 2, 3... is one and a same - equal to the light velocity.

Since MQs involved in RMQs have unchanged frequency, while EQs have, we may consider that both cases a. and b. shown in Fig. 2.34.A are allowed. This means that in the transverse section of the photon wavetrain the RMQs can be occupy both the central position and the boundary circumference, while in the same time they can be phase synchronized with the REQs, which will be arranged in a helical volume inside the boundary defined by RMQs.

2.10.4.2 Quantum wave configuration.

The propagated wave occupies a region around the propagated axes with gradually but finite drop of the energy at given radius. Knowing that the vectors E and H are orthogonal each other in any moment, we can build configuration of the photon wavetrain with its vectors of electrical f and magnetic field. It is easier in first to illustrate the node configuration for a standing wave, where the E and H vectors appears in fixed spatial positions. Fig. 2.35 shows only the central parts of such waves fore two cases **a.** not polarized wave, and **b.** - polarized wave. (Note: According to conclusion (A) in §2.10.4.1, the central part should be occupied by RMQs. This is not shown in Fig. 2.35 with a purpose of drawing simplification).

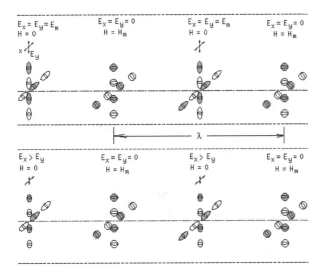

Fig. 2.35
Central part of standing quantum EM wave
(see the note in the above text)

The peripheral part of the quantum wave is also not shown for simplicity. In order to satisfy the boundary conditions, the elongated shape of the quasispheres in a radial direction gradually transverses to a spherical one (REQs gradually transfer to RMQs). The direction of wave propagation coincides with the horizontal axis. The spherical shapes in the standing nodes correspond to the vector of magnetic field H_m.

Fig. 2.36. illustrate part of the radial section of the wavetrain of a normal quantum wave, showing the instant arrangement of the RMQs and REQs. The spatial orientation and the SPM phase of RMQs and REQs define the orientation of the magnetic H and electrical E vectors..

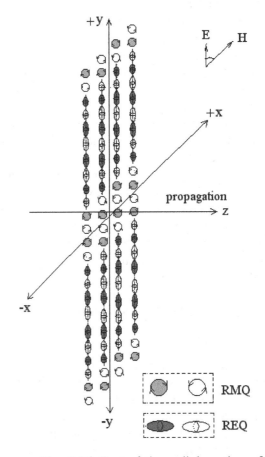

Fig. 2.36. Part of the radial section of the wavetrain of a normal quantum wave, showing the instant arrangement of the RMQs and REQs. The spatial orientation and the SPM phase of RMQs and REQs define the orientation of the magnetic H and electrical E vectors.

In an instant moment the electrical filed along the axis y and -y, for example is not at opposite direction but additive. This is because the phase difference of the SPM vectors of the involved stationary nodes in y and -y axes is equal to π. This condition is valid for any other diameter in any moment of the radial cross section of the photon wavetrain.

The spatial arrangement and phases of both types of running quasispheres allows explanation also of the photon polarization: linear (elliptical) and circular. In linear polarization, for example, the REQs along one diameter axis is larger than in the other perpendicular diameter. Right and left handed polarization also is explainable by the mutual rotation of the REQs and RMQs. The symmetrical components of REQs also provide better stability of the polarization state of the quantum wave.

In §2.10.2 it was shown that the CL space between charge particles provides interaction forces - effect caused by the synchronized neighbouring quasispheres. But the arrangement of the electrical quasispheres in the photonic wave is similar, only the opposite REQs are alternatively distributed across the wavetrain diameter. Consequently, we may expect an effect of attraction. In other words, the photonic wave containing positive and negative quasispheres should exhibit attractive Coulomb forces. This is the other unveiled factor that likely contributes to the integrity of the photon wavetrain. This will be further discussed later in this chapter.

2.10.4.3 Boundary conditions parameters of the quantum wave

A single photon can pass enormous distances in the Universe without any energy loss. In other words, its energy loss is practically zero. The classical wave function used in Quantum mechanics does not provide this feature of the photon.

Accepting the condition of magnetic quasispheres as an isolators, means that they provide a boundary condition of the photonic wave. Then the transverse dimension of the wavetrain should reach a limit value (this will be discussed in the next paragraph). From the case, shown in Fig. 2.37.b we see, that we may define two types of the light velocity: one helical and one linear. If accepting that a very small CL node excess will be propagated for one NRM cycle, then the light velocity becomes defined by the internode distance and the NRM period. In such case the helical velocity will appear larger but we must not forget that this is just a parameter of the wavetrain that could not exist independently.

The velocity of the quantum wave propagation appears as a linear velocity, while the path of the energy momentum propagation follow helical traces. Consequently, the helical light velocity should be defined for a definite radius of the wavetrain. Following the same logic, we will distinguish two types of wavelengths for a given quantum

wave: linear wavelength, referenced also as longitudinal and a helical wavelength. The helical wavelength, like the helical light velocity is not a measurable parameter, but analytical one. It is very convenient for the analysis, because it is aligned with the path of the momentum propagation. The helical wavelength should be defined for a definite radius from the central axes.

Unfolding the trace for one helical step we have: $L_h^2 = L_{lin}^2 + \pi^2 r^2$, where L_h and L_{lin} are respectively the helical and linear paths. Then the relation between the linear and helical wavelength is

$$\lambda_h = \sqrt{\lambda^2 + 4\pi^2 r^2} \qquad (2.20)$$

where: λ - is the linear wavelength

Now let use the concept of the REQ. **The REQs, appear to have energy above E_{cr2}, while the RMQs energy is normal or equal E_{cr2}. So the photon energy is carried by the running electrical quasispheres.**

Following the above considerations we may guess what is the configuration of the radial extent of the photon wavetrain. Fig. 2.38 illustrates this configuration in simplified way.

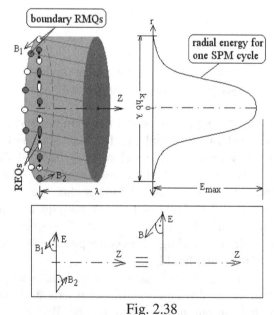

Fig. 2.38
Arrangement of REQs and RMQs in the wavetrain of a quantum wave (photon)

We must not forget that the single maximum shown in Fig. 2.38 is contributed from the REQs along the whole diameter integrated for one full SPM cycle of the stationary CL node of MQ type.

Figure 2.38 shows also the vectors of electrical filed E and magnetic field B. The equivalent magnetic filed vector is formed of multiple distributed vectors like B_1 and B_2, which keep the rotational helical features of the photon. The vector E and distributed vectors B_1, B_2 are equivalent to two vectors E and B perpendicular each other, with common origin lying in the axis Z. By squaring the E vector we can obtain the energy in any moment. **In such case the both vectors behaves in a same way as in the photon wave function used in the QED** (quantum electrodynamics), **while possessing at the same time boundary conditions**.

The photon wavetrain volume appears well isolated from the external CL space, but only for the quantum wave, defined by the photon frequency, which is a subharmonic of ν_{SPM} frequency. For a photon with a different subharmonic number $n > 1$ (or photon frequency ν_{SPM}/n) the wavetrain volume is transparent for photons with different subharmonic number.

It is useful to determine the helical wavelength at the boundary radius. If accepting that the transverse wavelength is a same as the longitudinal one, then $r = \lambda/2$ and the helical path will be $\sqrt{1 + \pi^2}$ times longer than the linear one. In Chapter 3, when discussing the electron and its quantum motion, it is shown that the boundary condition become satisfied at: $r = 0.6164\lambda_o$. In the analysis of §2.11.2.3 and §2.11.3 a conclusion is made that the transversal width is increased twice during the detection process, due to disappearing of the Coulomb forces, which keep the quantum wave integrity. So the full width of the detected wave becomes $2r = 2 \times 2 \times 0.6164\lambda = 2.4656\lambda$. This value is pretty close to the Airy disk diameter of the monochromatic wave pattern obtained by diffraction limited optics: 2.44λ. In such case the **coefficient for the boundary conditions is:**

$$k_{hb} = \sqrt{1 + 4\pi^2(0.6164^2)} = 4 \qquad (2.20.a)$$

The above calculated value should be consistent with the boundary conditions of the quantum motion of the electron. This is discussed in Chapter 3. Despite the accepted value of 4, we will

continue to use explicitly the boundary coefficient in all equations for two reasons: to show its involvement in the equations and to give a possibility for eventual correction of its value.

The linear light velocity expressed by the path of the first harmonic is:

$$c = \frac{\lambda_o}{t_o} = \lambda_o \nu_o = \frac{\lambda_{SPM}}{t_{SPM} k_{hb}} = \frac{c_h}{k_{hb}} \qquad (2.21)$$

where: $\nu_{SPM} = \nu_o$; $t_{SPM} = t_o$

ν_o and λ_o can be called respectively: fundamental frequency and fundamental wavelength.

c - is linear velocity;

c_h - is a helical light velocity.

In Chapter 3 we will show that, in the Earth local field, the fundamental frequency and wavelength are equal respectively to the Compton frequency and Compton wavelength. So in our calculations in the next chapters we will refer this SPM parameters to the Compton's parameters, which are known with a high degree of accuracy:

$$\nu_{SPM} = \nu_c \textbf{;} \quad t_{SPM} = t_c$$

Important note: Let us accept that λ_{SPMh} corresponds to the Compton wavelength but in the helical trace. While $\nu_{SPM} = \nu_c$, and $t_{SPM} = t_c$, then $\lambda_{SPMh} > \lambda_c$, because they are referenced to different velocities (helical and linear velocities).

In Fig. 2.38, the shape of the electrical quasisphere is shown as for not shrunk radial space. The photon is propagated in a cylindrical volume, whose boundary are defined by properly oriented peripheral magnetic quasispheres. The apparent SPM vectors of the REQs and RMQs operate at subharmonic of λ_{SPM}. The value of this subharmonic defines completely the photon energy. The running electrical quasispheres inside the cylindrical volume have their axes normal to the axis of the wave propagation. The opposite pairs are equally deformed. The gradient of their deformation, discussed in the next paragraph, should provide an energy density similar like the Lorenzian function, with an exception of the tails. The tails fall to zero at the radius, defined by the magnetic quasispheres, due to the mentioned above transverse wavelength compact effect. The ratio between the linear and transverse wavelength is a finite value, expressed by the factor k_{hb}.

2.10.4.4. Radial energy distribution of the quantum wave

It is evident that the integrity of the photonic wave is possible if the SPM vectors of all neighbouring nodes - magnetic and electric are synchronized. It has been illustrated in Fig. 2.37, that the line of synchronization of the magnetic quasispheres formes a helix centred along the axis of the wave propagation. Then the electrical nodes also must posses SPM phase synchronization along the helical trace. In the photonic wave, the degree of polarization of any EQ depends of its radial distance from Z axis.

Unfolding the helical trace of the energy propagation of one EQ node at a distance r from Z axes, for a time interval of $t_{SPM} = t_c$ we have.

$$\lambda_h(r) = \sqrt{\lambda_c^2 + 4\pi^2 r^2} \qquad (2.22)$$

Dividing the trace path by the time t_c we get the helical velocity.

$$\upsilon_h = \frac{1}{t_c}\sqrt{\lambda_c^2 + 4\pi^2 r^2} = v_c\sqrt{\lambda_c^2 + 4\pi^2 r^2} \qquad (2.23)$$

Eq. (2.23) provides the conditions for synchronization of the EQs along a helical trace with a radius r. Its plot is given in Fig. 2.39.

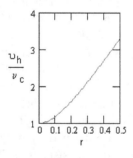

Fig. 2.39

The helical velocity at $r = 0$ approaches the value of the light velocity, c.

According to the boundary wave conditions, the REQs obtain minimum polarisation or practically are degenerated to magnetic quasispheres, at the boundary radius. Then the helical velocity should obtain a constant value, determined only by the harmonic number. It is logical to expect, however, that this change could not appear abruptly. It may influence also the shape of the radial velocity distribution.

We may find the shape of the radial velocity distribution by analysing the radial dependence of

the propagated momentum. For this purpose we may use the photon mass equivalence for the first harmonic wave.$m_1 = h\nu/c^2$, but referenced to a single node. Then the energy dependence from the radius could be described by the equation:

$$E = \left(\frac{1}{2}\frac{m_1}{n}\frac{1}{r^x}\right)\upsilon^2 = \left(\frac{1}{2}\frac{m_1}{n}v_c^2\right)(\lambda_c^2 + 4\pi^2 r^2) \qquad (2.24)$$

where: the factor $1/r^x$ gives the radial distribution. (In the next paragraph the photon mass distribution is discussed in details).

Fig. 2.40 shows the plots of Eq. (2.24), normalized to the product $\frac{1}{2}\frac{m_1}{n}v^2$, for the following values of x: 1, 1.5, 2, 3. The horizontal scale is one and a same for r and λ.

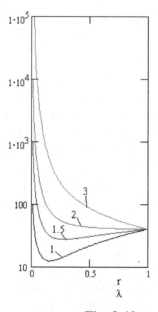

Fig. 2.40

The portion of the plots for x =1 and x=1.5 fall below the level common point at r = 1. This means too large negative (or reactive energies), so we may exclude this options. The plot for x = 3 is also not acceptable, because at radial distance of $r = \lambda/2$ the energy is still too high.

The plot for x = 2 only gives a reasonable energy distribution. Only this plot has a shape closer to Lorenzian shape of line. This corresponds to a distribution law proportional to inverse square of the distance. It is not difficult to guess, that this is caused by the interaction forces between the positive and negative electrical quasispheres, which simulate Coulomb like forces. We may likely conclude that: **The radial integrity of the quantum wave is kept by the Coulomb forces between the positive and negative d and complimentary synchronized EQs of the involved CL nodes.**

Now let make association between the radial energy distribution and the line shape for spontaneous emission. The spectral shape of line, corresponding to a spontaneous emission is given by the Lorenzian function. Expressing the Lorenzian function by the wavelength, and applying it for the first harmonic we get.

$$g(\Delta\lambda) = \frac{\delta\lambda}{2\pi c\left[\left(\frac{\Delta\lambda}{\lambda_c}\right)^2 + \left(\frac{\delta\lambda}{2\lambda_c}\right)^2\right]} \qquad (2.24)$$

where: $\Delta\lambda$ - is the wavelength change, $\delta\lambda$ is the spectral line width.

Fig. 2.40.A.a shows a plot of the radial energy distribution with the accepted inverse square law. Fig. 2.41.b gives the normalized Lorenzian line shape for different values of the spectral width, but in function of the transverse wavelength.

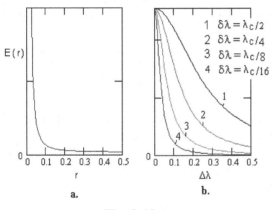

Fig. 2.40.A

Comparing the plots in Fig. 2.40.A a and b., we find that the energy distribution corresponds to a transverse width of $\lambda/16$. **This means that at distance $r = \lambda/16$, the node excess momentum falls approximately to half of its value.** Comparing to the detected line width this value appears narrower. But this line width is valid only for the quantum wave before the detection. During the detection process the Coulomb forces between EQs are destroyed and the line appears wider.

There is one important consideration about the Coulomb forces effect. The node distance should not been changed for two reasons: first one

- it could affect the synchronization conditions and second one - the space occupied by a quantum wave with longer wavelength is completely transparent for a shorter wavelength quantum wave. So it appears that the Coulomb forces interaction affects the angular momentum within the resonance cycle. This is equivalent to regarding the space of the electrical quasisphere as shrunk along its major axis. We will call this a **quasishrink effect of the CL space**.

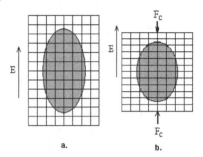

Fig. 2.40.B
Electrical quasisphere affected by a quasishrink effect of the CL space

This is illustrated graphically by Fig. 2.40.B, where: a. - shows the envelope of the quasisphere without shrinking, b. - shows the same quasisphere, shrunk by the Coulomb forces F_c.

2.10.5 Boundary conditions for the wave equation

The provided analysis about the quantum wave configuration allows to define boundary conditions for the E and H vectors, when used in the classical wave equation, for vacuum and air environment.

The radial trend of the electrical field has to fall to zero at the boundary defined by the boundary coefficient (2.20.a). **Practically, however, the EQ could not convert abruptly to MQ without some additional oscillations beyond the defined above radius. This is confirmed by the small concentric rings around the central maximum obtained by a diffraction limited optics.** The oscillating energy beyond the boundary level, could not be part of the photon momentum, because its energy is independent on the path it has passed before the detection. Consequently, it can be a reactive energy exchanged between the photon and the CL space. In other words, it is an energy borrowed by the CL space and used to provide a complete preservation of the photon momentum. During the detection process this energy is returned back to the CL space. Practically this is possible if the slightly deformed quasispheres in the boundary domain, (whose shape is closed to MQ shape), have proper phase differences of their SPM vectors.

From the above analysis, it becomes evident that when considering the energy distribution only, the E vector could be regarded as dropping to zero at the boundary radius. When analysing the intereferometric effects, however, the extended boundary radius for the reactive energy exchange should be considered.

The same boundary conditions are valid for the magnetic field vector H (or B), but having in mind, that it has a constant value. So, when we are interested of the energy only, the magnitude of the H vector obtains a constant value. In case of interferometric conditions it has a fluctuation component added to its constant value.

For simplified calculations, the E vector could be replaced with a constant value at the half maximum of the quantum wave.

The discussed above boundary conditions are valid for vacuum and air (with good approximation), but not for a transparent media (glasses and liquids). The CL space in this media is highly modulated by the EQs in the proton's proximity field, which are spatially arranged in a well defined order. The absolute frame analysis of the photon shape in this conditions is not convenient. The amazing fact, however is that the photon preserves its integrity and momentum, when passing through a homogeneous optical media. The polarization of the E vector although could be affected by the media properties, that are in fact properties of the internal CL space.

2.10.6 Propagation of quasiparticle waves (virtual particles).

Another important feature of the cosmic lattice, is that a wave is possible containing only positive or only negative EQs. In other words a wave propagation only by left handed or by right handed nodes is possible.

In all cases of neutral EM waves, including the photons, the energy is emitted due to a multiple oscillations of helical particle. In this case the energy is well mixed between the right handed and left handed nodes.

In case of short and strong aperiodic oscillations, caused usually by change of the external geometry of the helical structure, the picture is different. It will be shown in Chapter 6, that such kind of oscillations occurs when the neutron transforms to a proton or vice versa. The transformation in this case is accompanied with unlocking or locking of the near field of the elementary particles and respectively with birth or death of the far field electrical charge. The process is also accompanied my aperioding motion of helical structures, complimentary to a charge oscillation. The process is faster, than the CL relaxation time, so the invoked wave energy does not have a time to be equally distributed between both type of nodes. It propagates as wave exhibiting an electrical charge. We can call this type of wave a **quasiparticle wave, or virtual particle.** There are few important characteristics of this waves.

- they may reach energies, higher than the energy of the first harmonic,
- they are deflected by electrical and magnetic fields,
- their motion do not exhibit sharp quantum features like the moving electron

The first feature is possible, because the motion of this wave does not involve transmission of any intrinsic matter.

The second feature is obvious.

The third feature is explained by the difference between the quasiparticle wave and the quantum wave. While the quantum wave has an excellent isolation by the synchronized magnetic quasispheres, the quasiparticle wave does not have such one. The lack of such isolation excludes the transverse packing of the electrical charge, that is a major condition for a quantum wave formation. So the transverse radius of the wave does not have sharp boundary, but depends on the energy of the virtual particle. The quasiparticle wave behaves as a high energy electron or positron. In the process known as a positron moderation, the positive quasiparticle wave interacts with an electron, as a result of which, a low energy real positron particle is extracted. (see Chapter 6).

The quasiparticle waves, likely propagate with a velocity of light, but the BSM theory does not have enough experimental data about such conclusion.

The lack of quantum effect means a lack of synchronization of the running MQ's involved in the wave volume. But why such synchronization is missing? The neutral quantum wave has one important feature: the neighbouring positive and negative (left and right handed) quasispheres have complimetary interaction of their resonance momentums. The complimentary interaction is supported by the energy momentum carried by the quantum wave. In the quasiparticle wave, such interaction are missing. If the wave is with a positive charge, for example, every neighbouring quasisphere is a magnetic, and not synchronized to it's neighbour EQ's. In such way, the amount of the not synchronised magnetic quasispheres in the volume is quite large in comparison to the neutral wave. Then a formation of boundary magnetic quasispheres, synchronized to external SPM frequency is not possible. The strong quantum effect defined by the external boundary conditions is missing.

One question arises: How the integrity of the quasiparticle wave is kept? The possible explanation is the following.

The EQ eccentricity in the wave cross section has a radial dependence. If assuming, that it is similar to the stationary E-filed of the electron (see §3.4, Chapter 3), it has a maximum at some distance from the central axis of the wavetrain. This is illustrated in Fig. 2.41.A, where **a.** shows the radial E field of the electron with an energy of 1.51 eV (at confined motion with 3rd subharmonics, as discussed in Chapter 3) and **b.** shows the radial E field of a quasiparticle wave.

When the radial distance, r, approaches zero, the E-field also tends to zero. This is reasonable for the electron as discussed in Chapter 3. In the same chapter it become apparent that the interaction of the electron with a positive quasiparticle wave leads to emission of gamma wave or extraction of the internal positron (the latter according to BSM takes part in the phenomenon known as thermalization of high energy positrons, which are in fact quasiparticle waves or virtual particles). Then we

may conclude that the radial E-filed distribution of the quasiparticle wave has a similar configuration as the near E-field of the electron. In such case, the central axis zone of the quasiparticle wave should be occupied by MQ's. Then they could be synchronised by SPM frequency.

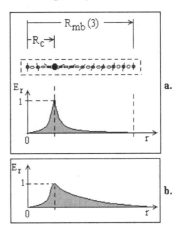

Fig. 2.41.A

(a) Electrical field of electron for confined motion with a third subharmonic

(b) Electrical field of QP wave (virtual particle) moving with a velocity of light

It is apparent that the wavetrain of the quasiparticle wave may have a central zone of MQ's which is a phase synchronized by the SPM frequency. This central zone could keep the integrity of the wavetrain. It could also serve as a quantum feature, assuring at the same time the propagation of the virtual charge with a velocity of light. This feature may appear as an energy modulation condition for waves with different energies. Despite the relative weakness of this quantum feature, it may play a role for the wavetrain integrity. A signature of such feature appears in the spectra of the quasiparticle waves, obtained from a β radioactive decay. Fig 2.41.B, shows spectra of β "particles" from decay of ^{64}Cu.

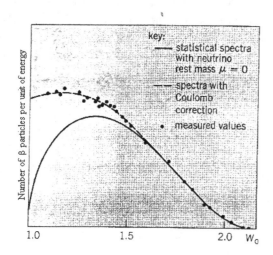

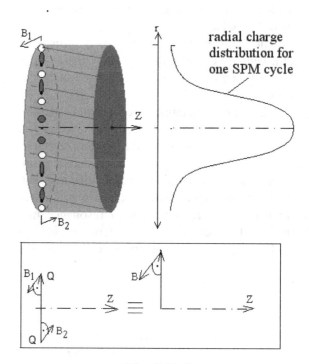

Fig. 2.41.B

Spectra of β decay of ^{64}Cu. The quantity W is the total beta-particle energy in the units of the electron rest energy. (C. P. Parker, 1993), courtesy of L. M. Langer et al., 1949)

Fig. 2.41.C

REQ and RMQ arrangement in a wavetrain of a QP wave (virtual particle)

From the analogy between the quasiparticle wave with the electron we see, that the radial arrangement is similar only a positive or a negative EQs. From the spectral plot in Fig. 2.41.B we also see, that the limit value of the negative QP wave approaches a quantum feature of energy modulation at 2.1 MeV. It is approximately 4 time the energy of the neutral first quantum wave (energy of the electron). This may lead to guess about the possible radial configuration of the QP wave. In the wavetrain cross section of a neutral quantum wave shown in Fig. 2.38, the E-field vector is contributed by the diametrically aligned EQs of both types. Fig. 2.41.C. (b) shows the possible arrangement of EQs for a case of quasiparticle wave, called also a virtual particle.

In the QP wave, only one type of aligned EQs exists (shown as dark ellipses in Fig. 2.41.C), while the boundary conditions are absent. One may see that EQs arrangement, corresponding to one sign charge, has similarity to a quantum wave (photon), but without boundary conditions. In such case it should carry electrical charge, while lacking a hardware helical structure. Such wave however could not exist as a stationary charge, which is an indispensable property only of the real elementary particle.

Let us pay attention on the similarity between the E-field of the electron shown in Fig. 2.41.A.(a) and E-field of the virtual particle shown in Fig. 2.41. A. (b). While the volume of the virtual particle wave does not contain boundary MQ's like the photon, it has quite a lot distributed MQ's. They may assure a quantum feature at SPM wavelength, that could allow keeping of the charge integrity. But this integrity may exist only if the wavelike structure is moving with the speed of light. The SG conditions that keep the charge unity however are missing, so this quasiparticle structure is most probably to be absorbed at once in some interaction process like the so called thermalization of β particle (this is discussed in Chapter 3). The absence of boundary conditions also does not put a limit on the radial extent. The EQs in Fig. 2.41.C are arranged along one diameter, perpendicular to the direction of propagation. We may accept that such configuration of virtual particle carry energy of 511 KeV. Although, there is not any restriction for another arrangement of similar structure but with two diametrical REQs alignment, perpendicular each other. For a first harmonics wavetrain the latter should carry energy of $2 \times 0.511 = 1.022 \text{ MeV}$. The right angle between the two cords might be an optimal arrangement corresponding to the observed maximum in Beta decay illustrated by Fig. 2.41.B. The cut-off about 2 MeV may correspond to four diametrical arrangements (with angles of 45 deg between them) carrying a total energy of $4 \times 0.511 = 2.044 \text{ MeV}$ The decreasing continuum of Beta spectra, shown in Fig. 2.41.B could be explained by the feature that the charge unity is not preserved (missing SG field as in the static charge particle). Then some decrease of the electrical charge may mimic a charge moving with a lower velocity.

Quasiparticle waves are emitted in the radioactive decays, where the proton-neutron (and neutron-proton) conversions are involved. Additional discussion about the process of their generation is presented in Chapter 6. (Virtual particles). Quasiparticle waves are virtual particles matching the Dirac idea. They could be born in pairs in some high energetic interactions. This happens in high energy collision experiments.

2.10.7. Gamma rays as a bunch of quantum waves

From the two basic requirements for the boundary conditions and the integrity of the quantum wave, it follows, that the SPM frequency of the CL space puts an upper limit for the wavelength (and frequency) of the quantum wave. This means that the described so far wavetrain configuration is valid for quantum waves not shorter than the first harmonic quantum wave (including the first one), but not for harmonics. But how to explain the gamma photon, whose energy can be much above the energy of the first harmonic quantum wave? The only possible configuration of the gamma photon is to be an entagled packet of SPM quantum waves.

The photon entaglement is proved to exist in the light produced by the lasers. The physical shape of the entagled photon could be inferred from the point of view of BSM considerations. The entagled photons could be consisted of two and more quantum waves.

Let us consider initially photon energy below the energy of the first harmonic. We may distinguish two types of entagled photons: parallel and serial. The serial entagled photons should be exactly with the same subharmonic number. The parallel entagled photons, when their number is above two, may contain photons with different quantum numbers. Let us pay attention especially on the last case. The physical conditions allowing the entaglement obviously are related to the condition assuring the keeping of the photon energy in a compact space, which however is not stationary but moving with the velocity of light. But for the stationary nodes this means a reduced interaction between the surrounding CL space and the photon space. The only possible way for this is to reduce the common boundary surface between the entagled photon and the surrounding space. According to these considerations, the possible configuration of the parallel entagled photon is a "rope" like structure. Such configuration is shown in Fig 2.42.

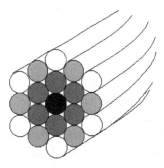

Fig. 2.42
Cross section of parallel entagled photon structure

From the shown configuration, we see, that the boundary structure of the entagled photon is smaller that the sum of the boundary structures of the individual photons. In the real case the individual circles in the cross section could be slightly separated. In such case the individual boundary conditions still could be satisfied, due to the reactive energy exchange, mentioned in §.2.10.5.

The parallel entagled photon could include individual photons with different wavelengths. In this case, they are twisted around the central core wavelength. This gives to the entagled photon not only compactness but additional stiffness. So they are difficult to be separated. The photon included in the centre of the structure is with the shortest wavelength. This is the core photon of the entagled structure. The core photon in the structure, shown in Fig. 2.42 is shaded black. The surrounding photons whose centre in the cross section is peripheral in respect to the central one posses larger wavelengths. The wavetrain cross sections of the photons with the same wavelengths, shown in Fig. 2.42 are shaded with a gray level of same colour. The common wavetrain structure could be regarded as a group.

The configuration shown in Fig. 2.42 does not cover the diversity of the entagled photons. The entagled photons are proved to exists in the laser light in the VIS and near IR range. According to BSM analysis, the photons in the ultra shot laser pulses (known as "femtosecond pulses") contain a large number of parallel entagled photons. So if entagled photons are able to exists in the visible and near IR range, they are even more likely to exist in the range near to the Compton wavelength.

Let us introduce an index of "basic order" corresponding to the number of radii passing through the centre of the wavetrain with equal wavelength. The basic order number of the configuration shown in Fig. 2.42 is 3. Let us suppose, that the basic order number reaches some limit, which is intrinsic of the CL space. The group in such case is completed. Then it could be possible to get a second order bunch comprised of such groups in a similar way, as the individual photons of the group. We may use the index "group order" in a similar way as the "basic order" The total energy of such entagled photon will be a sum of the energy of all contained individual photons. If the energy of the entagled photon is much larger than the first harmonic, the photon structure could be a parallel entagled photon containing bunches of groups. Now

we arrive to the possible configuration of the high energy gamma photon.

It is obvious, that the connection between the individual wavetrains within the group is stronger, than the connection between the groups, due to the different interspaces in the radial section. Then a parallel entagled photon of group order higher than one will exhibit one specific feature. The groups may be displaced along the wavetrain length. This displacement could be influenced by the CL relaxation time. Then the phase of the entagled photon regarded as a pulse of light could exhibit curled features in the front and back ends. Such features are observed in the ultra short laser pulses in the near IR region. Such pulses are achromatic. Their specific features show that parallel entagled photons are possible also in the much shorter quantum waves. This is one indirect indication that the gamma quant is a bunch of entagled photons.

We may summarize:
- **The gamma quant is a bunch of parallel entagled quantum waves, arranged in groups.**
- **The gamma quant is achromatic, comprised by individual photons that are harmonics and subharmonics of the SPM frequency.**
- **For high energy gamma quant, the central photon of any one of the group is a first harmonic wave.**
- **The lower energy gamma may contain only one group and the central photon may be a subharmonic of the SPM frequency**

2.10.8 Trapping mechanism of rectangular lattice inside the helical structures.

When a first order helical structure does not contain internal structure its modified rectangular lattice forms an axial trapping hole. Such configuration was discussed in §2.6.5 This type of lattice does not possesses SPM effect, and consequently does not propagate a magnetic field. The radial stripes, however are excellent propagator of the SG TP field of the helical shell. So we may consider, that the electrical field of this shell is propagated without losses and focused on the central hole. If the first order helical structure is built by right handed prisms, a node of left handed prism aligned to the axis will be attracted, then folded into a core of four prisms and trapped by the hole. If number of such nodes are folded and trapped, they will be connected together, forming a straight core structure, consisting of one central and six peripheral prisms. So we see, that the CL folded nodes can built a long core, that is not distinguished from the core the helical structures are built of.

The process of prism to prism interaction in the hole is enhanced due to two factors: 1) The gravitation of the central part of the trapping whole serves only to align the prism. Therefore the trapped node of prisms appears weightless in the centre of the hole, and only the twisted part interaction is effective. 2) The lattice configuration provides a focusing of the helical core handedness into the space of the trapping hole. This enhance the peripheral interaction between the focused SG field and the trapped prisms, of the folded node.

According to BSM theory the trapping mechanism shows its signature in some of the processes in the particle coliders. **The regeneration of κ_o^L from κ_o^S, for example, known as a CPT violation, is explained by BSM theory as a central core generation by the trapping mechanism. This is discussed in Chapter 6.**

The trapping mechanism is important effect, helping to explain some of the processes during the phase of particle crystallization. This phase precedes the birth of the galaxy.

2.11 Light velocity in CL space

2.11.1 Energy balance between CP and TP components of the NRM vector of CL node

When two systems of matter are involved in a common oscillation, the time duration of the oscillation process is very dependent of their ability to exchange equal energy momentum. In a common oscillation system, we may distinguish different subsystems by their interactions, even if they look physically inseparable. In such aspect we may provide analysis of CL node oscillations, as a process in which the following two subsystems are involved:

- Central Part (CP) of the prisms.
- Twisted Part (TP) of the prisms.

Now let applying this kind of separation to the CL nodes assuming that the prisms properties of CP and TP SG interactions are transferred to the CL node properties. Then we may assigned the mentioned properties to the NRM(CP) vector and NRM(TP) vector, respectively.

Let us find what are the main distinguishing properties of these two vectors. From the previously discussed "prisms to prisms" interactions (§2.5), we know, that the CP of the prism exhibits a low inertial factor, while the TP has a higher one. In the node oscillations, the CP is involved mostly in translational motions, while the TP in rotational motions. For the twisted prism model the volume of the TP is smaller than the volume of CP, however, the TP have the larger inertial interaction factor (see §2.4). The CL space is a highly spatially order system with a self-supporting balance. Then we may expect that for a normal (not disturbed) CL space the following balance is relevant:

$$[V_n(CP)] \times [F_n(CP)] = [[V_n(TP)] \times [F_n(TP)] \quad \textbf{(2.24A)}$$

where: V_n - is the node intrinsic matter volume of the corresponding part

F_n - is the node inertial factor of the corresponding part

For CL nodes from right and left handed prisms with defined dimensions the relation (2.24A) might be fulfilled at suitable node distance.

Another distinguishing feature between CP and TP interactions is the angular momentum of the oscillating node.

The NRM MQ has four bumps, indicating the directions of increased linear momenta. These momenta are same for the neighbouring right and left handed nodes.

If considering the linear momenta of NRM(CP) MQ along the positive and negative direction of any axis passing through the central point of MEQ, they are equal in the case of:

- the equivalent diametrically opposite motions during the resonance cycle
- the right and left handed nodes.

If considering the linear momenta of NRM(TP) MQ along the positive and the negative direction axes defined defined in a similar way, they are different for the cases of:

- the equivalent diametrically opposite motions during the resonance cycle
- right handed and left handed nodes

If considering the EQ nodes, the momentum magnitudes in the orthogonal axes are affected, but the NRM(CP) and NRM(TP) have similar features as for the MQ nodes.

In domains of MQ nodes, the above mentioned features contribute to the formation of magnetic protodomains. In domains of EQ nodes, the above mentioned features contribute to the propagation of EM waves. In this case the momentum possessed by EQs appears as an excess momentum. The SG forces acting on the node in such conditions are not conservative and the excess momentum is propagated by *abcd* axes of the CL nodes, which are interconnected.

While the CP component of NRM vector is hidden for electrical and magnetic property, it is not hidden for the inertial and gravitational mass properties of the matter. It is involved in the relativistic increase of the body mass, when approaching the light velocity This tissue is discussed in Chapter 10.

We may conclude:

- The electrical, magnetic, and electromagnetic fields in CL space are contributed by the interactions in which the twisted part of the prisms is involved. The central part of the prisms supports the CL structure integrity, while its features are hidden for these fields.

- A stationary EQ node involved in a quantum wave is not able to keep the excess angular momentum.

- The excess node momentum is propagated as EM field due to the not conservative force conditions. The energy is carried by the TP type of SG interactions.

- The central part of the prisms is directly involved in the gravitational and inertial mass property of the matter.

2.11.2 Momentum propagation of a CL node involved in the quantum wavetrain

2.11.2.1 Excess CL node energy and momentum

Let us accept that the CL node has inertial mass. Then we may provide a simplified analysis of the CL node dynamics, regarding it as a rotating mass point around a fixed point, connected to it by a massless rod. For analysis simplification, the trace curve of a MQ CL node oscillating at its proper resonance (NRM) frequency can be replaced by an equivalent circle. Then the angular momentum of the oscillating NRM vector for MQ node is given by Eq. (2.25).

$$L = m_n \omega_R r^2 \qquad (2.25)$$

where: L - is the node angular momentum at the resonance frequency.; m_n is the intrinsic inertial mass of the oscillating node; ω_R is the resonance frequency;

r - is the equivalent radius of the node trace per one resonance cycle

The inertial mass m_n, could be regarded as an average value of the intrinsic inertial mass of quite large number of MQ type of nodes. In such case it obtains very accurate value, depending only on the node distance. Consequently m_n **is a constant for a steady state CL space.**

When all CL nodes from a space domain have a normal ZPE, their angular momentum is constant. When a quantum wave is propagated through CL space any CL node may become MQ or EQ node for short time intervals. For this reason we must consider that a stationary node can be MQ or EQ only for intrinsically small time interval. The MQ should have a normal angular momentum. The EQ nodes involved in the quantum wave, obtain ZPE above E_{cr}, and their momentum is larger, than the normal one. The CL structure is self-preserved from imbalance, so the stationary EQ nodes transfer their **excess momentum** very fast to their neighbours in well organized spatial and time order. Therefore, the energy transfer could be expressed by the angular momentum change ΔL.

2.11.2.2 Energy and resonance period analysis for MQ and EQ type of node.

Note: In a global aspect, the CL space is interconnected. Then it contains two types of energy (discussed in Chapter 5 and called Zero Point Energy): a connection energy - called a Static type of ZPE (or ZPE-S) and kinetic (oscillation) energy - called a Dynamical type (or ZPE-D). In a normal not disturbed CL space the ZPE-S energy is hidden, while the ZPE-D is detectable and its existence is recognized by Quantum mechanics. In the following analysis the ZPE-D type is only considered.

Let us make some energy analysis of a single CL node. We may simplify the analysis by using a simple analogical model, whose parameters are able to represent approximately the dynamics of the CL node. Two classical models could be used: a three dimensional harmonic oscillator and a conical pendulum. Here the model of the conical pendulum is used, because it is more convenient for some illustrations for energy propagation in CL space.

The conical pendulum may have two types of motions: a circular and an elliptical one. **The circular motion of the pendulum can be associated with the NRM cycle of the MQ CL node, while the elliptical motion with the NRM cycle of the EQ CL node.** In order to select properly the parameters of the conical pendulum, we will use the equivalence between the CL node energy and the pendulum energy. More specifically, we will use the energy dependence on the CL node displacement from its equilibrium point.

The node energy in relative units, can be estimated by the return force curve given in Fig. 2.24. While the equation for this curve is pretty complex, it is replaced, by a fitted curve, given by the equation (2.26) and valid only for a displacement range of: $0.21 < r < 0.6$.

$$F_{ret} = (3.025 - 3.8269 e^{-r})^2 \qquad (2.26)$$

where: F_{ret} - is the normalized return force, and r is the displacement from the geometrical equilibrium.

Let us consider first the **MQ case, corresponding to a circular pendulum.** In the node displacement, the return force is aligned to the axis passing through the geometrical equilibrium point.

This force is equivalent to the force aligned with the cenrapetal acceleration. Then the tangential velocity for a circular motion is:

$$\upsilon^2 = \frac{rF_{ret}}{m_n}$$

The kinetic energy is:

$$E_K = \frac{m_n\upsilon^2}{2} = \frac{r(3.025 - 3.8269e^{-r})^2}{2} \qquad (2,26.b)$$

The plot of the node kinetic energy (in relative units) as a function of the displacement is shown in Fig. 2.43A. The amplitude of $r = 0.39$ (calculated by modelling the ZPE of CL node), corresponds to a displacement along xyz axes at normal ZPE.

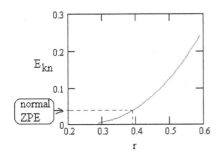

Fig. 2.43A
Kinetic energy of the oscillating CL node (in relative units) as a function of node displacement along *xyz* axis

The equivalent trace of the oscillating MQ node has a radius of $r = 0.39$, corresponding to a normal ZPE.

Fig. 2.43.B shows the pendulum with its parameters.

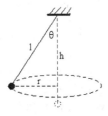

Fig. 2.43.B
Conical pendulum

The kinetic energy (E) for a circular pendulum (without friction losses), could be expressed by the potential energy, dependant on the height from the central position, but expressed by the displacement r.

$$E = mg(l - \sqrt{l^2 - r^2}) \qquad (2.26.c)$$

where: m - is the pendulum mass, g - is the Earth acceleration, l - is the arm length, r - is the displacement.

In order to fit the equation (2.26.c) to Eq. 2.26.b in a limited range of displacement, we normalize it to mg and introduce adjustable parameters a and b, for energy scaling and displacement, respectively.

$$E = [l - \sqrt{l^2 - r^2}]a - b \qquad (2.26.d)$$

where: a - is a scaling parameter, b is displacement parameter.

The Eq. (2.26.d) is suitable for simulation of the MQ node energy, considering a relative displacement up to 0.39 along anyone of the *xyz* axes.

The plot of Eq. (2.27.d) for $l = 0.53$, $a = 0.4$ and $b = 0.035$, for a range $0.2 < r < 0.5$, is shown in Fig. 2.43.C.

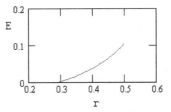

Fig. 2.43.C
CL node equivalent energy dependence on the displacement by using of the model of a circular conical pendulum)

In the case of EQ node, the equivalent trace of the pendulum obtains elliptical shape corresponding to a radius larger than r = 0.39. From the energy plot shown in Fig 2.43.A, it is evident that for equal deviations around 0.39, the slope has different steepness. Then an EQ node with a larger eccentricity of the NRM trace will have a larger energy. If considering not equivalent but real trace shape, this difference is even larger. This case could be analysed by the model of elliptical conical pendulum. There are two important parameters in this case to be taken into account: the node inertial mass and the trace length.

The separation of the inertial mass from the velocity is a difficult task, because the inertial mass may not be constant during the node cycle. For this reason, we will consider that the node inertial mass is constant, only when it is estimated for a full cycle and averaged on a large number of nodes.

The trace length is an important parameter, because it defines the duration of the cycle. Having in mind the influence between the neighbouring nodes, we may assume, that any node has a tendency to keep its cycle duration equal to the cycle duration of its neighbours, so this will provide a stable constant value for the period of NRM MQ. Then the following question is reasonable:

Is it possible the elongated trace of EQ node to have the same NRM frequency as the MQ node?

We will try to reply answer by analysing the oscillation of the elliptical conical pendulum. Initially we will estimate the period dependence on the displacement r. The periods for a circular conical pendulum $T_{con(cr)}$, and a planar pendulum T_{pl}, are given respectively by the equations (2.26.e) and (2.26.f)

$$T_{con(cr)} = 2\pi\sqrt{\frac{h}{g}} = \frac{2\pi}{\sqrt{g}}(l^2 - r^2)^{1/4} \qquad (2.26.e)$$

$$T_{pl} = 2\pi\sqrt{\frac{l}{g}}\left(1 + \frac{1}{4}\sin^2\left(\frac{\theta_1}{2}\right) + \frac{9}{64}\sin^2\left(\frac{\theta_1}{2}\right)\right) \qquad (2.26.f)$$

Using up or down arrow index, for annotation of the increasing or decreasing of the parameter, we have the following dependence from r and θ according to Eqs. (2.26.e) and (2.26.f):

Circular pendulum (MQ case): When: r/|\ T\|/ L_{tc} /|\ (26.g)
Planar pendulum: When: θ/|\ T/|\ L_{tp}/|\ (2.26.h)

where: L_{tc} and L_{tp} are the trajectory lengths for both types of pendulums, respectively

Now we have to determine how the period changes when the trace becomes elliptical (corresponding to conversion of MQ to EQ). This is a more complicated task, but we may simplify it by examining the following **two options**: a circular pendulum and a planar pendulum. The planar pendulum could be regarded as a degenerated circular pendulum for very large eccentricity. In such aspect, the mentioned two options appear as bounda-

ry cases, when changing the elipticity. We can formulate the task: What is the ratio between the periods of the planar and the conical motions for one and a same pendulum if the trajectory length in both cases is equal?

In order to get a simple expression, we will use the deviation angle θ, knowing that the increasing of the displacement r leads to increase of θ. Equating the trajectory lengths we get $\theta_1 = \pi\sin\theta_2$. where: θ_1 and θ_2 - are the angles of the planar and conical pendulum, respectively. The period of the planar pendulum for a large angle is determined by a sine series. Then we obtain the ratio between the periods for the planar and the conical pendulum.

$$\frac{T_{pl}}{T_{con}} = \frac{1}{\sqrt{\cos\theta_2}}\left(1 + \frac{1}{4}\sin^2\left(\frac{\pi\sin\theta_2}{2}\right) + \frac{9}{64}\sin^4\left(\frac{\pi\sin\theta_2}{2}\right)\right) \quad (2.26.i)$$

The source of the return force g disappears from the period ratio. The plot of the period ratio is shown in Fig. 2.43.D.

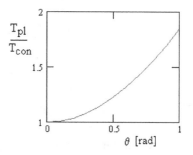

Fig. 2.43.D
Period ratio between planar and conical pendulum for equal trajectory lengths

Having in mind, that the planar pendulum is a degenerated elliptical conical pendulum, it is reasonable to expect, that the period ratio changes as a continuous function between the two pendulum cases. Consequently, the ratio between the periods of the elliptical conical pendulum and circular conical pendulum will be also a growing function.

Based on the provided analysis by the conical pendulum we arrive to the following conclusions for a EQ CL node.

The increase of the CL node energy causes increase of its NRM cycle, which means a decrease of its NRM (proper resonance) frequency.

Since EQ nodes have a larger oscillating energy than MQ node and the latter energy is fixed by the ZPE equalization, then EQ nodes will have a lower NRM (proper resonance) frequency than MQ nodes.

The single EQ node can be regarded as a carrier of a very small fraction of the electrical charge. The unit charge of any charge particle (electron, positron or any unstable particles) is a constant due to the SG forces of the particle that influence the number and eccentricity of its EQ's. At the same time, the SG field assures the equalization of the proper frequencies of the EQs and their synchronization (by SPM vector). In such way both, the constant unit charge and the charge integrity are assured.

For EQ and MQ CL node properties in a normal CL space environment we summarise the following conclusions:

- **The main distinguishing feature of the EQ is that it possesses an excess kinetic energy ZPER-D over the normal ZPE of MQ.**
- **The degree of the EQ eccentricity is determined by the node excess energy. The maximum value of the eccentricity is limited, due to the CL space self-supporting energy balance which resists to interactions leading to a break of CL structure (this intrinsic property of the CL space will become more apparent in the following chapters of BSM).**
- **The EQs forming the E-field of the elementary particles are stationary (in respect to the particle coordinate system). They are kept by the SG field of the particle.**
- **The EQs involved in the wavetrain of the quantum wave are of running type.**

While the above made analysis is simplified, it is evident that more complicated analysis of the 3D NRM and SPM vectors is needed. This is out of the scope of the present course of BSM theory. In the next chapters we may touch this problem again, because it is related to number of physical aspects: the integrity of the quantum wave, the integrity of the electrical charge around the elementary particle, the electron motion in quantum loops in electrical field and so on.

2.11.2.3 Excess momentum of EQs involved in a quantum wave. Quasishrink effect of CL space.

In the previous paragraph we found that for interactions in which the (CP) of the prisms are involved do not contribute to energy of electromagnetic type. Therefore, we will not take into account the CP interactions in the provided below analysis.

Let us consider a CL domain of normal CL space away from any charge particle and any external electrical and magnetic field. In this case the node inertial mass and the NRM frequency are constant, so we may express the MQ and EQ momentums by some equivalent common parameters.

Knowing the EQ distribution in the quantum wave configuration, we may introduce a constant that depends only on the distance of the EQ from the wavetrain axis. It is convenient to introduce multiplication factor for the MQ node radius of rotation in a form: $\sqrt{e/2}$ (where e is a linear eccentricity of the equivalent elliptical trajectory) for a reason, that will be explained below. Then the Eq. (2.25) for the angular momentum takes the form

$$L = m_n \omega_R \left(r \sqrt{\frac{e}{2}} \right)^2 \qquad (2.27)$$

From the quantum wave configuration, we know, that the eccentricity of the EQ is dependent on the radial distance from the central axis of the wavetrain. For a helical trajectory with constant radius, the eccentricity e is a constant. According to the above analysis, ω_R is a constant for any quantum wave. We assume also that the node inertial mass averaged for one resonance cycle is a constant. For a given radial distance, the excess node momentum, could be expressed as a change of the angular momentum. Then differentiating (2.27) on r we obtain the excess node momentum.

$$\Delta L = m_n (\omega_R e) r \qquad (2.27.a)$$

For a neutral quantum wave, r changes from some initial value r_o to the boundary radius r_b, at which the eccentricity e of EQ becomes zero (or it converts to a boundary MQs). Consequently, for $r = r_b$, ΔL becomes also zero.

From Eq. (2.27.a) we see, that the excess momentum is a product of a constant linear momentum $m_n r$, multiplied by the factor $\omega_R e$. Then the linear momentum for a constant radius is also a constant. Assuming a constant node inertial mass averaged for one resonance cycle, the velocity of the momentum transfer between the neighbouring nodes along one helical trajectory is also a constant.

We may express the equivalent excess momentum for the radial cross section of the wavetrain, when using the eccentricity e_{eq} corresponding to one equivalent radial distance r_{eq}. Then the equivalent excess momentum is:

$$\Delta L_{eq} = m_n(\omega_R e_{eq})r_{eq} \qquad (2.27.b)$$

If comparing the central point of the EQ node motion with the Keplerian motion of planets it is different. For the oscillating EQ node, the return forces along the major axis are larger than the minor one and the velocity change is much faster. The real trajectory shape contributes additionally to this effect. At the same time, the node trajectory could not obtain very large eccentricity, because the maximum and minimum radii of the node trace are restricted within a limited range. This restriction is imposed by the resistance of the CL structure to destruction. **Therefore, we may expect that maximum of the SG(TP) field of the prisms interactions occurs in a finite sector of the trace around the major semiaxis. Then the transfer of the excess momentum evidently takes place in that sector.** The NRM quasisphere is aligned to the *xyz* axes. **Then the transfer of excess momentum could be considered as a vector composed of components along *xyz* axes.** The actual momentum transfer, in fact is provided by the *abcd* axes during the resonance cycle, but this is not in contradiction with the above made considerations, using *xyz* coordinates.

There is one additional feature of the momentum transfer in the quantum wave. The helical trajectories (within the wavetrain) containing EQs with one and a same eccentricity, can be left handed or right handed. This obviously must be valid for a case of polarized and unpolarized quantum wave.

Let us find out what may determine the correct conditions for the excess momentum transfer?

The motion of every CL node, involved in the quantum wave, is characterised by both vectors NRM and SPM. The trajectories of these vectors are 3 dimensional, and consequently, they posses a handedness. The handedness momentum of the

SPM vector have much larger weighting factor. than the NRM momentum. Therefore, the SPM vector is responsible for keeping the wave handedness. The CL space provides equal conditions for propagation of left and right handed wave. Once the quantum wave is generated and the CL space in which it propagates is homogeneous, the handedness is preserved. This means that if the generated quantum wave has any kind of polarization it will be preserved.

The EQ node resonance trace and the excess momentum are illustrated in Fig. 2.44. The case **a.** shows the real NRM trajectory, where 1 - is the zone of the maximum momentum change, and 2 is the zone of maximum kinetic energy. The case **b.** shows the equivalent ellipse.

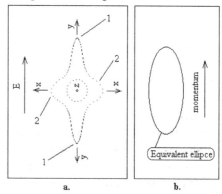

Fig. 2.44. EQ node momentum during the resonance cycle

The maximum linear momentum, shown in Fig. 2.44 has a direction of y axis. Let us assume that some external forces, having components only along $\pm y$ direction and acting always against the maximum linear momentum are applied, however, their magnitudes are smaller. In result of this, the vector of linear momentum will be affected as shown in Fig. 2.45, while the energy balance of the system must be preserved. The resulting linear momentum will have a direction at angle respectively to y and z axes (see the explanation below). This effect exactly appears in the neutral quantum wave, where the electrical quasispheres are affected by the Coulomb forces. We may call this a **quasishrink effect of CL space**. The term quasishrink is used, because it does not affect the node distance, but only the components of NRM vector. Fig. 2.45 illustrates how this effect changes the transfer momentum direction of the oscillating node.

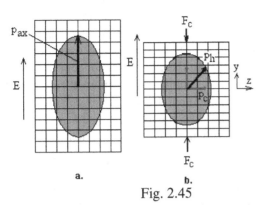

Fig. 2.45

Change of the transfer momentum direction
due to the **quasishrink effect** of the CL space

In case **a.** the equivalent momentum is
shown without the quasishrink effect. In case **b.**
the quasishrink effect of the space is provided by
the Coulomb forces F_c. The energy for this forces
is taken from the total momentum of the quantum
wave and more accurately from its twisting features. As a result of this, the apparent NRM momentum along axis y is reduced, but an equivalent
momentum P_h appears at angle in respect to the E
direction. In fact, if considering the point position
of the CL node in the wavetrain, the direction of P_h
coincides with the tangent of the helical trace passing through this point. One of the components of
P_h (vertical one) provides the balance between the
Coulomb forces F_c and the centripetal acceleration,
while the other one, P_c, provides the velocity for
the energy propagation in direction Z. Since conditions for not conserved angular momentum are valid for such CL node, the new component of the
linear momentum will be propagated between the
neighbouring nodes.

The induced Coulomb forces are moving
with the running EQs. They are responsible for
keeping a finite transverse width of the quantum
wave, in order to assure the boundary conditions.
They assure also the transversal compactness of the
quantum wave by narrowing the radial energy distribution, as was discussed in § 2.10.4.4. According
to that analysis, the most suitable width was estimated to be of the order of $\delta\lambda = \lambda/16$ (see Fig.
2.40.A). During the detection process, however,
the Coulomb forces are destroyed and the trans-
verse width appears as a $\Delta\lambda = \lambda/8$. This value

matches well with the relation between, c, μ_o, ε_o,
h, in the expressions, derived in the next paragraphs.

2.11.3 Light velocity equation and relation between the CL space parameters and the fundamental physical constants.

It was assumed in the previous paragraphs,
that the energy between neighbouring nodes is
transmitted per one resonance cycle. While the
components P_h and P_c are features of the resonance
cycle, they will be valid for all REQ nodes involved in the quantum wavetrain. This consideration facilitates the derivation of the light equation.
Having in mind the configuration of the quantum
wavetrain, we may also simplify the task by analysing the vector of the running EQ node.

Fig. 2.46 illustrates the orientation of the
running EQ, where: (a) - is a 3D view showing two
consecutive positions of the running EQ; (b) - is a
view showing the helical path and REQ in a perpendicular plane; (c) - a view of the two positions
of EQ in another plane. Any running EQ node at
distance R_h from the axis Z will have constant instant momentum components P_h and P_c. If considering consecutive time points, separated at time
distance of t_r, the REQ node will pass a distance d_n
on a curve along the helical trajectory HT.

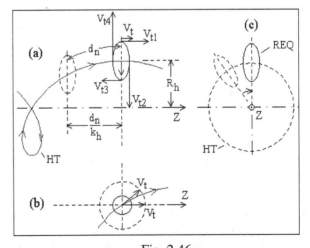

Fig. 2.46

Momentum propagation expressed by a running
EQ (REQ) through a helical trajectory HT

The long axis of the REQ is always normal to the axis of propagation Z, while its centre is at distance R_h from this axis. The vectors from V_{t1} to V_{t4} are the tangent momentum velocities of the oscillating node. The change of the angular momentum due to the Coulomb forces as discussed in §2.11.2.3 (having some radial gradient in the quantum wavetrain) provides a velocity component for moving the EQ in a helical trajectory instead of straightforward. If referencing to the laboratory rest frame the V_{t1} velocity will contain an advancing velocity component aligned to the direction of the quantum wave propagation. For this reason V_{t1} is shown larger than V_{t3}. The advancing velocity component could be translated to the central point of the quasisphere, because it is always parallel to the Z axis. View B shows that the velocity V_t has one and a same magnitude for the tangential axis and the Z axis, because of the circular symmetry of the REQ in this plane. We have a right to apply this consideration, because the momentum transfer occurs per one resonance cycle.

For the running EQ, the NRM vector carries an energy momentum along the helical trace, which we may call a helical momentum

$$p_h = m_n v_h \qquad (2.28)$$

where: p_h is a helical momentum (momentum along the helical trace), v_h is a helical component of velocity, m_n is a node inertial mass.

The defined helical momentum does not need to be compensated for a centripetal acceleration, because it is already compensating by the Coulomb forces and the lattice quasi shrink effect. The photon energy is comprised of all helical momentum carried by all REQs involved in the wavetrain. For simplification of the analysis, however, we may consider that the total photon energy is caries by an equivalent helical momentum of one equivalent REQ at equivalent radius from the central axis of the wavetrain.

Knowing that the integrity of the propagated quantum wave is preserved, we have to find the corresponding velocity, v_z, in a straight direction along Z axis. There is **one important consideration**: the excess momentum is propagated by the right and left handed nodes, which interact between themselves. In such case, the quantum wave momentum could not be considered as a sum from the right and left handed nodes momentums. The same

consideration is valid for the propagation velocity, of the photon. According to the present concept of Modern Physics this velocity carry the whole "photon mass". In order to comply to this consideration, but using the presented concept and having in mind the complimentary interactions between the right and left handed CL nodes, the propagation velocity could be regarded as a square root of the product from the right and left handed velocity contributions.

$$c = \sqrt{v_R v_L} \qquad (2.29)$$

where: c - is the light velocity (propagation velocity), v_R and v_L are respectively the linear velocity components contributed by the right and left handed nodes.

Considering the integrity of the whole E-field of the quantum wave, we can replace the radial energy distribution whose shape is shown in Fig. 2.38 by a rectangular function having the same area. Then the sum of the individual NRM vectors at one moment will be replaced by one equivalent NRM vector, corresponding to equivalent electrical quasisphere at a radial distance corresponding to the half maximum of the radial E-field. The equivalent helical path can be defined by a radial distance at half maximum equal to $\delta\lambda$ (see Fig. 2.38). As a result, we may relate the energy properties of the individual running EQ, but via some **equivalent REQ** located at equivalent radial distance from the central axis of the wavetrain.

The introduced equivalent REQ will carry the whole energy of the quantum wave (photon) while moving through an equivalent helical path centred around the direction of the quantum wave propagation.

Let us estimate the components v_R and v_L by the division of the equivalent helical path, on the time for this path. We may use the first harmonic wave, for simplicity, and to show later that the result is valid for all harmonics. Knowing that in the Earth local field $v_{SPM} = v_c$, we may use the Compton parameters for the SPM vector. Then the equivalent helical path for one SPM cycle is

$$\lambda_{he} = \lambda_c k_{he} \qquad (2.30)$$

$$k_{he} = \sqrt{1 + 4\pi^2 (\delta\lambda/\lambda_c)^2} \qquad (2.31)$$

where: k_{he} - is the coefficient for the equivalent helical path

Applying the Eq. (2.30) for the boundary radius we have:

$$\lambda_{hb} = \lambda_c k_{hb} \qquad (2.31.a)$$

where

$k_{hb} = 4$ - is the boundary coefficient according to Eq. (2.20.a):

Without confusing with the node distance of CL, that is a constant, in order to distinguish the calculated distances for both helical traces, we will use the distances:

d_{nb} - a node distance between RMQs lying in the boundary path (the helical path at the boundary radius, where MQs form the boundary conditions of the wavetrain)

d_{ne} - a node distance between REQs lying in a helical path corresponding to the equivalent path (the helical path at the equivalent radius valid for the equivalent REQ, defined above and carrying the total photon energy).

Fig. 2.47.a shows the resonance traces for magnetic and electrical quasisphere respectively. Fig. 2.47.b shows the same traces, but represented as equivalent circles (for simplification of the following analysis).

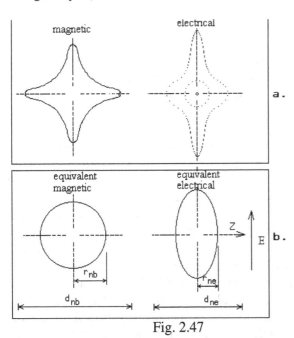

Fig. 2.47

Trace projections of resonance cycles (a.) and their equivalent presentations (b.)

Let us express the node distances d_{nb} and d_{ne} by their equivalent radii r_{db} and r_{de}, which are

shown in the same figure. We may accept, that the following ratio is valid:

$$\frac{r_{ne}}{d_{ne}} = \frac{r_{nb}}{d_{nb}} = k_{rd} \qquad (2.33)$$

Assuming that the energy is transferred between the neighbouring nodes per one resonance cycle we can write:

$$\lambda_{he} = N_{RQ} d_{ne} \qquad (2.34)$$

where: λ_{he} - the path of energy transfer for N_{RQ} cycles of NRM vector, N_{RQ} is the number of resonance cycles per one SPM EQ cycle.

For MQ CL node, the Compton time t_c is related to the NRM cycle time, t_r, according to the relation:

$$t_c = t_R N_{RQ} \qquad (2.35)$$

Then the resonance frequency is

$$\nu_R = \nu_c N_{RQ} \qquad (2.36)$$

where: N_{RQ} - is the number of resonance cycles in one SPM MQ cycle (must not be confused with cycles per second).

Evidently, the following relations are valid.

$$\omega_R = 2\pi\nu_R = \frac{2\pi}{t_R} = 2\pi\nu_c N_{RQ} \qquad (2.37)$$

The momentum velocity projection on the Z axis can be estimated by the path along Z axis per one SPM cycle, that is equal to λ_c. Therefore, the light velocity component, contributed by the CL nodes of one type handedness is:

$$\upsilon_R = \frac{\lambda_c}{t_c} = \frac{\lambda_{he}}{t_c k_{he}} = \frac{d_{ne}}{t_R k_{he}} = \frac{r_{ne}\omega_R}{2\pi k_{he} k_{rd}} \qquad (2.38)$$

If using not the first harmonic, but any subharmonic, the Eq. (2.37) gives the same result (because λ_c and t_c get multiplication by one and a same number). The velocity component, contributed by the CL nodes of other type handedness υ_L is a same as υ_R Then according to (2.29), the equation for the light velocity is

$$c = \sqrt{\frac{r_{ne}^2 \omega_R^2}{4\pi^2 k_{rd}^2 k_{he}^2}} = \frac{\omega_R d_{nb}}{2\pi k_{hb}} \qquad (2.39)$$

The Eq. (2.39) is not still a light velocity equation in a required, since it involved parameters that are not still determined. Although, it will help us to identify the relation between the well known parameters of the physical vacuum (permeability and permittivity) and the CL space parameters.

Substituting c in Eq. (2.39) with $(\mu_o \varepsilon_o)^{-1/2}$ and rasing on square we get

$$\mu_o \varepsilon_o = \frac{4\pi^2 k_{rd}^2 k_{he}^2}{r_{ne}^2 \omega_R^2} \quad \left(\frac{s^2}{m^2}\right) \tag{2.39}$$

The dimensions of the expression (2.39) are easily determined, having in mind the expression (2.37). Now the task is to find the expressions of the separate parameters of the product. The simple separation of the parameters in two terms could not give the correct result, because μ_o and ε_o may contain common parameters, which are eliminated in their product. However, we may guess what are the eliminated parameters, by examining the dimensions of μ_o and ε_o. Working in SI system, we can manipulate the dimensions of the $\mu_o \varepsilon_o$ product, by eliminating the common dimensions, until obtaining the dimensions of Eq. (2.39).

$$\mu_o \varepsilon_o \equiv \left(\frac{N}{A^2}\right)\left(\frac{A^2 s^2}{N m^2}\right) = (kg)\left(\frac{s^2}{kg \, m^2}\right) \tag{2.40}$$

Eliminated dimensions for μ_o: $\left(\frac{m}{A^2 s^2}\right)$ (2.41)

Eliminated dimensions for ε_o: $\left(\frac{A^2 s^2}{m}\right)$

Some eliminated parameters that are dimensionless are not directly apparent. Some other parameters, as the electron charge, for example, are defined at special conditions.

From dimensional expression (2.40) we see, that some mass should participate in μ_o and ε_o, while it is eliminated in their product. Let for this reason we multiply the nominator and denominator of Eq. (2.39) by the mass parameter m_n. The correctness of the obtained expression will be verified later. Then providing a proper grouping in brackets we get:

$$\mu_o \varepsilon_o = \left(\frac{4\pi k_{rd}^2 m_n}{N_{RQ}}\right)\left(\frac{k_{he}^2}{2\nu_c m_n r_{ne}^2 \omega_R}\right) \tag{2.42}$$

We will see later that m_n is a constant. Then all the parameters in the left bracket are constants, not dependent on the energy of the propagated wave. The term $4\pi/N_{RQ}$ also could be regarded as a solid angle corresponding to one resonance cycle. Its shape is shown in the figure Fig. 2.47.0.1. It has a non zero value because the NRM trace is not a planar curve, but a three dimensional one.

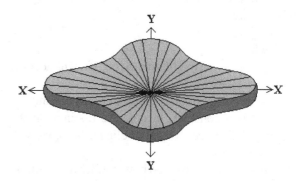

Fig. 2.47.0.1
Shape of the equivalent solid angle (exaggerated) defined by a single cycle of NRM vector

Therefore the left bracket in Eq. (2.50) appears to be a dynamical parameter of the NRM vector implemented in the SPM vector which forms the MQ. Consequently, the left bracket shows features indicating that it corresponds to the magnetic permeability of the physical vacuum μ_o. Then the right bracket should be the expression for the permittivity of the physical vacuum ε_0. The latter could be presented also as:

$$\varepsilon_0 = \frac{k_{he}^2}{2\nu_c (\Delta L)} \tag{2.43}$$

where: ΔL - is the angular momentum change of NRM vector, i. e. the momentum that carries the energy. This momentum multiplied by the **number of SPM cycles in all wavetrain** will give the total energy of the photon.

We have to find what is the reaction of the CL space to a disturbance pulse with infinite small duration. The response of such disturbance will be equal to the relaxation time constant. The relaxation time constant will likely define the transition envelope of the photon wave at the start. A more detailed discussion of the relaxation time constant, which is important but not investigated so far parameter is provided in §2.13.A. Its apriory accepted theoretical value is given by Eq. (2.44), while keeping in mind that further investigation of this important CL space parameter is necessary.

$$t_{CL} = \frac{(c)}{\nu_c} = 2.426 \times 10^{-12} \quad (sec) \tag{2.44}$$

where: (c) - is a dimensionless parameter equal to the light velocity in SI system.

Due to the space-time considerations of the relaxation time constant, the light velocity put in brackets is used as a dimensionless factor. The CL space relaxation constant is referred elsewhere in BSM also as space-time constant. This consideration is later used in Chapter 3 for definition of Dynamic CL pressure which is used successfully in Chapter 5 for derivation of the background temperature of CL space as a signature of the Zero Point Energy.

We will use the equivalence between the first harmonic energy (511 keV) and the electron mass in order to determine the m_n. parameter. In Chapter 3 the charge to mass equivalence principle also will be explained. Applying this principle, we can estimate the inertial node mass by the equation:

$$\Sigma m_n = \frac{h\nu_c}{c^2} \qquad (2.45)$$

For this reason we need to estimate the number of involved nodes, from the expression:
No of nodes = [(photon volume)/(node cell volume)] (2.46) The node cell volume should be determined from the boundary conditions: $d_{nb} = \lambda_{hb}/N_{RQ}$.

In order to estimate the volume we need the **wavetrain length. This is in fact a length**, that could be practically measured by Michelson interferometer with adjusted path length. The maximum path length at which the interferogram is still possible will provide this length that we may call a coherent length.

The coherent length l_{coh} and coherent τ_{coh} time are related by the simple relation:

$$l_{coh} = c\tau_{coh} \qquad (2.46.a)$$

The coherence time could be considered as the elapsed time, between the front and back end of the passing quantum wave, measured by a stationary observer.

The coherent times for a monochromatic thermal source and for lasers are very different. In our case we will consider only the first one. For best monochromatic thermal sources the coherent time is of the order of 10^{-8} sec.

Notes:
(a) The above definitions of a coherent length and time are for a not correlated single photon (its generation is not time and space correlated with

other photon). They should not be confused with the coherent length and time parameters of the lasers, where the photons are correlated by time (time instant of emission) and space (mutual spatial interactions).

(b). We should not confuse the coherence time with the detection time. The latter is much smaller, because, **the photoelectron in the detection process appears after all wavetrain energy is transferred to the detector.** The detection process in fact follows the end of the wavetrain.

If considering the CL space relaxation time constant as a transition time, the transition length l_{tr} of the wavetrain is:

$$l_{tr} = ct_{CL} \qquad (2.46.b)$$

The estimation of the coherent time or length for a single 511 keV gamma photon is a difficult task. For this reason we will use the CL pumping time for generation of this quantum wave. It is equal to the lifetime of the parapositronium 1'S$_0$ (p-Ps). (This is discussed in Chapter 3). Its value in vacuum is 125 psec. Here **we will assume that the pumping time is equal to the coherence time in this case.**

According to Eq. (2.46.a) and (2.46.b) the ratio between the coherent and transition length of the wavetrain is

$$k_d = \frac{125 \times 10^{-12}}{t_{CL}} = 51.52 \qquad (2.47)$$

The total wavetrain length is $t_{CL}k_dc$, while its cross sectional area, defined by the boundary radius of $\lambda/2$ is $\pi(\lambda_c/2)^2$ (0.5λ is used instead of 0.6164λ, because of possible EQ slope change near the boundary). Then the volume of the wavetrain can be expressed as:

$$V = \pi\left(\frac{\lambda_c}{2}\right)^2 t_{CL}k_dc = \frac{\pi}{4}\frac{(c)c^3k_d}{\nu_c^3} \qquad (2.47.a)$$

Substituting the volume from Eq. (2.47.a) in Eq. (2.46) and dividing the total mass of 511 keV by the number of nodes, we get the equation of the **node inertial mass** m_n, expressed by the CL parameters.

$$m_n = \frac{4h\nu_c k_{hb}^3}{\pi(c)c^2 N_{RQ}^3 k_d} \qquad (2.48)$$

The node inertial mass, m_n, could be regarded as an equivalent parameter. This is because the

CL node distance defines a linear dimensional boundary of the distance scale in CL space, so the inertial properties we are familiar with are valid only for larger distances than this one. The equivalent node inertial mass, however, is useful for finding the relation between the intrinsic CL space parameters and the fundamental physical constants.

The parameters μ_o and ε_o in system SI are estimated by using the Coulomb unit of the charge. However the terms of Eq. (2.42) are more convenient to be referenced to the charge of the electron. Then the following expressions are valid:

$$\mu_o = 4\pi \times 10^{-7} \quad \left(\frac{N}{A^2}\right) = \left(\frac{N}{C^2 s^2}\right) \quad \text{- permeability of free space}$$
defined in SI system
referenced to Coulomb unit

$$\mu_{oe} = \frac{\mu_o}{(q^2)} \qquad \text{- permeability of free space}$$
referenced to electron charge q (2.49)

$$\varepsilon_o = 3.854 \times 10^{-12} \qquad \text{- permittivity of free space}$$
defined in SI system referenced to Coulomb unit of charge

$$\varepsilon_{oe} = \varepsilon_o q^2 \quad \text{- permittivity of free space defined in SI system}$$
referenced to electron charge q (2.50)

Applying some substitutions in the second term of Eq. (2.42), and referencing to the electron charge we obtain:

$$\mu_{oe}\varepsilon_{oe} = \left(\frac{4\pi m_n k_{rd}^2}{N_{RQ}}\right)\left(\frac{N_{RQ}}{4\pi m_n c^2 k_{rd}^2}\right) \tag{2.50}$$

From the dimensional Eq. (2.41) we identify the eliminated parameters for μ_o

$$\left(\frac{m}{A^2 s^2}\right) \rightarrow \left(\frac{\lambda_c}{(qt_c)^2 t_c^2}\right) \tag{2.51}$$

The reason to use λ_c and t_c in the guessed parameters is that they are the basic parameters of the SPM effect, which is responsible for the constant light velocity. Multiplying the left term of the bracket of (2.50) by the eliminated parameters according to Eq. (2.51) and by q^2, according to Eq(2.49), we get the final equation for μ_o.

$$\mu_o = \frac{4\pi m_n k_{rd}^2 c v_c^3}{N_{RQ}} \tag{2.52}$$

In a similar way the final equation for ε_o is obtained.

$$\varepsilon_o = \frac{N_{RQ}}{4\pi m_n v_c^3 c^3 k_{rd}^2} \tag{2.53}$$

If knowing the factor k_{rd}, we can determine the parameter N_{RQ}, and consequently the resonance frequency of the CL node. The factor k_{rd}, given by Eq. (2.33) could be approximately estimated by the return forces plot of the node displacement, shown in Fig. 2.24 in Chapter 2. From this figure the node displacements along *abcd* and *xyz* axes are respectively: 0.2 and 0.4 values, normalised to the d_{abcd}, which is the node distance along one of the *abcd* axes. Then the average displacement, *r* is: $r \approx 0.5(0.4 + 0.2)d_{abcd} = 0.3 d_{abcd}$

$$k_{rd} = \frac{r}{d_{xyz}} = \frac{0.3 d_{abcd}}{2 d_{abcd}} = 0.15$$

For a value of $k_{rd} = 0.15$, we get the following results for the Cosmic Lattice:

$N_{RQ} = 0.88431155 \times 10^9$ - number of resonance (2.54)
cycles for one SPM cycle

$v_R = 1.092646 \times 10^{29}$ (Hz) - node resonance frequency (2.55)

$d_{nb} = 1.0975 \times 10^{-20}$ (m) - CL unite cell size (along xyz axes) at boundary (2.56)

$m_n = 6.94991 \times 10^{-66}$ (kg) - CL unit cell inertial mass (2.57)

Note: v_R is valid for *xyz* CL unit cell and node resonance frequency, while d_{nb} and m_n are defined for CL unit cells only. For approximate calculations, m_n could be considered valid for a single CL node, because any CL *xyz* cell includes sharing nodes from the neighbouring cell.

From eq. (2.48) we can directly express the Planck's constant by others fundamental constants and CL node parameters.

$$h = \frac{\pi(c)c^2 m_n N_{RQ}^3 k_d}{4 v_c k_{hb}^3} \quad \text{(N m s)} \tag{2.58}$$

where: (c) - is a light velocity as a dimensionsless factor.

k_d - is a dimensionsless factor given by Eq. (2.47)

The unit electron charge expressed by CL space parameters is:

$$q = \frac{N_{RQ}^2}{2v_c^2}\sqrt{\frac{c\alpha k_d}{2k_{rd}^2 k_{hb}^3}} \qquad [C] \qquad (2.58.a)$$

Summary and conclusions:
- **The introduced parameter of node inertial mass allows to find the relation between the CL space parameters and the fundamental physical constants.**
- **The CL node parameters: a node distance, a proper resonance time and an inertial mass are basic parameters defining the space-time properties of the CL space.**

2.12 Relation between the intrinsic and the inertial mass of the CL node.

The inertial factor defined by the Eq. (2.6.a) is a ratio between the interaction energy and average gravitational energy.

The CL node interaction energy is in fact the kinetic energy of the node oscillations. The node moment of inertia is $m_n r^2$, so the interaction energy is:

$$E_I = \frac{m_n r^2}{t_R^2} \qquad (2.59)$$

The average gravitational energy, can be expressed as a gravitational potential between two neighbouring node. In fact every node, regarded as a central one, has 4 neighbours (connected along the *abcd* axes), so we may consider that the central node interacts with 1/4 of every neighbouring nodes. In such case, the magnitude of the gravitational potential could be regarded as between two nodes at distance of d, defined for *abcd* axes. This potential is obtainable by integrating on a distance the SG forces between two nodes in a void space.

$$E_{SG} = \int \frac{G_{od}m_{no}^2}{x^3}dx = \frac{G_{od}m_{no}^2}{2d^2} \qquad (2.60)$$

where: G_{od} - is a gravitational constant in empty space between the two deferent types of intrinsic matter (in our case the two types of CL node); m_{no} - is the intrinsic mass of the node, (averaged between the right and left handed nodes); d - is the node distance along *abcd* axes.

Substituting (2.59) and (2.60) in Eq. (2.6.a) and having in mind that $d = d_{nb}/3$, we get the expression of the inertial node mass.

$$m_n = \frac{9I_F t_R^2 G_{od}m_{no}^2}{2d_{nb}^4 k_{rd}^2} \qquad (2.61)$$

where: I_F - an intrinsic inertial factor of CL node

The inertial factor is a function of the node shape, node distance d_{nb} and resonance time t_R.

From equation (2.61) we see that for CL spaces with different node distances, the parameters that may affect the inertial node mass are: I_F, t_R and d_{nb}. All other parameters are constants.

2.12.A. Planck's constant estimated by the parameters of the intrinsic matter and the CL space

Substituting m_n from (2.61) in (2.58), we get for the Planck's constant:

$$h = \frac{9\pi I_F N_{RQ}^4 G_{od}m_{no}^2 k_d}{8d_{nb}k_{hb}^6 k_{rd}^2} \qquad (2.62)$$

Note: The dimension of G_{od} is not equivalent to the dimension of G (universal gravitation constant). This is because G_{od} is involved in SG equation (2.1), where the distance participates in a cubic power instead of square. For this reason, in order to avoid any confusion in the analysis in BSM a SI systems of units is always used.

The Equation (2.62) could be useful for estimation of the quantum energy exchange between two different gravitational fields. This is a problem that is related to the General relativity. For this reason the inertial factor I_f, however, is necessary to be analysed. This is a complicated task, requiring number of unknowns, so it is not discussed in the present course of the BSM theory.

2.13 Physical meaning of the Planck's constant, by using the basic parameters of CL space.

The physical meaning of the Planck's constant appears more apparent if using the basic parameters of the CL space.

Analysing ε_o by Eq. (2.43) we see that the term ΔL has the same dimensions as the Planck's constant: (m²kg sec). Then the product $v_c\Delta L$ have a dimensions of energy. We see also, that the expression (2.58) for the Planck's constant contains v_c in the denominator. If multiplying this equation by the first SPM harmonic frequency ν_c, we obtain $h\nu_c/q = 511$ (KeV) (division on electron charge provides energy in (eV). This energy value is equivalent to a integral momentum change of the CL nodes, when a first harmonic wave is propagated. Following the same logic we may apply this for n-th subharmonic. **The frequency of the n-th subharmonic quantum wave (photon) is equal to the SPM frequency divided on n.** Then the photon energy is

$$E_{ph} = (first \text{ harmonic energy}) \times n = \frac{(first \text{ harmonic energy})}{\nu_c} \nu$$
(2.63)

where: ν - is the photon frequency

The Eq. (2.58) could be presented also as a torque referenced to SPM (Compton) frequency.

$$h = \frac{Torque}{\nu_c} \qquad \frac{kg \ m^2}{Hz} \qquad (2.64)$$

where:

$$Torque = \frac{\pi(c)c^2 m_n N_{RQ}^3 k_d}{4 k_{hb}^3} \qquad (2.65)$$

Consequently the Planck's constant could be regarded as a specific torque, measured at the SPM frequency. In such case it is expressed only by the CL space parameters.

The Equations (2.64) and (2.65) provide useful link for estimation of the Planck's constant by the Compton frequency, which is experimentally determined value of the SPM frequency. The Compton frequency (discovered by the great america physicists Compton) is simultaneously the first proper frequency of the oscillating electron.

The obtained expression of the Planck's constant gives a possibility to estimate not only the quantum wave features of CL space, but also to find its basic parameters: **the static and the dynamic pressure**. These two parameters are directly related to the following physical parameters and relations:

- the Neutonian (apparent) mass of the atomic particle and macrobodies in CL space

- the energy balance between ZPE of CL space and the minimum kinetic energy of the elementary particles

The determination of the static and dynamic pressure is discussed in Chapter 3.

We can summarize that:

- **The Planck's constant expresses the equivalent angular momentum change of all electrical quasispheres for a first harmonic quantum wave.**
- **The Planck constant can be measured as a specific torque resistance, at the SPM frequency.**

2.13.A. Zero Point Energy uniformity and CL space relaxation time constant

The measurable parameter of the ZPE-D according to BSM is the temperature estimated by the Cosmic Microwave Background (CMB). We may call it a CL background temperature. In Chapter 5, the relation between the background temperature, from one side, and the proton volume and the ideal gas constant, from the other is shown. The derived expression for the CL background temperature gives a value that differs only by 0.06 K from the "relict" temperature estimated as a blackbody temperature using the cosmic microwave background. In BSM analysis, the CL space-time constant is used in a sense of relaxation time, characterizing the space-time features of wave-like fluctuations responsible for ZPE uniformity of CL space. These fluctuations are obtained by the self-synchronization of the neighbouring CL nodes. Such fluctuations are called Zero Point Waves - a therm introduced in BSM. They are regarded as spontaneously created and destructed magnetic protodomains (because they are not closed loops) with a length equal to the Compton wavelength, λ_c. The very rarefied gas substances of low z-number elements and especially the Hydrogen, distributed in the deep space, obtain dynamical equilibrium with the ZPE-D. Then the ZPE-D of the space is estimated indirectly by the emission spectrum of these elements (in atomic and molecular form) and especially the Hydrogen. Without the existence of the zero point waves, the uniformity of the Cosmic background temperature is not possible to be explained. **It is quite logical to consider that the average time of recombination of the magnetic protodomains should be characterised by some relaxation time constant, which will be discussed below.**

Firstly, we will use a theoretical approach, considering that oscillating CL nodes could be presented as Phase Look Loop (PLL) oscillators connected by *abcd* axes. Let us suppose that such oscillators are not synchronized at some particular moment. The interconnected neighbouring CL nodes (oscillators) will obtain a self synchronization for one full cycle of SPM vector the period of which is given by the Compton's time $t_c = 8.0933 \times 10^{-21}(s)$. For this time the synchronization will be propagated by the speed of light at distance equal to the Compton wavelength. The light propagation, however is based on the CL node resonance frequency, expressed by NRM vector. The number of NRM cycles in one SPM cycle (Compton period) is given by NRQ which in §2.13.B is estimated to be of the order of $N_{RQ} = 0.884 \times 10^9$. For a stationary frame we may expect that the phase between the NRM and SPM frequency for the spontaneous synchronization may not match. Then we must multiply the Compton's time by the number of NRQ cycle, obtaining $t_c N_{RQ} = 7.157 \times 10^{-12}(s)$. The reciprocal value of this is a frequency of

$$1.397 \times 10^{11} (Hz). \hspace{2cm} (2.65.a)$$

One must keep also in mind that the SPM vector referenced to the direction of the phase propagation may have a left or right-hand rotation. Additionally, since the ZPE waves are based on a random process it should have some statistical distribution, but the phenomenon is not so simple in order to assign some known distribution law. We may only guess that the value given by Eq. (2.65.a) might be multiplied by the standard deviation of such a distribution.

There is one additional theoretical guess about the relaxation time constant of the CL space that fits well to many theoretical equations derived in BSM. Let us consider a virtual observer moving with a velocity of light. Its propagation properties are those of the photon. From the Modern Physics it is known that in such conditions the time stops to run. Then the passed distance also loses its meaning. In such conditions the value of the light velocity must be preserved, but the dimensions will be lost, so it could be considered as a <u>dimensionless factor</u>. In fact the motion of the oscillating electron (discussed in Chapter 3) with a velocity approaching the speed of light exhibit theoretically such features. Using the light velocity as a dimensionless factor, denoted as (c), we may define one theoretical value of the CL space relaxation time constant:

$$t_{CL} = \frac{(c)}{v_c} = 2.426 \times 10^{-12} \text{ sec} \hspace{1cm} (2.66)$$

where: (c) is a light velocity used as a dimensionless factor.

The value of the CL space relaxation time constant given by Eq. (2.66) is valid only in SI units. This does not mean that the relaxation constant is dependent on the system unit of measurement, which can be easily verified by checking the dimensions identity.

$$\frac{(c)}{v_c} \equiv \frac{(m/\text{sec})}{1/\text{sec}} = (m) \hspace{1cm} (2.66.a)$$

$$\frac{(c)}{v_c} \equiv \frac{(cm \times 100)/\text{sec}}{1/\text{sec}} = (cm)100 \equiv (m) \hspace{0.5cm} (2.66.b)$$

The dimensional equation (2.66.a) is for a SI measurement system, where, the length unit is 1 m. The equation (2.66.b) is for a measurement system, where the unit length is 1 cm. From the two equations we see that **the relaxation time could not be considered as an intrinsic parameter of the CL space related to its the space-time features.**

The reciprocal of the relaxation time constant can be regarded as a **relaxation frequency parameter**:

$$\frac{v_c}{(c)} = 4.12148 \times 10^{11} \quad [\text{Hz}] \hspace{1cm} (2.66.c)$$

If comparing the frequencies given by Eq. (2.65.a) and Eq. 2.66.c) we see that their ratio is about 0.339, which might be caused by the previously mentioned statistical distribution of unknown type.

The relaxation time constant and its reciprocal - the relaxation frequency are parameters of the zero point waves, which are responsible for equalization of the ZPE of dynamical type (discussed later in details). Then it is reasonable to consider that the relaxation time constant could provide some small contribution to the Johnson type of noise. It could be measured only at very low temperatures approaching the CL space background temperature of 2.72K. Such experiment is reported in Physics Review B by R. K. Koch, D. J. V. Harlingen and J. Clarke, (1982). Fig. 2.47.0.2 shows the plot of the measured spectral density of Johnson noise as a function of the frequency in a sample at temperatures 1.6K and 4.2K .

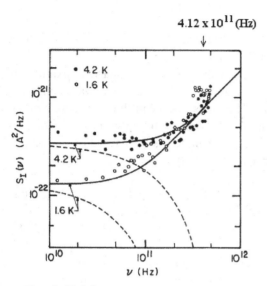

Fig. 2.47.0.2. Measured spectral density of current noise in shunt resistor of junction 2 at 4.2K (solid circles) and 1.6K (open circles). (Courtesy of Koch et al. (1982).

From the plot shown in Fig. 2.47.0.2. we see that the trend of the spectral density of the current noise points to the value 4.12×10^{11} (Hz), which is the theoretical value of the relaxation frequency parameter given by Eq. (2.66.c).

In Chapter 3, the relaxation frequency is used in the definition of the dynamical pressure of CL space, which allows derivation of theoretical expression of the CL space background temperature in Chapter 5. The derived constant is quite important parameter, participating in many physical phenomena.

The CL relaxation time constant and the Compton's wavelength could be considered as inseparable natural unites for time and space in CL space environment.

The above consideration agrees with the current vision in contemporary Physics, on which base the postulates in Modern Physics are defined. We must emphasize, however, that the theoretical guess of the relaxation time (or space-time) constant must not be considered as a final truth and further experiments are needed. The experiment of Koch et al. provides a value, which is relevant for the CL space in the sample, but the CL space relaxation time constant may depend on the refractive index of the medium.

2.13.B. Space-time relaxation of CL space domain

The CL space relaxation time and frequency have a feature of 1D dimension, since they are considered as a propagation of phase which takes place in a linear or curvilinear path. The question is, how much time will take the restoration of the CL node synchronization in a finite volume of macrodomain with dimensions much larger than the Compton's wavelength. It is reasonable to consider that the self-synchronization may start from many points or microvolumes, but the SPM handedness and phases could be random. Consequently, an effect of concurrence will occur. Additionally the motion of Earth and solar system through the galactic CL space, which is discussed in Chapter 10 and 12, must be also taken into account, since it may influence some priority in the effect of concurrence. Evidently, the restoration of the CL space in a 3-dimensional volume will be much larger that the theoretically derived relaxation time constant. This is important issue for the special applications about the control of the gravity and inertia, discussed in Chapter 13.

2.13.C. Characteristic frequencies related to the levels of matter organization

One important approach adopted in BSM is to determine the frequencies of identified basic interactions and operate with them. Operating by the frequency unit has advantage over operating by the length unit, since the frequency is preserved in the CL space domains with different refractive indices. For example, the photon energy estimated by frequency is independent from the refractive index of the medium in which it is emitted and propagated, while its wavelength is not.

It has been mentioned that the prisms have internal structure, so they are not in the lowest level of matter organization. This issue is discussed in details in Chapter 12, where it will be shown also why different levels of matter organization have different characteristic frequency. The lowest level has a highest frequency and every upper level has a lower one, defined by a common mode oscillations. One of the initial assumption in BSM was that the Planck's frequency is related to the lowest level of matter organization. It is a reciprocal to the

Planck's time defined by Max Planck by proper combination of physical constants. The obtained small value of this parameter have bogged the mind of the physicists for many years.

$$t_{pl} = \sqrt{\frac{Gh}{2\pi c^5}} = 5.39\times10^{-44} \ (\text{sec}) \qquad (2.67)$$

where: G - universal gravitational constant, h - Planck constant, c - velocity of light.

We have identified also the following characteristics frequencies for some upper level in the hierarchy of matter organization:

$$\nu_R = 1.092646\times10^{29} \ (\text{Hz}) \text{ - node resonance frequency } (2.55)$$

$$\nu_c = 1.23559 \times 10^{20} \ (\text{Hz}) \text{ - Compton frequency}$$

Note: The Compton frequency, as mentioned in the previous sections is the SPM frequency for MQ CL node in Earth gravitational field.

In Chapter 3 it will be shown that the first proper frequency of the oscillating electron is equal to the SPM frequency of MQ CL node.

Table 2.3 shows the value of the three identified frequencies, ν, and their reciprocal values (as "Period"), arranged in some order according to identified levels of matter organization (discussed also in Chapters 10 and 12). In order to plot them in a common scale, they can be expressed by their natural logarithms.

Levels of matter organization   **Table: 2.3**

Level x	Period (sec)	Frequency ν (Hz)	$\ln(\nu)$	Type of oscillation
0	5.39E-44	1.855E43	99.629	
1				
2	9.152E-30	1.0926E29	66.86	NRM
3	8.093E-21	1.236E20	46.26	SPM & Electron

Fig. 2.74.0.0 shows a plot of $\ln(\nu)$ versus the level of matter organization , x. The points of $\ln(\nu)$ are very close to a fitted robust line if one level of matter organization is missing. This level is identified as a level 1 in Table 2.3. The analysis presented in Chapter 12 are in favour of the existence of such level).

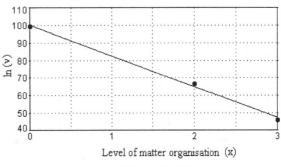

Fig. 2.47.0.0

The CL space exists in levels 2 and 3 of matter organisation, but not in level 0 or 1.

The very steep falling trend (having in mind the logarithmic scale) might be explained by the strong dependence of the inertial factor on level of matter organization, which is associated with different structural complexity. We see that the relation between the trend and the inertial factor (defined in Chapter 2) follows the rule:

large factor - large structural complexity - low frequency.

According to this rule, the smallest inertial factor should be attributed to the zero level of matter organization. In Chapter 12 (Alternative Cosmology) we will see that it could be attributed to a simplest material structure that possesses oscillation properties.

It has been mentioned in the previous analysis that the intrinsic matter should have some intrinsically small but finite time constant. In such aspect we may accept that:

- **The Planck's time is likely the mean value of the time constants of the two most fundamental and indivisible particles, from which the prisms are built.**

2.14. Basic measurable parameters of the CL space.

According to the considerations expressed in §2.6.4, the Cosmic Lattice is able to occupy a definite volume in empty space without need of boundary conditions.

Static pressure of CL space

When a complex helical structure is put in CL space, the CL nodes are displaced only by the volume of the first order helical structure (FOHS), because it contains an internal structure of RL type, which possesses much larger density and stiffness. When such structure is in motion, the CL nodes fold, deviate, pass, and restore their positions, so they passes through the stronger local field of the structure, but not through the volume of its FOHS. Consequently, for any complex helical structure, the CL space could exercise a pressure only on its FOHS's. We call this parameter a **Static CL pressure**. When the structure is in motion, it exhibits two types of interactions with the CL space: inertial and field type. The latter one is a magnetic field caused by the electrical field of the charge particle.

Dynamic pressure of CL space

The CL space exercises forces on the envelopes of the most external formation of FOHSs in form of ZPE waves or Zero Order Waves. According to the considerations in §2.13.A, the length of these waves is equal to the Compton wavelength λ_c. While this length is comparable with the length of the single coil structure of the electron, it could not exercise a pressure on it. For the proton and neutron, however, (the dimensions of which are determined in Chapter 6 and confirmed in later chapters) the ZPE waves will exercise a dynamical pressure. The reason, that these waves may exercise forces on the proton (neutron) envelope, but not on the individual FOHSs from which the proton (neutron) is built is that the length of the ZPE waves is much longer, than the radial section diameter of the FOHSs. As a result, the external shell of the proton feels a **dynamic pressure exercised by the smallest waves of the CL space - the Zero Point waves. This kind of CL pressure is dynamical with a characteristic frequency equal to the recombination frequency of the magnetic protodomains. (reciprocal to the CL relaxation constant).**

While the Static CL pressure gives a possibility to formulate the apparent mass of the elementary particles, the Dynamic CL pressure helps to estimate the energy equilibrium between the CL space and the atoms.

The Static and Dynamic pressures are expressible by the known physical constants. The derivation of the expressions for Static and Dynamical CL pressure is shown in Chapter 3, where original equation for the (Newtonian) mass of the elementary particles is also derived.

Background temperature of CL space

Another important parameter of the CL space is the Zero Point Energy of dynamical type (ZPE-D). Its measurable parameter is the CL space background temperature. According to BSM, the cosmic microwave background radiation is a signature of the background temperature of the deep space. The radiation is formed by the emission from atoms and molecules in the deep space, as they are in constant dynamical equilibrium with the ZPE of the space. The background temperature of the Earth local field can be calculated by using the universal gas constant, the static CL pressure and the proton dimensions. Its value is calculated in Chapter 5. It appears about 0.05 K lower than the measured CMB from the deep space, but this is reasonable (see the discussion in Chapter 5).

Basic parameters of the CL space

It is useful to show now some of the basic parameters of CL space, while some of them (Static, Dynamic and Partial pressure) will be derived in the following chapters. (Note: The number of equation derived in other chapter is shown in square brackets).

ν_R - proper resonance frequency of CL node

ν_{SPM} - SPM frequency, equal to the Compton frequency for Earth gravitational field ν_c)

N_{RQ} - number of resonance cycles per one SPM cycle

c - light velocity (as a propagation velocity of quantum wave in CL space)

P_S - Static CL pressure

P_D - dynamic pressure

m - apparent mass of elementary particle (in CL space only)

m_{ni} - node inertial mass

T_{BG} - background temperature parameter of ZPE

t_{CL} - relaxation time constant of CL space

h - Palnk's constant

q - unit electrical charge

Some of the derived basic equations (see Chapter 3) expressed directly by the CL parameters are the following:

The **Static CL pressure**, when using the SPM (Compton) frequency is:

$$P_S = \frac{g_e^2 h \nu_c^4 (1-\alpha^2)}{\pi \alpha^2 c^3} \quad \left[\frac{N}{m^2}\right] \qquad [(3.53)]$$

where: α - is the fine structure constant; g_e is the gyromagnetic factor of the electron (other parameters has been mentioned above)

The **Static CL pressure**, when using the CL resonance parameters is:

$$P_S = \frac{h g_e^2 (1-\alpha^2) \nu_R k_{hb}^3}{\pi \alpha^2 N_{RQ}^4 d_{nb}^3} \quad \left[\frac{N}{m^2}\right] \qquad [(3.54)]$$

where: d_{nb} - is the node distance estimated in *xyz* coordinates of CL structure

k_{hb} - is the quantum wave boundary condition factor, given by Eq. (2.20.a):

$$k_{hb} = \sqrt{1 + 4\pi^2(0.6164^2)} = 4 \quad,$$

where: 0.6164 - is a factor complying to the Rayleigh criterion for detection of light by diffraction limited optics.

The **Dynamic CL pressure** is:

$$P_D = \frac{g_e h \nu_c^3 \sqrt{1-\alpha^2}}{2\pi \alpha c^3} \qquad [(3.62)]$$

The Newtonian mass of any particle of helical structures in CL space is determined by the volume of its FOHSs. **The mass equation** [(3.57)] (derived in Chapter 3) allows to calculate the newtonian mass of the elementary particles, if the configuration of their helical structures is identified.

$$m = \frac{g_e^2 h \nu_c^4 (1-\alpha^2)}{\pi \alpha^2 c^5} V \quad [kg] \qquad [(3.57)]$$

where: V - is the volume of the FOHS's included in the particle

Note: If a first order positive structure is included in first order negative one, the external volume only should be considered.

The inertial mass of the oscillating node is:

$$m_{ni} = \frac{4h\nu_c k_{hb}^3}{\pi(c)c^2 N_{RQ}^3 k_d} \quad [kg] \qquad (2.73)$$

where: (c) - is the light velocity as a dimensionless factor

k_d - is a factor given by Eq. (2.47).

The light velocity by the resonance CL parameters is:

$$c = \frac{\omega_R d_{nb}}{2\pi k_{hb}} = \frac{\nu_R d_{nb}}{k_{hb}} \qquad (2.75)$$

where: ω_R - is the resonance angular frequency; d_{nb} and k_{hb} - are respectively the node distance and the boundary factor for a quantum wave.

The Zero Point Energy of dynamical type discussed above, has its measurable parameter: a **temperature background.** Its value for a deep space is provided by the Cosmic Microwave Background. In the local field the temperature background can be calculated. This is demonstrated in Chapter 5.

The current model of BSM theory, provides the following estimates for some of the CL space parameters:

$$N_{RQ} = 0.88431155 \times 10^9$$

$$\nu_R = 1.092646 \times 10^{29} \; [Hz] \qquad t_R = 9.152093 \times 10^{-30} \; [sec]$$

$$m_n = 6.94991 \times 10^{-66} \quad [kg]$$

$$P_S = 1.373581 \times 10^{26} \quad \left[\frac{N}{m^2}\right]$$

$$P_D = 2.025786 \times 10^3 \quad \left[\frac{N}{m^2 Hz}\right]$$

$$d_{nb} = 1.0975 \times 10^{-20} \quad [m] \quad \text{- node distance along } xyz \text{ axes}$$

$$d_{na} \approx d_{nb}/2 = 0.54876 \times 10^{-20} \; [m] \text{ node distance along } abcd \text{ axes}$$

2.15. Gravitational law in CL space

The gravitational law in CL space is the Newton's universal law of gravitation. Why the inverse cubic law if Super Gravitation in empty space becomes inverse square law of Newtonian gravitation in CL space?

The answer of this question is not simple enough, in order to be provided in this chapter. But some useful consideration, related to this aspect are the following:

a. A unit volume of cosmic lattice around a massive object has a specific weight.

b. For the first order structures, the cosmic lattice behaves as a real gas at constant temperature, defined by the ZPE. The volume occupied by the FOHSs of the particle displaces an equivalent volume of CL space

c. When an elementary particle, comprised of spatially arranged FOHSs, is in a gravitational field of a massive body, it feels an attractive force, which is a resultant of forces between the massive object and the FOHSs of the particle. These forces are SG forces propagated through *abcd* axes of the CL node and more specifically by the central parts of the prisms.

d. The lattice pressure around the massive object is slightly higher than in a CL space away from such objects. Since the ratio of the "volume/ number of prisms" of any macrobody is much smaller than the same ratio for FOHS, the effect of the CL space shrink from a macrobody is very weak.

The described above features lead to the following important conclusions:

(1) The gravitational forces defined by the Newton's law of universal gravitation are result of SG forces propagated in CL space by the abcd axes of CL nodes.

(2) The node resonance frequency and the Zero Order Waves serve as upper limit frequency attenuations of the SG forces propagation in CL space. Consequently, the Newtonian gravitation is not a primary fundamental law but derivable and dependent on the state of CL space.

3) The feature c. permits a clear explanation of the space curvature defined by the General Relativity. It also indicate that similar effect may exist in close proximity to the atomic nuclei (this is discussed elsewhere in BSM in connection to Lamb shift effect in atomic spectra)

The provided above logical considerations leads to a conclusion, that the Newtonian gravitation is a propagation of the Super Gravitation in conditions of CL space environment. While the propagation of the SG field between prisms that are not in motion could be quite fast its propagation through the oscillating CL nodes is likely affected by the oscillation period of NRM. This consideration is supported by the theoretical work of H. E.

Puthoff starting with the Planck's frequency and using one hypothesis of A. Saharov.

2.15.0. Mass - energy - charge equivalence principle.

Presently, the Newtonian mass (a mass, we are familiar with) is often is identified as equivalent to matter. According to the BSM theory, however, this is a serious misconception. The two fundamental particles embedded in the elementary particles and the CL space could be only considered as matter. The above-mentioned misconception leads to significant deviation of number of theoretical model (mostly based on mathematical physics) from reality.

The (Newtonian) mass is only attribute of the matter related to formations of fundamental material particles at proper level of matter organization. This attribute may disappear due to destruction, for example, as we will see in other chapters, but the fundamental particles and consequently the matter could never disappear.

For simplicity we may consider the Newtonian mass as an apparent mass (or simple a mass), and the intrinsic mass as a SG or intrinsic mass.

The principle of mass-energy-charge equivalence, discussed in the next section is valid only for the particles exhibiting apparent (Newtonian) mass in CL space.

2.15.1 Mass-energy equivalence

The mass - energy equivalence, according to BSM, uses the Einstein's equation $E = mc^2$, but with a remark, that the intrinsic matter does not disappear, when the apparent mass vanishes. Instead of that, the matter undergoes one of the two types of conversion:

- a Newtonian mass may vanish if a FOHS of one handedness is inserted inside of another FOHS of opposite handedness.

- a Newtonian mass may disappear if some FOHS is disintegrated into nodes or prisms (which may pass a huge distances in CL space until recombining in CL nodes - some types of neutrino).

The both processes are related with energy release, but the prisms are unchanged. The BSM theory shows that, there is not annihilation of the matter at all, not only at the temperature of the nuclear fusion but also in high energy cosmological phenomena.

2.15.2 Energy equivalence principle for the electrical charge and charge unit equality

2.15.2.1 Considerations and principles

The static electrical charge could be regarded as a kind of energy distributed in form of electrical field around the particle. Indeed, the electrical quasispheres around the particle contain larger energy, than the magnetic quasispheres. In Chapter 6 it will be discussed, that the neutron to proton conversion is related to creation of pair charges: one static and one dynamic as a quasiparticle wave. The proton gets mass deficiency, because its toroidal shape is twisted. In this process the internal rectangular lattices (RL) of all FOHSs get partially twisting, which leads to a small volume shrinkage. The energy equivalence of this volume shrinkage according to the mass equation is equivalent to the sum of the energies of the static charge and the quasiparticle wave. The both are reaction of CL space in order to preserve the energy balance.

Then applying the energy conservation law, the charge-energy equivalence principle can be formulated. Instead of universal formulation, which requires mentioning of lot of conditions, we can reference the principle to the neutron - proton conversion process (see details in Chapter 6).

- **The total energy of the created electrical charges, in the neutron to proton conversion in free CL space, is equal to the energy equivalence of the newtonian mass change.**

The term free CL space is used to emphasize that ideal conditions are considered in order to neglect the influence of external gravitational, electric and magnetic interactions. The formulated above principle allows to provide a logical explanation of the processes of the neutron-proton and proton-neutron conversions. (Details are given in Chapter 6).

As a consequence from the above conclusion it follows that a static (not moving) charge could exists only around a particle, possessing a matter. Having in mind the energy conservation law and the analysis of the electron oscillations in CL space (Chapter 3) we arrive to the following **conclusion:**

- **The electrical field energy of the electron (positron) is equal to its mass equivalent energy.**

This principle will become more apparent in Chapters 3 and 6. It will become apparent also that

- **The charge value of any kind of helical structure in CL space, is one and same, equal to the charge of the electron (unit charge equality principle)**

In fact the above principle is well known by the QED, but BSM is able to explain, why different size elementary particles have one and a same value of electrical charge. In Chapter 3 we will see, that the Planck's constant is an intrinsic feature of the electron structure. Based on such connection the expressions for the Static Dynamic and Partial CL pressure are derived. At the same time we see from from Eq. (2.58.a) (for the unite charge) that a number of space parameters participates but neither the Planck's constant or the physical dimensions of any particle are involved. Consequently:

- **the unit charge is intrinsic feature of CL space and does not depend on the mass of the elementary particle.**

The physical explanation of charge unity is provided in §2.15.2.2.

The matter does not annihilate, but we may use this therm for the electrical charge, regarded as an attribute of the helical structure in CL space.

When the process of creation or annihilation of electrical charge does not involves a particle destruction, the following rule is valid:

In CL space, electrical charges detectable in far field could be created or annihilated only in pairs.

The latter rule is a result of the intrinsic behaviour of the CL space. Knowing, that the electrical charge causes a creation of spatially arranged EQs around the FOHS, the sudden appearance of such domain in CL space, causes an opposite reaction. The space reacts by creation of opposite charge. The birth of electrical charge, for example, may be a result of: unlocking of near field (neutron - proton conversion); or exiting of some internal FOHS from the RL(T) hole of external one (in case of positron thermalization as described in Chapter 30). But this two cases do not exhaust all the possibilities. The processes related to particle destruction show quite more diversified reactions between the destructed helical structures and the CL space. This is due to the complicated interaction that takes place between the released internal RL structures and the CL space.

In case of FOHS destruction, it is possible one new born charge from destruction of FOHS to interact with one charge of not destructed FOHS (case of J/ψ and τ lepton decay are discussed in Chapter 6).

2.15.2.2 Physical explanation of the unit charge constancy.

The unit charge constancy and some features of the near locked field can be explain physically, when analysing the spatial configuration of the electrical field lines. Figure 2.47.A. illustrates the electrical field lines of a single coil FOHS, containing one internal RL(T) structure, which modulates the CL surrounding nodes converting them to EQ types oriented in lines, which are E-filed lines. Two views are shown. Such structure made by positive prisms with internal axial core of negative prisms, really exists. This is the positron.

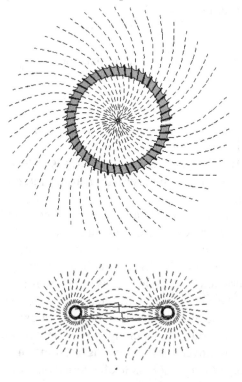

Fig. 2.47.A
Electrical filed lines of single coil FOHS

The internal RL(T) lattice of the positron is shown as gray shaded in the top view of Fig. 2.47.A. In the same view the E filed line alignment to the intercoil zones of the RL(T) is shown. Only the lines normal to the boundary of RL(T) will modulate the CL space. Lines exiting from the RL(T) at angles much smaller than 90° (not shown in the figure) will be locked by the SG (CP) field of the structure (including the internal RL(T)). In the bottom view of same figure we see, that lines, closed to the structure plane are connected between themselves, despite that the lines are result of EQs of same handedness. At first glance, the explanation of the proximity connected E lines seems to contradict the BSM explanation of the E field between charges of same polarity. Although the above discussed case is valid only for field lines generated by a single charge particle, whose RL(T)s are in synchronization. **In a case of separate charge particles the E fields of both particles are not synchronized, and the EQs of CL nodes between them could not get adequate synchronization. In the case, shown in Fig. 2.47.A, the CL node EQs in the proximity are synchronized by one and a same field, induced by the commonly synchronized internal RL(T).** The proximity synchronization is also facilitated by the strong SG(CP) field. **Due to these two features. the neighbouring quasispheres of opposite handedness get induced complimentary motion, and behave as an opposite quasispheres included in the normal E-field lines.**

At the same time the proximity connected (in the middle of the structure) lines exit and enter into the RL(T), so they are not open. Therefore, they could not be able to interact with external field lines created by another charge particle. In this case we consider, that these lines are locked by the SG(CP) field. The energy of the E-field is part of SG energy balance. But the SG field of the helical structure in CL space defines simultaneously two important parameters of this structure: the confined shape of the helical structure (the radius of FOHS envelope and the helical step) and the balance between the locked and unlocked E filed lines. Consequently:

The constant value of the electrical charge of single helical structure in CL space is a result of self regulating process in which a total SG en-

ergy balance is involved. **This balance includes the internal particle SG energy (of its RL(T) lattice) and the energy of the surrounding CL space (including ZPE and e-field energy).**

The above made conclusion helps to explain the following cases:

- the locked near field of the neutron

- the unit charge equality for helical structures with different size

- the locked near field between two single coil FOHS's in a superconducting state of the matter

The last case is discussed in the superconductivity state of the matter (Chapter 4.).

If the created E-filed lines are completely symmetrical, then the ability of the SG(CP) field to lock the whole charge in the near field is stronger. But if the structure is twisted, this ability is degraded. When a particle with locked E-filed is involved in an optimal confined motion, the electrical field could become unlocked. **This is the case with the moving neutron exhibiting a magnetic moment despite its neutrality when it is in rest.**

The explanation of the unit charge equality for structures with different sizes is illustrated by the Fig. 2.47.B. The figure shows a multiturn Second Order Helical Structure, comprising of four turns of FOHSs

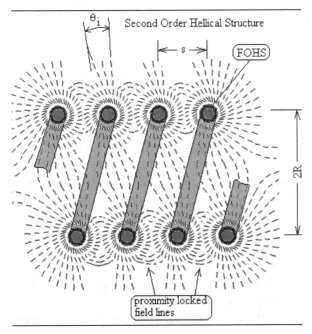

Fig. 2.47.B
E field lines of multiturn SOHS comprised

of 4 turns of FOHS

We see, that the multiturn SOHS can be regarded as a composed of single coils of FOHSs. The proximity intermediate space between the coils contains a large number of proximity connected lines. Adding more single coils makes the proximity SG(CP) field stronger and more lines are locked (proximity connected). Some of the escaped lines are curved by the SG filed. Only the lines that are within angle θ_i are able to escape and modulate the external CL space. They namely contribute to the detected external charge. The angle θ_i is one and a same for any intermediate coil. The angles of the E-field lines from the two ends have a similar configuration as the single coil structure. Adding more single coils affects the angle θ_i, making it narrower. Larger SG field also curves more lines and makes them locked in a near field. As a result of all this factors, the charge constancy is preserved. We may conclude, that:

The charge constancy is intrinsic feature of the CL space. It is self regulated by a complex dynamical balance between the CL space from one side, and the helical structure with its internal RL(T), from the other.

2.16 Confined motion of the helical structures in CL space.

Let us consider a single coil structure of type SH_1^2:-(+(-) shown in Fig. 2.17.a, moving in CL space under some electrical force. This structure could be regarded also as a cut toroid. We can consider now (and later will be proved) that the toroidal radius is much larger than the node spacing. The structure have internal RL(T), whose density is much larger, than CL density. Therefore, the CL nodes could not pass (even partially folded) through the much denser rectangular lattice, so they will be displaced. Then the motion could be regarded as a motion in a fluid. It is obvious that the screw type of motion will exhibit a smaller resistance. In this case, the main resistance is from the radial sectional area at the helix ends. We may call this type of motion a **confined motion.** A confined motion with peripheral speed equal to the light velocity is named an **optimal confined motion.** The axial velocity for optimal confined motion is named an **optimal confined velocity** and is much lower, than the peripheral one.

When moving with the optimal confined velocity, the electrical field of the helical structure becomes locked in some distance from the external shell, because the modulation properties of the RL(T) of the structure could not exceed the speed of light. The picture is similar like the electrical quasispheres in the first harmonic quantum wave. At this distance a boundary surface is formed. The quasispheres at the boundary surface and beyond it, are of magnetic type and are synchronised at SPM frequency of MQ type CL nodes. So the MQs at the boundary layer play roles of bearings for the moving helical structure. This is illustrated by Fig. 2.48. The helical circumference length is equal to the helical SPM wavelength λ_{SPM} of the magnetic (not disturbed) quasispheres. In such conditions the structure exhibits an optimal screw-like motion with less resistance. The boundary magnetic radius r_{mb}, for such motion is defined also by the SPM frequency of the magnetic quasispheres. In the next chapter we will see that the external shell of the electron appears to be a single coil of SOHS of the type shown in Fig. 2.48.

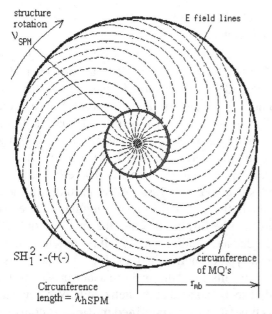

$SH_1^2 : -(+(-)$

Circunference length $= \lambda_{hSPM}$

circumference of MQ's

r_{mb}

Fig. 2.48. Electrical field in optimal confined motion of $SH_1^2 : -(+(-)$ helical structure

Second order structure with a helical shape also have well defined optimal confined velocity. The structure could move also with axial velocity larger than the optimal one (but always smaller than the speed of light, however, this is not completely screw type of motion. In the limiting case, when the linear velocity approaches the light velocity, the rotational motion tends to zero. In some conditions (accelerating by magnetic field) the light velocity limit may may even cause a rotation in a reversed direction.

Multiturn second order helical structures, as those shown in Fig. 2.10 and 2.12, also exhibit confine type of motion. Twisted toroidal structure as this shown in Fig. 2.14.b, will have also a confined motion, characterized by some equivalent step. The folded structure, shown in Fig. 2.14.a have also some equivalent step for confine motion. Both of these structures (from Fig. 2.14), however, do not exhibit so sharp features of confine motion as the structures with a helical shape.

The rotational velocity of the structure shown in 2.48 is a component of the helical velocity, while other component is axial. For axial velocity approaching the speed of light, the rotational velocity becomes zero. For electron accelerated to such velocity a Cherenkov - Vavilov type of radiation occurs. In this case the motion may cause generation of shock waves in CL space.

So far we have discussed helical structures as static combinations of simple structures. Dynamical combinations between some kind of these structures are also possible. They may interact due to their electrical and SG fields and may appear more or less as a stable oscillating system. Dynamical combinations between some structures are very stable, and may appear externally as neutrals, despite the fact, that they are composed of structures possessing equal opposite charges. All these combinations could be classify under the name **ordered helical systems of dynamic type.**

One important feature of the ordered helical systems is that they could be composed by structures, having different external shape and size, but possessing equal opposite charges.

We can summarize the dynamical features of the helical structures and ordered systems by definition of the following rules:

- **All kind of ordered systems, having external helicity or twisting, posses optimal confine velocity in CL space**
- **The effect of confined motion for particles with external helical shape is much stronger**
- **The optimal confine velocity of a second order helical structure in a lattice space is completely determined by the diameter of the helix, the helical step, and the speed of light.**
- **In a normal confine motion the peripheral velocity of the helical structure could not exceed the light velocity**
- **The lattice space is able to influence the helical step of some opened structures when they move with a higher velocity.**
- **Charged particles with different sizes, involved in common motion, may appear neutral in the far field, if the duration of the common motion cycle is shorter than the CL relaxation time.**

2.17 Basic CL space parameters and their connections to some fundamental properties of matter

The properties of the ordered helical structures of primordial matter in CL space, provide a clue for definition of the basic physical parameters and properties of the matter we are acquainted with: time, space, inertia, mass, light velocity, Zero Point Energy. Consequently the mentioned above basic parameters are a not arbitrary, but tightly connected to the property of the intrinsic matter. They permit to understand the controversial space-time feature of the physical vacuum.

Table 2.4 shows some known fundamental properties of the matter we are familiar with, and their connections to CL space parameters.
Table 2.4

Basic parameter	Defined by CL parameter
Space distance:	node distance, d_{nb}
CL primary time base:	CL node NRM period, t_R
CL secondary time base	CL node SPM period, t_c (Compton time)
Inertia:	node inertial mass, Eq. (2.61)
Light velocity:	quantum wave velocity, Eq. (2.39)
Particle mass (Neutonian mass)	Static CL pressure exercised on the FOHSs volume;
Background temperature: (CMB)	signature of ZPE of CL space (kinetic type of ZPE)

The first three parameters in the Table 2.4., d_{nb}, t_R and t_c, define the space time parameters of the physical vacuum. They are dependable on the background temperature, estimated by the Cosmic Microwave Background. In some laboratory experiments (discussed in Chapter 4 of BSM) this temperature could be changed within some limit, which leads to a partially change of the space-time parameters of the physical vacuum.

Chapter 3. Electron. Structure and physical parameters

The electron and positron appear to be the smallest stable elementary particles possessing an elementary charge. The electron is a compound system, consisting of three helical structures and possessing two proper frequencies (while the free positron possesses only one). Investigating the behaviour of the electron we may understand the complex interaction processes between helical structures in CL space. At the same time the electron may play a role of a test probe for estimation of the basic CL space parameters. Due to its complex structure it is referenced in some places as "electron system".

3.1 Electron structure and basic features.

3.1.1 Structure configuration

The electron is a compound helical structure of type: $H_1^2: -(+(-)$, so it is composed of three single coil helical structures.

The configuration of the electron was already shown in Fig. 2.13 a. In Fig. 3.1 a sketch of the electron is shown and the basic dimensions are denoted by letters. We will use this notations later in order to determine the physical dimensions of the electron components.

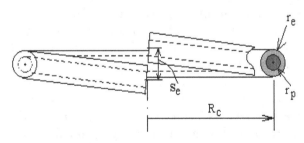

Fig. 3.1
Sketch of the electron's structure

The electron consists of an external negative shell and an internal positive shell with a core. Each shell includes a boundary helix, which encloses an internal RL(T) structure of same type prisms (matter and handedness). The positive shell with a core is the positron (Fig. 2.13.c). The intrinsic gravitation of the electron is not strong enough in order to keep the twisted SG field locked. So the

CL space disturbance propagates in a far field, i. e. the system exhibits a charge. The external E-filed lines were shown in Fig. 2.47A (p. 2-90) and discussed in §2.15.2.2. The electron has two internal rectangular lattices of twisted type (RL(T)). The external helical shell serves as a boundary of the negative RL(T), while the positron helical shell - a boundary of the positive RL(T). Due to a different helicity, the both RL(T) practically are not connected. This is illustrated by Fig. 3.1.A.

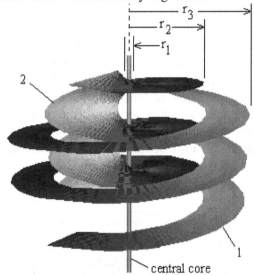

Fig. 3.1.A
Layers of both internal RL(T) structures of the electron

Figure 3.1.A shows only the radial layers connected to the boundary helical cores. The layer 1 (negative) is connected to the external shell, while the layer 2 (positive) - to the positron helical shell. The radial structure of the layers have been shown in Fig. 2.16.c and 2.17.a, where two radial layers are only shown for simplicity. The negative central core of the electron is positioned along the axis of the central hole of the internal positive layer. It is evident, that the interaction between SG(CP) of the two RL(T) is minimized due to the following two features:

- concentric symmetry
- helicity mismatch (between the right and left helicity) between both types of RL(T).

The shown configuration allows a free axial motion of the positron inside of the external negative shell. At the same time, the helicity mismatch and the self adjusted concentric symmetry, provide conditions of ideal bearing. A similar freedom and motion conditions possesses the negative core inside the hole of the positive RL(T) with a radius r_1.

In some conditions of extremely high velocity or operation in a low ZPE CL space domain, the electron may lose its internal positron and convert to a degenerated electron. The degenerated electron is shown respectively in (Fig. 2.13.b)

It is more difficult for the positron to loose its central core, because it is very thin (3 prisms diameters) and its interaction with the external CL space is much weaker. The degenerated electron or positron, however, preserve their internal RL(T), and consequently their dimensions and second order helicity. These features allows them to recombine again in a normal electron. If the central core is lost, it could be regenerated by the trapping mechanism (discussed in §2.8.2). The oscillation of the electron is illustrated by Fig. 3.2.

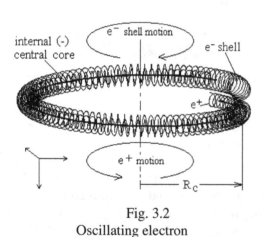

Fig. 3.2
Oscillating electron

We can distinguish two simple oscillating systems: "electron shell - positron", and "positron shell - central core". Therefore, the electron oscillates in a complex way. It is evident that every one simple system has its own proper resonance frequency.

3.1.2 Proper frequency of the oscillating system "electron shell - positron".

Let us analyse the system "electron shell - positron", in order to determine what kind of factors define the proper resonance frequency.

It is evident that the resonance frequency should depend on two types of interaction forces: the intrinsic gravitation between the electron and positron from one side, and the interaction between the EM field of the system and NRM and SPM vectors of CL space, from the other. The SG (CP) field between the electron shell and the positron is spatially structured by their RL(T). The both lattices although have opposite handedness and the radial stripes meet themselves at angle, which is in the range between 170 and 150 deg (see Fig. 3.1.A). This angle is determined by the radius to helical step ratio of the electron structure as a single coil SOHS. When the system oscillates with its proper frequency (Compton frequency), the individual vectors of both types of radial stripes meet themselves for a very short time. Having in mind, that the calculated xyz node distance of CL space was of the order of $d_{nb} = 1.0975 \times 10^{-20}$ (m), and the central axes of the electron is $2\pi R_c$, we estimate that a simple cell from a positive RL(T) could be aligned at a single cell of a negative RL(T) for a time no longer than 1×10^{-40} (s). The prism diameter is at least 12 times smaller than the minimal node distance of RL. So, the time during which the radial stripes between both RL(T) may appear aligned is extremely short and the SG interaction between both RL(T)s may not take place. This means that the SG field is not able to propagate between the two RL(T) lattices. **As a result of this, the gravitational mass of the positron with its RL(T) appears hidden for the external observer.** The positron E filed is propagated by the RL(T). When the positron is inside the electron, its positive field could not pass through the RL(T) of the electron shell, because of the different handedness. So **the E field of the positron in this case also appears hidden.**

Let us analyse now the SG field leaking between the nodes. If the electron overall shape, for example, was not a coil but a straight compound FOHS, then the leaking SG field would be different for a case when the positron is inside, and when part of it is outside of the electron's shell. But for the coil shape as shown in Fig. 3.1 the partially coming out positron core does not go away from the electron's shell. Hence, the returned forces for this kind of shape will be significantly reduced.

From the considerations, discussed above, we may accept that the resonance frequency of the electron depends mostly on the EM interaction with the Cosmic Lattice. This conclusion will become more apparent in the next chapters of BSM.

When the electron oscillates, a portion of a positive charge alternatively appears on both sides of the electron. The interaction of this charges with the lattice in fact influences the motion of the whole electron system.

The dependence of the electron resonance frequency from the CL space features at different value of ZPE complicates the analysis. However, when considering its motion in CL space of normal ZPE and constant node distance, the analysis is simplified, and the electron resonance frequency can be considered very stable. In this and following chapters, we will see that in all cases of photons generation and detection, the electron system is involved.

The quantum feature of the electron could not be explained if it is considered only as a passive system. The electron self-energy is a discussed topic, now, in Modern physics. Some quantum processes without such energy could not be explained. The BSM model of the electron shows that it has the ability to store energy. We can distinguish two different energy "reservoirs", capable to store kinetic energy. The first one is the oscillation energy between electron's negative RL(T) lattice and the internal positron. The second one (much smaller) is the oscillation energy between the positron positive RL(T) lattice and the central negative core.

The total kinetic energy of the oscillating electron interacts directly with the CL space environment by inertial and EM fields.

Experimental evidence exists about the ability of the electron to accumulate energy after it has been dumped. Such conditions are created in experiments observing the transition between the normal and the superconductivity state of the matter.

The superconductivity will be discussed in details in Chapter 4. Here only some features will be mentioned. In the superconductivity state of the matter, where CL domains possess a ZPE energy, the positron could come out and can be attracted externally to the electron shell. In conditions of low ZPE, the internal energy stored in the RL(T) can be dumped. When the ZPE is gradually rasing to a normal value, the electron system recombines, but it needs to restore its lost energy. So if the conduc-

tor temperature is elevating slowly, the electron heat capacity exhibits a peak. The peak, known as electronic specific heat coefficient is clearly observable in the experiments. Figure 3.3 demonstrates this feature for one type of superconductor (BSM interpretation).

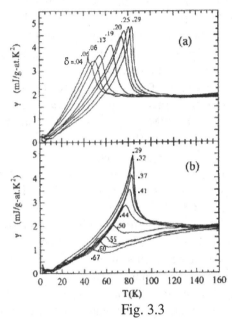

Fig. 3.3

Specific electronic heat coefficient as a function of the superconductor temperature (Courtesy of J. W. Loram et al., 1997)

According to BSM theory, however, not the whole energy of the specific heat goes for refilling the electron self energy. Part of it goes to refill the ZPE of the domains inside the conductor.

When the electron is in motion driven by external field, it oscillates and automatically keeps its internal energy at a nominal value. If the electron has a very low velocity (approaching zero), then the stored energy is still able to modulate the CL space, causing an electrical field. The stored energy provides momentum, which keeps the two subsystems in continuous motion. So this energy is very important factor, influencing the system behaviour in a CL space.

Analysing the dynamics of the electron-positron oscillations we see that: When the positron is inside of the electron shell, the field of its RL(TP) is completely shielded, so it could not exhibit a static charge. When it is oscillating, however, conditions for charge appearance occur periodically from both ends of the electron shell. The oscillating "electron - positron" system interacts with the CL nodes, which are congregated in magnetic protodomains, synchronised by the SPM frequency. In such aspect, the SPM frequency appears a quite strong factor. In Chapter 2 we saw, that it is responsible for the quantum waves. It is enough strong factor able to affect the oacillational motion of the "electron - positron" system. The most important common feature between the CL space parameters and the electron is:

- **The first proper frequency of the electron system (electron shell - positron) is equal to the CL node SPM frequency. In the Earth gravitational field these two frequencies are equal to the Compton frequency.**

- **The above feature means that the energy of the oscillating electron system is supplied directly by the CL space ZPE via resonance transfer. Consequently, in CL space environments, the energy reservoir of electron system is always filled up.**

Let us imagine that an electron is put in a CL space with constant spatial and time parameters, but away from any gravitational or EM field. We may call such a system a fundamental oscillator and may use one of its parameter, namely the first proper frequency as a stabilized frequency etalon. **Its period could serve as a time base** for investigation of the interaction between the helical structures and the CL space, so the CL space parameters to be estimated quantitatively. We may call the frequency of such electron **a fundamental frequency**. Despite the fact that it is not a primary frequency etalon (as we will see in Chapter 12) it is quite convenient for exploring the CL space parameters.

The proper frequency of such system is equal to the SPM frequency of the CL node, which is equal to the well known Compton frequency (valid for the Earth gravitational field).

$$\nu_{ep} = \nu_{SPM} = \nu_c \qquad (3.1)$$

where: ν_{ep} - is the proper frequency of the electron shell - positron, ν_c -is the Compton frequency

The relation (3.1) will become more apparent through the course of the BSM theory.

The Compton frequency is estimated by the Compton wavelength: $\nu_c = c/\lambda_c$ $\qquad$ (3.1.a)

where: λ_c - is a Compton wavelength, c - is the light velocity

It is known (from all physical courses) that the Compton wavelength is given by Eqs. (5.1):

$$\lambda_c = \frac{h}{m_e c} = 2.4263 \times 10^{-12} \quad \text{m} \qquad (3.1.b)$$

where: h - is the Plank's constant
$\qquad m_e$ - is the mass of the electron

The fundamental frequency appears more general parameter, than the Compton one. The fundamental frequency was defined for electron in CL space away from heavy objects, while the Compton frequency is measured in the Earth gravitational field. Secondly, the electron mass is involved in the determination of the Compton frequency. The mass of the charged particles may have mass deficiency, due to the charge potential field in CL space, in comparison with the neutral particle (as the neutron). This possibility, however, is not enough investigated in BSM theory and we will rely on the estimated Compton frequency. Despite the fact, that the Compton frequency is estimated in the Earth gravitational field, we will use it instead of the fundamental frequency. **For the purpose of our analysis we will accept that the above defined fundamental frequency is equal to the Compton frequency and the fundamental period is equal to the Compton time.**

$$\nu_o \approx \nu_c = 1.23559 \times 10^{-20} \quad \text{Hz} \qquad (3.2)$$
$$t_c = \frac{1}{\nu_o} \approx \frac{1}{\nu_c} = 8.0933 \times 10^{-21} \quad \text{sec} \qquad (3.3)$$

From the provided so far analysis, we may summarize the basic features of the electron:

- **In CL space with normal ZPE, the electron possesses internal stored energy. This energy keeps the oscillations of the electron subsystems.**

- **The electron obtains a proper resonance frequency equal to the SPM frequency of the CL space.**

- **The adjustment of the electron proper frequency to the SPM one, may provide explanation of one of the effects of the General relativity: the gravitational redshift of photons emitted in a strong gravitational field.**

3.1.3 Proper frequency of the oscillating system "positron-central core".

The proper resonance frequency of the "positron - central core" system appears different from the first proper frequency of the electron due to its different volume and core dimensions.

When the positron is free (outside of electron shell), its behaviour is similar to the oscillating system: "electron shell - positron" Its proper frequency, however, is different, due to its different external environments. The behaviour of the free positron will be discussed in §3.9.3. It will be shown, that the positron - core proper frequency is related to the Compton frequency by the simple expression:

$$\nu_{pc} = 2\nu_c \quad \text{- for free positron} \qquad (3.3.a)$$

where: ν_{pc} - is the proper frequency of the system "positron - central core".

Additional difference appears for oscillations with small and large amplitudes, when the positron is inside of the electron shell. These features are discussed in §3.5, §3.9.3 and in Chapter 4. The proper frequencies for smaller and larger amplitudes are the following:

$$\nu_{pc}' = 3\nu_c \quad \text{- for positron inside of the electron} \qquad (3.3.b)$$
$$\text{(small amplitudes)}$$

$$\nu_{pc} = 2\nu_c \quad \text{- for positron inside of the electron} \qquad (3.3.c)$$
$$\text{(large amplitudes)}$$

The proper frequency of the positron inside the electron in a case of large amplitudes is the same as the proper frequency of the free positron. This conclusion will become more evident in the course of the BSM theory.

3.2 Electron oscillations and lattice pumping effect leading to a photon emission. "Annihilation" or change of state of the matter.

We can distinguish two types of electron frequency oscillations: **weak** (small amplitudes of oscillation) and **strong** (large amplitudes).

Oscillations with weak amplitudes appear, when the electron is forced to move in the lattice space. The amplitude is much smaller than 180 deg deviation of the positron in comparison to the electron shell. It does not lead to generation of EM wave (photon). The oscillations have small amplitude, induced between the interaction of the electron proper frequency and the SPM frequencies of the CL nodes. The positron in the activated electron oscillates reversibly around the middle position. When a portion of the positron goes out it is not any more shielded by the electron external shell and portion of positive charges appear periodically in both sides. The invoked alternative field interacts with the external negative field created by the electron shell, while the latter interacts with the CL space. The electron in this case induces waves in CL space. The interaction between the induced waves and SPM frequency of CL domains exhibits a quantum effect. Its features are discussed later in this Chapter.

In the **strong amplitude oscillations**, the amplitude may reach 180 deg and over. So this type of oscillation may lead to a separation of the positron from electron system or recombination, as well. Such separation or recombination is always accompanied with absorption or emission of high energy photons. Oscillations with strong amplitudes appear in many observed physical phenomena: electron - positron "annihilation"; 1^1S_0 singlet of parapositronium activated by different methods: by X or gamma rays, by bremshtrahlung, by high energy electron or positrons and so on. It could be activated also by a collision of accelerated electron to a target or a collision with a high energy particle, including a quasiparticle wave. All these processes are accompanied with emission of two or three gamma photons, depending of the amount of activated energy. Let us analyse the dynamical behaviour of the energetically activated electron, leading to emission of two gamma photons.

We can analyse the example of interaction between a normal electron and a positron. They may have initial velocities or may start from a rest. In both cases they will have different potential energies. Let us assume that the potential energy (equal to the energy of activation) is equal to 511 keV. When the two systems accelerated by the attractive Coulomb forces approach each other, the external positron will be directed (by the interacted proximity electrical fields) to enter into the electron system and to replace the internal positron. Although its energy does not permit to expel the internal positron completely. As a result, the two positrons will start to oscillate in the inside hole of the electron shell. Initially the external positron will not hit the helical shell of the internal one, because of the repulsive fields around their edges, so some gap could exists. Once they start oscillations, this gap might be eliminated, because the positron shells around both ends are always inside of the electron RL(T) hole, where no space for stable EQ formation exists. Then the two positrons will oscillate as a single structure. As a result of this, the amount of the positive charge will oscillate alternatively at both sides, and they will interact with the external shell negative charge. In fact the electron system usually is not fixed in the space and both, the positron and the external shell will oscillate around a common equilibrium position. The periodically appeared positive charge and the moving negative charge will cause a lattice disturbance. This disturbance, however, will not be propagated far from the system, because the relative speed of the oscillating structures, as we will see later, is close to the speed of light. The oscillation energy from a single cycle is very small in order to overcome the SG forces and to escape from the system, so it is accumulated in the surrounding CL space. The oscillating system in such way provides some kind of energy of the surrounding CL nodes, increasing their ZPE. We may call this effect a **lattice pumping effect.** (A pumping effect will be also discussed later when explaining the photon emission process in atoms). As a result of this, a pumped energy becomes accumulated in both sides of the electron. Knowing, that the EQ only could handle an excess energy, the latter will produce a large number of EQs of both types (positive and negative). At the same time, the pumping is an energetic process and should have opposite reaction from the CL space. This means that, the CL space should have a saturation value for the number density of the generated EQs per unit volume. Consequently, the increased amount of both types EQ will continuously reduce the spatial modulation properties or E-fields of the oscillating system and the efficiency of the pumping process. Consequently, conditions for multiple oscillations with a gradually reduced amplitude exist. During the duration of this oscillation process the proper resonance frequency of the electron, however, is not changed, because it has enough stored internal energy. When

the pumping process falls below some critical level, the accumulated energy in the CL space domains from both sides of the electron, will be suddenly released as two quantum waves (photons). Note, that the pumping velocities of the electron positron shells had initial value of the linear light velocity. The pumping process in this case is optimal and completely symmetrical. The released two quantum waves have 180 deg direction and are orthogonally polarized. They are first harmonics of the SPM frequency, every one possessing an energy of 511 keV. Both emitted waves are orthogonally polarized because this is a condition for easier separation of the pumped energy from the both sides of the electron. The quantum waves are emitted when the effective strong type of oscillations are attenuated. They are neutral type waves, i. e. equally affecting the right and left handed CL nodes.The time of the oscillations and the finite time required for the energy mixing between the both types of CL nodes is obviously related. It is determined by the intrinsic property of the CL space and the electron. This time is known as a Positronium life time, and its value in vacuum is about 145 psec.

But what happens with the final state of the system? At the end of the oscillations the half of both positrons are equally out of the external electron shell. So the amount of the negative field lines from the electron external shell is equal to the amount of the positive field lines from the half of both positrons. Both types of the field lines are interconnected in proximity, so the far electrical field disappears. The obtained new structure is relatively stable and its mass is equal to the sum of the electron and positron masses. Such small neutral mass will appear undetectable. For this reason it seams that the electron and positron are **annihilated, but annihilation is a misinterpretation from a point of view of BSM space concept. The mass is an attribute of the structure of elementary particle in CL space and it is not equivalent to matter. The matter is composed of the building blocks - the prisms, which never annihilates. In the describe above particular effect, the combined structure of electron and positron only loose its far field electrical field, so it appears undetectable.**

3-6

3.3 Confined motion of the electron. Electron spin.

One important feature of the electron is its confine motion in the CL space. It is a result of the interaction between the SG field of the internal negative RL structure of the electron with the oscillating CL nodes of the surrounding CL space, in which magnetic lines (CL space energy circulated in closed loops) are created. The anomalous magnetic moment of the electron (discussed later) is such environments is an important factor for the preferred screw-like motion of the electron. The large R/r_e ratio also facilitate such type of motion.

As a result, the whole electron system rotates with a spin direction determined by its second order handedness. The positron system has the same second order helicity and handedness as the electron system. Therefore, the electron system and the free positron, both, always tend to perform a screw like type of motion. This type of motion we call a **"confined motion"**. The electron structure is moving and rotating like a screw. The efficiency of the confined motion depends on two factors: the motion velocity, from one side, and the momentum interaction between the proper frequency oscillations and the momentum of SPM vector of the stationary nodes, from the other. **In the case, when the tangential velocity of the electron is equal to the speed of light (linear), the motion is called an <u>optimal confine motion.</u>** The corresponding axial velocity of the electron is called an **optimal confined velocity**. For velocities below the optimal confine one, the electron motion is **completely screw like.** For velocities above the optimal confine one, the system exhibits a **quasi screw type** of motion.

In a completely screw like motion, all the points, lying on the central core pass through a common helical trajectory. In a quasi screw type of motion, every point of the central core has own helical trajectory. In both types of confine motion, no one point lying in the core curved axis could exceed the linear light velocity. **The axial electron velocity for an optimal confined motion is 2.187×10^6 (m/s), corresponding to an electron energy of 13.6 eV.**

The axial and tangential velocities for the two types of confine motions are illustrated in Fig. 5.4.,

where the electron is shown as a single coil, while the trajectory - by a dashed line.,

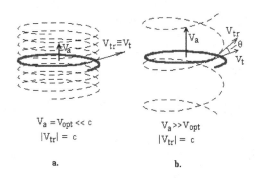

Fig. 3.4

Two types of confine motions of the electron. The electron is shown by a thick solid line and its trajectory by a dashed line

Figure 3.4.a illustrates a case with a complete screw type of motion at optimal velocity V_{opt}, while 3.4.b - a case with a quasi screw type of motion. For both cases the trajectory of the front end of the electron is shown (dashed lines), with the momentum position of the electron (thick solid line). The axial and tangential trace velocities are denoted as V_a and V_{tr}, and their ranges are shown below the drawing. The velocity vector V_t indicates the electron - positron oscillation.

Due to the interaction with the CL space, the oscillating system "positron - electron shell", induces a magnetic field. As a result, the whole electron system exhibits some small momentum with alternatively changing direction. This could be attributed to the electron spin. At condition of normal motion, the spin should have one preferred direction, determined by the conditions for motion with less resistance. Then a question may arise: What is the physical explanation of the $\mp \frac{h}{2}$ spin value assigned to the electron? To reply to this question, we have to distinguish between three cases:

- electron spin when the electron is in a motion around the proton
- electron spin flipping in EMR technique
- electron polarisation

In Chapters 6, 7 and 8 we will see that the proton has its own handedness. The electron trajectory appears as a closed loop curve (with a shape like a digit 8) around the proton, so it also has its own handedness. As a result of this, two combinations are possible: (1) the electron close loop curve and the proton are both with a same handedness. (2) They both are with different handedness. As a result of these, the Quantum Mechanical spin has two values, which are known as $\pm\frac{h}{2}$ spins.

In a conditions of normal motion of the free electron, the phase of the two oscillating systems "positron - electron shell" and "positron - central core" are automatically adjusted for a less resistance in the interactions with the CL nodes. The interaction energy of the first oscillating system is much larger than the second one. In some experiments like in EMR, the oscillating phases of the both system may be temporally affected (so called spin flipping)

In some motion cases, a collimated electron (positron) beam is striking a plane under angle. The reflected electrons (positrons) in this case exhibit a strong polarisation. This effect, however, is not related to the same type of Quantum Mechanical spin, which appears in the optical spectrum. According to the BSM, it is a result of the off-axial momentum obtained in the internal rectangular lattice of the oscillating internal positron during the impact. This effect is experimentally observed. The obtained momentum is preserved and appears quite strong, because the internal rectangular lattices contains a large intrinsic matter and the off-axis oscillation is a SG type of interaction. This kind of SG interaction through the RL(T) internal structure affects directly the external E-field of the electron, so the oscillation energy is transferred to the electron's electrical field. This affects the motion of the electron in a way that its behaviour becomes detectable. At the same time, this effect shows that the internal RL(T) has some freedom to oscillate. Such kind of oscillation may cause a minor change of the spatial geometry of the electron (more specifically the helical structure twisting) but the involved SG field can accumulate comparatively large energy. Consequently, the electron and the positron may have ability to store internal energy. This conclusion independently confirms the accepted feature of the electron to posses a selfenergy.

The discussed so far basic features of the electron are summarized below:
- **The electron exhibits a confined (screw type) motion in the lattice space**

- **The electron posses internal energy well. In CL domain of normal ZPE, the stored energy provides stable oscillations of the electron system components.**
- **The effect known as "annihilation" of electron and positron is in fact a damped oscillation of the compound system "electron - free positron", terminating with emission of two gamma photons at 511 KeV.**
- **The photon emission is a sudden release of the energy pumped in the surrounding CL space due to the self dumped electron oscillations. The released energy is propagated through the CL space as a quantum wave (photon).**
- **The quantum motion of the electron is a result of interaction between the compound oscillating momentum of the electron system from one side and the SPM vector of the surrounding CL space, from the other.**
- **The oscillating electron could be considered as a fundamental frequency etalon, if placed in CL space of normal ZPE, away from massive objects. Its frequency value in the Earth gravitational field is the Compton frequency. The fundamental frequency provides an absolute time base for analysis of processes at atomic level in a frame of absolute coordinates.**

3.4 Electrical field of the electron at confined motion

The electrical field of the electron is created by the SG (TP) forces of the internal RL(T). This forces form a highly ordered spatial field, which modulates the external CL space, causing a formation of electrical quasispheres in the surrounding CL space. The field is different for the cases of "static" (not moving) and "dynamical" (moving) electron. The static case is mostly theoretical, because the electrons always have some velocities.

In a case of "static" electron, the electrical field has a maximal radius, which is practically determined, by the surrounding noise level, defined by the noise of the "permittivity fluctuations" of the CL space.

In a case of "dynamical" electron, the situation is different, and very dependable on the electron velocity. In the confined motion the electron rotates, so its negative RL(T) structure can modulate the external CL space only to a range limited by the velocity of light. It is apparent that such modulation create aligned EQs. Since the electron is moving we may introduce running EQs, like in the analysis of the photon wavetrain in Chapter 2. In §2.11.2.2 and §2.11.3 we have concluded that the energy through aligned EQs is propagated with a helical light velocity, which is larger than the linear light velocity, c, propagated by MQs (This is possible due to the quantum properties of CL space defined by the bumps shape of MQ and EQ - SPM vector spends more time in the bumps). Keeping in mind the above considerations, it is reasonable to accept that the radial extensions of EQs around a confined moving electron may overpass the boundary exhibiting a spatial features, similar as *sinc* function (related to a spherical Bessel function of first kind) shown in Fig. 3.4.A.

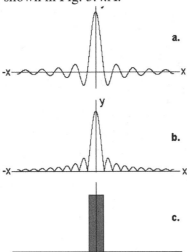

Fig. 3.4.A. Modulation of CL space by electron at confined motion. **a.** The curve portions above and below the horizontal axes corresponds to a same type EQs (negative for electron) but with SPM vectors having an opposite phase rotation; b: a curve with inverted parts of the curve in b. - providing the total E-filed energy of the moving electron

Fig. 3.4.A shows the EQs in a radial section of the field generated by the electron in confined motion. The points, where the sinc curve intercepts the horizontal axes are points where the EQs convert to MQs. The portion of the sinc curve below the axis x are in fact the same electrical charge but with SPM vectors at phase of π. For this reason the

curve b. showing these part above the axis x represents the total electrical field of the confined moving electron. In the real E-filed the sides of the sinc function falls rapidly below the Zero Point Energy (dynamical) of the CL space. For this purpose, we may consider an equivalent distribution, having a rectangular shape with an area equal to the whole area of the field shown in b. This equivalent field, shown in c. is characterised by equivalent radius R_{eq}.

Using the equivalent field will facilitate the further analysis, since we may show graphically the gradual conversion of the EQs to MQs only up to the first crossing point.

Figure 3.4.A illustrates the modulation of the surrounding CL nodes at one particular confined motion of the electron. The axial velocity in this case is equal $V_{ax} = \alpha c/2$, corresponding to energy of 3.4 eV. The tangential rotational velocity is equal to $c/2$, while the conversion of EQs to QMs becomes at radius $R = 2R_c$.

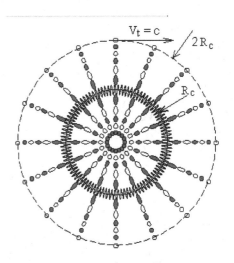

Fig. 3.4.A
CL nodes modulation at confined
motion of the electron at 3.41 eV.

Figure 3.5 shows the radial configuration of confined motion of electron with an axial velocity of $(\alpha c)/3$, corresponding to energy of 1.51 eV. A diametrical section *dy* with the orientation of the long axes of EQ is shown. These axes coincide with the E-field lines, shown by dashed lines. The radial dependence of the E-filed and the tangential

CL node momentum are shown in the bottom part of the figure.

If making analogy between the radial E field configuration of the electron and the field configuration of the radial section of the wavetrain of a quantum wave (see §2.11.3), we will find the following similar features:

- They, both, have boundary conditions provided by MQs.

- The E-field lines are aligned

From the other hand they have the following distinctive features:

- The integrity of the E-filed in the quantum wave is kept by the phase synchronized SPM vectors of EQs. Both types of EQs are equally affected (for a neutral wave) or complimentary affected (for a quasiparticle wave)

- The integrity of the stationary E field of the electron is kept by the SG(TP) field of its internal rectangular lattice.

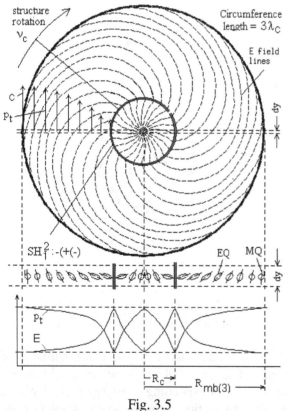

Fig. 3.5

Radial configuration of E field of the electron at confined velocity corresponding to energy of 1.51 eV)

Let us focus on the integrity feature of the E-filed for a quantum wave (photon) and electron field. In the case of quantum wave, the E-field integrity is kept by the alignment of the EQ in helical trajectories and we can distinguish two types of integrity: - along the helical trajectories and between them. Both types of integrity takes place in a cylindrical volume with a radius equal to the boundary radius and a volume length of λ_c. In the case of electron, the equivalent boundary circumference length is equal to λ_c, which is the same for a first harmonic quantum wave (photon with energy of 511 KeV), however, the axial length is much smaller, since it is defined by the helical step of the electron se, which is much smaller than λ_c. Such space, does not allows fulfilment of both types of E-field integrities, as in the quantum wave. Consequently only one type of integrity is possible, and this is the alignment of the EQ in field lines. For the far field the negative EQ's of the electron may repel each other, so they must be radially aligned, but the picture in the near field is different. All EQ of same type are synchronized by one and a same source - the internal RL(T) of the electron. Then in the proximity near field, the negative EQs are able to influence stronger the neighbouring nodes of opposite handedness (by abcd CL node axes). As a result, the neighbouring CL nodes of opposite handedness get some complementary motions, but their energy could not compensate the field of the negative EQ, whose source is the SG field of RL(T).

From the above considerations it is apparent that all E-field lines of the electron are connected to the internal RL(T), as shown in Fig. 3.5. The EQ polarisation is changing gradually from the strongest value, near the helical shell, to weakest one, near the boundary defined by MQs. The lines are bent and terminated at the boundary zone by MQ (only for a rotating electron). The field line intensity is proportional to the polarisation of EQ. Then one important feature emerges: **The electrical field lines of the electron are not strongly connected between themselves, and have a freedom for taking a proper space position.** This is very important effect, because when the electron is moving in a not homogeneous CL space, as in the metal crystals, its E-filed lines could automatically sense the lower resistant domains. This effect provides one very important feature: **a path sensing property of the moving electron.** The path sensing property is related to the NRM cycle, so it can operate faster than the quantum magnetic interaction, which is related to the SPM cycle.

The radial configuration shows, also, another important feature: **the angular frequency of the rotating electron in the optimal confined velocity is equal to the angular frequency of the SPM vector, associated to the MQ at the boundary zone.**

At the bottom part of Fig. 3.5 the radial dependence of the tangential CL node momentum of the SPM vectors is shown. The same momentum is illustrated also by arrows in the radial section. The shape of the curve presenting this momentum is determined by the orientations of the EQs. The long axis of EQ has a larger momentum than the shorter one. The shape of the radial dependence of the tangential node momentum is not calculated, but given as an example. Assuming that the node inertial mass is one a same for MQ and EQ, we may right:

$$p_t(R_c) = m_n c = m_n 2\pi R_c / t_c = m_n 2\pi R_c v_c$$

Fig. 3.6 shows the radial dependence of some field variables for the confined velocity in which the electron axial velocity is $\alpha c/3$ (the electron makes one turn for a time equal to $3t_c$.

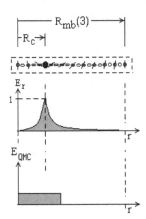

Fig. 3.6
E-field parameters of electron with a confined velocity of $\alpha c/3$.

The radial dependence of the equivalent E-filed intensity is shown as E_r and normalized to the maximum value at the radius equal to R_c. By re-

shaping the shaded area of E_r to a rectangle, we obtain the equivalent electrical radius, which should correspond to the Quantum mechanical radius of the electron RQM. This radius will be determined in §3.11.

We should not be surprised, that some of the internal E-field lines are shown as connected. In fact the E-field line intensities in the internal side of the radial section are expected to converge to zero (or MQ) at the centre. The reason for this comes from a consideration that the diametrically opposite EQs (in respect to the central axis) get opposite SPM phase synchronization from the RL(T).

So far, the radial section of the E-field in the confined motion of the electron was discussed. What are the boundary conditions and the field configuration in the axial section? The E-field configuration in the vicinity to the electron shell has been shown in Fig. 2.47.A, and it is given again in Fig. 3.7.

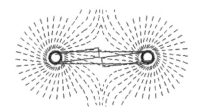

Fig. 3.7
Axial section of the proximity E-field lines of the electron

At confined motion, the boundary condition of the E-field in a radial plane (perpendicular to the motion axis) has a circular shape as shown in Fig. 3.5. The boundary section in an axial plane (passing through the axis of motion), however, is not circular but slightly elliptical, as shown in Fig. 3.8. It is evident, that the concentration of E-field is also not homogeneous and will have a different spatial configuration. As a result of this, the density of the terminating E-filed lines at the boundary section is not uniform as in the radial section. The boundary section is illustrated in Fig. 3.8, where the denser E-field line termination is presented by denser points.

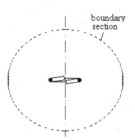

Fig. 3.8

Axial boundary section of the electron at confined velocity. The terminated E-filed line density is shown as a point density

From the radial and axial E-filed line configuration we see that at the confined motion, the boundary conditions zone has a shape of oblate spheroid whose axis coincide with the rotational axis of the moving electron. The density of the terminated E-filed lines is larger at the equator and lower at the poles. In such configuration, the E-field lines still posses strong guiding feature, which keeps the electron orientation in its screw type of motion. In the next paragraph we will see, that at motion with velocities lower, than the optimal one, the circumference length of the central section is equal to $n\lambda_{SPM\,MQ}$, where n is an integer and the boundary shape in the axial section of the E-field also approaches a sphere, so the magnetic radius of the moving electron approaches a sphere for $n > 1$.

Only for the region inside of the boundary surface, the E-field appears not uniform. For a "static" electron or moving electron with axial velocity higher, than the optimal one, the E-field configuration is different, and in many cases may appear to have a uniform spherical shape. The helical configuration, although, is always preserved.

We may summarize:
- **At optimal confined motion of the electron, the E-field is locked in a boundary surface, whose central sections has circumference length of** $\lambda_{SPM\,MQ}$
- **Inside the boundary surface the E-field possesses a helical configuration**
- **The boundary surface at confined motion with low *n* number has a shape of slightly oblate spheroid, with maximum density of**

terminated E-field lines in the equatorial region
- **At velocity much lower than the optimal one, the boundary surface is much larger and approaches the shape of sphere.**
- **The high efficiency confined motion of the electron is supported by its electrical field**
- **The moving electron possesses a path sensing property, due to the relative freedom of its electrical lines.**

3.5 Dynamical properties of the electron in confined motion

3.5.1 Oscillation properties at optimal confined velocity

For electron moving **with an optimal confined velocity** the following relation (3.6) is valid:

(rotational frequency of electron shell) = (electron-positron proper frequency) = (SPM MQ frequency)

We may say, that the motion with an optimal velocity is a motion at first harmonic of SPM (as for the quantum wave).

The second two terms of the expression (3.6) are always equal to the Compton frequency, independently from the electron velocity. This is so, because the positron edge is moving in the proximity E-filed of the external negative shell. This E-field, influenced by the proximity SG(CP) filed, is "carried" by the electron negative E-filed and is not affected by the electron velocity. Due to the proximity of SG(CP) filed, it has a constant value for any sub optimal velocity. The central negative core, also, oscillates in the negative E-filed in the proximity of SG(CP) filed.

The motion environments for both types of oscillating system are illustrated in Fig. 3.9

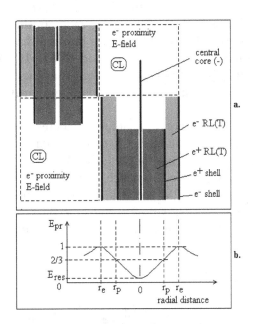

Fig. 3.9

Motion environments around the edges of the oscillating electron

Figure 3.9.a shows the motion environment of the positron and the central negative core. Figure 3.9.b shows the radial dependence of the negative proximity electrical field E_{pr} in the motion region, near the external electron shell. This field does not fall completely to zero in the central core axis, but to some residual value E_{res}. The oscillation conditions then are different for the positron shell and for the central core. If assuming a linear dependence of E_{pr} on the radial distance inside of r_e, the positron E-filed will interact with the negative proximity field, lying above $2/3E_r$, while the central core - with the field R_{res}. The interaction forces for the positron then are attractive but with a small cosine between the vectors. The interaction forces for the negative core are repulsive but with a cosine equal to zero. The electron and positron helical shell, however, has their own energy storage system RL(T), while the central core does not have such one.

The oscillation conditions of the **positron - central core system** will be discussed later in this chapter and in Chapter 4 (about the superconductivity). It will become evident that depending on the external conditions and the strength of the in-

voked oscillations, this system will get two different proper frequencies

(1). case: for small amplitude oscillations: $3v_c$

(2) case: for large amplitude oscillations and free positron: $2v_c$

Let us analyse now the first case. As a result of the simultaneous oscillations of both systems, the moving electron exhibits a 'hummer drill" effect, which facilitates the displacement (and simultaneous folding and unfolding) of the CL nodes. In this case, the oscillating internal positron with its core oscillating at third harmonic, provides a "hammering" effect, whose momentum is tangential to the helical trajectory. This effect, from one side, provides alternative component of the electron rotational motion, and from the other, contributes to the average velocity stabilization. The motion property of the electron due to the "hummer drill" effect are discussed in the next paragraph.

For a motion with a velocity lower than the optimal one, the direction of the oscillation is still tangential to the trajectory and the "hummer drill" effect is still working. For velocity above the optimal one, however, the momentum of oscillation is not tangential to the trajectory (see Fig. 3.4.b) and the effect is significantly reduced. The "hummer drill" effect is stronger at the optimal confined motion, less stronger in the range of lower velocities and negligible for higher velocities. In all cases however, it contributes to the screw type of motion of the electron, together with its electrical field.

3.5.2 Motion properties of the oscillating electron. Quantum motion.

We may denote the oscillation system as a e⁻/e⁺ system. From Fig. 3.9.a we see that the positron shell with its RL(T) oscillates in the negative proximity field of the electron, whose radial dependence is shown in Fig. 3.9.b.

The SPM EQ phases are synchronised along the E-field lines, while their long axis orientation depends on the radius. From the configuration shown in Fig. 3.5 and 3.6 we see, that the cosine between the tangential vector of the electron shell and EQ long axis approaches 90 deg. The same is valid for the positron motion inside of the electron. The EQ SPM momentum in this direction is reduced. In such case the electron moves and oscillates with less resistance. The hummer drill effect of the central core also contributes to the motion. Such motion conditions are valid only when the radial boundary circumference is equal to a whole number of Compton wavelengths. Only at such condition one full rotation of the external electron shell in CL space, contains whole number of its subsystem oscillations. It is obvious that similar motion conditions are valid for preferred selected velocities. We may call this type of electron motion a **quantum motion**.

Consequently, in conditions of quantum motion, we may refer the motion of the electron-positron system with its proper frequency directly to the SPM frequency of the CL space.

Both, the proper frequency of the e^-/e^+ system and the SPM frequency of the boundary have one and a same value equal to the Compton frequency.

Let us analyse the NRM and SPM of the CL stationary node during one cycle of SPM. The duration of the cycle is equal to the Compton time (period). During this time the direction of NRM and SPM vector changes in 3D space. Figure 3.10 illustrates the timing diagram of both vectors in 2D drawing, where **a.** - shows the spatial direction of the CL node NRM vector as sinusoids at step of $\pi/4$, **b.** shows the positions of the CL node SPM vector.

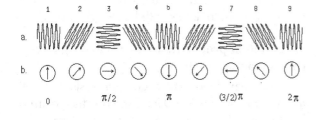

Fig. 3.10
NRM and SPM vector during one Compton period

The NRM(CP), shown in Fig. 3.10.a has one and a same momentum for the opposite direction of 180 deg. The NRM(TP), however has a different momentum due to the twisted parts of the prisms. For this reason the NRM(CP) appears, to have twice shorter cycle, than NRM(TP). The SPM vector, however, is determined by the NRM(TP). **In such aspect we may consider, that the NRM(CP)**

serves as a stroboscopic carrier of NRM(TP), providing in this way stronger quantum features of the CL space. Having in mind the fixed positions of the CL nodes and SPM MQ synchronization, it becomes evident that the quantum features are simultaneously temporal and spatial.**

The temporal quantum features of the NRM(T) of a stationary CL node are illustrated in Fig. 3.11, where: **a.** - shows the SPM vector phases, **b.** - the CL node resonance momentum, **c.** - the time phase of SPM vector in one spatial direction (denoted as positive), **d.** - the time phase of the SPM vector in a opposite spatial direction (denoted as negative)

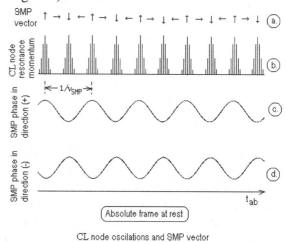

CL node oscilations and SMP vector

Fig. 3.11
Temporal quantum features of CL node vectors

Let us assume that the electron motion is invoked by external E-filed with a relatively small gradient, so the e^-/e^+ oscillation amplitude is small. Then the main disturbance of the CL space, due to the electron motion is from the electron E-field, while the disturbance effect of the e^-/e^+ oscillations on the CL space could be neglected). In order to analyse this motion we have to take into account the frame of the reference. For this reason we will use the concept of the virtual observer.

Let us imagine that a virtual observer, sitting in the front edge of the electron (external shell), observes the motion of the positron edge using its virtual fundamental clock. So its time base is the Compton time. Let us accept that the peripheral velocity of the electron is equal to a linear light velocity, and the PP (phase propagated) SMP vector is synchronized to appear in the same direction. If the SMP frequency is equal to the electron proper frequency, the virtual observer will see both vectors vibrating in phase. But this means that the SMP phase is propagated together with the virtual observer with velocity, which is strictly dependent on the node SMP frequency and node spacing. In this conditions, the virtual observer will move with velocity equal to a linear light velocity. The task of the virtual observer is to register the timing diagram of electron in absolute units. Knowing a priory that he moves with a light velocity he may prefer to reference the local velocities to the light velocity. Then the timing diagrams of the electron system oscillation from his point of view will look like this shown in Fig. 3.12.

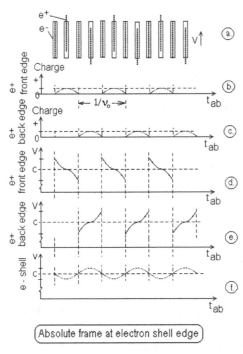

Fig. 3.12

Timing diagrams of electron system motion with simultaneous oscillations at first proper (Compton) frequency

The amplitudes of motion, velocity and oscillating charge are exaggerated in the drawings. In fact they are intrinsically small.

The electron is shown in **a.** as a cylindrical oscillating body for simplicity. It is assumed that the electron shell moves with light velocity and the velocity of the front and back positron ends are

shown respectively in **d.** and **e.** The sharp velocity spikes corresponds to a maximal interaction with the CL space. They are responsible for the synchronization of the electron oscillation with the SMP vectors of the nodes. The periodical appearance of the positive charge from both sides of electron interacts with the lattice space and is responsible for the phase locking conditions. As a result of the positive charge oscillations the external shell practically will not have exactly light velocity but will possess **a small AC (alternative) component around the DC component of the light velocity.** The value of this AC component is automatically self regulated due to the oscillation interactions between the electron and the CL space.

Equivalently we may consider that the electron always oscillates with its frequency v_o**, but meets the SMP vectors of the surrounding nodes with a correct phase.**

We may accept, that small oscillations of the whole electron system around the linear light velocity are possible. The interaction at this conditions with the CL space, however, is significant. As a result of this, the quantum effect is significant. The positron - central core oscillations also contributes to the "hummer drill" effect. Having in mind, that the proper frequency of the positron-core (for small amplitudes) is three times the e^-/e^+ proper frequency (Compton), the alternative electron motion will get a third harmonic in phase.

From the condition of phase synchronization between the e^-/e^+ proper frequency and the CL node SPM frequency (both equal to the Compton one), it follows that for one full turn of the electron the following relations are valid:

$$\lambda_{SPMMQ} = 2\pi R_{mb}$$

$$\lambda_c = 2\pi R_c$$

These are the same relations, assumed in §3.4 from the analogy with the first harmonic quantum wave. The second relation shows that the Compton wavelength can be considered as a path that any point lying on the external helical shell of the electron structure passes for one Compton period, when the tangential velocity of the rotating electron is equal to the light velocity. This relation is used in §3.6 for determination the physical dimensions of electron's material structure.

3.6 Dimensions of the electron.

The intrinsic mass of the electron is much smaller than the proton. Then we can assume that the electron does not shrink the lattice space (this will become evident later). Let us consider a confine motion of an electron with an optimal velocity. In this case, the peripheral part will move with a speed of light. We may consider however that the light velocity corresponds to the radius R instead of $f(R+r_e)$. This is acceptable (and will be evident later) because, from one side, the ratio r_e/R is small (0.0229), and from the other, the CL disturbance effect from the higher velocity at $(R+r_e)$ will be biased by the lower velocity at $(R-r_e)$. Then taking into account the screw like motion, the following relation is valid:

peripheral velocity: c   - path: $\sqrt{4\pi^2R^2 + (s_e)}$
axial velocity: υ_{op} - path: s

Then the axial velocity is:

$$\upsilon_{op} = \frac{cs_e}{\sqrt{4\pi^2R^2 + s_e^2}} \qquad (3.6)$$

Equation (3.6) gives the axial electron velocity for its optimal confined motion. This velocity is very specific and practically it appears in many cases related with electron motion. (for example the electron motion in the lowest stable orbits in atoms). Then our guess (which will be confirmed later) is: this is the velocity of the a_o orbit in the Bohr atom model, corresponding to energy of 13.6 eV.

$$\upsilon_{op} = \frac{q_o^{\;2}}{2h\varepsilon_o} = \alpha c \qquad (3.7)$$

where: q_o - is the electron charge,
     h - is the Plank's constant
     ε_o - is the permittivity of vacuum
     α - is the fine structure constant

We will prefer to express the velocity by the fine structure constant α, whose physical meaning will be revealed right away.

$$\alpha = \frac{\upsilon_{op}}{c} \qquad (3.8)$$

The fine structure constant is the ratio between the axial and tangential velocity of the electron at its optimal confine motion. (The tangential velocity at the optimal confine motion is equal to the linear light velocity).

We see that the fine structure constant, from one side is very basic parameter, and from the other, it helps to determine the electron dimensions. We see, also, that it is a dimensionless ratio of one and same parameter - velocity. Consequently, the fine structure constant will be not affected of eventual lattice space shrinkage (filed curvature).

Combining eqs. (3.6), (3.7) and (3.8) we get the step to radius ratio of the electron.

$$\frac{R}{s_e} = \frac{\sqrt{1 - \alpha^2}}{2\pi\alpha} = 21.809 \qquad (3.9)$$

From Eq. (3.9) we see, that the fine structure constant is completely determined by the radius R and the helical step s_e. It is convenient to express α directly by the ratio R/s_e.

$$\alpha = \frac{1}{2\pi}\left(\frac{R^2}{s_e^2} + \frac{1}{4\pi^2}\right)^{-1/2} \qquad (3.10)$$

In order to derive another equation about the electron, we will analyse its trajectory at optimal confined motion.

Having in mind the E-filed integrity we may express the path of one point of the central core, for example the frond edge, per one cycle time of the electron proper frequency:

$$path = 2\pi R = ct_c = c\frac{1}{\nu_c} \qquad (3.11)$$

Solving the system of (3.9) and (3.11), we get the values for R and s_e.

R = 3.86159 x 10^{-13} (m)
s = 1.77061 x 10^{-14} (m).

It is not a surprise, that the obtained value of R is exactly equal to the Compton radius, however, we obtained the value of the helical step, that is very important initial result. Returning to Fig. 3.1 and Fig. 3.2 we see that the helical step could not be less than $2r_e$, because the positron then could not be able to come partly out in order to make oscillation. It also could not be much larger and this will become evident in Chapter 6 (because then the negative muon will not be able to crash and decay to a positron if his step is too large. In most of the experiments performed by positive muons it is found that they could oscillate longitudinally until they decay to a positron). The helical step similarity between the muon and electron (positron) according to BSM is obvious. The electron (positron) could be regarded as a single coil structure of that of muon. From these considerations and from the stable appearance of the fine structure constant in the electron spectrometry, we may accept, that the edges of the external electron shell are just touching. This means that:

$$s_e \approx 2r_e \qquad (3.12)$$

According to the discussion given in §3.11.2, **the relation between s_e and r_e is more accurately given by the gyromagnetic factor.** This is a dimensionless physical factor, theoretically calculated and experimentally determined with very high accuracy (see Eq. (3.23.c). Then the small electron radius r_e is directly obtained from the relation:

$$s_e = g_e r_e = 2.002319 r_e \qquad (3.12.a)$$

The ratio between the electron and positron small radii is determined by the accepted ratio between the left and right handed prisms: 2/3. This ratio is further confirmed by quantum motion of the electron (discussed later in this Chapter), the fractional quantum Hall effect (discussed in Chapter 4, Superconductivity), and by the BSM interpretation of the τ particle decay. According to this ratio we have:

$$r_p/r_e = 2/3. \qquad (3.13)$$

Then the dimensions of the electron are the following:

R = 3.86159 x 10^{-13} (m)
r_e = 8.8428 x 10^{-15} (m)
r_p = 5.8952 x 10^{-15} (m)
s = 1.77061 x 10^{-14} (m)
R/r_e = 43.669
R/r_p = 65.5

The shown above dimensions define also the positron.

The thickness of the boundary helical core (assuming to be equal to 3 prisms diameter) is very small in comparison to the r_e and r_p. This is evident from the calculations in §2.14, where the CL node distance along *abcd* axes is found to be: $d_{na} \approx 0.872 \times 10^{-21}$ [*m*]. Having in mind that this distance is larger than the sum of the right handed and left handed prism lengths, the prisms diameter then is at least one order below this value. The helical core thickness is only three prism diameters, so it is negligible in comparison to r_p. Then, practically,

for many calculations, we may consider, that the internal RL(T) of the positron has external radius of r_p, and the internal RL(T) of electron is from r_p to r_e. Realistically some radial gap should exist between the electron internal RL(T) and the positron structure. This could be explained, if having in mind that the radial thickness of the electron's negative RL(T) is less than half of the radius r_e, so this shell it is not completed like the RL shells of the positron. Then its average radial node density is larger than this of the positron, and in the process of RL twisting, its internal radius may not shrink so much.

The RL(T) of the positron also has some finite internal radius. This is its internal hole radius in which the central core oscillates.

The internal holes of both RL(T) of the electrons are not affected, if the internal structure is lost. The probability of the positron to lose the central core is much lower. However, it may regenerate the lost negative core by the trapping mechanism.

The large ratio R/r_e and R/r_p favours the confine motion of the electron and positron in the lattice space.

From Eq. (3.10) and the above made calculations, we found that the fine structure constant α is completely determined by the ratio of R/s_e. But this ratio together with the radius r_e (according to the analysis in Chapter 2) are determined by the balances of forces between the internal RL(T), the bending helical core and the CL space forces (E - filed).

Consequently, we may conclude, that:
- **The embedded fine structure constant permits indirect estimations of the CL space parameters by using of the geometrical parameters of the electron**
- **The dimensions of the electron, are dependable on the CL space parameters.**

The second conclusion shows one very important feature of the electron. It may help to explain some effects in the General Relativity, for example, the red shift of a photon emitted in a strong gravitational field.

The above obtained dimensions appear very useful for solving the following tasks:

- estimation of CL space parameters: Static and Dynamic CL pressure

- derivation of theoretical equation for estimation of Newtonian mass (mass equation)

- derivation of theoretical equation for a background temperature of CL space

- calculation of the proton dimensions, including its substructures

In solving the above tasks, we will use the geometrical parameters of the electron. They could serve as a basic reference units.

It is useful to know, that any one of the geometrical parameters of electron can be expressed by the known physical constants. These expressions are given below.

$$R_c = \frac{c}{2\pi \nu_c}$$ Compton radius   (3.13.a)

$$s_e = \frac{\alpha c}{\nu_c \sqrt{1 - \alpha^2}}$$ helical step   (3.13.b)

$$r_e = s_e / g_e$$ small radius   (3.13.c)

$$r_p = \frac{2}{3} r_e$$ positron small radius

3.7 Interaction between the moving electron and the external electrical field

When the electron is forced to move by external electrical field it exhibits a confined motion. If the accelerating field possesses an axial symmetry and the electron has some initial velocity it will be accelerated by the field, but will preserve its straight trajectory. In the interaction process, the external field interacts directly with the electron. The interaction forces of the accelerating field, may be considered applied to the circumference at radius equal to its equivalent electrical radius R_{eq}. For a symmetrical field, the forces acts as a symmetrical "pull-up" forces and do not cause change of the straight line trajectory of the electron.

From the other hand, the interaction process between a moving electron and an external magnetic field is different.

3.8 Interaction between a moving electron and an external magnetic field

The process of interaction between a moving electron and external magnetic field is explained by the help of Fig. 3.13. In this figure the central plane of the radial section of the electron field is shown.

The electron axis is perpendicular to the drawing plane. The external magnetic field is presented by parallel lines with arrow pointing the field direction. The interaction takes place only in the circumference with radius R_{bm}. The magnetic field lines can include only magnetic quasispheres (not disturbed CL nodes), whose phases are synchronized. In the radial section of the electron field, the circumference at radius R_{bm}, only, include magnetic quasispheres. Due to the electron rotation in its screw-like motion, the effective forces from both sides of the axis OO' are different.

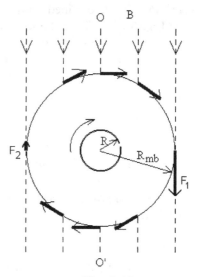

Fig. 3.13
Interaction forces between external magnetic field and the field of the moving electron

In the drawing presented in Fig. 3.13, only two interaction forces, F_1 and F_2, aligned with the direction of magnetic field B are shown, for simplicity. The right side of the electron single coil structure will get acceleration from the magnetic field, while the left side - deceleration. This will cause the electron to get angular momentum around the axis OO'. The electron containing kinetic energy will make a cyclotron curve in a counter clockwise direction. For stronger magnetic field, the filed lines are denser and the cyclotron radius will be smaller. If the direction of the magnetic field is reversed, the cyclotron rotation will be in a clockwise direction. The classical equation for the cyclotron radius is:

$$r = \frac{m_e v}{qB} \qquad (3.14)$$

where: m_e is the electron mass, v is the velocity, q is the electron charge and B is the magnetic field.

Accelerated electron makes a circle with an angular frequency ω named a cyclotron frequency.

$$\omega = \frac{v}{r} = \frac{qB}{m_e} \qquad (3.15)$$

In §3.4 and Fig. 3.8, it was shown, that the boundary conditions for electron with optimal confined velocity has a shape of oblate spheroid, and the density of the terminated E-filed lines is larger at its equator. The simplified presentation of the interaction mechanism, presented above is equally valid also for this case.

Fig. 3.14 shows the electron motion in quadrupole magnetic field. If the electron moves exactly in the centre of the field along the axis normal to the drawing plane, the field will exercise symmetrical forces on the magnetic boundary radius and it will not get deviation. If the electron is slightly of the central axis, it will get a helical trajectory around this axis. The shown type of magnetic filed is used in the synchrotron accelerators.

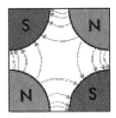

Fig. 3.14
Quadrupole magnetic field

3.9 Quantum motion at optimal and suboptimal velocities. Quantum velocities.

3.9.1 Quantum stabilised velocities and their corresponding energy levels

From the axial boundary section of the electron E-field with confined motion as shown in Fig. 3.8, we see that the E-field is restricted in a near spherical volume. The exact boundary conditions, i. e. the isolated magnetic quasispheres are valid

only for a part of the total E-field volume. This volume could be approximated with a cylindrical volume with a base approximately equal to the central section of the spherical volume with radius R_{mb} and small thickness. For the optimal confined motion, corresponding to energy 13.6 eV, the boundary condition is $2\pi R_{mb} = \lambda_{SPM\ MQ}$. The next possible boundary conditions, is fulfilled, when the boundary radius of the external surface is equal to $2\lambda_{SPM\ MQ}$. In this case the electron rotates with a twice lower frequency. Similar type of motion is possible if the circumference length is equal to n times $\lambda_{SPM\ MO}$. so the rotational frequency of the electron is respectively v_c/n. The n appears to be **a subharmonic number** in a similar way as in the quantum wave (photon).

Fig. 3.15 a. and b. shows the radial distribution of the equivalent E-field for rotational frequency of $v_c/3$. and $v_c/6$., respectively.

The electrical field parameters of the moving electron in both cases are shown in Fig. 3.15.

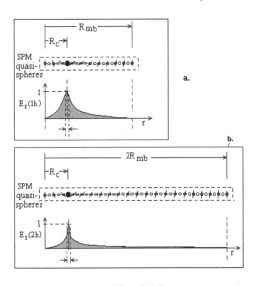

Fig. 3.15
Radial E-field and boundary conditions of the electron for confined motion with: a. - n=3; b. n=6

Having in mind, that the electrical charge is one and a same, we see from Fig. 3.15 that for a subharmonic number $n > 1$ the equivalent radial section of E-field will occupy an area of $(nR_c)^2$. Consequently, the energy density of any subharmonic normalized to the first one ($(n = 1)$ is inverse proportional to the subharmonic number. At the same time, the E-field density within the range of $0 < r < R_c$, do not decrease proportionally to n^2, because the field lines have different orientations, as shown in Fig. 3.7. This feature assures a stable proper resonance frequency of the electron as a self-oscillating system.

As a result of described above features, the **moving electron exhibits a quantum motion at preferable velocities.** This velocities define preferable kinetic energies of the electron.

Let us determine what are the resistive forces, which oppose the optimal confined motion at the quantum subharmonics. They are two:

- The CL resistance due to the rotation of the electron E-field;

- The CL resistance due to displaced CL nodes by the electron volume

The resistance from the rotated E-field is smaller at confined motion, because part of the radial circumference of the E-field is isolated. The isolation effect is only a partial, so the electron is still able to interact with some external electrical field.

The second resistive force, mentioned above, is from the displaced and folded CL nodes. (They do not pass through the RL(T) structures of the electron). Larger tangential velocity causes a larger number of folding and restoring nodes. This means a larger resistance of the CL space when the electron is accelerating (this issue is discussed in Chapter 10).

The tangential velocity of the electron external shell at different subharmonic numbers is:

$$\upsilon_t(n) = \frac{c}{n} = \frac{2\pi R_c}{t_c n} = \frac{2\pi R_c \nu_c}{n} \qquad (3.16)$$

where: n - is the subharmonic number

Knowing that the fine structure constant gives the relation between the tangential and axial velocity for an optimal confined motion, we can work directly with the axial velocity, expressed by the equation.

$$\upsilon(n) = \frac{\alpha c}{n} \qquad \text{- axial velocity} \qquad (3.16.a)$$

Eq. (3.16.a) is in fact the classical velocity of the electron as a function of n. Putting this velocity in the classical formulae $E = 0.5m\upsilon^2$ we get the cor-

responding energy level for such velocities in electron volts.

$$E_{ev} = \frac{1}{2}m_e[\upsilon(n)]^2\frac{1}{q} = \frac{1}{2}\frac{m_e\alpha^2 c^2}{n^2}\frac{1}{q} = \frac{1}{2}\frac{h\nu_c\alpha^2}{n^2}\frac{1}{q} \text{ (eV)} \quad (3.17)$$

where: m_e is the electron mass, q - is the electron charge

One useful expression, derived from (3.17) is the equation of the axial velocity. Its average value should not exceed the linear light velocity.

$$\upsilon(n) = \frac{\alpha}{n}\sqrt{\frac{h\nu_c}{m_e}} \qquad (3.17.a)$$

We may denote the preferred energy levels as **SPM subharmonic energy levels**, and the corresponding velocities - **SPM subharmonic velocities or quantum velocities of the electron,** knowing that they are referenced to the SPM frequency of the magnetic quasisphere. In other words they are the preferable quantum levels of interactions.

Table 3.1 shows the energetic and boundary parameters of the first six quantum levels denoted by n where: l_{mb} is the circumference length of the radial section, **v** is the axial velocity of the electron, E is the energy level in (eV).

Table 3.1

n	Effective boundary radius	l_{mb}	v [m/sec]	E [eV]
1	R_{mb}	$\lambda_{SPM\ MQ}$	2.187×10^6	13.6
2	$2R_{mb}$	$2\lambda_{SPM\ MQ}$	1.094×10^6	3.401
3	$3R_{mb}$	$3\lambda_{SPM\ MQ}$	7.292×10^5	1.5117
4	$4R_{mb}$	$4\lambda_{SPM\ MQ}$	5.469×10^5	0.8054
5	$5R_{mb}$	$5\lambda_{SPM\ MQ}$	4.375×10^5	0.544
6	$6R_{mb}$	$6\lambda_{SPM\ MQ}$	3.646×10^5	0.3779

The energy levels of the Bohr model of the hydrogen atom are given by Eq. (3.18).

$$E_n = \frac{-2\pi\varepsilon_o q^2}{a_o}\left(\frac{1}{n^2}\right) \qquad (3.18)$$

where: a_o is the Bohr radius

The derived Eq. (3.17) gives exactly the same energy levels as the Eq. (3.18). While the Eq. (3.18) is based on the atomic model, suggested by Bohr, the proposed by BSM equation (3.17) expresses directly the electron quantum behaviour in the CL space. Consequently, the quantum behaviour is intrinsic feature of the electron. Then it can appear in the combine motion between the electron and proton. In the Hydrogen atom, every possible subharmonic number of the electron quantum motion defines the bottom level of the series. They are the following:

Table 3.1

Subharmonic number	Lowest level in [ev]	Series name
1	13.6	Lyman
2	3.4	Balmer
3	1.51	Pashen
4	0.85	Bracket
5	0.544	Pfund
6	0.3779	Sixth

These levels, when considered as a quantum numbers are more stable than any other transitional levels, because of the complimentary interactions between the oscillating electron and the oscillating CL nodes. (The SPM frequency of the CL nodes are phase synchronized by the Zero Point Waves, which always exists in a normal CL space).

Similar motion conditions exist not only for the electrons in the Hydrogen atom but for any other atom. In the second case, however, the energy levels are modified due to the common positions of the protons in the nucleus, the stronger nuclear SG field and the orbital interactions.

3.9.2 First harmonic motion and Rydberg constant

The Rydberg constant (known also as Rydberg) is involved in the well known Rydberg-Ritz formula. It is a measurable parameter by the atomic spectroscopy. It may be expressed in wavenumbers, electron volts, or wavelength. The constant value has a very slow change from element to element. For the very heavy atoms, the Rydberg in wavenumbers is given by the equation:

$$R_\infty = \frac{m_e c \alpha^2}{2h} = 1.09737315 \times 10^7 \quad [\text{m}^{-1}] \qquad (3.19)$$

For the Hydrogen atom it is little bit smaller.

$$R_H = \left(\frac{m_e m_p}{m_e + m_p}\right) \frac{q^4}{8c\varepsilon_o^2 h^3} = 1.09677587 \times 10^7 \qquad (3.19.\text{a})$$

where: the term in the bracket is known as reduced electron mass

The Rydberg constant, according to BSM, is defined directly by the condition of the first harmonic quantum motion of the electron.

For a first harmonic motion, the electron energy in SI units is:

$$E = h\nu_c = hc\sigma \qquad (3.20)$$

where: $\sigma = 1/\lambda_c$ is the wavenumber

The quantum energy level according to Eq. (3.17) for $n = 1$ in SI units is given by Eq. (3.21), while in (eV) - by Eq. (3.21.a)

$$E = \frac{1}{2}\frac{h\nu_c \alpha^2}{1^2} = \frac{1}{2}(h\nu_c \alpha^2) \quad [\text{J}] \qquad (3.21)$$

$$E(eV) = \frac{1}{2}(h\nu_c \alpha^2)/q = 13.6057 \quad [\text{eV}]$$

Equating (3.20) and (3.21), and solving for σ, we get the value of Rydberg in wavenumbers.

$$\sigma = \frac{\nu_c \alpha^2}{2c} = 1.09737315 \times 10^7 \quad [\text{m}^{-1}] \qquad (3.21.\text{a})$$

If making a substitution $m_e = (h\nu_c)/c^2$ in Eq. (3.19) it converts to Eq. (3.21.a). Consequently:

- **The Rydberg constant corresponds to the electron's motion at first SPM harmonic (a case of optimal confined motion).**

The Rydberg constant, according to Eq. (3.21.a) (containing only CL space parameters) appears to be a parameter of the CL space. The fine structure constant is also a CL space parameter, but estimated by the electron parameters. In §3.11 it will be shown, that the electron parameters in fact are defined by the CL space parameters, because they determine the shape and dimensions of the electron. There is one very small contribution from the bending resistance forces of the helical core, that are not defined by the CL space parameters. This small contribution, in fact, gives the general relativistic deviation. Ignoring the latter one for now, we can make a conclusion, that:

- **The Rydberg constant is a CL space parameter**

In the table of fundamental constants, the Rydberg constant is given also in frequency units, and in energy units. In the latter case, when estimated in (eV) it corresponds to 13.6 eV - the energy of the electron optimal confined motion.

One question may arise: Why the accurate value obtained by Eq. (3.21.a) matches exactly the Rydberg for the massive element and not for the Hydrogen? The explanation is the following:

The Rydberg constant can be regarded as an energy parameter of the CL space. When a photon is emitted as a result of CL pumping, an exact equivalence exists between the pumped and the photon energy.

CL pumped energy = photon energy

For this reason the signature of the Rydberg constant appears in the atomic spectra. This gives a possibility for its experimental estimation. The pumping conditions in atoms are obtained by the circling of the electron around the much heavier nucleus. They both are not fixed in the CL space, but only by their masses. **So for the pumping effect of the stationary CL space (in our case the Earth local field) we have to consider their common motion.** For the much heavier nucleus, the comparative electron mass become intrinsically small. **Then the heavier nuclei could be considered as a fixed in CL space.** From the other hand, the Hydrogen nuclei is lighter, and could be not considered as fixed in the space. We see, that Rydberg constant approaches the maximum value at heaviest atom and is smaller for the Hydrogen. For this reason the reduced electron mass is used in the Eq. (3.19.a) (the bracket term). While the Rydberg is proportional to the photon energy it appears, that, **the CL pumping efficiency is highest, when the electron circle around a stationary fixed nucleus.**

The above made conclusion is confirmed, also by the Positronium transition $1^3S_1 - 2^3S_1$, discussed in §3.17.4.

Let us find the physical meaning of the electron reduced mass. The bracket term of the electron reduced mass, can be presented in a form:

$$\left(\frac{m_e m_p}{m_e + m_p}\right) = \eta m_e \qquad (3.21.b)$$

where:

$$\eta = \frac{m}{m_m + m} \quad \text{is the CL pumping efficiency} \quad (3.21.c)$$

m - is the mass of the heavier nucleus around which the electron is circling orbiting

Then the expression of the Rydberg constant takes a more general form, in which the CL pumping efficiency is explicitly involved.

$$R_y = \eta \frac{v_c \alpha^2}{2c} \qquad (3.21.d)$$

Equation (3.21.d) shows, that the Rydberg constant apart of the lattice parameters depends only on the pumping efficiency, determined by the involved masses.

The pumping efficiency for the Hydrogen atom is 0.9994557, while for the Positronium it is 0.5.

We may summarize that:

- **The electron exhibits a quantum motion due to the interaction between SPM MQ frequency and the proper frequency of the electron-positron system**
- **The quantum levels of the electron velocity are defined by kinetic energies at which the electron exhibits a screw-like type of motion with a less resistance**
- **The rotational electron frequencies for the quantum levels are subharmonics of SPM MQ frequency, including also the first harmonic.**
- **The quantum levels of the electron velocity define the bottom levels of the Hydrogen series and are relevant also for other atoms.**
- **The Rydberg constant is directly defined by the first harmonic quantum motion (optimal confined motion)**
- **The CL pumping effect obtains a maximum value, when the electron is orbiting in a fixed orbit around a heavier charge particle, for example the proton.**

3.9.3 Quantum properties of the positron system

We found that the oscillations with small amplitudes are relevant for the quantum motion of the electron at suboptimal velocities. The same amplitude conditions should be relevant for the positron quantum motion. When compared to the electron, the confined motion of the positron has some similarities and some differences.

The similarities are the following:

- the same type of boundary conditions

- a similar screw-like type of motion

The differences are the following:

- The external motion environment of the central oscillating core is different

- The proper frequency is different

At the first glance, the efficiency of the positron quantum motion could look much lower, in comparison to the electron, since the system has only one internal RL(T) and the intrinsic matter of the central core is much smaller. This, however, is partly compensated by the increased hummer drill effect, as we will see from the following analysis.

When the positron system oscillates inside the electron with small amplitudes, the central core oscillates in a slightly negative external field, as was shown in Fig. 3.9.b.

The oscillating conditions of the central core of the free positron, however, are different. Now, the external negative field is missing, and the gradient of the positive E-field falls to zero. For small amplitude oscillations of the free positron, we may accept that the central core oscillates in environment of MQ nodes. But the SPM frequency of the external and internal MQ's is one and a same. Then the oscillating central core will exhibit stronger hummer drill effect. This will partly compensate the efficiency of the free positron quantum interaction in comparison to the electron. The optimal interaction will be obtained at such positron rotation, at which the phase difference between the PP SPM vector and the proper frequency of the free positron is zero. To obtain this motion conditions we need to know the free positron proper frequency. Experiments with positrons provide confidence about its inertial mass, but this is not enough in order to obtain the proper frequency. The Planck's constant may appear different, when estimated by the electron and positron parameters (and this will become evident by the course of BSM.) So we will make initially some theoretical analysis, and then we will look for experimental confirmation.

Let us make comparison between the electron system and the free positron in order to find the conditions, when their central cores exhibit one and a same resistive momentum. Let us analyse initially the motion of both systems when they have one and a same tangential velocity of their external shells. We may use the classical equation for the proper frequency of oscillating system:

$$f = \frac{1}{2\pi}\sqrt{\frac{k}{m}} \qquad (3.21.c)$$

The inertial mass of the central core should contribute to m - parameter, while the repulsive forces between the negative core and external field should affect the spring constant k - parameter.

a) Inertial considerations of positron- core system

In the case of electron, the positron shell, moves in a CL domain with SPM momentum, which according to Eq(3.5) (see also Fig. 3.5) is given by: $k_{hb}p_t(R_{mb})$

In this case the central core is carried by the positron shell, and for small oscillations it does not feel the above SPM momentum. In case of free positron, however, the core will exhibit momentum $p_t(R_{mb})$. Then its inertial interaction appears lower, and it will behave as a lower inertial mass, that according to Eq. (3.21.c) will means a higher proper frequency.

b) k - parameter considerations

Here we have to consider two electrical components: the external component and the internal one - the latter is related to the trapping hole effect (discussed in §2.8.2).

In the case of internal positron, the negative core exhibits a slightly repulsive force from the external negative field of the electron (see Fig. 3.9). In the case of free positron, the external field is positive. The E-field interaction of the central core with that field is not so strong, since the core does not posses RL(T). However, it may obtain a small external attraction due to the positive external field. The internal trapping force, however, is not affected and predominates the external one. The SG(CP) forces are also unchanged. It is obvious, that the k parameter in the case of internal positron will be larger than for the free positron.

We see, that the factors m and k, both change in a same direction so: the proper frequency of the free positron is expected to be lower in comparison to the internal one. In fact we could not expect much change of the m factor, between both cases, because the SG(CP) is focused onto the central core is much stronger.

In order to find out what is the possible oscillating frequency of the free positron we will analyse the oscillations of the positronium, known as

Ps 1^3S_1 - 2^3S_1. It leads to emission of a photon at wavelength of 243 nm. This positronium is a result of common oscillation motion between a normal electron and a free positron. The energy of the emitted photon is 5.1 eV. The only possible quantum energy level transition for this value is

(13.6 eV - 3.4 eV)/2 = 5.1 eV.

It is not difficult to guess, that the electron participates in the oscillations with its optimal quantum velocity, corresponding to energy level of 13.6 eV. Then the positron energy is 3.4 eV. Obviously, this is a velocity with a stronger larger quantum effect. The oscillation process, leading to a photon emission, is analysed in more details in §3.17.3. The level difference is divided by two because the two masses are similar. This reduces the efficiency of the lattice pumping by a factor of two in comparison with the Hydrogen series, where the mass ratio of proton/electron is very large and the proton could be considered as a stationary body.

Other emissions from the above mentioned combination are not observed. It is reasonable to not expect another quantum energy levels, because the interaction properties of the central core in comparison to those of the positron shell are very week. The boundary conditions for the quantum motion of the electron and the free positron are similar. Then the energy level of 3.4 eV should correspond to the optimal confined motion of the free positron, which means that its oscillation frequency must be twice larger than the first proper frequency of the normal electron, or in other words - twice the Compton frequency.

$$\nu_{pc} = 2\nu_c \qquad (3.22.a)$$

where: ν_{pc} - is the proper frequency of the free positron system

The relation (3.22.a) is confirmed also by the fractional quantum Hall experiments, discussed in Chapter 4.

Consequently, the free positron exhibits an optimal quantum motion at 3.4 eV, due to the interaction between the CL node SPM and the proper frequency of the system.

3.9.4 Electron acceleration

For velocities higher than the optimal one, the electrons can be accelerated by two methods: by electrical field or by magnetic wave. Both types of acceleration exhibit distinctive features.

In the case of electrical field acceleration, (considering a symmetrical field), the electron is pulled by the electrical lines of the accelerating E-field. The accelerating force may be considered as applied at the electron equivalent radius. In this case, the stretched helical trajectory have the same handedness as the second order handedness of the electron.

In the case of wave type acceleration by magnetic field, the accelerated field interacts only with the electrons E-field lines terminated with a MQ's. For velocities much higher than the optimal one, these MQ's are arranged in very stretched helical trajectories, which tends to delay from the AC phase of the magnetic field. The alternative magnetic field is synchronized with the electron momentum velocity in order to not miss the phase of the electron proper frequency. At large velocity however, the quantum effect of the electron oscillation is small. Then the proper frequency could not be kept synchronized to the SMP MQ frequency of the accelerating field. As a result a squeezing effect may appear constantly between them. This effect, may reduce the reaction of the accelerated electron that will appear as an opposite magnetic field (as in the selfinduction). Due to reduced reaction, the acceleration effect appears more effective. Since the tangential velocity of the rotating electron is limited by the velocity of light, tt very high velocity the electron may rotate in a reversed direction.

From the provided simplified analysis it appears that a high energy (velocity) acceleration of an electron beam is more effective when provided by alternative magnetic field. This type of acceleration is used in the synchrotron accelerators.

3.10 Magnetic moment, gyromagnetic factor and Quantum mechanical spin of the moving electron

3.10.1. Magnetic moment

The confine motion of the electron means, that it rotates continuously. Consequently, its electrical field creates waves in the CL space. The waves accompany the moving electron as closed magnetic lines, formed of connected in loops magnetic protodomains. In uniform, not disturbed by other particles CL space, the magnetic lines are circles around the electron trace. The direction of the induced magnetic field is determined by the axial direction of the "screwing" electron.

For electron moving with a velocity closer to the optimal one, the motion behaviour is strongly influenced by the oscillation properties of the electron. When bundle of electron is moving with such velocity, the common synchronization effect is also very strong. This provides a strong modulation effect on the lattice space, appearing as an magnetic field.

Let us analyse the magnetic disturbance of the CL space from a single electron, moving with the optimal confine velocity, corresponding to energy of 13.6 eV. For one full turn of the external shell, the electron-positron system makes one cycle, whose period is the Compton time. The induced magnetic field in this conditions is characterized by the **electron magnetic moment**. It is given by the equation:

$$\mu_e = \frac{qh}{4\pi m_e}\left(1 + \frac{\alpha}{2\pi}\right) \quad [\text{A m}^2] \qquad (3.23)$$

where: q - is the electron charge, h - is a Plank constant, m_e is the electron mass and α - is the fine structure constant.

The electron magnetic moment is considered anomalous, so far, because of the second term in the brackets. But according to BSM model of the electron, α is completely determined by the electron radius R and step s_e (see Eq. 3.8). Consequently, this term shows the contribution of the helical step of the electron, due to the screw type effect. The electron motion at this velocity is affected stronger and the effect is detectable. **Therefore, the magnetic moment of the electron suggested by BSM should not be considered anomalous.**

Let us explain, why the magnetic moment is increased when α is larger. The dependence of α, on the ratio R/s_e was given by Eq. (3.8). For default value of α, this ratio is:

$R/s_e = 21.809$. Fig. 3.16 shows a plot of the fractional change of α, for R/s_e range from 21 to 23.

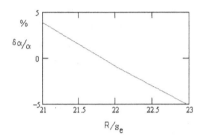

Fig. 3.16

R_c/s_e ratio as an estimation factor of the fine structure constant

We see, that for small range of R_c/s_e, the change of α is linear but not perfectly. The physical explanation of the R_c/s_e change and its direct effect on the magnetic moment is a following:

The electron parameters: R_c, r_e, s_e, are dependable on the forces balance, between the internal RL(T), external CL space and the helical core bending (see Eq. (2,8)). In this balance, the strong SG forces are involved including the electrical field created by RL(T) lattice of the electron's external shell whose energy is included in the SG energy balance. The balance causes adjustment of some of the parameters of the helical structures. For a SOHS, some parameters, such as the radius R_c and the step s_e are stronger affected than for the FOHS. (this was discussed in §2.8, Chapter 2). Therefore, if the CL space parameters are changed (for example the node distance), the twisting of the internal rectangular lattice will be affected. This twisting of RL controls the angle of the external E-field lines, emerging outside of the helical shell. Consequently, the magnetic moment is dependent on the degree of RL twisting and the helical step s_e of the SOHS.

The magnetic moment is a measurable parameter.

3.10.2 Gyromagnetic factor and Quantum mechanical spin

3.10.2.1. Gyromagnetic factor

While Quantum mechanics can not provide a reasonable classical explanation of the electron's spin. BSM model of the electron allows such explanation for a first time, as will be shown below.

Let us first show some of the known QM properties of the electron, which are related to the spin. One of them is the intrinsic magnetic moment of the electron, known as a Bohr magneton (because the Bohr atomic model is used for its derivation).

$$\mu_B = \frac{qh}{4\pi m_e} = 9.273 \times 10^{-24} \quad (J/T) \qquad (3.23.a)$$

where: q - electron charge, h - Planck's constant and m_e - mass of the electron.

According to QM model the expected magnetic moment of the electron must be $\mu_e = \pm(1/2)\mu_B$. However the measured value turns out to be about twice that.

$$\mu_e = \pm(1/2)g\mu_B \qquad (3.23.b)$$

where: the parameter g is a dimensionless coefficient, called gyromagnetic ratio or g-factor.

Accurate measurements of the g-factor are possible by using physical effects, such as Electron Spin Resonance and Lamb shifts. The accurate value of g-factor given by NIST is

$$g_e = -2.0023193$$

One experiment demonstrating the estimation of the g-factor based on an ESR is popular in universities physical laboratories. (for example an experiment provided by Leybold-Heraeues, based on a manuscript of Prof. H. K. Sheiner, Elensburg). The g-factor is estimated by ESR data according to equation:

$$g_e = \frac{h}{\mu_B B}f \qquad (3.23.b.1)$$

where: μ_B - is the Bohr magneton (theoretical value), h - is the Planck's constant. The ratio (f/B) is estimated directly from the experiment.

According to the theoretical treatment, the **g**-factor is expressed by a series containing only the fine structure constant as a physical parameter. In

the BSM model of electron the fine structure constant is an embedded parameter. The derived Eq. (3.13.b) shows that the helical step of the electron is expressible by the fine structure constant and CL space parameters. Therefore, it is not difficult to guess, that the dimensionsless g-factor is defined from the ratio between the helical step s_e and the small radius r_e of the electron.

$$g_e = s_e/r_e = 2.0023193 \qquad (3.23.c)$$

Consequently, the g-factor allows us to determine the small electron radius (as the helical step s_e was determined in §3.6).

One strong prove of the correctness of Eq. (3.23.c) is the calculation of the Static CL pressure using the derived in §3.13.3 equations (3.51) and (3.53). They both should give one and a same value. In Eq. (3.51) the volume of FOHS of the electron participate, while in Eq. (3.53) the g-factor is involved. When using g-factor given in (3.23.c) Eq. (3.51) and (3.53) give the exact same value up to the 10th significant digit. For any other value of g-factor the results departs significantly (in Eq. (3.53) g-factor is squared).

One may argue, that the relation (3.23.c) is not exactly a same for the free positron, because $r_p = (2/3)r_e$, while their inertial masses are equal. In case of positron, however, number of other factors should be also considered: the different proper frequency (defining different optimal confined velocity), the different physical dimensions, the different intrinsic matter densities of the right and left prisms and the different intrinsic time constants of the two substances of intrinsic matter.

3.10.2.2. Quantum Mechanical spin of electron

The electron spin is well known quantum mechanical property of the electron. Its signature is apparent in spectral lines and some experiments, for example, Electron Spin Resonance (ESR). The value of the spin is

$$S = \frac{h}{4\pi} = 5.2729 \times 10^{-35} \quad (J.s)$$

Let us analyse the Eq. (3.23.b.1) for measurement of the electron spin by ESR method. If putting dimensions of the involved parameters, the g-factor appears as a ratio of torques (Js/Js) or a ratio of energies (J/J). The torque ratio indicates that the QM spin has an angular momentum. It is similar as the dimensions of the Planck's constant, which physical meaning is revealed later in this chapter. In §3.5.1 it was shown how the oscillating electron interacts with the SPM vector. It is apparent that the magnetic field created by the confined motion (with a spatial parameters defined by the electron magnetic radius - see §3.11) will influence the phase of the electron oscillation with its first proper frequency (Compton). **Consequently, the proximity magnetic field may serve as a directional reference of the QM spin of electron.**

Then we may formulate the following physical properties of the QM spin of electron. For an optimal confined motion (first harmonic velocity) we may have the following two cases of QM spin:

S = +(1/2) h - The moving edge of the internal positron matches the phase of the SPM vector

S = -(1/2) h - The moving edge of the internal positron is in antiphase with the phase of the SPM vector

The proximity field SPM vector serves as a directional reference.

Physically, it is quite understandable how the BSM electron interacts with the SPM vector phase. The partial appearance of the edges of the internal positron beyond the edges of the external negative FOHS (with a frequency v_c) creates a pulsating field having the same frequency as the SPM vector of the CL nodes, so they may interact strongly. The distance between both edges of the electron structure (the helical step s_e) is much shorter than the Compton and magnetic radius of the electron (see §3.11). Then the pulsating magnetic field from both, the leading and trailing edges may get a phase lock to the external field SPM phase. This will correspond respectively QM spin value of the electron. This is a feature of the hummer drill effect of the electron confined motion as discussed in §3.5.2. While this effect is strongest at the optimal confined motion (13,6 eV), it is also valid for suboptimal confined motion with subharmonic numbers n = 2, 3 and so on, but with a degrading strength. This is possible due to the time quantum properties of the SPM quasisphere, which is discussed later in §3.12.2. Additional features clarifying the interaction related to QM spin are discussed in §7.7.1, Chapter 7.

When the electron is moving in closed orbit around the proximity field of the proton it gets one additional reference direction defined by the overall shape of the proton structure. This provides conditions for additional   splitting of the observed spectral lines in the strong Zeeman effect. This issue is discussed in Chapter 7.

Summary:

- **The magnetic moment of the electron is a parameter expressing its spatial and velocity stabilizing properties. The large magnetic moment of the electron (in comparison to the proton's one) is a result of its fast rotation in confined motion in comparison with the proton.**

- **In absence of magnetic field, the large magnetic moment of the electron assures its straight forward motion despite the displacement of the CL nodes. In environments of external magnetic field, the magnetic moment causes the electron to perform a motion in a cyclotron curve.**

- **The gyromagnetic factor is a parameter of the electron structure. It appears to be a ratio between the helical step of the electron and its radius.**

- **The oscillating structure of the electron with a first proper frequency equal to the SPM (Compton) frequency of the external magnetic field allows a strong quantum interactions with the CL space.**

- **The BSM model of the electron allows for a first time a classical explanation of its Quantum Mechanical spin.**

3.11 Quantum magnetic radius of the electron

The quantum features of the electron define its preferential velocities (energies), referred also as quantum velocities. Electron motions with such velocities exist not only in the Hydrogen, but in all atoms. In Hydrogen they appear explicitly, as the lowest level energy of the series. In other atoms, however, they do not appear explicitly, because their energy levels are added with the SG potential of the atomic structure. These potentials are included in the CL space pumping and photon emission. The electron motion with quantum velocities is involved in all emission and absorption spectral lines. For this reason, we will pay a special attention about the quantum conditions corresponding to suboptimal velocities (electron energies below 13.6 eV).

There is one value of the electron radius, that fits well to the spectroscopic data. It is known as a Quantum Mechanical radius of the electron, R_{QM}. It is related to the Compton radius by the factor of $\sqrt{3}$

$$R_{QM} = \sqrt{3}R_c = 1.732R_c = 6.688 \times 10^{-13} \text{ (m)} \qquad (3.24)$$

In the following analysis we will derive the equivalent quantum radius of the electron. This is the radius, corresponding to the equivalent radial field, shown in Fig. 3.6. It is evident, that this radius depends on the subharmonic number.

Let us derive, in first, the equivalent radius for the first harmonic, corresponding to energy of 13.6 eV.

A. Case: Quantum radius at first harmonic (at optimal confined velocity $\upsilon = \alpha c$):

In CL space with a normal ZPE (not superconducting state), the quantum magnetic field Φ_o is given by the relation

$$\Phi_o = \frac{h}{q} = 4.135 \times 10^{-15} \qquad Wb \qquad (3.25)$$

The quantum magnetic strength H_o is:

$$H_o = \frac{B_o}{\mu_o} = \frac{\Phi_o}{\mu_o S} = \frac{h}{q\mu_o} \frac{1}{S} \qquad \frac{A}{m} \qquad (3.26)$$

where: B_o is the quantum magnetic inductance, μ_o is the permeability of free space (cosmic lattice), S is the surface area through which the magnetic flux flows.

From the E-field configuration of the electron having a confined velocity discussed in §3.4, it becomes evident that when the subharmonic number increases, the shape of the boundary surface approaches a sphere. In a first approximation we may accept, that the surface has a spherical shape also for the first harmonic motion. Since the proximity E-field in the range of $0 < r < R_c$ is different (as shown §3.9.1), we may consider that the E-filed is enclosed in a spherical volume with a radius of $k_{hb}R_c$, where k_{hb} is a correction factor. Here we assume that it is the same factor, related to the boundary conditions of the quantum wave (defined by Eq. (2.20.a) in §2.10.4.3., for which consistent results were derived for $(\varepsilon_o, \mu_o, q, h)$ in §2.11 and §2.12). The corresponding surface of this volume is

$S = 4\pi(k_{hb}R_c)^2$. Substituting this surface in Eq. (3.26) we obtain

$$H_o = \frac{h}{q_o\mu_o 4\pi(k_{hb}R_c)^2} \quad \left[\frac{A}{m}\right] \qquad (2.27)$$

Eq. (3.27) having dimensions of [A/m], can be regarded as a quantum magnetic strength.

The magnetic moment of the electron was given by Eq. (3.23). It has dimensions [Am^2]. The electrical field of the screwing electron generates disturbance in the CL space - a magnetic field. The disturbance volume is obviously external to the helical structure, having a similar shape but with $r_{eq} > r_e$. Due to the larger R/s$_e$ ratio, we can express the volume of the electron structure as a torus volume with a larger radius r_{eq} and a smaller one r_e. $V_{eq} = 2\pi^2 R_c(r_{eq}^2 - r_e^2)$ The magnetic moment of the electron is given by Eq. (3.27.a).

$$\mu_e = \frac{qh}{4\pi m_e}\left(1 + \frac{\alpha}{2\pi}\right)$$

Then dividing the magnetic moment on the volume V_{eq} we obtain

$$\frac{\mu_e}{V_{eq}} = \frac{q_o h}{4\pi m_e} \frac{1}{2\pi^2 R_c(r_{eq}^2 - r_e^2)}\left(1 + \frac{\alpha}{2\pi}\right) \quad [\text{A/m}] \qquad (3.28)$$

Eq. (3.28) has the same dimensions as Eq. (3.27), and expresses also the quantum magnetic strength H$_o$.

Equating (3.27) and (3.28) and solving for r_{eq}, we get the quantum equivalent radius for the first harmonic.

$$r_{eq} = \left[\frac{q^2\mu_o R_c k_{hb}^2}{2\pi^2 m_e}\left(1 + \frac{\alpha}{2\pi}\right) + r_e^2\right]^{1/2} \qquad (3.29)$$

For $k_{hb} = 4$ (as defined by Eq. (2.20.a) we have $(r_{qe} = 1.057 \times 10^{-13})$ m.

The equivalent quantum field for the optimal confined motion of the electron is a torus with a R_c but with $r_{eq} > r_e$. The real equivalent quantum radius then is:

$$R_{eq} = R_c + r_{eq}. \qquad (3.30)$$

B case: Quantum radii at subharmonic numbers $1 \le n \le 3$.

In a subharmonic motion, the flux Φ_o is the same but the generated magnetic inductance B is changed because the boundary surface s is larger.

The boundary radius at *n-th* subharmonic is: $R_{mb}(n) = nR_ck_{hb}$. Then the boundary surface is:

$$S(n) = 4\pi(nR_ck_{hb})^2$$

Substituting the expression of $S(n)$ in (3.26) and processing in a similar way as the previous case, we get the equivalent radius as a function of the subharmonic number *n*.

$$r_{qe}(n) = \left[\frac{q^2\mu_o n^2 R_c k_{hb}^2}{2\pi^2 m_e}\left(1 + \frac{\alpha}{2\pi}\right) + r_e^2\right]^{1/2} \quad (3.31)$$

The quantum radius is always smaller, than the boundary radius. In order to simplify the calculations, it could be regarded as an equivalent radius for idealised E-field with a square shape of its radial distribution (see Fig. 3.6). This is convenient for field calculations.

The values of the equivalent radius r_{qe}, the boundary radius R_{mb}, and the relevant energy *E* for few subharmonics are shown in Table 3.3.

Table 3.3

n	E [eV]	r_{eq} [m]	$R_{mb}(n)$ [m]	$(R_c+r_{eq})/R_{QM}$
1	13.6	1.057E-13	1.544E-12	0.736
2	3.401	2.1087E-13	3.089E-12	0.893
3	1.51	3.161E-13	4.634E-12	1.05
4		4.215E-13		

In the derivation of Eq. (3.31) we have assumed that the equivalent radius is symmetrical around the radius R_c. Then the condition $r_{eq} \le R_c$ must be satisfied. From Table 3.3 we see, that this condition is satisfied up to the third subharmonic. The fourth subharmonic does not satisfies this condition, so Eq. (3.31) is valid for $1 \le n \le 3$ only. **For these cases, the equivalent quantum radius is still equal to** $(R_c + r_{eq})$.

The last column of Table 3.3 shows the ratio between calculated equivalent quantum radius and the accepted quantum radius R_{QM}, used in Quantum mechanics, according to Eq. (3.24). We see that this ratio approaches unity at n =3. This quantum number is the most populated one. Then the calculated equivalent quantum radius appears consistent with the observed spectral data.

C case: Quantum radii at subharmonic numbers $n > 3$.

For subharmonics larger than 3, the equivalent quantum field will not have the same radius R_c, but larger, as shown in Fig. 3.17.

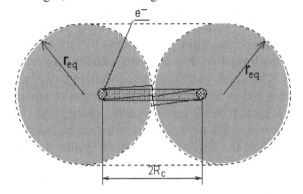

Fig. 3.17. Equivalent quantum radius of the electron for subharmonics numbers > 3.

In this case the volume V can be expressed as: $V_{eq} = 2\pi^2(r_{eq}^3 - R_c r_e^2)$. Neglecting for simplicity the small helical factor $\alpha/2\pi$, the derived radius r_{eq} in this case is:

$$r_{eq}(n) = \left[\frac{q^2\mu_o n^2 R_c^2 k_{hb}^2}{2\pi^2 m_e} + R_c r_e^2\right]^{1/3} \quad (3.32)$$

The equivalent quantum radius is:

$$R_{eq} = 2r_{eq} \quad \text{for } n > 3 \quad (3.33)$$

Table 3.4 shows the equivalent and boundary radii for subharmonics numbers $n > 3$.

Table 3.4

n	E [eV]	r_{eq} [m]	$R_{eq}(n)$ [m]	$R_{mb}(n)$ [m]
4	0.85	4.092E-13	8.184E-13	6.178E-12
5	0.544	4.748E-13	9.496E-13	7.72E-12
6	0.377	5.361E-13	1.072E-12	9.268E-12

We see that at smaller velocity the quantum radius is larger. The velocity for the sixth subharmonic is 3.646×10^5 m/sec. (see Table 3.1). According to BSM, the electron motion around the proton is characterized by large velocities up to velocity corresponding to 13.6 eV. The motion of free electrons in metals however is characterised by velocities smaller than this.

It is interesting to derive the electron equivalent quantum radius for a small velocity, because, it will help to unveil the interaction effect of the electrons with the atoms in solids. This interaction plays important role in understanding the resistivity of the conductors.

D. Case: Equivalent quantum radius of the electron moving with a small velocity

The average electron velocity in copper according to the drift theory is 3.54×10^{-5} m/sec. This is much smaller, than the sixth subharmonic velocity. Obviously the expected equivalent radius will have much larger value. Here we must open a bracket that this is an equivalent velocity since the electrons paths obviously are not straight, but nevertheless, the drift velocity is much smaller than the velocities corresponding to quantum numbers up to $n = 6$, for example. According to Eq. 3.32, the volume of the electron structure at such small velocity becomes insignificant and we can ignore it. The configuration of the E-field lines for a very small velocity is also changed. The density of the terminated lines at the boundary conditions becomes more uniform. The helicity, however, is preserved. Having in mind all this considerations, we may accept that the E-field occupies a spherical volume. Then neglecting for simplicity the small helical factor $\alpha/2\pi$, as in Eq. (3.32), the energetic equivalent volume can be expressed by the equation:

$$V_{eq} = \frac{4}{3}\pi R_{eq}^3 \qquad (3.34)$$

We can not apply a similar approach for calculation of R_{eq} as the previous cases, because the boundary conditions for a large volume does not work. Instead of that, we will consider the change of the magnetic flux Φ due to the slower electron rotation in comparison to its rotation at first harmonic. Consequently, now we will reference the magnetic flux surface s to the surface corresponding to the equivalent quantum radius at first harmonic.

$$S = 4\pi^2 R_c r_{eq1}$$

where: $r_{eq1} = 1.057 \times 10^{-13}$ is the equivalent radius for a first harmonic motion

For a small velocity case, the rotating speed of the electron will be smaller, but the fundamental period (first proper frequency) is unchanged. Let us assume that the positron makes **n** cycles for a full electron turn. Then the dependence of **n** on velocity υ according to Eq. (3.16) is:

$$n = \frac{\alpha c}{\upsilon} \qquad (3.35)$$

The lattice twisting will be **n** times smaller, and so the magnetic flux also:

$$\Phi = \frac{h}{nq} \qquad (3.36)$$

The magnetic strength then is:

$$H = \frac{\Phi}{\mu_o S} = \frac{h\upsilon}{\alpha c q \mu_o} \frac{1}{4\pi^2 R_c r_{eq1}} \qquad (3.37)$$

The volume expressed by Eq. (3.34) is the equivalent volume for the quantum interaction. Dividing the magnetic moment by this volume, we get:

$$\frac{\mu_e}{V_{eq}} = \frac{qh}{4\pi m_e} \frac{1}{\frac{4}{3}\pi R_{eq}^3} = H \qquad (3.38)$$

Solving (3.37) and (3.38) for R_{eq}, we get the equivalent quantum radius as a function of the velocity (for $\upsilon \ll \alpha c$).

$$r_m = \left[\frac{3\alpha c q^2 \mu_o R_c r_{eq1}}{4\pi m_e \upsilon} \right]^{1/3} \qquad (3.39)$$

The graphical plot of Eq. (3.39) for a velocity range between 10^{-6} **m** and 1 **m** is shown in Fig. 3.18, where the velocity is in a log scale.

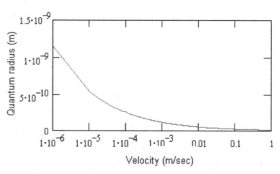

Fig. 3.18. Quantum radius of the electron at low velocities

The graphical plot of Eq. (3.39) shows significant increase of the magnetic radius of the electron at very low velocities. Such velocities exist in the metal conductors.

Example: The average electron velocity in copper according to the drift theory is 3.54×10^{-5} m/sec. Then the corresponding magnetic radius according to Eq. (3.39) is 2.77×10^{-10} (m). This is comparable to the gaps between the atoms. The magnetic field of the moving electron, obviously interacts with the proximity fields of protons and neutrons of the atomic nuclei. This could explain the ohmic resistance. It is well known that the ohmic resistance was not able to be explained so far my a classical approach. Quantum mechanics gives explanation by the wavefunctions. While this could not be directly translated as a real picture, the mathematical explanation of Quantum mechanics is in agreement with the presented BSM model, which provides an explanation by a classical mechanical approach.

The quantum radius for small velocities, can be expressed also by the electron kinetic energy. For this purpose, the velocity in Eq. (3.39) can be substituted by the expression (3.40), where, the energy is in eV.

$$ \upsilon = \sqrt{\frac{2E_{ev}q}{m_e}} \qquad (3.40) $$

The equivalent quantum radius as a function of the kinetic energy is:

$$ R_{eq} = \left[\frac{3\alpha c \mu_o R_c r_{eq1} \sqrt{q}}{4h\sqrt{2E_{ev}m_e}} \right]^{1/3} \quad \text{for } E_{ev} \ll 1.36 \text{ eV} \quad (3.41) $$

where: E_{ev} - is the electron kinetic energy in (eV).

The quantum radius dependence on the velocity is very important feature of the moving electron. From one hand it helps to analyse the orbital motion conditions of the electrons in the atoms. In this aspect, the equations (3.29) and (3.31) are relevant. From the other hand, when applied for metals, the quantum radius helps to understand the interaction of the free electrons with the atomic nuclei of the metal lattice. In this case the equations (3.39) and (3.41) are relevant.

We may summarize the analysis by the following conclusions:

- **At optimal confine velocity corresponding to energy of 13.6 eV, the transverse equivalent quantum radius is the smallest one.**
- **The dependence of the electron quantum radius on the velocity helps to understand the orbital motion of the electron around the proton and the ohmic resistance in metals.**

3.11.A Relativistic motion of the electron. Relativistic gamma factor and quantum efficiency.

So far, the quantum motion of the electron for optimal and suboptimal velocities was discussed.

The electron motion with a velocity larger than optimal one also exhibits a quantum feature, but the quantum effect is weaker. For correct physical analysis of the electron behaviour, two relativistic factors are necessary to be considered: the relativistic gamma factor and the quantum efficiency. The first one is well known from the relativistic theory. The second one is not considered so far, but it is very important for the correct estimation of the electron behaviour.

In a case of relativistic motion, according to the special relativity, the Lorentz transformation is used, where the gamma factor is given by

$$ \gamma = (1 - V^2/c^2)^{-1/2} \qquad (3.42.A) $$

In the next paragraph the same gamma factor will be derived based on the electron motion behaviour.

3.11.A.1 Quantum efficiency of moving electron

(A) Quantum efficiency at suboptimal velocity

The quantum effect in this case is strong, so it is enough to derive expressions as a function of the subharmonic number.

Quantum efficiency dependence on the boundary conditions

The surface of the boundary conditions is proportional to square of the magnetic radius, while the latter is defined by the subharmonic number. The smallest boundary surface corresponds to the first harmonic whose quantum efficiency is maximal. Then the quantum efficiency dependence on the boundary conditions should be inverse proportional to the subharmonic number.

$$ \eta_{BC} = 1/n , $$

Quantum efficiency dependence on the "hummer drill effect" of moving and oscillating electron

This problem is discussed in Chapter 4 in relation with the Fractional quantum Hall experiments. It is shown that the efficiency is inverse proportional to the subharmonic number.

$$ \eta_{HD} = 1/n $$

The total quantum efficiency is a product of both types efficiencies. This is **the quantum efficiency for a suboptimal velocity motion.**

$$\eta = 1/n^2 \qquad\qquad (3.42.B)$$

The quantum efficiency affects the width of emitted spectral line. Then comparing the linewidhts (normalized to the wavelength) from the different series of the Hydrogen atom, we may test the validity of Eq. (3.42.B). The Lyman series should contain the narrowest spectral lines.

(B) Quantum efficiency at superoptimal velocity, (velocity above the optimal one but lower, than the relativistic velocities)

In this case the quantum efficiency is determined by the efficiency of the "hummer-drill effect". The analysis in this case is similar as the hummer-drill effect analysis in the Quantum fractional Hall effect, discussed in Chapter 4.

(C) Quantum efficiency at relativistic velocities.

The decreased efficiency in its case is caused by the large cosine between the vector of trace velocity and the vector of the first proper frequency of the electron (electron-positron oscillations).

Figure 3.18.A shows the confined electron motion for both cases: a. - for an optimal velocity and b. - for a superoptimal (relativistic) one. The electron structure is shown by a thick curve, while the front edge traces are shown by a dashed line. The trace projection of the front edge on a plane normal to the axial velocity V is a circle, shown below the helical trace. The selected points from the trace are denoted in the circle projection by the same letters with primes ('').

We will consider the following two time intervals:

T_R - a rotational cycle time for one full rotation of the electron

t_c - a period of the electron oscillations with its first proper frequency; it is equal to the Compton time

This two time intervals determine two different properties:

- **The rotational cycle time determines the wavelength of the generating wave (also the magnetic radius) - the magnetic interaction with the CL space**
- **The proper frequency time determines the quantum interaction with the CL space (a phase match with a SPM vector propagated**

with a velocity of light and the hummer-drill effect of the central core oscillations)

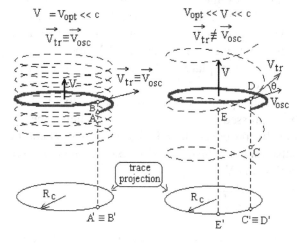

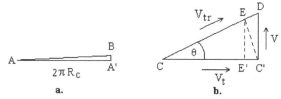

3.18.A
Electron motion with an optimal (a.)
and relativistic (b.) velocity

It is reasonable to consider that **the first proper frequency of the oscillating electron does not depend on its confined motion velocity,** so its oscillating period (equal to the Compton's time) is unchanged: $t_c = 1/\nu_c = const$.

For the case a. (optimal confined motion), the axial velocity is $V_{opt} = \alpha c$ and we have $T_R = t_c$.

Let us consider the case b., which is valid for relativistic velocities $V_{opt} < V < c$. If point C is our initial reference point, the electron will complete one full oscillating cycle (with the Compton frequency) in point E, earlier than the completion of its full rotational cycle in point D, so $T_R > t_c$

One important feature of the quantum interaction with CL space at the optimal confined motion is the angle between the proper oscillating frequency vector V_{osc} and the vector V_{tr}, tangential to the helical trace. While the angle θ between these two vectors is zero at optimal confined motion, it has a finite value at velocities larger than αc. It approaches 90 deg when the axial velocity V approaches the speed of light. Accepting the quantum interaction at the optimal confined motion as an unity, the quantum efficiency at any other relativistic velocity can be expressed by

$$\eta = \cos\theta$$

Using the unfolded trace for case b. we may express the path of the front edge of the electron as a result of two velocity vectors: - by the velocity vectors V_{tr} (tangential to the trace) and by the axial velocity vector V. Both velocities acts simultaneously for a time T_R.

$$C'D = V T_R$$
$$CD = V_{tr} T_R$$

Then

$$\tan\theta = V/V_{tr} \qquad (3.42.C)$$

The light velocity puts a limit on V_{tr} velocity:

$$V_{tr}^2 = c^2 - V^2$$

Substituting V_{tr} in Eq. (3.42.c) and expressing *tan* by *cos* function, we obtain the expression for the quantum efficiency at relativistic velocity:

$$\eta = (1 - V^2/c^2)^{1/2} \qquad (42.D)$$

The plot of the quantum efficiency vs velocity is shown in Fig. 3.18.B.

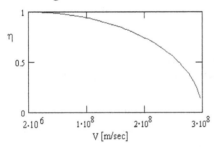

Fig. 3.18.B Quantum efficiency of the electron at relativistic velocities

From the point of view of the physical explanation, the trend of the quantum efficiency is quite reasonable. We see that it appears as inverse function of the relativistic gamma factor.

3.11.A.2 Relativistic gamma factor

According to the special relativity, the gamma factor is equal to the ratio between the relativistic and the nonrelativistic momentum: $\gamma = p_{rel}/\mathrm{p}$. It is also equal to the ratio between the relativistic and not relativistic time: $\gamma = T_{rel}/T$.

According to the physical analysis of the electron motion in the previous paragraph, the cor-responding two time periods are: $T = t_c$ and $T_{rel} = T_R$. Then the gamma factor is

$$\Upsilon = \frac{T_R}{t_c} = \frac{1}{\cos\theta} = \frac{1}{\eta} \qquad (3.42.E)$$

When expressed by c and V, the gamma factor is:

$$\gamma = (1 - V^2/c^2)^{-1/2} \qquad (3.42.F)$$

Conclusion: The quantum efficiency is an inverse function of the relativistic gamma factor. This should be taken into account, when estimating the physical properties of the real particles exhibiting a short lifetime. Some processes may obtain quite different physical explanation. This is valid in full for the muon lifetime, for example, leading to a different explanation of the factors that influence its decay (discussed in Chapter 6).

3.12 Quantum loops and orbits

So far we analysed the quantum motion of the electron system in an open trajectory. This does not exhausts all the possibility of the electron quantum motion. One special case still exists. This is the quantum motion of the electron in a closed loop. The term quantum loop is more universal, meaning a closed trajectory. A typical case is the electron - positron oscillation trajectories. The term quantum orbit is more suitable for the electron motion around the proton. In this case the proton can serve as a frame of reference, due to its large mass.

3.12.1 Quantum loop conditions

Now we will analyse the motion of a normal electron, from a point of view of a stationary frame. When performing quantum motion as repeatable loops, some portion of the electron orbit may have equipotential paths. The magnetic field created by this motion also could tend to extend the length of these paths. In this conditions the electron - positron system oscillates with small amplitudes. Then we have the following **proper frequencies:**

$$\nu_{ep} = \nu_c \text{ - proper frequency of electron - positron} \quad (3.43.a)$$
$$\nu_{pc} = 3\nu_c \text{ - proper frequency of the internal positron-core} \quad (3.43.b)$$

Let us consider two cases of motion:
- a very slow motion approaching zero velocity

- a motion corresponding to energies from 0.3 to 13.6 ev

In the first case, the velocity of both ends of the positron are equal and symmetrical in respect to the stationary CL nodes.

In the second case, the velocities of both ends of the positron are not symmetrical in respect to the stationary CL nodes.

If trying to reference the period of the v_{pc} oscillations to the stationary nodes, we will find the following features:

(a) the electron system is displaced axially due to the second order helicity by distance equal to the step s_e.

(b) the internal positron system is carried by the electron shell

(c) the path that the positron system is carried depends on the velocity of the electron shell

If we compare the feature (a) with the boundary conditions of MQs, we will see that: The displacement due to the step s_e does not have a symmetrical counterpart in the boundary conditions. **Consequently, it will cause a phase difference between the proper frequency of the electron-positron system and the SPM frequency of the stationary CL nodes.**

Let us estimate this phase delay for electron motion with an optimal confined velocity (E = 13.6 eV). For this purpose, we will present the single coil of the electron as an unfold helix, shown in Fig. 3.19.

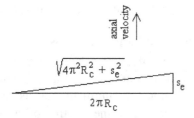

Fig. 3.19   The electron as unfolded single coil helix

The hypothenuse of the triangle can be regarded as a path of the front edge of the electron, while the kathet s_e is its axial displacement. If s_e approaches zero, the discrepancy between the oscillations with the proper frequency and the CL node oscillations will disappear. The both frequencies are equal to the Compton frequency. For some finite velocity in the range $0 < v \leq \alpha c$, the phase dif-

ference due to the discrepancy should be proportional to s_e. We need to reference the phase difference to the full revolution. Then it could be expressed as a ratio between the kathet s_e and the hypotenuse (the other option - the ratio between the two kathets does not provide consistent result later). So the fractional phase difference is:

$$\frac{\Delta\varphi}{2\pi} = \frac{s_e}{\sqrt{4\pi^2 R_c^2 + s_e^2}} = 7.2973531\times10^{-3} = \alpha \quad (3.43.c)$$

The fractional phase difference, defined in such way appears equal to the fine structure constant. We can refer to it as a **phase difference per one turn.**

Then we can define a quantum loop, which is an important feature of the electron circulating in a closed loop trajectory:

The quantum loop is a closed loop trajectory, whose length corresponds to a whole number of carrier oscillations.

Let us take into account only the discrete velocity values of the electron, corresponding to the preferable quantum velocities. These velocities are defined by the subharmonic number. We must also keep in mind that the oscillating electron has two proper frequencies v_c and $3v_c$. The triangle shown in Fig. 3.19 could be considered as a path of the front edge of the electron at optical confined motion (referred also as a first harmonic). At any other subharmonic, the triangle is similar but with sides divided on the subharmonic number. Then the phase difference given by Eq. (3.43.c) appears not for one turn of the electron system, but for one proper cycle (Compton time). Then for a quantum motion at n subharmonic, the phase difference accumulated per one turn of the electron system is:

$$\frac{\Delta\varphi}{2\pi}(n) = n\alpha \quad (3.43.d)$$

3.12.2 Quantum loops and orbits for electron with an optimal confined velocity. Embedded signature of the fine structure constant.

Let us find the path length at which the quantum loop condition for the electron system is fulfilled. The electron system possesses two proper frequencies and we must check the quantum loop conditions for both of them. It is reasonable to look for a path length defined by some CL space parameter. One of this parameter is the Compton wavelength $\lambda_c = \lambda_{SPM}$

If an electron possessing a first harmonic velocity travels in a closed loop with length λ_c, the number of turns N_T is:

$$N_T = \lambda_c/s_e = 137.03234 \qquad (3.43.d)$$

The value of N_T could be regarded as a condition for a phase repetition for two consecutive passes through a chosen point in the loop, keeping in mind a confined (screw-like) motion of the electron. The trace length of $\lambda_c = 2.4263 \times 10^{-12}$ (m), however, is quite small, when comparing to the Bohr orbit length of $2\pi a_0 = 3.3249187 \times 10^{-10}$ (m). Therefore, we may look for a phase repetition conditions at a larger loop length. From Eq. (3.43.d) we see that N_T is close to $1/\alpha = 137.036$ and this seams not occasional. Then we may substitute N_T in Eq. (3.43.d) by $1/\alpha$ and multiply the expression by λ_c. The latter is a CL space parameter, from one side (a distance that the SPM phase propagates for one SPM cycle) and from the other - the circumference length of the electron structure. In such case we obtain:

$$N_T \lambda_c \approx \frac{1}{\alpha}\lambda_c = 3.24918460 \times 10^{-10} \qquad (3.43.e)$$

We see that the obtained value of Eq. (3.43.e) having dimensions of length is equal to the Bohr orbit length given by CODATA 98 up to the 9th significant digit.

$$a_0 = 3.24918460 \times 10^{-10} \text{ (m)} \qquad (3.43.f)$$

where: $a_0 = 0.52917721 \times 10^{-10}$ (m) - is the radius of the Bohr atomic model of hydrogen.

The term λ_c/α of the expression (3.43.e) is not something new. The important fact, however, is the way of its derivation related with the suggested physical model of the electron. The obtained loop length appears equal to the orbit length of the Bohr atom, defined by the Bohr atomic radius, a_0. The latter is one of the basic parameters used in Quantum mechanics. From the BSM point of view, however, the physical meaning of this parameter appears different.

According to BSM concept, the well known parameter a_0 used as a radius in the Bohr model, appears defined only by the quantum motion conditions of the electron moving in a closed loop with an optimal confined velocity corresponding to an electron energy of 13.6

eV. Then the main characteristic parameter of the quantum loop is not its shape, but its length.

The identity of Equations (3.43.e) and (3.43.f) also indicates that **the signature of the fine structure constant is embedded in the quantum loop.**

Now we may use the new obtained meaning about the quantum loop associated with the Bohr orbit, and more specifically the orbital length $2\pi a_0$. For a motion with an optimal confined velocity, the number of electron turns in the quantum orbit is equal to the orbital length divided by the helix step (s_e).

$$\frac{2\pi a_0}{s_e} = \frac{\lambda_c}{\alpha s_e} = 18778.365 \text{ (turns)} \qquad (3.43.g)$$

Let us find at what number of complete orbital cycles (for orbit length of $2\pi a_0$) the phase repetition of the first and second proper frequencies of the electron is satisfied (in other words the smallest number of orbital cycles containing whole number of two frequency cycles). The analysis of the confined motion of the electron in Chapter 3 and 4 of BSM indicates that its secondary proper frequency is three times higher than the first one (the first one is equal to the Compton frequency). Equation (3.43.g) shows that the residual number of first proper frequency cycles is close to 1/3. If assuming that it is exactly 1/3 (due to a not very accurate determination of the involved physical parameters), then the condition for phase repetition of both frequency cycles will be met for three orbital cycles. The whole number of turns then should be $(3\lambda_c)/(\alpha s_e)$ Substituting s_e by its expression given by Eq. (3.13.b) and knowing that $v_c/c = \lambda_c$ we get

$$\frac{3(1-\alpha^2)^{1/2}}{\alpha^2} \text{ (turns)} \qquad (a)$$

We have ignored so far the relativistic correction, but for accurate estimation it should be taken into account. The relativistic gamma factor for the electron velocity of $v_{ax} = \alpha c$ is $\Upsilon = (1-\alpha^2)^{-1/2}$. Multiplying the above expression by the obtained gamma factor we get.

$$3/\alpha^2 = integer \text{ (turns)} \qquad (b)$$

　　The validity of obtained expression (a) and (b) could be tested by the following simple procedure: calculating these expressions by using the best experimental value of α, rounding the result to the closer integer (satisfying the condition for two consecutive phase repetitions) and recalculating the corresponding value of α. The rounded integer (a whole number of turns) could be correct only if the recalculated value is in the range of the accuracy of the experimentally determined α. Let us use the recommended value of experimentally measured α according to CODATA 98.

$$\alpha = 7.2973525(27) \times 10^{-3} \quad \text{(CODATA98)}$$

where, the uncertainty error is denoted by the digits in the brackets.

　　The calculated values of α from Eq. (a) and (b) exceeds quite a bit the uncertainty value of experimentally determined α given by the CODATA 98. Consequently, the condition for phase repetitions of the two proper frequencies is not fulfilled for three orbital cycles with total trace length of $3 \times 2\pi a_0$. Therefore, we may search for the next smallest number of orbital cycles in which the phase repetition conditions are satisfied. It stands to reason that the approximate value of the orbital cycles could be about 137 $(1/\alpha)$. Then if not considering relativistic correction, the corresponding number of electron turns is $(1-\alpha^2)/\alpha^3$. When applying a relativistic correction (multiplying by the estimated above gamma factor for the kinetic energy of 13.6 eV) the number of the electron turns becomes $1/\alpha^3$. The phase repetition conditions will be satisfied if this number is integer. Substituting α by its value from CODATA 98 we get: $1/\alpha^3 = 2573380.57$

　　It is interesting to mention, that the closest integer value of 2573380 is obtained by Michael Wales, using a completely different method for analysis of the electron behavior (See Michael Wales book "Quantum theory; Alternative perspectives", www.fervor.demon.co.uk).

　　We may use one additional consideration, for validation of the above obtained number. The number of turns multiplied by the time for one turn (the Compton time) will give the total time on the orbit (or the lifetime of the excited state, according to the Quantum Mechanics terminology). If accepting that the total number of turns are 2573380 then we obtain a lifetime of 2.0827×10^{-14} (s), that appears to be at least two order smaller than the estimated lifetime for some excited states of the atomic hydrogen.

　　Following the above analysis we may check for phase repetition at $1/\alpha^4$ turns. The participation of α at power of four is in agreement also with the following consideration: In the analysis of the vibrational mode of the molecular hydrogen, an excellent match between the developed model and observed spectra (section 9.7.5 in Chapter 9 of BSM) is obtained if the fine structure constant participates at a power of four. In such case we may accept that the phase repetition conditions is satisfied for a number of turns given by the closest integer in Eq. (3.43.i).

$$1/\alpha^4 = integer \qquad (3.43.h)$$

　　Using the CODATA value of α we obtain $1/\alpha^4 = 352645779.39$. Rounding to the closest integer we obtain an expression for the theoretical value of α (if its experimental estimation is accurate enough).

$$\alpha = (352645779)^{-1/4} = 7.2973525 \times 10^{-3} \qquad (3.43.i)$$

　　The small difference of the theoretically obtained value of α from the experimental one could be caused by an experimental error. One of the methods for accurate experimental estimation of α is based on the measurement of the Josephson constant, K_J. Its connection to α is given by the expression

$$K_J = \frac{2}{c}\left(\frac{2\alpha}{\mu_0 m_e \lambda_c}\right)^{1/2}$$

where: μ_0 - is the permeability of vacuum,

m_e - is the electron mass, c - is the light velocity, λ_c - is the Compton wavelength.

The accuracy of α according to this method depends mostly on the accuracy of the Josephson constant measurement, because all other parameters are accurately known. The recommended value for this constant according to CODATA 98 is $K_J = 483597.898(19) \times 10^9$ (Hz/V). If replacing α in the above expression of K_J with the value obtained by Eq. (3.43.i) we will get the value of K_J that is in the uncertainty range given by the CODATA 98.

The conclusion that the orbital time duration may depends only on α is reinforced also by the consideration that the Compton wavelength, λ_c, was initially involved in the analysis (Eq. (3.43.d), (3.43.e), (3.43.f)), but it disappeared in the derived Eq. (3.43.i). Consequently, the phase repetition condition is satisfied not only for the two proper frequencies of the electron, but also for the SPM frequency of the CL nodes included in the quantum orbit (λ_c is the propagated with a speed of light phase of the SPM vector for one SPM cycle of the CL node (SPM frequency = Compton frequency)).

In §3.5 it was described, that the central core is moving in the CL zone of magnetic quasispheres (MQ's). When the quantum loop condition is satisfied, the phase of core motion appears as repeatable in respect to the stationary CL nodes. The arrangement of MQs along the orbit trace will have a helical shape. This is illustrated in Fig. 3.20.

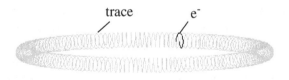

trace   e⁻

Fig. 3.20

Electron motion in quantum loop. MQ trace is shown by green line with a shape of close loop helix. The momentary position of the electron structure is shown as a black single turn

It is evident that the motion of the electron system is like a screwing in a helical curve. The

MQs along this helical curve possess a strictly determined spatial order.

There are following important features of the electron motion in such conditions:

(1) The phase difference between the stationary MQs along helical curve and the oscillating central core is zero for any point in the curve.

(2) In the absence of external electrical and magnetic field, there is not a phase mismatch between the electron first proper frequency and the SPM (Compton) frequency of the CL space. In other words, there is not a disturbing interaction caused by the CL space environments.

(3) In presence of external electrical or magnetic field up to some limit, the electron orbit could exhibit self adjusted properties. Above this limit, however, its quantum conditions are disturbed. (The signature of this is that the corresponding spectral lines become split - Stark and Zeeman effects).

The second feature, is valid only in the absence of external electrical field. The "near field" of the electrical field of the proton, for example, exhibits spatial configuration. In such conditions, the above feature becomes valid only for the boundary orbit. For other orbits, the total phase sum is preserved, but continuous phase difference appears, as a running phase in the closed helical curve. This causes a phase shift in the helical loop of MQs and **creation of magnetic line**. This effect will be additionally discussed in Chapter 7.

One question here may arise: The MQs oscillate with SPM frequency equal to the Compton one, while the positron - core frequency is three time higher? How the phase can be kept close to zero in this case?. The explanation is in the SPM vector quasisphere. From the spatial characteristics of SPM quasispheres it is evident that the bumps are much narrower, than the sinusoids. From the temporal point of view, however, they are much wider, because, the SPM vector spends much more time in the bumps. Fig. 3.21 illustrates the interaction process between the oscillating central core and one of the MQ bumps, unfolded in time. The time diagram should be considered in a frame travelling with the moving electron

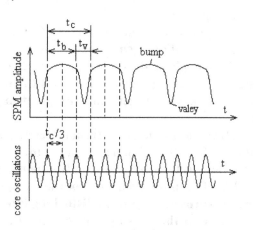

Fig. 3.21
Interaction between central core and SPM
vector of stationary MQ nodes

The interaction between the two oscillations with different but constant ratio of their frequencies is possible due to the ratio $t_b/t_v > 1$, where t_b is a bump time and t_v is a valley time of the SPM vector.

The provided concept of the fine structure constant embedded in the quantum loop, which defines the quantum orbit, agrees well with the analysis of Balmer series in Hydrogen, provided in Chapter 7, where α is involved in the **orbital time of the electron** (time duration of the electron circling in one quantum orbit).

3.12.2.A. Quantum orbits and time duration for a stable orbit

It is apparent from the provided analysis that a stable quantum loop is defined by the repeatable motion of oscillating electron. The shape of such loop, however, is determined by external conditions. Such conditions may exist in the following two cases:

 - a quantum loop obtained between particle with equal but opposite charges and same mass, as in the case of positronium (see Fig. 3.25).

 - a quantum loop obtained between opposite charged particles but with different masses (a hydrogen atom as a most simple case and other atoms and ions as more complex cases).

In both options the quantum loops are repeatable and we may consider that any quantum orbit is formed of whole number of quantum loops.

A single quantum orbit could contain one or few serially connected quantum loops (in both cases the condition for phases repetition is preserved). It is obvious that the shape of the quantum orbit is defined by the proximity field configuration of the proton (or protons). The new space concept of BSM allows unveiling not only the electron structure but also the physical shape of the proton with its proximity electrical field (Chapters 6 and 7 of BSM). The shape of any possible quantum orbit is strictly defined by the geometrical parameters of the proton.

Let us consider now the induced magnetic field of the electron motion in a quantum orbit by using the electron magnetic radius. The magnetic radius of the electron moving with different sub-harmonic numbers n was analyzed in §3.1. Its value for $n = 1$ (a kinetic energy of 13.6 eV) matches the estimated magnetic radius corresponding to the magnetic moment of the electron. For larger numbers (decreased electron energy), however, the magnetic radius shows an increase. The physical explanation by BSM is that at decreased rate of the electron rotation its SG field of the twisted internal RL structure is able to modulate the surrounding CL space up to a larger radius until the rotating modulation of the circumference reaches the speed of light. Keeping in mind that the circumference of the electron is equal to the Compton wavelength (with a first order approximation) the circumference length of the boundary (defined by the rotation rate) should be a whole number of Compton wavelengths. Then the integer number of the Compton wavelengths corresponds to integer sub-harmonic number. In such case, the orbiting electron with optimal or sub-optimal velocity could not cause external magnetic field beyond some distance from the nucleus. This provides boundary conditions for the atoms, if accepting that in any quantum orbit the electron is moving with optimal or sub-optimal confined velocity (integer sub-harmonic number). Here we must open a bracket that the higher energy levels in heavier elements come not from a larger electron velocity but from the shrunk CL space affected by the accumulated protons and neutrons. Such CL space domain is pumped to larger energy levels in comparison to the CL space surrounding the hydrogen atom.

The existence of the SG law changes significantly the picture of the orbiting electron in a proximity field of the proton. In Chapter 7 of BSM an analysis of Balmer model of Hydrogen atom is developed based on the BSM concept of the electron and proton and the SG law influence on the orbital electron motion in the proximity to the proton. It appears that the limiting orbit has a length of $2\pi a_0$ (where a_o - is the Bohr radius) while all other quantum orbits are inferior. This conclusion is valid not only for the Balmer series in Hydrogen but also for all possible quantum orbits in different atoms, if they are able to provide line spectra. Therefore, the obtained physical model of Hydrogen puts a light for solving the boundary conditions problem of the electron orbits in the atoms.

Time duration for a stable orbit (lifetime of excited state).

The following analysis could be valid only for free CL space. This exclude the quasishrunk space in proximity to the proton, or so called Bohr surface as defined in Chapter 7.

Keeping in mind the screw-like confined motion of the electron, the axial and tangential velocities will be inverse proportional to the subharmonic number. Then the condition for phase repetitions for a motion with a subharmonic number n will be satisfied for n times smaller number of electron turns, or the quantum orbit will be n times smaller. It is reasonable to consider that the first and second proper frequencies of the electron are stable and not dependent on the subharmonic numbers. Then for estimation of the time duration of the orbit (the lifetime of excited state) it is more convenient to use the number of the cycles of the first proper frequency of the electron. It is equal to the number of electron turns for $n = 1$. In such way we arrive to the conclusion:

(a) If conditions for stable quantum orbit are defined only by the phase repetition conditions and the whole number of Compton wavelengths, the time duration (lifetime) of the orbiting electron does not depend on the subharmonic number of its motion.

(b) If (a) is valid, the lifetime of the excited state will be equal to the product of the total number of the first proper frequency electron cycles

(according to Eq. (3.43.h)) and the Compton time (the time for one electron cycle with the first proper frequency).

According to condition (b) the theoretical lifetime for an excited state of the hydrogen is

$$\tau = t_c/\alpha^4 = \lambda_c/(c\alpha^4) = 2.85407 \times 10^{-12} \quad (s) \quad (3.43.k)$$

where: $t_c = 1/\nu_c$ - is the Compton time.

Note: The obtained Eq. (3.43.k) does not take into account the possible modification of the surrounding space in a close proximity to the proton (a quantum quasishrunk space as discussed in Chapter 7). Such modification (a slight shrinkage changes the size of the Compton wavelength, or in other words a space curvature takes place in proximity to the protons and neutrons) may cause aliasing for the phase repetition conditions of the proximity CL zone that will lead to significantly departure of the real lifetime from the value given by Eq. (3.43.k). The CL space modification in the proximity field becomes apparent in the analysis of Balmer series orbits, provided in Chapter 7. For elements with more than one electron, the mutual orbital interactions also may lead to increase of the real lifetime.

3.12.3 Quantum loops and orbits, for electron with any suboptimal quantum velocity

The analysis so far was done for an optimal confined motion - first harmonic quantum motion. Let us see how the quantum loop condition is satisfied for motions with subharmonics.

If the electron is moving with a second subharmonic, its velocity is two times slower. The positron - core system will make the same number of oscillations for twice shorter path. Consequently, the same conditions for a quantum orbit are satisfied also for a twice shorter orbit. For quantum motion with n subharmonic, the quantum loop will be n times shorter. This conclusion is evident also from Eq. (3.43.d). Then the length of the quantum orbit, L_{qo}, can be expressed by the equation:

$$L_{qo}(n) = \frac{2\pi a_o}{n} = \frac{\lambda_c}{\alpha n} \quad (3.43.j)$$

where: n - is the subharmonic number

In a similar way as we used the term ***subharmonic*** for the quantum motion of the electron, we may use it again for the quantum loop. Then the first harmonic quantum loop corresponds to electron motion with energy 13.6 eV, the second harmonic quantum loop - energy of 3.4 eV and so on.

From energy point of view the orbital shape is is not important feature of the quantum loops. This is quite important conclusion that answers one important question: Why the Bohr atomic model and the Quantum mechanical models (operating only with energy levels) provide accurate values of the energy levels.

The quantum loops are very important features of the electron motion around the proton in the atoms. When discussing the Hydrogen orbits in Chapter 7, we will see, that they are folded 3D curves.

The quantum orbits play important role, also, between the atomic connections in the molecules. In this aspect additional combinations of the quantum loops are possible: **Two or more quantum loops can be serially connected, permitting quantum orbits with larger physical dimensions.** Such orbits are possible in the atomic nuclei and between atoms.

Summary:
- **MQs in the helical trajectory of the electron motion in closed loop are connected in closed magnetic lines.**
- **The central core trace in the quantum loop is a helix of aligned MQs.**
- **The quantum orbits are formed of closed loop electron trajectories, containing whole number of central core oscillation periods**
- **In a case of absence of external electrical field, there is not distributed phase shift between aligned MQs and the CL space MQs.**
- **The characteristic parameter of the quantum orbit is the orbital length, while its shape is defined by the proximity E-field of the proton (or protons).**
- **The quantum orbits are possible for the first harmonic and any subharmonic quantum velocities. Consequently, the attribute *n-subharmonic* defines completely the energy level of the quantum orbit.**
- **The length of the n-subharmonic orbit is n times shorter, than the length of the first har-**

monic (corresponding to the optimal confined velocity).
- **Subharmonic quantum loops are able to be connected in series, forming a common quantum orbit.**

3.13 Estimation of basic CL space parameters by the parameters of the electron system. Derivation of the mass equation.

3.13.1 Physical interpretation of inertial mass ratio

In Chapter 6 the similarity between the electron and muon (and positron and muon) is discussed. The muon is a second order helical structure whose central radius is the same as the electron radius R_c. The evidence for this comes from the fact that the muon can oscillate longitudinally and when it crashes, only a single coil could be left from one of its ends. All other portion of the muon helical structure together with its internal lattice is disintegrated and could be detected as one type of neutrino. When providing a physical interpretation of the mass and magnetic moment we arrive to the conclusion that the muon has 206.7 more windings than the electron system. Then the volume ratio of their FOHSs is also equal to this value. It follows, that the inertial mass of the muon is equal to the inertial mass of the electron multiplied by their volume ratio that is 206.7. This can be expressed by the equation:

$$\frac{\mu_e}{\mu_\mu} = \frac{m_\mu}{m_e} = 206.76 = \frac{206.76}{1}\frac{\text{windings}}{\text{winding}} \qquad (3.44)$$

From Eq. (3.44) it follows that there is a direct proportionality between the total volume of the FOHSs embedded in the elementary particle and it mass.

The equivalence of ratio between the mass and magnetic moment is valid only for particles having similar helical structures. The same ratio, for example, is not valid between the electron and proton or neutron. The latter two particles are formed of higher orders helical structures. They also have confined motion, but due to a equivalent high order helicity. However, all helical structures exhibiting confined motion, contain FOHSs

From the considerations discussed above, the following conclusions can be made:

The inertial mass of any helical structure, exhibiting confined motion, could be expressed by the electron mass multiplied by the volume ratio between the helical structure of consideration and the helical structure of the electron.

3.13.2 Relation between CL node displacement from FOHS and the Broglie wavelength

Accepting the apparent mass of the electron as unity, we will derive equation that relates its mass to the cosmic lattice parameters.

The mass to magnetic moment ratio is valid for similar structures, such as the electron (positron) and the muon. Similar expression between the electron and proton is not valid, because their shapes are different. However, there is some similarity in their motion behaviour in CL space. This is their confined motion and we will use this feature in the following analysis.

It is well known fact that the elementary particles exhibit a wavelike motion with wavelength determined by the Broglie's equation:

$$\lambda = \frac{h}{m\upsilon} \qquad (3.44.a)$$

where: λ - is the wavelength of the wave like motion, m - is the particle mass, υ - is the particle velocity

Now we will provide a physical interpretation of this equation from the point of view of BSM. It was pointed out that the confined motion of the proton and neutron is due to their equivalent helical step. The confined motion means that the particle rotates. Consequently, there is some periodicity of the particle interaction with the CL structure. This periodicity will depend on the particle mass, the motion velocity and the ability of the particle to twist the CL lattice. All this parameters are contained in the Broglie equation. The important feature of this equation is that the mass is involved, and this will give us a key for derivation of the inertial mass equation valid for the elementary particles. Let us use the Broglie's equation for the electron, in the case when $\upsilon = c$, and make some manipulations, as shown in Eq. (3.44.b).

$$\lambda = \frac{h}{m_e c} = \frac{h}{m_e c}\frac{c}{c} = \frac{hc}{h\nu_c} = \frac{c}{\nu_c} \qquad (3.44.b)$$

Now using the corresponding physical dimensions in SI for the parameters in Eqs. (3.44.a) and (3.44.b) we get:

$$wavelength = \frac{\text{N m sec}}{\text{kg m sec}^{-1}} = \frac{\text{N m}}{\text{kg m sec}^{-2}} = \frac{torque}{force} \qquad (3.45)$$

From Eq. (3.44.a) we see that λ becomes the wavelength of the SPM frequency (Compton frequency in Earth local field) $\nu_{SPM\,MQ}$ when $\upsilon = c$.

From dimensional interpretation of Eqs. (3.44.a) and (3.44.b), shown as Eq. (3.45), we see that the Broglie wavelength can be expressed as a ratio of torque over force that moves the particle. The torque is a result from the particle helicity.

The waves invoked by particle having confined motion in CL space could be regarded as a dynamical disturbance of the CL space surrounding the particle. The wavelength of this disturbance is equal to the torque that the particle exercise on the CL structure under the pushing force.

The inertial mass of elementary particle can be regarded as a static lattice disturbance causing a lattice displacement. Interpolating the Broglie expression for a motion with a velocity of light without taking into account the relativistic mass change, provides the inertial mass of the particle.

Having in mind that the optimal confined motion of the electron is completely determined by its geometry and its first proper frequency, we can make the following general conclusions:

- **The electron could serve as an inertial mass unit in CL space.**
- **The mass of the electron can be expressed by its 3-D geometry, the Compton frequency and the CL space parameters.**

3.13.3 Static CL pressure and apparent (Newtonian) mass of the helical structures

It has been already mentioned, that the Compton frequency is a value of the SPM frequency at Earth local filed. Let us express the electron inertial mass using Eq. (3.44.a), when $\upsilon = c$ and applying some manipulation of the dimensions.

$$m_e = \frac{h}{\lambda c} \qquad (3.46)$$

$$\frac{Nm\sec}{nm\sec^{-1}}\frac{m^2}{m^2} = \frac{N}{m^2}\frac{m^3}{m^2\sec^{-2}} = \frac{(\text{pressure})\text{x}(\text{ref. volume})}{(\text{light velocity})^2} \quad (3.47)$$

Eq. (3.47) is a dimensional expression of Eq. (3.46). From Eq. (3.46) we see, that the inertial mass can be expressed by the parameters shown in the brackets. Then the **equation for the inertial mass** of helical structure exhibiting confined motion, will take a form given by (3.48).

$$m = \frac{P_S}{c^2}V_{H(SI)} \quad (3.48)$$

where: P_S - is the cosmic lattice static pressure exercised on the external shell of FOHS, $V_{H(SI)}$ - is the FOHS volume referenced to the measuring system (SI in this case), c - is the light velocity

Eq. (3.48) permits the formulation of the inertial mass of an elementary particle in CL space:

- **The inertial mass of particle in CL space is proportional to the static CL pressure and the volume of FOHS's contained in the particle, and inverse proportional to the square of the velocity of light.**

The inertial mass of helical structure with a second order helicity is equal to its gravitational mass. So we may refer it as a **Newtonian mass (or Newton's mass)**. The Newtonian mass of any elementary particle is different than its intrinsic mass. **The Newtonian mass does not take into account the amount of the intrinsic matter inside of its FOHS.**

The pressure P_S is called static, because the CL nodes are constantly displaced by the volume of the FOHSs of the elementary particle. This volume is occupied by RL(T), which is so dens that even partly folded CL nodes could not pass through. The electron contains only a single coil of FOHS (external negative and internal positive). Knowing the total volume occupied by the RL(T) we can estimate the static pressure, P_S, by applying Eq. (3.48)

$$P_S = \frac{m_e c^2}{V_{e(SI)}} \qquad \frac{N}{m^2} \quad (3.49)$$

where: $V_{e(SI)}$ is the electron volume, expressed in units of SI

From Einstein mass - energy equation we have:

$$m_e c^2 = h\nu_o = 511 \text{ KeV/c}^2 \quad (3.50)$$

Then the static CL pressure, can be expressed also by Eq. (3.51).

$$P_S = \frac{h\nu_o}{V_{e(SI)}} \qquad \left[\frac{N}{m^2}\right] \quad (3.51)$$

Checking the dimensional correctness of Eq. (3.51) we get:

$$\text{pressure} = \frac{N}{m^2}\left(\frac{m\sec}{m\sec}\right) = \frac{Nm\sec}{m^3}\frac{1}{\sec} = \frac{Nm\sec}{m^3}Hz \quad (3.52)$$

The accepted in §3.6 relation $s_e = 2r_e$. matches well with all the calculations, physical considerations and models developed by BSM. Having in mind the relation between R_c and s_e, given by Eq. (3.9), we may express the static pressure only by the CL parameters. We have two options for this purpose: by the SPM (Copmton) frequency or by the resonance frequency:

The **static CL pressure**, when   using the SPM (Compton) frequency is:

$$P_S = \frac{h\nu_c^4 g_e^2(1-\alpha^2)}{\pi\alpha^2 c^3} = 1.37358\times10^{26} \qquad \left[\frac{N}{m^2}\right] \quad (3.53)$$

where: α - is the fine structure constant, g_e - is the electron gyromagnetic factor

The **static CL pressure**, expressed by the CL node parameters is:

$$P_S = \frac{hg_e^2(1-\alpha^2)\nu_R k_{hb}^3}{\pi\alpha^2 N_{RQ}^4 d_{nb}^3} \qquad \left[\frac{N}{m^2}\right] \quad (3.54)$$

where: d_{nb} - is the node distance for a not disturbed CL space; ν_R - is the node resonance frequency; N_{RQ} - is the number of resonance cycles for one SPM MQ cycle

k_{hb} - is the quantum wave boundary condition factor, given by Eq. (2.20.a) as derived in Chapter 2.

$$k_{hb} = \sqrt{1 + 4\pi^2(0.6164^2)} = 4 \qquad [(2.20.a)]$$

where: 0.6164 - is a factor complying to the Rayleigh criterion (see §2.10.4.3, Chapter 2).

The ratio m_e/V_e in Eq. (3.49) could be regarded as a mass density of the electron. A single coil from muon has the same mass density. The pion and kaon structures could be also referenced to this value. Comparing Eq. (3.49) and (3.53) we see, that the mass density of the electron is:

$$\rho_e = \frac{m_e}{V_e} = \frac{g_e^2 h\nu_c^4(1-\alpha^2)}{\pi\alpha^2 c^5} = 1.528315\times10^9 \left[\frac{kg}{m^3}\right] \quad (3.55)$$

Then the Static CL pressure can be expressed

$$P_S = \rho_e c^2 \qquad (3.56)$$

The expression (3.56), is quite convenient especially in the analysis of the inertial features of the particles and macro systems in CL space and their relativistic features. Such analysis is presented in Chapter 10.

Eqs. (3.51), (3.53) and (3.54) express the Static Pressure of CL space by using known physical constants or CL space parameters estimated by BSM analysis. They all provide one and a same value:

$$P_S = 1.373581 \times 10^{26} \qquad [N/m^2]$$

We might be surprised, in a first gland, that P_S is so large. If estimating also the total force exercised on the electron surface S_e we will find that it is also quite large. But this is an area where large energy interactions take place. The interactions involving the CL static pressure, however, are static and we can not feel them. We can detect them only when a change of the FOHS takes place. Two types of changes exist for the electron system: (a) separation of the positive (internal) FOHS from the negative negative (external) FOHS; or (b) destruction of anyone of both FOHSs of the electron. This issue is discussed in Chapter 6, where some high energy experiments from particle coliders are analysed from a new point of view.

Substituting the value of P_S in Eq. (3.48) and knowing the volume of the FOHS involved in the elementary particle, we can calculate its apparent mass in CL space, referred as a Newtonian mass according to BSM.

Mass definition:

- **The Newtonian mass of any helical system in CL space exhibiting confined motion could be determined by the fundamental parameters h, ν_o, c, and the total volume of its first order helical structures.**

By substituting P_S from one of Eq. (3.51)., (3.53) or (3.54) into Eq. (3.48) we obtain the mass

equation, estimated by the CL space parameters and the FOHS volume.

$$m = \frac{g_e^2 h v_c^4 (1 - \alpha^2)}{\pi \alpha^2 c^5} V \quad [kg] \qquad (3.57)$$

where: V_{HS} is the volume of the FOHS

Note: The mass equation (3.57) in this form is valid only for negative FOHS's. For positive FOHS's the volume, the proper frequency and the tangential to axial velocity ratio are different. This requires a use of correction factor (see §3.14).

- **When the mass equation is applied for a positive FOHS the right side of the mass equation should get a multiplication factor of 2.25.**

It is evident from Eq. (3.48) that **the mass of any helical structure, is determined by the volume of all of its FOHS's, which are in direct contact with the surrounding CL space.** The electron system contains only one coil of combined (positive inside a negative) FOHS, but only the external negative FOHS is in a direct contact with the surrounding CL space. Consequently, it is a suitable mass unit for estimation the mass of more complex structures. Sometimes another task is more useful - determination of the dimensions, when the mass ratio is known. In this case, another form of the mass equation is more suitable. Substituting (3.51) in (3.48) and introducing the volume normalisation factor K_v **the Newtonian mass equation** takes a form:

$$m = \frac{h v_c}{c^2} K_V \quad [kg] \qquad (3.58)$$

where:

$$K_V = \frac{V_{H(SI)}}{V_{e(SI)}} = \frac{V_H}{V} \qquad (3.59)$$

K_V is a ratio between the total volume of all FOHS's of the particle (with mass m) and the volume of the FOHS of electron.

Note: The mass equation (3.58) is valid for negative FOHS's.

For Newtonian mass of positive FOHS's, we must use the positron estimate of the Plank's constant and the positron's proper frequency. In the next paragraph (§3.14) it is shown, that the product of both parameters is

$$h' \nu_{pc} = \left(\frac{9}{8} h\right)(2 \nu_c) = 2.25 h \nu_c \qquad (3.59.a)$$

Consequently, when applying the mass equation for a positive FOHS the factor 2.25 should be used in the nominator.

The Eq. (3.58) provides results, consistent with the practically estimated masses of the following particles: proton, neutron, pion, muon. They all have second order helicity. The experimentally estimated masses of the kaons are not consistent with the calculated masses by the mass equation. The kaon is strait FOHS, but this is not the main reason. The reason is the following:

The mass of the kaon is not correctly estimated in the experiments in the particle accelerators, because it possesses active jet during its lifetime. This jet is from destructing internal RL(R) or RL(T) structures which provides a reactive force with a velocity vector coinciding with the direction of kaon's motion. Since this effect, discovered by BSM was not envisioned so far, the kaon mass appears largely overestimated (see the calculations in Chapter 6). The pulsar theory presented in Chapter 12 also confirms the evidence of jets from the kaon disintegration in CL space.

The mass equation is valid for any single particle up to the size of the proton (neutron). However it is not exactly valid for the atomic nuclei, larger, than Hydrogen. When the number of protons and neutrons, forming the atomic nuclei, increases, a mass deficiency effect appears due to the shrinkage of the CL space around the nuclei from the SG(CP) forces. This is a General Relativity micro-effect at atomic level, envisioned by BSM analysis and discussed elsewhere. In such case the atomic mass appears slightly smaller than the sum of the neutrons and protons masses. Such mass difference is known as a nuclear bonding energy.

Summary notes:

- **The static pressure of cosmic lattice is the pressure exercised on the surface of the first order helical structure (FOHS)**
- **The electron is a convenient helical structure for estimation of the CL static pressure.**
- **The apparent mass of a first order helical structure is equal to the product of the static pressure and the structure volume, divided by the square of the light velocity**

- **The mass of any elementary particle is completely determined by the total volume of its FOHS's.**
- **The inertial mass of the elementary particle is completely defined by its helical structures and the CL space parameters**

3.13.4. Physical nature of inertia and inertial mass

The inertia can be regarded as an effect preventing the helical structures to get infinite acceleration. This is a result of increased interaction between the field of the internal RL(T) and the oscillating nodes of the CL space.

Any kind of elementary particle is comprised of helical structures. Any type of helical structure is build of FOHS that may be curled into second and third order helical structure. Only FOHS contains internal RL(R) or RL(T). They both are much denser than the CL structure. Therefore, the the CL nodes could not penetrate inside the FOHS. When the particle moves in CL space, the surrounding CL nodes are partly folded, displaced and temporally inserted among the normal CL nodes surrounding the particle. The motion of any FOHS through the CL space causes continuous folding and unfolding of CL nodes. **Therefore, the folded and displaced CL nodes form CL pressure, which can be regarded as a partial pressure.**

From the point of view of the moving FOHS, the static pressure is a scalar, while the Partial pressure could be associated with the velocity vector of the particle motion in respect to a stationary CL space. Consequently, the Partial CL pressure can be involved in the definition of the inertia for any particle comprised of helical structures. In such aspect, the inertial mass could be expressed by the equivalent interaction energy in CL space: $E = mc^2$. The gravitational mass is measurable only if gravitational interaction exists. It also have equivalent energy in CL space. A normal CL space assures equivalence between the inertial and gravitational mass for all types of helical structures.

The inertial properties of particles and macrobodies beyond the Newtons law of inertia are discussed in Chapter 10.

3.14. Free positron. Newtonian mass and Planck's constant estimated by its motion in CL space.

It is experimentally known fact that the masses of the electron and positron are exactly equal. From the analysis of the positronium in §3.17.3 and the Fractional Quantum Hall Experiments (FQHE) in §4.48 it is concluded that the proper frequency of the free positron is different than the proper frequency of the internal one. While the internal one has a proper frequency of $3\nu_c$, the proper frequency of the free positron (directly immersed in CL space) is $\nu_{pc} = 2\nu_c$, so having in mind the volume ratio of the electron - positron $K_V = r_{pc}^2/r_e^2 = 4/9$ we may express the positron mass for a case of a free positron.

$$m_{pos} = \frac{h'\nu_{pc}}{c^2}\frac{4}{9}$$

Equalising the positron and electron masses we get:

$$h' = (9/8)h$$

where; h' - is the Plank's constant estimated by the positron parameters; ν_{pc} - is the proper frequency of the free positron, equal to twice the Compton frequency

Then for the free positron the following product becomes valid:

$$h'\nu_{pc} = 2.25h\nu_c \qquad (3.59.a)$$

Eq. (3.59.a) shows, that when the mass equation is applied for the positive FOHS, the product $h'\nu_{pc}$ is valid, or the equation should get a multiplication factor of 2.25.

3.15 Dynamic pressure of CL space

In Chapter 2 it was discussed that the background uniformity of the CL space is maintained by ZPE (or zero order) waves. These wave are responsible for spontaneous creation of magnetic protodomains, whose concentration is a constant parameter. In all these effects the CL relaxation space-time constant is involved. Its accepted theoretical value was discussed in Chapter 2, §2.13B, where it was compared to the experimental results from Johnson noise estimated at temperatures closer to absolute zero. The reciprocal value of the theoretically accepted space-time constant has a dimension of frequency. It could be called a relax-

ation quasifrequency, because it is not defined by exact periodical motion. It is given by the equation:

$$\nu_{CL} = \frac{1}{t_{CL}} = \frac{\nu_c}{(c)} = 4.12148\times10^{11} \quad [Hz] \qquad (3.60)$$

where: (c) - is a light velocity as a dimensionless factor

Defined in this way, we may use the relaxation quasifrequency only in one and a same measuring system - the SI system.

In a similar way as the static pressure, given by Eq. (3.51), we may define a dynamic pressure, which however is referenced to the relaxation quasifrequency of CL space, given by Eq. (3.60). The dynamic pressure is caused by the zero point waves, characterized by wavetrain length equal or multiple of λ_c. **Consequently, they may envelope around the electron or positron, but could not penetrate inside the FOHS volume.** So they may exercise forces on the envelope of the helical structures. For this reason the surface of the external electron shell will be used for a reference. The dimensions of this pressure should be: $\left[\frac{N}{m^2Hz}\right]$.

Then the the Dynamical CL pressure is:

$$P_D = \frac{h}{c}\frac{\nu_c}{S_{e(SI)}} = 2.025786\times10^3 \quad \left[\frac{N}{m^2Hz}\right] \qquad (3.61)$$

where: $S_{e(SI)}$ - is the surface of the electron's external shell envelope

The correct dimensions of Eq. (3.61) appear when the light velocity participates with its dimensions. For this reason the brackets used in Eq. (3.60) are not used in Eq. (3.61).

The dimension of the Eq. (3.61) is a "pressure unit per frequency". For this reason it is called Dynamical pressure. When applied to the envelope of a helical structure it exercises bouncing forces with a frequency given by Eq, (3.60).

The Dynamical pressure is a pure CL space parameter as the Static pressure. It is expressible by the known physical constants:

$$P_D = \frac{g_e h\nu_c^3\sqrt{1-\alpha^2}}{2\pi\alpha c^3} \qquad (3.62)$$

where: g_e - is the electron gyromagnetic factor

Eq. (3.62) gives exactly the same value as Eq. (3.61).

Note: The Dynamical CL pressure, is equally applicable for a negative and positive external shells, and is not influenced by the type of the internal structures. The Static CL pressure, however, has different value for negative and positive FOHS's, and this should be taken into account, when applying the mass equation. The latter conclusion is confirmed by the calculations about the physical dimensions of the proton and its substructures.

The Dynamical CL pressure provides a way for indirect estimation of the ZPE by the measurement of the behaviour of an atom, which is in equilibrium conditions with the surrounding CL space. This approach is used in Chapter 5 for calculation of the background temperature of deep CL space. It corresponds to the experimentally determined parameter known as a Cosmic Microwave Background.

3.16 Scattering experiments for electron and positron from the point of view of the BSM theory.

Presently, it is well known that a large discrepancy exists between the Compton radius of the electron and the electron radius determined by the scattering experiments (scattering radius). While the Compton radius is 3.86×10^{-13} (m), the scattering experiments give a value about 1×10^{-16} (m) This problem is solved by the BSM theory.

The "electron - electron" scattering model is developed by C. Moller (1932) and the process is known as a Moller scattering. The electron - positron scattering equation is derived by H. J. Babha and the process is known as a Babha scattering. Later modifications, based on the Dirac theory are applied involving correction for the QM spin. Some improvements are also contributed by Scott, 1951; Barber, 1953; Ashkin, 1954 and others. The Moller and Babha equations has been corrected, but the basic assumption is not changed. The basic assumption for both types of scattering is that the electron and positron are regarded as a point-like particles possessing a charge. The scattering models take into account the kinetic energy of both particles and allow to determine the angular distribution and the differential scattering cross section. From such data one can determine the size of the electron with a priory accepted shape and features.

Let us consider the Babha scattering model, for example. The following parameters are taken into account: electron (positron) mass, electrical charge, velocity, spherical radius, two spin parameters (+h and -h).

Fig. 3.22 shows an angular distribution of the scattering events for Babha scattering at 29 GeV (D. Bender et all., (1984). In the same figure, the theoretical curve with Monte Carlo simulation is shown.

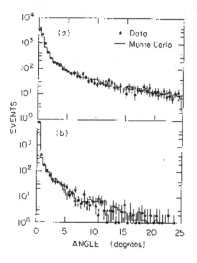

Fig. 3.22

Angular distribution of scattering events for Babha scattering at 29 GeV (D. Bender et all, 1984)

The vertical axis is in a logarithmic scale because the peak is very sharp. When assuming a spherical shape, the data of the scattering experiment lead to a result, that the radius of the sphere is very small - of the order of 1×10^{-16} *(m)*. The existing so far theories are not able to explain the huge discrepancy between the Compton radius $R_c = 3.86159 \times 10^{-13}$ *(m)*, and the scattering one.

From the point of view of the BSM, the discrepancy between the Compton and scattering radius of the electron, comes mainly from the assumption that the electron does not possess a structure. In both, the Moller and Babha scattering models, the following factors are not taken into account:

a. The form factor: a sphere is assumed, instead of single coil of first order helical structure.

b. The confined motion in CL space

c. The oscillation properties of the electron subsystems

d. The possibility for different rotational phase at the moment of meeting in the high energy collision

e. The Super Gravitation between the helical structures in close proximity

f. The distributed charge appearance in a close encounter

It is evident that if applying a scattering model with all above factors taken into account, the output result could be quite different.

Fig. 2.23 illustrates the scattering process according to the Moller and Babha assumptions - case **a.**, and BSM - case **b.**.

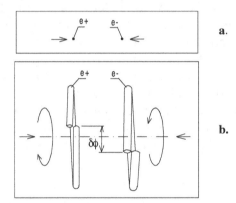

Fig. 3.23

Electron positron scattering according to: a. Babha model; b. Twisted prisms theory

Fig. 3.23.b illustrates the orientation, spin direction and the phase difference $\delta\phi$ in the moment of meeting.

In Babha model, the spin momentum has only two values ($+h/2$ and $-h/2$). The both values may express correctly the quantum energy, but only for motions with suboptimal velocities. In the Moller and Babha scattering, the velocities are much higher, so the quantum motion effect, according to BSM is significantly reduced. Then it is not correct to use the same spin momentum as in the low energy motion.

3.17 Positronium

The positronium is a state of temporally stable dynamical system between electron and positron keeping in mind the complexity of the dynamical properties of their helical structures. In some cases a virtual (quasiparticle) can be involved, which increases the number of possible combinations. In the end of oscillating process a photon is emitted.

A pretty large number of combinations are possible, but we can list here a few of them, for which experimental evidence exists:

  - $Ps1^1S_o$ state

  - $Ps1^3S_1$ triplet

  - Ps $1^3S_1 - 2^3S_1$ singlet

3.17.1 $Ps1^1S_o$ state

This state involves oscillations between a normal electron and a free positron. The free positron is directed toward the internal positron, and both positron start to oscillate as a common system in the electron shell. The oscillation process provides an energy pumping of the surrounding CL space (in close proximity), which terminates with emission of two polarized gamma photons of 511 KeV, propagated in opposite direction. Note that there is not matter disappearance in this case and even the Newtonian mass is preserved, but the obtained particle is so small and neutral that it is practically undetectable. The obtained common helical structure is comprised of the electron shell and two halves of the positron shells. The opposite E-fields are locked in the proximity (by the SG field) and the final (quiet) system appears as a neutral.

3.17.2 $Ps1^3S_1$ triplet

This state is usually activated when a positive β particle from a radioactive decay begins to oscillate with a normal electron system. The positive β particle is a virtual particle (quasiparticle wave), possessing a positive charge of running EQs moving as a quantum wave. This type of wave does not have strong boundary conditions and behaves as an electrical charge. The radial dimension of this wave is a function of its energy. Smaller energy means larger radius. When the quasiparticle wave meets the electron system, their electrical fields interact and cause multiple repeatable oscillations of the electron - positron system. In a such process, an energy pumping effect of the surrounding CL space occurs. During the pumping process the energy of the quasiparticle, that have been distributed only among the positive EQs, redistributes between the positive and negative EQ'. As a result of this, the positive charge is gradually consumed, and its energy is converted to a pumped CL space energy. The latter finally is released as 3 gamma quants, if the Beta particle energy is less than 511 keV. The spectrum of 3 gamma emission is a continuous. Here one question arises: Why 3 gamma quants are emitted?

The explanation is the following:

The most energetic quantum wave is the first harmonic wave with energy of 511 keV. According to the boundary conditions, only subharmonics wave are possible. This condition put a limit on the spectrum continuity in the vicinity of the first harmonic. The quasiparticle wave, however, may posses any value of energy, that do not coincides with the subharmonics quantum conditions. **Such energy could not be presented as sum of two subharmonics quantum wave, but with sum of three subharmonics**. Consequently, the emission of 3 gamma wave is cause by the reaction of the CL space, due to its quantum constraints.

The described above process is valid for a vacuum or air conditions.When the electron is in solids, the process is modified. The process known as a positron thermalisation belongs to this category. It is discussed in §3.17.5

3.17.3 Ps $1^3S_1 - 2^3S_1$ state

This is a positronium that terminates with an emission of a single photon at 243 nm. One of the experiments in which the above state is activated is provided by Mills, Berko and Canter, (1975).

The transition $1^3S_1 - 2^3S_1$ is obtained by the following way. By moderation of Beta particles from radioactive decay of ^{58}Co, using MgO covered gold foil converter, slow positrons are obtained. These positrons strike MgO covered gold foil converter and then magnetically guided by 150 cm long curved solenoid, they strike a copper plate. The copper plate is faced to microwave cavity operating around 8860 MHz. When the microwave

(RF) is off, a Ps with a lifetime of 1.1 ms decays in 3 gamma photons. When it is on, emission at 243 nm is detected, in first, and after delay of 1.13 ms a 3 gamma photons are detected.

The explanation of the emission from the point of view of BSM is a following:

The particles obtained by the moderation process are hardware positrons.

Case A. The RF is off.

The slow positrons striking the copper plate are combined to oscillating pair of normal electron - free positron. The oscillation process invokes a CL space pumping and the external E-fields of the normal electron and the positron become gradually consumed. Approaching a neutral field and possessing kinetic energy, they escape easily from the copper plate and enter in the cavity. Here they continue to oscillate with partially lost energy. Their residual energy is lower than 511 keV, because part of it has been exhausted for escaping from the copper plate. As a result of this, the oscillation process terminates with emission of 3 gamma particles. From the mass point of view, the final system is consisted of two positrons inside of the electron shell with a total mass of 1.2 MeV. The positive E - filed of the positron is locked in proximity with the electron field and the particle appears as a neutral. Such particle is very difficult for detection.

Case B. The RF is on.

After the escaping of the the electron - positron system from the copper plate, the process in this case is different. The frequency of the RF field is suitable for creation of curved loops for both the electron and the positron. They are suitably folded to match the interaction between the moving charge and the magnetic field. As a result of this, both carriers, having still enough energy, do not move directly one to another. In this type of oscillations, supported by the combination of magnetic and RF field, quantum conditions are created, in which the carriers adjust their velocities to 13.6 eV and 3.4 eV. The quantum interaction with the CL space allows them to stay longer in this condition. At the same time, the started pumping process continuously degrades the quantum motion in the loops. At some point, they lose the motion in the orbits. Then the CL pumped energy escapes as a photon. The both carriers now become involved in direct interaction. After 1.1 microsecond pumping time, the oscillations are gradually suppressed and the accumulated pumped energy is emitted as 3 gamma photons. It is evident that the free positron does not have enough energy to expel the internal positron from the electron system. So the final system is again neutral, comprised of one electron shell and two positrons inside.

The RF frequency 8625 MHz appears as an optimal oscillation frequency of the loop. The loop, **however may contain a large number of serially connected first harmonics quantum loops.** This is easily verified by the corresponding period and the known velocity (corresponding to 13.6 eV and 3.4 eV). When approaching this frequency the emission efficiency for 243 nm photons is improved. This curve, reference by the authors as a line is shown in Fig. 3.24.

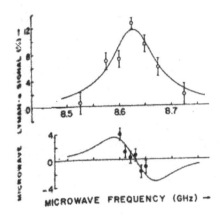

Fig. 3.24

The observed Lyman signal S (open circles) and logarithmic first-difference signal S* (solid circles) as a function of microwave frequency (Courtesy of A.P. Mills, Jr. et al.)

The subharmonic number of the carriers motion in the loops can be easily determined from the photon energy. <u>The only possible combination is:</u> $(13.6 - 3.4)/2 = 5.1$ eV. This means that:

- The quantum motion of the electron corresponds to its first SPM harmonic - optimal confined motion.

- The quantum motion of the free positron satisfies simultaneously two quantum conditions: a second subharmonic of SPM frequency, and a forth subharmonic of its proper frequency. **This is one additional confirmation, that the proper frequency of the free positron is twice the**

Compton frequency. If it was $3\nu_c$ as the internal positron, such combination could not be possible.

- The lattice pumping effect is a result of the energy difference between the two quantum loops, divided in two. The factor of two means 50% pumping efficiency, according to the pumping efficiency Eq. (3.21.c):. $\eta = m_e/(m_e + m_e)$.

The axial velocity of the electron is twice the axial velocity of the positron. This condition perhaps makes the lattice pumping effect possible. While the electron motion with optimal velocity is most stable, the positron motion is additionally stabilised by the mentioned above two quantum conditions.

In the same experiment, the dependence of the "line" from the RF power is investigated for different power levels. The increase of RF power causes the fitted "line" to move up, giving an impression of broadening, that the authors are not able to explain. According to BSM interpretation, this effect is completely logical. Its possible explanation is illustrated by Fig. 3.25 where simple illustrations of the single quantum loops are shown without pretending of their exact shape. In the provided experiment with RF frequency of 8625 MHz, the real orbits lengths should be equal to multiple number of first harmonics quantum loops.

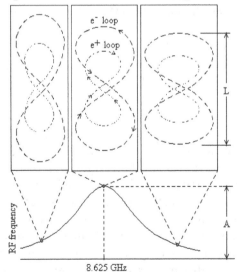

Fig. 3.25. Possible loops of electron and positron motion in $1^3S_1 - 2^3S_1$ state

The RF field in TM010 microwave cavity is parallel to the incident slow positron velocity. Then the formed positroniums also could be prefferentially aligned to this field. The RF frequency then will determine the oscillation period of the loop. The carrier velocity in the loop, is fixed by the quantum motion conditions. Then the RF frequency can tune the length of the loop. There is a constant magnetic field of 50 G, however, which fixes the curvature of the trace by the cyclotron radius. This means, that the loop is properly aligned as shown in the Fig. 3.25. Then the tuning of RF frequency could make the curve length L shorter or larger, but one optimal value of L should exists. This is the frequency of 8.625 GHz. At this frequency the amplitude of the A parameter of the lorenzian shape is estimated at 11.4%. At different RF power levels of 0.13 mW, 0.41 mW, and 2.0 mW., the corresponding A parameter is 5.3%, 12.3%, and 16.3%. The A parameter is increased, because the larger RF power in the cavity is possible to bias the week magnetic field, increasing in such way the range of L variation.

Summary:

- **The $1^3S_1 - 2^3S_1$ transition gives an indirect confirmation, that the proper frequency of the free positron is twice the Compton frequency.**
- **The pumping energy between quantum motions in loops with different velocities is equal to the carrier energy difference multiplied by the pumping efficiency.**

3.17.5 Positron "thermalisation"

In the process, known as a "positron thermalisation", a thin plate of proper metal cut at proper crystal plane is radiated by positive Beta particles (quasiparticle waves). The quasiparticle wave enters into oscillations with a free normal electron of the plate, forming an oscillating system. The CL space inside the sample, however, is different, than the free space (vacuum). Due to the influence of the proton's fields, and the motion of the formed oscillating system in CL space environment with stiffness gradient, the **internal positron may come out** at much smaller energy of the Betta particle. So a two possibilities may exist in this case:

- the thermalised beam is comprised of particle positrons (this option is more logical)

- the thermalised beam include both: particle positrons, and quasiparticle waves with reduced energies.

Chapter 4. Superconductive state of the matter.

4.0 CL space inside a solid body

The CL space of astronomical bodies and small bodies are discussed in Chapter 10. Here only some features of CL space inside a small body could be mentioned. Any solid body placed in the Earth gravitational field is immersed in the CL space of the Earth, which is in fact a CL space of the Milky way galaxy, modulated by the Earth gravitational field (this issue is discussed in Chapters 10 and 12). The atomic matter of the solid body may partially distort the Earth CL space inside of the solid body volume. The distortion is very weak, because at the level of the elementary particles the Newtonian type of gravitational field is quite smaller in comparison to the SG field between the nodes of the CL space structure.

The distortion in fact depends on the intrinsic matter density included in the volume of the solid body. The intrinsic matter density can be expressed by the **number of prisms** included in a **unite volume, using a properly selected length scale.** If using the parameters of the electron's structure (revealed in Chapter 3), the above highlighted parameters are defined. At the same time, the physical parameters of the electron structure as a single coil SOHS could be easily referenced to any helical structure, which is embedded in the elementary particles. The protons and neutrons differ only by their overall shape, since they have one and a same internal structures (comprised of helical structures). It is shown in Chapter 8 (and in the Atlas of Atomic Nuclear structures) that the protons and neutrons in the atomic nuclei are arrange in a strict order, while the valence protons have some limited freedom of their spatial positions. From the revealed order it becomes apparent that in a micro-scale range the density of the intrinsic matter exhibits a complex spatial gradient even inside a homogeneous solid body. Since the intrinsic matter density depends on the chosen scale, we may formulate the following scale ranges:

Case (1): a scale comparable to the proton core envelope thickness, which is equal to:

$$2(R_c + r_p) = 0.0074 \times 10^{-10} \quad (m)$$

Case (2): a scale comparable to the proton width, estimated as:

$$W_p = 0.195 \times 10^{-10} \quad (m)$$

Case (3): a scale comparable to the average internuclear distance in solids: 1 to 3 *Angstroms* $(1 Angstrom = 1 \times 10^{-10} \quad (m))$.

The scale parameters in cases (1) and (2) are determined in Chapter 6.

The average internuclear distance in case (3) depends on number of factors: a pure metal, an alloy or a chemical composition.

The superconductivity is known as a first and a second type.

(a) The pure first type of superconductivity appears in metals in a solid aggregate state

(b) The second type of superconductivity appears in solids of alloys and chemical compositions

The properties of both type superconductivity are well known. We will try to provide some physical explanation from a BSM point of view of the matter in a CL space environment.

The atoms in metals in a solid aggregation state are closer than the nonmetals in a same state. So the metals will exhibit a stronger modulation of the CL space in proximity to the nuclei when having in mind the SG forces between nuclei. The specific gravity of the element in a solid state could serve as a reference parameter of this type of modulation. At the same time, solids with a same or similar specific gravity may have different arrangement of the atoms in the crystal, for example: the structure of the metals and the alloys. While the atoms in the metal crystal are more uniformly spatially distributed, those in the alloys are not. This provides conditions for a stronger SG gradient inside the alloys, which means a stronger modulation of the CL space (or CL space spatial non uniformity).

Now if associating the both types of superconductivity with the CL space modulation inside the solid body we may distinguish two cases:

(a) The CL space modulation is not dependent on the particular element, so this type of modulation could be excluded from the superconductivity considerations.

(b) The CL space modulation may depend on the number of the hadrons (proton or neutron) in the nucleus and their arrangement. **Consequently**

the case (b) may determine which pure metal may exhibit a first type superconductivity.

(c) The CL space modulation depends on the nonuniformity in the crystal structure configuration. The nonuniformity is smaller for pure metals and larger for alloys, doped metals or chemical compositions. **Consequently the case (c) might be involved in the conditions for second type superconductivity.**

The proof of the above made conclusions will become apparent from the analysis presented in this chapter.

4.1. Normal and superconductive operational mode of CL node.

The superconductivity, according to BSM, is a state of the solid matter, directly related to the dynamical type of the zero point energy (ZPE-D) of the CL space inside of the solid body. When the conductor is cooled to a very low temperature, the CL domains inside its volume may get lower ZPE-D than the normal one. This condition is dependable on the internal structure of the material. When approaching the absolute zero we see that different metals have different superconductive temperature. Some metals as gold, silver, copper, do not exhibit superconductivity. The nonuniformity of CL space in the solids means that CL structure exhibits stiffness gradients.

The CL nodes of domains with different stiffness have different value of their return forces and consequently different energy wells. This difference affects also their proper resonance frequency. At normal temperature all domains have a normal ZPE

When the body temperature drops to a very low level the domains with a lower stiffness get a lower value of their ZPE-D in comparison to domains with a higher stiffness. This is much more relevant for the second type or high temperature superconductors, for which the structural mass gradient is much larger, than for the pure metals. The stiffness is very sensitive to the node distance. Small differences in the node distance (or node density) leads to larger stiffness differences, due to the inverse cubic dependence of the SG forces on the CL node distance.

The return force dependence and the energy diagram of the CL node, was presented in Chapter 2, but the <u>superconductive (**SC**) state</u> was not discussed in details. Here we present the same diagram, while emphasizing on the SC state.

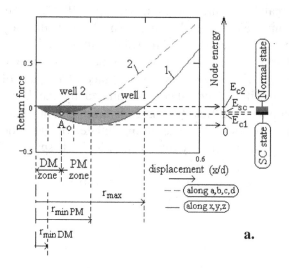

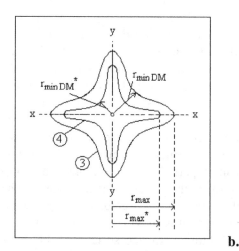

Fig. 4.1
a. CL node return forces and CL energy diagram as a function of displacement,
b. trace projection of the CL node SPM cycle for two different values of CL node energy

Figure 4.5. a. shows the return forces and the energy diagram as a function of the node displacement with identification of the working energy ranges for a normal and a SC state. Fig. 4.1.b shows a 2D projection of the CL node SPM cycle. The node displacement corresponding to the bumps of the SPM quasisphere is denoted as r_{max}, while the displacement corresponding to the dimples - by r_{min}.

The full well capacity of the node energy is artificially separated into *energy well 1* and *energy well 2*. The *energy well 1* is spatially allocated around *xyz* axes, while the *energy well 2* is allocated around *abcd* axes. The point A_0 is the lowest point of the *energy well 1*. The energy scale shown in the right side, is not linear, but showing only the order of the energy levels. The diagram indicates the displacements, corresponding to the r_{max} and r_{min} of the node trace. The radius r_{max} is determined by the curve 1, while r_{min} - by curve 2. We see that both radii are dependent on the level of ZPE down to the critical point A_0, which corresponds to the critical energy level E_{c1}. The level E_{c2} corresponds to a normal ZPE. The radius r_{max} is always defined by the curve 1 for the case of normal and subnormal energy. The same rule, however, is not valid for r_{min}. When the ZPE begins to decrease from E_{c2}, the radius r_{min} initially is determined by the sector of the curve 2 lying in the right side of p. A_0. However, when the energy approaches E_{c1}, the smaller radius may flip to the left side of curve 1. This flipping may occur as a result of an external field influence.

The flipping of r_{min} at a very low energy level may flip the phase of the SPM frequency of the affected domain by a value of π. Then the magnetic filed of this domain could act repulsively to the external magnetic field.

The mentioned effect has direct impact on the external magnetic filed, which tries to penetrate the sample. For r_{min} operated in the left side of p. A_0, the conductor material appears diamagnetic, while in the right side of A_0 it may still appear paramagnetic.

In order to distinguish the node operation of this two cases related with r_{min}, we denote them as a **DM** (diamagnetic) and **PM** (paramagnetic) type of operation. Therefore, p. A_0 separates the diagram in two zones: a **DM zone** and a **PM zone**.

The separation of the node operation in a normal and a superconductor state is a function of the CL node energy. In fact the normal state should correspond only to one value of the energy level denoted as E_{c2}. The SC state, however, does not begin just below E_{c2}, but is closer to E_{c1}. So between the normal and the SC state, some transition zone with finite energy range should exist. The width of this zone is dependable to some extent on the conductor crystal structure and the atomic mass gradient.

The diamagnetic state is a result of not stabilised resonance frequency. In such case both, the NRM and the SPM vectors are affected. The magnetic filed is very severe disturbed, because it depends on the SPM phase synchronization.

Let us see how the trace shape changes when the ZPE-D decreases. When approaching SC energy point, the decrease of r_{min} becomes faster than the r_{max}. This changes the shape of the trace from 3 to 4 as shown in the Fig. 4.1. At the same time, the NRM and SPM quasispheres are also affected. Their bumps become sharper. The size of the EQ, however, appears restricted by the lower ZPE. As a result of this change, the electrical field around the FOHS may appear stronger, but localised in a smaller space.

Let us see the SPM behaviour in a CL space domain, when the ZPE of this domain decreases. At E_{c2} level the SPM frequency is stabilised by the own internal resonance frequency stabilization mechanism. Below this point, this mechanism is lost, but the stabilization could be supported by the external magnetic field, which is still able to penetrate. This becomes apparent in the FQHE experiments, where an induced quantum effect takes place (BSM interpretation). The point E_{c1} corresponds to a bottom critical energy level. **Above this level the spatial rotation of the NRM vector covers the solid angle of 4π, so the SPM vector is defined. Below the E_{c1} level, however, the NRM vector is not able to cover the full angle of 4π. In this case the SPM vector is not defined for three-dimensional space, but it could still exist in a two-dimensional one.**

The following conclusion could be made:

- **The point A_0 corresponding to level E_{c1} is very characteristic point for switching the SPM vector behaviour.**

Some signatures of the SPM vectors behaviour in two-dimensional space, including the characteristic point A_0 from the energy diagram, are apparent in the Quantum Hall Effect (QHE) experiments, in which very thin samples are put in cryogenic temperature close to the absolute zero.

The SC energy zone, E_{sc} is between levels E_{c1} and E_{c2} of the CL node energy diagram shown in Fig. 4.1.a.

$E_{C1} < E_{SC} < E_{C2}$

In order to explain the superconductivity we must consider the CL domains located near the surface and operated in a PM mode. Only in a such mode the SC carrier could interact with the external CL space and create a strong magnetic filed. Then the following question arises: **What kind of interaction is able to keep the (internal) CL nodes located near the surface of the superconductor in PM type of operation? The answer is: The interaction of the strong magnetic field outside of the superconductor (where the SPM vector is stabilised) with the oscillating SC carriers. The expelled magnetic lines partially penetrate below the surface of the superconductor, where PM zone is created. This is the penetration depth of the superconductor.**

The SC carriers are modified versions of the electron system. They are discussed in §4.2.

The penetration depth has an exponential shape, because it depends on a temperature gradient, between the external and internal CL space domains. The electron and its modified SC configuration, has own self energy in their internal RL(T). So they are automatically attracted in the surface region, where they can exhibit a quantum motion. They are able to keep their normal level of self energy by participating in a quantum motion under control of the external stabilised SPM frequency. The external magnetic field could not pass through the bulk region of the semiconductor, occupied mostly by domains operated in a DM mode. The SPM vectors of these domains (whose frequency is not stabilized) "bounce" the penetrating magnetic lines. This is the **Messner effect.** If applying, however, external filed above some **critical level**, the SC state is destroyed. **In this case the CL nodes, which have operated in the DM zone, forcibly are pressed to operate in the PM zone. The mediator in this interaction is performed by the SC carriers. They are able to sense the external SMP vector by their oscillations and their affected motion can interact stronger with the CL domains.** If the external field is removed the opposite transition takes place, because the bulk of the superconductor has a lower ZPE.

If we consider, that the energies in both energy wells (well 1 and well 2) are reduced proportionally, then we may accept that N_{RQ} parameter is not affected. This means that the change of the NRM frequency will be in the same direction as the SPM frequency change. We don't know in which direction the SPM frequency change when the temperature change, but we may use the analogy with the conical pendulum of circular type. The angular frequency of such pendulum is equal to $(g/h)^{-1/2}$, where g - is the Earth acceleration and h is the pendulum height. From Fig. 4.1 we see that the decreasing of the CL node energy is equivalent to decreasing the amplitude of such pendulum, which means an increase of h. Therefore, the pendulum frequency decreases when the amplitude decreases. We cannot expect that the CL node distance of the sample changes with temperature, otherwise the mass of the cooled body must depend on the temperature, which is never observed. Then the light velocity (determined by the node distance for one NRM cycle) will change, which means that the sample refractive index will be changed. From this analogy follows that the SPM frequency of the sample will decrease when the absolute temperature is decreasing. The relation may not be linear proportional, so we will use an arrow for direction of parameter change with the temperature T.

$$T\backslash\!\!/ \quad \nu^*_{SPM} \quad \backslash\!\!/ \quad c^*_{SPM} \quad \backslash\!\!/ \quad n_i/\!\!\backslash \qquad (4.1)$$

where: ν^*_{SPM}, c^*_{SPM} and n_i are respectively the SPM frequency, light velocity and the refractive index of the sample at cryogenic temperature.

From the analysis of the QHE experiments we will see, that the feedback leading to the NRM frequency stabilization is negative. It is the same type of feedback operated in a normal state CL space (possessing a normal ZPE-D) Consequently, the feedback leading to a NRM frequency stabilization is of negative type and corresponds to a PM zone, the zone at the right side of point A_0 (Fig. 4.1.a). The slop sign is obviously related to the sign of the feedback, which provides the NRM frequency stabilization. The slop of curve 2 from the left side of A_0 changes the sign. Then the feedback for the nodes operated in the DM zone will become positive. This conclusion is supported by the results from the fractional quantum Hall effect experiments.

4.2 The electron system configurations in superconductive environments

4.2.1 Electron system in SC state environment. Carriers in SC state of the matter.

The very distinctive features of the electron from the proton and neutron is that it is composed of three separate helical structures and possesses internal energy, which is like an energy buffer for its oscillations. We may expect that in domains of normal ZPE the electron always oscillates with small amplitudes. The oscillations create a small alternative magnetic field. This field is directly related to the quantum motion of the electron. The same field, also, assures the symmetrical oscillations of the internal structures in respect to the external one. In other words, this field keeps the internal positron inside the electron's shell. Then we may expect that if the magnetic field is disturbed as a result of abnormal CL space conditions, the electron system could be reconfigured. <u>The most probable reconfiguration is the exiting of the positron out of the electron shell.</u> In the DM zone of SC state, however, the electrical field is stronger and the exiting positron may attach itself externally to the electron shell.

Another mechanism may also favour the exiting of the positron. The normal electron system contains self energy in its internal RL(T)'s. This energy keeps a proper frequencies of the electron system (mostly the first one which is equal to the Compton's frequency) for a finite time. At very low temperature of the superconductor material, the ZPE gradients between domains with different node density are increased. Then small differences between the SPM frequencies of different neighbouring domain may take place. If the moving electron is forced to pass through such domains (by external fields) it may loose the quantum synchronization with the SPM vectors of these domains. Then an effect of low frequency phase difference may occur between the electron proper oscillations and the SPM frequencies of these domains. At large phase biasing, the positron may occur outside of the electron shell. When the oscillations change from small to large amplitudes, the proper frequency of the "positron-central core" system changes respectively from $3v_c$ to $2v_c$ (this is confirmed by

the FQHE experiment analysis) and this may also contribute to above underlined effect. Such process may happen, when the electron is in domains of SC state of matter and operated in PM zone of the energy diagram, shown in Fig. 4.1.a. The conditions for such phase biasing leading to exiting of the positron from electron system are shown in Fig. 9.6

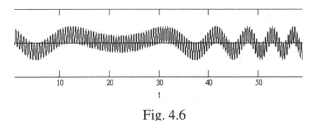

Fig. 4.6

Phase biasing between the electron's oscillations and the SPM frequency of a CL node domain with a lower ZPE energy leading to exiting of the positron out of the electron's shell (out of the negative helical structure of e⁻)

The black horizontal line shows the central position of the oscillating positron. We see, that there is number of points at which the positron appears almost outside of the electron's shell. The simulation was provided, by using two close frequencies, one stable for the electron proper frequency and another one with a slight frequency sweep, simulating a motion of the electron in a CL domain having a gradient of the CL node proper resonance frequency.

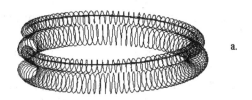

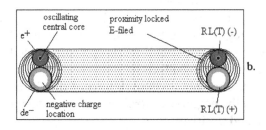

Fig. 4.7. Superconducting electron system and the external proximity locked field. A small gap (not shown) exists between both structures (see the considerations in §6.4.3)

　　　　　　4-5

Fig. 4.7.a, and b. shows respectively the configuration of the SC electron and the proximity locked external E-field. In fact, the two helical structures (positive and negative) could not touch each other, due to the different handedness of the SG fields generated by their internal RL(T) structures. This feature is confirmed in the analysis provided in Chapter 6.

The superconductors of first type are usually metals. However, heavy metals like gold, silver and copper do not exhibit superconductivity. The reason for this is that their crystal structure is pretty uniform, and they have a larger hadron (protons and neutrons) density, which means a larger Intrinsic Matter density. In such conditions, ZPE gradient is relatively small. From the other side, some compounds of heavy and not heavy metals may have CL domains with a larger difference between their stiffness. Then such domains may get a larger ZPE gradient in a comparatively high (in respect to the low temperature superconductivity) temperature. In this conditions, some domains might be in a normal state, while others in a SC state. In this case, the superconductor will have a channel structure. As a result of this, the resistance does not fall so sharply. The Messner effect, however could work even at this conditions, because the magnetic lines need closed paths in order to pass through. A II type superconductors are characterised by such features.

We may conclude that: **The SPM frequency gradient in a CL space domain is an important condition for appearance of SC electrons.**

The SC electron distinctive features from the normal electron are the following:

a. The external electrical field is locked in a proximity. In the far field the SC electron appears as a neutral particle.

b. The negative charge is hidden in the internal region of the twisted mono rectangular lattice of the electron shell. It has openings to the CL space only from both ends.

c. The only oscillation part of the SC electron is the negative central core of the externally attached positron

When analysing the FQHE experiments we will see, how the SC electron gives its signature as fractional charges of 1/3, 2/3, 4/3, 5/3. The conclusion that the charge appears hidden in the far field

is supported also by some experimental observations. J.D.F. Franklin et al. (1995), for example, surprisingly discover in their QHE experiment that the quasiparticles can tunnel through a barrier.

The SG mass of the positron with its internal lattice appears hidden, when it is inside the electron shell, because the lack of coupling between their SG(CP) lattices, due to the oscillations. Once the positron is external and in a proximity to the electron shell, the motion is absent and the SG(CP) forces are active. They force the E-field lines to get closer and to be connected in proximity. Despite the different external shells radii, the E-fields of the electron and positron are exactly equivalent, because the angular density of their field lines is equivalent. (The Intrinsic Matter difference between the right and left handed prisms is compensated by the CL space). In this conditions the SC electron appears as a neutral in the far field. The hole inside the electron RL(T), however is open and available for the CL nodes. **As a result of this, the existing before the reconfiguration negative charge is transferred to this hole.** The location of this charge is shown in Fig. 4.7.

There are few reasons for the conclusion that the negative charge of the electron is transferred into the hole. The first one comes from the analysis of the experimental behaviour of the electron during the refurbishing process. If the charge happens to disappear, the CL space should react with an emission of a virtual particle as in the case of proton - neutron conversion (in which case the reaction of the CL space creates a β particle). **Effect of** β particle emission, however is not detected when the matter approaches a superconductive state. The second reason comes from the fact that the SC electron is able to be controlled by the electrical potential in the superconductor and it creates a magnetic field outside of the superconductor. Consequently, it still possesses a guiding feature from the internal negative charge, which is open to the external E-field from both sides of the hole.

In the junction between the superconductor and conductor, the SC electron undergoes refurbishing to a normal electron. **This is the reason for the appearance of the Josephson resistance.** If the charge was missing, the CL space should react as in the neutron - proton conversion, by emission of a β particle. **Such effect is not observed.**

There is a direct observational evidence that the SC electron exhibits a tunnelling effect, passing through a barrier (see quasiparticle tunnelling through a barrier observed by J .D. F. Franklin et all., 1995).

The motion behaviour of the SC electron is different than the normal one. Its interaction with the atoms is greatly reduced. At the same time, it still has two important features assuring its quantum type of motion: The first one is the guiding feature of the hidden negative charge, open in both ends of the electron shell. The second one is the oscillating feature of the central core, which assures the quantum motion of the whole system while interacting with the magnetic field of the external space. In SC state, the SC electron naturally prefers to move in zones near the surface of the superconductor, where the central core interacts with the external magnetic field, which possesses a normal ZPE. We may expect, that the ZPE of this zone of superconductor falls exponentially from the surface to the bulk. Such zone really exists and it is well known under name **a penetration depth.** The internal energy of the positron RL(T) is preserved and serves to support the oscillation of the central core. During a full cycle of the oscillation, both ends of the negative core move periodically inside the electron shell hole, occupied now by the negative charge (see Fig. 4.7). Therefore, the conditions for the central core motion are similar as in the normal electron. **Then we may expect, that the proper frequency of SC electron is the same as the positron-core system in the normal electron.** We will see in the next paragraphs, that this is confirmed by the fractional charge experiments.

The SC electron is able to keep its integrity not only in domains in SC state. It can be temporally stable in domains possessing a normal ZPE. Then the motion of the SC electron in a normal ZPE domain will exhibit some resistance, which has to be overcame before the SC electron converts back to a normal electron. This conclusion is confirmed by the experimental data (BSM interpretation). Figure 4.8. shows the temperature dependence of the resistivity of II type superconductor Ba-La-Cu-O for different concentrations of Ba and La with a measurable parameter - resistivity, ρ.

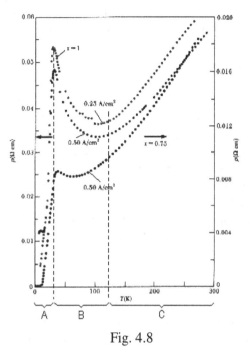

Fig. 4.8

Temperature dependence of the resistivity, ρ, for IInd type superconductor based on Ba-La-Cu-O

The division of the temperature scale into ranges A, B, C on Fig. 4.8 is made by BSM interpretation. The range A provides temperature conditions for the CL domains in SC energy state. In the range B, the CL domains approach the normal ZPE level, but the conversion of the SC electron to a normal one is not completed. The SC electrons may temporally survive, when passing through some domains with a normal ZPE. However, their oscillation properties are not optimised for these zones, so they may feel an increased resistance. At the same time, the abundance of the normal electron is small, while the SC electrons are still the predominating fraction of the total electron gas. As a result of this, the measured resistance arises with the temperature. For some concentrations of Ba and La, the domains with a near critical ZPE exist at higher temperature and some SC electrons are still attracted to them. Evidently the balance between the normal domains, the SC domains and the normal electrons leads to a smooth resistance change in the range B. In the temperature range C, all CL domains are with a normal ZPE. The resistance of the SC electron for the back conversion to a normal electron comes from the SG(CP) forces between the degenerated electron and the positron. These forces could be compensated only by the increase of ZPE of the CL nodes. Then more and more CL domains operate in a PM zone. The increased ZPE also removes the restriction on the eccentricity of the EQ quasispheres. In such conditions the electrical and magnetic interactions can overcome the SG(CP) forces. All this factors provide a <u>hysteresis effect</u> in the direction of SC to normal electron conversion process. The hysteresis effect is an important factor, keeping the stability of the SC electron in the SC state. Without such effect the surviving of the SC electron, especially for the IInd type of superconductors would not be possible.

- **The integrity and stability of the SC electron in SC state of the matter is kept by an hysteresis effect.**

The carriers of the SC state are not comprised only of single SC electrons. The BSM analysis of the Fractional Quantum Hall experiments (FQHE) unveils the signatures of another configurations of the electron system: - **stacked SC electrons**. A configuration of two stacked SC electrons is shown in Fig 4.9.

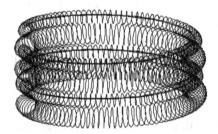

Fig. 4.9

Two SC electrons stacked together. The small gaps between the stacked helical structures are not shown (see the considerations in §6.4.3)

The stacked SC system also appears as externally neutral. But it has one distinctive feature in comparison to the single SC electron. In the single SC electron, only one end of the central core oscillates in a domain of a hidden electrical charge. The core and the hidden charges are negative, so the partially exciting core is repelled back. This keeps the stability of such oscillating system. For a stacked SC system comprised of two SC electrons one of the hidden negative charges appears between the two central cores. For this reason, the two cores oscillate synchronously and the system exhibits one proper frequency. However this frequency is different than the single SC electron, due to the different oscillation conditions of the cores. In a similar way three or more stacked SC electrons are possible. They have different proper frequencies the signatures of which appear in the FQHE experiments, when analysed by BSM. In a system of **n** stacked SC electrons, **(n-1)** positrons are between electron shells. So we may expect that the proper frequency of the stacked SC electron changes with **n**.

The single and stacked SC electrons are the real carriers in superconductivity. One question arises: How the SC electron creates so strong external magnetic field, while its charge is hidden? The answer is:

When in motion, the proximity locked field of the single or multiple stacked SC electron is able to generate a magnetic field. This feature is similar to the case of moving neutron. The neutron has a proximity E-filed (generated by RL(T) of its external FOHS) which is in the near field by the SG forces, but generates a magnetic moment when the neuron is in motion. The SPM frequency in CL domains with a low ZPE is different than in CL domains with a normal ZPE. The SC electron quantum motion tends to follow the SMP frequency of the local domains through which it moves. Simultaneously its oscillations are influenced by the external magnetic field via direct frequency synchronization. This is because the magnetic field generated by the single SC electron (which secondary proper frequency is unchanged) interacts with the external magnetic field. This field partially penetrates below the surface of the superconductor, where the SPM vector has a gradient. As a result of the interaction of the SC electron with the strong external magnetic field it tends to escape in a zone with SPM gradient closer to the external CL space - the surface of the superconductor of Ist type. This zone is a transition one between the internal CL space with a different SMP frequency and the external one with the normal SPM frequency. The transition zone corresponds to a penetration depth - a well known parameter of the superconductor of Ist type.

The inertia of the SC electron is strongly dependent on the intensity of its proximity field. In a superconductive state of the matter **the "energy hungry" CL domains do not provide resistance but rather exhibit a complimentary behaviour (oscillating feature) which permits the SC electrons to move almost without resistance.**

Having in mind that the resistance that SC electrons exhibit is very low, the driving voltage is also low. This means that the SC electrons may also operate at large subharmonic numbers. In this case their magnetic radius is larger (as for the normal electron discussed in Chapter 2). So it is quite reasonable to expect that the SC electrons have common phase synchronization. This feature also explains the appearance of the magnetic field outside of the volume of the superconductor, while the SC carriers are at some finite distance.

4.2.2 Proper frequencies of the stacked SC electrons.

In the previous paragraph the reason for the stacked SC electrons proper frequency change was mentioned. The negative central core of any consecutive combination of stacked SC electron will get an additional pushing force from the interaction with the hidden negative charge. This will decrease the oscillation period, which means an increase of the proper frequency. It is reasonable to expect that the proper frequency of the system will increase with the number. Here some theoretical considerations will be presented.

Let us make an analogy with a Quantum harmonic oscillator. Its angular frequency is given by the expression

$$\omega = \sqrt{k/m} \qquad (4.2)$$

where:

k - [N/m] - is a force constant

m [kg] - is a mass of oscillating body

The combined negative core of the stacked SC electrons oscillates in environment of the combined stacked internal negative charge. The number of combined charges increases with the stack number. Then we may expect an increase of the spring constant of the common oscillating negative cores, which will means an increase of the oscillating frequency with the stack number. However, there are some specific features arising from the specifics of the SG field, from one hand

and the configuration of the electron structure, from the other.

(a) in a single CS electron, only one end of the central negative core oscillates in a repulsive negative field of the hidden negative charge (inside the FOHS of the electron).

(b) in 2, 3 or more SC electrons, both ends of the central negative core oscillate in repulsive negative fields.

(c) the radial SG forces focused on the oscillating central core are of opposite handedness and are also well balanced, due to the radial symmetry.

Features (a) and (b) mean that the trend of the proper frequency change for a single SC electron will be different than for *i* stacked electron, where $i > 1$.

Feature (c) means that the SG mass of the stacked central cores will not influence the trend of the proper frequency change for n-stacked CS electrons.

Now let consider the force constant *k* used in Eq. (4.2). This equation is valid for a Newtonian gravitation $F = (Gm_1m_2)/r^2$. The force constant regarded as a first derivative is

$$k = 2(Gm_1m_2)/r^3 = 2(F/r) \quad \text{[N/m]} \qquad (4.3)$$

For stacked SC electrons we must use the inverse cubic SG law. The analogous force constant for the SG law will be:

$$k_{IG} = 3(G_0 m_{o1} m_{o2})/r^4 = 3F_{IG}/r^2 \qquad (4.4)$$

F_{IG} changes with the distance much faster than F. When comparing (4.3) and (4.4), we may accept that $F_{IG} \sim F^2$, from which follows that $k_{IG} \sim k^2$

Then by analogy with Eq. (4.2) the proper frequency of a SC carrier made by *i* stacked SC electrons will be proportional to i according to the expression:

$$\nu_{pr} \sim \sqrt{k_{IG}} \sim k \sim i$$

Let us use for a reference the proper frequency of the single SC electron (equal to the secondary proper frequency of the normal electron), for which only one end of the internal negative core oscillates in a domain of a hidden negative charge. For any additional stacked SC electron both ends oscillate in such domains, consequently we may have a multiplying factor of 2 for $i > 1$. Since one single SC electron is embedded in every SC electrons with $i > 1$, the combine internal core will oscillate with a single frequency that could be a combination of the single SC electron frequency and the frequency of the stacked fraction for $i > 1$. Having in mind that the combined internal negative core oscillates in a highly non-linear environments (the return forces are inverse proportional to the cube of displacement (SG law) permits to expect that the obtained proper frequency is a combination of the above mentioned two frequencies. Since the oscillation energy increases with the number of stack, we may accept that the resultant frequency is a sum of the two frequencies. If expressing it as a normalized to the proper frequency of a single SC electron, it will appear as a dimensionless number, denoted as $\tilde{v}_{pr}$

$$\tilde{v}_{pr} = 2i + 1 \qquad (4.5)$$

In the analysis of FQHE it will become apparent that the proper frequency of the single SC electron is three times the first proper frequency of the normal electron, while the latter is equal to the SPM frequency of the CL nodes with a normal ZPE-D. Compton frequency (discussed in Chapter 3). Then Eq. (4.5) can be expressed as a ratio between the proper frequency of the SC electrons and the SPM frequency of the normal CL space.

$$v_{pr}/v_{SPM} = 2i + 1 \qquad (4.6)$$

where: i - is a number of the stacked single SC electrons in the SC carrier

The ratio given by Eq. (4.6) is inverse proportional to the parameter known as a filling factor, introduced by Landau. It is usually denoted as $1/v$ but so far it has not been considered as a parameter related to some frequency. The relation between the filling factor and parameters of the superconductive carriers, revealed by BSM as stacked SC electrons is shown in Table 4.1.

Table 4.1

Filling factor	Carrier type	v_{pr}/v_{SPM}
1/3 - single SC electron		3
1/5 - two stacked SC electrons		5
1/7 - three stacked SC electrons		7
1/9 - four stacked SC electrons		9
1/11 - five stacked SC electrons		11

where: v_{pr} is the proper frequency of the stacked SC electron and v_{SPM} is the SPM frequency of the CL node with a normal ZPE of dynamic type.

4.3. Integer and Fractional Quantum Hall Effect experiments.

All experiments of IQHE and FQHE experiments are performed at very low temperature, close to the absolute zero and by applying a strong magnetic field. The sample is usually a very thin GaAs/AlGaAs heterostructure (denoted as a two-dimensional), grown in thin layers atop a suitable substrate. The applied magnetic field is normal to the plane of the two-dimensional sample. The typical experimental setup and the measured parameters are illustrated by Fig. 4.10.

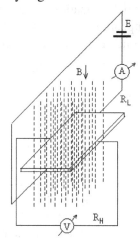

Fig. 4.10
A typical experimental setup for IQHE and FQHE

The charge carriers are driven by the electrical field applied in X direction. Due to their interaction with the magnetic field, they are deflected and a potential is accumulated in the Y direction. The accumulated potential continues until the created electrostatic force balances the magnetic force on the charge carriers. The equation of balance is: $qv_d B = q E_H$. Then the generated Hall voltage is:

$$U_H = E_H d = v_d B d \qquad (4.7)$$

where: E_H is the generated Hall potential per unit length, d - is the width of the strap, B - is the magnetic field, v_d - is the drift velocity.

Using the kinetic theory, the drift velocity is obtainable. For the semiconductor materials it depends on the temperature.

By changing the strength of the magnetic field, the Hall voltage and the longitudinal resistance are changed. The parameters of interest are the Hall resistance R_H and the longitudinal resistance R_L. They are obtainable if the carrier concentration in the sample is known (a routine technique).

The classical Hall resistance is given by the simple equation,

$$\frac{U_H}{I_d} = R = \frac{B}{nqt} \qquad (4.8)$$

where: q is a unit charge, n - is the number of carriers, and t - is the thickness of the sample.

The sample thickness is usually known, so the above equation can be normalized to t. The number of carriers, n, for the semiconductors depends on the temperature, but for a constant temperature Eq. (4.7) gives a linear dependence of the Hall resistance (R) on the magnetic induction (B).

The quantum Hall effect (QHE) is discovered by Klaus and Klitzing in 1980 by observing a two-dimensional system at very low temperature and strong magnetic field. In integer and fractional QHE experiments with very thin samples at very low temperature, the Hall resistance dependence on the magnetic field B departs from linearity in some particular points. At these points the Hall resistance appears as plateau. The plateaus are centred about integer or value of the Hall resistance, which is named a Von Klitzing constant, R_K, and appears to be exactly equal to the ratio between the Planck's constant and the square of elementary charge.

$$R_K = \frac{h}{q^2} = 25813 \ \Omega \qquad (4.9)$$

Therefore, the two-dimensional sample exhibits specific conductivity given by

$$\rho = I\left(\frac{e^2}{h}\right), \qquad (4.9.a)$$
where I is a small integer.

Additional experiments of similar arrangement showed plateaus not only for integer, but for fractional values of I, so the effect was referenced as a Fractional Quantum Hall Effect (FQHE). FQHE was observed, as a big surprise, in number of experiments. In order to explain the effect, the concept of degeneracy of the Lanndau level is used. This degeneracy defines a "filling factor" denoted as ν.

$$\nu = nh/qB \qquad (4.9.b)$$

Then the Hall resistance is modified to the form

$$R_H = \frac{h}{q^2}\frac{1}{\nu} \qquad (4.9.c)$$

Fig. 4.11 shows the observed Hall and longitudinal resistance for integer and fractional quantum Hall effect. (Summury article by J. P. Eisenstein and H. L. Stormer (1990).

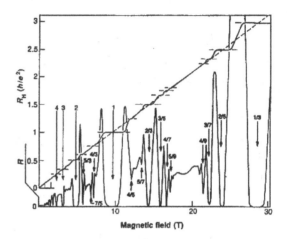

Fig. 4.11

Integer and Fractional quantum Hall effect (courtesy of J. P. Eisenstein and H. L. Stormer (1990)) RH - Hall resistance; R - longitudinal resistance (curve with peaks)

From Fig. 4.11 we see, that the Hall resistance R_H obtains plateau for $R_H = R_K$ and for other values, which could be interpreted as integer or fractional charges. They correspond to the Landau filling factors, denoted as ν and defined by Eq. (4.9.c).

Figure 4.12 shows some other data by D. Tsui et all (1982), where ρ_{xy} denotes the Hall resistance and ρ_{xx} the longitudinal resistance, respectively.

The plots show, also, the temperature dependence on both resistances that is very useful information for the BSM interpretation.

For IQHE the observed plateaus correspond to the filling factors ν, which are small integers. For FQHE, the plateaus appear for filling factors of: 1/3, 2/3, 4/3, 1/5, 4/5, 1/7 an so on. One of the existing and accepted so far theory is proposed by Robert B. Laughlin. The Laughlin's theory explains the integer and fractional charges as a manybody wavefunction using Landau levels.

However, besides the fractional values shown above, additional values are also observed: 5/2, 9/2, 11/2 (M.P. Lilly et all., 1998). They cannot be explained by the Laughlin quasiparticle theory.

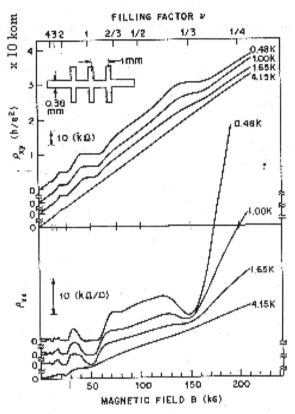

Fig. 4.12
QHE and FQHE (courtesy by D. Tsui et all.)

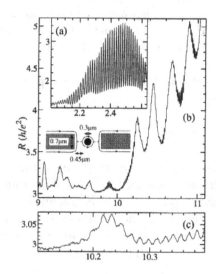

Fig. 4.13
Aharonov-Bohm oscillations for integer and fractional QHE (J.D.F. Franklin et all., (1995)

The accepted from some authors explanation of the Aharonov-Bohm oscillation, as a charge spin separation of the electron in hollon and spinon is also very controversial.

The surprising behaviour of Hall resistance at very low temperature and magnetic field, are subject of extensive and controversial discussions. At temperature below 100 mK and magnetic field approaching zero, the Hall voltage shows large departures from the classical expected value. Figure 4.14 shows such observations, provided by C. Ford et all. (1988).

Some other strange phenomena are observed by M. Lilly et all. (1998). Investigating the quasiparticle effects for filling factors of 5/2, 7/2, 9/2, 11/2, 13/2, they observed strong peaks in the longitudinal resistance with not smooth peak top. Then by simply changing the direction of the current through the sample, they observed strong anisotropy of the magnitude of the peaks. Such phenomena has been briefly mentioned by H. L. Stormer et all. (1993) as a puzzling behaviour.

In 1997 two groups, Israeli (R. de-Picciotto et all. (1997)) and French (L. Saminadayar et all, (1997)) reported a shot noise observation corresponding to $1/3e^-$ fractional charges in two-dimensional structures at temperature near to absolute zero. The observed shot noise also puts a doubt about the quasiparticle nature of $1/3e^-$ charge.

Lately in some experiments, Aharonov-Bohm oscillations are clearly observed in the longitudinal resistance for integer and fractional charges. Figure 4.13 shows such data, provided by J. D. F. Franklin et all. 1995. The authors, also, **discovered, that the quasiparticles can tunnel through a barrier**

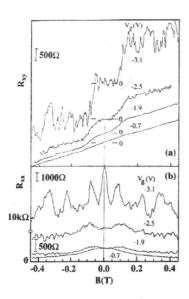

Fig. 4.14
The quenching of the Hall effect near B=0 at T,100 mK
(C. Ford et all. (1988)

When the FQHE has been examined using a method of surface-acoustic-wave propagation, it is observed a strange anomaly of attenuation at $\nu = 1/2$ (E. L. Willet et all. (1990)). The observed phenomenon appeared only at very low temperatures and disappeared for temperatures above 700 mK.

Experiments related to the fractional Landau levels but using the cyclotron behaviour of the "quasiparticles" has been provided by Kennedy et all. (1977). They observed a shift in the cyclotron resonance frequency, concomitant with a drastic line width narrowing.

The presented above data are small fraction of the experiments in this area. The BSM theory provides a new explanation for all of the above mentioned phenomena.

4.4 QHE and FQHE experiments as examples of active control of the light velocity in the sample

The analysis of the QHE and FQHE from the point of view of BSM concept about space leads to very consistent results if accepting that the electron preserves its internal proper frequency, ν_c, (equal to the Compton's frequency in a normal CL space) in a CL space domain with a lower ZPE-D.

4.4.1 General considerations

The quantum Hall effect according to the BSM is very useful phenomenon for investigation of the CL node dynamics at lower ZPE. Our goal is to understand the CL node behaviour at low ZPE. The analysis will provide a valuable information about:

- the operation of the SPM vector
- the quantum conditions of CL space
- the oscillation properties of the oscillating electron system
- the carriers of the superconductive state of the matter
- the refractive index properties of the super-conductors
- the possibility to invoke quantum states and to control the refractive index in the 2D sample (light velocity) in discrete levels

The environment conditions of QHE experiments are very important factors. The most typical conditions at which the experiments are provided are following:

- The sample is at very low (cryogenic) temperature, close to the absolute zero
- The sample is a two-dimensional, i. e. it contains deposited layer with a very small thickness (in the QHE experiments the samples are usually referred as 2D structures)
- A very strong magnetic field is applied perpendicular to the sample plane

The plane orientation of the layers of 2D sample, in respect to the magnetic field and direction of carrier motion, enhances the condition of their interaction with the magnetic field. Due to the thin multilayer configuration, the refractive index in the layer cross section is not uniform. This reduces the strength of the quantum effect based on the boundary condition of magnetic quasispheres. The influence of another quantum effect, however, described as a "hummer - drill" (HD) effect, becomes stronger. This is a frequency interaction between the simultaneously moving and oscillating carrier (SC electron) and the CL space parameters of the external field, which penetrates into the sample as a strong magnetic field. We distinguish two types of interactions due to this effect:

- direct interaction between the moving carriers and the external magnetic field

- interaction between the moving carriers and the quantum conditions in the sample.

Let us pay attention, in first, about the direct interaction between the applied magnetic field and the carriers in the sample. The observation of Aharonov-Bohm oscillations in the longitudinal resistance (see Fig. 3.14, courtesy of J. D. F. Franklin et all (1995)), indicate **that the carriers are likely moving in eshellons.** In this case their quantum motions are synchronised and they are able to be detected. The second benefit from understanding the carrier grouping in eshellons is that their quantum motion could be influenced directly by the strong magnetic field. At the same time, the sample SPM frequency is not stabilized. So we have the following oscillators:

- a stabilised SPM frequency of external space, providing the strong magnetic field

- SC single electrons, with preserved proper frequency (kept by its internal energy)

- SC stacked electrons (with a different but stable proper frequency)

- a not stabilized SPM frequency of the sample CL space

Now, let see what happens, when changing the magnetic field *B*. The resistance between the plateaus follows the law of the classical Hall effect. When changing the magnetic field, more or less electrons are forced to deviate from the longitudinal path due to the direct interaction with the external space SPM vector. If the sample was at normal temperature, its SPM vector may create some resistance, but then its frequency is stabilised. If the sample is in a SC state, the node SPM frequency is not stabilized. In this case the number of the magnetic protodomains is decreased. The SPM frequency of the individual domains or nodes may have a phase and frequency dispersion. Then trying to resist the carrier motion, their central frequency could be dragged by the forced motion of the carriers. In such case, by changing the magnetic induction *B* we are able to tune the SPM frequency of the sample. In this process the **SC carriers serve as a mediator between the external space SPM frequency and the SPM frequency of the sample.**

They provide the necessary connection between the SPM frequency of the external and sample CL spaces. If the sample SPM frequency becomes exactly tuned to a harmonic or subharmonic of the external one, (corrected by the sample's refractive index), then the carriers become involved in the quantum motion conditions provided by both spaces - the external one and the internal one of the sample. In a case of exact tuning, the strength of the SPM frequency interaction is increased. As a result of this, the sample quantum effect obtains a feedback from the quantum motion of the SC carriers until the phase difference between the external and sample SPM oscillations becomes zero. The sample SPM frequency gets a strong frequency and phase synchronization from the SPM frequency of external CL space. We will call this condition a **synchronized quantum effect.** The synchronized quantum effect appears for the following three types of the carriers:

- normal electron,

- SC electron

- stacked SC electrons (2, 3, 4, and more stacked SC electrons)

For any one of the above SC carriers, the quantum synchronization effect takes place when the sample SPM frequency is either equal or a subharmonic of the proper frequency of the carrier. Such synchronization may occur for every one of the above mentioned SC carriers.

Additionally to the above described effect of phase lock between the sample and external SPM frequencies, there is an effect of frequency holding effect in this phase lock due to a Doppler shift. It will be discussed later in this chapter.

The theoretical considerations in §4.1 and Eq. (4.1) show that the decreasing of the sample ZPE-D leads to decrease of the SPM frequency of the sample CL space domain. We may accept that the CL node distance in the cooled sample does not change. Otherwise the Static CL pressure of the sample CL space would change that will lead to change of the sample weight, but such effect is not observed. Then denoted the physical parameters in the sample by * we have:

$$n_i = \frac{c}{c^*} = \frac{\nu_{SPM}}{\nu_{SPM}^*} \qquad (4.10)$$

where: n_i - is an abnormal refractive index of the sample at very low temperature, c - light velocity and ν_{SPM} - SPM frequency (equal to Compton's frequency in Earth local field); * - denotes the parameters of the sample CL space at very low temperature

Let us denote the confined motion of the normal electron or SC electron by the subharmonic number N. For $N^* = 1$ we have and optimal confined motion and for $N^* > 1$ - a confined motion with a subharmonic number N. Then in a SPM frequencies phase lock condition the following expression will be valid

$$\nu_{SPM}^* = \nu_{SPM}/(N^*) \qquad (4.11)$$

Let us consider, for example, that the applied magnetic field corresponds exactly to a filling factor $\nu = 1$. Then if $N^* = 1$ the normal electron is moving with the first harmonic (an optimal confined motion) and the sample SPM frequency becomes equal to the external SPM frequency. At $N^* = 1, 2, 3, 4$ the sample SPM frequency will be respectively 1,2,3,4 times lower. The normal electron will rotate respectively 1,2,3,4 times slower and it quantum magnetic field will appear respectively 1,2,3,4 time lower. Then the Von Klitzing constant RK defined by Eq. (4.9) will exhibit the same ratio 1,2,3,4. In QHE experiment the signature of the 1,2,3,4 ratio will appear as plateau centred at the same values 1,2,3,4 of the measured Hall resistance as shown in Fig. 4.11.

Consequently, the subharmonic number for the normal electron in QHE appears as Landau levels of generacy.

Let us consider now the SC electrons. According to the theoretical considerations in §4.22, the single SC electron has a proper frequency of three times the SPM frequency and *i* stacked SC electron has a proper frequency as shown in Table 4.1. When a SC electron of any type is moving by its optimal confined velocity (defined by the sample CL space parameters), and a phase lock effect occurs between the sample and external CL space SPM frequency, the following expression will be valid

$$\nu_{SPM}^* = \nu_{SPM}p_{pr} \qquad (4.12)$$

where: p_{rp} is the ratio between the proper frequency of the SC electron and the sample SPM frequency

The SPM frequencies locked by SC electrons comprised of 1,2,3,4 single SC electron moving with optimal confined velocity will correspond to plateaus of 1/3,1/5,1/7,1/9 according to Table 4.1.

The SC electron may also have a confined motion at subharmonics, as the normal electron. Then Eq. (4.11) will be valid. Therefore, combining (4.11) and (4.12) we for the SC electrons

$$\nu_{SPM}^* = \frac{\nu_{SPM}p_{pr}}{N^*} \qquad (4.13)$$

For a normal electron $p_{pr} = 1$, so Eq. (4.13) appears to be universally valid for normal and SC electrons in QHE and FQHE experiments.

When analysing data shown in Fig. 4.11 we see that the ratio $p_{pr}/(N^*)$ corresponds to $1/\nu$. From the other hand, the sample refractive index at locked frequencies is

$$n_{iL} = \frac{\nu_{SPM}}{\nu_{SPM}^*} = \frac{N^*}{p_{pr}} \qquad (4.14)$$

Then the sample refractive index at any locked SPM frequency of the sample is equal to the Landau level, ν.

$$n_{iL} = \nu \qquad (4.15)$$

Now let show that the locked values of the sample refractive index are discrete value of the sample refractive index dependence on the magnetic field B.

The first derivative of the sample refractive index of the sample velocity of light is

$$\frac{\Delta n_i}{\Delta c^*} = \frac{c}{(c^*)^2} \qquad (4.16)$$

Since c* is proportional to ν_{SPM}^*, we have
$$c^* = c\frac{p_{pr}}{N^*} \qquad (4.17)$$

Combining (4.16) and (4.17) with (4.1) and having in mind that $p_{pr}/N^* = 1/\nu$ after simple manipulations we obtain: $n_i = \nu$. Since n_i is now a continuous function it should be expressed by the line connecting the value set of ν. In Fig. 411 we see that the Hall resistance, R_H, depends linearly on B and according to Eq. (4.9.c) the parameter $1/\nu$ is linear dependent on B. Then we will have

$$1/n_i = 1/\nu = R_H\frac{q^2}{h} \qquad (4.18)$$

From the analysis of the QHE and FQHE experiments we will see that the following relation is valid

$$n_i p_{pr} / N^* = 1 \qquad (4.19)$$

The expressions (4.14) and (4.15) are valid for the plateau location in the Hall resistance plot, while (4.18) is for the slope of the plot. They, however, do not show the conditions defining the plateau width.

From QHE we see, that the plateau width for the filling factor set 1,2,3,4, and 1/3, 2/3, 3/3, 4/3, 5/3 falls pretty fast with the subharmonic number n^*. The plateau width dependence on the subharmonic number is a result of an effect that we may call a **holding effect.** The main parameter characterising the holding effect, which is directly measurable in QHE experiments is the locking range. For a particular filing factor, this is the range of B for which R_H is a constant. In other words this is the plateau width in unites of B. holding effect characterizes the locking range of the quantum effect. The locking range depends on the subharmonic number at which the carrier is involved (referenced to the sample SPM frequency). It is largest for N = 1 and falls exponentially when N is increasing. It can be illustrated as a holding force between two sliding sinusoids possessing a weight. If the sinusoids have the same period, the holding force is strongest. When the periods are dissimilar the holding force falls rapidly. **We will see from the experimental data analysis that the range of the holding effect depends on the type of involved SC carrier and its subharmonic number.** It can be described by a holding function, for which the subharmonic number of the involved carrier is an its argument.

The plateau width can be explained if finding a mechanism, which tends to keep the obtained quantum feature of the sample despite the low ZPE level. Such mechanism should provide conditions for keeping the mentioned above feedback when the change of the magnetic flux tries to force the sample SPM frequency to exit from its quantum conditions. The mechanism could be based on a Doppler shift. The latter would not appear between the carrier motion and the applied magnetic field, but between the proper frequency of the carrier and the quantum conditions of the sample.

From the experiment of J.D.F. Franklin et all., (1995), we see, that the carriers are moving in eshellons. We may expect, that the bunch of this eshellons, has a velocity dispersion. Then they will exhibit a range of Doppler shift dispersion. But from number of experiments we see, that the plateau width, if explaining by a Doppler shift due to the axial carrier velocity, will require much larger Doppler shift, than the average drift velocity. In the specific conditions of the QHE experiments, the Doppler effect should be analysed from the point of view of the direct interaction between the PP SPM of the applied magnetic field, propagated in the sample and the **hummer drill effect (HD)** of the carrier motion.

All different carriers, mentioned above, moves as rotating rings exhibiting a confined screw-like motion. Each of them possesses an individual proper frequency, and interact with the internal and external quantum conditions, by the HD effect. As a result of this, its rotational velocity obtains an alternative frequency component (denoted as AC component). This component provides a reference condition for exact comparison between the sample and external SPM frequency with an accuracy up to a fraction of the phase difference. If both frequencies are not exactly equal, a continuous running phase will exist between them. This running phase, regarded as a frequency difference, namely, can contribute a Doppler shift. Having in mind, that electron tangential velocity at the optimal confined motion is equal to the light velocity, the running phase may give a very large Doppler shift. It will become evident from the experiments, that the quantum synchronisation effect occurs always for electron tangential velocity in the vicinity of the light velocity estimated by the external space parameters. This is valid for all types of mentioned above carriers.

We must keep in mind that when changing the magnetic flux, we actually scan the sample SPM frequency, not directly, but by the help of the moving carriers. The local feedback between the sample SPM vector and the carriers provides conditions for a local sample quantum effect. So when trying to push the sample frequency to exit from the quantum hold conditions, the carriers take the frequency discrepancy on themselves, by changing and adjusting the mentioned above running phase. The existing quantum conditions in the external space and in the sample convert the running phase into a Doppler shift. **In such conditions we see, that the Hall effect shows the same resistance, not only for exact frequency value, defined by Eq. (4.12), but in the vicinity of this value, as well. For this reason a quantum plateau in the Hall resistance is observed.**

The direct interaction between the magnetic field and the rotating electron due to the <u>HD effect</u> is illustrated by Fig. 4.15.

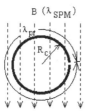

Fig. 4.15. Direct interaction between the external magnetic field and the rotating electron

It is convenient to estimate the influence of the external and internal quantum conditions on the tangential velocity of the rotating carrier by using the Doppler shift formula for relativistic velocity. For this purpose we will use the tangential velocity of the rotating electron and all SC carriers:

$$v_t = 2\pi R_c v_r \qquad (4.19)$$

where: v_t - is the tangential velocity and v_r - is the angular frequency of the rotating electron.

It is evident, that the Doppler shift will contain a pretty large contribution from the difference between the light velocities in the sample CL space and in the external CL space. Let us estimate this contribution by using the quantum wave refractive index of the sample, defined by the ratio between the two light velocities, respectively for sample and external CL space. It will be shown from the QHE experiments, that the rotational frequency of the carriers is always in the vicinity of v_{SPM}/n_i.

Let us estimate the frequency range that the relativistic Doppler shift could provide for the different combinations, between N and p_{pr} parameters, involved in Eq. (4.14).

We can express the fractional change of the sample SPM frequency by using the Doppler shift formula in which the rotational velocities of the SC carriers are involved, estimated by the CL space quantum conditions of the sample and external CL space.

$$\Delta_1 = \frac{\Delta v_{SPM}{}^*}{v_{SPM}{}^*} = \sqrt{\frac{v_{int} - v_{ext}}{v_{int} + v_{ext}}} \qquad (4.20)$$

where: v_{int} - is the tangential velocity confined with the internal quantum interactions (sample CL space), v_{ext} - is the tangential velocity from the direct interaction with the applied magnetic field (external CL space).

The velocity v_{int} estimated by the sample CL space parameters is:

$$v_{int} = (c^* p_{pr})/N^* . \qquad (4.21)$$

The velocity v_{ext} (defined by the direct type of frequency interactions) estimated by the external CL space parameters is:

$$v_{ext} = v_{SPM}/n_i \qquad (4.22)$$

Substituting (4.21) and (4.22) in Eq. (4.20), and after simple manipulations we obtain the relation between the plateau width and the parameters N^* and p_{pr} of the SC carriers.

$$\Delta_1 = \sqrt{\frac{|1 - N^*/p_{pr}|}{1 + N^*/p_{pr}}} \qquad (4.23)$$

Note: Both v_{int} and v_{ext} are estimated by the external space parameters.

The nominator under square root is put in modulus, because the velocity difference can take either positive or minus sign depending on the tuning direction of $v_{SPM}{}^*$ by B, from one side, and from the different ratio N^*/p_{pr}, from the other.

According to Eq. (4.23), the plateau width for one and a same type of carrier will depend on N^* and will decrease when N^* is growing. However, the plateau width is not completely determined by Eq. (4.23). From the experiments we see, that the plateau width falls pretty sharp with N^*. <u>The plateau width does not depend on a single factor. If introducing a holding function, defining the plateau width, it should take into account the following two factors:</u>

(1) the Doppler shift, defined by Eq. (4.23)

(2) The strength dependence of the holding effect on magnetic field B

(3) the strength of the direct interactions

Let us determine the contribution from factor (2). The observed plateaux appear at different values of B. While B influence the velocity of the SC carriers, it can not define their directions. Therefore, the velocity direction will depend on statistical uncertainty. Then if the number of involved carriers is m the statistical error will be $\sqrt{m}$. If m is linear dependent on B than the statistical error will be proportional to $\sqrt{B}$. Consequently, we may condole:

$$\Delta_2 = \sqrt{B} \qquad (4.24)$$

The width of plateau contributed by the factor (2) will be proportional to $\sqrt{B}$.

The strength of the direct interactions (factor (3)) can be estimated by the analysis of the HD effect by using the concept of the drag momentum. In a first approximation, the drag momentum could be simulated as a normalised momentum difference between two sliding sinusoids with a frequency ratio, corresponding to a subharmonic number. Analytically, such condition is expressed by the equation:

$$\int_0^{2\pi} [\sin(x) - \sin(n^*x + \varphi)]dx = \frac{\cos(2\pi n^* + \varphi) - \cos(\varphi)}{n^*} \quad (4.25)$$

The solution (4.25) is not defined for integer n^*, but is defined for $(N^* + \varepsilon)$, where ε is small enough (for example 0.001). Then we may obtain the solutions for consecutive $(N^* + \varepsilon)$, and normalise them to the value of $N^* = 1$. The obtained function fits excellently to a simple function:

$$y = 1/N^* = \eta_{HD} \qquad (4.26)$$

The Eq. (4.26) could be regarded as a quantum efficiency of the HD effect, referenced to the external SPM field. It defines the contribution of the third factor.

The contributions from all three factors, presented by Eq. (4.20, (4.24) and (4.26) act simultaneous, so the total frequency shift should be equal to their product. **Consequently, we obtain the expression for the plateau width, Δ, as a total shift** between the sample and external SPM frequencies

$$\Delta = \frac{k}{N^*}\sqrt{B\frac{|1 - N^*/p_{pr}|}{1 + N^*/p_{pr}}} \qquad (4.27)$$

where: k - is a coefficient closer to unity, but dependent on individual experiment parameters, such as the homogeneity of the magnetic field, for example. If its dimensions are in $(T^{1/2})$, then Δ has the same dimensions as B.

Eq. (4.27) gives closer values of calculated plateau width for single and stacked SC electrons, for all Landau filling factors different than one. In the following analysis of QHE and FQHE data it will be shown that the plateau for $v = 1$ is contributed by any SC carrier with a motion parameters

$p_{pr} = N^*$, so the contribution of the individual SC carriers for this plateau cannot be estimated.

Eq. (4.27) may give an approximated value of the plateau, because we used an idealised case of momentum difference between two sinusoidal sources for the direct interaction between carriers and the magnetic field. While the carrier proper frequency is of sinusoidal type, the bumps of the CL node magnetic quasisphere are not sinusoidal. This may affect the interaction function given by the simple Eq. (4.26).

Additional small Doppler shift, which may influence the quantum stabilization, might be contributed by the velocity dispersion between the carrier eshellons. This Doppler shift is convolved with the major Doppler shift, because the eshellon carriers participate in both interactions simultaneously. This will give a comb like structure of the combined Doppler shift. There is also a threshold level for development of synchronised quantum effect. If the coefficient k is large enough and the threshold level is inside the comb like Doppler shift, oscillations appear in the longitudinal resistance. This oscillations are known as Aharonov-Bohm oscillations. They are experimentally observed (see Fig. 4.13). In many QHE experiments, these type of oscillations are not observable, because, the lack of special conditions necessary to enhance this effect to a level of detection.

The Aharonov-Bohm oscillations, even undetectable, may play a role in the plateau formation. When changing the magnetic flux, the quantum synchronization effect begins to work before the central point of the plateau is reached. This could be explained by the velocity distribution of the carriers participating in the eshellons. Let us suppose, that all electrons in one eshellon have synchronised rotational frequency due to their local HD effect and the interaction with the local protomagnetic domains. Such eshellon may have much stronger direct interaction with the magnetic field. When the Doppler shift from the combined factors satisfies the Eq (4.27), a synchronised quantum effect will be obtained.

The frequency range at which the sample is synchronised is usually lower than its SPM frequency at normal ZPE (not cooled conditions). This is due to the resonance frequency decrease when the temperature approaches the SC state.

This explains why all plateau widths exhibit the same temperature dependence. This is clearly shown by the experimental data provided by D. Tsui et all (1982) (see Fig. 4.12).

As a result of the low temperature SPM frequency shift, the sample's quantum features are also shifted.

The SPM frequency defines the light velocity in the sample. If all quantum features are shifted as a result of cooling, **the sample CL parameters can be estimated if the signature of the optimal confined motion of the electron is identified.** Analysing the FQHE experiment data it may appear that the quantum refractive index of the sample appears with a value smaller than unity for some value of B. This does not automatically means that the velocity of light in the sample is larger than the light velocity in the external CL space, since the refractive index at cryogenic temperature has a lower value than in a normal temperature.

Comparing the quantum motion of the electron in normal ZPE, derived by hydrogen series, with this in the QHE experiments, we see that the allowed levels correspond to a consecutive subharmonic numbers of: 1, 2, 3, 4, 5. This provides additional confidence and confirmation about the subharmonic set used to describe the quantum motion of the electron.

One additional properties of the carrier motion is apparent from the behaviour of the longitudinal resistance in around the plateau. It is related to the carrier mobility in a confined motion. At plateau centre, the Doppler shift is zero and the strength of the confined motion due to the internal quantum effect has a maximum value, which gives a maximal value of the longitudinal conductance. This means that the sample CL space permits the carriers to move with a minimum energy loss.

From the provided analysis we see, that the quantum Hall experiments give a possibility to scan the SPM frequency of the 2D sample at very low temperatures, by changing the applied magnetic field. The involved interaction processes are illustrated schematically in Fig. 4.16.

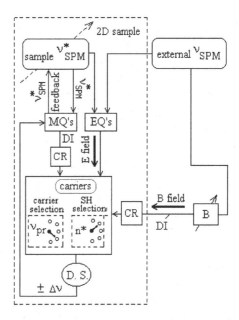

Fig. 4.16

Scanning of the sample SPM frequency in QHE

DI - direct interaction; CR - carrier rotation,

D.S. - Doppler shift; SH - subharmonic (defining the carrier velocity)

CR - carriers participating simultaneously in the applied electrical filed in the 2D sample and the applied magnetic field perpendicular to the sample plane

Summarizing the synchronised quantum effect, it is useful to emphasize the following features:

• **The permitted stabilization frequencies for the induced quantum effect are determined by the combinations between the proper frequency of the involved carrier** v_{SPM}^***, and the subharmonic number** $N*$ **at which the carrier performs a confined motion in the condition of the induced quantum effect.**

• **The quantum wave velocity (respectively the refractive index) is defined only for the points of the induced quantum effect.**

• **In conditions of induced quantum effect, the first derivative of the longitudinal conductance as a function of the Doppler shift is negative.**

4.4.2 Signature of the electron system in the Integer quantum Hall effect

The first proper frequency for the normal electron is equal to the SPM frequency, so we have $p_{pr} = 1$. The largest plateau observed in all QHE experiments is for $v = 1$. Since the sample is in air or vacuum, according to Eq. 4.13 we have $N* = 1$. Consequently for the integer QHE, the plateau for $v = 1$ is contributed by the normal electron, moving with an optimal confined velocity. The plateau width is contributed by the Doppler shift according to Eq. (4.27). It provides a holding effect for the Hall resistance R_H in the following way: When the applied magnetic field B is near the left side edge of the plateau, less carriers are involved in the quantum stabilization effect. Approaching the plateau centre, more carriers are involved. This is consistent with the change of the longitudinal resistance (see Fig. 4.11). If estimating the electron velocity by the number of passed nodes per one proper cycle (Compton time), it will appear closer to the drift velocity of the sample at normal temperature, corrected by the effect of the temperature change of the sample refractive index. Therefore, the plateau for $v = 1$ is contributed by eshellon of electrons moving with an optimal confined velocity. In other words, the SPM frequency of the sample is locked to the external space SPM frequency by the normal electrons (moving with an optimal confined velocity) as intermediate system.

If the applied magnetic field approaches a value for a Landau level 2, the resistance known as a Von Klitzing constant becomes twice smaller. Assuming that the electron charge is always a constant, it follows from Eq. (4.9) that the CL space twisting (generating a magnetic field) is twice smaller. This will be valid if the angular velocity of the rotating (screwing) electron is twice slower. This corresponds to a confined motion with a second subharmonics, or $N* = 2$. The physical process is similar for other subharmonics numbers.

Consequently we may conclude:

- **The quantum signature of the normal electron is provided by the QHE plateaus. When estimated by the sample CL space parameters, the quantum motion corresponds to the first harmonic motion, but when estimated by the external CL space parameters it cor-**

responds to a particular subharmonic motion. The subharmonic number is equal to the so called Landau level (denoted also as a filling factor v**).**

Let us use the experimental data plot provided by J. Eisenstein and H. Stormer (1990) and shown in Fig. 4.11. The identification of the motion parameters of the normal electron is provided in Table 4.2. The plateaus for the integer QHE are provided by the normal electron.

Signature of the normal electron

$p_{pr} = v_{pr}/v_{SPM} = 1$ **Table 4.2**

v	$N*$	$1/n_i$	B [T]
1	1	1	9.59
2	2	1/2	4.79
3	3	1/3	3.14
4	4	1/4	2.37

where: v_{pr} - proper frequency of the electron, v - Landau filling factor, $N*$ - subharmonic number valid for the sample CL space parameters, n_i - refractive index of the sample, B - applied magnetic field.

4.4.3 Signature of the SC electron in the FQHE

The single SC electron has a proper frequency 3 times larger than the Compton frequency: $v_{pr} = 3v_{SPM}$. The SC electron gets an optimal confined motion (defined by the sample SPM parameters) at $N* = 1$. This corresponds to a level $v = 1/3$. (this means a higher $v_{SPM}*$ but at cryogenic temperature its value is lower than at the normal one).

At level corresponding to $v = 2/3$, the SC electron is moving as a second subharmonic (estimated by the external CL space parameters, while corrected by the sample refractive index n_i.) In a similar way the levels 3/3, 4/3, 5/3, 6/3 correspond to quantum motion at 3, 4, 5, 6 subharmonics.

The locking conditions for the single SC electron, according to Eq. (4.15) are given by Table 4.3, where: $N*$ - is the subharmonic number of its confined motion. The last column is the tangential velocity, v_t, estimated by the external space parameters. The frequency ratio between the proper frequency of the SC electron and external SPM frequency v_{SPM} is $p_{pr} = 3$.

Signature of the single SC electron

$p_{pr} = 3$ **Table 4.3**

ν	$N*$	$1/n_i$	B [T]
1/3	1	3/1	28.7
2/3	2	3/2	14.3
3/3	3	3/3	9.59
4/3	4	3/4	7.14
5/3	5	3/5	5.69
6/3	6	3/6	4.79

We see, that the plateaus at filling factors 1 and 2 are contributed by both, the normal electron and the single SC electron.

4.4.4 Signature of the stacked SC electrons

The proper frequencies of the stacked electrons were given in Table 4.1. The signatures of the two and three stacked electrons are shown respectively in Table 4.4, and Table 4.5 The same source of the experimental data is used.

Signature of two stacked SC electrons

$p_{pr} = 5$ **Table 4.4**

ν	$N*$	$1/n_i$	B [T]
1/5	1	5/1	
2/5	2	5/2	23.9
3/5	3	5/3	15.9
4/5	4	5/4	11.96
5/5	5	5/5	9.59
6/5	6	5/6	
7/5	7	5/7	6.77

Signature of three stacked SC electrons

$p_{pr} = 7$ **Table 4.5**

ν	$N*$	$1/n_i$	B [T]
1/7	1		
2/7	2		
3/7	3	7/3	22.3
4/7	4	7/4	16.7
5/7	5	7/5	13.36

4.4.5 Scanning the sample SPM frequency and invoking a synchronised quantum effect

Without the presence of the electron and its SC configurations, the scanning of the low energy SPM frequency of the sample and the synchronised quantum effect would not be possible.

Fig 4.16 shows a plot of the inverse refractive index of the sample $1/n_i = 1/\nu$ (which is also equal to the ratio $\nu*_{SPM}/\nu_{SPM}$)as a function of the magnetic inductance B. The data are taken from the Tables 4.2 to 4.5. The carriers are shown as square dots and are connected with a line, showing a perfect linear trend. The notations by the filling factor are shown below the line, while the notations of the identified carriers - above the line.

While the SPM ratio could be considered as a continuous function, the refractive index is defined only for these discrete values, for which a synchronized quantum effect is possible.

One surprising result of the synchronised quantum effect is that, the refractive index of the sample for $\nu < 1$ becomes smaller than unity, but one must not forget that it is referenced to the refractive index at cryogenic temperature which is higher then at the normal temperature.

It is apparent from the presented plot in Fig. 4.16 that the change of the magnetic field causes a scanning of the sample SPM frequency in some particular range. The synchronized quantum effect, however, appear only when R_K is constant - at plateaus. These are discrete points in the SPM frequency scale. The quantum light velocity and the refractive index that the sample obtains are defined only for these points.

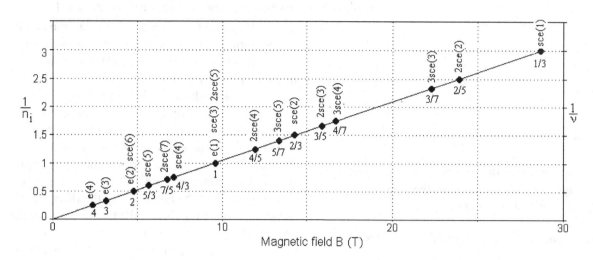

Fig. 4.17

Plot of $1/n_i = 1/\nu$ as a function of the magnetic field B using the integer and fractional quantum Hall data provided by J. P. Eisenstein and H. L. Stormer (1990) (see Fig. 4.11). The carrier identifications are shown above the plot. The Landau filling factors are shown below the plot.

ν^*_{SPM} - sample SPM frequency (tunable)

ν_{SPM} - SPM frequency of external CL space

N^* - subharmonic number of the SC carrier

$e(N^*)$ - normal electron moving with N^* subharmonic

$sce(N^*)$ - SC electron moving with N^* subharmonic

$2sce(N^*)$ - 2 stacked SC electrons with N^* subharm

$3sce(N^*)$ - 3 stacked SC electrons with N^* subharm

4.4.6 The plateau width as a signature of direct interaction between the moving carriers and the quantum conditions

Using the data plot in Fig. 4.11 the position and width of the plateaus for different Landau level are measured. The plateau width is calculated by using the derived Eq. (4.27) assuming $k = 1$ and the results are compared. They are shown in Table 4.6, where the following notations are used:

> normal electron: e $\nu_{pr} = 1$
> single SC electron: SC e⁻ $\nu_{pr} = 3$
> double SC electron: 2SC e⁻ $\nu_{pr} = 5$
> N^* - subharmonic number
> $\nu = n_i$ - filling factor
> B - magnetic field
> ΔB_{dat} - plateau width (by the data plot in Fig. 4.11)
> ΔB_{calc} - plateau width calculated by Eq. (4.27)

Calculated plateau widths   **Table 4.6**

SC carrier	N^*	$\nu = n_i$	B	ΔB_{dat}	ΔB_{calc}
e⁻	1	1	9.59	1.15	
e⁻	2	2	4.79	0.52	0.63
e⁻	3	3	3.14	0.31	0.418
e⁻	4	4	2.37		
SC e⁻	1	1/3	28.7	3.82	3.79
SC e⁻	2	2/3	14.3	1.0	0.85
SC e⁻	3	3/3			
SC e⁻	4	4/3	7.14	0.29	0.252
SC e⁻	5	5/3	5.69	0.24	0.239
2SC e⁻	1	1/5	23.9	1.8	1.6
2SC e⁻	2	2/5	15.9	0.5	0.66
2SC e⁻	3	3/5	11.9	0.39	0.29

The difference between ΔB_{dat} and ΔB_{calc} is a result of:

- the values of ΔB_{dat} are taken from the plot in Fig. 4.11

- the plateau widths may contain a statistical experimental error (they may include for example, the time constant of scanning magnetic field), since this parameter has not been of a primary concern in the experimental setup.

4.4.7. Theoretical simulation of the holding effect in the SPM frequency synchronization

We may consider the plot $1/n_i = f(B)$ as a continuous function, interpolated for a discrete value corresponding to the plateaus. The plot shown in Fig. 4.17 fits to a robust line with a slope $k_{sl} = 0.10445$. If plotting data from different experiments, we will get the same linear dependence of $(1/v)$ on B, but the coefficient k_{sl} may vary. Then the following relation is obviously valid:

$$\frac{1}{v} = \frac{qB}{nh} = k_{sl}B \qquad (4.28)$$

Then

$$k_{sl} = \frac{q}{nh} \qquad (4.29)$$

where n - is the number of carriers. We see from Eq. (4.29) that the slop of $1/n_i = f(B)$ depends only on the number of carriers, while the unit quantum flux (h/q) is unchanged. There is no evidence, again, that a fractional charge could exist. This is in agreement with the accepted rule, for the charge unity (discussed in previous chapters).

There is one additional outcome from the function expressed by Eq. (4.28). Since $1/v = 1/n_i = v^*_{SPM}/v_{SPM}$, the plot of Eq. (4.28) is very similar to the theoretical plot of the frequency of the conical pendulum as a function of the relative displacement x, which was discussed in $2.9.3 and given by Eq. (2.17.f). (Resonance frequency stabilization, Chapter 2). The general form of the model equation is:

$$f = [a[(L(x))^2 - (x - 0.23)^2]^{-0.5}]^{0.5} - b \qquad (4.30)$$

where: f - is the resonance frequency, $L(x)$ - is the energy dump function, x - equivalent node displacement in the PM zone, 0.23 - offset factor for the operating in the PM zone, a - adjustable span coefficient, b - adjustable frequency offset. The plot of the Eq. (2.17.f) is shown again in Fig. 4.19

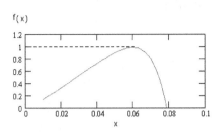

Fig. 4.19

Node resonance frequency with ZPE stabilization by the conical pendulum model

Comparing the plot in Fig. 17 and the linear range of the plot in Fig. 4.17 we see, that they are similar with the following correspondence:

B -> x ;   (1/v) -> f.

This similarity shows, that the change of the magnetic field B corresponds to a linear change of the displacement x. In order to use the conical pendulum model for investigation the holding effect of the resonance frequency behaviour (proportional to $\sqrt{B}$), we have to correct the model by taking a square root of f. Then the simulation equation for the resonance frequency dependence on x will take a form:

$$\frac{\Delta v_{SPM}^*}{v_{SPM}} = \left\{[a[(L(x))^2 - (x - 0.23)^2]^{-0.5}]^{0.5} - b\right\}^{0.5} \qquad (4.31)$$

$x = k_x B$, where k_x - is a constant of proportionality.

Equation (4.31) may be used for approximative investigation of the CL node resonance frequency of the sample.

We may show, that a resonance stabilization could appear as a plateau, by using the Eq. (4.30) or (4.31). For this purpose we may simulate the dump energy as a gaussian function. It will correspond to the energy of some carrier when getting direct quantum interaction by the external CL space. The plots of the selected dump energy function and the fractional change of the SPM frequency ratio are shown respectively in Fig. 4.20 a, b. The model does not simulate the Doppler shift, and does not provide a features for the plateau centring. But it demonstrate the mutual parameter behaviour. It also shows, that the frequency is very sensitive to the energy dumping.

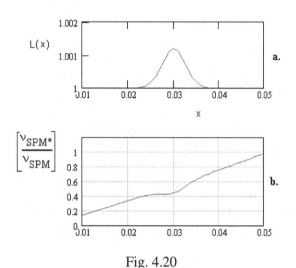

Fig. 4.20
Plateau simulation by the conical pendulum model

In the theoretical treatment of the QHE and the provided so far data analysis, an assumption was made, that the quantum motion of the carrier corresponds to a first harmonic confined motion estimated by the sample CL space parameters. If this was not true, the dependence of the observed states on magnetic field B would not be arranged in a line, as the plots, shown in Fig. 4.17, but spread in an area.

There is one additional confirmation that the dependence of the sample SPM vector from the filling factors and from the magnetic field B is aligned in a continuous curve. It is obtained, when investigating the shift and the width of the cyclotron resonance frequency as a function of the filling factor. The first successful measurements are provided by T. A. Kennedy et all.,(1977). Figure 4.21 shows experimental data of the spectral width and shift of the cyclotron frequency as a function of the filling factor at different values of the magnetic field.

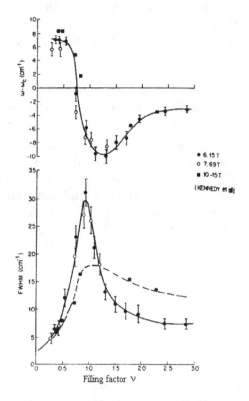

Fig. 4.21
Spectral width and shift of the cyclotron frequency in function of the filling factor (T. A. Kennedy et all. (1977)

The shift of the cyclotron frequency vs the filling factor, according to BSM, means that the SPM frequency of the sample is changed. The maximum of the cyclotron width vs the filling factor could be explained by the tendency of the different carriers to move with a common velocity, estimated as a number of passed nodes per unit time. Someone may argue, that the state $v = 1$ is contributes by few types of carriers: a normal electron, a single sc electron(3), a two stacked sc electron(3), and so on. However, the peak of the resonance line has a finite width. This means, that different combinations of carriers with different subharmonics numbers have a tendency to move with similar velocity. This is completely logical, if we estimate the velocity as a number of passed CL nodes per unit time. The CL node distance of the low ZPE channels is not changed by the temperature and the inertial property estimated as CL node displacement due to the FOHS motion through the CL space should be a similar. So we may conclude, that:

In a domain with a low ZPE, the carriers tend to move with velocities closer to the optimal confined velocity of the electron.

The above conclusion is important for understanding the superconductive state of the matter.

4.4.8 Signature of the SPM vector behaviour around the critical energy level E_{c2}.

It was discussed in §4.1, that the critical energy level E_{c1}, corresponding to p. A_0 of energy well 2, can be detected by the behaviour of the SPM vector. Below the E_{c1} energy point, the SPM vector could not get 4π spatial rotation. This change of the SPM vector behaviour can be detected as disturbed quantum behaviour of the carriers. In some of the QHE experiments, this effect is observed and called "quenching of the Hall effect". Its signature is a plateau for $B = 0$ at very low temperature. The experimental data given in Fig. 4.14 (C. Ford et all. (1988) demonstrate this effect. The plateau appears only for temperature below 100 mK and for B approaching zero. The plateau and the close region around the plateau, both, appear shifted from the trend of the classical Hall effect. At the same time, the longitudinal resistance sharply arises, indicating a disappearance of any internal quantum effect. The strong temperature dependence of this effect is completely consistent with our theoretical considerations.

4.4.9 Signature of the positron in FQHE

The positron is comprised of a positive FOHS with an internal RL(T) in the central hole of which a negative core oscillates. According to the mass equation, and from the 1^3S_1 -2^3S_1 positronium, the proper frequency of the free positron is twice the Compton frequency. Following the same logic as for the SC electron, **the positron signature in FQHE should be proportional to 1/2 state.** Such states are observed in number of FQHE experiments.

For observation of 1/2 or multiple of 1/2 state, one specific condition is important. The signatures of such states appear at much lower absolute temperature than the other fractional levels. In the experiment provided by M. P. Lilly et all.

(1998), they appear at sample temperature below 150 mK. The dependence of the longitudinal resistance on the magnetic field B at different temperatures is shown in Fig. 4.23.a., b.

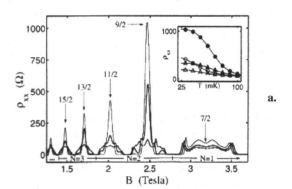

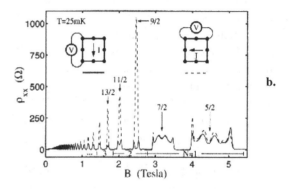

Fig. 4.23
Courtesy of M. P. Lilly et all. (1998)

The 1/2 (and multiple) states are distinguished from the integer and fractional states by a number of specific features:

- Strong peaks are observed only in the longitudinal resistance

- The observed set follows the order: 3/2, 5/2, 7/2, 9/2, 11/2,13/2,15/2.

- The plateaus of Hall resistance after 5/2 are missing

- Strong peak features appear for values above 7/2

- The peak amplitude between 7/2 and 9/2 changes with a jump and then gradually decreases.

- Some anisotropy is observed when changing the direction of the current in the sample (see Fig. 4.23 b.)

Before reaching this low temperature (during cooling) the sample has been passed through ZPE levels (suitable for creating of SC electrons according to BSM). Let us focus on the data plot shown in Fig. 4.23 a. At very low ZPE level, according to the theoretical diagram given in Fig. 4.1, the small radius r_{min} approaches the point A. The absolute polarisation of the EQ is limited by the low ZPE. Then the proximity E-field becomes weaker and the positron could be separated from the electron shell. Due to the lower E field it could not be attracted inside of the electron. Therefore, the environment allows existence of free positrons. Since they are quantum oscillating systems, they are able to participate in a direct interactional motion due to the external quantum field. The positron charge, however is not hidden as the SC electron and may interact much stronger with some distant CL domains, which have comparatively higher ZPE. As a result of this it gets much stronger resistance. We found in the presented analysis that for all carriers, exhibiting plateau in Hall resistance, the direct interaction between them and the external SPM frequency is given by the simple relation: $v_{pr} = v_{SPM}/n_i$. This means one rotation cycle per one v_{SPM}/n_i cycle. The n_i is additionally lowered by the much lower temperature (in comparison to QHE). Then the above condition could be satisfied only for higher axial velocity. The both factors: the larger resistance and the twice higher proper frequency (than the normal electron), are obstacles for such motion. Then the possible motion of the positron is that it may oscillate not once but number of cycles per one rotational period of v_{SPM}. **Then the rotational frequency of the positron should be a subharmonic of** v_{SPM}/n_i. Here again we have to keep in mind that n_i is defined only, when an internal quantum effect occurs.

One specific characteristic of 1/2 states is that the nominators are odd. This means that only the odd subharmonics are presented. But this is different in comparison to the integer and fractional QHE. The explanation of this new feature will become evident when analysing the direct interaction between the applied magnetic field and the rotating positron, interacting at subharmonic. Fig 4.24 illustrates the direct interaction due to a HD effect for three cases:

(a) - the carrier interacts at $v_{pr} = v_{SPM}/n_i$

(b) - the carrier interact at $4v_{pr} = v_{SPM}/n_i$

(c) - the carrier interact at $5v_{pr} = v_{SPM}/n_i$

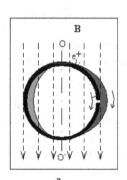

a.

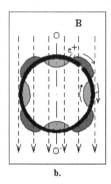 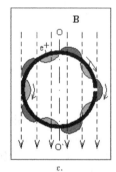

b.   c.

Fig. 4.24
Direct interaction between the carrier and the magnetic field for three cases (case b. does not work)

The direction of magnetic field is indicated by dashed lines with arrows and could be considered as a direction of phase propagated SPM vector. The rotational direction of the positron ring structure is shown by arrows. The oscillations due to the HD effect are presented as a sinusoid around the ring. The dark shaded area of the sinosoid indicates a direction coinciding with the ring rotation, while the light shaded area is in the opposite direction. The screw type of motion due to a second order helical step of the positron structure assures axial motion in one direction. Due to the homogeneity of the magnetic field and carrier motion orientation, the latter takes rotational energy from the field, that is transferred to axial motion energy. The rotational energy is obtained as a result of interaction between forces from the HD effect and the phase propagating SPM vector of the applied magnetic field. In order to estimate the resulting force, an axis OO' is drawn vertically through the centre of the positron ring. Then the interaction can be easily estimated by the balance of the forces between left and right side of the axis OO'. The phases between PP SPM vector of B and the proper oscillations are self adjusted due to the inertial moment of the carrier from the screw type of motion.

In case **a.** the right side forces are clearly predominant over the left side ones. This is the HD type of interaction for the FQHE, discussed in the previous paragraph.

In case **b.** the right side forces are equivalent to the left forces. Consequently the net effect of the ring rotation is zero. Therefore, this is a not working case. The rotation frequency is a subharmonic number four. The result is the same for all even subharmonics

In case **c.** the right side forces are predominant. So this is a working case. The rotation frequency is the fifth subharmonic number of the carrier proper frequency. Similar interactions are valid for other odd subharmonics.

Consequently, we may conclude:

When the direct HD interaction between the phase propagated SPM vector of the magnetic field and the oscillating carriers takes place in subharmonics, the even subharmonics are excluded due to the equal balance between the right and left-side forces.

Figure 4.23.a and b. shows two features of the peaks:

(1) - the peak maximum has an optimum values between 7/2 and 17/2;

(2) - the peaks are strongly dependant on the sample temperature

The first feature is explainable by the previous analysis. If the number of subharmonics is large, the difference between the right and the left forces will be decreased. In fact the interactional forces are not sinusoidal but possessing higher harmonics, because of the sharp bumps of the SPM MQ. The fall of the ρ_{xx} amplitude in the side of lower subharmonics motion number, shown in Fig. 4.23.b., could be a result of the magnetic radius dependence on carrier velocity. At lower subharmonics the carrier motion is closer to the optimal one

and the magnetic radius is smaller. Then the HD interaction also could become smaller.

The observed phenomena of the large peak change appears in a temperature range of the sample from 25mK to 100 mK. The strong dependence of the peaks from the temperature is due to the node operation at very low ZPE levels, close to E_{c1}. In order to illustrate the node operation in such conditions, the return forces - energy diagram is shown zoomed in Fig. 4.25.

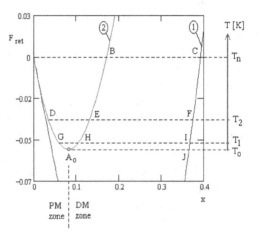

Fig. 4.25

Return forces - energy diagram by absolute temperature scale

Instead of the energy levels, the corresponding temperatures are indicated. The temperature T_n corresponds to a normal ZPE. The FQHE from electrons and SC electrons are observed in a range between T_1 and T_2. The positron signature is observed in a range between T_0 and T_1.

The small radius r_{min} for the range between T_1 and T_2 is determined by the sector HE. For operation in the range between T_0 and T_2, the ZPE is so small that the carrier electrical field (generated by the carrier RL(T)) will be very weak in order r_{min} to operate in HE sector (the particle RL(T) could not cause enough EQ polarisation). So the r_{min} is shrink and begins to operate in GD sector. The GD sector, however has a different sign of the slop, in comparison to the HE sector. **This affects the frequency stabilization of the synchronised quantum effect.** In such case, the oscillating carriers (only positrons) can still interact with the applied external field, but the induced quantum effect cannot get frequency stabilization because the feedback appears positive in a contrary to QHE and FQHE, where it is negative. In such conditions an increase of the longitudinal resistance occurs instead of increase as in the QHE and FQHE. The missing plateaus in Hall resistance indicate a lack of frequency stabilization effect.

The operation of r_{min} in the HE sector corresponds to a node operation in the PM mode, while in GD sector - to the DM mode. For lower subharmonic numbers as 5/2 and 7/2 some CL domains may operate in PM mode and others in DM mode. For this reason the both features appear; a large positive pulse (DM mode) and a small deep in the middle (PM mode). At lower temperature, more CL domains flips to DM mode.

Now let explain the last feature observed in this type of experiments - the anisotropy. When the SC electron is decayed into a positron and a degenerated electron, only the positron is an oscillation system. The degenerated electron could not participate in a quantum motion. The electrical charge of the separated degenerated electron again appears external, so it can move due to the external electrical field, but the motion is not of quantum type. From the other hand, the carriers in the external conductors (wires, power supply) are normal electrons. So in the contact points between the conductor and superconductor, the positrons and electrons has to be reconfigured into normal electrons. However, they have obtained different spatial distribution in the sample. When the direction of the current through the sample is changed, without changing *B* and temperature, the different spatial distribution appears in conflict with the required one for refurbishing **This gives an effect of anisotropy**. The anisotropy effect is also mentioned by H. L. Stormer et all, (1993).

The Landau level of 1/2 in a two-dimensional sample at lower temperature is experimentally confirmed also by investigation the propagation of surface acoustic waves (R. L. Willet et all. (1990). Some data from this experiment are presented in Fig. 4.26.

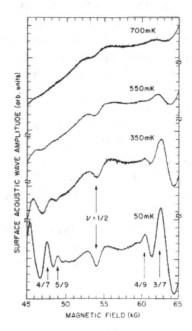

Fig. 4.26

Surface acoustic waves amplitude vs magnetic field at four different temperatures at 700 MHz (courtesy of R. L. Willet et al.)

The valley in the longitudinal resistance appears only at very low temperature. This is an indication of the SC electron decay into a free positron and a degenerated electron.

The signature of the free positron gives a possibility to investigate the node resonance frequency at very low temperatures. The relation between the sample and external SPM frequency is similar as those given by Eq. (4.12), but with some difference about the direct interaction

$$\nu_{SPM}* = \frac{\nu_{SPM}P_{pr}}{N'} \qquad (4.31)$$

where: N' - is a subharmonic of the proper frequency, but indicating a different type of interaction (without SPM frequency stabilization effect)

A different notation of N' instead of $N*$ is used in Eq. (4.31) in order to notate a different type of interaction. The difference is the following:

By analogy to Eq. (4.14) and (4.15) we have:

$$n_{iL} = \frac{\nu_{SPM}}{\nu*_{SPM}} = \frac{N'}{P_{pr}} \quad \text{and} \quad n_{iL} = \nu$$

The expression $1/n_i = 1/v = f(B)$ is also valid.

Figure 4.27 illustrates the plot of this expression by using the data of M. P. Lilly et all. (1998)., shown in Fig. 4.23.b. The N' subharmonic numbers are identified in a similar way as in the FQHE experiments. The points fits excellent to a robust line.

$1 - \dfrac{1}{v} = f(B)$ $2 - n_i = \sqrt{\dfrac{1}{v}} = \sqrt{f(B)}$

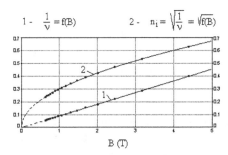

B (T)

Fig. 4.27

Plot of sample refractive index and (p_{pr}/n) ratio in function of magnetic field, B

We see that the positron signature gives additional possibility to investigate the SPM frequency at very low ZPE.

4.4.10 Summary and conclusions

- **The QHE gives a possibility to investigate the CL node operation at low ZPE-D**
- **The analysis of the QHE and FQHE experiments confirm the structure and the oscillating properties of the electron unveiled in previous chapters:**

    - the first proper frequency of the electron (electron shell - positron) is equal to the Compton frequency v_c (SPM frequency in the local Earth field).

    - The proper frequency of the free positron is $2v_c$

    - The proper frequency of the single SC electron, is equal to $3v_c$, and is the same as the proper frequency of the internal positron (a positron inside the electron.

- **The SC electrons are the carriers in the superconductive state of the matter**

4.5 More about the superconductivity

Understanding the superconductive properties of a two-dimensional system permits to extrapolate some results for a three-dimensional (bulk)

superconductor. The generated magnetic field by the SC carriers now is defined by the 3D sample. It is not difficult to guess that different type of SC carriers will tend to move in eshellons with one and a same velocity. This is the velocity of the optimal confined motion of the normal electron, corresponding to Landau level of 1. But now it is contributed by the SC electrons for which the expression (4.19) is valid: $n_i p_{pr}/N^* = 1$. In other words:

In a bulk superconductor the SC carriers tend to move with a velocity corresponding to the velocity of the normal electron at optimal confined motion.

Relying on the carrier behaviour in QHE, we may try to explain also one of the observed effect in the superconductors - the effect of long lasting current loop.

It is well known fact that if a current is induced and allowed to flow in a closed loop inside the superconductor, it continues to flow infinitely. The necessary conditions for such state are a low temperature support and a lack of external magnetic field, which may disturb the current flow.

From the QHE experiment analysis we see, that the sample could get refractive index below unity, when CL domains with enough low ZPE are created. In §4.2.1 it was discussed why the SC carriers are localised in the zone of penetration depth of the superconductor. In this zone the ZPE gradient is large. So it may always happens that some CL domains in this zone obtain a quantum refractive index equal to this of the external space (unity in air and vacuum).

$$n_i = n_{ext} = 1 \quad \text{or} \quad c^* = c \qquad (4.32)$$

The depth from the surface at which the condition (43.2) is satisfied determines the penetration depth. This is illustrated in Fig. 4.28.

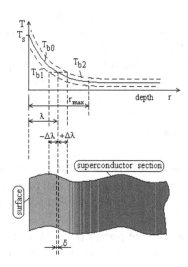

Fig.4.28
Penetration depth in function of temperature gradient and bulk temperature

The bottom part of the figure presents a transversal section of the superconductor, where the darker area corresponds to a lower ZPE and the lighter one - to a higher ZPE of the internal domains. The temperature dependence of the penetration depth, referenced to the surface at different bulk temperatures, is shown at the top of the figure. The shape of the curve near the surface is exponential. The relation between the three bulk temperatures is $T_{b1} < T_{b0} < T_{b2}$. The penetration depth denoted as λ, corresponds to a point of the curve in the exponential part. This gives a possibility the condition (4.32) to be satisfied for range of temperatures. In such case the penetration depth will vary with values $\pm\Delta\lambda$. The dimension of the CL domains, for which the condition (4.32) is satisfied determines the channel width δ. The channel may have not straight shape due to the hadron density nonuniformity. It however obtains a finite width, despite the exponential ZPE gradient, due to the tendency of the MQ's to congregate in magnetic protodomains. The exponential gradient of ZPE-D provides a direction of the induced magnetic field of the moving carrier, which tends to escape the superconductor. So **the motion of the carriers is accompanied with external magnetic field.**

It became evident, from the QHE experiments, that the carriers of the SC state are single or stacked SC electrons and they have a tendency to move with velocity corresponding to the optimal confined velocity of the electron. This means, that the single and the stacked electrons are moving with the same velocity. Due to their common interaction they moves in eshellons, as we saw by the Aharonov - Bohm oscillations. In such case, their proper oscillations are phase synchronised. This gives a strong magnetic filed, that is directed to the external space. The interaction of the moving eshellons with the sample nodes is small due to the hidden charges, but their guiding properties are preserved. At the same time, the carriers carry a huge SG mass in their internal rectangular lattices. If we take this mass into account and apply the mass energy balance principle, we will see, that it could balance a large SG energy. **So, we may speculate that the energy of the external magnetic field balances part of the Super Gravitational energy of the superconductive carriers, which otherwise is unbalanced in the superconductor volume due to the lower ZPE-D of its CL space.**

Let us suppose, that the current flow velocity is deviated by some internal factor. This is equivalent to deviation of B from the point of the synchronised quantum effect in the QHE experiments. The large energy balance of the system "moving carriers - external magnetic filed" has enough momentum in order to return the velocity to its previous value. The stabilization mechanism involves the internal quantum effect and the automatic self adjustment of the penetration depth.

So we see, that **the induced current in the superconductor is able to flow infinitely, due to the involvement of a large SG energy.** The necessary conditions for this effect are only the lower temperature to be in a finite range and the lack of external opposing magnetic field or energy dumping.

The properties of the low energy CL domains in the superconductor will become more apparent, when the reader is acquainted with the atomic nuclear structure in Chapters 6 ,7 and 8 and how this structure influences the CL space in the superconductor.

Chapter 5. Energy of the Physical Vacuum. Hidden space energy.

The analysis of CL node dynamics in Chapter 2 and 4 of BSM leads to the conclusion that CL space contains distributed Zero Point Energy (ZPE), which is of two types: a static energy (ZPE-S) and a dynamic energy (ZPE-D).

ZPE-S is the energy that keeps the CL nodes at proper internode distance. This energy is quite large, since it involves the superstrong SG field between the two types of nodes, formed respectively by the two types of prisms. An important parameter for this field is derived in Chapter 9 and denoted as C_{SG}, while the possible physical origin of the superstrong SG field is further discussed in Chapter 12 indicating that SG field is likely related to the super-high Planck's frequency given by the expression:

$$f_{PL} = \sqrt{\frac{2\pi c^5}{hG}} = 1.855 \times 10^{43} \ \ (Hz) \qquad (2.0)$$

ZPE-D is a the vibrational energy of the permanently oscillating CL nodes. Its intrinsic parameters are the node resonance frequency and the SPM (Spatial Precession Mode) frequency. The first one is related to the velocity of light, while the second one to its constancy in an uniform space medium. (SPM frequency is equal to the Compton frequency, valid for the Earth gravitational field (General Relativistic effect)). The relation of these two parameters to the permeability (μ_0) and permittivity (ε_0) of free space defines the light velocity by the well known equation $c = (\varepsilon_0 \mu_0)^{-1/2}$. In fact the constant value of ε_0 and μ_0 are result of the common synchronization of the oscillating CL nodes. This synchronization is assured by permanently existed spatially and time recombined Zero Point waves, which can be regarded as a magnetic protodomains. So ZPE-D is directly related to the zero point waves as discussed in Chapter 2.

ZPE-S is related to the Static pressure, while ZPE-D - to the Dynamic pressure of the CL space as discussed in §3.13 Chapter 3 of BSM. The Static pressure is also related to the Newtonian mass of the elementary particles, while the Dynamic pressure is related to the magnetic and electrical interactions. Presently, the contemporary physics identifies a zero point energy related only to the magnetic and electrical interactions. This corresponds to ZPE-D according to BSM concept.

ZPE-S is out of vision in contemporary physics, due to the currently adopted concept about space. In fact this is the energy behind the nuclear atomic energy, but presently it is not envisioned as a space energy. While the nuclear binding energy is correctly expressed by the Einstein equation $E = mc^2$, the real source of the nuclear atomic energy and its physical presence in space has not been properly understood so far.

If referenced per unite space volume, ZPE-S is much larger than ZPE-D. The huge difference between them could be understood if using an analogy with the energy of the ocean waves. ZPE-D is analogous to the energy of the ocean waves carried by a layer with a small thickness in comparison to its depth and normalized to unit surface area. The ZPE-S is analogous to the potential energy of a column of water under the same unit area. If a zunamy wave is invoked by an Earth quake, for example, the released energy is enormous and it transfers to strong tides and waves when approaching the cost. This energy analogically corresponds to the hidden ZPE-S energy of the physical vacuum. ZPE-S is responsible for the constant internode distance of the CL structure, in which the SG law is directly involved. (this specifics of the SG law is discussed in Chapter 12). BSM theory predicts that it could be accessed by some particular interactions between elementary particles and the ZPE-D energy. Since the physical vacuum structure has ability to keep a constant ZPE-D, the extracted portion of ZPE-D is refilled by ZPE-S, which is unlimited. Some particle physics experiments provided in the particle accelerators show a signature of accessing the ZPE-S. In the collision of accelerated high energy particles some jets often appear with energy much larger than the input energy of the colliding particles. These phenomena known as Regge resonances appear as "infinities" (from energetic point of view) in the Feynman's diagrams. Contemporary modern physics does not have explanation about the origin of this energy.

5.1. Zero point energy of CL space and its relation to the Cosmic Background Radiation

According to BSM theory, the Cosmic Background Radiation, known also as a relict radiation, is not a simple signature of the Universe evolution. It originates from every point of CL space, so it carries information about the distributed space energy and more specifically the ZPE-D. The deep space contains rarefied gas in a state of cold cosmic plasma. In such environment, the gas molecules or atoms are in dynamic equilibrium with the surrounding CL nodes. This equilibrium involves absorption of zero order waves (ZPE waves) from the gas molecules or atoms and emission of photons, while the total energy balance of the system of the gas particles and the surrounding CL space is preserved. The most abundant interstellar gas is an atomic and molecular Hydrogen. This fact facilitates the estimation of the ZPE and permits to determine one important parameter of the proton - the volume of its envelope. This parameter is later used in Chapter 6 of BSM for obtaining the physical dimensions of the proton with its substructures by cross calculations with other experimentally determined parameters.

5.2. Derivation of expressions about CL space background temperature.

The following method is based on two fundamental expressions: The first one is the well-known in the Classical termodynamics equation of the ideal gas and the second one is the Dynamical pressure of CL space, derived in Chapter 3. The ideal gas equation is:

$$R_{ig} = \frac{PV_\mu}{T} \qquad (5.1)$$

where: $R_{ig} = 8.31451 \; (J \; kmol^{-1})$ - is the universal gas constant; V_μ is the molar volume at absolute temperature T.

The ideal gas constant is an experimentally measured physical parameter for a gas in conditions when it behaves as an ideal gas.

The CL space dynamical pressure, P_D, is given by Eq (3.61) (Chapter 3 of BSM).

$$P_D = \frac{h\nu_c}{cS_e} = \frac{h\nu_c}{4\pi^2 cR_c r_e} \qquad \left[\frac{N}{m^2 Hz}\right] \qquad (5.2)$$

where: h - is a Plank's constant, ν_c - is a Compton frequency, c - is the velocity of light, R_c is the Compton radius, r_e is the small electrons structure radius.

In conditions of dynamical equilibrium, the hydrogen atom gets momentum from the ZPE waves in a form of dynamical pressure on the proton envelope. The surface of the proton envelope, considered as a torus, is given by the envelope of the proton circumference $2\pi(R_c + r_p)$ and its axial length L_{pc}.

$$S_p = 2\pi(R + r_p)L_{pc} \qquad (5.3)$$

The dynamical force exercised by the CL space on the proton surface is:

$$F_D = P_D S_p = \frac{h\nu_c(R_c + r_p)L_{pc}}{2\pi cR_c r_e} \qquad (5.4)$$

The pressure unit in SI system is $[N/m^2]$. This means that the resultant total force should be referenced to a unit surface of $1 \; m^2$. In such case, the pressure can be regarded as a sum of bouncing individual forces on large number of protons, while the resultant force is reference to a virtual wall with an area of $S_W = 1 \; m^2$.

The number of protons in one molar volume of atomic hydrogen is given by the Avogadro number N_A. Then the resultant force on a virtual wall from N_A number of protons is $\Sigma F_D = N_A F_D$. Normalizing the resultant force to a virtual wall with an unite area of $1 \; m^2$, we get the normalized value of the exercised pressure:

$$P = \frac{\Sigma F_D}{S_W} = \frac{N_A F_D}{S_W} = \frac{N_A h\nu_c(R_c + r_p)L_{pc}}{2\pi cR_c r_e} \qquad (5.5)$$

Let us consider a quantity of one *mol* of neutral Hydrogen atoms in a deep space. This could be regarded as a normal CL space environment, in which the atoms are in a dynamical equilibrium with ZPE-D. This equilibrium could be estimated by the dynamical CL pressure exercised on the proton. It has been mentioned that the dynamical CL pressure is caused by the zero point waves, responsible for equalization of ZPE; this means a background temperature uniformity of CL space. The dynamical equilibrium means that the energy obtained by the hydrogen atom will be equal to the energy radiated back into space. Obviously, the radiated energy should be contributed by energy level transitions in the atomic and molecular hydrogen. It is performed by small amount of atoms distributed in the space, so the optical depth is quite large. This conditions allow us to consider that the hydrogen distributed in deep space behaves as an ideal gas. The distance between atoms is large enough in order to eliminate the collision effect. The background temperature is also very low, so we may consider that the photon energy exchange between the atoms in such environments is negligible. In such conditions the probability of the hydrogen electron to be in a ground state is high. But the electron could never stop its motion in the quantum orbits. So it will have a continuous interaction with the CL space by its magnetic moment. This means that the Hydrogen atom will have some finite velocity different than zero. The physical effect of such motion is some small but finite pressure. In order to estimate this pressure we need to define a finite volume. Such volume could be the molar volume. It could be defined as:

$$V_\mu = V_H N_A \qquad (5.6)$$

where: N_A - is the Avogadro number

V_H - is the Hydrogen volume, considered as a neutral in the interactions with the zero point waves (related to ZPE-D).

The interacting volume should be some volume around the proton core where the interaction takes place. It is very probable this to be the volume enclosed by the Bohr surface, so in the outside volume the atom should behave as a neutral (the system of proton and orbiting electron appears externally neutral). Then comparing such described system of Hydrogen (possessing the mentioned hydrogen volume connected to the Avogadro number) with a similar volume defined for a single neutron, we see that they both exhibit the following common features:

- they appear neutral in the far field
- in the near field they exhibit magnetic field
- the proximity electrical field of the neutron is locked by the SG(CP) forces due to the symmetrical spatial configuration
- the proximity field of the proton in the Hydrogen is locked inside the Bohr surface due to the proximity coupling with the electrical field of the orbiting electron in a quantum quasishrunk space (see Chapters 7 and 9 of BSM).

The above features provide possibility to replace the magnetic interaction (with CL space) of the moving neutral Hydrogen by the magnetic in-

teraction of the neutron. So we can use some of the neutron's parameters and more specifically its magnetic moment as a dynamical interaction with the CL space.

Let us examine firstly, could the following relation be correct: $V_H / V_p = m_p / m_e$, where V_H is the above defined volume, V_p - is the volume of the proton envelop, m_p and m_e - the proton's and electron's masses. From well known relation $m_p / m_e = \mu_e / \mu_p$, where μ_e and μ_p are respectively the magnetic moments of the electron and the proton we arrive to $V_H / V_p = \mu_e / \mu_p$. While the left side of this relation is a volume ratio between a neutral (hydrogen) and a charged particle (proton), the right side is a magnetic moment ratio between two charge particles (electron and proton). According to the above mentioned considerations for neutrality of the hydrogen, we may replace the magnetic moment of the proton μ_p with the magnetic moment of the neutron μ_n.

$$\frac{V_H}{V_P} = \frac{\mu_e}{\mu_n} \qquad (5.6.a)$$

The envelope volume of the proton structure (whose surface is expressed by Eq. (5.2)) is:

$$V_p = \pi(R_c + r_p)^2 L_{pc} \qquad (5.6.b)$$

Combining Eqs. (5.6), (5.6.a) and (5.6.b) we may express the interaction molar volume of the hydrogen as:

$$V_\mu = \left(\frac{\mu_e}{\mu_n}\right)\pi(R + r_p)^2 L_{pc} N_A \qquad (5.7)$$

Substituting (5.5) and (5.7) in Eq. (5.1) we obtain the equation of the CL space background temperature.

$$T = \frac{N_A^2 h \nu_c (R_c + r_p)^3 L_{pc}^2}{S_W} \cdot \frac{1}{2cR_c r_e R_{ig}}\left(\frac{\mu_e}{\mu_n}\right) \quad [K] \qquad (5.8)$$

where: $S_W = 1$ (m^2)   - is a reference wall area

The proton core length L_{pc}, obtained directly from Eq. (5.8) is:

$$L_{pc} = \frac{1}{N_A}\left(\frac{2cR_c r_e T S_W R_{ig} \mu_n}{h \nu_c (R_c + r_p)^3 \mu_e}\right)^{1/2} \quad [m] \qquad (5.9)$$

The measured background temperature by COBE experiment is:

$$T_{exp} = 2.726 \pm 0.01 \quad [K] \qquad (5.10)$$

Then from Eq.(5.7) we obtain

$L_{pc} = 1.6429 \times 10^{-10}$ (m), but this is still approximate value. In §6.12.2.1 (Chapter 6 of BSM), we use this value and by strobing with other experimental data we obtained the accurate value for the proton core length

$$L_{pc} = 1.6277 \times 10^{-10} \quad [m] \qquad (5.11)$$

This value is extensively used in number of expressions, especially in Chapters 9 and 10 of BSM. It matches quite well the theoretical results and the experimental data.

The calculated background temperature for L_{pc} according to (5.11) is:

$$T = 2.6758 \quad [K] \qquad (5.12)$$

The difference between the estimated temperature by BSM and the experimentally measured one is only 0.05K.

The CMB (cosmic microwave background) temperature is measured by a satellite looking in a deep space, while the universal gas constant is measured in Earth conditions. Some difference may exist between the ZPE-D of the deep space and the Earth local field that could be a result of the Earth gravitation influence on the CL density. This is kind of General relativistic effect.

The concept applied for the Hydrogen in fact should be valid for any other simple molecule, because, the zero point waves have a very short wavetrain. Therefore, in conditions of dynamical equilibrium in a deep space, a large number of molecules could be involved. The resultant spectrum obtained as a summation of their radiation may have an envelope approaching the theoretical curve of the blackbody radiation the maximum of which is the estimated temperature of 2.72K

5.3. CL space background temperature expressed by the parameters of CL space.

The deep space background temperature is a pure CL space parameter existing in both conditions: a deep space and in a vicinity of a massive objects as well. The first option indicates that it could be expressed directly by some of the intrinsic parameters of CL space. In order to obtain such expression we must replace the proton and electron geometrical parameters in Eq. (5.6) with pure CL space parameters.

The proton length L_{pc} could be substituted by some length parameter of the quantum orbit. In Chapter 3 of BSM (§3.12.3) it has been shown that the trace length of a quantum orbits is defined by the equation [(3.43.j)].

$$L_q(n) = \frac{2\pi a_o}{n} = \frac{\lambda_c}{n\alpha} = \frac{c}{n\alpha v_c} \qquad [(3.43.j)]$$

where: n - is the quantum number, defined by the subharmonic number defining the confined velocity motion of electron in CL space.

For a second subharmonic we have:

$$L_q(2) = 1.66246 \times 10^{-10} \quad [m] \qquad (5.13)$$

This value differs from L_{pc} only by 2%, so L_{pc} in Eq. (5.8) could be substituted. The parameters of the electron: R_c, r_e and r_p can be expressed by the fine structure constant, α, the velocity of light c, and the electron gyromagnetic factor g_e, as shown in §3.6, Chapter 3. Then we arrive to an equation, in which the background temperature is expressed only by the CL space parameters and magnetic moment ratio μ_e/μ_n :

$$T = \frac{N_A^2 hc^2 \left(3g_e\sqrt{1-\alpha^2} + 4\pi\alpha\right)^3}{864\alpha^3 v_c^2 \pi^2 g_e (1-\alpha^2) R_{ig}} \frac{\mu_e}{\mu_n} \qquad (5.14)$$

The magnetic moment ratio μ_e/μ_n could be also considered defined by the CL space.

The provided analysis is correlated with the calculated mass budget of the proton and cross-validated with the eta-particle mass, and the high energy collision resonances (1.7778 GeV, 1.44 GeV, 80 GeV and 91.18 GeV) (See Chapter 6 of BSM).

5.4. Considerations for breakdown of the equivalence principle at internode range distance.

At first glance it seams that ZPE-D could be estimated if using the classical equation $E = 0.5I\omega^2$, where the moment of inertia, I, could be estimated from the parameters of the oscillating CL node. This equation, however, implies the presumption that the Newtonian gravitational mass is equal to the inertial mass (equivalent principle). Extensive analysis of experiments (not presented here) from a point of view of BSM, however, indicates that equivalence principle breaks down when approaching the internode distance. This is understandable if considering that the CL space could not exercise a static pressure on a particle object whose size is comparable with the internode distance. If the boundary structure of the elementary particle, for example is broken, the internal structure is released as rectangular nodes of six prisms. Such extremely small but superdens particle could not feel the CL pressure and will have an enormous penetration capability. It probably corresponds to one kind of the neutrino particles.

5.5. Hidden space energy

The second type, ZPE-S, is embedded in the connections between the CL nodes. The alternative CL nodes are connected by their abcd axes, in which the SG law is directly involved. In a normal non-disturbed CL space they are well balanced When an elementary particle is immersed, the CL space exercises strong SG forces on its impenetrable volume of the First Order Helical Structures (FOHS). The static energy from this pressure is related to the Newtonian mass by the Einstein equation $E = mc^2$. This pressure called a Static CL pressure is estimated in Chapter 3 by analysis in which the the unveiled structure of the electron is used. Its estimated value is:

$$P_S = 1.3736 \times 10^{26} \text{ (N}/m) \text{ - Static CL pressure}$$

While the obtained value of the Static CL Pressure is very large, one must take into account that it could be exercised only on the volume of the FOHS, since it contains a more dens internal lattice. For the electron, this volume, V_e, is calculated by its identified physical dimensions as a cut toroid with a large radius R_c - (Compton Radius) and a small radius $r_e = 8.8428 \times 10^{-15} (m)$.

$$V_e = 2\pi^2 R_c r_e^2 = 5.96 \times 10^{-40} \text{ (m}^3)$$

According to the mass equation (3.48) derived in Chapter 3, the mass of electron is:

$$m = (P_S V_e)/c^2 = 9.109 \times 10^{-31} \ (kg)$$

Using Einstein equation $E = mc^2$ we have:

$$E = P_S V_e = 8.187 \times 10^{-14} (J) \equiv 511 (KeV)$$

Scaling this energy to 1 cubic meter we obtain the value of ZPE-S energy in system SI:

$$E_S = 1.3736 \times 10^{26} \text{ (J)}$$

How such enormous energy is hidden in space? In fact ZPE-S is composed of two energies related respectively to the left and right-handed CL nodes, which are behind the positive and negative charge. In a non-disturbed CL space, both energies

are in accurate balance, so they appear hidden for ordinary EM and gravitational interactions. It is evident from Einstein equation that ZPE-S is accessible if the mass is changed. This in fact is the binding nuclear energy. The energy from the nuclear power stations is a result of changing the nuclear binding energy. This involves a micro-effect of General Relativity, discussed in Chapter 13.

5.5. Summary

The derived expression about the CL space background temperature is a CL space parameter related to the Dynamic type of the ZPE. It corresponds to the estimated blackbody temperature of the Cosmic Microwave Background (CMB).

Eq. (5.8) connects many experimentally measured physical constants. It provides also a relation between the ideal gas constant and the CMB temperature.

The revealed relation between the CMB temperature and ZPE of the CL space, provides a physical meaning of the universal gas constant and the Boltzman constant, as parameters of CL space.

The ZPE-S is the primary source of the nuclear energy. It is directly involved in the definition of the Newtonian mass of the elementary particles.

Chapter 6. Structure of the elementary particles.

6.1 Atomic and subatomic particles. Conversion processes.

6.1.1 End products of the crystalization process

The process of particle crystallization is discussed in Chapter 12 of BSM. The helical structures are formed in an unique crystallization process taking place in a hidden phase of a galaxy recycling, the scenario of which is presented in Chapter 12. It precedes the birth of the new galaxy accompanied by a new CL space creation and expansion of the new crystalized particles. The crystallization process is possible only before the galaxy birth, so in the new environments the following crystalized particles are initially available:

- Protoneutron;
- Open helical structures
- Electrons
- RL nodes and free prisms

The identification of the atomic and subatomic particles as helical structures has been provided in §2.7.A.2., Chapter 2. The shape of the proton was shown in Fig. 2.14.a in Chapter 2, which is shown below.

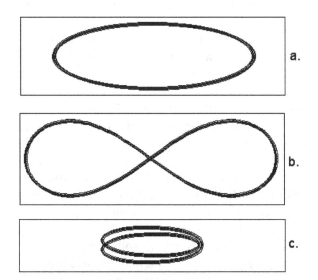

Fig. 2.14 Three conformations of one and a same higher order helical structure: a. - torus; b. - twisted torus (proton); c. folded torus (neutron)

The crystalized two basic particles, involved in the atoms are the **electron** (discussed in Chapter 3 and 4) and the **protoneutron, shown in** Fig. 2.14.a.

The electron is obtained from crashed SOHS's. In the new born CL space, the opened SOHS's undergo a decay chain reaction similar like the *pion - muon - electron* decay. Some high collision processes between hadrons (including proton and neutron) may cause a destruction, leading also to electrons (and positrons) as final products.

The protoneutron has a toroidal overall shape, which however is not stable in a CL space.

The unstable free protoneutron in CL space converts to a proton or neutron. The proton is a twisted torus having a shape of a Hippoped curve (Fig. 2.14.b), while the neutron is a double twisted torus (Fig. 2.14.c)

The free neutron is also unstable in CL space. Its lifetime is about 12 min. It may convert to a proton or in combination with a proton may form a Deuteron (presented in Fig. 2.15.c, Chapter 2, and shown below).

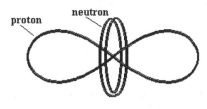

Fig. 2.15.C
Shape of Deuteron

The internal structure of the protoneutron is a same as the proton and neutron. It has been described in Chapter 2, §2.7.A.2.2 (see also Fig. 2.15.A and 2.15.B. Here we give again the table with the atomic and subatomic structures. Their shape has been shown in Chapter 2 from Fig. 2.9 to Fig. 2.14. The column RL refers to the type of the internal structure of Rectangular Lattice for the most external helical shell of the helical structure.

Table 6.1

Name	Notation	External helix	RL structure	Internal helix
electron system	e^-	H_1^1:-(+(-)	T	e^-
positron	e^+	H_1^1:+(-)	T	H_1^0:(-)
degenerated electron		H_1^1:-( )	T	missing positron
degenerated positron		H_1^1:+(-)	T	missing (-) core
pion (+)	π^+	CTH_m^2:+(-)		common type
pion(-)	π^-	CTH_m^2:-(+(-)	PT	common type
Kaon	K_L^0	TTH_m^1:-(+(-)	R	common type
protoneutron		TH_1^3:+(-)		# pair pions, kaon
proton	p	TTH_1^3:+(-)		# pair pions, kaon
neutron	n	FTH_1^3:+(-)		# pair pions, kaon

Notes:

FOHS - first order helical structure

SOHS - second order helical structure

For other type see the notation in Chapter 2.

For RL column: R - radial; T - twisted; PT - partly twisted

The term **common type** means, that inside of a negative FOHS there is a positive FOHS with its internal central core (negative).

6.1.2 Twisting process of an open helical structure

During the crystalisation process, the internal lattices of the FOHS is a RL type without twisting. When a free and open helical structure (not connected in a loop) is in CL space, its internal RL creates a strong SG(TP) field. The TP of SGSG field interacts dynamically with the CL nodes, causing formations of EQ SPM. At the same time, the FOHS is under static pressure of the CL space. As a result, the whole internal RL gets twisting and the RL unit cell is distorted to a shape of rhomboid. This leads to decrease of the small radius until the balance of the forces is obtained. It is a kind of energy balance in CL space including the hidden SG energy, in which the energy of the new born electrical charge (from EQ SPM) is also involved. The latter is a result of the dynamical interaction be-

tween the twisted SG field of the internal RL(T) structure and the surrounding CL nodes. The radial shrinkage means a volume shrinkage, because the length of the FOHS is preserved. The length is preserved due to the much larger stiffness of the internal layers of RL (see §2.8.2.A. in Chapter 2) in comparison to the CL structure. The volume shrinkage according to the mass equation means a decrease of the Newtonian mass of the particle. The energy equivalence from this mass change is equal to the created charge energy plus the energy of the quasiparticle wave, which is emitted as a result of preservation of the SG energy balance of the CL space domain. At the same time, the RL and FOHS twisting is accompanied by some relative rotation of both ends of the helical structure.

The preservation of the energy balance is dictated by the ability of CL space to keep a constant ZPE-D (the vibrational energy of CL nodes). For this reason, the CL space reacts to the new born electrical charge by emission of opposite charge as a quasiparticle wave. The latter simulating a charge particle obtain a handedness opposite to the handedness defined by the twisting of the real structure. While this effect is known as a CPT violation in contemporary Physics, it is a normal reaction of the CL space, according to the BSM concept. Using some analogy with a propeller rotational motion in water, it is like the reaction of the water caused by the propeller rotation.

The opposite handedness is a normal reaction of the CL space on the effect of twisting of the helical structure.

6.1.3 Protoneutron internal structure

The internal structure of the protoneutron, proton and neutron is one and a same. The external shell is a CH_m^2:+(-) structure serving as an envelope of internal proton space. At the same time, the external shell have a second order step, which is larger enough in order to allow the CL space to occupy the internal space of the proton. The internal space is occupied also by the following structures:

- CH_m^2:+(-)   - internal positive pion
- CH_m^2:-(+(-)   - internal negative pion
- SH_m^1:-(+(-)   - internal kaon

Thee pair of internal pions have a closed loop SOHS's with a same second order helicity. The internal kaon is a FOHS with a shape of torus. (The correct number of the pions (and kaon) contained inside the proton (neutron) are obtained later in this chapter).

6.1.4 Protoneutron conversion

The following conversion processes are possible:

- protoneutron -> proton   (1)
- protoneutron -> neutron   (2)
- proton <===> neutron   (3)

The first and third cases are related to a birth or death of a far field (detectable) electrical charge associated with the particle and emission of an opposite charge as a quasiparticle wave. The quasiparticle waves from a new born positive charge of the proton will behave as a negative charge. The proton also could undergo a transition by additional folding to get a shape shown in Fig. 2.14.c - the normal neutron. In this case, the detectable charge will be lost and the process will be accompanied with emission of a negative quasiparticle wave.

In other words, the neutron could be obtained from the protoneutron if folded twice or from the proton, if folded additionally. The processes of charge birth and death are symmetrical.

Note: The double folded protoneutron appears again as a neutral in the far field. The reason for this is that the SG field symmetry of the neutron is larger than for the proton. We may test this by applying the rule described in Chapter 2. As result of this the charge obtained by the twisting of RL appears locked in a near field. The appearance of magnetic moment, when the neutron is in a confined motion, serves as a proof of the concept that some residual proximity charge is locked in the near field, but undetectable in the far field.

The proton to neutron and neutron to proton conversions are experimentally known as a decay processes. They naturally occur in the processes of radioactivity. In this case, however, the conversion is accompanied also with additional gamma radiation because of the interaction the energy of which has been involved in the binding nuclear energy. The binding energy is a General Relativistic effect of the CL space shrinkage from the matter contained in the atomic nuclei.

The table 6.2 shows the discussed processes of particle conversion according to the BSM theory and the contemporary nuclear physics.

Particle conversions   **Table 6.2**

By BSM theory	By Nuclear physics
$TH_1{}^3{:}{+}{(-)} \to TTH_1{}^3{:}{+}{(-)} + \beta^-$	(1)
$TTH_1{}^3{:}{+}{(-)} \to FTH_1{}^3{:}{+}{(-)} + \beta^+ + E_{UL}$ $p \to n + \beta^+ + \nu$	(2)
$FTH_1{}^3{:}{+}{(-)} \to TTH_1{}^3{:}{+}{(-)} + \beta^- + E_{UL}$ $n \to p + \beta^- + \tilde{\nu}$	(3)

The protoneutron -> proton to -> neutron conversion takes place just after the birth of a new galaxy (a scenario is presented in Chapter 12). So this type of reaction is experimentally unknown. The second and third conversions are experimentally known respectively as a positron beta decay of the proton, and a negatron beta decay of the neutron.

According to the BSM, the above mentioned conversions are not related to emission of any hardware particle, but only of quasiparticle waves. When the reaction of the neutron - proton decay has been discovered, a hypothetical neutrino particle has been introduced in order to explain the missing mass which is energy equivalent according to Einstein equation $E = mc^2$. (Note: According to BSM concept the mass is not equivalent to matter. This issue is very important and discussed later). According to BSM, the mass deficiency is completely explainable by the reaction of the CL space. So no need of a neutrino particle in this case is necessary. The energy contributed as a "neutrino" is dissipated in the CL space as zero point waves, which are not detectable. (This explains the problem with the "missing Solar neutrinos").

The presented concept of the particle conversion does not mean that the neutrino does not exist. The measured neutrinos from space, from the Sun and in the high energy particle colliders are real hardware particles. However, according to BSM, they originate from processes, which are different than the proton - neutron conversion. They will be discussed in one of the next paragraphs.

The internal pions and kaon appear as backbone structures of the proton and neutron. They are composed of closed loop FOHS's, but possessing a different stiffness. The internal kaon appears as a most stiff structure, followed by the pions. The higher order helical structure has a lower overall stiffness (responsible for the overall shape). As result of that the external helical shell of the proton and the pions get twisting even as a closed loop structures. The internal kaon could be considered not twisted, and consequently not contributing a detectable charge (it may have only a weak proximity E-field). The opposite charges of the pions are locked in a proximity, however, they support the symmetrical configuration of the pions inside the proton. The positive charge of the proton is contributed by the external helical structure (shell), which has the lowest level of stiffness.

6.2 Virtual particles (waves)

Let us take for example the conversion of the normal neutron to a proton. The neutron is folded torus with a shape shown in Fig. 2.14.c. In this case, the Super Gravitation is able to lock the lattice disturbance into the near field and the neutron appears a neutral in the far field. Once the folded structure begins to unfold, the Super gravitational field will lose its previous symmetry. Some zones with large shear potentials will appear, which facilitates the escape of the locked field. The escape process however is not smooth, because it is related to creation of a large number of EQs in the far field. This means a creation of lot of energy in the surrounding CL space. This is kind of CL pumping process that appears in a very short time. The energy is induced by the internal RL, which gets twisting. Due to the short time duration of the process, the released energy does not have a time to be equally distributed between the positive and negative EQs. In some particular moment it escapes the region as an opposite charge wave. This kind of wave has an internal aligned MQs in its centre, which keeps the integrity of the propagated wavetrain. However, it does not possesses a boundary conditions, as the neutral quantum wave. So it moves with a speed of light, but could lose part of its energy. The wave appears and behaves as a high energy electrical charge due to its propagation with a velocity of light.

All betta "particles" from the radioactive decay are virtual particles. In the process, known as moderation, they interact with a real electron. As a result of that, low energy charge particles are obtained, which are real particles. When obtaining slow positrons, the latter has been part of the electron (internal positron).

In other words we may say that the neutron to proton (and proton to neutron) conversion is a process of unlocking (or locking) the near field and birth (or death) of electrical charge. It is accompanied with emission of specific waves behaving as high energetic charge particles with a charge opposite to the born one. The mass deficiency (or excess) is a result of a new energy balance between the new born electrical charges and the CL space.

6.3 Neutrino particles

Assuming that the neutrino particle originate from destructed helical structures, the following formations are most likely possible:

(1) partly folded nodes comprised of two axially aligned prisms and four in the periphery

(2) one or more stacked nodes of 4 prisms each

The first type of neutrino could originate from the RL structures embedded in the pions, muons, and kaons which are released when the proton is broken. The second type could originate from the helical boundary envelope of the FOHS, which holds the internal RL structure.

Both of the above types of neutrino may have a large penetration capability passing even through the proton and neutrons, but not through their internal RL structures. The probability to hit the FOHS is quite small because they posses a large spin and despite the intrinsically small inertial factor, they still could be guided by the oscillating CL nodes and the proximity field of the RL(T) of FOHS's. Consequently, the interaction capability of these types of neutrinos are intrinsically small in comparison to the elementary particles.

Note: The above envisioned neutrino particles are real particle of matter possessing intrinsic mass but not a Newtonian mass. They may not correspond to all definitions of neutrino particles currently used Modern physics, some of which are not even real particles (for example the "neutrino" from the proton beta decay).

6.4 High energy particle collision

High energy particle collisions are created in the particle accelerators. High energy collisions or other forms leading to a destruction of elementary particles are likely possible in some cosmological events.

6.4.1 Braking of the proton and neutron

The proton and neutron has stable structures in a CL space environment. The electrical field of the external helical shell of the proton and the closed proximity field of the neutron provide a guiding property helping to avoid the collision with other particles. In such way, the proton and neutron structure are preserved from damage.

When the proton, however, is involved in a high energy collision with another particle, and the collision energy is above some threshold level it hardware structure brakes. One of the most probable brake, if the hardware destruction is not so large, is a cutting of the folded torus in one place only. In this case the following optional products are possible:

  - the proton is cut, but the external shell is not broken
  - the proton is cut and the external shell is broken
  - the external shell only is broken without cutting and damaging the internal structures
  - the external shell is broken plus one of the pions is cut
  - other options

Cutting the proton in one place only.

In this case the, external helical shell and all internal structures obtain one cut. The internal structures then are free to comes out of the proton enclosure. This type of breaking gives the richest information about the internal structure of the proton. One very useful feature for identification of the physical structure of the released particles is the following:

- all particles, whose mass is estimated with great accuracy have been loop-shaped internal structures, cut in one place only.

Using this feature it is not difficult to guess that the following particles has been loop-shaped internal structure with an overall shape of torus or curled torus:

  - separately identified internal pions (positive and negative), an internal neutral kaon.

  - a combination of pair pions (having an opposite charge and the central kaon). This combination corresponds to so called **eta particle**.

The eta particle contains two helices (positive and negative pions and one central kaon. It is a neutral since the charges are compensated (the internal kaon is neutral when its FOHS is straight) and has oscillation possibility but a short lifetime, since the FOHS are not stable in this case.

6.4.2 Identification of internal structures of the proton (neutron)

Table 6.3

Name	Notation	Structure	RL	Comment
Kaon (internal)	TTH_m^1:-(+(-)		R	internal
Kaon	K_L^0	SH_m^1:-(+(-)	R	external broken (1)
Kaon	K_S^0	SH_m^1:-(+( )	R	external broken (1)
Kaon(-)	K^-	SH_m^2:-( )	T	external broken (1)
Kaon(+)	K^+	SH_m^2:+( )	T	external broken (1)
pion (+)	π^+	CTH_m^2:+(-)	PT	internal pion
pion(-)	π^-	CTH_m^2:-(+(-)	PT	internal pion
pion (+)	π^+	SH_m^2:+(-)	PT	external pion
pion(-)	π^-	SH_m^2:-(+(-)	PT	external pion

Notations: R - radial; T - twisted; PT - partially twisted

Kaons.

The internal kaon is a TTH_m^1:-(+(-) structure. It is a common type of structure, i. e. inside of the negative FOHS there is a positive FOHS with central negative core. When the proton structure is cut in one place, the kaon comes out and become a "long lived" kaon K_L^0 with a lifetime of 5.8×10^{-8} (s). It is an open FHOS with a straight shape. **Moving through CL space, such structure can easily loose its internal core. In such case it K_L^0 is converted to a "short lived" K_S^0. Without internal core, the stiffness of K_S^0 is much lower.**

For this reason it has a shorter lifetime of 0.89×10^{-10} (s) in comparison to K_L^0.

K_L^0 can be regenerated from K_S^0 if passing of K_S^0 through a localised field of EQs. This is experimentally verified, for example, by passing of K_S^0 through a Deuterium gas (BSM interpretation of the experiment). The internal positive FOHS of the kaon has a positive internal RL. Due to the focusing features of the radial SG(TP) it exhibits attracting property for the partially folded negative CL node which are of opposite handedness. Some of these nodes are attracted and trapped in the hole, where they quickly loose their spin. Then the prisms of the trapped nodes are stacked by the SG forces. The accumulation of additional such nodes leads to creation of a central core with a thickness equal to 3 prism's diameters. The obtained central core is the same as the original one. **The kaon regeneration experiments confirm the trapping whole effect, which was envisioned in the process of particle crystalization (more details about this process are provided in Chapter 12).** One question may arise: Why the regeneration occurs, when the K_S^0 is passed through Deuterium? The answer is: The kaon needs partly folding nodes. In CL space free of atoms, the kaon moves in homogeneous environment. The nodes may appear displaced and partially folded but away from the central whole. The Deuterium, however, has a stronger local field with some inflection points. The most probable location of these points is between the proton and the neutron. The space around these points could be not homogeneous. Passing through this spots may cause some of the partially folded nodes to enter the trapping hole.

This mechanisms, referred by BSM as a trapping mechanism has played very important role in the phase of the particle crystalisation.

Why K_L^0 and K_S^0 behave as neutral particles?

Initially the internal RL are almost untwisted. Consequently its internal RL does not generate external electrical field. In order to get such field its RLs have to get twisting. A few options are possible for the twisting process:

     a) Twisting of the whole structure K_L^0

     b) Twisting of the whole structure of K_S^0

     c) Separation of K_S^0 into K- and K+ followed by twisting of the latter two, but preserving the straight shape

     d) Additional twisting of K^- and K^+ and converting to charged pions.

The options (a) and (b) are less probable, than losing the central core and converting to K_S^0.

The option (b) is the most probable one. It leads to charged kaons.

While the neutral and charged kaons are still FOHS's, the twisting option (d) convert them to SOHS's. The internal RL of the SOHS get more twisting than the FOHS.

Why the neutral K_S^0 kaon is temporally stable?

The straight shape FOHS gives some temporally stability. In the straight shape, **the radial stripes** of internal positive RL of $SH_m^2{:}+(-)$ and the external (to it) negative RL of $SH_m^2{:}-(\ )$ **are both aligned**, because they are not twisted. **This alignment is kept by the SG(CP) of the prisms, from which the RL nodes are made.** This state, however, could be easy disturbed, once the straight kaon is bent. The conditions of bending are dictated by the CL space environment. In case of high velocity motion of the kaon in a uniform CL space, **the time life appears extended, because of the confine interaction, dependence of the velocity.** Ones the velocity is below some level, the interaction strength is reduced and the kaon undergoes not reversible twisting that converts it to a charged pion. The internal structure $SH_m^2{:}+(-)$ may come out before the twisting, and then the both structures may undergo twisting according to options (c) or (d). The twisting of internal RL in all cases leads to appearance of external charge in CL space (generation of EQs). The charge of twisted structure is always equal to one elementary charge, because the far field is controlled by the SG(CP) forces. The twisting case (c) leads to charged kaons K^+ and K^-. They are still straight FOHS's, but possessing a twisted RL(T). The finite life time of the charged pions is kept by the straight symmetry of the proximity locked E-lines between the turns. The symmetry however could be easy broken. At high speed motion in uniform CL space, the symmetry could be preserved for a longer time, because the proximity locked E-lines may participate in the interaction (as for the case of neutron, exhibiting a magnetic moment). The lifetime of the charged kaons is about 1.2×10^{-10} (s).

One important feature of all free kaons is, that they have a low inertial factor due to their FOHS shape. **The confined motion of a straight FOHS is different than the confined motion of a SOHS.**

The masses of the short lived particles are estimated by their traces in magnetic field and their penetration capability. **In both cases the straight kaons will give a false indication that they are much heavier. The kaons looks heavier than the pions because they do not have second order helicity.** The neutral kaons will slide faster through the lattice and will have less interaction. The charge kaons also will exhibit less interaction with the lattice for the same reason. This behaviour of the kaons leads to overestimation of their Newtonian mass.

The estimation of the inertial mass of the kaon, without taking into account its different inertial factor, leads to wrong Newtonian mass value. As a result of this, the proton (neutron) mass balance from the masses of the substructures appears not adequate.

In the derivation of the mass equation, it was emphasized that only a SOHS (formed of FOHS with a second order helicity) gives an accurate estimate of the inertial mass, by direct application of the equation. This however does not mean that the mass equation is not applicable for kaons. If a proper correction, corresponding to the inertial factor is used, the mass equation gives a correct result. The inertial correction factor of the kaon is taken into account in the BSM model about the proton mass balance.

6.4.3 Pions and their decay.

The position of both pions inside the proton shell has been demonstrated in Chapter 2, Fig. 2.15.A and 2.15.B. Their internal RL are partly twisted as mentioned above. In this case they still have internal tension. By the BSM model, the internal pions are estimated to have about 294 windings if the radius of their second order helical structure is 2/3 R_c. **The negative pion is of combined type.** Inside of its negative FOHS there is a positive FOHS, identical to the positive pion. The configu-

ration is similar as for the electron system (the positron is inside the electron shell).

If the proton is cut without breaking of the external positive shell, it is more probable a positive pion to come out. Most of the pion and muon experiments are provided with positive pions. When the pion gets out it can live in this condition only for a short time - about 2.6×10^{-8} (s). This is the time necessary for the internal RL to get modification and the structure - additional twisting. The life time, depends also of its velocity. The latter condition is essential for its longer lifetime, when in motion. When the velocity drops below some critical level, it undergoes additional twisting. The additional twisting of the internal RL structure affects also the second order step and radius of the helical structure. **After completion of the twisting the pion obtains a second order radius and step equal to the radius R_c and step s_e of the electron. The obtain structure is the muon.**

The muon is more stable than the pion because the helical structure possesses a confined curvature, determined by the CL parameters. The curvature of the FOHS is the same as the electron (positron). This means that the force balance Eq. (2.8) is satisfied for CL space. The muon has a longer lifetime ($2.2 \ (\mu s)$) and possesses a larger penetration capability. It is able to penetrate deeper in a solid material with a metal structure, like iron, for example. Its penetration capability is due to its screw type of motion. The muon has one specific feature: the penetration path through material with high but uniform density (for example iron) is larger than in material with smaller one. This strange at first glance feature gets explanation, when analysing the decay process $\pi \rightarrow \mu \rightarrow e$, according to BSM interpretation.

Explanation of muon - electron decay. "burning effect" of RL structure destruction.

From the physical point of view there are two cases of the muon decay caused by different environment conditions:

- decay due to extensive longitudinal oscillations

- decay due to a slow motion (below the optimal confined velocity)

The reason for the decay in both cases, however is one and a same: when the neighbouring turns gets too closer, the SG(CP) forces get interference, which leads to destruction of the RL(T) structure. Similar conditions could never happen for any substructure of the normal proton or neutron. The interference between the RL(T) of the neigbouring turns of FOHS is illustrated by Fig. 6.1, which shows the radial section of the RL node structure of the most external layers of two neigbouring turns of the FOHS (the RL nodes are shown enlarged, while they are much smaller). In the normal proton, the external shell and the internal pions are multiturn SOHS's. The distance a is kept constant by the neutral internal kaon, which is a FOHS. <u>In such configuration the neutral kaon plays a role of a backbone of the proton (neutron) opposing any shrinkage or stretching.</u> The external shell is twisted and possesses electrical field. The pions are partially twisted, but it is enough in order to posses electrical field (controlled to unity charge by SG field). In fact both factors - the neutral kaon and the electrical fields of other substructures keep the neigbouring turns of the proton (neutron) SOHS's apart. In such case a constant finite gap between neigbouring SOHS turns is guaranteed. This gap is denoted as a, in Fig. 6.1.

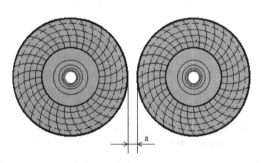

Fig. 6.1
Sectional view of neigbouring turns of SOHS showing the external RL(T) (RL node size is not in scale). a - finite size gap

The gap a is large enough to allow existence of CL space between the neighbouring turns of SOHS. The CL space attenuates enough the external propagation of SG(CP) forces between the neigbouring turns. In such case their internal RP(T) structures are free from interference. If the gap a is decreased below some critical value, the SG(CP) forces from the radial stripes may interfere and change their position. Then the accurate balance,

between the RL forces and the core bending opposing forces could be disturbed. Such disturbance could propagate quite fast through the RL structure leading to crash of the helical core of the FOHS. The crash is not momentary for the whole structure, but could propagate preferentially in one direction due to the helicity. It is similar to a fire in a fire guide cable used in the detonation devices. If the destruction is initiated from one end due to a high energy collision, it is very probable the RL structure to form a jet like a rocket propulsion engine. We me refer this as a **"burning effect"** of the RL structure destruction. Due to the reaction force from the CL interaction and the helicity, the whole "burning" structure will get a spin momentum. The jet of such "burning" structure contains RL nodes getting individual linear and spin momentum, after their separation from RL(T). **These RL node configurations are particle neutrinos.** These particles posses enormous energy and are able to penetrate deep in the matter. They are distinguished by the folded CL nodes by the number of prisms in the nodes (6 for RL and 4 for CL). Practically they could not be refurbished into CL nodes, without interaction with a matter, made of helical structure formation. The particle neutrino is the smallest particle formation, possessing linear and spin momentum. The latter two features keep their partly folded shape. In such conditions, they are able to pass freely through the protons and neutrons, even between the turns of the SOHS, but not through the FOHS's. **In some experiments the signature of these particles is detected as a "neutral current" - a term used in the Electroweak theory.** Individual neutrino particle from the space, however is much more difficult to be detected. The detection probability is increased when it loses some fraction of its momentum, due to the weak interaction with the CL space. (The angle between the prisms in the partially folded nodes of RL(T) may change, which will lead to some increase of their interaction with oscillating CL nodes of the normal CL space). In such interaction, they may transfer a fraction of their energy to the oscillating CL nodes. Then the probability of interaction with the protons and neutrons from the atomic nuclei is increased. Such particles are able to leave a trace in a large volume of suitable selected liquid. Neutrino detectors rely on this principle.

Let us focus now on the muon decay, which according to the BSM is a process of destruction of its multiturn helical structure. The main feature of the destruction is the "burning" of the muon internal RL(T) (one positive for the positive muon, and positive and negative for the negative muon). The RL(T) "burning" provides a rotational velocity which, drives the screw-like motion. We may distinguish more than one option of destruction depending on external environments - the media through which the muon decay. It could be a vacuum or some media. If the medium is not uniform the muon helical structure may obtain longitudinal oscillations in which case its helical structure may crash due to a collision between the neigbouring turns, preliminary before continuously "burning" of the whole internal RL(T). The both ends of the muon structure, however a free of such collisional destruction. One complete (leftover) turn (coil) is always left, inheriting the structure momentum. The leftover turn is an electron or positron (depending on the sign of muon charge). This conclusion is supported by the experiments of muon decay, when a longitudinal "polarization" is invoked. The direction of emitted positron (or electron) is related to the phase of the muon's "polarisation".

We see, that the main reason for starting the decay process of the muon is the decrease of the distance *a* below some critical value. Such conditions may appear, also, when the muon passes through CL space with different node density.

In a less dense space environment, the muon can oscillate longitudinally. When the oscillations are large enough, the muon structure could break due to the collisions between neigbouring coils. When the screw type of motion is performed in a denser but uniform space, like in iron, the longitudinal oscillations are suppressed, by the stronger interaction with the CL space in such environment. This keeps the muon from destruction.

It is well known fact, that the decay of the muon exhibits a CPT violation. The obtained electron (positron) posses an opposite spin. The decay mechanism, according to BSM, however, is free of any physical contradiction. Two decay options, according to BSM are possible: 1) - a positive quasiparticle wave; 2) - a real positron

1) case: The handedness of the quasiparticle wave is determined by the reaction of the CL space. The process is similar as the proton - neutron conversion.

2) case: During the violent process of breaking the muon structure, when only one single coil (positrons) is left, it is quite probable the central core to be lost. Then the positron (one coil structure) will take the confine shape with second order helicity determined not of the central core, but by its RL(T) internal structure, that is opposite. The degenerated positron, however, could be regenerated to a normal one by using its trapping hole mechanism to build a negative core. The process is similar as the regeneration of the K_0^L from K_0^S.

6.4.4 Experimental evidence for muon shape

There are few experimental evidences about the adopted by BSM muon configuration, based on a similarity with the electron. They are the following:

- the reciprocal equivalence between the mass and magnetic moment ratio between electron and muon
- similar "anomalous" magnetic moments
- the close values of the g-factors for electron and muon.

6.4.4.1 Cross relation between the Newtonian mass and magnetic moment of muon and electron

The provided so far conclusions are consistent with the experimentally determined relation between the masses and magnetic moments of the electron and muon.

$$\frac{\mu_e}{\mu_\mu} = \frac{m_\mu}{m_e} = 206.77 \qquad (6.1)$$

If considering a negative muon and electron, their structures are of the same type, having the same confined radius and second order step. The only difference is the number of turns. So Eq. (6.16.1) automatically gives the number of turns of the muon.

The larger radii of the electron and muon are both equal to the Compton radius R_c, and the small radii are also equal.

$$R_\mu = R_c \qquad\qquad r_\mu = r_e \qquad (6.2.a)$$

6.4.4.2 Newtonian mass change caused by the twisting (shrinking) of FOHS. Conversion of pion to muon, as a typical example.

The fact that the number of turns of the muon, (according to Eq. (6.1) is not an integer, indicates that the structure has been twisted during the conversion process from pion to muon. In this process the internal rectangular lattice cell is distorted and the volume of the FOHS is shrunk. In the twisting process, the obtained stiffness in the radial section of FOHS could not get higher than the stiffness in the vicinity of the internal radius of the RL layers. So the absolute shrinkage of the layers close to the central core is negligible. As a result of this, the length of the FOHS is preserved. It is equal to the length of the central core.

- **The twisting of RL causes only a radial shrinking, while the axial length of the FOHS is preserved.**

This is very important feature. It indicates, that the **length of the FOHS of the internal pion is equal to the length of FOHS of the muon.** So multiplying the Compton radius R_c by the number of turn according to Eq. (6.1) we get the FOHS length of the pion, L_π

$$L_\pi = 2\pi R_c \frac{\mu_e}{\mu_\mu} = 5.0168 \times 10^{-10} \qquad (6.2)$$

The obtained pion length L_π will be used later for accurate determination of the proton dimensions and unveiling its internal structure.

According to the mass equation, the shrinkage of FOHS volume causes a reduction of the Newtonian mass. Then knowing the mass ratio between the pion and muon, we can determine the volume shrink factor. But according to above made conclusion, we can determine directly the volume shrink factor, k_V, and the radius shrink factor, k_S.

$$\frac{m_\pi}{m_\mu} = \frac{\pi r_\pi^2 L}{\pi r_\mu^2 L} = \frac{105.7}{109.6} = 0.964416 = k_V$$

$$k_S = \frac{r_\mu}{r_\pi} = \frac{r_e}{r_\pi} = 0.87007 \qquad k_{rst} = \frac{1}{k_S} = 1.14933 \qquad (6.3)$$

where: r_π and r_μ are the radii respectively of the pion and muon FOHS.

The factor k_s is a shrink coefficient, while k_{rst} is the restore coefficient (to be used for convenience).

From the calculations in §6.12.2 it appears that the internal pion possesses a negligible twisting in comparison to the muon, while the latter is identical to the twisting of electron (positron). Consequently, the obtained shrink factor is valid also for the electron (positron).

The factor k_s (and its reciprocal) appears quite useful when applying the mass equation, because the latter operates with the volume of FOHS's. It gives a possibility to restore the radius and then the volume of the internal pions from the parameters of the measured pions.

6.4.4.3 Muon g-factor

For the electron system we found, that the g-factor is equal to the ratio between the helical step and the small electron radius r_e. The same relation is valid for the muon. The g - factors for electron and muon are experimentally determined with very high accuracy. According to NIST data, they have the following value.

$$g_e = -2.0023193 \pm 0.82 \times 10^{-8} \qquad \text{electron g-factor (by NIST)}$$

$$g_\mu = 2.0023318 \pm 0.13 \times 10^{-10} \qquad \text{muon g-factor} \qquad \text{(by NIST)}$$

T. Coffin et al. (1962) measured the muon g-factor in different targets: Cu, Pb, CH2, CHBr2. He observed variation of the g-factor depending on the target.

We see that the muon g-factor may vary depending on the density of the matter it penetrates. This indicates that the second order step may exhibit a slight change. In solids with different density, the internal CL space should also have slight variation of node density. This will influence the helical step of the passing muon.

It is interesting to see the difference between the positive and negative muon g-factors (Baley (1979).

$$(g_{\mu+} - g_{\mu-})/g_{average} = -(2.6 \pm 1.6) \times 10^{-8} \qquad (6.4)$$

The negative muon contains turns like the normal electron system, while the positive muon - turns like the free positron. In the first case the second order step is determined by the stiffness of - RL(T-) and the negative central core, both opposing the stiffness of +RL(T+) that react against step increasing. In the second case, the negative central core only contribute to the step increasing, while the +RL(T+) structure opposes to this. The difference is small, because the g-factors are usually measured during a confined motion of the muon, in which case the CL space is a third strong factor, which keeps the helical step at its initial value. The provided considerations give explanation of the slight non equality of the steps given by (6.4).

6.4.4.4 Quantum motion of the muon

In analogy with the electron system, the quantum motion of the muon is characterised by the interaction between the muon proper frequency and the SPM vector of CL space.

The negative muon is a structure similar as the electron system, while the positive muon is similar to the positron. They are distinguished only by the number of turns. Since muons are larger structures than the electron (positron) they should have lower proper frequency. Let us speculate to determine the proper frequency of the negative muon, using the classical expression for the proper frequency and the similarity of its structure to the electron:

$$\nu = \frac{1}{2\pi}\sqrt{\frac{k}{m_{in}}} \text{, where } k \text{ is a stiffness and}$$
m_{in} is the intrinsic mass.

For both particles (positive and negative muon) the intrinsic mass should not affect the stiffness k because the lack of connection between the internal RL(T) structures. Then the only factor is the external charge, which is the same as the electron one. Then the difference is only in the intrinsic mass of same type. If m_{in} is the intrinsic mass of electron and n is the number of turns of the muon, the latter will possess n times larger intrinsic mass or nm_{in}. Then using the electron intrinsic mass as unit we can express the muon proper frequency by the muon turns and the electron proper frequency:

$$\nu_\mu = \frac{\nu_c}{\sqrt{n}} = \nu_c\sqrt{\frac{\mu_\mu}{\mu_e}} = \nu_c\sqrt{\frac{m_e}{m_\mu}} = \frac{\nu_c}{14.379} \quad (6.5)$$

The muon proper frequency is lower than the electron one. In such case the quantum motion of the muon will have two major differences, in comparison to the electron.

- The muon proper frequency is lower than the SPM frequency of CL space

- The optimal quantum feature should be at velocity 14.379 times larger than the electron optimal velocity. The quantum efficiency, however, should be lower. The reasons for this are the difference between the proper and SPM frequencies, from one side, and the disturbed conditions of the boundary MQ effect, from the other. For one proper cycle of muon, its external shell passes much longer distance than the electron.

Despite of the differences, the quantum features of the muon at superoptimal velocity have some similarity to the electron quantum features. The hummer-drill effect could be enhanced, while the boundary conditions of MQ isolation are deteriorated. Then it is quite logical to accept, that the confined motion exhibit much more resistance from CL space. At the same time, the quantum interaction, even relatively weak, is able to keep the muon substructure oscillations. These oscillations play additional role of keeping the low stiffness structure from contraction due to the SG forces. If the velocity falls below some critical level, the SG forces between neighbouring turns may predominate the confined motion interaction, causing the structure to shrink. Then the distance a between the neigbouring turns could fall below the critical level and the destruction process described above will begin. Obviously **the muon decay will begin at a particular critical velocity, which depends on the CL space parameters. Inside of solids, these parameters are different than in vacuum (a different refractive index), so the critical velocity for solids will be different than in the vacuum.** The critical velocity in iron for example, could be much lower, than in the vacuum. Additional factor in this case could be also the influence of the iron magnetic domain, causing enhanced conditions for the muon confined motion.

The muon is unstable hardware particle and its lifetime, according to BSM, is not defined by a relativistic time dilation, but by the CL space environment. **The apparent "time dilation" according to BSM is a result of velocity thermalization process. This means, that the muon superoptimal velocity is continuously decreased until reaching the critical one.**

In Chapter 3, we found that the relativistic gamma factor, γ, and the quantum efficiency, η, derived for superoptimal velocities of the electron are inverse functions.

$$\gamma = \left(1 - V^2/c^2\right)^{-1/2} \qquad \eta = \left(1 - V^2/c^2\right)^{1/2}$$

The same functions should be valid also for the muon. The "time dilation" is obtained by multiplying the non relativistic lifetime by the gamma factor. But if considering a velocity thermalization process, the delayed time is a result of the longer time for the thermalization. It could be obtained by dividing the nonrelativistic lifetime by the quantum efficiency η. The result is exactly the same as the "time dilation", however, the physical process is quite different.

6.4.4.5 Pion and muon lifetime and connections to the Fermi coupling constant

While the enlarged "relativistic time" of the muon decay is a result of a thermalization process, the non relativistic muon life time of 2.2 usec is an intrinsic feature. According to the electroweak theory, without taking into account the anomalous corrections, the electroweak coupling constant, known also as a Fermi coupling constant, is related to the muon lifetime according to the expression:

$$\tau^{-1} = \Gamma = \frac{G_F^2 m_\mu^5}{192\pi^3} \qquad (6.6)$$

where: τ - is a muon lifetime, G_F - is a Fermi coupling constant.

The Fermi coupling constant is estimated also by the parameters of the nuclear β decay. According to R. Feynman and Gell-Mann (1958), the β decay and the muon decay are related to one **and a same physical process**. The strong argument for that, according to him, is the close value of the Fermi constant, determined experimentally from the both processes (within 2%).

The BSM looks for a physical process, which determines the finite lifetime of the pion. The following **hypotheses** is considered.

The pion lifetime is the time duration of the conversion process from pion to muon

Some features of the nuclear β decay from a BSM point of view has been mentioned in §2.10.6,

Chapter 2. It involves neutron to proton or proton to neutron conversion, which was described elsewhere. In this process, the RL structure of the FOHS's is only modified, by getting more or less twisting, accompanied by a Newtonian mass change (due to a slight change of the small radius of the helical structures, as a result of the obtained twisting). If the physical process that determines the pion lifetime is the RL twisting, then it will correspond to the pion to muon helical structure conversion. The conversion should take a finite time because it is related to the following processes:

- The external effect of the twisting involves a change of the FOHS volume, and the second order step and radius. The CL space reacts on this changes

- The twisting involves rearrangement of the RL nodes between the neigbouring turns characterised with "flipping" of the tangential aligned prisms to different neigbouring RL nodes. The speed of the "flipping" may depend on some oscillation frequency of the tangential nodes, that could keep a constant speed of the twisting.

- The twisting, likely, may start from one end of the pion structure and propagate until reaching the other. Such process may influence the velocity of the spin momentum of the structure. According to the calculations in §6.12.2.2 the second order step ratio between pion (Eq. (6.63)) and muon (the latter is equal to the step of electron) is 31.27.

According to suggested hypotheses, the internal structure is not able to oscillate, during the conversion process. The pion velocity is usually large but its confined velocity is also large due to the large second order step, as mentioned above. Then at the end of pion - muon conversion, the new born muon will posses superoptimal velocity. The velocity of the refurbished helical structure is decreasing to the level of the optimal quantum velocity. In such conditions, the internal core is able to oscillate, and the quantum motion is possible. This feature, namely permits detection of the muon's magnetic moment and spin. The critical velocity of the muon, at which the "burning effect" begins, is below the optimum quantum velocity for this structure.

6.4.4.6 Trace signature of the pion-muon-electron (positron) decay

If the physical process defining the muon lifetime is the RL twisting, then the lifetime should be equal to the duration of the "burning" effect. One may try to find some signatures about this process by analysing the pion-muon- electron traces. Figure 6.2 shows such traces in bubble chamber. (Courtesy of Lawrence Berkeley Laboratory, University of California, Photographic services. Adopted from Physics for scientist & engineers with Modern physics, R. A. Serway, 1992).

Analysing only the muon and electron traces, we observe the following features:

- The muon trace is included between two abrupt changes of the velocity vectors. The trace between the beginning and the end is smooth and with a constant curvature.

- The electron trace possesses a shape of Arhimed spiral, ending with a smaller radius. The curvature change from the beginning appears as a constant.

According to BSM, the "burning" effect of the muon RL(T) is included in the muon trace.

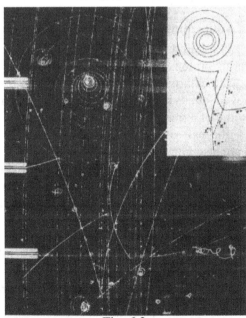

Fig. 6.2

(The plot is adopted from Physics for scientist & engineers with Modern Physics, R. A. Serway, 1992)).

According to the analysis in the previous paragraph, the muon possesses an optimal quantum velocity immediately before starting the "burning" process. The axial velocity in this case is 14.379 times larger (if this factor is correct), than the axial optimal velocity of the electron. The tangential velocity, however, is limited by the velocity of light. Then the muon confined motion will be similar as the electron motion with a particular subharmonic number. When the "burning" process reaches the last turn of the muon helical structure its oscillating proper frequency will get a value as the normal electron. Then the obtained electron, which has inherited energy from the decayed muon will get an optimal confined motion. This corresponds to the sharp kink of the muon curve, as shown in Fig. 6.2. The following curve of Arhimed-like spiral is the trace of the electron as an end product from the destruction of the negative muon (or a positron from a positive muon).

We see that the spin momentum is provided by the SG energy of the "burning" RL(T) structure as a result of its interaction with the CL space. This is not the full SG energy of the destroyed RL structure, but only its immediate signature. A vast amount of the SG energy is carried out by the RL nodes as neutrino particles. This type of neutrino, however is undetectable in such type of experiments. In some particular experiments a fraction of the released energy from the muon decay is detected as a **"neutral weak current"** (see A. Benveniti et al. (1974). This term is adopted and used by the Electroweak theory.

6.5 Neutral pion π^o

The **neutral pion** is formed by pair of charged pions, cut in one place, when they had a chance to come out together from the broken or destroyed proton shell. In this case the turns of the negative pion are between the turns of the positive one. In such configuration the SG(CP) field is able to lock their electrical fields in proximity, and they are not detectable. When exited, such system is able to oscillate. In this case it can pump the CL space. The oscillation process is terminated with an emission of two gamma photons. (The gamma photons are formed of parallel entaglement first harmonic quantum waves). Such radiation is experimentally detected. This is a gamma decay mode of the neutral pion.

6.6 Ifinities in the Feynman diagrams.

When using the Feynman diagrams for $e^- + e^+$ reaction with small energies the problem of infinities does not exists. The sum of input energy is equal to the sum of output energy. The picture is different when using the Feynman diagram for hadron-hadron collisions. In this case the S-matrix cannot be calculated as the output results is an infinity. (The only option in this case is to look for Regge resonances - short time lived energies). The infinities are not explained so far.

The explanation from the point of view of the BSM theory is simple. In the case of $(e^- + e^+)$ interaction, for example, there is not any process of destruction of helical structures. But in a hadron to hadron collision involving for example a break of a proton, a destruction of helical structures takes place. The stiffness of the proton structure is enormous in comparison to the stiffness of atoms and chemical compositions. In order to look for Regge resonances, experiments with high energy collisions are usually used. When first order of structure is broken, the internal RL occurs in CL space. Such structure (less or more twisted), possesses a large amount of Super Gravitational energy with no boundary, so its destruction may lead to a temporal but strong quasiparticles, which may dissipate in "showers", "pairs" and particle neutrinos. The refurbishment of the RL to a CL causes a vulnerable process in CL space. It may involve also energy from the space itself as a result of SG energy fluctuations. Some similar processes may exist in the nuclear explosion. In order to restore the equilibrium of ZPE, a huge energy may come from the space.

We see, that the destruction of the FOHS and the release of the SG energy embedded in the internal RL structure, is a kind of strong disturbance of CL space invoking of large CL energy fluctuations. The obtained energy is manifested as showers, gamma radiation, charge pairs, neutrinos and so on. It bursts the momentum of the hardware particles, such as pions, kaons, muons, electrons, positrons and their possible combinations. This is the energy that causing the "infinities in the Feynman's diagram".

6.7 Eta particle

In some cases of the proton destruction, only the external shell $TTH_1{}^3$:+(-) could be destroyed, while the other substructures are only cut. Then the obtained particle is comprised of three combined helical structures: one central kaon with a pair of muons of opposite charge, centred around it with a possibility to oscillate:

$K_L{}^0$ $SH_m{}^1$:-(+(-) - cut torus

π^+ $CH_m{}^2$:+(-) - cut curled torus

π^- $CH_m{}^2$:-(+(-) - cut curled torus

The identified oscillating structure corresponds to the observed neutral **Eta particle** η^0 with estimated mass of 547 MeV/c^2 and lifetime of 3.5×10^{-8} (s).

The E-fields of the pions are locked in proximity and do not appear in the far field. The kaon is kept not twisted and does not possesses E-filed. Consequently, the particle having oscillating possibility appears as a neutral, but for a very short time, because the symmetry could be broken easily after the oscillating energy is exhausted.

6.8. "Antiproton"

Let us suppose that in some collisions of accelerating protons we get a product in which:

- the external positive shell of the proton is destroyed, but its internal negative core (as the negative core of a free positron) is intact
- the pair of internal pions and the central kaon are also intact

Then the obtained particle has an overall shape of the proton but with a negative charge. This structure corresponds to the so called "antiproton". So the "antiproton" is consisted of:

         $TTH_m{}^1$:(-)   (twisted torus)

$K_L{}^0$ $TTH_m{}^1$:-(+(-)   (twisted torus)

π^- $CTTH_m{}^2$:-(+(-)   (curled twisted torus)

π^+ $CTTH_m{}^2$:(+(-)   (curled twisted torus)

When compared with the proton, the obtained compound of helical structures has one positive helical structure less. Consequently the negative charge will dominate, so it will have an unite negative charge.

We see that the antiproton structure is quite different than the proton one. While the external shell of the proton contains more that 1000 turns made of FOHS (the estimation is discussed in §6.12), the antiproton is pretty "nude".

One of the characteristic features of the "antiproton" is that it is unstable at low velocities. The antiproton is stored in traps, where it is in continuous motion, interacting only with the guiding magnetic field. If the antiproton is out of this condition and in touch with a normal matter (protons, neutrons), it decays. The process is known as annihilation, but in fact the Newtonian mass may also disappear. The matter never annihilates.

Note: while the "annihilation" energy is estimated by the Einstein equation $E = mc^2$, this energy does not take into account the energy released from the destruction of RL(T), which is much larger. This is the energy of hypothetical "quarks" and "bozons" which is much larger than the "annihilation" energy estimated by the Einstein equation.

Despite the enormous efforts for obtaining an antihydrogen, the results are not very promising. The obtained "atom" could exists only about a second in conditions of high velocity motion. Additionally to the different structure, we have to emphasize, also, another obstacle for obtaining a stable antihydrogen. The positron is not identical to the electron. The electron is a three body oscillating system possessing two proper frequencies. The positron is a two body system possessing only one proper frequency. The latter may not possess stable quantum orbits like the electron. For this reason the efforts for obtaining a stable antihydrogen are predestined to fail.

Here and in the following analysis it becomes apparent that the matter is not equivalent on antimatter. The equivalence is now a big misconception. It resulted from particle physics built on a wrong concept of the physical vacuum.

6.9 High energy particle collision

Conditions for high energy collisions are created in the high energy accelerators. Pair particles possessing charge or magnetic moment are accelerated to very high velocities and directed to collide.

In the high energy colliders today energies up to few hundred GeV are possible. A large variety of the colliding processes and reactions exists, but we will concentrate only on few aspects of the high energy collisions.

6.9.1 High energy collision between electron and positron (e^+e^- high energy reaction).

Let us analyse the collision processes for the most simple pair of structures with opposite charges - the electron and the free positron, if their collision energy is small (this process was described in Chapter 3). During their collision, the quantum features of the CL space are able to control their oscillations. Guided by the electrical fields, the free positron enter correctly in the electron shell. As a result of this, a process of multiple oscillation and CL pumping develops, that is terminated by emission of two or three gamma photons. The annihilation occurs for the electrical charges, but not for the matter of the helical structures. The helical structures with their internal RL(T) are not disturbed. The condition for entering in such oscillations is the both particles to approach each other with an optimal confined velocity. The corresponding energy for the electron is 13.6 eV.

If however the both particles are accelerated to energies above few GeV, their velocities are much above their optimal confined values. The quantum efficiency at such high velocity is negligible. The electrical guiding features fail at such conditions and the particle collide striking directly their helical structures. The interaction process is illustrated by Fig. 6.3.

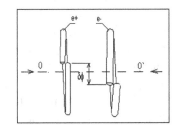

Fig. 6.3

High energy collision between electron and free positron

If the applied energy exceeds the gravitational energy equivalence of the structure, the latter will be disintegrated. So the gravitational energy equivalence put a threshold limit, below which the particle could not be disintegrated. In the collision process, very often one particle is only disintegrated, while the other takes part of the applied kinetic energy. For stronger energy collisions the two particles are disintegrated. So for successful disintegration of the particles, the provided centre of mass energy (a term used in particle collision) should be larger, than the threshold level.

The disintegration of FOHS is marked by the following processes:

- Reaction of the CL space, by generation of quasiparticle wave, if only one of the particle possessing RL(T) is disintegrated

- Slow refurbishing of the released RL(T), in which case it is able to simulate a high energy charged particle. In this case it could be misidentified as high energy pion or muon. The integrity of high energy RL node ensemble could be kept due to the following three features: much larger stiffness in comparison to CL space; high frequency spin of the nodes; and larger intrinsic node mass in comparison to CL node (factor 3:2).

- Fast energy dissipation in form of neutrino particles. They may pass a enormous distance without interaction. The RL nodes of a same type are very difficult to be integrated in the CL space.

- Gamma radiation as a secondary effect from RL destruction

The disintegration process involves also some elastic energy exchange. The fact that the axes of the two cones do not coincide, supports this conclusion. This is explainable if both particles do not meet with an exactly zero phase, as illustrated in Fig. 6.3. The elastic energy exchange, however, makes possible the appearance of a Regge resonance in an energy range below the applied centre of mass energy. Obviously some of the energy is escaped from detection. Some not detectable energy could be spent for "shaking" the local CL space.

In the next section some experimental results from e+e- collision are presented.

6.9.2. Experimental data about e⁺e⁻ high energy collision

The process described in the previous section may provide a number of resonances. The Regge resonance corresponding to the energy of 1.778 GeV is identified as a "tau lepton". Fig. 6.4 show a data obtained by R. Balest et al., (1993)

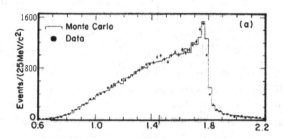

Fig. 6.4
Courtesy of R. Balest et al. (1993)

Figure 6.5 clearly shows the resonance estimated by C. Eduards (1982) to be at 1.44 GeV (+0.02 - 0015). The text under the figure is from the author. (the interpretation according to BSM is different).

The resonance at 1.44 GeV has been observed and discussed from a large number of experimenters. Good data are published by C. Edwards et al, (1982), F. J. Gilman et al., (1985), M. Procario et al. (1993), D. Bortoletto et al. (1993).

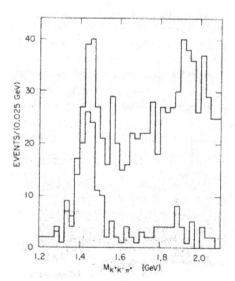

Fig. 6.5
K⁺K⁻p⁰ invariant-mass distributions
for events consistent with $J/\varphi \to \gamma \, K^+K^-p^0$
Courtesy of C. Eduards (1982)

Another interesting result is presented by large group of investigators (J. Z. Bay et al, (1996)). They observed a small third track at different angle. The sketch with the text shown in Fig. 6.6 is from their paper.

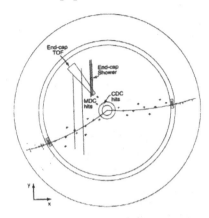

FIG. 9. A $\tau^+\tau^-$ candidate event in x-y projection, with evidence for a low-angle, third charged track, as indicated by the four hits in the CDC, followed by hits in layers 1 and 2 of the MDC; the associated hits in the end-cap TOF and shower counters further corroborate the presence of a third charged track.

Fig. 6.6
Courtesy of J. Z. Bay et al, (1996)
Note: The notation Fig. 9 with the text is f rom the paper of J. Z. Bay et al, (1996)

Another resonance obtained by e^+e^- collision is at 3.1 GeV/c^2. It is known as **J/psi particle**. It is reported firstly by G. S. Augustin et al (1974) and J.J. Aubert et al. (1974).

A large group of authors (M. Procario et al. (1993)) report experiment about tau decay (e^+e^- collision) in which, they investigate the two gamma photons. They found a well defined resonance at 0.14 GeV/c^2 . Some of their results are shown in Fig. 6.7. and Fig. 6.8.

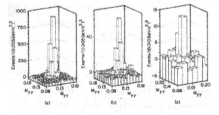

Fig. 6.7
Invariant mass of one pair of photons (randomly selected) vs that of a second pair (Courtesy of M. Procario et al. 1993)

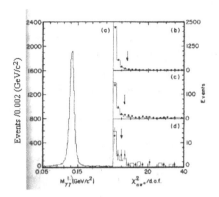

Fig. 6.8
Reconstruction of $\pi^o \rightarrow \gamma\gamma$ decays in lepton-tagged event samples (Courtesy of M. Procario et al. 1993)

Recent experimental results about high energy colliding heavy ions has been discussed in a special seminar "A new State of Matter: Results from the CERN Lead-Beam Programme", on 10 February 2000. Web site:

http://webcast.cern.ch?Archive/2000/2000-02-10/.

Most of the experiments are based on colliding of *Pb* ions with very high energy. According to BSM interpretation, a large destruction of protons and neutrons (also electrons) and their internal RL(T) and RL(R) structures takes place. One good indicator of this type of destruction is the fibrous shape of the recorded events. The reader of BSM is advised to visit the following Web site:

http://cern.ch/CERN/Announcements/2000/ NewStateMatter/Slides/slide02.html

Please observe slides 05 and 55, where the fibrous structure in the massive destruction process is evident. In the very last slide 97 of one of the presentations, some of the conclusions are following:

The measured J/ψ suppression pattern:
- Rules out the available conventional models
- Provides evidence for a change of state of matter

According to BSM the smallest particles obtained by this type of destruction are to be the two types of prisms. We barely could expect, that the prisms itself could be destructed in a such process. Despite the assumption of internal prisms structure by BSM, it could be change only in quite different process involving enormous pressure of highly dense intrinsic matter. This is a subject of discussion in Chapter 12 of BSM.

6.9.3 SG energy embedded in the helical structures and "electroweak forces".

The BSM is able to provide complete explanation of the Newtonian mass of the elementary particle, its internal material structure and physical dimensions. The forces of nature described by the BSM appear naturally unified and there is no reason for their separation into "four forces of nature".

The BSM theory does not use any concept of the Quantum chromodynamics, quark models, and the particle classification in fermions, leptons, and bosons. The BSM accepts to use the term "hadron" for a particle possessing a material structure, in order to be distinguished from a quasiparticle. Despite the different interpretation between particle physics and BSM, the necessary credit should be acknowledged to the particle physicists for the useful experimental data. In such aspect, the BSM concentrates on some measurable parameters and features, for which a physical explanation is possible. Some useful parameters, related to the electroweak theory will be used. They are following:

- The Fermi coupling constant: G_F
- "Tau mass" and the resonance at .44 GeV
- "Masses" of the W+- and Z bosons
- experimental evidence of "neutral currents"

The above mentioned parameters reflect real physical processes or features. Some of the parameters are intentionally put in quotation marks, because, they do not have the same physical meaning according to BSM. Although, in correct physical interpretation, they become very useful for unveiling the real structures of the elementary particles and the nature of the observed phenomena.

6.9.3.1 Embedded SG energy

The opened FOHS's are only temporally stable in CL space, so they are characterised by a finite lifetime. If, however, we try to destroy them during their life time, we have to apply high energy. The stability of the FOHS's is kept by a balance of forces according to Eq. (2.8). The free FOHS's possess two distinguished states, determined by the two distinctive states of their internal RL: - not twisted and twisted state.

For FOHS with not twisted RL (for example: the neutral kaons), the external E - field is locked by the SG(CP) forces, and the equation of the force balance includes:

Shrink forces:
- SG(CP) forces of RL
Opposing forces:
- helical core bending

The SG(CP) forces of the RL balance the helical core bending and provide locking forces for the external E - field.

For FOHS with a twisted internal RL(T) (for example, the charged pions), possessing external E - field (unlocked), the force balance includes:

Shrink forces
- SG(CP) forces of RL(T)
- SG(TP) forces of RL(T)
Opposing forces:
- helical core bending
- External E - field (unlocked) - CL reaction

For a stable structure like the proton (neutron), the external shell is a priory twisted during the crystalization-formation process, but the SG(CP) forces from all contained substructures are still able to lock and unlock the E - filed generating by the external shell, depending on the overall shape of the whole structure.

Let us analyse the embedded SG energy for the simple helical structures as the charge pion and kaon.

It is evident from the force balance, that the state of the internal RL has a signature, which appears in CL space. The helical core forces opposing the RL forces are quite strong, because, the prisms are stacked along their length, and the spaces are much shorter, than the node spaces of RL structure. In order to disintegrate the helical structure, one must apply a destruction energy much larger than the energy for refurbishing the RL or RL(T). At the same time, the number of the prisms forming the helical core of the FOHS is negligible in comparison to the prisms forming the internal RL. So when the structure is destroyed, only the signature of the RL or RL(T) appears detectable. Their signatures are specified, because they have well ordered spatial structures of RL nodes. The lack of boundary envelope is the reason for their very short lifetime, but during this time they are able to enter in strong interactions with the CL space. It is easier to detect their specific signature as electrical charge interaction, than the dissipation of the RL nodes as a neutrino particles. In many cases the released RL(T) can simulate real particles as pions, muons, electrons, while at the same time they are virtual particles (quasiparticle waves) or particle neutrinos. It is clear why the applied energy measured by a centre of mass parameter has to be larger than the energy of the detectable RL signature: some fraction of the applied energy is spent for breaking the helical core of the FOHS's. At the same time a much larger part of the energy embedded in RL structure is released as powerful jets. These jets in fact are the observed Regge resonances, which contribute to the "infinities of the Feynman diagram. These infinities has been unexplainable enigma so far. BSM provides a clear explanation for a first time. The following sections will provide enough arguments for this conclusion.

6.9.3.2 Signatures of RL(R) and RL(T) destruction

In the process of destruction, some helical structures have an internal RL(T) (pions and external helical shell envelope), others have RL(T) (the internal kaon. At the same time the degree of twisting for the internal and external (released) pions may be slightly different. It is reasonable to expect that the embedded energy in RL(R) should be different than RL(T).

A. Destruction of helical structures with RL(T)

Identification signature: two jets of opposite charges originated from one point.

The RL(T) exhibits an electrical charge before the break-up of the helical envelope. The charge has an unit value due to the regulation process of SG(CP) forces. When the envelope is destroyed, the precise spatial order of SG(CP) forces is disturbed, and the regulation process is disturbed. Then the RL(T) is able to generate a much larger charge (number of unit charges). The CL space reacts by generation of equivalent opposite quasiparticle charges propagated as a waves. The destruction process due to the collision is much faster, than the muon "burning" in the pion-muon-electron decay. So it takes place in a small space re-

gion. The point of the beginning of destruction becomes origin of two jets: a positive and a negative one. This is a specific signature of such process. RL(R) and RL(T) destructions show some differences in their signatures, which will be discussed later.

BSM theory identifies the following particles or resonances as destruction processes carrying the above described signature:

(1) τ lepton decay
(2) Resonance at 1.44 GeV
(3) Z vector bozon

(1). Lepton decay

The BSM uses the energy equivalence instead of mass, because, the detected energy is not a signature of a new born particle. It is a signature of the destruction energy of existing particle.

The τ lepton with energy equivalence of 1.7778 GeV is a destruction signature of +RL(T+) of the positron.

According to BSM interpretation, the "mass" of the "tau lepton" corresponds to the energy spent for full disintegration of the degenerated electron. The charge equivalence energy is negligible in comparison to this value. So this is the energy necessary for disintegration of the RL(T) There are few reasons, for this conclusion:

It is very probable, that the high energy electron has lost its internal positron during the acceleration. The most probable moment for this to happen is, when the accelerating electron exceeds the level of the synchrotron radiation.

(2) Resonance at 1.44 GeV

The 1.44 GeV equivalence energy is a destruction signature of -RL(T-) of the electron shell.

There is one experimentally observed resonance at 3.1 GeV/c^2, known as J/psi particle. The sum from the 1.44 and 1.7778 energies is 3.218 GeV. This value is larger, than the J/psi energy by **0.118** GeV/c^2. Could be some explanation about that?

Procarrio et al. (1993) measure a peak at **0.14** GeV/c2 by estimation of two gamma radiation in the "tau' decay.   This value is pretty close to the above mentioned difference.

According to BSM, the measured energy of 0.14 GeV is a signature of the destruction energy of the central core. It may look that the central core should have a negligible destruction energy due to the negligible number of prisms, in comparison to RL. However the 7 prisms in the core thickness section are stacked along their length and are much closer, than the smallest distance between the prisms in the RL structure. Taking into account the inverse cubic SG law, it is not difficult to estimate, that one single core node has few thousand times larger destruction energy than a single RL node. Additionally the central core is enclosed with layers having smaller number of interface nodes. This increases the probability some internal layers to be cached by the core. This could explain why the directly measured value of 0.14 GeV is larger than the difference of 0.118 GeV. At the same time, in most of the cases of +RL(T+) destruction, the central core could be inside of this structure, contributing to the total energy of 1.7778 GeV.

One more confirmation of our hypotheses comes from the experiment reported by J. Z. Bay et al. (1996). They clearly observe a third charge track. Fig. 6.6 (adopted from their paper) illustrates the third track. This track according to BSM is a **signature of the central core**. More accurately, this could be central core with some of the internal RL(T) layers.

(3) Z vector bozon

The Z vector boson with energy equivalence of 92.37 GeV is a destruction signature of the negative kaon. The latter is a straight FOHS with -RL(T-).

The prove of the presented in this paragraph conclusions is given in the next paragraph, where some theoretical treatment is given.

B. Destruction of helical structure with RL(R)

Identification signature: two separated jets of opposite charges

The RL structure in this case is not twisted and does not possesses a charge before the destruction. After the helical core get broken, the RL undergoes a twisting process, in which it obtains a charge. This is different than the destruction of RL(T) where the RL is preliminary twisted. The transition from radial to twisted shape of RL now appears after the helical core destruction and takes a finite time duration. **The slower twisting process does not lead to adequate CL reaction accompanied with generation of a virtual particle.** In such

condition the detected jet is a signature of the RL, getting twisting and obtaining a charge. The positive and negative jets are from physically different RLs. For this reason they are spatially separated. This is the main distinctive feature of this process, when comparing to the process in case A.

BSM theory identifies the following processes with the described above signature.

W^+ and W^- vector bozons with energy equivalence of 80.396 GeV.

The W^+ vector boson is a destruction signature of a not-twisted positive kaon (not possessing a far field charge). It is comprised of a straight FOHS with a +RL(R) internal lattice.

The W^- vector boson is a destruction signature of a not-twisted negative kaon. It is comprised of a straight FOHS with a -RL(R).

6.9.3.3 Tau energy equivalence as an unit of destruction energy for RL(T)

A careful analysis of the experimentally observed energy 1.7778GeV (considered so far as a **tau particle**) and the observed energy of 1.44 Gev could be regarded **as unit destruction energies**, respectively for a +RL(T) and -RL(T) embedded in a single turn of a SOHS. In other words the +RL(T) is the embedded energy in the normal positron, while -RL(T) is the embedded energy of the degenerated electron the structure envelope of which was shown in Fig. 2.13.b. The unit destruction energies are convenient for identification of all helical structures embedded in the proton (neutron). These units are given in Table 6.4

Units of destruction energy for RL(T)   **Table 6.4**

Destruction energy	Type of lattice	Particle	FOHS length
1.7778 GeV	+RL(T+)	e^+	R_c
1.44 GeV	-(RL(T-)	de^-	R_c

where: de^- - degenerated electron (without internal positron

The unit destruction energy could be used for estimation of the destruction energy of the subatomic substructures: pions and kaons, if knowing their FOHS lengths. Together with the radius shrink factor k_S (or restore factor), they could be used for restoration of the volume of the RL(R), (the volume they possesses before the proton break) and consequently the estimation of the New-tonian mass by applying the mass equation (the length of the FOHS, is preserved, because the most dens part of the layers is not changed). Some small correction, should be necessary, if using the shrink factor k_S for the kaon, which is a straight FOHS. When analysing the twisting (shrink) process of the FOHS and SOHS, we see, that:

The straight FOHS could not exhibit so large shrinkage (twisting) as the SOHS. Beyond some level, the twisting of RL leads to a conversion of the FOHS into a SOHS. The shrink factor k_S is estimated by pion to muon conversion. They both are SOHS's. Consequently, the shrink factor for the K$^+$ and K$^-$ kaons should be a little bit different than K_S.

6.9.3.4 Considerations about the decay options of pions and kaon

In Chapter 5, the proton core length was estimated by using the CL background temperature 2.72K. Later in this chapter we will obtain a pretty accurate value of the proton's dimensions by cross calculations using experimental data for pion and muon (the obtained value of the proton core length is $L_{pc} = 1.6277 \times 10^{-10}$ m). The kaon is in the centre of the proton structure, so it has the same overall shape and length. Before cutting the proton, the external (negative) FOHS and internal (positive) FOHS, both have internal RL(R) type of lattice. Until the kaon FOHS is straight it is still a neutral long lived kaon κ_L^0. If it loses its central core it converts to a short lived kaon κ_L^S. The process of kaon lattice conversion from RL(R) to RL(T) takes a finite time duration after the proton's cut. In one option both straight FOHSs of the kaon are initially separated and then twisted, while still they are straight. The obtained particles are the charge kaons K+ and K- Further twisting converts them to pions, which a SOHSs with a helical step. They have a confined motion and consequently a definite lifetime. Further twisting convert them to muons, which have better confined motion and longer lifetime. In other decay option a destruction process of the kaon RL structure may start earlier even before separating of both straight FOHSs. Since the RL destruction is accompanied with the released of the embedded SG energy, the process is characterised

by strong jets. These jets provides strong reactive force to the disintegrating structure. This effect is not revealed so far in the particles physics, so the mass of the kaon is significantly overestimated.

Considerations about the embedded SG energies in pions

Note: Since the FOHS of the internal pion is longer than the kaon, the embedded SG energy in the pion will be also larger.

The ratio of the pion to kaon FOHS length according to the calculations in §6.12.2.3 is 3.08. The destruction signatures for the pions according to those calculations are respectively 297.5 GeV for -RL(T-), and 367.3 GeV for +RL(T+). The required centre of mass energy should be at least twice larger. However, sharp resonances as from the kaons, may not be observed, due to the following reasons:

- The kaon is a straight FOHS and the reactive force from the destruction jet is aligned with the structure axis. The cosine of the reaction force is unity and it could not lead to a structure break-up in pieces

- The jet in pion is not aligned to the structure central axis. The reaction force will cause a structure rotation and will support the confined motion.

- Then for a combined helical structure like the negative pion, the destruction process of the RL will be different for the external (negative) and internal (positive) FOHS. A jet from the external one may appear, while the internal one is inside. The reverse option is not possible.

6.9.4 Theoretical analysis of the embedded SG energy in the elementary particle

The released RL after helical core breaking has a large initial potential energy. The particle colliders provide a possibility this energy to be estimated experimentally, by measuring the energy contributed by the jets. The obtained energy usually is converted to mass equivalence by the Einstein formulae $E = mc^2$ and assigned to some hypothetical particle. In our analysis we will use just the measured energy.

When the helical core of FOHS is destroyed, the RL undergoes a volume expansion. The volume expansion is anysotropic due to the different radial and tangential stiffness of RL(R) and RL(T). Both types of RL have a different behaviour.

6.9.4.1 Theoretical consideration

The obvious indications about the large amount of the embedded energy comes from the particle collision experiments. These are the Regge resonances discovered a few decades ago. They provide infinities in the Feynman diagrams, for which the particle physics theories failed to provide adequate explanation. One simple example is the estimated "mass" of the top quark, which is a decay product from a proton or neutron. The top quark "mass "is estimated as 174 GeV/c^2, while the proton mass is only 0.938 GeV/c^2. This mystery is a long time problem in particle physics. However, it clearly indicates the existence of large embedded SG energy. One could make analogy with the energy stored in the bullet before firing.

Why the energy of RL structure is so high?
- the RL node contains 6 prism in comparison to 4 prisms of CL node
- the node distance of in the RL structure is smaller than in CL one
- in both cases the embedded energy is defined by the Supergravitational law. The SG forces are inverse proportional to the cube of the distance

When the large energy is embedded in the helical structure? It is embedded during the phase of particle crystallization. According to process described in Chapter 12, the crystallization phase takes place in a gravitational lattice of mixed type, which have a smaller node distance in comparison to the CL space node distance. Such dens lattice will have a larger static pressure that the static pressure of CL space. Then for theoretical analysis we may use the concept of the SG pressure enclosed in the internal RL structure of the helical structures from which the elementary particles are built.

6.9.4.1.1 Embedded SG energy in the FOHSs of the elementary particles.

The energy embedded in the FOHS of the elementary particles could be estimated by the destruction energy. The theoretical problem can be easier treated in a classical void space, which however is not achievable environments for experiments. In this case the results will be useful for CL space structure in order to find out the energy balance between the two types of nodes. From the oth-

er hand, experimental estimation of the destruction energy is possible only in CL space by high energy particle collision experiments. However, one consideration appears to be very useful for solving this problem. When the FOHS appears in CL space and is free to twist it twist until a balance is obtained between following forces: SG forces of the RL structure, SG forces of the helical boundary core and the static CL pressure forces.

A. SG energy estimate assuming empty space environment

The embedded energy will be equal to the destruction energy estimated by the work for expansion of the RL structure from its initial volume to an infinite one. The destruction energy can be estimated by integration of the RL pressure, P_{SG}, (defined by SG forces) on the expanding volume.

$$E_D = \int P_{SG} dV \quad [\text{N m}] \qquad (6.7)$$

$$P_{SG} = F_{SG}/A = \frac{G_{OS} m_0^2}{r^3} \frac{1}{6r^2} = \frac{G_{OS} m_0^2}{6 V^{5/3}}$$

where: r - is the length of the cubical RL cell, A -is the surface area and V is its volume.

Let us assume firstly an isotropic expansion of the unite cell with an initial volume V_0. Solving (6.7) we get

$$E_{RL} = \frac{G_{OS} m_0^2}{4 V_0^{2/3}} \qquad (6.8)$$

Let us apply Eq. (6.8) for both types of RL, keeping in mind that SG constant G_{OS} is one and a same, while the RL node SG masses denoted as m_{0R} and m_{0L} are different. Since the prisms length ratio is $L_R/L_L = 3/2$ we have a volume ratio $V_{0R}/V_{0L} = 27/8$. Making a ratio between both types of E_{RL} we obtain

$$\frac{E_{RL-L}}{E_{RL-R}} = \frac{m_{0L}^2}{m_{0R}^2} \left(\frac{27}{8}\right)^{2/3} = 2.25 \frac{m_{0L}^2}{m_{0R}^2}$$

For energy balance between both types of RL structures the above ratio should be equal to one, so we obtain

$$m_{0R}^2 = 2.25 m_{0L}^2 \qquad (6.9)$$

The dimension of the CL pressure defining the Newtonian gravitation is a *force/area* [N/m^2]. The gravitational force is proportional to the product of the involved masses (square of the same type masses in our case). According to the derived mass equation in §3.13.3, Chapter 3, the mass of the elementary particle is proportional to the CL pressure. While the mass equation was referenced to the external electron volume it was discussed, that when applied for positive particles, a correction factor of 2.25 should be used. The factor 2.25 was discussed in Chapter 2, 3 and 4 and will be used later in this Chapter. Consequently:

The derived factor of 2.25 appears to be equal to the correction factor for estimation of the Newtonian mass of a positive elementary particle.

In the next paragraph it will be shown, that the initial energy balance is satisfied for a volume ratio of 0.8. **This corresponds to the accepted ratio between the electron and positron small radii:** $r_p/r_e = 2/3$.

B. SG energy estimated by particle destruction in CL space

The destruction energy in this case will be different because additional forces are involved - the forces responsible for the Static pressure of the CL space. They will work against the expansion of the released RL structure. Consequently, the released energy of the FOHS destruction will be the difference between the embedded SG energy and the Static energy of the CL space calculated in Chapter 5.

This case is valid for the high energy particle collision experiments. The observed high energy jets are from destruction of the FOHSs of the elementary particles. We must emphasize one important thing:

The embedded energies in +RL(R) and -RL(R) of a single combined helical structure are equal. The embedded energies in +RL(T) and -RL(T) of a single combined helical structure are different.

As mentioned before, the reason is that in the case of RL(T) the Static CL pressure is also involved in the total energy balance

The envisioned energy equivalence between +RL(R) and -RL(R) is of grate importance and we have experimental confirmation of this. This is the mass-energy equivalence between W+ and W- bosons. It is equivalent to the embedded SG energy of +(RL(R) and -RL(R) - from the untwisted straight positive and negative FOHSs of the kaon.

$$E_{D\ RL(R)} = E_{D\ RL(L)} = E_{W+} = E_{W-} = 80.396\ \text{(GeV)}\quad (6.10)$$

In other words this is the **initial energy balance, obtained during the phase of crystalization.**

Consequently, we may conclude that:

- **The kaon inside of the proton, possesses configuration preserved from the time of crystalization.**

The above conclusion fits excellently to the envisioned process of particle crystallization discussed in Chapter 12 and partly in Chapter 2.

The twisting of the RL structure affects the initially embedded SG balance, so the released energy of the pair particles of similar FOHS but opposite RL(T) will appear different. The reason for this is the length and intrinsic mass difference between the right and left-handed prisms. Experimental data supporting this conclusion are the identified energies from destruction of the -RL(T) and +RL(T) structures of the electron, as shown in Table 6.4. They are discussed later in this Chapter.

Summarizing the analysis we may conclude:

- **The individual FOHSs (positive and negative) of a combined FOHS have an equal embedded SG energy if their internal RL structures are not twisted**
- **A definite volume ratio exists for which the embedded SG energies of -RL(R) and +RL(R) are both equal. It is defined by the length and intrinsic mass differences of both type prisms.**
- **The RL structures of the internal kaon are preserved untwisted, since its toroidal shape is preserved. The short lived neutral kaon has also untwisted RL(R) types structures. Its signatures are the observed energies of W+ and W- bosons.**

6.9.4.1.2 Initial SG energy balance between external -RL(R) and internal +RL(R) structures

Let us analyse the case of electron structure. It will be valid for all subatomic structures as the small radii for all FOHS of same type are equal. The centre of radial section of the electron compound FOHS (external negative and internal positive with internal negative core) is a symmetrical point coinciding with the central negative core. For every single turn of the SOHS of the external negative FOHS corresponds, corresponds a single turn of the internal positive FOHS. So we can compare only single turns. The configuration of their RL(R) structures has been discussed in Chapter 2 (see Fig. 2.16 and 2.17).

Fig. 6.10.a shows a part of a radial section of untwisted combined FOHS. The external negative FOHS contains a single layer -RL(R), while the internal positive FOHS contains +RL(R) consisting of multiple sublayers. We must emphasize one important geometrical feature:

A. The radial prisms are connected without gaps, while the gaps between tangentially aligned prisms increases linearly from the internal to the external radius of any layer.

This feature will give us a possibility to estimate the dependence of the SG pressure of any RL sublayer on the sublayer radius. We see that the tangential distance between RL nodes within one sublayer depends linearly on the radius.

<u>Having in mind the inverse cubic SG law, we can express the SG forces by their inverse cubic roots. Such expression is convenient, since it is a linear function of the distance.</u>

Fig 6.10.b and c. show the plots of the inverse cubic root forces respectively for +RL(R) and -RL(R) (thick lines). They are in own scale, referenced to the own unit distance (the length of the right and the left prism, respectively). We may normalised the inverse cubic forces to the minimal node distance, which is twice the prisms' length (valid for the internal radius of any sublayer) and also to the product $G_{os}m_0^2$, where G_{os} is the SG constant between objects of same substance and m_0 is the intrinsic mass.

The normalised SG energy in a single layer is

$$E_i = \int P_i dV . \qquad (6.11)$$

where: P - is the layer SG pressure, V - is a layer volume, i - layer index

Since the tangential distance change linearly with the radius, as seen in Fig. 6.10.a, the inverse cube of the SG force will change also linearly as shown in Fig. 6.10.b.

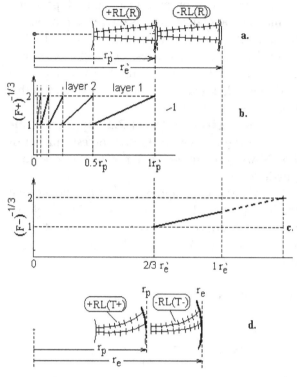

Fig. 6.10

a. Radial section showing the RL nodes of an external -RL and internal +RL layer

b. Tangential inverse cubic SG force (thick line) as a function of layer radius for +RL layer

c. Tangential inverse cubic SG force (thick line) as a function of layer radius for the external -RL layer

d. Twisted internal +RL(T) layer and external -RL(T) layer of compound FOHS (containing a positive FOHS inside of a negative FOHS)

For any sublayer of +RL(R) we have:

$$dE = (F(r)/A(r))dV \qquad (6.11)$$

where: F(r) is the radial component of SG force, $A(r) = \pi r^2 L$ is a cylindrical volume with a length L equal to the FOHS length

From (6.10.b) the inverse cubic SG forces for sublayers 1 and 2 are:

$$(F)_1^{-1/3} = 2r(G_0 m^2{}_0)^{-1/3} \qquad (F_2)^{-1/3} = 2r(G_0 m^2{}_0)^{-1/3}$$

In order to integrate Eq. (6.11) we must express F and A functions of volume, knowing that $V = r^3$. The range of integration in volume scale (normalised to 2π) is:

For $0.5 < r < 1$ we have $0.25 < V < 1$

For $0.025 < r < 0.5$ we have $0.0625 < V < 0.25$

Normalising to the parameters that are common constants for the two sublayers we obtain

$$E_{1n} = \int_{0.25}^{1} \frac{1}{2V^{2/3}}dV \qquad E_{1n} = \int_{0.0625}^{0.25(1)} \frac{1}{4V^{2/3}}dV$$

$$E_{1n} = 0.55506 \quad E_{2n} = 0.17483 \quad E_{2n}/E_{1n} = 0.31498$$

Following the geometrical rule that the thickness of every internal layer is a have of the thickness of the external one, the normalized energy sum of all sublayers (a total SG energy of RL(R) is

$$E_L = 0.55506 \sum_{i=1}^{l} 0.31498^{i} \qquad (6.12)$$

The sum E_L (index L stands for lefthanded RL nodes) diverge very fast for $i > 6$, while from geometrical considerations i may approach 20. The contribution of the sublayers with a large i-number is insignificant. Then we have: $E_L = 0.81028$

The SG energy of the negative RL layer can be expressed in a similar way for a correctly defined range in own length scale

For $0.5 < r < 0.75$ we have $0.25 < V < 0.421875$

$$E_R = \int_{0.25}^{0.421875} \frac{1}{V^{2/3}}dV = 0.3601184 \qquad (6.13)$$

The SG energies E_L and E_R are based on SG forces estimated in own length scale, and normalized to the product $G_{0S}m^2_0$. Consequently, the ratio between E_R and E_L will give the ratio between the SG masses of the two types RL nodes.

$$E_L/E_R = 2.25 = (m_{0L}/m_{0R})^2$$

$$m^2_R/m^2_L = 2.25 \qquad (6.14)$$

The derived Eq. (6.14) is the same as Eq. (6.9) - a ratio between the right and left handed RL nodes. It is also the ratio between the two types of prisms and CL nodes. We must keep in mind that this corresponds to the ratio $r_p/r_e = 2/3$ of the electron structure.

In the next paragraph we will see that Eq. (6.14) plays a role of SG mass asymmetrical factor when the SG energy is estimated by CL space units. Then the equality $E_L m^2_L = E_R m^2_R$ means that during the phase of particle crystalization, the SG energy of untwisted +RL(R) is equal to the SG energy of -untwisted -RL(R). While the SG energies are balanced during the phase of crystalization, they are not balanced when the helical structure occurs in CL space environment. In order to restore the balance it undergoes twisting in which the following conversion takes place:

right-handed RL(R) -> -RL(T)

left-handed RL(R) -> +RL(T)

In the new balance, however, the forces of the helical boundary are involved. We obtain information about the embedded SG energy only if the FOHS is disintegrated. This happens in the high energy particle collision experiments. In such experiments stable and unstable (short lived) particles are involved. Electron, positron, proton and neutron are stable elementary particles. This means, from a BSM point of view, that their helical structures are stable in CL space environment (a correct energy balance). The unstable particles appear only if some of the stable particles helical structures are broken.

The total SG energy balance of any particle or helical structure assures also a constant charge unity of the particle. When the structure of the stable particle is damaged, the released RL(T) (possessing a large embedded SG energy) may disturb also the mechanism of the charge unity. In such case, the released energy might be quite big and can mimic a real pion or muon, while in fact it is a virtual charge wave in CL space moving with a speed of light.

Released SG energies take place in high energy particle collisions, such as: collision between the electron and free positron (e+ e- reaction) showing two released energies: 1.7778 GeV and 1.44 GeV. The first one is from +RL(T) and the second one is from -RL(T). Another example of such process is the energy of 91,18 GeV, known in particle physics as a Z bozon particle. According to BSM, this is the SG energy of the -RL(T) structure of the short lived negative kaon.

Conclusions:

- **For untwisted combined FOHS, consisted of internal positive and external negative FOHS, the embedded SG energies of their internal RL structures are equal**
- **The insignificant energy contribution from the most internal layers provides a useful tolerance for termination of the process of RL crystallization.**

- **The twisting of a combined helical structure (which takes place in CL space) change the initial energy balance**
- **Untwisted RL(R) is preserved in the internal kaon and in the short lived neutral kaon (after the proton or neutron break)**
- **The electron structure with its internal positron is an example of a single coil compound helical structure in which both types of RL structures are twisted (+RL(T) and -RL(T)).**

6.9.4.1.3. Geometry of RL(T) in twisted FOHS

When the helical structure occur in CL space environment, its internal RL(R) converts to RL(T). In this case the volume of the FOHS slightly change, which affects its Newtonian mass and appearance of charge. We distinguish:

FOHS of $K_O^S \rightarrow \pi$: RL(R) -> RL(T)   (charge appearance)   (1)
FOHS of $\pi \rightarrow \mu$: RL(T) ->RL(T)   (k_S - measurable)   (2)

The twisting in the first case is smaller than in the second one. We have derived the radius shrink factor k_S for the second case (§6.4.3).

The shrinkage of the twisted FOHS affects mainly the geometry of the radial stripes of the internal RL structure. This is illustrated in Fig. 6.11.

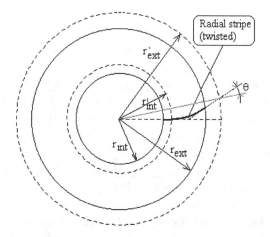

Fig. 6.11. Change of radii of the most external RL sublayer and the geometry of the radial stripes for twisted FOHS. The radial section before twisting is shown by dashed lines. θ - is the pitch angle of the twisted radial stripe

Knowing the radius shrinkage coefficient k_S for pion-muon conversion $\pi \rightarrow \mu$, and keeping in mind that the length of the radial stripe is un-

changed, one may determine the angle θ, by using geometry and assuming a constant curvature of the radial stripe or a parabolic curvature. This angle may play an important role in the magnetic interaction of the helical structures with an external magnetic field.

6.9.4.2.1 Asymmetrical reaction of the CL space

The selected ratio between prisms length 2:3 in fact predetermines the other parameters of the prism as the ratio between their intrinsic masses. In order to validate the selected ratio, BSM provides extensive cross-validating analyses with comparison with some experimental data. From the other hand this ratio is important for correct crystallization condition in the phase of particle crystallization, as discussed in Chapter 12.

So far we used the length or SG mass of the prisms or RL nodes referenced to one of both types - right or lefthanded prism (RL node). In CL space environment, however, we need only one reference for the length and for the Newtonian mass, since they are the measurable parameters in the experiments and observations.

A. Equilibrium node distance for CL space

Let us express the SG forces between pair of right handed and pair of left handed CL nodes, but for the own system length unit defined by the prism length. We can call them unit forces.

$$F_{R1} = \frac{G_{os} m_R^2}{1_R^3} \qquad\qquad F_{L1} = \frac{G_{os} m_L^2}{1_L^3}$$

where: F_R and F_L are respectively the right and left handed unit forces; 1_R and 1_L are respectively their unit vectors, proportional respectively to the right and left-handed prism's lengths.

We cannot compare both forces directly, as they are referenced to different scales. However, we may compare the change of their SG forces as a function of a distance change $\Delta F/\Delta L$, measured in their own systems (referenced to right and lefthanded prism scales). At the same time, we may express the distance change by a third scale, referenced to the unit of the CL space, which must be a common equivalent length unit.

Fig. 6.12 illustrates the units in the three scale systems. The CL unit is denoted by x. Keeping the distance between the node pairs the same, and referencing the unit forces, not to own length unit, but to the point X, we have the relations:

$$F_{RX} > F_{R1} \quad \text{and} \quad F_{LX} < F_{L1}$$

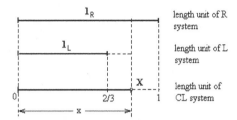

Fig. 6.12 Cross reference of the length unit scales

Applying the SG law for the ratios between left and right-handed forces we get:

$$\frac{F_{RX}}{F_{R1}} = \frac{G_{os}m_R^2\frac{1}{\frac{x}{3}}}{G_{os}m_R^2\frac{1}{(1_{CL})^3}} = \frac{1}{x^3} > 1 \qquad (6.25)$$

$$\frac{F_{RX}}{F_{L1}} = \frac{G_{os}m_L^2\frac{1}{\frac{x}{3}}}{G_{os}m_L^2\frac{1}{\left(\frac{2}{3}(1)_{CL}\right)^3}} = \left(\frac{2}{3}\right)^3\frac{1}{x^3} > 1 \qquad (6.26)$$

In order to find the equilibrium point, we make the product of both ratios equal to unity.

$$\frac{F_{RX}F_{LX}}{F_{R1}F_{L1}} = 1 \quad \text{or} \quad \frac{1}{x^3}\frac{8}{27}\frac{1}{x^3} = 1$$

Then for the equilibrium distance x, we get the values, representing the unit length correction factors:

$$x = 0.8165 \times 1_R = 1_{CL} \text{ - relation between CL and R} \quad (6.27)$$

$$x = 1.2247 \times 1_L = 1_{CL} \text{ - relation between CL and L} \quad (6.28)$$

Let us denote:

$K_{CL} = 1.2247$ -CL space correction factor for length **(6.28.a)**

We can call the unit length 1_{CL} - an equilibrium node distance for CL space, and the factor K_{CL} - a CL space correction factor for length, referenced to the longer prism

B. SG mass asymmetrical factor of the CL space

By analogy to the equilibrium node distance, we can define an equilibrium CL node mass, denoted as m_{CL}. The SG law in CL system between same type of nodes at fixed distance of 1_{CL} could be written as

$$F = G_{os}\frac{m_{CL}^2}{1_{CL}^3} = G_{os}m_{CL}^2 \qquad (6.29)$$

We may introduce an asymmetrical factor denoted as k_m and express the forces between two pair of nodes of one and a same type, but referenced to the CL system units. We will use the square root of forces for convenience (as a square root of Eq. 6.29).

$$\sqrt{F_R^{CL}} = \sqrt{G_{os}}m_{CL}\frac{1}{k_m} \quad \text{- for pair of R nodes} \quad (6.30)$$

$$\sqrt{F_L^{CL}} = \sqrt{G_{os}}m_{CL}k_m \quad \text{- for pair of L nodes} \quad (6.31)$$

The Eqs. (6.30) and (6.31) must be valid not only for R (right-handed) and L (left-handed) CL nodes but also for R and L types of helical structures in a CL space environment.

The product of the above two expression is

$$F^{CL} = \sqrt{F_R^{CL}F_L^{CL}} = G_{os}m_{CL}^2 \qquad (6.32)$$

We have a right to use a square root of the product of both forces, if their interactions are interconnected. So the physical meaning of the expression (6.32) is a common interaction. (The square root rule for the interconnected interaction were also used in the derivation of the light equation in Chapter 2.)

The Eq. (6.32) is equivalent to Eq. (6.29). The introduced factor k_m is eliminated in Eq. (6.32). This is in case of equal interactions for R and L, discussed in the previous paragraph or, said in other words - a symmetrical behaviour of CL space.

The factor k_m should persist for a not symmetrical behaviour on R and L. We can derive the value of this factor by equalising the pair forces referenced to the 1_{CL} unit distance, but with own intrinsic masses, properly multiplied or divided on the factor k_m. The factor k_m in this case expresses the asymmetry of the SG interaction between R and L in a CL space environment. For equivalence between R and L forces in CL space we must refer the SG forces to the unite length of CL space 1_{CL}:

$$G_{os} \frac{m_R^2 1/k_m^2}{1_{CL}^3} = G_{os} \frac{m_L^2 k_m^2}{1_{CL}^3} \qquad (6.33)$$

According to Eq. (6.9) and (6.14) we have $m_R^2/m_L^2 = 2.25$, from where

$$k_m = \sqrt{2.25} = 1.5 \qquad (6.34)$$

Eq. (6.33) shows that when estimating CL parameters proportional to SG forces, the square of the mass asymmetrical factor k_m should be properly used.

Conclusions:

- **The SG mass asymmetrical factor k_m is involved in the destruction process of RL(T) in CL space environment. Consequently, its signature must appear in the analysis of some experiments in particle physics.**

6.9.4.2.2 SG mass asymmetrical factor k_m in the mass equation for particles of positive FOHS

In the derivation of the Newtonian mass equation in Chapter 3, the CL static pressure was referenced to the electron, whose external shell is accepted to be from right hand prisms (negative). In this case the mass equation does not contain a correction factor for SG mass asymmetry. However, when applied for a positive particle (helical structure) the square of the mass correction factor k_m should be used. Then one question arises: Why the square of k_m must be used?

The SF forces are proportional to the square of the SG masses and consequently to k_m^2. The Static CL pressure is a result of SG forces. Since the Newtonian mass is proportional to the Static CL pressure (see Eq. (3.48)), it should be proportional to k_m^2.

Therefore the factor k_m^2 must be taken into account for the positive helical structures, such as the positron, the positive pion and muon. This factor is used in the mass balance of the proton (neutron) provided later in this Chapter, which is cross-validated with number of experimental data.

6.9.4.3. Identification of the particle structure by the embedded SG energy released in the particle collision experiments

In Chapter 3 we derived the mass equation of the elementary particle by using the volume of their FOHSs. In the theoretical analysis presented in this chapter we find out that the volume of the FOHS is involved also in the embedded SG energies. This is very important issue, since it give us a possibility to make a cross correlated calculation between the well known masses (newtonian) of the particles and some accurately measured energies from particle collision experiments. For this purpose we will use also the following factors derived in the previous sections:

$k_V = 0.9644$ - volume shrink factor for conversion of pion to muon (see §6.44.1)

$k_{CL} = 1.2247$ - CL space correction factor for length (§6.9.4.2.1)

$k_m = (m_{0L}/m_{0L})^{1/2} = 1.5$ - SG mass assymmetrical factor in CL space (Eq. 6.34)

From a pure geometrical considerations it is evident that the degree of twisting changes not only the FOHS volume, which is proportional to the particle mass according to the mass equation. It changes also the step of the obtained SOHS. In such way, when the straight FOHS of the kaon get twisting, it becomes a SOHS with a defined helical step (pion). Additional twisting converts the pion to muon, which have a smaller step of its SOHS - equal to the helical step of the electron (positron). Consequently, we have the relation:

Degree of twisting /l\ Mass \l/ Helical step \l/   (6.35)

The twisting of the electron and the positron is the same as the twisting of the muon. The relation (6.35) and factors, k_V, k_{CL} and k_m used in a correct way permit to identify the released SG energies and to find out from what type of helical structures the are by using data from particle collision experiments. The main criteria will be the equivalence of the SG energy from both types of RL(R) structures.

We must keep in mind that only FOHS with free ends can undergo a twisting. Those connected in loops, like the internal central kaon and pions, cannot undergo such process, so they preserve their untwisted or partially twisted internal RL structures, obtained during the particle crystallization. This means that they preserve also their initial volume which define their inertial mass. This is a very useful feature, which when combined with the estimated SG energy from the particle collision experiments will permit to restore the whole picture of the internal structure of the proton (neutron). Figure 6.13 illustrates how an unite volume of a FOHS is changed when undergoing twisting in CL space. In the same figure the correspondence between some subatomic particles and the type of their helical structures is shown..

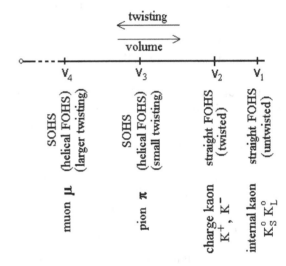

Fig. 6.13 Change of unit volume of FOHS with the degree of twisting

From the muon to pion decay we have
$$V_4/V_3 = k_V = 0.9644$$
The FOHSs of the electron (external negative and internal - positive) have the same volume shrinkage from the twisting like the muon. Consequently, the SG energy embedded in the electron's FOHSs can be used as a reference unite in a same way that the FOHSs volumes were used as a reference unite for the Newtonian mass.

From the high energy collision of electron and positron in which energies of 1.7778 GeV (tau particle) and 1.44 GeV are measured, we have:

$$\frac{1.7778 \, k_{CL}^3}{1.44 \, k_m^2} = 1.008 \qquad (6.36)$$

Eq. (6.36) indicates that the energy 1.7778 GeV (known so far as a tau particle) and the energy 1.44 GeV fulfil the SG energy balance between -RL(R) and +RL(R) of untwisted FOHS.

Consequently:

(a) 1.7778 GeV - is the SG energy released from the destruction of a +RL(T) embedded in a single coil of a positive FOHS - the positron

(b) 1.44 GeV - is the SG energy released from the destruction of the -RL(T) embedded in a single coil of a negative FOHS - the degenerated electron

From the conclusions in §6.9.4.1.2 it follows that the SG energies of -RL(R) and +RL(R) in untwisted pair of FOHS are equal. Therefore, we must look for two identical energies obtained in particle collision experiments. Such energies are for W^- and W^+ bosons (80.396 GeV). Obviously they will come from untwisted FOHSs. The only such structures are the helical structures of the central kaon. This is additionally verified by a mass equation balance of the helical structures comprised in the proton (neutron).

80.396 GeV - SG energy in -RL(R) of central kaon
80.396 GeV - SG energy in +RL(R) of central kaon

Let us use now the experimental data of Z boson corresponding to energy of 91.187 GeV. From the analysis in §6.9.4.1 it was evident that for one and a same FOHS the SG energy when it is twisted is larger than when it is untwisted. Then our analysis identifies the the 91.187 GeV as an energy released from a twisted negative straight helical structure of the kaon. Checking the initial balance we get

$$\frac{80.396 k_m^2 k_V}{91.187 \; k_{CL}^3} = 1.041 \qquad (6.37)$$

Eq, (6.37) shows 4% difference in the energy balance because the volume shrink factor k_V was derived from the pion-muon conversion. From Fig. 6.13 we see that the twisting of kaon, while it is still a straight FOHS is smaller than the twisting of the pion and muon, so the unite volume is different. So we must correct slightly this factor for the case of Z boson energy.

$$\frac{80.396 k_m^2 0.926}{91.187 \; k_{CL}^3} = 1.0000 \qquad (6.38)$$

where: $0.926 = V_4 / V_2$

Using the obtained volume shrink factor of 0.926 and the conclusions (a) and (b) we may estimate the energy released from destruction of the positive straight FOHS of the kaon, by using the corresponding relation:

91.187   -   1.44
x   -   1.7778

We must correct the proportionality by a factor of $0.926 / 0.9644$, since a correction volume shrink factor was used in Eq. (6.38), so the result is:

$$X = \frac{91.187 \times 1.7778}{1.44} \frac{0.926}{0.9644} = 108.09 \text{ GeV} \qquad (6.39)$$

The calculated energy of 108 GeV must be the energy of the searched X boson. A strong evidence for resonance at 105 GeV is found in the CERN document titled " Search for Higgs bosons: Preliminary combined results using EP data collected at energies up to 202 GeV" (CERN-EP-2000-055, April 25, 2000). The document is posted in Internet. Fig .6.16 shows the resonance as "reconstruction Mass mH [GeV/c²] (corresponding to Fig. 4 of the document).

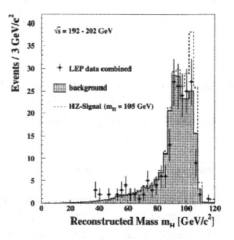

Fig. 6.16

(from technical document, published by CERN, 2000)
Paper text: "*Fig. 4 : LEP-combined distribution of the reconstructed SM Higgs boson mass...The figure displays the data (dots with error bars), the predicted SM background (shaded histogram) and the prediction for a Higgs boson of 105 GeV/c^2 mass (dashed histogram)....*"

Another indication of resonance at 105 GeV is found from the paper presented by W. M. Yao at FERMILABAB -Conf-99/100-E, and posted also in Internet.

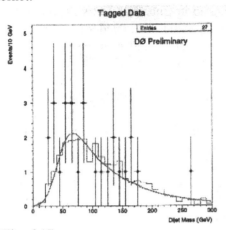

Fig. 6.17

Paper text: "*Fig.1. The measured two-jets mass distribution along with background ...*"

The paper shows a plot of the parameter "two jet mass distribution", in which slight evidence of the 105 GeV resonance is apparent. The statistical noise is large due to the small number of obtained events. Fig. 6.17 shows the plot of the above mentioned parameter (corresponding to the Fig. 1. (right) of the paper)

In the experimental data shown in Fig. 6.16 and 6.17 it is apparent that the energy of the 105 GeV resonance is very weak in comparison to the resonance at 91.188 GeV, so its peak position may be shifted in the energy scale, while the real value could be 108 GeV. Why it is so weak? There very reasonable considerations for this.

The destruction of the straight external negative FOHS with its -RL(T) (91.187 GeV)   may take place not only when both structures are separated, but also when they are still combined. The destruction get has a finite spatial angle as shown in Fig. The same condition cannot happen for the +(RL(T) structure while it is still inside of the negative FOHS. Therefore the peak of 108 GeV resonance can be much smaller.

Additionally, the central negative core could catch part of positive internal RL(T) layers, forming a different jet. This case will be similar as the observed 3-rd track of high energy e+e- collision, where the third track is from some layers of

+RL(T), catch the central negative core (see Fig. 6.6 (courtesy of J. Z. Bay et al, (1996)).

One additional experimental proof for the presented here concept is the observed asymmetry of Z boson "mass". It is known as forward-backward charge asymmetry" (see Abreu, (1994)). More accurately this is asymmetry between the right and left wing of the resonance energy. The $w^\pm$ boson "masses" does not posses such asymmetry. Experimental data about Z boson "mass" asymmetry, referenced as forward-backward charge asymmetry is shown in Fig. 6.18 (Abreu, (1994))..

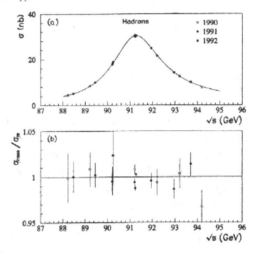

Fig. 6.18

BSM interpretation: The asymmetry between left and right wing. Courtesy of Abreu, (1994). Source paper text: *Hadronic cross section from 1990, 1991 and 1992 data. In (a) the data are shown together with the result of 5-parameter fit. Plot (b) shows the ratio of the measurements to the best fit value*

The BSM explanation of the asymmetry is the following: The destruction energy of RL(T) as a function of degree of twisting exhibits a definite trend. The twisting process, involving large SG energies, may exhibit small aperiodic oscillations of the twisting angle θ. (especially valid for K^- and K^+, which are straight FOHSs). The destruction process may be initiated during these oscillations. The variation of θ means variation of the volume of the FOHS, which affects the SG pressure and consequently leads to variation of the released destruction energy. Then the observed Regge resonance will be a weighted function of number of individual destructions for a Gaussian type of deviations centred around one mean value. At the same time, the SG pressure is not linearly dependent on the degree of twisting. This explains, why the right-hand wing of the Z boson "mass" is higher than the left hand one

Table 6.5 shows some of the identified SG energies and the type of the helical structures they are embedded in.

SG energy and corresponding structure　　**Table 6.5**

SG energy (GeV)	Structure	"Particle"	From
1.7778	+RL(T)	tau	+FOHS of e+
1.44	-RL(T)		-FOHS of e-
80.396	+RL(R)	W+ boson	+FOHS of kaon
80.396	-RL(R)	W- boson	-FOHS of kaon
91.187	-RL(T)	Z boson	-FOHS of kaon
108 (calc.)	+RL(T)	105 GeV res.	+FOHS of kaon

Conclusions:

• **The high energy Regge resonances in particle collision experiments are signatures of embedded SG energies in the structure of elementary particles. The presented theoretical approach allows identification of the particular helical structures, corresponding to the released and experimentally estimated SG energies.**

6.9.4.4 The real newtonian mass of the kaon

The kaon is a big enigma in particle physics. The BSM succeeded to explain the mysterious features of the kaon, after unveiling one of its puzzles. It is related to the mass of the kaons K_L^0, K_S^0, K^+, K^- estimated from the particle collision experiments. The mass suggested by the present interpretation did not not fit to BSM model. Extensive BSM analysis on the kaon interaction led to the conclusion that the provided mass is highly overestimated. The reason is the following: The cut kaon in all of its modifications is a straight FOHS and during its lifetime it has a jet of disintegrating RL structure, which provides a propulsion reaction force. While this effect is not envisioned so far by the particle physicists, the inertial mass of all kaon modifications are overestimated significantly.

Using the derived mass equation and masses of all helical structures embedded in the proton

(neutron) the correct rest mass of the internal kaon is estimated later in this chapter.

The effect of kaon disintegration appears to exist also in a cosmological scale. When pulsars are investigated in Chapter 12, it was found that they contain a bundle of aligned kaons obtained from destructed protons and neutrons in the wombs of the stars due to an enormous gravitational pressure. From one hand, the kaon bundle exhibits a super-strong anisotropic SG field, causing a superstrong magnetic field. From the other hand it possesses also a destruction effect and consequently a huge propulsion forces, which causes the gigantic motion of the pulsar through the home galaxy. The presented pulsar theory in §12.B.6.4.3 shows that the rate of RL release per unit CL space volume is constant ($\Lambda / V = $ const).

The destruction of a single kaon have some similar feature related to disintegration of the FOHS with the internal RL. The destruction of the RL structures starts after the cut kaon exits from the cut proton (neutron). The destruction is in a form of a single jet providing a strong thrust force of the structure. The aligned SG forces may influence the spatial angle of the jet fuse and also the fuse cone volume. Since they kaon structure continuously decreases, the effect may mimic a passive particle with an initial birth velocity. Fig. 6.21 illustrates a destruction of a kaon, showing the jet fuse in its "burning" end.

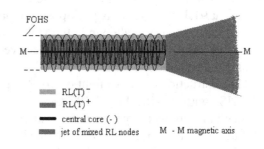

▦	RL(T)⁻
▦	RL(T)⁺
━	central core (-)
▦	jet of mixed RL nodes

M - M magnetic axis

Fig. 6.21
Destruction of the straight FOHS of a kaon showing the jet in its "burning" end

It is apparent, that:
If the kaon is considered as a passive particle with a finite lifetime its rest mass will be significantly overestimated.

The kaon jet may provide not only incorrect experimental masses of kaons's structure modifications but also for the eta particle and short lived "antiproton" (they both possess a cut kaon), (see §6.12.2.4).

Question: Why the experimental masses of pions and muons appear correctly?

Answer: Both particles posses a second order helicity. In the process of RL(T) destruction, the debit of released RL nodes depends on the following three parameters: linear velocity, spin rate and angle between the jet and the structure axis. Then the peripheral part of the high velocity jet containing packet RL(T) structure may reach the light velocity, but the flow debit could be below the parameter $\Lambda/V = const$. In such conditions a confined motion of the particle is performed, permitting to detect its correct newtonian mass. For the muon, the angle between the jet and the structure axis is close to 90 deg, which provides a better conditions for confined motion. For this reason the muon lifetime (the burn time of its helical structure) is about two order of magnitude larger in comparison to the pion.

6.9.4.5 Fermi coupling constant and effective mixing ratio for "leptons"- signatures of the helical structure twisting

The Fermi coupling constant is one of the basic parameters of the electroweak theory. It has been recognized, from particle physicists that the radioactive Beta decay and the muon lifetime are related to one and a same physical process, which signature is the Fermi coupling constant. BSM theory provides a new insight about this constant. The Beta decay as we see is related to the conversion of a proton to neutron. In this process there is not destruction of any helical structure. Only twisting is involved. Then the similar process is involved in the conversion of the pion to muon. In the latter case, the conversion process evidently affects the lifetime of the muon, as this was discussed in §6.4.4.5.

In the following analysis we will give a simple mathematical proof for the relation between the Fermi coupling constant and the twisting of the helical structures.

A. Twisting SG energy estimated by the embedded SG energies

The twisting equivalent energy, E_T, is the energy difference between the SG energy of $RL(T)$ and $RL(R)$, for one and a same structure

$$E_T = E_{D(T)} - E_{D(R)} \quad \text{GeV}$$

Using the SG energy data for kaon from Table 6.5 (we will use the resonance at 105 GeV)

$$E_T^+ = 108 - 80.396 = 27.6 \quad \text{GeV} \quad \text{for } K^+ \quad (6.47)$$
$$E_T^- = 91.187 - 80.396 = 10.79 \quad \text{GeV} \quad \text{for } K^- \quad (6.48)$$

Note: The calculations are for single structures only, i.e. the negative structure does not contain internal positive one. According to BSM, this corresponds to the experimental conditions for the bosons mass estimation.

Other basic parameter in Electroweak theory is the effective mixing parameter $\sin^2\theta_{eff}^{lept}$ for leptons.

B. Twisting SG energy estimated by the Fermi coupling constant

The Fermi coupling constant come from Enrico Fermi theory. It is usually normalised to the mass of the involved nucleon. So for a Beta decay or pion to muon decay it is normalized respectively to the proton mass or muon mass. Using experimental data from a Beta decay of O^{14} (provided by Bromley et al., 1955) Feynman and Gell-Mann (1958) estimated the Fermi coupling constant as $(1.01 \pm 0.01) \times 10^{-5}/m_p^2$, where m_p is the proton mass. Substituting $m_p = 0.9383 \, \text{GeV}$, we get the value of $1.1358 \times 10^{-5} \, [\text{GeV}]^{-2}$. This constant now is more accurately determined by the lifetime of muon decay. The currently accepted value of the Fermi coupling constant according to NIST data is:

$$G_F = 1.16639 \times 10^{-5} \quad [\text{GeV}]^{-2} \qquad (6.49)$$

Expressing Eq. (6.49) in units of GeV, we get the equivalent SG twisting energy for a neutron-proton conversion, $E_{T(p)}$.

$$E_{T(p)} = [1.16639 \times 10^{-5}]^{-1/2} = 292.8 \quad \text{GeV} \quad (6.50)$$

Now the Fermi coupling constant is more accurately measured from the pion-muon conversion, by estimation of the muon lifetime according the expression suggested by W. Marciano and A. Sirlin, (1988). This expression relates the muon lifetime with the masses of muon, electron and W boson, the fine structure constant are the Fermi coupling constant.

The NIST value of the Fermi coupling constant is determined by the measuring the muon lifetime that according to BSM is related to the twisting in the pion-muon conversion. The NIST value is given as normalised to the proton's mass, m_p^2. Let us renormalise it to the muon mass by a proper use of the mass ratio between the proton and muon $m_p/m_\mu = 0.9383/0.1057$. Then we get the average SG twisting energy for the pion-muon conversion, mentioned above.

$$E_{T(\mu)} = \left[G_F \frac{m_p^2}{m_\mu^2} \right]^{-1/2} = 32.98 \quad [\text{GeV}] \qquad (6.51)$$

The twisting energy of K^+ and K^- FOHS are given by Eqs (6.47) and (6.48), but they are shorter than the FOHSs of the pions. This feature, recognized by BSM analysis, was not recognised so far by particle physicists, since the reaction of kaon-pion-muon is very rear, so it gives a very low muon production in comparison to the pion-muon reaction. In §6.11 it was found that the ratio between the pion to kaon FOHS lengths is 3.082. Dividing (6.51) on this factor we get the twisting SG energy of the kaon, that appears to be quite close to the value of $E_T^- = 10.79$ GeV, obtained by Eq. (6.48).

$$32.98/3.082 = 10.7 \quad (\text{GeV}) \qquad (6.52)$$

The fractional error between the 10.79 GeV (from the Fermi coupling constant) and the twisting energy of 10.7 GeV (from the Z boson - W boson energy (mass) difference is only 0.8%. <u>This is a confident proof of the physical effect of twisting of an open FOHS in CL space.</u>

C. Effective mixing ratio as a signature of FOHS twisting

The effective mixing ratio for leptons is another parameter, which is related to the twisted of a FOHS, according to BSM. Let us use the effective mixing parameter for "leptons", to which category the electron belongs. We may rely on this parame-

ter, as it is experimentally determined, from on-resonance observations at LEP (Large Electron Positron collider at CERN, Switzerland) and SLC (Stanford Linear Collider, USA). The best averaged value for this parameter, according to G. Degrassi et al. (1997) is:

$$\sin^2\theta_{eff}^{lept} = 0.23165 \pm 0.00024 \qquad \theta_{eff}^{lept} = 28.77^0 \quad (6.53)$$

Let us denote the ratio between W-boson and Z-boson SG energies (or their mass equivalent) as $80.396/91.187 = \cos\theta$. Then we have

$$\sin^2\theta = 0.22267 \qquad \theta_{eff} = 28.15^0$$

We see that the calculated value of θ_{eff} from kaon twisting is closer to θ_{eff}^{lept}, valid for electron with a fractional error of 2%.

Summary:
- **The Fermi coupling constant is estimation of the equivalent SG twisting energy**
- **The nuclear β decay, the pion - muon decay, and the neutral kaon - charge kaons decay are related with one common phenomena - twisting of their FOHS's.**
- **The twisting of the FOHS does not change its length**

6.10 Physical explanation of the Newtonian gravitation and inertia of the elementary particles as interactions between SG(CP) forces of the RL(T) structures from one side and the CL space from the other.

6.10.1 Relation between RL state and SG forces balance

From the provided analysis and the interpretation of the experimental data, we found that the destruction energies of +RL(R) and -RL(R) are balanced for a radius ratio $r_e/r_p = 2/3$. When the FOHS is twisted, the radius ratio is preserved, but the destruction energies of +RL(T) and -RL(T) are different. The criterion for a stable helical structure requires a balance of the SG forces according to Eq. (2.8). In this equation the opposing bending forces of the helical envelope are also included. In CL space, the interaction forces between RL(TP) and CL node (TP) forces are also included, but they are comparatively small and could be ignored. We may accept that:

- The degree of twisting is determined by the balance of all SG forces according to Eq. (2.8).
- The SG forces of stable helical structures in CL space are not detectable in the normal interaction processes

The first conclusion means that the free FOHS gets twisting until the force balance according to Eq. (2.8) is fulfilled.

The second conclusion is valid for a normal processes of interaction, in which the stable helical structure does not change its twisting and shape. This exclude the high energy interaction as the particle destruction and high energy scattering (including Babha and Moller scattering). **The normal processes of interactions involve:**
- Neutonian gravitational mass
- Neutonian inertia
- Electric field
- Magnetic field

6.10.2 Neutonian gravitational and inertial mass of a helical structure

Let us analyse the possible interactions between the SG(CP) forces of a helical structure and the nodes of CL space. For these purpose, the following two cases will be considered.

Case A: a single particle formed of external and internal FOHS's is away from any other helical structure or gravitational filed

Case B: two identical particles formed of FOHS's are in a close distance, but away from any external gravitational field.

A typical example for the case A is the single electron, when it is away from other particles possessing a newtonian mass. Let us consider in first that the electron is in rest. Both RL(T) structures carry superdens intrinsic masses (and SG energies), but their SG(CP) forces are accurately balanced due to the symmetry. **As a result of such balance, they individual SG fields do not interact with the CL space nodes, so they appear as hidden.** The SG(TP) forces, however, are not balanced (due to the prism to prism interaction, discussed in Chapter 2), and they contribute to the E field of CL space, characterised by an unit charge. **Now let consider the case when the same particle is in motion** with a not-relativistic velocity. It now possesses a kinetic energy. **The universal energy conservation principle requires that the energy balance in the SG system should be preserved.** The electrical charge, contributed by the SG(TP) is a same, so the SG(TP) interaction with the CL space is a constant. The only left option for restoring the energy balance is a possible interaction between SG(CP) forces and the nodes of the CL space. This means that SG(CP) field for the moving structure is not any more hidden for the CL space. The generated disturbance from the twisted SG(CP) of the most external RL layer, is involved in this type of interaction. **This effect is manifested as a Newtonian inertia of the helical structure**.

In case B, we have the same two particles but in a close distance. In this case, some influence could appear between the SG(CP) of the structures, so the accurate balance between the SG(CP) forces of the structure and the CL space should be disturbed. In case of combined helical structure (internal positive and external negative) possessing RL(T)s, the major contribution for the disturbed balance is from the external shell. Due to the disturbed balance, a forces of attraction appears between the two particles, whose sum could be expressed by a common attraction force. The propagation media for these forces is the CL space between them. It becomes apparent, that the common attraction force is propagated by the (CP) of the prisms forming the CL nodes. We may express the common force as a summation of the individual forces from every point of the first particle, exhibiting SG interaction from the second one within subtended angle. If the distance between the two particle changes, the subtended spatial angle also changes with the inverse square root of the distance. In this interaction the CL space nodes are involved, and they are always outside of the FOHS volume. This is completely consistent with the definition of the mass equation, valid for the newtonian mass. So we arrive to the conclusion, that the described above interaction is a Newtonian gravitation, known as a "universal gravitational law".

We see that in the Newtonian gravitation the CL nodes are directly involved, however, their intrinsic masses are hidden. The newtonian mass is proportional to the static pressure, while the latter has some connection to the SG forces.

Conclusions:

- **The Newtonian gravitation is a result of disturbed balance between the SG(CP) of two close helical structures in a CL space environment. Its manifestation is the Universal gravitational law, derived By Isaac Newton.**
- **The Newtonian gravitation depends only on the volume of external helical shell of FOHS**

The second conclusion provides explanation of the **effect of the hidden neutonian mass.** This effect appears for the internal positron, and the positive internal structure of the negative muon and pion.

6.10.3 Radial gap between the two different types of helical structures in a combined helical structure

Let us consider again the structure of the electron. For proper oscillation of its two helical structures, two radial gaps are needed:

- between the -RL(T) and the positron helical structure

- between the central negative core and the internal radius of the most internal +RL layer.

It is experimentally verified, that electrons and positrons may obtain a strong polarization after striking a plane surface at oblique angle. The strong polarisation, according to BSM, is indication for the existence of the mentioned above gaps.

The mechanisms assuring the both gaps are different, because the +RL and -RL structures are different.

A. Internal gap of a helical structure with internal +RL(T) structure

The internal gap of +RL(R) structure is determined by two factors:

- finite length of the prism

- an anisotropy of SG propagation through the prism

The second factor is an intrinsic feature of the intrinsic matter, manifested in the prisms formation. It means, that the axial prism vector of SG force is larger, than the radial one. (It is to be discussed in the last Chapter of BSM).

The finite length of the prisms means a finite number of layers for +RL structure. In the following analysis it is shown, that the number of layers could be determined if we know the gap between

the prisms in CL space. In Chapter 3, the node distance along xyz axes is calculated as $d_{nb} = 1.1 \times 10^{-20}$ m. This is in CL unit length scale. Applying the correction factor of 0.8165 for a positive prism (lefthanded assumed) we get a value of 0.8981E-20 m. The CL node distance along abcd axes is approximately half of xyz distance, so it is 0.449E-20 m. The unknown parameter for CL space is the "gap to node distance ratio" for abcd axes. (The gap is between the neighbouring prisms, which are of different type). Let us assume, for example, that the gap is equal to the sum of right and left-handed prism lengths. When referenced to CL scale unit, it is equal to $2 \times 0.8165 = 1.633$. Now referencing to the positive prism scale it is $1.633 \times 0.8981 \times 10^{-20} = 1.4665 \times 10^{-20}$ .m. This value is half of the CL node distance (along abcd). Now expressing the CL node distance by the positive prism scale we obtain: $2(L_L + L_R) = 2(L_L + 1.5L_L) = 5L_L$. So we get: $L_L = 1.4665 \times 10^{-20}/5 = 0.2933 \times 10^{-20}$ m. Then the minimal node distance in the +RL(R) layer is twice this value or: 0.5866×10^{-20} m. Theoretically the smaller internal radius of the most internal layer is equal to the minimal node distance. This put the **theoretical limit of the internal hole radius** to be also 0.5866×10^{-20} m. From the other hand, knowing the positron small radius, r_p, and applying the rule of the half radius layer thickness we may express the hole radius by the equation:

$$r_{hole} = r_p \left[1 - \sum_{i=1}^{l} \frac{1}{2^i} \right] \qquad (6.52.a)$$

where: $i = 1$ in this equation corresponds to the most external layer of +RL(R).

The bar plot of Eq. (6.52.a) $15 < l < 22$ is shown in Fig. 6.19. In the same plot the theoretical limit of the radius is shown as a dashed line.

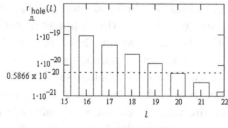

Fig. 6.19

The theoretical limit coincides with the value 0.5866, with amassing accuracy for number of layers equal to 20.

Note: The obtained number of layer is dependent on the gap between the prisms of neighbouring CL node gaps. It also does not take into account the SG anisotropy, mentioned above. However, in §6.9.4.1 it was concluded that the contribution of the most internal RL sublayer to the SG energy balance is very small, so a small difference in a needed number of layers is tolerance in the SG energy balance according to Eq. (2.8). Then in the real case, the number of layers may be smaller, assuring a larger hole radius.

Radial node distance affected by the finite length to radius ratio of the prism.

Due to the finite thickness (diameter) of the prisms, and the radial dependence of the tangential stiffness within one layer, the radial scale exhibits a small deviation from the linearity. This is illustrated by Fig. 6.20, where: **a.** shows RL nodes in a radial section near the small layer radius; **b.** shows a similar section near the large radius; **c.** shows the enlarge scale of nodes in case a.

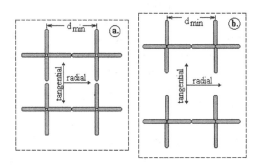

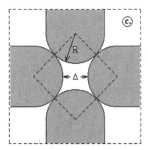

Fig. 6.20

Radial displacement due to a finite prism thickness

It is obvious, that the deviation from linearity is a function of L/R ratio of the prism. (In Chapter 12 it will be shown, that this ratio is a constant for all prisms belonging to one galaxy, but may have slight variation between the prisms of different galaxies).

The slight non linear radial scale leads to a slight increase of the central ("trapping") hole. Knowing the radius r_p and the number of layers, the radius of the hole could be calculated for a chosen L/R ratio. Approximate calculations for $L/R = 12$ and number of +RL(R) layers = 20, leads to a result, $r_{hole} \approx 1.6$ % of r_p.

B. Radial gaps in combined helical structures with RL(T).

After the twisting, the radial gaps of the combined helical structures is increased. This conclusion follows from the following features, demonstrated by the experimental data:

- the strong polarization of the electron and the free positron after striking a plane at oblique angle

- the difference in the two main decay reactions of K_S^0 with highest fraction (Γ_i/Γ) :

The first feature is well known by the experiments. The second feature is related to the decay reactions:

Mode	Fraction (Γ_i/Γ)	
$K_S^0 \rightarrow \pi^+\pi^-$	68.61 %	(6.53)
$K_S^0 \rightarrow \pi^0\pi^0$	31.39 %	(6.54)

In the reaction (6.54) both pions are pair of separated untwisted structures - one negative and one positive with internal structures, respectively - RL(R) and +RL(R). Consequently, they have been untwisted before and during the time of separation. The difference in the fraction parameter for both reactions could be explained by the differences in the radial gaps. Despite the fact that the end product of the reaction (6.53) are charged pions, the fraction parameter for this reaction is larger. This means that the separation in this case is easier. A possible reason could be a larger radial gap. This gap corresponds to the gap between the internal radius of -RL(T) and external radius of +RL(T). In the case of the electron and the negative muon, this gaps is even larger, due to the additional twisting.

There are **two additional features that may lead to an increased gap between the electron shell and the positron:**

(a) the structural difference of +RL from -RL:

(b) the reduced layer thickness of -RL

The feature a) is characterized with:

- the intrinsic matter of +RL(T) structure is in radial range of $0 \approx r_{hole} < r < r_p$

- the intrinsic matter of -RL(T) is in the radial range of $r_p < r < r_e$

In the first case, the radial size could be kept more compact due to the additional diametrical SG forces. In the second case, such forces are absent due to the larger hole.

The feature (b) means that the tangential stiffness of a -RL layer changes in smaller range in comparison to the tangential stiffness of a +RL layer. Consequently, the centre of mass of the SG matter density in -RL(T) will be displaced toward the larger layer radius in comparison to the +RL(T). This assures an additional gap increase between the electron shell and the positron.

We may summarise:

- **The radial gaps between the substructures of the combined negative helical structure assures conditions for oscillations without internal energy loss**
- **The central positions of the oscillating structures are kept by the radial configuration of the SG field**
- **The oscillation motion is frictionless**
- **The radial gaps for the electron system and the negative muon are equal. (The same rule is valid for the free positron and the positive muon).**

6.11. Physical dimensions and structure of the proton and neutron.

The protoneutron (an initial product from the particle crystalization), the proton and the neutron are distinguished only by the external shape and the degree of twisting of their substructures. If disregarding their common substructure, these three particles are closed loops, but with different shapes. Let us use the term **"proton core"** for the body of this loop. We may summarize:

- **The proton core is the body of the closed loop, from which any one of the three particles, protoneutron, proton and neutron, is formed.**
- **Referencing to the proton core is useful, when disregarding the internal substructure of the particle**
- **The length of the proton core is one and a same for the mentioned three particles, while its overall shape is different.**

It is obvious that the spatial configuration of the proton and neutron is defined by the following parameters:

- length of the proton core
- thickness of the proton core
- the plane projection of the twisted shape
- external twisting angle of the loop

In the next paragraph we will see that the determination of the dimensions of the proton core leads simultaneously to determination of some dimensional parameters of its substructure components.

6.12 Proton

6.12.1 Structure and shape

The proton is a complex helical structure composed of external structure $TTH_1^3{:}{+}({-})$, referenced as an external shell, which enclose internal structures of pions and a central kaon. The internal structure of the proton was shown in Fig. 2.15.A and 2.15.B. The protoneutron and neutron have one and a same internal structure, which differs only by the degree of twisting. The shape of the proton with exploded view of its external shell is shown in Fig. 6.22.

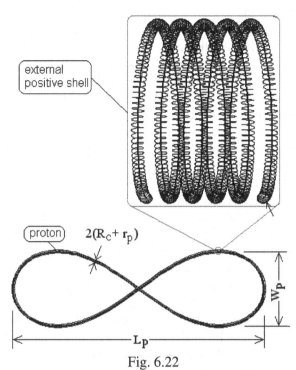

Fig. 6.22
Proton overall shape with exploded view of its external positive shell (without the internal pions and kaon)

The central axis of the folded proton core does not lye in a plane. For this reason we can call it a quasiplane. In order to use a simple mathematical expression, however, the quasiplane could be considered as a plane. The modelling of spring structure, possessing some degree of stiffness, shows that the most suitable mathematical function for the shape of the proton is the Hippopede curve. In polar coordinates this curve is given by Eq. (6.54.a), while in Decart coordinates - by Eq. (6.54.b)

$$r^2 = b^2(1 - a^2(\sin(\theta))^2) \qquad (6.54.a)$$

$$(x^2 + y^2)^2 - b^2[x^2 + (1 - a^2)y^2] = 0 \qquad (6.54.b)$$

The parameter a is related to the width to length ratio of the curve according to the equation:

$$\frac{1}{2\sqrt{a}} = \frac{W_p}{L_p} \qquad (6.56)$$

where: W_p - is the width of the curve
L_p - is the length of the curve

Based on a mechanical spring modelling (not presented here) the most suitable value of this factor is $a = \sqrt{3}$. The length and width of the curve for this value of a are physical parameters of the proton shape. They will be further referenced as a proton length and width. The polar plot of Eq. (6.54.a) for the accepted value of a is shown in Fig. (6.23), where the parameter b denotes the half of the proton's length.

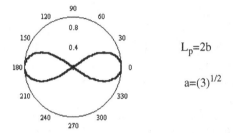

Fig. 6.23
Proton shape by Hippoped curve

We may call the two portions of the curve **proton clubs**. There are two important points in the plane of proton clubs, corresponding to the maximal vertical width of the hippoped curve. They are located on the horizontal axis at distance of **0.6455b** from the curve centre. Each of these points could be regarded as a centre of mass of the proton club. We may call them **locuses**. The ratio between the length L_p, and width L_w of the curve is:

$$W_p/L_p = 0.28867 \quad \text{for } a = \sqrt{3} \qquad (6.56)$$

We shall adopt this factor of a, because it matches very well to all equations in which the proton shape is involved.

The curvilinear length of the Hippoped curve can be found by a numerical integration in polar coordinates:

$$L_{pc} = 2L_p \int_0^{\theta_m} \sqrt{(1 - a^2(\sin\theta)^2) + \frac{a^4(\sin\theta)^2(\cos\theta)^2}{1 - a^2(\sin\theta)^2}} \, d\theta \quad (6.57)$$

where: L_{pc} - is the length of the curve, equal to the proton core length;

$\theta_m = 0.19590657\pi$ - is the adjusted angle limit for the integration (in order to include the whole curve sector)

$L_p = 2b$ - is the proton length;

b - is the outscribed radius

Notice: The upper limit of integral in the Eq. (6.57) depends on the *a* parameter and should be properly adjusted for any different value of *a* than $\sqrt{3}$.

By using the shape of the Hippoped curve, for $a = \sqrt{3}$, the proton dimensions can be completely determined if knowing the curvelinear proton core length L_{pc}. This parameter was approximately determined, in Chapter 5, by using the equation of the CL space background temperature and its experimental estimate by the Cosmic Microwave Background. (The latter phenomena has a different interpretation according to BSM - it is not a Relict radiation, but a CL space background radiation).

6.12.2 Cross calculation method for accurate determination of the proton dimensions

6.12.2.1 Accurate determination of the proton core length

In §6.9 it was proved that:

The twisting of the FOHS does not change its length. In other words, the length of the FOHS of pion and muon are equal each other.

Relying on this feature we can determine L_{pc} very accurately, by using properly selected data from particle collision experiments involved a proton destruction and applying the mass equation, derived in Chapter 3. For correct application of the mass equation, we have to use correctly the FOHS radius shrink factor $k_S = 0.87007$, or its reciprocal value k_R, and the mass correction factor for CL space $k_m^2 = 2.25$ for the positive helical structures. The true value of L_{pc} could be calculated by solving a system of mass equations. They should satisfy the following mass budgets, for which quite accurate experimental data exist

   - Correct mass budget of the proton

   - Correct mass budget of the eta particle

   - Correct stopping power ratio between the antiproton and proton

   Additional criteria used

   - The number of second order turns of the internal pions should be an integer, because it is a curled toroid

   - The calculated length of the proton core, should match closely the Cosmic Microwave Background experimental data

All of the masses, mentioned above are Neutonian masses.

Note: The mass of the Kaon determined in the particle collision experiments is overestimated because the thrust force from kaon jet (destruction of its RL structure) has not been discovered so far and consequently not taken into account. (The existence of such jet is confirmed also by the pulsar theory provided in Chapter 12). So when using the mass equation for compound particle containing a stable kaon a proper correction factor should be used for the kaon. Fortunately this correction factor affects the mass budgets of the proton, the eta particle and the antiproton/proton stopping power ratio in a different way. This provides the opportunity for pretty accurate determination of the mass correction factor of the stable kaon.

The parameter L_{pc} would be determined quite accurately, if we knew the helical radius of the internal pion and its second order step. Based on the provided so far analysis, we may choose a guessed value for the helical radius and test the equations of the mass budgets.

The helical radius is the radius of the projection of the pion SOHS in a plane normal to the proton core. In Chapter 3 it was mentioned, that during the phase of the crystallization, the internal cylindrical space enclose by the external proton shell likely contains RL(R) type of structure. (This is to be more extensively discussed in Chapter 12. Then the equation (2.8) of SG forces balance could be satisfied for a value close to 2/3 of the radial distance between the external positive shell and the internal Kaon. So the corresponding internal pion radius, R_π, can be expressed as

$$R_\pi = \frac{2}{3}(R_c - r_p - k_R r_e) + k_R r_e \qquad (6.57)$$

The length of FOHS in one turn of the internal pion is: $\sqrt{4\pi^2 R_\pi^2 + S_\pi^2}$ so the total length of its FOHS is:

$$n\sqrt{4\pi^2 R_\pi^2 + s_\pi^2} \qquad (6.58)$$

where, s_π is the second order helical step of the internal pion, and n is the number of its turns inside the proton.

One turn of muon is equivalent to the electron (positron). The number of turns in the muon are given by:

$$\frac{\mu_e}{\mu_\mu} = \frac{m_\mu}{m_e} = 206.76 = \frac{206.76}{1}\frac{\text{turns}}{\text{turn}}$$

The FOHS length of one turn of the muon and the FOHS length of the electron (positron) are equal. Then the total length of the FOHS in the muon is

$$\left(\frac{\mu_e}{\mu_\mu}\right)^2 \sqrt{4\pi^2 R_c^2 + s_e^2} \qquad (6.59)$$

Using the equality of the FOHS length of pion and muon, we equalise the expressions (6.58) and (6.59). Then solving for s_π we get:

$$s_\pi = \frac{1}{n}\left[\left(\frac{\mu_e}{\mu_\mu}\right)^2 (4\pi^2 R_c^2 + s_e^2) - 4n^2\pi^2 R_\pi^2\right]^{1/2} \qquad (6.60)$$

The total length of the pion as a SOHS inside the proton is a same as the proton core length L_{pc}. So $L_{pc} = s_\pi n$ or

$$L_{pc} = \left[\left(\frac{\mu_e}{\mu_\mu}\right)^2 (4\pi^2 R_c^2 + s_e^2) - 4n^2\pi^2 R_\pi^2\right]^{1/2} \qquad (6.61)$$

In Chapter 5, the approximate value of the proton core length was determined as: $L_{pc} \approx 1.6429 \times 10^{-10}$ m (based on BSM obtained expression, in which the experimentally obtained temperature of 2.72K by the Cosmic Microwave Background was used). Giving a range of numbers of turns n, we get a number of values L_{pc} around the approximately determined above one. These values are shown in Fig. 6.24 as a bar plot.

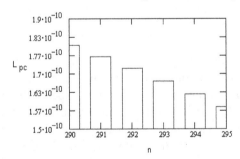

Fig. 6.24
Proton core length versus the possible number of pion turns, n

The closest value of L_{PC} to 1.6429×10^{-10} (m) is for $n = 294$.

$$L_{pc}(294) = 1.62772 \times 10^{-10} \quad \text{(m)} \qquad (6.62)$$

By using of Eq. 5.8 (Chapter 5), the recalculated ZPE temperature corresponding to this value of L_{pc} is $T = 2.676K$. This differs from experimentally estimated temperature based on the CMB background of COBE (2.726) only by 0.05K.

Now using Eq. (6.61) we determine the accurate value of the proton length. **This is the length of the proton along its long axes, without taking into account the thickness of the external shell:**

$$L_p = 0.667 \times 10^{-10} \text{ (m) - proton length} \qquad (6.62.a)$$

6.12.2.2 Some useful dimensions

Accepting n = 294, we get the following additional dimensions for the internal pion:

$$s_\pi = 5.5365 \times 10^{-13} \text{ (m) pion's second order step} \qquad (6.63)$$

$$\frac{\sqrt{\left(2\pi\frac{2}{3}R_c\right)^2 + s_\pi^2}}{s_\pi} = 3.088 \qquad (6.64)$$

ratio of one turn length to a second order step size of the internal pion

$$\frac{2r_p}{L_{pc}} = 0.72435 \times 10^{-4} \qquad (6.64.a)$$

ratio of the proton envelope thickness to core length

6.12.2.3 Volume ratio (between the proton's envelope volume and the total volume of its FOHSs)

The proton envelope is the surface enclosing its external positive shell. The volume of the proton envelope participates in the ZPE balance in which the CL space dynamical pressure and the CL space-time constant are both involved (see Chapter 5.).

Let $R_{Vp} = V_{env}/V_{FOHS}$ to be the volume ratio between the proton envelope and the total volume of its FOHSs. We may apply the mass equation for the proton and use R_{Vp} as a volume ratio. In order to eliminate the proton mass dependence on the electrical charge, we will use the mass of the neutron instead of proton one.

$$m_n = \frac{h\nu_c \pi (R_c + r_p)^2 L_{pc}}{c^2 \quad 2\pi^2 R_c r_e^2} \qquad (6.65)$$

From Eq. (6.65) the factor R_{Vp} is directly determined.

$$R_{Vp} = \frac{h\nu_c (R_c + r_p)^2 L_{pc}}{c^2 \quad 2\pi R_c r_e^2}\frac{1}{m_n} = 71.72 \qquad (6.66)$$

If the volume enclosed by the proton envelope was not accessible to the CL nodes, the factor R_{Vp} should be equal to unity. The obtained value, however, indicates, that the internal volume of the proton is accessible to CL nodes. Consequently:

The CL space penetrates inside of the proton (neutron) envelope, but not inside of its FOHSs.

6.12.2.4 Verification of the dimensions by the newtonian mass equation

The general form of the newtonian inertial mass equation for a single helical structure is (see Chapter 3)

$$m = \frac{P_S}{c^2} V_{H(SI)} \quad \text{[kg]} \qquad [(3.48)]$$

where: P_S is the CL static pressure and V_H(SI) is the FOHS volume (in SI units).

It has been noted that the equation in this shape is valid for FOHS with an external negative shell. For FOHS with an external positive shell, additional multiplication factor of 2.25 must be used.

For a compound structure as a proton or neutron containing a number of helical structures the mass equation is simply the sum of masses of all their FOHSs' they contain.

$$m_{tot} = \Sigma m = \frac{P_S}{c^2} \Sigma V_{H(SI)} \qquad (6.66.a)$$

The expression (6.66.a) can be applied for number of compound particles composed of FOHS's whose mass is experimentally estimated. Having in mind that the proton (neutron) substructures are: pions and kaon, it is not difficult to guess the possible combinations of them. Therefore, operating by the volumes of one and a same type of substructures and taking into account their possible modifications, the volumes of their FOHSs could be estimated. Then the mass of the compound particle can be calculated by Eq. (6.66.a) and compared to the properly identified experimentally measured value.

The correct charge balance for any compound particle will allow determination of the correct number of pion pairs.

A. Total mass equation applied for the proton (neutron)

When applying the mass equation for the proton, the volume V_H should be the total sum of all FOHS's., or $v_H = v_{tot}$. Having the volume ratio R_{Vp}, determined by Eq.(6.66), we can obtain the total volume of all FOHS's, V_{tot}

$$V_{tot} = \frac{V_{env}}{R_{Vp}} = \frac{\pi(R_c + r_p)^2 L_{pc}}{R_{Vp}} = 1.0959 \times 10^{-36} \ (\text{m}^3) \ (6.67)$$

According to the mass equation (3.48) the sum of the newtonian masses of the separate FOHS' is equivalent to the sum of their volumes, multiplied by the factor P_S/c^2. If we normalise the mass budget equation to this factor, we get the equation of the volume budget. In the case for the helical structures with external positive shell the mass correction factor of 2.25 should be applied for their volumes. The FOHS's volumes of the proton substructures are:

$$V_{\pi^-} = \pi(k_{RST} r_e)^2 L_{pc} \quad \text{FOHS volume of } \pi^- \quad (6.68)$$

$$V_{\pi^+}(cor) = 2.25\pi(k_{RST} r_p)^2 L_{pc} \quad \begin{array}{l}\text{corrected FOHS volume} \\ \text{of } \pi^+ \quad (6.69)\end{array}$$

$$V_{K0} = \pi(k_{RST} r_2)^2 L_{pc} \quad \text{volume of the neutral kaon} \quad (6.70)$$

$$V_{pext} = 2.25(2\pi^2 R_c r_p^2)n_t \quad \begin{array}{l}(+) \text{ corrected volume of} \ (6.71) \\ \text{the external shell}\end{array}$$

$$V_{\pi^-} = V_{\pi^+}(cor) \quad \text{- corrected volume equivalence}$$

where: n_t - is the number of turns in the positive external shell

$k_{RST} = 1.1493$ - radius restoration factor for twisted FOHS (see§6.4.4.4.2)

After the proton break-up, the pions and kaons are cut in one place only and their FOHSs have undergo a fast twisting in which process the internal RL(R) are modified into (RL(T) and their FOHS volume get a slight decrease. The factor k_{rst} allows to restore the radius of FOHS envelope before the twisting, and consequently to obtain its former volume. This factor has been estimated by the twisting obtained after pion-muon conversion and referenced to the small radius r_p of the positron. The internal pions have some twisting and consequently anyone of them possesses a charge, but they are locked in proximity and commonly neutralized. The internal kaon, however, is   not twisted (this is evident from the mass difference between $w^\pm$ and Z boson, which are destruction energies respectively of untwisted and twisted RL structures of the kaon). Consequently, the internal kaon possessing RL(R) could not have an external negative charge. Then the positive charge of the proton can be created only from the external positive helical structure, which forms the proton envelope, as shown in Fig. 6.22. This means that the pions, which are twisted and having charges, have to balance themselves, so they must be in pairs. Then we have to check options with one pair, two pairs and so on.

The FOHS volume of the external positive shell is expressed by the volume of a single turn, multiplied by the number of turns n_t. The volume of the single turn of the external positive shell is equal to the free positron volume.

The total volume of all FOHSs involved in the proton can be expressed as a sum of corrected (from twisting) volumes of their substructures:

$$V_{tot} = (V_{\pi -} + V_{\pi -})n_{pair} + V_{K0} + V_{pext} \qquad (6.72)$$

where: v_π, v_{KO} and v_{pext} are respectively FOHS volume of pion, kaon, and external proton shell

n_{pair} - is the possible number of pion pairs

The value of V_{tot} by Eq. (6.67) matches the value obtained by Eq. (6.72) for $n_{pair} = 1$ (one pair of charged pions) and one straight negative FOHS with external RL(R-) and internal RL(R+). The same value of pair pions is in agreement with the results for eta particle and antiproton/proton stopping power, presented in Table 6.8.

B. Mass correction factor for the neutral kaon

The experimentally determined mass of K_L^0 (and all other kaons modifications) is overestimated because the existing jet of cut kaon is not taken into account (see §6.10.4 and Fig. 6.21). Substituting $v_{H(SI)}$ by v_{K0} given by Eq. (6.70) we obtain a value of K_L^0 mass, that converted in MeV is: 45.286 MeV. This is about 11 times smaller than the experimentally provided value by NIST: 497.672 MeV. So the necessary correction factor is:

$k_{cor} = 10.99$

The same correction factor could be obtained simply by the equality of the ratios between newtonian masses and volumes of FOHS's:

$$\frac{m_{K0EX}/k_{cor}}{m_{\pi \pm}} = \frac{V_{K0}}{V_\pi}$$

where: $m_{\pi \pm}$ = 139.5699 MeV according to NIST data

C. Mass equation applied for the eta particle

The structure of the eta particle was discussed in §6.7 It contains one neutral kaon, and pair pions with opposite charges. All these structures are cut in one place. In this case the kaon possesses active jet and its real mass should be the experimental estimated mass of K_L^0 divided by correction factor k_{cor}. Consequently, when estimating the total volume of all FOHSs in the proton (neutron) the participated kaon K_L^0 volume should be divided by this factor:

$$V_\eta = V_{\pi -} + V_{\pi +}(cor) + V_{K0}/k_{cor} \qquad (6.73)$$

where: k_{cor} - correction factor of the experimental estimated kaon mass due to the jet (see §6.10.4 and Fig. 6.21)

The Eq. (6.73) is given for one pion pair, but it could be checked for different number of pairs.

D. Mass equation applied for the "antiproton"

The structure of the "antiproton" was discussed in §6.8. It contains one neutral kaon, a pair of pions and an envelope of a negative helical structures, which has been an internal core of the broken external positive helical structure. All they are still closed helical structures. We also envision that the overall shape of the antiproton might be additionally twisted until the full balance of the SG forces in CL space according Eq. (2.8) is fulfilled. In this case the internal kaon become more twisted and may contribute to the negative charge of the antiproton. Since the external negative envelope in such structure cannot contain RL structure, it could not feel the Static CL pressure and consequently will not contribute to the antiproton's mass. The corrected volume budget is:

$$V_{\bar{p}} = V_{\pi -} + V_{\pi +} + V_{K0} \qquad (6.74)$$

6.12.2.5 Calculated results

According to the mass equation (3.48) the ratio of the corrected volumes from different FOHS's is equal to the ratio of their newtonian masses.

Then for the eta particle and the antiproton we have the relations

$$\frac{m_\eta}{m_p} = \frac{V_\eta}{V_{env}} \quad \text{for eta particle}$$

The fractional error of the calculated mass of the eta particle is 3.5 %. The model provides additional parameters as : number of turns of the internal pion, step to radius ratio, pion to kaon FOHS length ratio. The value of the last parameter is 3.082. The most important parameter is the the proton core length, from where the proton length is determined. This parameter is verified and cross validated in Chapter 9.

The results of the composition and physical dimensions of the proton (neutron) structures are the following:

Internal pions: one negative and one positive; the negative one is a compound structure of external negative and internal positive FOHSs

Internal pion parameters:
Number of turns: 294

Ratio between FOHS length of a single turn and the size of the second order step: 3.088

External proton shell
Step to radius ratio of the SOHS: 0.3501

Pion to kaon FOHS length ratio: 3.0822

Volume ratio of internal RL(T):
Between external positive shell and
a positive kaon: ~ 30

Between pion and kaon same type of structures ~ 3.

Conclusion:
The mass equation applied for the proton (neutron) indicates that it is comprised of one not twisted central kaon (with external negative shell), two pions with opposite charges and external positive shell with internal structure of RL(T) type. The latter provides the positive charge of the proton.

6.12.2.7 External dimensions of the proton structure

Relying on the accepted shape of the proton a a Hippoped curve with parameter $a = \sqrt{3}$, one may determine the basic parameters of its dimensions: a proton's length; a proton's width and a core thickness.

The core length was L_{pc} determined in §6.12.2.1 (see Eq. (6.62). The length of the Hippoped curve, corresponding, to L_{pc} is given by the integral:

$$L_{pc} = 2L_p \int_0^{\theta_{max}} \sqrt{\left(1 - a^2(\sin\theta)^2\right) + \frac{a^4(\sin\theta)^2(\cos\theta)^2}{1 - a^2(\sin\theta)^2}}\, d\theta \quad (6.75)$$

where: $\theta_{max} = 0.19590657\pi$ - is a limit angle of integration

L_p is the proton length;
The upper boundary of the integration θ_{max} corresponds to the disectrice angle in [rad] at the origin of the curve.

The proton length L_p is determined from Eq. (6.75). The proton width is determined by the width of the Hippoped curve for $a = \sqrt{3}$. The proton core thickness is: $2(R_c + r_p)$.

The calculated proton dimensions are:

$$L_p = 0.667 \times 10^{-10} \quad \text{m} \quad \text{- proton length} \quad (6.76)$$

$$W_p = 0.19253 \times 10^{-10} \quad \text{m} \quad \text{- proton width} \quad (6.77)$$

$$2(R_c + r_p) = 7.9411 \times 10^{-13} \quad \text{m - core thickness} \quad (6.78)$$

Note: The proton length and width are referenced to the central axis of the proton core.

6.12.3 Experimental confirmation about the theoretically determined proton dimensions

So far one experimental confirmation about the length of the proton core was used: the proton core length, cross calculated by the ZPE and CMB temperature. In Chapter 8 the spatial rule of the protons and neutrons arrangement in the atomic nuclei is revealed. As a result an Atlas of Atomic Nuclear Structures (ANS) for all stable elements is built. Additional verification of the proton external shape and dimensions are also possible, by using of:

(a) The Atlas of Atomic Nuclear Structures

(b) The electron orbits around the protons in the atomic nuclei. (See the the orbital quasiplanes and Balmer series model in Chapter 7);

(c) The atomic nuclear configurations in the chemical compositions (Chapter 9);

(d) the internuclear distances, determined by X - ray crystalography (Chapter 8)

The derived above proton's shape and dimensions are verifiable by using data from the above mentioned cases.

In chapter 9, for example, the proton length appears apparent from the 3D structure of the Chlorine molecule with a covalent bonds. The only needed parameter is the internuclear distance. The latter is known by the X - ray crystalography.

One example for the proton length verification by using the Nuclear Atlas is the following:

Let use the experimentally determined radius of the Neon atom, while a spherical shape has been assumed. Looking at the configuration of the Ne atom in the Atomic Atlas we see that the pole to pole distance is just equal to the proton length. The electrons orbits in ground state does not contribute to the atomic radius. The atomic shape of Ne, in fact, is not sphere but oblate spheroid. In order to be able to use the estimated radius for Ne atom as a sphere, we will equate the volumes of sphere and oblated spheroid.

$$\frac{4\pi a^2 b}{3} = \frac{4\pi R_s^3}{3} \qquad (6.79)$$

where: a and b are the large and small oblated spheroid radii, and R_s is a radius of the sphere

From the physical model of Ne nucleus, as shown in the Atlas ANS, we see that it could be inscribed in an oblate spheroid with a radius ratio: $a/b = 1.805$. Substituting a in Eq. (6.79) and solving for b we obtain

$$b = 0.67455 R_s. \qquad (6.80)$$

Let use the radius of Ne provided by the Michigan Institute of Technology: http://wulff.mit.edu/pt

$R_s = 0.51 \times 10^{-10}$ (m)

From Eq. (6.80) we get: $b = 0.344 \times 10^{-10}$ (m)

Then the proton length is directly obtained as twice the smaller radius b of the spheroid envelope

$$L_p = 2b = 0.688 \times 10^{-10} \text{ (m)} \qquad (6.81)$$

The proton length calculated by the Ne radius appears larger than the theoretical value (Eq. 6.76)

by 3% only, but this difference could be explained by the applied method of measurement in which the electron orbits in the equatorial region of the Ne atom may contribute to the estimated dimensions. (The spatial position of the electron orbits of the inertial gazes become apparent in Chapter 8).

The most strict verification of the proton length is obtained in Chapter 9 where the parameters L_p and W_p are involved in number of equations whose result show excellent agreement with some spectroscopic data of the Hydrogen molecules.

6.13 Neutron

Figure 6.25 shows the overall shape of the neutron. Its external shell and internal structures are the same as the proton.

Fig. 6.25
Neutron overall shape

The external shape of the neutron is characterized by a folding node of the core. There are minimal distance between the two subloops at the position of the folding node so the external helical shells of the subloops do not touch each other. (This is partly evident from the drawing shown in Fig. 6.25)

The structural difference between the proton and neutron is only the degree of twisting of all helical structures they comprised of. The neutron possesses a electrical field, however, it is locked in proximity by the SG field. The main reason for the locking of the electrical field is the overall shape of the neutron core. Due to the small thickness to core length ratio, the core shape is almost equivalent to two torus structures. Applying the graphical rule for charge of particle defined in §2.10.1.A. we will see that it could appear as a neutral particle in a far field.

6.13.1 Neutron. Envelope shape and confined motion

Some of the features of the neutron has been already discussed in §6.1.4 The conversions between the neutron and proton with an emission of a quasiparticle was also discussed in §6.2.

The free neutron is unstable and converts to a proton, because its intrinsic gravitation is not able to overcome the core elasticity in the folded position. (Although when it is combined with a proton, it stays in the proton saddle, and the supergravitational field of the system is able to keep it in this position. This combination is a Deuteron). The free neutron has some equivalent helicity and exhibits a confined motion. The interaction with the CL space in such motion preserves the overall shape and prolongs in such way its lifetime. For this reason, the more energetic neutron has a longer life time. One additional factor keeping the neutron from conversion into a proton is the lattice resistance for creation or disappearance of the electrical charge (far electrical field).

There is one particular feature of the neutron confine motion. When a monohromatic beam of neutrons passes through a crystal with parallel walls under Brewster angle, the outcome signal is separated in two beams in the same plane, but symmetrically deviated. The effect is known as Borman effect. The conditions of confined moving neutron when entering the crystal surface are illustrated by Fig. 6.26.

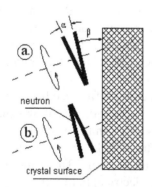

Fig. 6.26

The full explanation of the Borman effect is out of the scope of BSM, but some considerations will be pointed out.

a. When in motion, the locked positive field of the neutron external shell becomes unlock. Then it exhibits a magnetic field and consequently it may have a confined motion. The individual neutrons in the beam obtain common phase synchronization, which is connected to its rotation (like the beam of electrons)

b. The angle of opening α depends on the neutron velocity

c. The Brewster angle β also depends on the neutron velocity

d. There are two special cases in the moment when the neutron strikes the surface of the crystal at oblique angle, shown as case **a** and **b** in Fig. 6.26. The phase difference between them is π. For both cases the folding node of the neutron core lies in the drawing plane

e. When the folding node of the neutron core does not lye in the drawing plane, conditions for pendellosing effect occur.

6.13.2 Neutron-proton conversion and virtual particles

In the neutron to proton conversion process a positive charge appears as a far field and a negative wave is propagated as a virtual particle. In the proton to neutron conversion, the positive charge must disappear, so it propagates as a virtual particle. In both cases the propagated virtual charge particle could be considered as a reaction of the CL space to the unlocked (or locked) electrical field. The physical explanation of the process is the following:

In the case of photon emission from oscillating electron (discussed in Chapter 3) or atomic and molecular spectra (Chapter 7 and 9), the lattice pumping is a result of multiple periodical oscillations. The pumped energy is well distributed between positive and negative nodes. In the case of proton/neutron conversion **the pumping is from aperiodical motion.** If the time duration of this motion is shorter than the CL space relaxation time, the pumped energy can not be equally distributed between the opposite CL nodes. Then this energy propagates through the CL space in a different way. The obtained wave affects the opposite nodes in a non symmetrical way, exhibiting a feature of a moving charge. The configuration of the quasiparticle wave is little bit different than the quantum wave. It was described in §2.10.6, Chapter 2. Such wave could be deviated by electrical and magnetic field. **The major distinctive feature by which the virtual particle wave could be identified is its velocity. It is equal to the light velocity. It may be considered as a virtual electron or positron.**

The virtual particle wave, however, does not posses external boundary conditions and the quantum features are not so strong, as those of the neutral quantum wave. For this reason we may accept, that the charge dissipate gradually. In such conditions, the rule of the charge unity may not be valid.

The proton-neutron or neutron-proton conversion is accessible for investigation by experiments. Apart of the newtonian mass change and the creation of quasiparticle wave, the reaction of CL space includes additional energy, identified so far as a neutrino particle. Let us analyse the conversion of the free neutron to proton known as a Beta decay. In this process the neutron loses 1.294 MeV of its newtonian mass. But the unit charge of electron is 511 KeV. So the balance is:

1.294 MeV = 511 KeV(e+) + 511 KeV(ve-) + 272 KeV

According to the BSM interpretation, the 511 KeV(e^+) is a mass equivalent energy of the new born positive charge of the proton. It reflects the volume change of all FOHS's as a result of the twisting. The 511 KeV(ve-) is the mass-equivalent energy of the virtual electron and **272 KeV is a newtonian mass deficiency of the proton due to the changed FOHSs volume from the twisting.** This energy so far, has been wrongly contributed to a neutrino. According to BSM, however, this mass deficiency is the necessary energy to fulfil the balance of the SG forces for the twisted helical structure in CL space, according to Eq. (2.8). In a few words, the mass deficiency is an adjusting of the SG potential for a proton in order to appear as a charge particle in CL space.

- **The energy of 272 KeV, related with the neutron/proton conversion is a signature of an adjusting total balance energy in CL space.**

The physical explanation of the mass deficiency effect is the following.

The neutron does not have electrical far field. This means that the positive and negative nodes in the far field are not biased and their SMP vectors participate equally in the propagation of the super gravitational field through the prisms core, forming the gravitational mass of the neutron. When its structure is folded to a proton, the negative nodes of external surrounding field are engaged in a new complimentary motion of the SMP vector in a preferable direction aligned with the electrical lines of the unlocked positive charge. But the gravitational field is contributed by the equal distributed SMP vectors on the node quasisphere. The propagation of the SG field by the SMP vectors of the CL nodes involved in the electrical field is affected. As a result of this, the particle suffer some portion of its gravitational mass. The proton is a very large structure in comparison to the CL node spacing and a small percentage of the nodes is affected. The neutron and proton both are symmetrical structures, but during the conversion process they pass through a not symmetrical shape. The birth or death of a charge has a finite time duration and during this process the following conditions are relevant:

(a) The neutron possesses a local gravitational field with a spatial orientation determined by the folded node position. So every neutron has a definite orientation in respect to the laboratory frame

(b) The spatial orientation of the obtained proton is correlated to the spatial orientation of the neutron

(c) The emission of the quasiparticle wave is correlated to the neutron - proton orientation

(d) The spatial direction of the proton momentum, obtained as a result of the conversion is correlated with the initial neutron orientation and the direction of the emitted quasiparticle wave.

The feature (d) is a measurable parameters in the experiments. It is known as an "electron neutrino angular correlation effect." The neutrino for the Beta decay has been introduced with the purpose for conservation of the energy and angular momentum. It has never been directly observed in a Beta decay process (and especially of free neutron), but only calculated by the energy balance. The energy balance, according to BSM interpretation is preserved, when taking into account the reaction of the CL space. Consequently:

- **According to BSM, the β decay does not involve a neutrino or antineutrino particle.**

The process of neutron proton conversion in the Beta decay in atoms is similar, but the energy range of the virtual particles is larger. The atomic nuclear is a complex structure of neutrons and protons, arranged in a well defined spatial order, which could affect the energy of the emitted virtual particle.

It is well known fact that the average energy value for positive Beta particles is slightly higher than the negative Beta particles. The BSM explanation for this difference is the following: In the emission of positive Beta (virtual) particles, the structure is changing from unfolded (proton) to folded (neutron) shape. In the final moment the intrinsic gravitation start to attract the structure parts and accelerate the folding process. In the opposite case, the unfolded structure does not stop exactly at the proton shape but overpasses it aperiodically. An aperiodical motion in the first case is not possible. During the aperiodical motion some portion of the energy could be transferred to the nuclear system. **This energy will be missing from the energy of the emitted negative Beta (virtual) particle.**

The BSM explanation of the mass deficiency effect in the neutron/proton conversion may provide understanding of the discrepancy between the expected neutrino flux from the Sun and the measured one.

6.13.3 Neutron in Deuteron

We found that one may distinguish the proton from the neutron by their overall shape. While the free neutron is not stable in CL space, it is quite stable when combined with a proton in a configuration shown in Fig. 6.27 This is the shape of a Deuteron. .

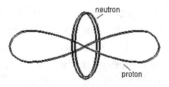

Fig. 6.27
Deuteron nucleus

The combination effect of intrinsic gravitational forces of the proton and neutron make the structure very stable. The stabilizing effect of the proton structure on the neutron is obvious and do

not need explanation. The Deuteron is one of the basic components in the atomic nuclear structures of most elements.

6.13.4 Neutrons in Tritium

The tritium is formed of one proton and two neutrons over its saddle. The stabilizing effect for the neutron is similar as in the Deuteron, but the Tritium is not a stable element because:
- The two neutrons could not be very close due to the repulsion of the locked electrical field in a very close distance
- The SG filed from the two neutrons affect the partially the proton shape invoking untwisting force. This may affect the distribution of the electrical field lines and finally could lead to losing of one neutron.

When the Tritium is connected to other structures in the atomic nuclei, however, the proton shape is less affected and the Tritiunm structure appears quite stable.

6.13.5 Neutron magnetic moment

One specific feature of the neutron is the possession of magnetic moment despite of its charge neutrality. The magnetic moments of the basic atomic particles are given in Table 6.2.

Table 6.2

Particle or combinations	magnetic moment (JT^{-1})
neutron	$\mu_n = -0.96623707 \times 10^{-26}$
proton	$\mu_p = 1.41060761 \times 10^{-26}$
electron	$\mu_e = 9.28477 \times 10^{-24}$
deuteron	0.43305×10^{-26}
tritium	1.5045×10^{-26}

The existence of the neutron magnetic moment has been already explained so far. **But why the neutron magnetic moment is opposite to those of the proton?**

The twisting of the FOHS changes slightly the pitch angle of the radial stripes. This may effect the magnetic moment of the particle in its confined motion. For a compound particle comprised of number of helical structures, like the proton, the pitch angle should be referenced to the axis of particle confined motion. The neutron is closer to torus and its magnetic moment is defined by the average prisms pitch angle of its shell. <u>When converted to a proton the pitch angle changes</u>. If the proton twisting was with the same handedness as its external shell, the magnetic moment will be larger but with a same sign. The fact that the magnetic moment changes the sign in the neutron-proton conversion means that the proton gets opposite higher order handedness. But the handedness of the proton twisting will be determined by the condition: which type of helical structure stiffness will predominate (right or left handed)? Let us remember that the lower order helical structure has a larger stiffness. All pions and kaon have negative central cores. The external shell of the proton (neutron) also possesses a negative internal core. Additionally, the negative pion and the central kaon have a negative external shells. Then the negative structure stiffness will predominate. Following the rule that a core of left handed prisms will favour a left handed twisting, we may conclude, that the twisted proton will have a higher order handedness opposite to the handedness of its external shelf. This explains the sign change of the magnetic moment. In fact the handedness of the proton twisting is predetermined by the protoneutron structure. According to the described considerations we obtain:

- the higher order handedness of the proton corresponds to the handedness of the negative prism

- the second order handedness of the electron and positron corresponds to the handedness of the negative prism

In Chapter 8 we will see, that the atomic nuclear structure possesses a handedness, corresponding to the proton handedness. The structure analysis of the simple molecules also shows that the handedness of the atoms is propagated to the molecules. Then we may consider that the handedness of a complex molecule like the DNA corresponds to the handedness of the atoms, and consequently to the proton. It is proved that the DNA possesses a right-hand helicity. Then the proton should have the same handedness. From this follows that the negative prism is right-handed.

- **From the analysis of the above considerations it follows that the negative prism is a right-handed.**

The author of BSM is not 100% sure about the above conclusion, which must be additionally

verified. For this reason in many cases a positive or negative attribute is used instead of the more accurate left-handed or right-handed one.

6.14. Nuclear binding energy

According to Modern Physics, the nuclear binding energy is estimated by the mass difference between the algebraic sum of the masses of the protons and neutrons from the which the particular atomic nuclei is built and its apparent mass. The apparent mass is always smaller, so the binding energy is estimated by the Einstein equation $E = (\Delta m)c^2$, where Δm is the mass difference.

The BSM explanation of the binding energy is the following:

The elementary particle contain an Intrinsic mass due to the large number of prisms. This Intrinsic matter influence the surrounding CL space by a slight shrunk of the CL node distance. This effect is in fact general Relativity, introduced by Einstein, but envisioned so far only for massive objects. Now after this effect is unveiled, we may a find a way for theoretical calculation of the nuclear binding energy. In this section we will provide such calculations between the proton and neutron in the Deuteron and for obtaining the separation distance between the two neutrons over the protons saddle for the Tritium. The obtained results will be compared to the corresponding binding energy known from nuclear physics data.

In order to provide the necessary calculations, we use a parameter called a Supergravitational factor C_{SG}, define as a product of the SG constant and the square of the Intrinsic masses of the proton. It ca be considered as a therm participated in the inverse cubic SG law.

$$C_{SG} = G_0 m_0^2 \qquad (6.82)$$

where: G_0 - SG constant in CL space units, m_0 - Intrinsic mass of the proton (or neutron)

The factor C_{SG}, is derived in §9.7.3, Chapter 9, by the analysis of the vibrational oscillations of the H2 molecules. It is given by the expression:

$$C_{SG} = (2h\nu_c + h\nu_c \alpha^2)(L_q(1) + 0.6455 L_p)^2 \quad [(9.24)]$$

$$C_{SG} = 5.26508 \times 10^{-33} \left[\frac{m^5}{kg\ s^3} \right]$$

The value of C_{SG} is additionally validated by matching of derived expressions for the energy levels of diatomic molecules with their vibrational-rotational spectra (Chapter 9 and Appendix9-1).

6.14.1 Binding energy between the proton and neutron in the Deuteron system.

In Chapter 2 §2.9.6.B it was accepted apriory, that the energy balance between both type of intrinsic energies in CL space (associated with the central part of the twisted prism model and its peripheral part) is expressed by the equation:

$$E_{SG}(TP) = 2\alpha E_{SG}(CP) \qquad (6.83)$$

where: α - is the fine structure constant

The approach used for calculation of the binding energy between the proton and neutron in the deuteron nucleus is illustrated by Fig. 6.28.

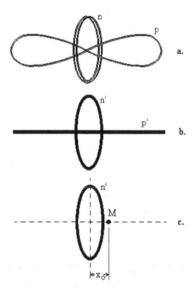

Fig. 6.28

Approach used for calculation of the binding energy between the proton and neutron in the Deuteron system

Fig. 6.28 a. shows the real configuration of the Deuteron.

Fig. 6.28 b. provides an equivalent model, where the neutron is presented as a mass ring and the proton as a mass bar, both possessing the same intrinsic mass (the intrinsic masses of proton and neutrons are equal). The length of the bar is an equivalent length of the proton, presented in this way.

Fig. 6.28 c. provides additional simplification by presenting the bar as a mass point with a

same intrinsic mass M residing on axis x, but at proper distance x_o from the ring..

We may estimate the binding energy by applying a classical method for estimation of a gravitational potential. For this purpose we will estimate the work for moving of the proton from its initial position to.

Consideration (1): The binding energy is equal to the disintegration energy when separating the proton and neutron of deuterium atom by moving the proton from neutron along the axis of symmetry x.

In the real situation, when a large number of deuterons may exists in a finite space volume, the integration to infinity is not possible. However, we will show later that even an integration up to a small distance, B, (comparable to the size of the proton L_{pc}) provides a good estimation of the disintegration energy, since the inverse cubic SG forces fall quite fast with the distance.

One question may arise: Why the mass point M is not in the middle of the ring, but displaced at distance x_0?

The possible answer of this question is: When the bar is pulling out of the ring along the x axis, the **SG(TP) interactions are involved**. In this interactions, the pull-out energy for any of differential mass point residing on the right side of the bar is not compensated by insertion of a symmetrical point from the left side. Then we accept considerations (1) and (2), which will become more evident later

The process of bar axial removing could be considered as removing of the two half of the bar in opposite direction.

Consideration (2): Let us consider the mass and ring respectively as a linear and curvilinear massive structures. Then their masses are completely defined by their length.

In the following mathematical model, the mass point M and the one dimensional bar are both regarded as a sum of equal number of small differential masses. Then applying consideration (1), the disintegration of the mass point in case c. starting from some initial distance x_0 is equivalent to disintegration of the bar in case b.

The mathematical approach for derivation the distance x_0 is simplified if regarding the ring and the bar as a one dimensional structures according to the consideration (2):

For derivation of the disintegration energy, the gravitational potential is initially derived. The process is clarified by Fig. 6.29.

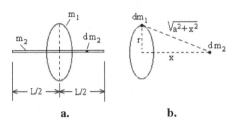

a.   **b.**

Fig. 6.29

Fig. 6.29 shows a point of the bar with differential mass dm_2. Using a classical approach, we derive the gravitational field of unit mass point from the bar as a function of distance x. (a trivial classical problem but for inverse cubic dependence of the attractive force)

$$g_x = \frac{dF}{dm_2} = -\frac{G_o m_1 x}{(x^2 + r^2)^2} \qquad (6.84)$$

where: g_x is the axial component of the gravitational field; r is the neutron radius, G_o is the intrinsic gravitational constant, x - is the distance of the point dm_2 from the ring centre, r - is the ring radius, m_1 is the ring mass

The neutron radius is a double folded torus (protoneutron), so expressed by the proton core length it is:

$$r = L_{pc}/(4\pi) \qquad (6.84.a)$$

Let us use x as a running parameter for integration. According to considerations (2) and (1) we have:

$$dm_2 = \frac{m_2}{L}dx \quad \text{and} \quad dF = g_x \frac{m_2}{L}dx$$

$$F = \int_{-L/2}^{L/2} \left| g_x \frac{m_2}{L} \right| dx = 2 \int_0^{L/2} \left| g_x \frac{m_2}{L} \right| dx$$

where: F is the axial SG force for one point of the bar with mass dm_2; L - is the mass bar length

The solution is:

$$F_x = \frac{2 G_o m^2 k_{cor}}{L}\left(\frac{1}{2r^2} - \frac{2}{L^2 + 4r^2} \right) \qquad (6.85)$$

where: $m^2 = m_1 m_2$

k_{cor} - is a correction factor, that will help to find how the final solution depends on L, for which the possible range is known.

For the mass point M at distance x_o (case c. of Fig. 6.28) we have:

$$F_x = G_o m^2 \left(\frac{x}{(x_o^2 + r^2)^2} \right) \qquad (6.86)$$

Equating Eq. (6.85) and (6.86) we get expression in which the parameters $G_o m^2$ are eliminated but its analytical solution for x_0 is difficult. For this reason we make a function of differences of the therms

$$f(x_o) = \frac{x_o}{(x_o^2 + r^2)^2} - \frac{2 k_{cor}}{L}\left(\frac{1}{2r^2} - \frac{2}{L^2 + 4r^2} \right) \qquad (6.87)$$

It is reasonable to expect, that L will be in the range of L_p or L_{pc}. For $L = L_p$ and $k_{cor} = 1.9246$, the solution is

$$x_o = 0.0747 \times 10^{-10} \ \text{m} \qquad (6.87.a)$$

The function $f(x_o)$ falls very sharply for this value. The plot is shown in Fig. 6.30.

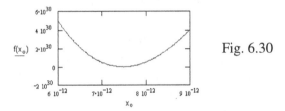

Fig. 6.30

For different L in the above mentioned range, the curve $f(x_0)$ still exhibits a well defined minimum. The position of the minimums on the x scale are not very dependent on the introduced parameter k_{cor}. **For the defined range of L, the second term does not influence the x_o parameter.** Then the binding energy could be determined by using the simple case c. of Fig. 6.28, where the ring mass is the intrinsic neutron mass and the point mass M is the proton one.

The estimation has to be in a CL space environments, so let try to use the SG(TP) type of energy. Having in mind the relation (6.83) and using the field expression (6.84), which is valid for this case we get:

$$E_{SG}(TP) = (2\alpha)2 \int_{x_0}^{\infty} C_{SG} \frac{x}{(x^2 + r^2)^2} dx \qquad (6.88)$$

where: ∞ in fact is some finite value little bit larger than L_{pc} (because the SG forces fall quite fast with the distance).

The factor 2 in front of the integral comes from the two arm branches (along *abcd* axes) of the CL space cell unit. The same factor of 2 has been used in Eq. 9.13 in §9.7.3 for derivation of C_{IG} factor.

The binding energy is obtained for initial value of $x = x_o$. Substituting Eq (6.87.a) into (6.88) and dividing on electron charge *q* we obtain the **binding energy** in eV.

$$E_{SG}(TP) = \frac{2\alpha C_{SG}}{(x_o^2 + r^2)q} = 2.145 \times 10^6 \ (eV) \qquad (6.89)$$

where: *r* is given by Eq. (6.84.a)

The experimentally measured value is 2.22457×10^6 eV. Consequently, the theoretical derived binding energy is quite close, which is one additional validation for the correctness of the unveiled structures of the proton and neutron.

6.14.2 Estimation of the distance between the neutrons in the nucleus of Tritium.

We will show a method for estimation of the binding energy between the nuclei of Tritium and Deuteron, from which the approximate value of the space between the two neutrons will be obtained. The Tritium is a three body system with configuration shown in Fig. 6.31.

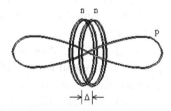

Fig. 6.31. Nuclear configuration of Tritium

The experimentally determined binding energy is:

$$E_T = 8.4818 \ MeV \ .$$

According to the analysis in the previous paragraph, we may regard it as a disintegration energy. We may assume apriory, that the distance Δ between the two neutrons is a few times smaller than the proton length. At the same time, they will be kept apart by the proximity electrical fields of their

external shells. So we may regard the two neutrons as separate systems. Since the electrical field energy is part of the SG field energy, we may consider that the total disintegration energy is a sum of two partial disintegration energies.

The disintegration energy between the proton and neutron system could be obtained in a similar way as for deuteron, but considering a ring with a twice larger intrinsic mass. We shall use again the theoretically derived Eq. (6.89). but multiplied by factor of two (because two neutrons SG masses are involved instead of one). So the first partial disintegration energy is:

$$E'_{SG}(TP) = \frac{4\alpha C_{SG}}{(x_o^2 + r^2)q} = 4.29 \times 10^6 \ (eV) \qquad (6.90)$$

The second partial disintegration energy then is:

$8.4818 - 4.29 = 4.1918 \ MeV$

The obtained energy of 4.1918 MeV will serve to determine Δ. We may simplify the task by regarding the two neutrons as two mass points at distance Δ. Let us consider the following possible options:

(1) - only SG(TP) fields are involved in the separation energy

(2) - only SG(CP) fields are involved in the separation energy

-(3) - both SG(TP) and SG(CP) fields are involved

The energy expressions for the both types SG fields are obtained by integration the SG field from Δ to infinity and using the relation (6.83). So we may assign the energy 4.1918 MeV either to (CP) or (TP) and see what reasonable value of Δ we may obtain.

$$\frac{C_{SG}}{2\Delta^2 q} = 4.1918 \ (MeV) \ \text{ if SG(CP) energy - then use} \quad (6.91)$$

$$\frac{\alpha C_{SG}}{\Delta^2 q} = 4.1918 \ (MeV) \ \text{ if SG(TP) energy - then use} \quad (6.92)$$

The calculated distance in angstroms for both cases is respectively:

when considering $E'_{SG}(CP)$ field:　　0.885 A

when considering $E'_{SG}(TP)$ field:　　0.107 A

The distance 0.888 A is not acceptable, because it is larger than the proton length. Consequently, only the $E_{SG}(TP)$ field is involved in the balance, so the distance is:

$$\Delta = 0.107 \times 10^{-10} \text{ m} . \qquad (6.93)$$

The neutron separation distance is about 16% of the proton length.

The obtained result shows one important fact: despite its neutrality the two neutrons are well separated by the SG(TP) field. This is one additional confirmation, that the neutron has a proximity E-field, which is kept locked by the SG(CP) forces, but is effective at close proximity. This concept is in agreement with the neutron's possession of magnetic moment.

6.14.3 Role of the SG(TP) field in the spatial order of hadrons in the atomic nucleus

The calculated binding energy between the proton and neutron in Deuteron was estimated if the disintegration takes place along the proton long axis and the neutron's equivalent plane is normal to this axis. The analysis shows, that the SG(TP) field is only involved. The result for the separation between the neutrons in Tritium also indicates the involvement of the same type of field and interactions. Consequently the SG(TP) fields and the interactions between the CL space and the internal RL(T) of the FOHSs are responsible for the spatial order of the neutrons and protons in the atomic nuclei. This also means that in any combination of protons and neutrons, the distance between their helical structures is always larger than some critical distance. This restriction preserves these helical structures and consequently the stable elementary particles from destruction (see §6.4.3 and Fig. 6.1, Chapter 6). This effect is so strong that the structures of the hadrons (proton or neutron) are separated at safely distance in the atomic nuclei. Obviously, this condition is fulfilled even in the process of nuclear synthesis in stars. Otherwise the FOHSs could crush, which means that an enormous flux of hardware neutrino would be observed from the Sun (while we have a lack of expected neutrino flux).

The SG(TP) fields allows also the protons to have a limited freedom mainly in the polar atomic plane. This feature will be discussed in next chapters.

6.15 Neutrino particle classification

The neutrino firstly was introduced in order to explain the mass-energy balance in the controversial Beta decay. Later additional types of neutrino were introduced for the muon to electron decay. Later, some types of neutrino has been experimentally detected from the space, from the Sun and from the nuclear colliders. Presently, number of neutrino types exist with their modifications and the neutrino concept is not solved in satisfactory level by Modern physics. The unsolved solar neutrino problem is just one example.

According to BSM, the accepted neutrino in the Beta decay does not have nothing common with a real neutrino particle, which must possess a matter. In order to mach the terminology we may call it a "virtual neutrino", having in mind the above comment. In some processes, both types of neutrino are obtained simultaneously. Table 6.3 lists some types of neutrinos and their relation to some decay processes.

Types of neutrinos according to BSM **Table 6.3**

Type	From process
neutrino from partly folded RL node	FOHS destruction
neutrino from folded RL node	FOHS destruction (also from $\mu \to e$)
"virtual" neutrino	$n \to p$; $\pi \to \mu$

The signature of the partly folded RL nodes in the experiments might be the electroweak current, introduced by the Electroweak theory. The neutrinos, detected by the neutrino detectors are from the folded RL nodes.

In $n \to p$ process (neutron - proton conversion), only a twisting is involved. The energy that is detected by the obtained proton momentum is not a quantum wave and is dissipated in CL space (contributing to the ZPE ocean). It comes from the Newtonian mass change, due to the shrinkage of the FOHS's. In contemporary physics this energy is wrongly accepted as antineutrino particle.

In the $\pi \to \mu \to e$ decay, there is a process of twisting, followed by FOHS disintegration. So the "virtual" and real neutrinos appear almost simultaneously. This is the confusing feature, as a result of which the "virtual " neutrino is misidentified as a real neutrino. The neutrino detectors can detect only real neutrinos. At the present time, their threshold level of detection is higher than the energy of the antineutrino from the inverse Beta decay.

The first experiment, considered as a neutrino confirmation is provided by F. Reines, C. Cowan et al. They measured the "energy-mass" difference in the reaction $p \to n$ and considered the missing part to be a signature of the neutrino. Therefore, it must be stated that direct neutrino particles in this case are not measured. (see Cl. Cowan, Jr, F. Reines et al. (1956). This signature is considered so far as a low energy neutrino. From the BSM theory, we already know that the low energy "neutrino" is a signature of the Newtonian mass difference due to the untwisting, which causes a slight volume change of all FOHS's. That's why BSM denoted it as a virtual neutrino, keeping in mind, that it does not have anything common with the real neutrino, which will possess matter.

The processes of obtaining of high Z number elements in the womb of a star are characterized with a large number of $n \to p$ type of reaction (the nuclear build-up process is described in Chapter 8). The energy of the "virtual" neutrino from this reaction is dissipated in the star gravitational field, so the "virtual" neutrino expected from our Sun is missing in the neutrino detectors used in the Earth.

Apart from the virtual neutrino, real neutrino particles (possessing matter) are also emitted from the womb of the stars, but they are result of quite different process - a destruction of neutrons and protons into straight helical structures - kaon type bundles, which cause a superstrong magnetic field due to the strong aligned SG forces. (see §10.14 in Chapter 10 and §12.B.6.4 in Chapter 12 of BSM). In this process some FOHSs are also destroyed and the released RL nodes escape as real neutrino particles possessing a matter. Such neutrinos possesses a high penetration capability, so the neutrino detectors identify them as high energy neutrino particles. For example, **a neutrino flow emitted by the supernova SN1987A has been detected** by IBM team (Irvine-Michigan-Brookhaven) Collaboration. The observation was led by Fred Reines. **This indicates that in the supernova a processes of massive destruction of FOHS's leading to a formation of a huge kaon nucleus takes place.** In a cosmic scale such process may involve an unimaginable enormous energy.

Chapter 7. Hydrogen atom

The Hydrogen atom is a dynamical system comprised of one proton and one electron. The internal structures and individual properties of both particles were discussed extensively in the previous Chapters 3 and 6. The analysis of the dynamical properties of the Hydrogen atom, is very useful for understanding the structure and dynamical properties of other atoms.

7.1 Proton as a nucleus of the Hydrogen atom

The proton is a twisted closed loop helical structure (with an plane projection like the digit 8) enclosing a pair of internal pions and one kaon. Its notation according to Table 2.1 of Chapter 2 TTH_1^3:+(-). Its internal structure was shown in Fig. 2.15.B. The proton shape (illustrated in Fig. 6.22 of Chapter 6) is shown again in the figure below.

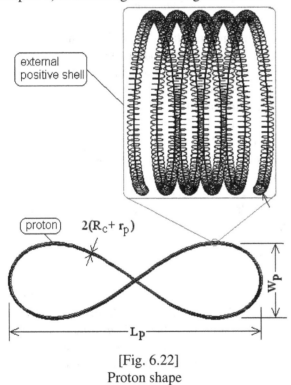

[Fig. 6.22]
Proton shape

The proton core is a three dimensional curve, whose plane projection is given by the Hippoped curve at parameter $a = \sqrt{3}$. The analytical expressions of the Hippoped curve (in polar and Decart system) were given in Chapter 6 (Eq. 6.54.a and 6.54.b). The plot of this curve with some specific dimensions, is shown in Fig. 7.1. We may call the

two portions of the curve **proton clubs**. While the real proton clubs do not lie in a plane but in a slightly curved surface, we may call this surface a quasiplane. At the same time, we may often treat it as a plane because the core thickness to length ratio of the proton (also for the neutron) is quite small:

$$(2(R_c + r_p))/L_{pc} = 0.0048$$

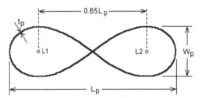

Fig. 7.1
Projection of the proton core on a 2D plane

The dimensions of the proton core, determined in §6.12.2.6 are following:

$$L_p = 0.667 \times 10^{-10} \quad \text{(m)} \quad \text{- proton length} \quad (6.76)$$

$$W_p = 0.19253 \times 10^{-10} \quad \text{(m)} \quad \text{- proton width} \quad (6.77)$$

$$t_p = 2(R_c + r_p) = 7.8411 \times 10^{-13} \text{ (m) - core} \quad (6.78)$$
$$\text{thickness}$$
$$L_{pc} = 1.62772 \times 10^{-10} \quad \text{(m)} \quad \text{- core length}$$

The ratio between the length and width of the curve for the accepted parameter $a = \sqrt{3}$ is 3.4643. There are two characteristic points in the proton quasiplane, shown in Fig. 7.1 as L_1 and L_2. They are located on the horizontal axis passing through the geometrical centre and corresponds to a maximal vertical width. The distance between them is *0.648L_p*. These points, called **locuses**, are characteristic points for the distributed proximity electrical field of the proton.

7.2 Bohr surface of the Hydrogen atom

7.2.1 Proton electrical field

The proton core structure was described in details in Chapter 6. The positive charge is contributed by the RL(T) of the external shell. Having in mind the finite dimensions and shape of the proton core, the electrical charge in the closed field is distributed over the external shell. The unit electrical charge is a result of the SG energy balance between the RL(T) structures of the proton's helical structures and the surrounding CL space. In the far field the electrical lines appear radial to the geometrical centre of the Hippoped curve and their density distribution simulates a point charge. In the near field, however, the electrical lines are curved with a spatial configuration determined by the proton core shape. Consequently, when approaching the proton from the far field, some boundary range should exist, beyond which the straight radial electrical lines are converted to curve lines. We may approximate this boundary range with an equivalent surface. Outside of this surface, the proton will look like a point charge particle, possessing the Newtonian mass of the proton. So it should be a closed surface. Inside of this surface the proton's electrical field will not converge to a point charge but to the proton core (Note: the proton core should not be confused with its internal kaon core. The proton core is the external enclosure of the twisted helical structure of the proton, which thickness is $2(R_c + r_p)$).

7.2.2 Relation between the BSM model of the Hydrogen atom and Bohr model

The Quantum mechanics is successfully built on the concept of the Bohr model of the Hydrogen atom. According to BSM theory, the Quantum Mechanical (QM) models of the Hydrogen and other atoms are good **mathematical models,** providing very useful quantum features. However, they could not serve as physical models of the atomic structure. The main discrepancy comes from the absence of CL space parameters and the structural features of the elementary particles in the QM models. The goal of the BSM theory is to provide exact physical model of the Hydrogen and other atoms. One useful parameter, that the BSM model will use from the Bohr model is the Bohr radius, denoted as a_o. The proper use of this parameter will provide an useful bridge between the quantum model of the atoms and the BSM physical models.

In §3.12.2 the parameter a_o was determined from the conditions defining the quantum orbit length (Eq. (3.43.f) and (3.43.g). At the same time a_o is determined by the Bohr model of the Hydrogen atom.

$$a_o = \frac{h^2 \varepsilon_o}{\pi m_e q^2} \tag{7.1}$$

Then we have:

$$\frac{h^2 \varepsilon_o}{\pi m_e q^2} = a_o = \frac{\lambda_c}{2\pi\alpha} \tag{7.2}$$

The left side relation is from the Bohr model of Hydrogen, while the right side is related to the quantum orbit equivalent radius for electron quantum motion with first harmonic velocity. Based on the a_o parameter, the Eq. (7.1) provides very useful relation between the CL space parameters, the Plank constant and the unit charge. Multiplying the nominators of relations (7.2) by 2π, we obtain relation valid for the first harmonic quantum orbit, whose length is equal to $2\pi a_o$.

$$\frac{2h^2 \varepsilon_o}{m_e q^2} = \frac{\lambda_c}{\alpha} = 2\pi a_o = const \tag{7.3}$$

The ground state orbit in Bohr model defines a shperical surface around the point-like proton with radius a_o. The proton shape according to BSM is quite different, and the ground state orbit will define a surface different than sphere. The equivalence of Eq. (7.2) and (7.3) allows us to use the spherical surface area, defined by the radius a_o, as a modified characteristic parameter of the quantum orbit condition. In such case, the radius a_o provides a bridge between the Bohr and the BSM model of the Hydrogen atom. In order to preserve (at least approximately) the relation to the Bohr model, we will make the area of the equivalence surface of the BSM model to be equal to the area of the spherical surface defined by the Bohr radius. We may call the equivalence surface, used in the BSM model, a **Bohr surface,** in order to emphasize the relation to the Bohr radius a_o. Then the right-hand part of the relation (7.3) could serve as one of the definition parameters of this surface. It is written separately as Eq. (7.3.a).

$$2\pi a_o = \frac{\lambda_c}{\alpha} = const \tag{7.3.a}$$

The parameter $\lambda_c = \lambda_{SPM}$ is valid for free CL space not disturbed by E-field. So it is satisfied outside of the boundary region, which defines the Bohr surface, but we may consider that it begins to be satisfied at the boundary region.

The other definition parameter of the Bohr surface is dictated by the shape of the proton or more accurately said - its proximity electrical field. **The latter possesses a spatial configuration defined by the proton shape and enclosed inside of the Bohr surface in a case of orbiting electron.**

In case of single proton without an electron (a positive ion) its electrical field breaks the Bohr surface. In this case the proximity field distribution is still preserved, but outside the Bohr surface the electrical lines are rearranged. So from outside, the proton (ion) appears as a point charge.

Summary:

- **The Bohr surface is an equivalent boundary surface around the proton with a shape of ellipsoid.**
- **The area of the Bohr surface is equal to the area of the sphere with a radius a_o.**
- **The spatial orientation of the Bohr surface is defined by the spatial orientation of the proton core**
- **The Bohr surface around the proton is implicitly defined by the CL space parameters λ_{SPM} and α.**
- **In a case of orbiting electron, the proximity E-field is enclosed inside the Bohr surface.**
- **In case of single proton (or ion), the E-field breaks the Bohr surface. The external E-filed lines in this case are radial to the geometrical centre and simulate a pointlike charge particle.**
- **The expression (7.3) requires only a constant trace length of the quantum orbit, while giving a freedom of its shape. This means that the shape of the Bohr surface could be also modified.**

The position of the Bohr surface around the proton is illustrated by two sections, as shown in Fig. 7.2.

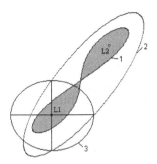

Fig. 7.2

Bohr surface around the proton

1 - proton core, 2 - Bohr surface section along the proton length, 3 - Bohr surface section across the proton width, passing through one of the locuses

The quasiplanes of the proton clubs are shown shaded. The two locuses are denoted as L_1 and L_2.

In order to determine the ellipsoid axes of the Bohr surface we need to know the E-filed lines distribution inside the surface. Their spatial distribution is determined by the twisting characteristic angle θ_w of the external proton shell and the overall shape of the proton. It is more convenient to present a section of equipotential surfaces inside the Bohr surface. The curved E-filed lines should intercept these surfaces at right angle. A possible configuration of the equipotential surfaces in a section passing through one of the locuses is shown in Fig. 7.3.

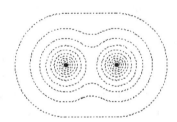

Fig. 7.3

Section of equipotential surfaces inside the Bohr surface

One possible way to determine the Bohr surface parameters is to fit quantum orbits around the proton core, taking into account the equipotential surfaces in all possible sections. At first glance, this is a complicated task. However, we know the shape and length of the proton core and the length of the quantum orbits. Then we could make models of quantum orbits around the proton core with possible shapes. One helpful rule in the search for the correct shape is the following:

- **the motion in the quantum orbit should be characterized with a minimum energy loss.**

    This condition is fulfilled, if:

- **the trace of the quantum orbit intercepts the E-filed lines inside the Bohr surface at one and a same angle.**

    In the following analysis we will see, that the second condition may not be fulfilled for the whole orbit trace, but for a larger or smaller section of the orbit. The strength of the quantum effect is dependable of this condition. It will become evident, also, that:

- **The quantum effect becomes stronger when the interception angle approaches the twisting angle:** $\theta_w = \theta_{eff}^{lept} = 28.762$ deg **(discussed in Chapter 6).**

    The criterion for the correctness of the quantum orbit shape will be the electron energy, corresponding to the quantum number. The electron energies for the consecutive quantum orbits will correspond to the energy difference between the neigbouring spectral lines in one series. Such model is developed for the Balmer series of the Hydrogen atom. It is presented in §7.8.

7.3 Coulomb force inside the Bohr surface

It is evident, that the distributed positive charge of the proton inside the Bohr surface will provide different interaction conditions between the electron and proton, when the electron is inside of this surface. The inverse square dependence of the Coulomb forces from distance is not any more aplicable in such conditions. In order to find the modification of the Coulomb force law inside the Bohr surface, we will make analogy with the optical radiation. For this purpose we will take example of a point light source and a point like detector as illustrated in Fig. 7.4. The source and detector, both have the same angle of view. The point source illuminates a screen, the distance from which is fixed and shown as r_{ref}. Let us consider that the screen is made of micro corner cube tape, having the property to reflect the rays at the same incident angle.

This will closely simulate a feature of E-field lines coupling between the point like electron and the distributed E-filed lines of the proton. The case *a*, *b*, *c*, corresponds to three different distances of the detector from the screen. The illuminated area is denoted as A_s and the pick up area by detector - as A_d.

In case **a.** the distance between the detector and the screen is larger than r_{ref}. This corresponds to an electron outside of the Bohr surface. It cannot pick up all the electrical field lines of the proton. The picked up signal is inverse proportional to the square of the distance r, normalized to r_{ref}. This is the classical law of Coulomb forces.

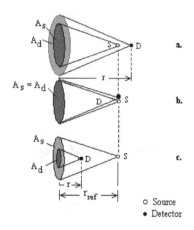

Fig. 7.4

In case **b.** the detector field of view covers exactly the illuminated surface A_s. This corresponds to an electron positioned at the Bohr surface. **All E-filed lines of the electrons are connected to all E-field lines of the proton. This is possible if the electron velocity is not very large.**

In case **c.**, the picked up signal is proportional to the square of the distance r, referenced to r_{ref}. This dependence is valid only for the range $0 < r \le r_{ref}$. This case corresponds to an electron inside the Bohr surface. **In such conditions the electron's E-field lines are not able to be interconnected to all E-field lines of the proton, because the proton's E-field lines emanates from a comparatively large proton core envelope surface.**

Note: Here we must open a bracket that the effect of not inerconnected E-field lines between the proton and the electron is compensated by the faster motion of the electron (as we will see in the Balmer series model presented in §7.8). This means that the faster moving electron will sweep more electrical field lines for a unite time, which is below the relacsation time of the CL space for the proximity E-field region of the proton. As a result, the dynamical system of the proton and orbiting electron will appear as a neutral in the far field.

The above considerations could not be valid at very close distance to the proton core, because the effect of the twisting SG field in the vicinity of the proton and electron external shell will predominate. This may give some increased repulsion at the very close distance. Such effect will assure a safe minimal gap between the hardware helical structures of both particles, which is a very important feature for keeping their internal RL(T) structures from destruction (see §6.4.3 Chapter 6).

According to the provided analysis the Coulomb force inside the Bohr surface could be proportional to the term: $\frac{q^2}{4\pi\varepsilon}r^2$. Outside of the Bohr surface the atom is neutral for all series, including the Balmer one. This means that the whole charge of the proton participates in this term. For the circular sector of the orbit around the proton core, however, we may consider that the electron interacts only with the half of the E-field lines, so the corresponding force will become proportional to $\frac{q^2}{2\pi\varepsilon_o}r^2$.

The shown above term should be normalized to a reference distance, corresponding to r_{ref} in the analogous optical example. It is not difficult to guess that this reference distance is the Bohr radius, a_o. It connects the CL space and the electron system parameters: h, ε_o, q, m_e, according to Eq. (7.1). Normalising to this distance, we get the equation of Coulomb force between the proton core and the electron in the circular section of the orbit, which is inside the Bohr surface.

$$F_C = \frac{q(q/2)}{4\pi\varepsilon_o a_o^2}\left(\frac{r}{a_o}\right)^2 \qquad (7.3.b)$$

where: r - is the distance between the proton core and the electron in the circular part of the orbit.

7.4 Orbital planes for the Hydrogen series.

We will consider here only the orbits, which are related to emission or absorption of photons.

The possible orbits are three dimensional curves and in fact could not define a plane surface, but we may define an equivalent surface, so the average distance of it from all orbital points (for small time intervals) to be a zero. Then such surface will have a twisted shape, so we may call it **an orbital quasiplane**. One orbital quasiplane is defined by one orbit, but a large number of possible orbits for one electron may have a common orbital quasiplane. One spectral series of the Hydrogen atom, for example, corresponds to set of orbits with common orbital quasiplane. **The limit of the series corresponds to the largest orbit of the set, called a boundary orbit.** It will be shown, by the model of the Balmer series, than the number of orbits in the series is limited. The electron kinetic energy in the boundary orbit can be determined by the limit energy of the corresponding spectral series. Knowing the length of the boundary orbit as a quantum orbit, the quantum velocity can be determined. **In such way it can be verified that the boundary orbit is a quantum orbit.** Consequently, we may determine the length of the boundary orbit for any one of the series, using the condition of the quantum orbit.

The equation of the quantum orbit trace length was derived in §3.12.3 (Eq. (3.43.i)

$$L_{qo}(n) = \frac{2\pi a_o}{n} = \frac{\lambda_c}{\alpha n} \qquad [(3.43.i)]$$

where: n is the subharmonic number of the quantum orbit

The positions of the orbits are referenced to the proton core geometry. Then it is more convenient to use the ratio between the quantum orbit trace length and the core length of the proton. For a quantum orbit corresponding to a first harmonic electron velocity, this ratio is:

$$L_{pc}/L_{qo}(1) = 2.042878 \approx 2 \qquad (7.4)$$

The ratio (7.4) is very close to a whole number and we may use integers, for convenience, neglecting the small fractions. In §3.12.3 it was mentioned, that subharmonic quantum loops are able to be connected in series, forming in this way a common quantum orbit. We may call such orbit a **serial quantum orbit**. Table 7.1 shows the ratio calculated for different subharmonic numbers, *n*, and for both types of orbits: single and serial. The ratio is rounded to integer or close fractional numbers for convenience.

Possible quantum orbits according to the approximate ratio $L_{pc}/L_{qo}(n)$ Table 7.1

n	single quantum orbit	serial quantum orbit comprising:				
		2 loops	3 loops	4 loops	5 loops	6 loops
1	**2**					
2	**1**	2	3	4		
3	1/3	2/3	1	4/3	5/3	2
4	1/4	2/4	3/4	1	5/4	6/4
5	1/5	2/5	3/5	4/5	1	6/5
6	1/6	2/6	3/6	4/6	5/6	1

A different subharmonic number means a different quantum velocity of the electron. The selection rule for a proper orbit is additionally influenced by the SG forces. For this reason a model involving the balance between all forces is necessary. Such model is developed for the Balmer series. Based on this model and additional considerations of orbits in other atoms, the orbital quasiplanes of Lyman and Balmer series are identified with a high degree of confidence. The boundary orbit for the Lyman series corresponds to a ratio 2, while for the Balmer series - to a ratio 1 (according to Table 7.1).

Figure 7.5 shows the position of the boundary orbits of the Lyman and Balmer series referenced to the proton shape. They define also the orbital quasiplanes.

The boundary orbits for the higher order Hydrogen series of spectral lines may occupy the same orbit as the Balmer or the Lyman series. Below the boundary orbit however, the higher order series may have serially connected quantum orbits (the latter option is not enough investigated by BSM). (For atoms with a higher Z number, the Lyman series quansiplane becomes less accessible and the Bohr surface becomes distorted).

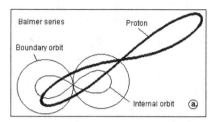

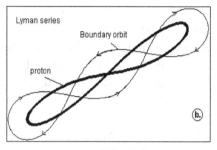

Fig. 7.5

We see that the orbital quasiplanes of Lyman and Balmer series are quite distinguishable one from another. The orbital quasiplane determines the positions of many quantum orbits, but the boundary orbit is the largest one. It is reasonable to accept that the boundary orbits of all possible quasiplanes are inside the Bohr surface, so in all this cases the Hydrogen atom appears as a neutral. The electron may change also the orbital quasiplane if getting or losing a large amount of energy due to some elastic collision of the Hydrogen with another molecule. The probability of quasiplane change in a spontaneous emission however is much lower than changing of the quantum orbit in the same orbital quasiplane. We may consider, that in the process of ionization, the lost electron has been in one of the possible quasiplane. It is reasonable to consider that atoms with Z >1 may also have conditions for different orbital quasiplanes as the Hydrogen. However, the possible quasiplanes are dependent on the arrangement of the protons and neutrons in the nucleus, as this will be shown in Chapter 8 and the Atlas of the atomic nuclear structures. In any case, however, the ionization is possible.

Figure 7.6 illustrates a possible shape of the boundary orbit for higher series of the Hydrogen spectra.

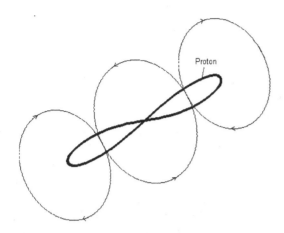

Fig. 7.6

Possible shape of a boundary orbit for higher order series of the Hydrogen spectra

The existence of more than one boundary orbit, could be explained also by the flexibility of E-field refurbishment that may modify the shape of the Bohr surface. We may assume that the Bohr surface have a constant area of $4\pi a_o^2$ but a flexible shape, depending on the working orbits and the interconnection of the proton to other protons in the atomic nucleus. Below the Bohr surface, the E-field possesses a spatial structure, confined to the proton shape and the characteristic twisting angle θ_w of the E-field line emerging from the proton's core. The circulated electron in the orbit intercepts most of the E-field lines at constant angle. In the Bohr surface region this condition (intercepting E-filed line angle) is disturbed. At the same time, the electron possesses a finite momentum even at the Bohr surface (this is shown later in the Balmer model). Consequently, the electron is able to escape from the boundary orbit, if the momentum is large enough.

Below the boundary orbits are all orbits contributing to the series, terminated with the ground state orbit.

- **According to BSM, every series has its own ground state orbit, which is the shortest one.**

This conclusion will be demonstrated by the Balmer series model.

The quantum conditions defining the stable orbits are discussed below.

We may use the term **orbital** for all orbits in one series. Although we have to keep in mind that it does not correspond to the term orbitals used in the quantum mechanics (where they are defined by the wave function). We see also that the orbital quasiplanes are curved but open surfaces. The electron transitions between any two orbits are in one orbital quasiplane. The passing of the electron from one to another orbital quasiplane requires special conditions and is less probable.

From the considerations presented so far and from the further analysis we can formulate the following physical rules for the orbits:

- **The orbits for all spectral line series are inside the Bohr surface**
- **Any orbital quasiplane, related with a photon emission or absorption, intercepts one or two proton clubs**
- **The boundary orbits approach the zero level potential**

7.5 Effect of the orbiting electron on the atomic motion in CL space.

It is evident that the electron trajectories reside on the orbital quasiplane. The latter is an open surface and could not affect the static pressure of the CL space exercised on the protons and neutrons, and consequently - the atomic mass. Although the orbital momentums of the electrons could affect the atomic motion in the lattice space, causing a spin rotation. For a simple physical analogy, the orbital's twisted quasiplanes behave as fins of mechanical object, causing a rotation of this object when moving in a fluid.

Let us consider a neutral Hydrogen atom, moving with a constant velocity. The most probable orbit is the ground state. The electron has its own momentum that defines the orbital momentum. But in order to keep this momentum when the Hydrogen is moving, the interaction with the CL lattice should be minimal. Then the Hydrogen has to rotate with some confined spin because the orbital shape is twisted. The motion behaviour of such atom or molecule will simulate a "flying bird" with a simultaneous rotational motion. As a result of such motion, the electron could make transitions between very close orbits. So it may pump the CL space with very low energy that could be periodically emitted as a low energy photon. The input energy for such emission may come from the equalization of the zero point energy of the CL space. Such radiation will contribute to the Cosmic Microwave Background, corresponding to the temperature of 2.72 K. It might be contributed not only by the Hydrogen atoms and molecules, but from other atoms and molecules, as well.

From the detailed atomic nuclear structure discussed in Chapter 8, we will see that many aspects of Hydrogen orbital structure are preserved in the atoms with a higher Z number.

Summary
- **The orbital momentum affects the proton confined motion in CL space**

7.6 Quantum motion of the electron in electrical field. Quasishrunk CL space.

The ionization energy of the Hydrogen atom is 13.6 eV corresponding to the optimal velocity of the electron. Consequently, all orbital velocities of the electron are of suboptimal type. The quantum motion for such velocities was analysed in §3.9. The analysis, was provided for CL space without external electrical field. In Hydrogen atom, however, all orbits are inside the Bohr surface, where the electrical field has a specific spatial configuration, defined by the proximity field of the proton (and the proximity locked field of the neutrons in Deuteron).

Let us denote a CL space without electrical and magnetic field (other than the electron's own fields) as a **free CL space** and the CL space with an external E-filed - as a **E-field CL space**. Then we may distinguish two cases of the electron confined motion:

   (a) electron motion in a **free CL space**
   (b) electron motion in a **E-field CL space**

The case (a) was analysed in Chapter 3. The quantum motion of the electron is defined by the CL space parameters. Among them are the Compton wavelength, λ_{SPM}, which is defined by the SPM frequency and the light velocity. For a free CL space we have $\nu_{SPM} = \nu_c = \nu_e$, where, ν_e is the first proper frequency of the electron (electron shell - internal positron). **At the end of §2.11.2.2 it was discussed, that the stationary EQ of the CL nodes may posses a higher resonance frequency,**

than the MQ node. This automatically means that they will have a higher SPM frequency (a higher SPM frequency of a CL space domain with same node distance means a shorter SPM cycle and a shorter propagated SPM phase, i. e. **a shorter SPM wavelength** λ_{SPM}' in comparison to the free CL space).

$$\nu_{SPM}' > \nu_{SPM} \quad \text{or} \quad \lambda_{SPM}' < \lambda_{SPM} \tag{7.5}$$

where: the prime sign denotes the corresponding parameter in the E-field CL space.

It is apparent from Fig. 7.5 that the Balmer quasiplane is much less twisted than the Lyman one. This makes Balmer orbits more convenient for analysis. The electron orbits for Balmer series occupy the range between the proton core and the boundary orbit. Around the proton core, their traces tend to follow the equipotential curves as illustrated in Fig. 7.3.

According to the derived rule for the interception angle between the orbital trace and the E-field lines (see §7.22), it follows that there is a tendency of keeping a constant value of this angle with variations within a limited angular range. Consequently, the condition (7.5) will be more or less valid for the motion in any one orbit below the boundary one. At the same time, the electron proper frequency ν_e is unchanged, because the electron system possesses own internal energy. While the E-field does not affect the CL space node distance, it affects the SPM wavelength, making it shorter. But the SPM wavelength is a specific quantum parameter of the CL space influencing the light propagation and the electron quantum motion. So if the electron velocity is estimated by the node distance, the quantum velocity appears smaller (following the shorter λ_{SPM}'). In the Balmer orbits model presented in the next paragraph, it is accepted that in the E-field CL space inside the Bohr surface, λ_{SPM}' changes linearly with the radius of the circular part of the orbit. In this case, the obtained results of the model are optimal. The figure of merit is the shape of the calculated energies corresponding to the Balmer series spectra. The linear dependence may be a result not only of the proton E-filed configuration below the Bohr surface, but also of the magnetic field lines caused by the electron motion and oscillation.

The reduced value (shrinkage) of λ_{SPM}' and the orbital length dependence on it gives a

possibility the space below the boundary orbit to contain a larger number of orbits.

The shrinkage of the λ_{SPM} inside the Bohr surface complicates the analysis, because the quantum scale becomes different. **In order to solve this problem, we may consider, that the quantum space is quasishrunk.** The term quasi is used, because the CL node distance is not changed and the proton's dimensions - also, but the shrinkage is valid only for the quantum conditions. In order to keep this into account, we have to translate the necessary parameters to the scale of the quasishrunk quantum space, i. e. to λ_{SPM}' (the prime is used to denote the shrunk value of the parameter). The field forces and inertial momentum also have to be referenced to this scale. **In such case, the inertial mass of the electron referenced to λ_{SPM}' scale will be affected.** When analysing the motion in the circular part of the orbit, **the apparent inertial mass will appear different**, because the electron intercepts smaller number of nodes per λ_{SPM}'. We may test a linear or a quadratic dependence of the apparent inertial mass in function of orbit length (or distance from the proton core). The quadratic dependence, which is a symmetrical function of the Coulomb force inside the Bohr surface provides better results in the Balmer model.

The quasihrunk quantum space affects not only the electron motion but the quantum waves as well. The internal space inside the Bohr surface behaves as an optical media with gradual index change. In such way, it affects the propagation of the quantum waves in the X-ray range. This behaviour is discussed in Chapter 8. Consequently, we may accept that the space is characterised with a gradual refractive index. This refractive index is valid only for electron motion at proper orbits and for incident quantum wave falling at proper incident angle.

The static pressure is from all direction forces exercised on the FOHSs, valid also for the electron. According to this formulation, the static pressure in E-field CL space should not be changed, because the average node distance is unchanged. Then the electron's parameters are also preserved. This is valid also for the fine structure constant, estimated as a ratio between the tangential and axial velocity of the electron.

Summary:
- **In E-field CL space, the SPM wavelength along the equipotential curves is reduced**
- **The electron performing a quantum motion in equipotential curve exhibits increased apparent inertial mass, if referencing its motion to the quasishrunk quantum space**
- **The refractive index of the quantum quasishrunk space is valid for electron motion at proper orbit and for incident quantum waves falling at proper angle in respect to the proton club quasiplane.**

7.7 Quantum orbit conditions for orbits inside the Bohr surface.

7.7.1 Quantum conditions, related to the orbital length

The confined electron motion in a closed loop trajectory was discussed in §3.12.1. The conditions of the phase repetition of the two proper frequencies and their match to the phase of propagated SPM vector were analysed. As a result a definition of a quantum loop was derived and formulated as:

The quantum loop is a closed loop trajectory of the electron moving with a confined velocity. The loop trajectory length is defined by the condition of whole number of carrier oscillations.

The expression (3.43.e), derived in a original way showed the relation between the BSM model of the electron confined motion and the Bohr model of Hydrogen. Apart of the unveiled quantum conditions, one of the most important conclusion was that the quantum orbit is defined only by the trace length, but not by its shape.

By the way, the phase match is a necessary but not enough condition in order to define the orbital time, which is known as a lifetime of the exited state. The BSM analysis unveils a second condition causing the dropping of the electron to a lower quantum orbit after a finite time known as a "lifetime of the excited state." It was found that the orbiting electron creates magnetic lines with very short wavelengths approaching the Compton's wavelength $\lambda_c = 2.426 \times 10^{-12}$ (m). The latter puts a limit for the length of the magnetic line, so this effect is referred later as a "short magnetic line conditions" (described in the next section). While the electron possesses two proper frequencies, the corresponding oscillating structures are influenced differently by the created magnetic field, so conditions for conflict occurs after a definite number of electron's rotation, which is strobed by the short magnetic line conditions. This defines the lifetime of the electron, if its motion is not disturbed by external fields. This lifetime corresponds to the spontaneous emission of a photon. The effect of the limited number

The result for a phase repetition of oscillating electron was derived for a free CL space in §3.12.1. Similar conditions are valid also for the boundary range of the Bohr surface. If the Compton's length below the Bohr surface decreases gradually, as discussed in the previous paragraph, the conditions defining the quantum loop and the orbital time (lifetime) are still preserved, because the SPM frequency in that region also changes with the same rate. It is only necessary the cosine between the negative core oscillation, and the E-filed lines of the proton to have enough small dispersion around one mean value defined by the orbital position. Keeping in mind, that the E-field lines (enclosed in the internal volume defined by the Bohr surface) are subordinated by the angle θ_w, it is apparent that the condition for quantum orbits could be satisfied for a finite spatial range inside the Bohr surface.

Consequently, the quantum orbit conditions may be valid for a large number of orbits, below the boundary one.

It will become evident from the analysis later, that the above mentioned conditions are valid for all orbits and orbital transitions that provide the spectral line series of the Hydrogen atom.

7.7.2. A minimal length of the magnetic line (effect of a short magnetic line) affecting the possible quantum orbits.

If the above conclusion is correct, an additional quantum condition is necessary in order to provide an individual orbit separation, corresponding to the different quantum energy levels.

- **The condition for orbit separation is provided by the magnetic line, aligned with the spin axis of the orbiting electron.**
- **The above condition is contributed by the spin rotation of the orbiting electron.**

When the electron provides a repeatable motion in a quantum loop, its spin rotation is an important attribute of the motion. The velocity vector of the rotating electron shell is normal to the orbital trace, so the magnetic lines induced by the spinning electron appear parallel to the orbital trajectory. Having in mind the radial E-field distribution (see Fig. 3.6, Chapter 3) and the quantum magnetic radius, we may distinguish **two separate bundles of magnetic lines from the spinning electron: peripheral and axial.** The peripheral one is related to the quantum magnetic radius and will have a shape of hollow tube around the electron shell. The rotating electron provides a large concentration of magnetic field lines passing through the axis of its rotation. This will cause deterioration of the external E-field of the proton in a narrow zone centered around the electron orbital trace. **Consequently, we may consider that the axially aligned field of the moving electron creates a path with MQs, having the same SPM frequency as the CL space outside of the Bohr surface, where the E-field is missing.** At the same time, the magnetic lines in the peripheral field occupy a larger volume and do not lead to disturbance of the proton's E-field. As a result of this analysis, we arrive to two important conclusions:

- **The peripheral magnetic lines from the spinning electron interact with the proximity E-field of the proton.**
- **The axial magnetic lines occupy a small volume space, while the proton's E-field in this space is deteriorated. The CL nodes in this space are of MQ type with a same SPM frequency as the external free CL space.**
- **The λ_{SPM} in the peripheral space along the orbital trajectory is shrunk, while in the axial space it is the same as in the free CL space.**

Having in mind, that the magnetic line is a loop of zero order SPM waves, it is close to the mind, that its length should contain a whole number of λ_{SPM}. For a large size magnetic lines this condition is quite easy to be satisfied. The satisfaction of this condition becomes problematic for magnetic lines with a size smaller than the trace of the boundary orbit. As a result a stroboscopic effect takes place. Therefore, this could be regarded as a quantum condition related to the magnetic lines. We may call this quantum condition a **short magnetic line condition**.

- **The short magnetic line condition, provides quantum conditions for the quantum orbit to appear as a set of allowable individual orbits. It is based on the assumption, that the length of the magnetic line loop should contain a whole number of Compton wavelengths.**
- **The short magnetic line condition is valid for the peripheral and axial magnetic lines, created by the spinning electron.**

It is evident, that the individual orbits corresponding to one series of spectral lines fulfil simultaneously two quantum conditions: the quantum loop condition and the short magnetic line condition. At the same time the short magnetic line condition is valid for the peripheral and axial magnetic lines from the spinning electron. Analysing the Balmer model we will see, that **the first one determines the orbits separation, while the second one defines the finite time of the electron on a particular orbit.**

Figure 7.7 illustrates the short axial magnetic line condition for one particular orbit in Balmer series.

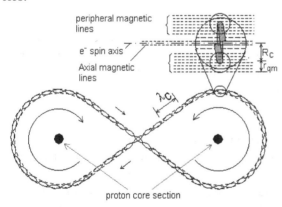

Fig. 7.7

Stable orbit defined by the short magnetic line condition. The peripheral and axial magnetic lines are from the spinning electron

The shown sinosoids along the orbital trace indicate the whole number of longitudinal $\lambda_{SPM}{}'$, in the proton's proximity E-field CL space. They fulfil the short magnetic line condition for the peripheral magnetic lines, which are induced by the

spinning (around its proper axis) electron. This condition defines the orbital separation in the Balmer series. In the same figure, the momentary position of the orbiting electron and its exploding view are also shown. The spatial configuration and density of the peripheral magnetic lines are defined by the quantum magnetic radius r_{mq}. The r_{mq} radius for Balmer series is defined by the second subharmonic quantum velocity. While the short magnetic line condition from the peripheral lines defines the orbit separation, the same condition for the axial lines defines the total time duration of the individual orbit. This will become evident by the analysis of the Balmer series model in §7.8.

The shape of the orbit shown in Fig. 7.7 is idealised. The shape of the real orbit could may differ in a way, that the section around the proton core may not be a perfect circular and the sections between the circular parts may not be straight lines. Despite of accepted simplification, the idealised orbital shapes of the orbits lead to consistent results of the model.

7.7.3 *Summary for quantum orbits:*

- **In the proximity E-field of the proton enclosed inside the Bohr surface, the SPM frequency depends on the distance from the proton, so the Compton wavelength $\lambda_{SPM}{}'$ appears shrunk. We may call such space a quantum shrunk CL space.**
- **For electron moving in a quantum quasishrunk space with a particular quantum velocity, the quantum orbit separates into a number of individual quantum orbits due to the short magnetic line condition strobed by the shrunk Compton wavelength $\lambda_{SPM}{}'$.**
- **In a shrunk quantum space, the quantum velocities for different orbits appear almost equal (because they are referenced to the internal $\lambda_{SPM}{}'$), while they are different when referenced to the length scale of the external CL space.**
- **The individual orbits have different mean distance to the proton's centre of mass. As a result, the energy levels of the individual orbits are different due to the strong influence of the SG field of the proton.**

- **A quantum velocity with a definite subharmonic number defines one set of orbits with one lower level orbit as a ground state**
- **The different energy levels for a electron motion with one quantum velocity correspond to the closely spaced spectral lines - known as line series in the atomic spectra.**
- **The time duration of any orbit (a lifetime of excited state) is defined by the conditions of magnetic field conditions mismatch and phase match between the two proper frequencies of the electron and the phase of propagated SPM vector.**

7.7.4. Electron orbits contributing to the sharp spectral lines in the series

The spectral series of the Hydrogen atom are measured with high accuracy. It is well known feature, that when approaching the energy limit of the series, the lines become less distinguishable and finally converts to continuum. The quantum mechanical model gives explanation of this effect by accepting infinite number of closed spaced energy levels. The BSM model, however, leads to a different conclusion:

- **The obtained continuum is not from infinite number of levels, but from deteriorated quantum conditions. Such conditions cause some energy variation of the emitted quantum waves that is detected as widen spectral line shape.**

Let us take for example the Balmer series. Not all orbits below the boundary one contribute to the sharp spectral lines. There is a range below the boundary orbit, where the quantum orbit conditions in E-filed CL space are not well fulfilled. This is illustrated by Fig. 7.9, showing the Balmer orbital quasiplane in a section perpendicular to the proton's quasiplane and passing through the locus point of one of its clubs. The equaipotential lines of the proximity E-field are shown by dashed lines, while the idealized trace of the limit orbits by a solid line (as two solid circles). All dimensions are in scale. Two possible boundary orbits are shown: 2 and 3. The both have one and a same length, but the orbit 3 is more probable for the escaping electron, while the orbit 2 still passes through the proton's club.

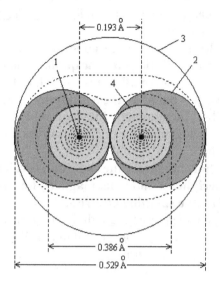

Fig. 7.9

Boundary and limit orbits in the Balmer orbital quasiplane: 1 - proton core; 2 and 3 - boundary orbits, 4 - limit orbit. The internal region associated with the generation of sharp line series is shown by a green (or dark grey) colour, while the external region for smeared spectral lines with a lighter grey colour. The equipotential E-field lines are shown by dashed lines, while the idealized limit orbit - by solid line (as two circles).

The space inside the limit orbit 4 (green or dark area), is occupied by orbits contributing the Balmer series spectral lines. In this region, a large section of the orbital trace coincides or follows the equipotential curves. (The E-field interception angle of the equipotential surfaces inside the Bohr surface are not exactly at 90° due to the characteristic twisting angle θ_w). In the region between the limit and boundary orbit (light grey area) the mentioned above condition is not fulfilled and the quantum orbital conditions are deteriorated. This causes an increase of the line width and appearance of continuum.

Figure 7.10 shows a shape of orbit from Balmer series in the region corresponding to the sharp spectral lines, together with the E-filed lines around the proton core. The E-filed lines are shown as normal to the proton's core in the surrounding circular regions, for drawing convenience, but in the real case the angle is not exactly $\pi/2$, due to the twisting angle θ_w, discussed in Chapter 6.

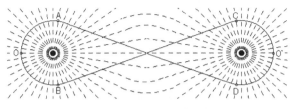

Fig. 7.10

Idealized orbital shape for a quantum orbit belonging to Balmer series in the region corresponding to sharp spectral lines.

7.8 Model of the Balmer series

Note: This is an example model with mostly qualitative output results. The quantitative results may not be considered as final, because the model have more than one adjustable parameters.

7.8.1 Purpose and general considerations

The purpose of the model is to provide some verification about the correctness of the quantum orbits concept, developed in the previous paragraphs. The model is aproximative, because it contains some unknown or partially known parameters, so there are more than one adjustable parameters.

The known parameters are:

(a) proton core and proton width

(b) shape and length of the boundary orbit

(c) shape and length of the limit orbit

(d) approximate shape and length of the ground state orbit

(e) the quantum conditions discussed in the previous section

The unknown parameter is:

(g) the inverse power degree of the leaking (in CL space) SG forces between the proton and electron structures

Partially unknown parameters:

(h) the parameters of the quantum quasishrunk space inside the Bohr surface

The **figure of merit** is the correct shape of the curve presenting the calculated by the model energy levels of the Balmer series.

It is evident, that the attraction SG forces between the proton core and the electron affect the motion of electron. These forces appear as leakage SG forces through CL space, so they are not any

more proportional to the inverse cub of the distance (like in a case of pure void space). The modified SG law through CL space should appear in higher inverse order. (see Fig. 2.8 and the discussion of feature 7 in §2.6.1). In our case we will simulate the SG low through CL space by using the Newtonian mass of electron as unit mass and the Newtonian gravitational constant. The attraction SG force is expected to appear with a large inverse power than 3, because the leaking SG field in CL space falls faster with the distance, than the SG field in pure empty space. Applying the defined above figure of merit we may obtain the degree of the SG law valid for the distance range limited by the Bohr surface.

The determination of the parameters of the quantum quasishrunk space is more controversial. The space inside the Bohr surface is characterised by:

- two different regions, as shown in Fig. 7.8, the region of spectral series and the region of the continuum

- the region of spectral line could be divided into following zones: two zones of the circular orbit trace around the proton core and one middle zone between them.

We may simplify the problem if deriving parameters based on the orbit lengths and the proton dimensions. For this reason we use idealised shape of the orbits, estimating the quantum quasishrunk factors for the orbits which lengths are known.

In order to express the orbital dependence on λ_{SPM}', it is necessary to introduce a quantum quasishrink factor. If assuming a linear dependence (that will be confirmed by the results) it is more convenient to define a quasishrink ratio, k_{qs}. It is equivalent to consider, that k_{qs} is defined as a ratio between the λ_{SPM} at Bohr surface (corresponding to a free CL space) and λ_{SPM}' at the Balmer Ground State (GS) orbit. Once determined, we may reference the quasishrunk ratio to the Bohr surface, where the parameters of the CL space and the electron are defined by the known physical constants. The reciprocal of the quasishring ratio is equal to the gradient refractive index. The existence of this index around the proton club will be discussed in Chapter 8 in connection to the X-ray properties of solids.

- **The quantum quasihrink refractive index is reciprocal to the quasishrink ratio. It could be denoted as** n_{qs}

$$n_{qs} = 1/k_{qs}$$

The Bohr surface could not be considered as a surface with a stable shape. The electron, when orbiting in different quasiplanes, may cause a different deformation of the Bohr surface. The physical constants, like h, q, m_e, ν_c are valid for the space outside of the Bohr surface. In order to use them we have to translate some of the Balmer model parameters to the Bohr surface. In many cases it is more convenient to use the Bohr radius or the length of the Bohr orbit.

7.8.1.A. Aproximative determination of the quasishrink ratio for Balmer series

The quantum orbits contributed to the Balmer series lie on the Balmer quasiplane (now considered as plane for a simplicity) and occupy the internal circle regions, shown in Fig. 9.7. It is reasonable to accept a linear dependence of the quasishrunk SPM wavelength λ_{SPM}' on a distance between the orbital trace and the proton core. Then the approximate value of the quasishrink factor could be obtained by the ratio between the Bohr orbit length $2\pi a_o$ and the shortest orbit. The shortest orbits is the ground state (GS) orbit. One factor restricting the orbital length is the finite distance between the two proton cores in the Balmer orbital plane. Having in mind the requirement for safety margin between closely spaced FOHSs discussed in Chapter 6, it is reasonable to accept the magnetic radius of the electron and proton as factor defining the minimal distance. This condition is illustrated in Fig. 7.11, where r_{eq} is the magnetic equivalent radius of the electron for the second subharmonic (see §3.11 and Table 3.3).

The magnetic radius for the second subharmonic was given in Table 3.3: $r_{eq} = 2.109 \times 10^{-12} (m)$. Then:

$$r_o = (R_c + r_p) + (R_c + r_{eq}) = 9.89 \times 10^{-13} \quad (m)$$

The length of the idealized orbits in the orbital range: GS orbit - limit orbit, is given by the Eq. (7.6):

$$L_{orb} = 2[d\sqrt{1 - 4r^2/d^2} + r(\pi + 2\operatorname{asin}(2r/d))] \quad (7.6)$$

where: r- is the orbit distance in the circular part from the centre of the proton core; d - is equal to the proton width (given by Eq. 6.77)

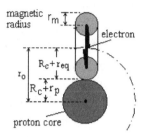

Fig 7.11

Definition conditions for r_o. corresponding to the Balmer GS orbit

The Balmer GS orbit length, obtained by the conditions of finite magnetic radius is 0.45018 A (1 Angstrom = 10^{-10} m). The length of the Balmer boundary orbit is: $2\pi a_o = 3.3249187 \times 10^{-10}$ (m)

Then the aproximative mean value of the quasishrink ratio, k_{qs}, could be defined as ratio between both orbits:

$$k_{qs} = 7.385 . \quad (7.7)$$

The quasishrink ratio gives a possibility to define the change of λ_{SPM}' in the Balmer orbital plane as a function of distance from the proton core. This is used in the next paragraph.

7.8.2 Concept of the Balmer series model

The concept of the model is based on the energy calculation of the possible quantum orbits, related to the spectral lines of the Balmer series (without the continuum near the limit). These orbits cover the range between the Balmer Ground State (GS) orbit and the limit orbit (denoted as 4 in Fig. 7.8). The orbit positions are illustrated by Fig. 7.13. Their shapes are idealized for convenience. Every orbit contains two circular sectors around the proton cores connected with tangent lines, passing through the proton club locus. The trace length of such geometrically simplified orbits is given by Eq. (7.6).

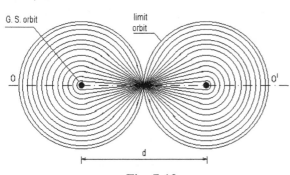

Fig. 7.13
Idealized orbits for Balmer series

Complying the short magnetic line quantum condition for the peripheral magnetic lines defined by the quantum magnetic radius of the electron, the length between the neighbouring orbits will differ by λ_{SPM}'.

The energy level of all orbits can be estimated by applying a balance of forces for the motion of the electron in the circular sector of the orbit. The electron velocity in any orbit depends on the super gravitational force, F_{SG}, the internal Coulomb force, F_C, and the inertial force from the apparent inertial mass. The balance of forces for this region is given by Eq. (7.8), from where the electron velocity is expressed by Eq. (7.9).

$$F_C + F_{SG} = m\frac{v^2}{r} \qquad (7.8)$$

$$v = \sqrt{\frac{r}{m}(F_C + F_{SG})} \qquad (7.9)$$

The Coulomb forces inside of the Bohr surface has been presented by Eq. (7.3.b) where the argument r is counted from the radius of the GS orbit. In order to use later the quantum numbers as adopted by the Quantum Mechanics, we will use a shifting parameter r_{GS}. From geometrical considerations we may consider that when the electron circles around one proton club, it interacts only with the half of the proton charge. Using Eq. (7.3.b) and applying these considerations we arrive to Eq. (7.10) for the Coulomb force inside the Bohr surface.

$$F_C = \frac{q(q/2)}{4\pi\varepsilon_o a_o^2}\left(\frac{r - r_{GS}}{a_o - r_{GS}}\right)^2 \qquad (7.10)$$

where: r - is a running parameter - the distance of the orbit interception point with the OO' axes from the proton core centre (in absolute units)

r_{GS} - is the distance of the Balmer G.S orbit interception point with the OO' axis from the proton core centre (see Fig 7.9). In order to satisfy the quantum condition, r_{GS} is very close to r_o, but at distance not larger than one λ_{SPM}'.

(q/2) - is the proton charge in the proximity E-field affecting the electron motion at point O.

In order to apply the quantum condition for orbit separation, we have to use λ_{SPM}', but it depends on the argument r. For this reason it is more convenient, to accept a constant λ_{SPM}, referenced to the distance of the boundary orbit and to correct the argument in the expressions of the SG force, the Coulommb force and the inertial force. It is equivalent to work in units of quantum quasishrink space. Then we can use directly the quantum number of the orbit.

The **inertial mass law in a quasishrunk space** is a controversial problem, not investigated enough. It was discussed in §7.6. A set of laws are tested in the model. The best results are obtained for a square law dependence, when the curve shape is a mirror image of Coulomb law inside the Bohr surface (the mirror axis is parallel to the horizontal axis). So the electron inertial mass dependence on the distance in the quasishrink space is simulated by the Eq. (7.11).

$$m = m_e n_i^2 - m_e(n_i^2 - 1)\left(\frac{r - r_{GS}}{a_o - r_{GS}}\right)^2 \qquad (7.11)$$

where: n_i is the quasishrink index of the zone around the proton core, assuming that λ_{SPM}' is a linear function of the argument r.

The simulation of the SG forces through the CL space was discussed in Chapter 2 §2.6.1, feature 7. The SG forces between the proton core and the electron are presented as a higher degree inverse power law between a mass point and a mass bar. The intrinsic (SG) mass of the electron is used as a unite mass point, while the proton core - as a bar of such mass points. One single coil of the external positive shell of the proton core contains approximately the same intrinsic mass as the electron. Then the mass of the proton core can be expressed as number of N electron masses. Then the differential gravitational filed, dg, is given by the Eq. (7.12)

$$dg = \frac{GM}{L((r^2+x^2)^{0.5})^P}\cos(\alpha)dx = \frac{GNm}{L((r^2+x^2)^{0.5})^P(r^2+x^2)^{0.5}}dx$$

$$(7.12)$$

where: L is the length of the mass bar; M is its intrinsic mass; m - is the point mass for which g is estimated; N is the number of mass points, from which the bar is consisted, r - is a distance; x - is a running parameter for integration; and P is the degree of the inverse power law.

The integration on x gives the super gravitational field, from which the gravitational force is expressed. The tuning of the model requires precise adjusting of the power degree. For this reason a numerical integration is preferable.

The bar length is proportional to the number of mass point. So it is more convenient to replace N by a length of the bar, L, in order to have one and a same units of distance. The parameter L could be expressed as a fraction of L_{pc} and the model could be tested for different L. Then we arrive to the expression of the SG force that leaks through CL space in the range between the electron and nearby proton core.

$$F_{IG} = \frac{2rGm^2}{L}\int_0^{L/2}\frac{1}{(r^2+x^2)^p}dx \qquad (7.13)$$

From the Eq. (7.13) we see that the factor p after the integration will corresponds to a power law of degree P, according to the expression (7.14)

$$P = \frac{p-0.5}{0.5} \qquad (7.14)$$

Replacing the value of F_C and F_{IG} in Eq. (7.8) we get the velocity in function of distance r. In all equations the converted mass is included by its expression given by Eq. (7.11). Now we need to connect the forces balance condition with the quantum condition of the orbit separation based on the whole number of λ_{SPM}' for any orbit length. For this reason the Mathcad program st_w_qn.mcd is used. The length of the orbit as a function of the distance r is determined by Eq. (7.6). In this point of the model, we have two options for applying the short magnetic line condition:

    (a) - for the peripheral magnetic lines
    (b) - for the axial magnetic lines

Note: The above two options are for the magnetic lines induced by the spinning electron, not by its orbital motion)

For option (a), assuming a linear dependence of λ_{SPM}' on the radius r in the circular zone and referencing to the $\lambda_{SPM} = \lambda_c$ for the boundary orbit, we have:

$$\lambda_{SPM}' = \lambda_c\left[1 - \left(1 - \frac{1}{k_{qs}}\right)\frac{a_o-r}{a_o-r_o}\right] \qquad (7.15)$$

The curvature of the line connecting the calculated energy levels from the model output is very dependable on the quasishrink ratio k_{qs}. Its value estimated in the previous paragraph is 7.385, however, the model shows better results with a value $k_{qs} = 7.728$. Taking into account the very aproximative method for estimation of this parameter such small departure is acceptable.

For option (b) the SPM wavelength is the same as for the external CL space λ_{SPM}.

The options (a) and (b) separate the model into two similar branches. Let us investigate the option (a), since it is related to the quantum orbit separation.

The length of the quantum orbit expressed as a whole number of λ_{SMP}' is given by Eq. (7.16).

$$L_q = (k_{min} - 2 + n)\lambda_{SPM}' \qquad (7.16)$$

where: k_{min} is the number of wavelengths for the nearest to the core orbit, but not closer than r_o; **n is a principal quantum number; the factor 2 is for matching the orbit number to the quantum mechanical principal quantum number for Balmer series**.

Substituting (7.16) in (7.15) and equating the result with the Eq. (7.6), we arrive to Eq. (7.17).

$$2\left[d\sqrt{1-4\frac{r^2}{d^2}}+r\left(\pi+2\operatorname{asin}\left(2\frac{r}{d}\right)\right)\right] = L_q(n)\lambda_{SPM}'(r) \qquad (7.17)$$

Giving consecutive numbers for n starting from 2, the corresponding distance r is determined as a discrete value function $r(n)$. The function $r(n)$ is fitted to a curve.

$$r(n) = a + bn^c \qquad (7.18)$$

where: $a = 0.00396279$; $b = 0.0051841$; $c = 0.7947166$

The curve fitting plot and the residuals are shown in Fig. 7.14.

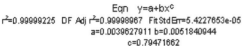

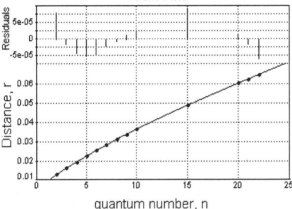

Fig. 7.14

Substituting the argument *r(n)* in all terms of the Eq. 7.9, **we obtain the electron velocity as a function of the quantum number**. The plotted curve of this discrete value function is shown in Fig. 7.15.

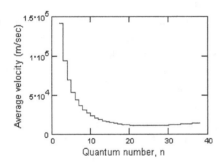

Fig. 7.15

Orbital electron velocity as a function of quantum number

All spectral lines of the Balmer series are between 3.4 eV and 0 eV. We see that the velocity in the GS orbit with $n = 2$ is 1.09×10^6 (m/s). It corresponds to energy level of 3.4 eV. The limit orbit appears at $n = 39$. It corresponds to energy level of 0 eV, according to Bohr model, but we see that the velocity is not zero. This is very important result, for a limiting orbits, which is an enigmatic problem for the Bohr atomic model and all QM models of atoms. The plot in Fig. 7.15 shows, that the velocity is decreasing with the quantum number for the

range $2 < n < 26$ and then slightly increased for $n > 26$. We may call the first region a **region of velocity inversion**.

The velocity curve is referenced to the GS velocity. This velocity is determined as a second subharmonic velocity corrected by the quasishrink index n_{qs}, which from the other hand is inverse proportional to the quasishrink ratio k_{qs}

$$V(2) = \frac{\alpha c}{2 n_{qs}} \quad \text{(m)} \qquad (7.19)$$

The model velocity equation (7.9) should give the same value for $n = 2$ as Eq. (7.18). For this reason only the parameter *p* in Eq. (7.11) is tuned. The corresponding degree of the inverse power low of SG forces, is obtained by Eq. (7.14). The plot of SG forces together with the plot of the Coulomb forces are both shown in Fig. 7.16.

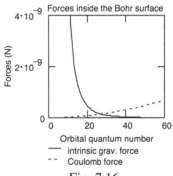

Fig. 7.16

SG and Coulomb forces as a function of the orbital quantum number (plotted as continuous curves)

It is evident that the slight increase of the electron velocity for $n > 26$, as shown in Fig. 7.15, is contributed by the increased Coulomb forces, as shown in Fig. 7.16.

Figure 7.17 shows plots of the Coulomb forces for two cases: $1 - F_c(2) = 0$; $2 - F_c(2) > 0$ ($n = 2$ corresponds to the Balmer GS orbit).

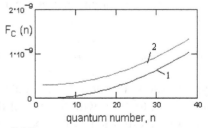

Fig. 7.17. Coulomb forces as a function of the quantum number (plotted as continuous curves)

Keeping in mind that the GS orbit is determined by the finite distance r_o, (see Fig. 7.11), it becomes apparent, why Coulomb force for $n = 2$ may not start from zero, but from some finite value. For this reason, the plot 2 is more probable. It does not affect significantly the curve shape of the energy level fitting, but may affect slightly the total orbital time, discussed in §7.8.3

The inertial mass dependence on the quantum number influences the shape of the velocity and the energy levels of the series. The best fitting result is obtained for a second order inertial mass dependence, given by Eq. (7.11). Expressed as a function of quantum numbers, the plot of Eq. (7.11) is a mirror image of the Coulomb forces expressed by Eq.(7.10). The plot of the inertial mass expression (7.11) is shown in Fig. 7.18.

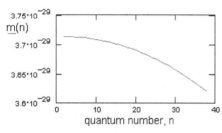

Fig. 7.18

Newtonian inertial mass of the electron
in the E-field quasishrunk space inside
the Bohr surface (plotted as a continuous curve)

Using the obtained expressions of the inertial mass and velocity as functions of the quantum number, we may express the electron kinetic energy in eV for any one of the quantum orbits by using of the well known classical equation:

$$E_k(n) = 0.5m(n)\upsilon^2(n)\frac{1}{q} \qquad (7.20)$$

The energy levels, according to the Quantum mechanics, are the potential energies but referenced to the limit orbit. So we have:

$$E_p(n) = 0 - E_k(n) = -(0.5m(n))\upsilon^2(n)\frac{1}{q} \qquad (7.21)$$

Fig. 7.19 shows the plot of the calculated energy levels, $E_p(n)$, together with the plot of the energy levels, $E_b(n)$, estimated by the spectral data.

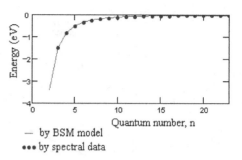

— by BSM model
••• by spectral data

Fig. 7.19

Calculated and experimental energy levels
for Balmer series

The shape of the calculated energies fits quite well to the levels from the experimental data (quantum mechanical levels). We also see, that we have referenced the model velocity only for $n = 2$, but the obtained energies cover the range of the Balmer series. The small discrepancy between the calculated and the experimental data might be contributed by:

- using idealised shape of the quantum orbit, as shown in Fig. (7.13)

- The SG law through LC space is different than in the empty space. In the former case, the leaked SG forces for a close distance are inverse proportional to a distance at power larger than 3. The power index is also dependable on the absolute distance value.

The Balmer model output parameters for the best fit, shown in Fig. 7.19 are following:

k_{qs} = 7.728 - a quasishrink ratio
Quantum orbits: 37 ($2 < n < 38$)
P = 5.474 - degree of inverse power SG low
through CL space

7.8.2.A Discussion:

The quantum efficiency for pumping the CL space is not considered in the Balmer model. When investigating the molecular vibrational spectra in Chapter 9, we will see that the quantum efficiency affects the CL space pumping. Then in the Balmer model, the quantum efficiency will be a hidden parameter. This might be the reason, why a velocity minimum appears at $n = 24$. It could be explained by the shape of the orbit. At $n = 24$ the orbital shape approaches the Hippoped curve with a parameter $a = \sqrt{3}$ and the distance between the locuses of the Hippoped curve - closer to the distance d (see Fig. 7.13). The quantum efficiency at such shape of the orbit might be a maximum.

The BSM model provides energy levels consistent with the levels obtained by the optical spectrum (Fig. 7.19). So the velocity concept may be considered as a correct parameter, including the quantum efficiency as a hidden one. Then we may calculate the quantum magnetic radius, by Eq. (3.39) from Chapter 3. For velocity value of 1.079E4 m/sec corresponding to $n = 24$ we obtain for the small magnetic radius:

$$(r_{eq} = 5.194 \times 10^{-13}) \ m \ .$$

Then the external magnetic radius is:

$$(R_c + r_{eq}) < 1.2 \times 10^{-12} \ m$$

The quantum magnetic radius of the quantum orbits with lower quantum number is even smaller. **Consequently, the quantum magnetic radii for all orbits of the series are inside the Bohr surface.**

Note: The quantum magnetic radius is estimated by the analysis in Chapter 3, where a repeatable motion in a quantum loop is not taken into account.

The Balmer model unveils one specific feature of the orbiting electron. When the electron drops to a lower orbit, despite the fact that it obtains a larger velocity, its potential energy is lower due to the SG forces. When such transition occurs, the energy difference will be emitted to the external space as a quantum EM wave - a photon. **Consequently, the emitted photon carries a portion of the SG energy, between the electron and the proton.**

The above conclusion is of great importance, because it is valid for the energy levels of all atoms. **In fact the SG energy contribution increases with the Z number of the atomic element.**

The electron's geometrical parameters, valid for free CL space have been used without change in the model. Consequently, the fine structure constant, which appears as embedded parameter of the electron geometry is also not changed inside the Bohr surface. Both proper frequencies of the electron as a system are also unchanged. These results obtained for the volume of the spectral line orbits should be also valid for the total volume inside the Bohr surface.

Investigating the separate contributions for the shape of the Balmer plot, shown in Fig. 7.19, we may see that the change of inertial mass could not affect significantly the output result. The main contributors are the SG field and E-field. The E-field in fact is controlled by the SG field due to the charge unity mechanism. Consequently:

- **the energy of the orbiting electron is defined mainly by the SG energy of the system**.

This feature may be considered valid also for the heavier atoms where the SG energy of the nucleus contributes to the energy of emitted or absorbed photons. (In such aspect the inertial mass contributes only a small fraction by the cenrapetal acceleration force).

The above conclusion is one of the major distinctive parameters between the BSM model and the Bohr model of the Hydrogen atom. This leads to the following major distinctions between both models:

Summary:

- **Major distinctions between Bohr model of Hydrogen and BSM model:**

    In the Bohr model, the orbit with a length of $2\pi a_o$ **is the most internal orbit.**

    In the BSM model, the orbit with a length of $2\pi a_o$ **is the most external possible orbit**

- **The spectral line positions in the series carry signatures of: the SG filed, the E-field, and the electron inertial mass inside the Bohr surface**

- **The resolvable spectral lines are in the range of velocity inversion**

- **The magnetic radius of the electron for all quantum orbits could not appear outside of the Bohr surface.**

- **The two proper frequencies of the electron as a system are not affected by the properties of the space volume enclosed by the Bohr surface.**

7.8.3 Orbital time

From the concept of CL space pumping and Balmer series model it is apparent that the CL space is pumped during orbital circling of the electron, after the electron falls to a lower orbit, the pumped energy is emitted as a photon (quantum wave). The well defined energy of the photon indicates that the electron makes a whole number of orbital cycles. This is in agreement with the relation between conditions of whole number of cycles discussed in §3.12, Chapter 3. This relation is shown in the following table:

Phase repetition condition	e- rotations	internal positron-core cycles
short time	18778.3(3)	56335
long time	56335	2573380

The phase repetition time could be considered as a necessary but not enough condition for the finite time of the electron on orbit. In Quantum mechanics (QM) this time is known as a lifetime of the excited state. It is a constant for a spontaneous emission, while it is shorter for stimulated emission used in lasers. The Quantum mechanics does not provide explanation of the physical mechanism behind the lifetime of the excited state. BSM theory provides such explanation for a first time.

One of the factors that defining the limit time duration of the orbit, according to BSM, is reduced to a possible number of full orbital cycles according to the considerations of the phase repetition. Another factor that may influence the possible number of orbits could be related to the short magnetic line condition. This condition is valid simultaneously for the axial and peripheral magnetic lines created by the rotational and orbital motion of the electron. While the axial lines are related to the Compton length λ_{SPM}, valid for the external free CL space, the peripheral lines are related to the shrunk Compton wavelength λ_{SPM}' valid for the orbital space inside the Bohr surface. Obviously, both conditions comes in conflict after some finite time from the beginning (when the electron start circling in the particular orbit). This could be inferred from the analysis of the relation between the CL space relaxation constant and orbital time, applied for the following two cases: axial and peripheral magnetic lines.

For the axial magnetic lines the CL space constant is t_{cl}. For the peripheral magnetic lines the quasishrunk CL space constant follows the same dependence on the distance from the proton core as the parameter λ_{SPM}' given by Eq. (7.15).

The orbital time from the Balmer model is obtained by division of the orbital length (Eq. 7.16) by the orbital velocity (plotted in Fig. 7.15).

$$t_{orb}(n) = \frac{L_{orb}(n)}{V(n)} \qquad (7.23)$$

Two plots of Eq. (7.23) for short magnetic line conditions applied for axial and peripheral magnetic lines are shown in Fig. 7.20. The time scale is multiplied by factor of 3 for a reason explained later.

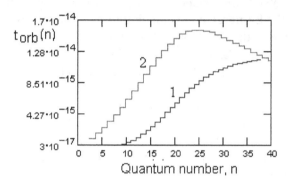

Fig. 7.20

Trend of the short magnetic line conditions as a function of quantum number
1- for the peripheral magnetic lines
2- for axial magnetic lines
The quantum numbers are in the scale
of curve 2, related to the quantum number of the orbit.

We see from the plot that the trends of the two curves are different. While their relative position might be influenced by the properly determined quasishring index, their trends will be always different. But this difference means that a conflict between the two types of magnetic line conditions may occur for some finite time of the electron on the particular orbit. The quantum features of the CL space and the electron system oscillation may make the conflict to occur in the time when the oscillation passes through the initial phase. Then the electron falls to a lower orbit. The most probable lower orbit is the ground state orbit of the series.

The provided considerations have to been combined with the conditions of the phase repetition between the two proper frequencies of the electron and the CL node Compton frequency. This issue has been discussed in §3.12.2.A.

The developed concept is valid for a single Hydrogen atom only. The obtained lifetime should not be confused with the cases of long lifetimes for some atoms or molecules. In the latter cases quite long lifetimes may result from different mechanisms involving complex interactions between multiple orbits.

Summary:
- **The spontaneous life time is defined by the mechanism of the short magnetic lines, strobed with the effect of the phase match conditions between the two electron proper frequencies and the Compton wavelength of the CL nodes**
- **The orbital time could not be shorter than one orbital cycle of the electron**
- **The finite lifetime is a result of conflict between axial and peripheral short magnetic line conditions, developed for a finite time of the electron motion in the proton E-field.**

7.9. Photon emission and absorption. Physical explanation of the uncertainty principle.

Let us analyse the electron motion in one orbit of Balmer series above the GS orbit. The induced peripheral magnetic lines by the spinning electron are in the region inside the Bohr surface. During the stable motion of the electron, the Hydrogen appears neutral outside the Bohr surface. Consequently, the electron momentum is able to neutralize the distributed E-field lines inside the Bohr surface. The CL space interaction will balance this momentum, so it will contain a balanced energy. This energy is distributed in the orbital trace, formed by the electron quantum magnetic radius (see Fig. 7.7). **The energy, kept in the volume swapped by the magnetic radius is sustainable due to the shorter refreshing cycle, supported by the orbiting electron.** Due to the finite orbital time, determined by the short magnetic line condition conflict (described in the previous paragraph), the motion in a given orbit may be terminated in two possible ways:

(a) the electron falls to a lower orbit

(b) the electron jumps to higher orbit, if the proton club space has received external energy in a proper time.

Photon emission. The case (a) (mentioned above) is related to the emission of photon. The termination of electron motion in the current orbit terminates the process of the energy pumping of the surrounding CL space volume (converting of MQs to EQs. The excess energy is a difference between the electron energies in the two orbits. This energy distributed in the former trace is in conflict with the proton E-field inside the Bohr surface. If regarding the excess energy volume as running EQ's they are pushed in direction to less intensive SG field (for the line series, the SG field predominates the E-field). The running EQ's carrying an excess energy above the ZPE, are refurbished into a quantum wave (photon) with a wavelength corresponding to the total excess energy. The most probable transition to a lower orbit is the transition to the GS orbit of the series. For this orbit the conditions of stable long lived quantum orbit are always present. For any other transition to an orbit higher than the GS, energy higher than the GS state is needed (This happens more often in atoms with more than one electron in the external valence shell (discussed in Chapter 8).

Photon absorption. In the case (b), an electron in a lower orbit may jump to a higher one, if obtaining an energy in a proper time, depending on how long the electron has been on this orbit. This is a process of absorption. The absorption of photon, also is not a sudden process. It obviously involves number of orbital cycles. It is well known from Quantum mechanics and the experiments, that the total energy of the photon is transmitted to one electron. But how the energy of the quantum wave wavetrain occupying a much larger volume than the volume enclosed in the Bohr surface, is shrunk into the latter in order to affect the electron motion (if it is in a proper orbit)? The only possible explanation is to accept that the combination of the space inside of the Bohr surface and the orbiting electron posses a feature, we may call an **energy dumping effect** (in analogy with a dumping effect for mechanical waves, propagated in a stiff media). If considering a solid optical detector, the size of the quantum wave usually covers many atoms with their Bohr surfaces. In order to start the dumping effect, however, the electron has to be on suitable orbit for a suitable time. Then the energy dumping effect selectively starts for an atom, having an electron at quantum orbit satisfying the above mentioned conditions. Once the dumping have started, the whole energy of the quantum wave is sucked, contributing its energy only to this electron. The process is not simple and may involve number of nonlinear factors inside the Bohr surface. The intuition for a possible nonlinear factors comes close to mind if analysing the experiment described by L. J. Wang et al. (2000).

CL space pumping. If the electron in a GS orbit, for example, gets some energy from absorbed photon, it jumps to a proper higher orbit, but stays here a finite time and returns back, most probably to the GS orbit. In this case the same obtained energy is reemitted. We may regard the process as a CL space pumping, a process discussed in previous chapters.

In §3.17.3 (Chapter 3) the case of the positronium transition Ps $1^3S_1 - 2^3S_1$ was analysed. The energy of emitted photon was found to correspond to $(13.6 - 3.4)/2 = 5.1 \, eV$. In this case, both, the electron and the positron are oscillate and pump the surrounding CL space, which terminates with an emission of a photon with energy 5.1 eV. This indicate that the oscillation is referenced to the common centre of mass in respect to the fixed CL nodes of the laboratory frame. In fact the common centre of mass is not fixed in CL space, but oscillating. For this reason the difference between 13.6 and 3.4 eV is additionally divided by two. When the CL space in this case is regarded as defining the frame of reference, the classical explanation of this effect is understandable.

Some very low energy photons from the atomic spectra also may get physical explanation, when considering the CL space interaction. In the case of Hydrogen atom, the proton mass is much larger than the electron one, and the proton could be considered as a carrier of the local frame. During the photon emission, the Hydrogen atom gets a kick in an opposite direction. Due to the orbital quasiplane twisting shape, the Hydrogen gets a simultaneous spin momentum. This momentum may cause an emission of another quantum wave, with much lower energy. This may explain the Hydrogen emission at 21 cm coming from the space.

Heisenberg uncertainty principle. The emission and absorption processes are able to provide explanation of the Hisenberg uncertainty principle, applied to the electron motion in the atoms. It is evident that emission and absorption are not sudden physical processes. They have a finite time duration involving many orbital cycles. The receiving system will absorb the energy of the photon at the end of such process. A particular atom even in a superfast detector will get the whole energy of the photon only when the electron quits the orbit (or when the corresponding atom is ionized).

The photon emission is a reversed process. The pumped CL space energy is emitted at the moment, when the electron drops to a lower orbit. Consequently, it could not be associated with any instant position of the orbiting electron. This provides a clear explanation for the Heisenberg uncertainty principle. At the same time we see that the arbitrary use of the uncertainty principle is not justified and may lead to very unrealistic speculations.

7.10 Fine structure line splitting

In Chapter3 the Quantum mechanical spin of the electron having a torque value of $\pm h$ was explained by a classical way. QM spin is valid for electron moving in quantum loops, therefore it is valid for quantum orbits. Let us consider the quantum orbits of Balmer series in Hydrogen atom. The two value of QM spin will allow two energy values for every orbit, however, they will be more apparent in the first harmonic and lower subharmonic number, since the quantum interactions is stronger at first harmonic (corresponding to 13.6 eV) and decreases with the increase of the subharmonic number. This is observed in the spectral lines known as a fine structure splitting. When applying a strong external magnetic field, however, the line splitting is more than two - a phenomenon known as a strong Zeeman effect. The BSM model provides a classical explanation for this: The quantum orbits passes through the proton club, which has a handedness. It is twisted in a definite direction (let accept a righthanded). Then the twisting defines a reference direction for the electron orbital motion. We may distinguish two opposite direction of the electron motion in respect to the twisted proton. This will give twice more combination for the possible energy levels, which signature is observed as additional splitting of the spectral lines. This feature is valid not only for the Hydrogen, but also for all elements. The strong Zeeman splitting is clearly observed in the single valence elements of the first group of the Periodic table, since they have a single quantum orbit in the external shell. For other elements additional energy levels appear due to magnetic interactions between the quantum orbits. Additional discussion about Zeeman effect is presented later in §7.14.

7.11 Pauli exclusion principle. Magnetic fields inside the Bohr surface.

According to Pauli exclusion principle one orbit could be occupy by no more than two electrons. When the orbit is occupied by two electrons, they have opposite spins.

These conditions are reasonable for all kind of orbits passing through the proton club. In BSM concept, the opposite spins means that both electrons circle the same orbit but in opposite directions. In Balmer orbits they pass simultaneously through the proton clubs near the locus. However they do not collide because:

- the guiding role of the magnetic field and the repulsion of the E-field of the electrons

- the orbits are pretty close, but not exactly the same due to the interaction between the axial rotation of the electron (around its proper axis) and the helicity of the proton.

The instant positions of the two electrons are symmetrically synchronized, i.e. in any arbitrary moment their positions are symmetrical in respect to the proton club. The instant symmetry in one particular moment is illustrated by Fig. 7.21. This symmetry is supported by the magnetic field from the orbiting electrons.

Fig. 7.21

Instant position symmetry of two electrons with opposite QM spin, occupying one quantum orbit of Balmer series

It is evident, why the two electrons cannot possess one and a same spin. The opposite QM spins allow their magnetic fields from the orbital motion to be mutual compensated with a symmetry referenced to the proton club. The magnetic lines of these fields are normal to the orbital plane. We may call this field an **orbital magnetic field**. This field has a different configuration than the magnetic field induced by rotational momentum of the electron around its own axis. In the vicinity of the electron, the magnetic lines of both fields are normal each other and do not interfere. The orbital magnetic field, however is created in a E-field CL space, inside the Bohr surface, where the λ_{SPM}' is different than the λ_{SPM} of the external CL space. For this reason the created magnetic lines may not be able to escape outside of the Bohr surface, so they are closed inside. This feature supports the fact that the orbiting electron does not exhibit external magnetic field. This is valid not only for a neutral atom, but also for pair electrons in a common orbit of a negative ion. In the far field the negative ion possesses completely symmetrical charge feature as the positive one, despite the different dynamics inside the Bohr surface.

From the above analysis it becomes evident that similar conditions for more than two electrons are not possible. Said in a simple way, two electrons "complete" the orbit (according to the Pauli exclusion principle), because no more than two orbits of opposite Quantum Mechanical spin are possible.

Summary:

- **Physical meaning of the Pauli exclusion principle: two electrons with opposite QM spins posses two individual symmetrical orbits with one and a same orbital quantum conditions. Such conditions are not possible for more than two electrons, because the magnetic field symmetry is disturbed.**

- **The orbital magnetic field from one or pair electrons in the orbit is enclosed inside of the Bohr surface, due to the different SPM frequency of the E-field in this region from the external free CL space.**

- **The negative ions does not exhibit external magnetic field despite the second orbiting electron, for the same reason.**

7.12 Superfine spectral line structure

During the photon emission, the excess energy kept so far inside the Bohr surface gets a fast escape as an emitted photon. The local gravitational field of the proton serves as a reference frame. It is reasonable to consider that the proton exhibits a reaction force during the moment of the photon shot. Due to its large inertial mass it gets a slight kick. Its helicity and twisted orbital shape of the circling electron, converts part of the kick momentum to a nuclear rotation. The emitted quantum wave have a finite wavetrain length and consequently a finite emission time. So the kick effect is able to influence the photon emission, providing in such way a small frequency shift, as a red Doppler shift. At the same time, the proton has preserved a small fraction of energy as a rotational momentum. At some particular moment this rotation may become in conflict with the orbiting electron. The atom could free this energy only as an emission of a low energy quantum wave. The energy of the emitted in this case photon, however, depends also on the current status of the QM spin of the electron. The signature of this dependence is **the superfine spectral line structure**.

For atoms with a higher Z number, the superfine structure may have more splittings due to the effect of orbital interactions. The latter is discussed in Chapter 8.

7.13 Lamb shift

In Quantum electrodynamics (QED), the Lamb shift is known as a displacement of the GS (ground state) level from its position, estimated by the difference between the expected energy level and the real one. This is observable from Hydrogen to higher Z number of elements. The Lamb shift increases with the Z number.

Let us consider the Balmer series. The orbital shape and dimensions of all orbits in the series with exception of the GS orbit are defined only by the quantum conditions. Only for the GS orbit additional conditions appear for termination of the quantum loop condition. It is related to the quantum magnetic radius interference with the proton core (see §7.8.1.A and Fig. 7.11). So it is very reasonable the lowest orbit quantum condition (corresponding to the GS) to appear slightly displaced.

The orbit deformation causes a slight shift of the quantum position, estimated by the Quantum mechanical model. The deviation from the exact quantum value becomes observable, because this orbit is closer to the proton core. In this range, the SG forces are stronger and are proportional of higher inverse power degree of the distance from the proton's core. For a similar reasons, the orbit deformation and the quantum shift for GS orbits in atoms with higher Z numbers is larger. It is also evident, that the Lamb shift may appear only for GS orbits near the proton core. This condition is valid not only for the Lyman and Balmer GS orbits in the Hydrogen, but also for the corresponding similar orbits in the heavier elements.

7.14 Zeeman and Stark effects.

The Stark effect is a spectral line splitting as a result of applied electrical field. The Zeeman effect is a spectral line splitting as a result of applied magnetic field. In fact the Zeeman effect could be also a line shifting. The detection effect may provide a signature of line splitting as a result of the following conditions:

- detection of photons from different atoms
- consecutive photon detection from one and same atom but with different orientations in respect to the applied field

In order to explain the physical process, we will use the term line shifting. There are two major differences between both effects. In the Zeeman effect, two different types of shifting are observed: for a small and for a large intensity magnetic field. The Stark effect does not exhibit such phenomena. These differences helps to identify the physical process.

In the Stark effect, the applied electrical field deforms the shape of the Bohr surface. This may influence the position of the quantum orbits. The orbital energy level of the quantum orbit is dependent on its position, because the strong gradient of the SG field. The gradient is also larger for orbits with low quantum numbers.

In the Zeeman effect, the applied magnetic field could not influence the Bohr surface. The Bohr surface, by definition, is generated by the static E-field of the proton and should not be affected by an external magnetic field. **The applied**

magnetic field, however, may generate magnetic lines inside of the Bohr surface, influencing in such way the orbital quantum conditions. The penetrating magnetic lines may obtain loops closely to the magnetic fields generated by the orbiting electron. Having in mind both quantum conditions, defining the quantum orbit (formulated in § 7.7.3), it is apparent that they can be effected differently by weak and by strong external magnetic field. For this reason we have two types of Zeeman effects:

(a) Zeeman effect caused by low intensity magnetic field

(b) Zeeman effect caused by high intensity magnetic field

The low intensity magnetic field may not influence the short magnetic line quantum condition, related to the axial magnetic lines of the electron. This field is in a tube-like volume with a very small thickness (smaller than the Compton radius). But it may affect the quantum loop condition, of the peripheral magnetic lines, which occupy much larger volume (still inside the Bohr surface).

The higher intensity magnetic field may affect both types of magnetic line quantum conditions. Having in mind, that the SPM frequencies of the applied magnetic field and the axial magnetic lines of the electron are equal, the stronger field may provide a different type of line shifts in comparison to the weaker one.

7.15 Cross validation of the Hippoped curve concept, used for the shape of the proton and the quantum orbits.

Here we will summarize, briefly, the cross calculations and validations, some of which are used so far and others - given in the next Chapters. The knowledge of the shape and dimensions of the proton, neutron, electron, and the quantum orbits, is very useful for understanding the structure of the atomic nuclei and their physical and chemical properties.

A. Shape and dimensions of the electron as an oscillating system of three helical structures with internal rectangular lattice (twisted)

- Static and Dynamic CL pressure expressed by the electron volume and surface, involving the Compton radius (wavelength) Plank constant, light velocity and fine structure constant (Chapter 3)

- Magnetic radius of the electron, calculated by the magnetic moment, the Compton radius and number of other physical constants (Chapter 3)

- Relation between the electron static charge and the virtual particle waves in the beta decay (virtual electron and positron) (Chapter 3 & 6)

- X-ray properties of the electron (Chapter 3)

- Electron system modifications and proper frequencies validation by experimental data of FQHE (Chapter 4)

- CL space pumping and proper frequency validation by the Positronium (Chapter 3)

- Electron system modification in Superconductivity state of the matter (Chapter 4)

- Internal gravitational lattice structure validation by analysis of the released SG energies in particle collision experiments (1.7778 GHz, 1.44 GHz, 80.396 GeV, 91.187 GeV) (Chapter 6)

- electron and muon magnetic moments, mass ratio and their physical meaning (Chapter 3 & 6)

- derivation of relativistic gamma factor by the dimensions and property of the moving and oscillation electron (Chapter 3)

- electron confined motion and quantum velocities. Quantum orbits (Chapter 3 & 7)

B. Proton shape and physical dimensions

- Cross validation of the physical dimensions of the proton by participation of its parameter LPC in Eq. (5.9) (Chapter 5)

- Matching the mass ratio of the pion to muon; the magnetic moment ratio between electron and muon; mass balance equation of the proton (involving all pions and kaon); mass balance equation of eta particle; relation between the Newtonian mass change due to FOHS twisting and the electroweak parameters $\theta_{eff}^{lept} = 28.762^0$ and Fermi coupling constant; internal FOHS destruction energy ratio between right and left handed structures, by tau lepton equivalent mass energy at 1.7778 GeV and the resonance energy at 1.44 GeV; the destruction energy of untwisted K+ and K- (kaons) matching to W+/- bosons; destruction energy of twisted K- by Z boson; prediction of the destructive energy of twisted K+ at 108 GeV; physical explanation of the relation between the muon lifetime, Fermi coupling constant and pion muon electron decay

- Matching the proton dimension to the Balmer model

- Matching the proton dimension to nuclear size estimated by internuclear distance in some molecules (Chapter 9)

C. Neutron shape and dimensions

The dimensions of the neutron are obtained directly from the proton, because the external difference is only in their shapes. While the proton is a torus twisted in a shape of a hippoped curve, the neutron is a folded torus with a shape of a double loops (with some small gap between the two loops).

D. Proton and quantum orbit dimensions

- matching the proton and quantum orbit dimensions for atoms in molecules (Chapter 9)

- matching the dimensions calculated for H_2 molecule ortho state and cross validating them by data from optical molecular spectra and photoelectron spectroscopy (Chapter 9).

Notice

In all cross validations only accurate physical constants and reliable experimental data are used. All sources of experimental data are referenced. The CL space concept of the physical vacuum helps to explain the relation between the electric, magnetic and gravitational fields. Based on it BSM provides explanation of the fundamental rules adopted in Quantum mechanics and General and Special relativity. However, the dimensions and structure of the atomic and subatomic particle appear different than the existing so far models and theories. The BSM models, however agrees quite well with the experimental data and the observations from different fields of natural science.

Chapter 8. Atomic nuclear structures

Appendix: Atlas of atomic nuclear structures.

The Periodical Table of the elements, discovered by Dimitrii Mendeleev provides a signature of the spatial arrangement of the protons and neutrons in the atomic nuclei. (a conclusion apparent at the end of this chapter).

The recent technological achievements in many applied fields of natural science, such as: nanotechnology, new materials, transmutation of elements (including the cold fusion), biomolecules, put a dough about the long time belief that the QM models of atoms describe completely the physical reality. One of the major problem is that the QM models do not provide information about the structural features of the atomic nuclei (how the protons and neutrons are spatially arranged).

8.1 The Quantum Mechanical models of the atoms are only mathematical models

While the Bohr planetary model of the hydrogen provides the energy level in agreement with the spectroscopic observation, it has a number of problems. The following questions have never get satisfactory explanations:

- why the orbiting electron does not radiate energy and does not fall on the nucleus

- what is the physical mechanism behind the quantum levels

- what defines the finite lifetime of the excited states

The QM models of the atoms are based on the Bohr atomic model, but upgraded additionally with a number of adopted rules. The fact that the number of the protons in the neutral atom is equal to the number of the electrons is used by the Quantum mechanics as a major distinction feature between the elements, while additional rules are adopted for arrangement of the orbiting electrons in shells and subshells. (Among such rules are the Hunds rule, the Pauli exclusion principle and a completed shell.). The fitting of the elements to the Periodic table and their valences is based on adopted rules. The major benefits from the QM concept about the atoms is the possibility to calculate the atomic spectra as transitions between quantum levels. The definitions of the quantum levels is based on wave-functions based on additionally adopted rules. Most of the rules are derived empirically without a logical physical explanation, while number of necessary constants are determined experimentally.

The nuclear radius, according the QM models is of the order of 10^{-15} (m), while the size of the orbits is at least 10^4 times larger. Then a logical question arises: How the protons enclosed in so small volume are able to define extremely well the position and shapes of all possible orbits, especially for heavier elements? The Quantum Mechanics is not able to provide a logical answer of this question.

One of the strongest QM argument for the planetary atomic models comes from the scattering experiments. The Rutherford scattering experiment is cited as a classical one. **However, the output results from the scattering experiments are strongly dependable on the apriory adopted considerations and postulates. This put a serious doubt about the validity of the results.** Not only the underlying structure of the physical vacuum is ignored, considering a void space, but any kind of structure of the elementary particle is completely excluded. In order to obtain the nuclear size (or proton and neutron dimensions), for example, an interpretation scattering model is used. The directly measured experimental result is the angular distribution of the scattering. **The Atomic nuclear size is derived by admitting that it has a spherical shape. But an extended toroidal shape or a twisted torus with the same volume will provide a similar angular distribution.**

From the point of view of the BSM, it appears that the obtained results may differ significantly from the reality because the following considerations are not taken into account:

- the atomic nucleus is considered a spherical with a uniform density

- the structure of the physical vacuum is not taken into account

- the structure of the elementary particles involved in the scattering experiment is not taken into account

Additional discussion on this issue is presented later in §8.9.

One of the major problems for the Quantum mechanics is that it cannot unveil the physical structure of the atomic nuclei. QM models operate only by energy levels. Another major problem is the lack of logical explanation for many adopted rules. For justification of this, the orthodox supporters of the QM concept claim that the human logic fails. Such unjustified conservative claim is an obstacle in the search of the physical structures of the atomic nuclei and causes enlarging the gap between the adopted theoretical models and the experimental results in number of applied fields.

From the point of view of BSM theory, **the QM models of atoms are not real but mathematical models only.** Why the Heisenberg uncertainty principle is one of the most essential rule in Quantum mechanics? Because the Quantum mechanics is based on adopted concept of the physical vacuum as a void space, despite the facts that it must possess a well defined physical properties. While the space is considered as a void, the process of the photon emission or absorption has to be directly related to the position of the electron in the orbit. This unsolved problem (and other related) led to adoption of the uncertainty principle suggested by Heisenberg. While this principle is useful up to some point with the adopted vacuum concept, it will appear obsolete in a new correct concept about the physical vacuum. In the new vacuum concept on which the BSM theory is based, the photon emission does not need to be associated with an exact position of the orbiting electron. The emission is a result of the CL space pumping obtained by the multiple orbital cycles of the electron. This opens a possibility for a classical analysis of the QM phenomena without using of the Heisenberg uncertainty principle. Such new kind of analysis was shown in Chapter 7 for the hydrogen atom. Additional examples of such analysis are provided in Chapter 9 for estimation of the vibrational energy levels of simple molecules. Although the derived method for simple diatomic molecule gives approximate values, it permits to obtain a valuable information not only about the proton and neutron arrangement in the nuclei but also about the connections of the atoms in molecules.

The QM models could not provide information about the real nuclear structure because the following features are missing:

- **the material structure of the CL space environment with its Zero Point Energy;**

- **the complex spatial structure of the atomic particles;**

- **the structure and the oscillation properties of the electron;**

- **the SG forces inside the Bohr surface;**

- **the distributed proximity E-field of the proton inside the Bohr surface and the modified Coulomb law**

The Super Gravitational law, as seen from the total course of BSM, is the most fundamental physical law behind the known fields: Newtonian gravitation, electrical and magnetic fields. It is directly involved in the nuclear binding energy and is apparent in the vibrational properties of the atoms in the molecules (as shown in Chapter 9). The missing SG field in the QM models is compensated by number of adopted rules, which appears unlogical.

8.2 BSM concept about the atomic structure

According to BSM, the proton has a shape of Hippoped curve with a parameter $a = \sqrt{3}$ and a length of 0.667×10^{-10} m. The proton core thickness is 7.84×10^{-13} m, while its cut length is 1.627×10^{-10} (m). The proton is only a twisted shape of the protoneutron - a primary particle as a result of the particle crystallization process, which has preceded the birth of the galaxy (as discussed in Chapter 12). While the particle structure is stable, the protoneutron shape is a torus, which is not stable in CL space. It has two options; a twisted shape like the digit 8, which is a proton, or a double folding shape, which is a neutron. So the proton and neutron contain one and a same types of helical structures (torus and curled toruses). The difference between them is only in their overall shape. All FOHSs have internal RL latices that are able to modulate the CL nodes of the surrounded space by their SG field. In the shape of the neutron, the external modulation from the RL SG field appears symmetrical and forms a proximity field that is locked by the SG field. In the shape of the proton, however, the external modulation from the RL SG field is not symmetrical, so it appears unlocked and creates a positive charge (positive EQs) in a surrounding CL space domain. In both cases, the charge in the proximity is distributed around the external helical structure shell of the proton (neutron). This feature allows suitable conditions for the electrons to create quantum orbits. Therefore, the orbital trace of the electron is strictly defined by the proximity E-field of the proton, which may be slightly modified by a neutron if the latter is over the proton's saddle.

The electron is a single coil helical structure whose external diameter is only slightly larger than the proton thickness $2(R_c + r_e) = 7.9 \times 10^{-13}$ m.

The BSM theory was able to unveil the atomic structure of all elements as spatially ordered systems, composed of two basic particles: proton and neutron. The mutual spatial positions of the protons and neutrons in the atomic nucleus is one of the most important atomic feature for every element. Their arrangement is defined by the shape of the proton and neutron, the SG fields and the distributed proximity E-field around the proton core, inside of the Bohr surface (see the definition of the Bohr surface in Chapter 7). Such nuclear structure defines separate orbital quasiplanes for every proton. The electrons are subordinated to occupy the orbital quasiplanes inside the Bohr surface of the protons. The individual Bohr surfaces from closely connected protons may combine in one integrated Bohr surface for the whole atom. The electron orbits of the neutral atom are always inside of the integrated Bohr surface.

8.3 Atlas of Atomic Nuclear Structures (ANS)

The nuclear configuration for the most abundant isotopes are shown in an appendix titled: Atlas of Atomic Nuclear Structures (ANS). In order to simplify the drawings of the complex three dimensional nuclear shapes, two dimensional sketches are used, where the basic elements (protons and neutrons) are shown by properly adopted graphical symbols. Figure 8.1 shows these symbols and their use for a graphical presentation of some of the most simple elements: hydrogen deuteron, tritium and helium.

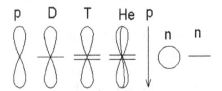

Fig. 8.1

Symbols for atomic nuclear structures

In Fig. 8.1, the following annotations are used:

p - Proton
D - Deuteron
T - tritium
He - helium
n - neutron

8.3.1 Building rules of the atomic nuclei, related to Z number

Using the Z number as an argument, the build-up trend is investigated and the following rules are discovered. They are logical consequences of a build-up process, governed entirely by the shape of the proton and neutron and the SG and proximity electrical fields. They form a complex of forces which are discussed in the next section.

8.3.2 Natural forces defining the build-up trend of the atomic nuclei of the elements

- **External gravitational pressure**
- **Proton - proton repulsion**
- **Neutron - proton near field interaction in the moment of neutron insertion over the proton saddle**
- **SG (Super gravitational) attraction between protons and neutrons**
- **Dynamical interaction between the internal RL(T) of the protons and neutrons and the surrounding CL space**
- **Space locking of the proton electrical field by the orbiting electron and the SG field (for a neutral atom only)**

The shown above complex allows formulation of the following rules in the build-up tendency of the atomic nuclei

- **Chain building principle: a consecutive building of complex structures by starting from a simple basic one and bonding additional basic structures**
- **End product as a particular element: a compact shell structure, formed of basic elements (protons and neutrons), satisfying the balance between their attractive SG fields and the repulsive E-fields.**
- **A natural selection of stable atomic structures in which the natural process of radioactive decay is involved**

- **The radioactive decay is a result of disturbed nuclear balance between the SG and E fields, due to the stable hardware structure of the protons and neutrons.**
- **The disturbed nuclear balance in the radioactive elements likely contain modes of oscillations between the Supergravitational and electrical fields, which are related to rate of the radioactive decay, known as a half-time.**

8.3.4 The formulated rules must comply to the following:

- The **periodicity in the unveiled atomic nuclear structures must fit to the periodicity of the Periodic table (Mendeleev's periodical law).**
- **Must comply to the Hund's rule**
- **Must comply to the Pauli exclusion principle.**
- **Must comply to the principal and secondary valence properties of the element**
- **Must contain signatures apparent from the X-ray properties of the elements.**

8.3.5 Useful data for unveiling the nuclear structure

- **X-ray properties of the elements in a solid state.**
- **Laue back-reflection patterns**
- **Relation between the nuclear bonding energy and the X ray spectra of the elements**
- **Oxidation number by experimental results. Principal and secondary valencies.**
- **Ionization potential dependence on Z number**
- **Considerations for orbital interactions and pairing between the electrons from different orbitals**
- **Radioactive decay of unstable isotopes**
- **Optical atomic spectra**
- **Photoelectron spectra of molecules**
- **Nuclear magnetic resonance of the elements**
- **Nuclear configuration and VSEPR model for chemical compounds**
- **Vibrational properties of the atoms in the molecules in a gas phase.**

8.3.6 Type of bonds in the atomic nuclear structure

The protons and neutrons in the nuclear configuration are kept together by different types of bonds, in which SG and electrical fields are involved. SG bonds are annotated as G (gravitational) bonds. The abbreviated notations of these bonds are given in Table 8.1. The SG filed

Bonds in the atomic nuclear structure   *Table 8.1*

Bond notation	Description
GB	Gravitational bond by SG forces
GBpa	polar attached GB
GBpc	polar clamped GB
GBclp	(proton) club proximity GB
GBnp	neutron to proton GB
EB	electron bond (weak bond)

8.3.7 Basic rules in the process, leading to build-up of stable isotopes.

The factors defining the build-up process of the atomic nuclei are the shapes of the proton and neutron and their proximity fields. The neutron shape is stabilized when it is inserted in the proton's saddle. To be inserted over the proton, the folded neutron has to overcome initially the repulsion forces between its proximity near E-filed (locked by the SG field) and the proton E-field. The obtained composite structure is a **Deuteron.** The next dens composite structure is the **tritium** - two neutrons over the proton saddle. The next dens structure, which in fact is a denser one is the **He** nucleus. It can be considered as built of two deuterons. The accumulated SG matter causes, also, a small shrinkage of surrounding CL space. This affects the total detectable Newtonian mass, which is the BSM explanation for the nuclear binding energy. **He** nucleus possesses the largest SG field density in comparison to other composite structures from protons and neutrons, consequently a larger binding energy. Therefore, it is quite reasonable to consider that a He nucleus should be embedded in the central part of the nucleus of any stable element. In this chapter will show a strong proof that the atomic nuclei are built of the mentioned above basic elements, whose spatial arrangement is dictated by natural geometrical rules with balanced SG and electrical fields. The **basic elements** are listed below in an order of increasing matter density, corresponding to increasing binding energy (increasing shrinkage of the surrounding CL space)

proton (or neutron)
deuteron
tritium
helium

Following the above considerations and those in §8.3.4 and 8.3.5, it was found that the trend of congregation into larger atomic nuclei leads initially to building of polar structure, in which the He nucleus is in the middle. After such composite structure is completed to some level at which the equatorial radius is larger than the polar one, the build-up trend leads to connection of such structures in their polar regions. As a result of this, the build-up trend leads to a polar-connected chain structure. The envelope of the completed composite structures and their interconnections into a polar-chain composite structure are illustrated in Fig. 8.2.

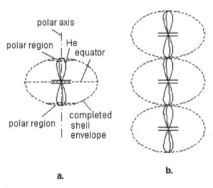

Fig.8.2

a. - a composite structure with a polar axial symmetry; b. - a composite polar-chain structure

The detailed analysis of the build-up trend leads to discovering the following rules:

(**1**) The build-up trend, following the increase of Z number (number of protons in the nucleus), leads to a building of congregation of polar-chain structure. The first level of the congregation is a completed polar structure containing one He nucleus in the centre. The second level of the congregation is a chain connection of two completed polar structures in their polar region, so their polar axes are aligned.

This trend appears valid up to three polar structures (see Fig.8.2.b). The third level of the congregation is related with a build-up process, in which basic components are attached in the equatorial regions of the composite polar-chain structure. In such case the atomic nucleus approaches a more spherical shape. The first level of the congregation is completed at Ar. The build-up phases for the second and the third levels overlap. The third level of the congregation is a build-up of equatorial shells, so the nuclear shape tends to approach a more spherical form.

(**2**) The building of the atomic nuclei is based on a process of bonding of basic elements (<u>proton, Deuteron, tritium, Helium</u>). Once bound, every basic element is kept in that position **due to the SG fields involved in the synchronization between the internal RL(T) of all involved FOHSs. Such type of synchronization automatically controls the balance between the attractive SG forces and the repulsive electrostatic forces between the bound elements.** This type of binding is named a gravitational (G. binding, or bond). Four different types of G. bonds are distinguished: GBpa, GBpc, GBclp, GBnp (see Table 8.1).

(**3**) The main binding mechanism, responsible for the atomic build-up is the polar gravitational binding. It appears in two types: - polar attached GBpa, and polar clamped, GBcp

(**4**) Every GBpa bound proton (Deuteron or tritium) has some rotational freedom of motion in the polar plane. Its freedom in the equatorial plane is more restricted, because of the flat geometrical shape of proton and the SG forces. Neglecting the equatorial freedom, the proton position may be defined by the angle between the nuclear polar axis and the long axis of the proton.

(**5**) Polar region problem: The polar electron circles around the core of all Gpa bound protons (Deuterons or tritium). This is a place of increased SG field density and concentrated E-field lines. The protons contributes to this SG field concentration, while the neutrons over the proton saddle, pull up the field to the periphery. For this reason, a correct numbers of protons and neutrons are necessary for formation of stable elements, known as stable isotopes. Unstable isotopes converts to more stable ones by β⁻ or β⁺ radioactive decay, which provides the necessary ratio between the number of protons and neutrons and their proper positions.

(6) While the positions of electronic orbits is defined by the spatial arrangement of the building elements in the nucleus, the Hund's rule appears indirectly related to the spatial positions of the building elements. It was found that the Hund's rule exhibits four types of appearance: two types of orbital pairing and two types of QM spin pairing. The different types of pairing possess different **pairing strengths**. Additional requirements for formation of a stable isotope is the spatial arrangement of the protons to fulfil the Hund's rule according to which the largest possible pairing strength appears as a dominant

(7) The proton shell build-up trend matches the corresponding raw of the Periodic table.

(8) For Z- numbers larger than 23, more that one branch of the growing trend are possible. Some branches may pass through a zone of unstable isotopes and reach a stable isotope zone again.

(9) The single polar structure is completed at the Argon atom. The additional nuclear growing by bindings in both polar regions continues to the group 10 of the Periodic table, where it is interrupted. The likely reason is a disturbed balance between the Super gravitational and the electrical fields of the external protons. Group 11 is from another branch that has been grown in one pole only, but has been unstable. From groups 11 to 18 this branch is stable and the growing trend is completed at the group of noble gases.

(10) The growing mechanism for the atoms after Argon, could be based not only on a binding of basic structures but also on an attachment of smaller nuclear structures, either with completed or not completed shells. The attachment is also at the polar region, but conditions for a clamp type of binding are required (GBpc). This may reduce the probability of such growing.

(11) For the groups between 11 and 18 up to the element Rn, the external atomic structures defining the chemical valences are similar. The Lantanides and Actinides have different external structures, not matching the structures of the mentioned above groups.

(12) For the groups 11 to 17, another weak binding mechanism, named electron binding (EB), interferes with the main build-up mechanism, based on

a polar binding. The EB binding is based on interactions between the quantum orbitals and is related to the Hund's rules. The EB bound protons are excluded from the chemical bonds. Under influence of a strong chemical reagent, however, they could be unbound.

(13) The weak EB bonds convert to strong GBclp bonds at group 18. In such case, the bound protons could not participate in any chemical reaction.

(14) A chain of three polar structure is completed at Xe atom and the additional chain growing is interrupted (the aspect ratio of the polar to equatorial radius becomes unacceptably large). Then the tendency of equatorial growing becomes predominant, leading to build-up of the Lantanides. (In the Periodic table, the Lantanides are shown after Ba, but the position of the two polar protons in Ba are preserved in all Lantanides).

(15) The Lantanides grow by GBclp binding of deuterons (proton, tritium) to the equatorial region of the two ends polar structures. They periodically become EB bound in pairs due to an orbital pairing (Hund's rule). Their EB bonds behave differently due to the different spatial positions of the protons. Such bound protons (D, T) become underlying shells in the next following elements, but they can not be converted to strong GB bonds. They are source of alpha particles in the alpha radioactive decay.

(16) The nuclear structures from Hf to Rn grow by GBcpr binding of deuterons (tritium) to the equators of the middle and two end polar structures, forming two shells (2 x 8) with a rotational helical symmetry.

(17) The Actinidies grow by GBclp binding of mainly tritium (proton, deuteron) to the equators, forming another two shells between these of Hf - Rn.

(18) All atomic nuclei exhibit a higher order helicity possessing the same handedness as the proton.

Additional less general features will be discussed in the next paragraph.

8.3.8 Discussion about the basic rules

Figure 8.3 shows mechanical mock-ups for illustration of the building process.

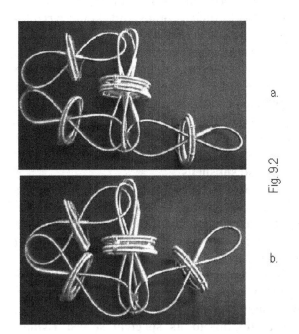

Fig. 9.2

a.

b.

Fig. 8.3
Mock-ups illustrating the spatial
aggregation by binding of basic structures

Polar angle of the polar bound proton and its range of freedom

The central structure to which Deuterons are bound by GBpa is a He atom. The GBpa binding allows a limited freedom of rotational motion in a limited angle (about 90 deg) in the polar plane as illustrated in the mock-ups shown in Fig. 8.3. The freedom of motion in a plane parallel to the equator is much more restricted - only a few degrees. It is evident that the freedom of motion is restricted by the shape of the hardware structures and the E-field inside the Bohr surface. The electrical interaction, for example, will not allow the free ends of protons to be so close as shown in Fig. 8.3 **a**. In one case of Hund's pairing the free ends could approach the case shown in **a.**, but little bit apart, in order to have a room for both circling electrons to pass between the proton's clubs. In case.**b** a BGclp type of bond is shown. This type of bond exists in the internal proton shells and in the external shell of the noble gases.

Neglecting the equatorial freedom, **the position of the proton could be defined by the angle between nuclear polar axis and the long axis of**

the bound proton. This angle may be defined as a **polar angle**, characterised by an angular range and boundary conditions.

- **The polar angle of any GBpa proton could never become zero.**
- **The range of the polar angular freedom of a GBpa proton is restricted by the structural configuration and by pairing with protons, bound to the opposite pole of the nucleus.**

The polar angle and its range are very characteristic parameters of the valence protons. The range of the polar angular freedom of the GBpa protons is different for different elements.

The four types of the gravitational bonds and one type of the electronic bond are shown in Fig. 8.4

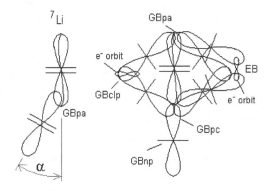

Fig. 8.4
The structure in the left side is ^{7}Li nucleus, while the structure in the right side is a part of atomic nucleus. The annotation of the different types of bonds is according to Table 8.1
α - is the polar angle of a valence proton (deuteron)

The electronic bond (EB) corresponds to one type of Hund's rule electron pairing. The common electron orbits for EB and GBclp bonds are also shown. The electron orbital plane for EB is closer to the polar plane of the atom, while the electron orbital plane of GBclp is closer to the equatorial plane of the atom.

It is evident from Fig. 8.4 that when using the Hippoped curve symbols, the drawing become very complicated for elements with a high Z number. For this reason, in the Atlas of Atomic Nuclear Structures (ANS) drawings, only the He symbol is used in the centre, while the protons are shown by the simplified symbols given in Fig. 8.1. Even in this way, the drawings of the heavier atoms become overcrowded. In such case, additional drawing simplifications are used as the following:

- more complex nuclear structures are shown by three views (projections); the middle projection is a polar section; the top and bottom projections are polar views.

- any polar view shows the peripheral structures from its side only and without showing the helicity

- in the polar view, the angles between the symbolic vectors and the lengths of the vectors may deviate from the real perspective view, for a reason of a drawing clarity

- completed and known internal structures (like Ar), when participating in element with higher Z number, may be shown as a dashed shaped ovals or polygons (close to the structure envelope)

- the nuclear helicity is not shown in the drawings

- additional symbols for electronic bonds (EB), corresponding to Hund's rule are also used.

Electron pairing and Hund's rule.

In §7.11, the magnetic field related to the electron magnetic radius was discussed for the single proton. It was emphasized, that it is kept inside the Bohr surface. When more than one protons are connected to the polar region of the He nucleus, their individual Bohr surfaces become combined into one **integrated Bohr surface**. This surface now closes a common volume. In such case, it is possible the magnetic fields related to the magnetic radii of the electrons to interfere. The electrons are still in their orbits, but their QM spins become mutual dependent. This affects the position and orientation of the orbits of the different protons and consequently their positions (having in mind, that the proton has some angular freedom in the polar nuclear plane.)

There are number of differences between the QM model and BSM model that reflect also the used terminology:

(a) the electrons in QM model circulate around a common nucleus, containing all protons and neutrons

(b) the electrons in BSM model are strongly connected to their protons and do not circulate around the whole nuclei

One or more electrons according to BSM may pass to other protons only in special cases like: an Auger effect; a radioactive decay, a mutual alignment of valence protons bound to one pole.

(c) According to QM model, the position of the atomic element in the Periodic table is directly related to the electronic orbits, defined by shells and subshells

(d) The electronic shell and subshell configurations, according to BSM, do not have the spatial arrangement as those accepted by the QM model, however, the ionization energies are the same.

(e) According to BSM, the orbital quasiplanes for the valence protons in the atoms are of similar type as the orbitals in the Hydrogen, corresponding to its spectral series: Lyman, Balmer, Paschen, Bracket etc. In the heavier atoms, the Lyman type series is significantly restricted by the spatial positions of the protons and neutrons, while the Balmer type series is always available.

Despite the above differences, the BSM model provides excellent match of the atomic elements to the Periodic table with clearly defined building rules. The BSM model provides also clear physical explanations of the Pauli exclusion principle and the Hund's rule. The Pauli exclusion principle, discussed in §7.11 for the Hydrogen atom is valid in a similar way for all atoms.

The Hund's rule according to BSM model appears in two options, each one possessing two suboptions:

- **an orbital pairing**
- **an electron pairing**

The orbital pairing subdivides additionally into:

- **an orbital pairing for protons attached to a common pole**

- **an orbital pairing for protons, attached to different poles**

The electron pairing is valid for two cases of the proton bondings: EB and GBclp.

The different options of the Hund's rule are illustrated by Fig. 8.5.

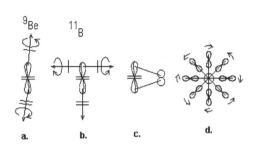

Fig. 8.5
The Hund's rule appears to be of four
different types (symbolic notations are used)

The pairing in case **a.** and **b.** is a complimentary alignment of two orbits, due to their common interactions. The interactions are result of the magnetic fields induced by the electron magnetic radius of the orbiting electrons, but kept inside the integrated Bohr surface. The arrows in Fig. 8.5 show the rotational directions of the electrons.

The electron pairing in case **c.** has a different physical aspect. In this case, the free ends of the protons are approached and their electrons circle in a common orbit, which passes through their clubs. (The electron traces, however, are not intercepting). The electrons must have opposite QM spins. This type of pairing provides an electron bond (EB) between protons connected to the opposite poles. The protons with EB bond are excluded from the principal valence of the element. This type of pairing corresponds to a Hund's rule applied for groups 11 to 17 of the periodic table.

The electron pairing in case **d.** corresponds to a completed proton shell. The paired electrons pass through a pair of GBclp bound proton clubs. The orbital planes in this case are perpendicular to the orbital planes of EB bonds. The drawing plane for the case in Fig. 8.5.d coincides with the equatorial plane of the atom. The orbital planes are slightly tilted in respect to the equatorial planes due to the atomic helicity. The neighbouring orbits are paired due to their proximity. This type of pairing will be additionally discussed later.

Another feature for the cases shown in **a.** and **b.** is that both electrons may have one and a same quantum number. At first glance, it may look that such condition is in conflict with the Pauli exclu-

sion principle, but in fact it is not. The principle is formulated in accordance to the quantum mechanical model, where the electron kinetic energy and transition moment are associated with the emission of photon. According to BSM theory however, the photon emission is associated with a pumped CL space energy in a region inside of the Bohr surface. Such physical effect (revealed by BSM) is present prior to the electron transition to lower orbit (lower energy level), so it affects the environments defining the conditions of quantum orbits. Let us admit that both electrons of the pairing in case **a.** and **b.** have one and same quantum numbers. They should have, however, different QM spins due to the orbital interaction. When one electron losing an energy falls in a lower orbit, the other electron will profit of this and will take a portion of this energy before the photon is emitted. Consequently, the second electron now will have different kinetic energy. So the Pauli exclusion principle is still valid. In order to avoid some misinterpretation, when considering the BSM model, the Pauli exclusion principle could be reformulated as: **Simultaneous emission of two photons from one atom, corresponding to one and a same transition is not possible.**

Many isotopes are possible, but most of them are not stable for different reasons. The alkali elements start always with a polar bonding. The element ^{5}Li (form by He and one D) is not a stable isotope, but ^{6}Li is a stable one. The reason is that the SG attraction between He and D is not strong enough in order to keep the GBpa bond between them. Adding one more neutron over the proton makes the SG forces stronger. But how the SG forces are able to overcome the E-field repulsion between the protons? The SG field manages to synchronise the high oscillations of the internal positive RL(T) structures of the protons (the high frequency oscillation modes of the SG field between SG matter are discussed in Chapter 12). While SG defines also the E-field of the protons (neutrons), their proximity E-fields are likely displaced towards the club of the bound protons. All polar clamped proton clubs then posses a total of one elementary charge, balancing the elementary charge of the polar electron. In such case, the charge balance between the protons and electrons (for a neutral atom) is preserved.

Binding nuclear energy and a polar region problem

Let us consider the building of the heavier atom as continuous growing process, by adding new protons (deuterons). Since the SG mass increases, the SG field becomes stronger. The increased SG field causes a CL space shrinkage, that is small, but still detectable, because it affects directly the CL static pressure and consequently the Newtonian mass. The detectable parameter in this case is the nuclear binding energy.

- **The nuclear binding energy is estimated by the Newtonian mass difference between the sum of the masses of the embedded protons and neutrons and the apparent atomic mass**

The CL space shrinkage in fact is a General Relativistic effect discovered by Albert Einstein, but BSM provides its physical explanation by a classical way.

For elements up to Z-number of 56 (Ba), every new proton (deuteron or tritium) attaches to the polar region by GBpa or GBpc type of bond. The increasing SG field is able to control the proximity E-field distribution in the polar region, but within some limit, because conditions for orbiting polar electron deteriorate. If this limit is exceeded, one polar bound proton must be detached, so the balance to be restored. The observed effect is a radioactive decay. We may call this a **polar region problem**. This problem may be solved partly by adding a Deuteron or a tritium instead of a proton. The added neutrons redistribute the SG field (which affects the proximity E-field) towards the nuclear periphery. When adding only protons, the E-field concentration in the polar region is stronger, than adding Deuterons or tritium. The polar region problem unveils when analysing the trend of the bonding energy as a function of proton and neutron numbers and the X-ray properties. This is discussed in §8.6

Most unstable isotopes exhibit a polar region problem. The stable elements have well balanced ratio between protons from one side and Deuteron or Tritti from the other. In the heavier nuclei, the polar region problem requires more neutrons in order to rarefy the field. For this reason the ratio between neutrons and protons gradually increases with the Z number. While the tritium is unstable as

a element, it appears stable when included in the nuclear structure of heavier elements. So protons with one or two neutrons provide a stabilizing effect. More than 2 neutrons over one proton however are unstable, and if such structure eventually appears, it may lead to a radioactive decay by a neutron emission. In noble gases additional nuclear stabilizing option appears by adding neutrons over the equatorial GBpcl type of bonds (see Fig. 8.4). The noble gas nucleus ^{40}Ar, for example, has four such neutrons over its equatorial external GBpcl bonds

8.3.9. Atomic nuclear build-up trends

Let us describe briefly the build-up trends, following the Z increase in the Periodic table. We will see, that this trend is characterised by a shell structure made of consecutively added protons. In most cases, the protons are with one or two neutrons over their saddle, so in the build-up process, protons, deuterons and tritium are added. Since the Z-number is defined by the positive charge, we may mention protons, knowing that deuterons and tritium elements are added when Z-number is increasing.

The He nucleus must be always in the middle, since it is the most dens basic element. The first shell of the protons (deuterons) is built complying the rule of orbital pairing (one type of Hund's rule). The free proton clubs define the principal valence of the element. From group 15 to 17 the element valence is determined by the free proton clubs that are not EB bound (third type Hund's rule). In Nitrogen atom there is only one equatorial pairing, forming a weak EB bond. The bound protons are excluded from the principal valence, so the Nitrogen has a principal valence of 3. At the Oxygen element, the four protons obtain conditions for equatorial pairing. Instead of two weak EB, much stronger GBclp bonds are obtain, due to the perfect symmetry. The evidence of such conclusion comes from two different types of experimental data: the Z dependence on the first ionization potential (see §8.5 and Fig. 8.16) and the photoelectron spectra of molecules. (The latter will be discussed in Chapter 9). At element F, the symmetry is broken and the EB type of equatorial bonding is restored. At element Ne, all proton clubs get GBclp type of bond and are excluded from any valence. At the same time, Ne gets a more compact structure than any previous atom and the local lattice space gets a relatively larger shrinkage. This affects the nuclear mass according to a relativistic considerations, explained in Chapter 10. The structure of Ne nucleus gets a completed rotational polar symmetry at azimuthal step of 90 degrees. The transformation of GBclp to EB bonds is related to orbital transformation too, as shown in Fig. 8.5. The orbital planes for EB are closer to the polar plane, while the orbital planes for GBclp are closer to the equatorial plane. In fact the orbital planes of the equatorial GBclp bonds at Ne intercepts the equatorial plane of the atom at small angles.

The second shell of protons (mostly deuterons), starting in Na and ended at Ar, follows a similar build-up process. The equatorial structure is completed at Ar with an increased CL shrinkage effect (mass deficiency equivalent to a nuclear binding energy). The deuterons in the second completed shell are very close to the deuterons of the first one. The most abundant isotope ^{40}Ar gets four neutrons symmetrically disposed over its external GBclp bonds (in the equatorial region). The Ar structure gets also a completed rotational polar symmetry with an alternatively arranged symmetrical structural features (at azimuthal steps of 45 deg).

The next shell starting from the element K grows initially by GBpa bound deuterons in the two polar regions symmetrically up to ^{39}Co. The Cr nucleus is an exception from the ordinary growing trend. This is evident, when examining the oxidation states (O state - 70%; +2 state - 21%; +3 state - 38%; +4 state -3%). The lack of +1 oxidation state means that there is not a free valence proton. The possible nuclear configuration, shown in the ANS is a four equatorial EB bound structure of pair deuterons with He nucleus, which is polar connected to Ar nuclear structure. The three polar electrons in the two He structures have parallel orbits in a strong SG fields. So they may give a particular signature in the electronic spectra, which could be misinterpreted as a signature of a single electron in the external shell.

The build-up tendency is restored for a symmetrical polar growing at the element ^{55}Mn. This tendency is kept till Ni or Cu. It is controversial at which of both elements a jump to Ne like structure occurs. The correct assignment requires more detailed investigation of the differences between the physical and chemical properties of the neighbouring elements. The ^{59}Cu atom, however, reasonably possesses a Ne like structure, but with EB type of bonds instead of GBclp bonds.

Following the growing trend, the stable isotope ^{63}Cu has one valence deuteron (bound by GBpc to the Ne like structure. The stable isotope ^{64}Cu has an one valence tritium instead of deuteron. In Zn, two valence protons are GBpc bound to the modified Ne structure. At Ga atom, the weaker EB bonds of the Ne like structures are converted to much stronger GBclp bonds. The bound in such way protons are excluded from the valence. The build-up tendency after Zn, continues with bondings, attached only to the poles of the Ne - like structure. The growing trend provides valances similar like the previous raw of the Periodic table, until the element Br. In the noble gas Kr, all previous EB bonds are converted to GBplc type. The orbit plane of the two electrons with opposite spins now is almost a parallel to the equatorial plane with a slight tilt due to the nuclear helicity. The external shell and a second Ar like structure are completed at the noble gas Kr. Both Ar-like structures are attached in their polar regions by GBpc bonds. The clamp attachment slightly changes their equatorial radii. The nucleus of Kr, however, contains four more neutrons than the two separate Ar nuclei. They compensate the polar region problem, mentioned above.

The 5th raw of the Periodic table (Rb - Xe) follows generally a similar build-up tendency, but some differences appear. One of them is that the axial polar D (Deuteron) begins to get replaced by a He nucleus. The first replacement is in Tc, but for this element only one polar D is replaced by a He nucleus. This asymmetry is likely the reason for the instability (radioactivity) of the element Tc. The build-up trend for Ru and Rh adds GBplc deuterons with EB equatorial bonds. In Pd all equatorial bonds become of GBclp type so the bound protons are excluded from the valence. Then for For $Z = 47$ a deuteron can be bound only to the Ar polar structure. This is Ag. From Ag to Xe the process is similar.

The six raw begins at Cs with a polar GBpc attached deuteron, but after Ba, the polar binding tendency is intercepted from the Lantanides build-up trend. The initial tendency is restored after all Lantanides are built, however, the bonds are not any more attached to the poles but to the equatorial points. So the growing tendency here has been changed from a polar chain to an equatorial growing. This is dictated by the balancing conditions between the complex shape configuration, held by the proximity SG field (leaking in CL space) and the distributed electrical fields of the protons.

All Lantanides grow by GBclp type attachments of protons (deuterons) between the equatorial GBclp clubs of the two end polar structures. The attachment position initially follows the pairing rule, shown in Fig. 8.4.b., but only up to two protons, diametrically positioned. Then a third proton (deuteron), at Pr, appears attached in the equatorial region of the opposite Ar structure and makes an EB bond with one of the first two attached protons. At Nd a new proton (deuteron) is attached in a similar way and makes a second EB bond. At Gd four EB bonds appear, making the structure completely symmetrical with a rotational symmetry of 90 deg. The tendency of growing after Gd continues with EB bonds repeated at any second proton (deuteron). This keeps the principal valence between 3 and 4 (keeping in mind that two of the valences are contributed by the polar attached deuterons).

The structure of Gd is shown in Fig. 8.6.

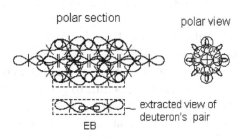

polar section   polar view

_ extracted view of
deuteron's pair

EB

Fig. 8.6
Gd nucleus, EB - electronic bond

In the left projection, only the central elements, closer to the drawing plane are shown. Below the polar section, the positions of EB bound deurons are shown. These EB bound deuterons are in these positions also in the higher Z-number

atomic nuclei. The proton clubs are too apart and their EB bonds could not be converted to stronger GB bonds in the further build-up trend. These bonds, however, could be broken in some chemical reactions, especially for the elements of the raw 6 of the Periodic table after Hf. The higher oxidation numbers obtained for Au in cases of strong reagents could be explained by this effect. In Lantanides, some of the EB bonds could be also broken by strong chemical reagents. The EB bound peripheral deuterons in the Lantanides are well axially aligned and their positions are preserved in the elements of higher Z number.

One of the most important feature of the peripheral deuterons in the Lantanides is that they provide the He nuclei in the alpha radioactive decay of the heavier elements.

The above made conclusion is apparent from the spatially oriented positions of the pair EB bonded deuterons, shown in the bottom of Fig. 8.6. They may undergo transmutation into He if some fluctuation of the SG field takes place.

Some artificially built elements, by nuclear collision processes, may have very deformed or different structures. For this reason they have very short decay halftime (of the order of msec and sec).

The different build-up process of the Lantanides determines their properties, which are distinctive from the lower Z-number rows of the Periodic table. One interesting feature is that the nuclear structure of Hf looks more completed than the nuclear structure of Lu. At Hf, all eight equatorial pairs of Deuterons are EB bound. One feature, that may look strange, is that Hf exhibits only oxidation number of four. This corresponds to a break-up of two symmetrically positioned EB bound proton's pairs. After Hf, the growing tendency is characterized by attachment of GBclp bound protons (deuterons, tritium) to the equatorial proton clubs of the middle and one end polar structure. At Pt one polar deuteron disappears. At Hg the two polar deuterons disappear. They do not appear again and at Rn the external shell becomes GBclp bound. One interesting growing transition appears between Au and Hg. Despite their neighbouring positions in the Periodic table, their nuclear structures are quite different. The weak EB bonds of the Lantanides structures could be broken up to Rn and even at Rn. This may explain some chemical compositions of Rn.

The polar section of the atomic nuclear structure of Rn is shown in Fig. 8.7. Only the central elements, parallel to the drawing plane are shown.

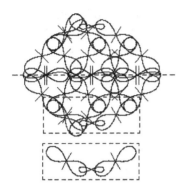

Fig. 8.7

Sectional view of Rn nucleus. For a 45° rotation around pp axis the sectional view is a mirror image

Below the sectional view of Rn nucleus a pair of deuterons is shown. Such deuteron pair is the source of the He nuclei obtained during alpha decay.

The seventh raw of the Periodical table starts again with polar GBpa bonds for Fr and Ra, but it is discontinued for the build-up trend of the Actinides. The growing process in the Actinides is different than the Lantanides. The Actinides grow by elements (deuterons, protons, tritium) GBclp bound to the adjacent two equatorial proton clubs in the equatorial space left over from the previous 6-th raw. The growing trend of the Actinides and the following transuranium elements tends to form a more spherical shape of the nuclear structure.

8.4 Experimental data supporting the unveiled atomic nuclear structures

8.4.1 The polar region effect and a proton to neutron ratio

The polar region effect can be investigated if making a plot of edge levels of mass attenuation coefficient of X-ray data as a function of Z-number, then fit to a polynomial and observe the residuals. The residuals, obtained in this way, look differently for the internal and the external shells. Careful investigation provides indications that: the neutrons are added in the periphery; the neutrons and the protons affect differently the internal and

the external shells. Such plot for K edge levels of x-ray mass attenuation coefficients towards the Z-number is shown in Fig. 8.8.

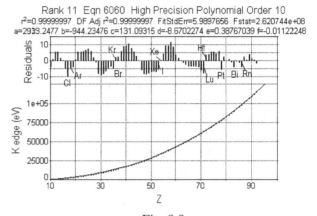

Rank 11 Eqn 6060 High Precision Polynomial Order 10
r^2=0.99999997 DF Adj r^2=0.99999997 FitStdErr=5.9897656 Fstat=2.620744e+08
a=2933.2477 b=-944.23476 c=131.09315 d=-8.6702274 e=0.38767039 f=-0.01122248

Fig. 8.8

The plot starts from Z = 10, instead of Z = 0, because the residual fluctuation depends also on Z number. The statistical error, however, is of random origin and is easier identifiable. For $z > 10$ it is significantly reduced. Analysing the residuals, we see larger positive jumps in the transitions from the group 17 to the group 18. This is due to the compactifying the external shell, when all eight proton clubs get GBclp type of binding. The SG field is pull out to the equatorial region, causing a decrease of the CL shrinkage in the polar regions. Another two interesting features are observed. The element Pt gets a negative jump. This is a result of losing one polar Deuteron. The next atom Au, although, gets a positive jump. **This shows that its external shell containing 8 protons (tritium for ^{197}Au) gets GBclp type of bonds and therefore it is completed. This explains why Au is a noble metal**. The Au atom still has one valence proton that plays an important role in the solid and liquid aggregate states. Otherwise it would be a noble gas. Examining the oxidation number we see that it has a large first oxidation state, a very small 3rd and insignificant 2nd one. The 3rd and 2nd states come from braking one of the weak EB bonds from the underlying EB bound deuterons in the Lantanides shell. The Lantanides type of bonds, cannot be converted to GB bonds, because the distance between the proton clubs is large. Examining the spatial nuclear structure we see that these bonds are below the external shells and could be broken in very special conditions. From the change of the growing tendency between Au and Hg one interesting conclusion could be made:

Let us suppose that the valence deuteron of Au is removed but the nuclear structure is not refurbished. Then it will have the same number of protons like Pt, but its nuclear configuration will define different chemical and physical properties. Its properties could be rather similar to those of the noble gases. We are not sure, however, how stable will be this element.

8.4.2. X ray properties of the elements

It has been considered so far (in the contemporary physics), that the X - crystalography is applicable for studying the structure of the solid crystals made of atoms, but not the atoms themselves. This concept is a result of the consideration that the atomic nucleus is a very small sphere. The proton radius, for example, has been considered to be of the order of 1×10^{-15} (m). The BSM theory, however, unveils that the proton has a shape of a Hippoped curve (close to the shape of figure 8) with external dimensions of 0.667×0.1925 A , with a core thickness of 7.841×10^{-13} (m), where $A = Anstrom = 1 \times 10^{-10}$ (m). Then the nuclear structure is also possible to be studied by a X rays technique including the X ray crystalography.

8.4.2.1 X ray transmission of solids as a function of X-ray energy

A narrow beam of monoenergetic X-ray with an incident intensity I_o, penetrating a layer of material with a thickness t and a density r, emerges with an intensity I given by the exponential attenuation law. $I = I_o \exp((\mu/\rho)\rho t)$. The grouped parameter ρt is called a mass thickness, while μ/ρ is known as a mass attenuation coefficient. The following relation is valid between these two parameters

$$\mu/\rho = \frac{\ln(I_o/I)}{\rho t} \quad (cm^2/g) \qquad (9.1)$$

The mass attenuation coefficients are different for every element. A good reference source, where the X-ray data are summarized in tables and plots is provided J. H. Hubbel & S. M. Seitzer, (1998). They are presented usually as tables and

plots of μ/ρ as a function of the x-ray energy, together with experimental data. The experimental data usually use the total attenuation. They are collected from a large number of published experiments, provided from the beginning of 20th century. Between the most characteristic features of the plots are the absorption edges, characterized by a sharp change of μ/ρ at some particular energy. This is illustrated by Fig. 8.9 showing the experimentally measured total attenuation of $_{29}Cu$ (included in the data source given by J. H. Hubbell, 1971).

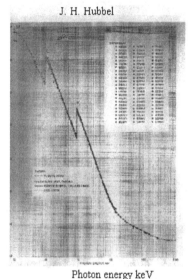

J. H. Hubbel

Photon energy keV

Fig. 8.9 Experimentally measured total attenuation for $_{29}Cu$

It is evident that the edges are signatures of the shell structure of the atomic nucleus.

The detection of the atomic shell structure by X-ray methods cannot get a logical explanation if using the QM atomic models, because the following questions cannot get a satisfactory answer: If the atomic nuclear is of the order of 1×10^{-15} (m) (as accepted by QM atomic models), how the internal structure will be detected with wavelengths which are ten thousand times larger? According to the QM model of the atom, the hardware structure of the atoms must be determined by the electron orbits. But how they are kept in very ordered positions orbiting with a relativistic velocities without radiation of energy? How do they resist to any external stress and the penetrating X-rays? These are simple physical questions for which the QM atomic models do not provide answers.

BSM models are free of such problems. The explanation of the transmission behaviour of the solids, according to BSM, is the following:

The observed edge zones of the mass attenuation coefficient are caused by the refracted and reflected X-rays from the quasishrunk CL space inside the Bohr surfaces. Inside of this surface the shell arrangement of the protons & neutrons (mostly deuterons) causes CL zones with a different degree of shrinkage and a complex shape. The additional effect of Brag reflection, which is defined by the atomic arrangement in the crystal structure will provide additional signature, which will be superimpose on the above mentioned effect.

The quasishrunk space exists inside the Bohr surface, due to the E-field spatial distribution (discussed in §7.6, Chapter 7.). This space namely behaves as a gradient optics for the x-rays.

In order to show the edge effect, a simple analogy is made by using an example of two concentric optical layers with different refractive indices. For this reason an optical ray tracing is provided, as illustrated by Fig. 8.10. The layers are shown as lenses having concentric radiuses, but different indices of refraction. The internal lens L1 has $n = 2$ corresponding to a more shrunk CL space, while the external L2 has $n = 1.5$, corresponding to a less shrunk CL space. An optical ray tracing program is used and the result is evaluated by annulus scanning aperture.

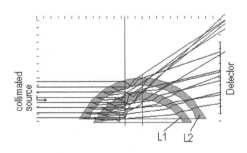

Fig. 8.10

Ray tracing is provided for a case of one and two lenses. The results are shown in Fig. 8.11, where radiometric analysis with a scanning annular aperture for a case of two lenses is also shown.

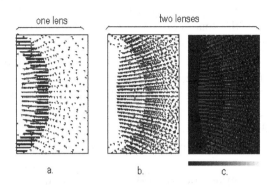

Fig. 8.11

a. is a spot diagram in a case of one lens only;
b. is a spot diagram for a case of two lenses and
c. is a race tracing radiometric analysis with a scanning annular aperture for a case of two lenses
The appearance of curved edges is due to the effect of Fresnel and Brag types of reflections

In a case of one lens, a deep valley appears. In the case of two shperical lenses, the appearance of one sharp edge is evident. In the atomic case, the shells will behave as optical layers with a gradual index, but the edge effect could appear in a similar way. If changing the wavelength, the index of refraction will change and the edge will be shifted respectively. However, this is still a spatial edge effect. To explain the edge effect of the mass attenuation coefficient, additional consideration must be taken into account. This is the atomic positions inside the solid cell, where the atomic polar axes may have different orientations, but following a strictly defined order. Then in the case of symmetrical angular distribution (as in case of one lens), the transmitted total flux will have symmetrical behaviour around the wavelength of the maximum transmission. In the case of asymmetrical angular distribution, (corresponding to the example with two lenses), the transmission as a function of the wavelength will have also asymmetrical appearance. Then the analogy with two or more lenses will be valid for the external shell of the nuclear building tendency, which gets a completion at Ar. The shell, however, begins to obtain a focusing effect when getting EB bonding. In the optical example, this shell could be associated with the lens L2. The lens L1 will be associated with the He nucleus.

The analogy could give only a rough impression of the picture inside the solid crystal. In the real case the spatial distribution of the rays is quite more complex because of number of factors: a gradient index change, the proton shape and the spatial positions of the atoms in the solid crystal and so on.

Investigating carefully the X-ray transmission and the characteristic X lines associated with the shells (next paragraph) we may conclude:

For X-ray range between 1 KeV and 100 MeV:

K edge corresponds to the shrunk CL space around the proton (deuteron) shell, which shell gets a completion at Ar atom

L edge corresponds to the shrunk CL space around the proton (deuteron) shell, which shell gets a completion at Kr.

M edge corresponds to the shrunk CL space around the proton shell, which shell gets a completion at Xe.

N edge corresponds to the shrunk CL space around the equatorial proton shell, beginning at Ta.

For X-ray range between 100 eV and 1 KeV:

The edge frequency depends on the crystal geometry and the spatial positions of the valence protons

Discussion about **additional features:**

The top of the L edge begins to appear slightly doubled from Cu. This could be explained by the CL shrinkage, which appears for Cu as a result of EB type of bindings. The doubling increases continuously, because the proton arrangement around the two end polar structures and the middle one becomes different.

The N edge never gets the sharpness of the previous edges. This is an equatorial proton shell, following different build-up principle. At Rn two similar equatorial shells exists, but the Actinides protons grow between them.

8.4.2.2. Characteristic lines of X ray spectrum

The spectrum of the characteristic x-ray lines is strongly related to the transmission edge position. Lot of experimental data are accumulated mainly in the first half of 20th century. Following the energy increase order, the lines are annotated as

a, b, g and so on. For many elements the a line is measured as a doublet. The energy condition for appearance of these lines requires the activation energy to be above a specific threshold value equal to the line quantum energy. This energy is little bit below the transmission edge energy. Ordering the line set by the wavelength, the a line appears with a longest wavelength, and strongest intensity. **However, one very interesting effect exists with the following features: 1) A monochromatic X-ray radiation (activation energy) applied in a broad spectral (energy) range leads to emission of spectral lines with stable wavelengths; 2) The increasing of the activation energy leads to intensity increase of the characteristic lines with a preserved intensity ratio between them.**

Similar effect does not exists in UV, VIS and IR spectral range.

At the transmission edge energy, the intensity of the characteristic lines begins to grow much faster. This effect was observed and extensively discussed by D. Webster (1916). In his and other discussions in that time, an energy storing mechanism is thought to have place. This energy storage mechanism works for the time the activation energy is present. The reason for such conclusion is the independence of the line wavelength from the wavelength of the activating monochromatic energy. At the same time the increase of the activation energy does not show discontinuity. In Webster's experiment the Rh lines: a (doublet), b and g, are studied for a pretty large range of the activation energy (from 20 to 32 KeV for a line). The plot of the a and b lines with the fluorescence radiation is shown in Fig. 8.12.

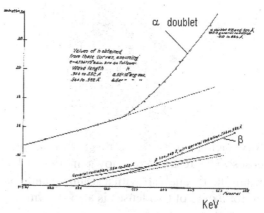

Fig. 8.12 (Courtesy of D. Webster (1916)

The beginning of sharp intensity growing is carefully investigated by Webster, and found to be 23.3 KeV +/- 1%. This value is very close to the edge energy for Rh, that has a value of 23.22 KeV.

The physics of the energy storage mechanism has not been explained so far. BSM analysis arrived to a reasonable conclusion about the possible mechanism. It is discussed in the next section.

8.4.2.3 Characteristic x-lines and energy storage mechanism according to BSM

For UV, VIS and IR spectral range, the photon emission and absorption is a result of electronic transitions between different quantum orbits. So this is an atomic type of photon emission and absorption process. The **weak electron oscillations** serve only to support the confined type of motion, as it was discussed in Chapter 3 and 7.

The mechanisms of X-ray photon emission and absorption is quite different than the atomic type. The photon emission in X-ray region is due to the **strong type of electron oscillations** (previously discussed in Chapter 3). From the analysis of the experiments we may conclude that:

- **The electron can emit X-ray photons during its orbital motion around the proton club.**
- **The X-ray radiation is not an electronic transition process between the shells. The emitted electron does not change the quantum orbit**

A possible explanation of the X -ray activation of orbiting electrons: The X-rays are able to penetrate inside the quasishrunk CL space (which is inside the Bohr surface) due to their shorter wavelengths. The characteristics lines are located around the transmission edges. For well defined sharp edges they are stronger. This means that the activation radiation is able to penetrate in the shrunk CL space and may affect the electron confined motion in the quantum orbit. Then a special interaction process may occur, when **the electron orbital length is an integer number of radiation wavelength (condition A)**. When this condition is fulfilled for the current electron orbit, the activation energy could affect the SPM vector of the space swapped by the electron magnetic radius. Then the electron has to fulfil simultaneously two quantum conditions: (1) keeping the same quantum number (from the short magnetic line condition (see section 7.7.1, Chapter 7); and (2) - meet the new modulation of SPM vector. The first condition is related to the positron-core oscillations, while the second one - to the electron spin rotation. The energy causing the discrepancy between the two quantum conditions is much larger in comparison to the quantum time conditions in CL space without external energy. **So this energy will cause strong oscillations of the electron structure (oscillations between the electron shell and its internal positron).** From Chapter 3 we know, that such oscillations lead to emission of X rays.

X-ray energy storage mechanism

There is a second feature. The continuous range of the activating energy means that the described process should be satisfied for some limited but continuous range of wavelengths. The possible explanation, is that the different wavelengths are distributed in the quasishrunk space around the proton core. The Balmer model in Chapter 7 shows good results for linear dependence of the quasishrink factor from the distance r (between the electron orbit in the circular part of the orbit and the proton core). The refractive index is a reciprocal to the quasishrink factor and also is gradual (but not linear). In conditions of gradual refractive index, the applied monochromatic radiation may be converted to standing waves, occupying volume with a finite thickness around the proton core. **Then the condition A, could be fulfilled for a volume adjacent to the quantum orbit, but from the side to the proton core.** These waves could be able to transfer their energy to the space volume of the current quantum orbit, due to the lattice stiffness gradient. The energy transfer is from larger to smaller stiffness. In such case, a finite range of wavelengths could be able to transfer their energy to the electron.

This kind of energy transfer could be possible due to the non-linear gradient of the SG field around the proton core. We may call this effect a **X-ray Energy Transfer Mechanism (XETM).**

The spatial position of the XETM in respect to the proton's core is shown in Fig. 8.13, where a section across the quasishrunk space around the proton club is shown.

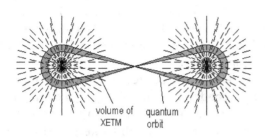

Fig. 8.13

Orbital range with conditions of standing waves
The range is around the GBclp bound protons

The volume, where the XETM could function is shown by a gray level. It is internal in respect to the quantum orbit trace. The quasishrunk CL space is around the two GBclp bound proton clubs. The position of the current electron orbit is also shown.

One question arises: Does this process possess a finite time constant, in order to be considered as an energy storage mechanism?

The available experimental data does not allowed a definite answer, but the possible time constant may be related with the modified CL time constant for the range of the quasishrunk space, where the effect may take place.

Characteristic line splitting

One important feature found in the experiment is related to the characteristic line splitting. Only doublets are observed. The author of BSM is not aware of observed triplets or multiplets. To explain this feature, having in mind the X-ray emission mechanism, we will return to Fig. 8.5.d. The drawing plane coincides with the equatorial atomic plane and shows the orbits around the GBclp bound proton clubs of a completed proton shell. (Every internal shell is completed). As a result of the orbit proximity, they could influence each other. The lowest energy they could get according to the Hund's rule is when they are paired. The arrows show the electron motion for such pairing. The applied radiation penetrates inside the Bohr surface of each proton guided by the proton helicity. While the proton helicity is one and a same, the electron motion in the neighbouring orbitals is opposite, because of the orbital interaction. For this reason the

line is doubled. Then the following conclusion could be made:

The fact that the line doublet appears clear and well separated when applying external radiation means that:

- all orbits in the shell obtain the same principal quantum numbers

- half of the orbits posses a right handed and another half - a left handed QM spin

- the above two conditions are valid only for completed proton shells with GBclp type of bonds in the equatorial region

From the above made conclusions, it follows also, that the simultaneously observed different lines come from different atoms. It also means, **that the lines could be split only in doublets.** This latter statement is in agreement with the observational data, but in contradiction with the QM methods for calculating the characteristic lines, that is based on electronic transition between the shells.

There is a lack of experimental data showing close triplets or multiplets, while a lot of doublets are observed. The multiplets if existing should appear very close, like the doublets. They should be not confused with the lines from other quantum orbits. The elements after La, although, could exhibit more complex line structures, because the condition for absence of additional equatorial bond is disturbed. In such case W, for example, shows a larger number of lines, but only one doublet is identified (A. Compton, 1916).

The pairing effect makes the characteristic lines to appear quite strong. Using this feature and XETM effect, X-ray lasers could be built.

Discussion:

It is a question if the *a* characteristic line is emitted from the lowest or higher level orbit of the series. When the activation energy scans from lower to higher energies, the order of the line appearance is *a*, *b*, *g*. Then the *a* line should correspond to the most external, from this set of three orbits. But then its strength will be strongly dependent on the temperature. The question could get an answer by suitable temperature data set.

8.4.2.4 Laue patterns

The Laue patterns are diffracted images, when a solid structure is radiated by collimated X -

rays. The solid structure should be carefully cut on order to match the particular crystal plane. Such images of pure elements are of particular importance. The interpretation of the results is complicated, due to the large CL shrink factor around the proton and neutron cores. But the polar symmetry is helpful in order to unveil some structural features. For this reason, the back-reflection and transmission Laue patterns are useful.

In the back-reflection patterns, the most intensive spots are due to the external shell protons (deuterons), whereas some weak patterns are from the nearest internal shell. The polar axis of the atoms may have different orientation even in one cell. In a proper orientation of the crystal to the collimated beam, when the polar axes of group of atoms become normal to the image plane, some spots in the Laue pattern image will carry an angular signature of the of the azimuthal arranged protons (deuterons).

From the Lauer Atlas, by Eduard Preuss et al., containing Lauer pattern for some elements, one may find that:

- The rotational polar symmetry of 45 deg (corresponding to 8 protons in a polar view) appears for: Ga(010), In(100), In(001), -Sn(001), W(100), Np(001), Np(100).

- For Mg the "45 deg pattern" is missing and this is a good sign, because it does not have an 8-proton shelf.

For elements with Z number larger than 12, the "45 deg" pattern may appear, because only one proton in a "45 deg" position could be able to contribute to the Laue pattern (due to the different atomic positions in the neighbouring crystal cells).

The transmission Lauer patterns are also useful. Figure 8.14 shows a transmission pattern of Fe under stress, provided by Wadlund (1938). The "45 deg" feature has a strong appearance. The origin of the radial lines has been in discussion in the time of the experiment, but not reasonable explanation is obtained. Now, having in mind the BSM model, the stronger appearance of the radial stripes in the transmission pattern are easily explainable. The stress may cause deformation of the Bohr surfaces, whereas the polar symmetry is preserved.

Fig. 8.14
Laue photograph of compressed iron crystal. Courtesy of Wadlund (1938)

Another quite informative experiment is provided by G. Preston et al. (1939). They measured the transmission pattern of aluminium, in cases of polychromatic and monochromatic X-ray beam, at room and at elevated temperatures. The images are shown in Fig. 8.15.

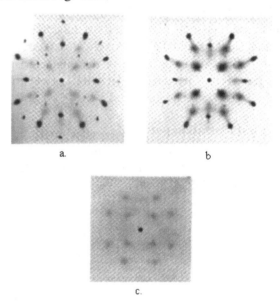

a.　　　　　　　　b.

c.

Fig. 8.15. Aluminium, X-rays parallel to [110] vertical a. room temperature; b., 500° C; c. monochrome Ag Ka radiation. Courtesy of G. Preston et al. (1939)

The spots whose appearance is stronger at elevated temperature have been a subject of hot discussions at the time of the experiment, but without good explanation. According to the BSM, these spots are due to the characteristic line radiation. At elevated temperature, the spots are quite more intensive and diffused than at the room temperature. At monochromatic radiation, the spots become smaller but well defined. This corresponds to an increased contribution from the characteristic lines. The fixed and symmetrical locations of the spots are in good agreement with the BSM atomic model and the concept of the X-ray radiation, described in the previous paragraph.

The X-ray crystalography technique could be very useful for confirmation of the unveiled atomic structures. The diffraction image is a result of two major factors: the focusing property of the CL space around the proton clubs and the Bragg conditions caused by the repeatable surfaces. However, it is necessary to distinguish the spots contributed by the proton shells from those provided be the atomic positions in the crystal. A proper orientation of the crystal is necessary in order to make the polar axes of group of atoms normal to the detector plane. One very useful feature would be the X-ray source to be monochromatic with an option of variation of the wavelength and the beam intensity. In such case, the spots contributed by the completed proton shells could be selectively observed and identified by the appearance of the characteristic line radiation. The azimuthal positions of electron orbitals also could be identified.

8.5 First ionization potential

The trend of the first ionization potential as a function of Z number for the first 21 elements from the Periodic table is shown in Fig. 8.16.

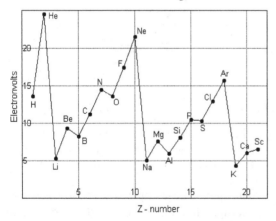

Fig. 8.16
First ionization potential of elements
for $1 < Z < 21$

The rising trend for the elements in one raw of the Periodic table is explainable by the increas-

ing number of the protons (deuterons) in the external shell. The sharp potential drops for Li, Na and K elements indicate the beginnings of new shells. Proton shells are compactified at Ne, Ar, Kr and Xe. At the same time, all electronic orbits in the completed shell become closer to the central region and their orbital planes are almost parallel to the equatorial nuclear plane. The orbital planes of the polar electrons are also parallel to the equatorial plane. For this reason, a removing of electron from the inert gas up to Rn requires a larger ionization energy. For Rn, the first ionization potential is much smaller in comparison with the noble gases with lower Z numbers, because the electron might be extracted from the weaker EB bonds corresponding to the Lantanide's built shell.

A less strong decrease of the ionization trend appears between Zn and Ga. This indicates that the EB bonds of the four previously bound protons are converted to GBclp bonds. The attached structure becomes identical to a Ne structure. Consequently, the protons of this structure are completely excluded from the valence of the element. A similar trends exist between Cd - In and Hg - Tl. This means that a similar effect takes place.

In the rising trend of the ionization energy within the first few rows of the Periodic table, the following anomalies are observed: Reversing of the potential trend between: Be - B; Mg - Al; N - O; P - S; As - Se. The Be - B trend indicates, that Be has two bound protons at one and a same polar region, corresponding to one type of Hund's rule pairing. (The pairing in fact is between the electron orbital planes, but they influence the proton position in the nucleus). The next bound proton for B is attached to the opposite pole, so it slightly disturbs the proton pairing obtained at Be. This facilitates the removing of the electron from that proton (for B).

A smaller reversal jump occurs between N - O and P - S. This might be also a conversion of EB to GBclp bonds. The possible explanation of the ionization potential drop is a following:

Let us analyse the magnetic field orientations, considering the plane of the electronic orbits (orbitals), for simplicity. In Nitrogen, the plane of the three valence orbitals (determined by the valence proton) are (approximately) normal to the equatorial plane. The planes of the two electronic orbits from the equatorial EB bonds are also approximately normal to the equatorial plane. Due to the proton twisting, both orbits are not completely normal between themselves, so some weak interaction could exist between them. In the Oxygen atom, the two GBclp are completely symmetrical and the valence protons also. Then the corresponding orbital planes are also symmetrical. The equivalent plane of the orbits of the GBclp could be considered exactly parallel to the equatorial plane. The equivalent planes of the two symmetrical valence orbits (at the proton's free clubs), however, appear always at angle in respect to the equatorial plane, so the interaction effect should be smaller. This means, that the valence electron should be ionised by a smaller energy. The possible configuration of Oxygen atom is illustrated in Fig. 8.16.B by two views: A and B.

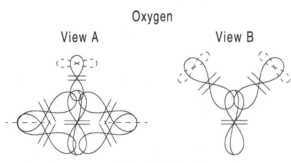

Oxygen

View A   View B

Fig. 8.16.B

The analysis of the photoelectron spectrum of the atomic oxygen, according to BSM, is also in agreement with the shown configuration (especially the autoionization features of the oxygen obtained by excitation with the strong neon line at 73.6 nm).

The same effect should be valid for the atoms of the same group: S, Se Te. For the group of halogens (F, Cl, Br, I), the equatorial bonds are three. Three EB bonds are able to generate closed magnetic lines around the polar axes. So the equatorial proton pairing is by EB bonds. This means a larger ionization potential, according to the above analysis. This is also in agreement with the plot shown in Fig. 8.16.

8.6. Atoms at different aggregate state of the matter

In a gas phase of the matter the atoms have a full freedom of motion, while the external protons (deuterons, tritium) has a limited freedom of motion relative to the atomic nuclear core. In the liquid phase the average distance between atoms (molecules) is a constant, but the protons still have some freedom. It is known that so called hydrogen bonds play a role in the liquid water. If taking into account the limited angular freedom of the valence protons (deuterons) and the orientational freedom of the molecules this is completely understandable for the BSM atomic models.

In solids **the valence protons (Deuterons, tritium) of the nucleus lose their freedom of motion, because they provide the interconnections between the neighbouring atoms in the crystal. This connection is different for the insulators and for the metals. In both cases the atoms are arranged in a crystal structure.**

The amorphous solids (like glasses) may contain not oriented domains, while the individual domain may contain a well defined crystal structure. The size of domains will depend on the rate of temperature decrease from a liquid to a solid phase.

The crystal structure of the insulators is different than this of the metals.

In insulators **all valence protons of the nucleus from one crystal cell are connected to the valence protons of the nucleus from the neighbouring cells by EB.** As a result of this, the solid state of insulator will not contain free electrons. The distances between the atoms in this way are usually larger and the Super Gravitation between the bound protons (deuterons) is smaller. In the space between atoms, the CL space may exhibit a local variation of its parameters, influenced by the atomic structures. This makes a suitable condition for free passing of EM waves and light, but with a reduced light velocity in comparison to the vacuum due to the slightly changed proper resonance frequency of the CL nodes inside the solid body.

Metals When analysing the differences between the external proton shells of the metals and insulators (for single elements only) we see that:

(1) There is a larger abundance of GBclp in the equatorial peripheral range for the metals, than for the insulators

(2) The GBclp of the completed polar structure Ar is always uncovered by valence protons

(3) The number of completed Ar structures increases (up to 3) with the row number of the Periodic table.

(4) The number of metals in the transition elements of the Periodic table gradually increases. So the boundary between the metals and semiconductors in the Periodic table is not a vertical line but a diagonal: Zn In Pb (Bi).

It is evident, that the number of the equatorial GBclp bonds play a role in the crystal cell of the metals. These bonds possess larger SG fields and may attract and align the neighbouring atoms. This conclusion is experimentally confirmed by the BSM interpretation of the observed gold crystal structures by T. Kawasaki et al. (2000). For this purpose the group of Kawasaki used a new developed transmission electron microscope with a resolution of 0.6 A (0.6×10^{-10} m).

Figure 8.17 shows the transmitted pattern for 15 nm thick Au film by successive vacuum deposition onto Ag layer (200 A thick). The comments for Fig. 2 (from the paper) are from the authors of the article. The images (a) and (b) correspond to two different crystal planes.

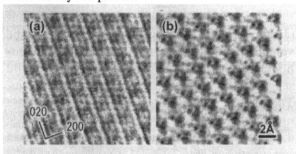

FIG. 2. (200) lattice fringe ($d = 2.04$ Å) images of a Au(001) thin film: (a) 1/3 spacing fringes formed from ($\overline{2}00$) and (400) reflections; and (b) fringes formed from many high-order reflections. The Fourier transform of the micrograph extends to ~0.5 Å

Fig. 8.17
Courtesy of T. Kawasaki et al. (2000)
BSM interpretation: Crystal (lattice) images of two crystal planes of Au

While the Kawasaki's group interpretation relies on the QM concept, the interpretation according to BSM concept is different.

According to the QM concept: The image is a fringe pattern contributed by the electron clouds

According to the BSM concept: The fine structures of the observed pattern are contributed by the electronic configuration, which comply to the nuclear structure, because all the electrons are around GB type of bonds. Consequently, the fine structure details of the observed pattern carry a signature of the atomic nucleus of the Au (gold) atom.

The shape and the configuration of Au atom, according to BSM is shown in Fig. 8.18, where **a.** is the polar section and **b.** is one polar view. In the polar section, only the protons intercepted by the section plane are shown for clarity.

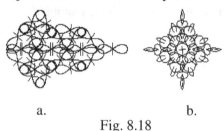

a.   b.
Fig. 8.18
Au nucleus: a. - polar section; b. polar view

The single proton (Deuteron) in the right side of the view **a.** is the valence one. The valence electron (not shown) is orbiting around the free proton's club. From Fig. 8.18 we see, that the shape and dimensions of Au nucleus are completely defined, when knowing the proton dimensions. This gives a possibility to explore the possible arrangement of the Au atoms in different crystal planes using a real scale. Fig. 8.19 and Fig. 8.20 provide the possible synthesized images corresponding to the crystal planes for the images in Fig. 8.17. The scales are in dimensions referenced to the proton's length.

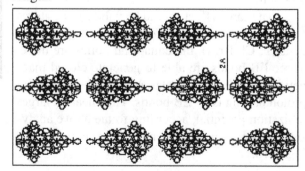

Fig. 8.19
Synthesized image of Au in one section of the crystal plane

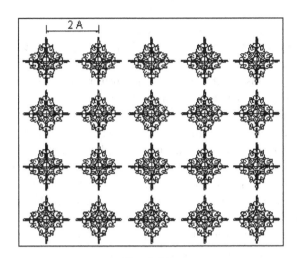

Fig. 8.20

Synthesized image of Au in another section
of the crystal plane

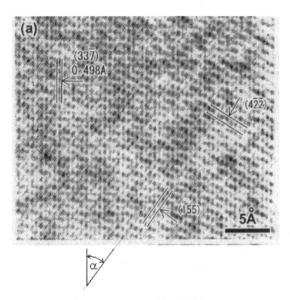

Fig. 8.21
Electron micrograph of Au(111)
(Courtesy of T. Kawasaki et all. (2000)
Note: The angle α below the image is
is added by the author of BSM

neighbouring atoms. Then we again come to a conclusion, that pure SG type of binding may exit between the atoms in solids.

Comparing the synthesized images to the experimentally observed, we see, that the dimensions between atoms and patterns match quite well. From the image in Fig, 8.19 we see, that the valence protons are adjacent, and consequently connected by EB bond. In the opposite side of the nucleus, however, EB type of bonds are not possible and the possible connection is only by SG forces or this is a kind of SG bond between the separate atoms. This corresponds to one type of the Wan Der Wall forces. The alignment between neighbouring rows is possible only by the SG forces between GBclp bonds in the equatorial region. The softness of the pure gold is a result of the SG forces between the adjacent gold nuclei, which provide much larger freedom and switching between different nuclei.

Fig. 8.21 shows another image with a different resolution from the same authors (T. Kawasaki et all. (2000)).The angle α between the line patterns is not 45 deg, but 40 deg. The same angle appears also in Fig. 8.17. This feature indicates, that the cell of four adjacent atoms does not has a rectangular shape, as shown in the synthesized image of Fig. 8.20. The cell is slightly rhomboidal. The only possible nuclear feature that may cause the rhomboidal shape of the cell is the nuclear twisting along the polar axis. This twisting, however, could influence the shape of the cell only by the interaction between the equatorial GBclp bonds of the

The GB bonds between atoms are effective, if the atoms are at enough close distance. The electronic orbits of EB bonds probably counteract for such proximity keeping the atoms apart, even if their polar axes are aligned. The EB bonds start from the group 11 of the Periodic table and their number progressively increases in the group of halogens. The boundary of the metals - semiconductors however is not vertical from the rows 4 to 6, but follows a diagonal, because the elements of every consecutive row after number 4 have one more additional Ar like structure with their GBclp bonds.

One additional question should be replied for the metals: *How the "gas" of free electrons is formed in the metals?*

When the atoms are separated, like in the gas phase, every electron is orbiting around its proton.

In the solid phase of the metals, some electrons obviously should be permanently free due to some physical mechanism. Using again the example with the Au crystal structure, we see, that every four proton pairs with EB are in close distance. The polar proton (deuteron) have some freedom of angular motion. By small tilting of this proton, it is possible every four pair of EB bound proton to form a **cluster of closely spaced protons**. The increased SG field of this cluster might be able to synchronize the dynamical interactions between the internal RL(T)'s of the proton shells and to control the common charge of the cluster to 2 unit charges, for example. Then this volume will be served by one pair of electrons with a different QM spin. We may call this hypothetical effect a **free electrons effect**. As a result, six electrons will become free, contributing to electron gas formation. In other metals, the free electrons effect is also possible, when more than two valence electrons are in proximity. The effect is possible not only for valence electrons, but also for EB bound protons (of one atom). The pairs of EB bound protons are not necessary to be unbounded. In proper orientation of the atomic nuclei they may become close enough in order to form a proton cluster and to allow a free electron effect. This conclusion could be confirmed by the example with the electrical conductivity for Fe and Cu. Their possible cell configurations can be found using the configurations of Fe and Cu given in the Atlas of ANS. Both, Fe and Cu, have almost the same mass density, but very different electrical conductivity. At the same time, Fe has more valence protons, than the Cu (only one). The copper, however, has four EB proton pairs, while the iron does not have such pairs.

From the above made analysis it becomes apparent that the metallic feature of the elements appears only when they are in a solid aggregate state.

In semiconductors, **the external protons are closer in comparison to the** *isolators***, but not closer enough to form clusters like in the** *metals***.** When the applied field exceeds some level, some electrons could migrate from a local atomic orbitals to the orbitals of the neighbouring atom.

The hadrons involved in the nuclei (protons and neutrons) are immersed in a CL space. The average internuclear distance, normalized to the nuclear size, is a largest one for the insulators, a smaller one for the semiconductors, and a smallest one for the metals. The non spherical shape of the atoms and the internuclear bonds explain the orientation of the atomic nuclei, or atoms, in order to provide different crystal planes. This is apparent by their Laue patterns. Different metals have different hadron concentration, which modulates the CL space to a different level. So the CL space inside the solids exhibit a spatial nonuniformity even for a crystal structure from one and a same element. Such uniformity may increase significantly if the solid body contains impurities. The obtained in a such way local nonuniformities play a significant role in the superconductivity state of the matter (this was mentioned in Chapter 4). Compounds of different elements, mostly metals may get larger local nonuniformity of the internal CL space. They are of increased interest in the area of high temperature superconductors.

8.7. Nuclear magnetic resonance applied for atomic element

The Nuclear Magnetic Resonance (NMR) technique is applicable for the single atomic elements and for chemical compounds. In NMR technique, the atoms are in a very strong homogeneous permanent (DC) magnetic filed. A RF radiation is applied in a direction perpendicular to the DC field and a signal is detected by a coil the axis of which is usually perpendicular to the DC and RF field. The RF field causes the individual atom to rotate, but then it interacts with the strong magnetic field. As a result of this interaction, the whole atomic structure obtains oscillating motion with a precessional component due to the interaction of its close proximity fields (possessing a helical feature) with the external fields.

It is obvious, that the orbital planes should plane important role in the precessional motion. This means that the electrons participate in the complex motion by transitions between allowed energy levels. The contribution of the external valence protons with their orbiting electrons should be stronger as the relative electron speed of these electrons compared to the nuclear electrons (referenced also to a local rest frame) is larger. By detecting the signature of the precessional motion, a spectrum is obtained, known as a NMR spectrum. When a more precise scanning is performed, the signature of the electron transitions appears. This technique is known also as an Electron Paramagnetic Resonance (EPR).

The NMR can be applied for elements and for molecules (chemical compounds). When applied for elements, one specific feature deserves to be mentioned. **The detection efficiency is optimal, when the RF filed is applied at 54.4^0 to the strong magnetic field (instead of at 90 deg). (This feature indicates that the atomic nucleus possesses a helicity, which is an indication that the proton itself possesses some helicity).**

The angle of 54.4^0 also provides indication about the twisting feature of the helical structures see §6.9.4.1.3, Chapter 6).

8.8. Giant resonance

Presently heavy atoms mostly are used for experiments related with the nuclear giant resonance. The BSM explanation of this effect is quite straightforward. The effect is caused by the vibrational motion of the neutron, which is over the proton saddle. Therefore, following the Z-trend of the Periodic Table, the Giant resonance will become apparent for the deuteron and its signature will become more complex with the increase of the Z-number. The intensity of this type of resonance is quite high because the large neutron mass is involved. The resonance peaks in the heavier atoms are contributed by the separate groups of the Deuterons and tritium.

8.9 Scattering experiments

The shape of the atomic nucleus according to the QM atomic model is close to sphere with dimensions much smaller, than the electron orbits. This assumption came initially from the interpretation of the Rutherford's experiment, provided in 1906. In this experiment α particles (He nucleus) from radioactive source strike one or pair of golden foils at normal incidence. The angles of their deviation from the initial direction are measured by large photographic plate behind the foils. Most of the alpha particles passed clear through, unaffected and undiverted, recording themselves on the photographic plate behind. There were, however, some particles that were scattered even through large angles. Since the gold foil was about two thousand atoms thick, and since most alpha particles passed

were not deviated, it would seem that the atoms were mostly empty space. Since some alpha particles were deflected sharply it has been interpreted that somewhere in the atom must be a massive, positively charged region, capable of turning back the positively charged alpha particles. Rutherford concluded, that the atomic nucleus is very small but containing all mass and the total positive charge. He admitted that the atom has a planetary model with a mass concentrated in the small spherical nucleus, while the electron circles around the nucleus as planets. In the contemporary physics this concept has not been change too much since the time of the Rutherford's experiment.

The data analysis from the Rutherford's experiments is based on the following considerations:

A. QM considerations: The following assumptions are adopted:

- the nucleus has a simple spherical structure and its positive charges are located in an extremely small volume inside the sphere. (later the charges are considered distributed in the nuclear volume);

- the alpha particle does not have a structure;

- the inertial mass and the Coulomb force (according to inverse square law) are unchanged at small distances (below the Bohr radius)

Applying these considerations the Rutherford experiment leads to a nuclear size of the order of 10^{-15} (m).

According to the BSM, the following considerations, should be used for correct interpretation of the data from the same experiment:

B. BSM considerations: The following features must be taken into account:

- the complex shape of the atom of Au and the distributed E-field inside the Bohr surfaces

- the crystal structure formed by the Au atoms

- the shape of the He nucleus and its helicity

- the complex structure of the proximity E-fields around the individual atoms

Additionally, inside the golden foil the CL space is not uniform but spatially modulated by the hadrons building the atomic nuclei, so it may form preferable path channels for the He nuclei

Using the real atomic models of BSM, it is evident that the interaction mechanism is much more complicated and any simple model for data interpretation is not adequate. For such reason the used so far interpretation of the scattering experiments may lead to a result quite distant from reality. One simple example is if considering the toruses (also folded or twisted) instead of sphere with a similar core thickness like the sphere. The experimental results from scattering of such structure will be similar.

Figure 8.22 helps to provide some qualitative insight about the interactions between He nuclei and the golden foil. The drawing shows only one layer of the crystal lattice structure of Au, corresponding to the left panel of Fig. 8.17 (T. Kawasaki et all. (2000)).

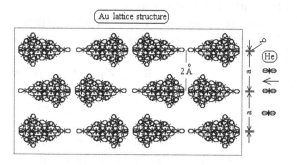

Fig. 8.22 Illustration of the Rutherfords scattering showing the atomic nuclei with their structure (Au thin foil irradiated by α particles, which are He nuclei.)

The dimensions of He and Au nuclei are given in one and a same scale. The width of the "pass through" zone is annotated by *a*. Due to the helical interactions between the alpha particles and atomic nucleus (by their fields), the atomic crystal provides a guiding of the alpha particles. For this reason a small amount of alpha particles are deflected. The channelling structure of the metal crystal will affect tremendously the propagation of the particles in such type of scattering experiments. The data interpretation of the Rutherford's experiment according to BSM theory is quite different.

8.10 Three dimensional view of the atomic nucleus

Providing an accurate three dimensional view of any atomic nucleus is not so simple, due to the nuclear helicity. Fig. 8.23 and 8.24 show two views of a mechanical mock-up made by steel springs. The upper part of the structure in Fig. 8.23 corresponds to Ar atom (only four neutrons over equatorial GBclp are missing). The lower part is a modified He structure with EB bonds. The same figure shows, also how the lower structure is clamped to the upper Ar-like structure by GBpc bonds. If one additional Deuteron is bound in the bottom of the whole structure, the shape of Cu atom will be obtained.

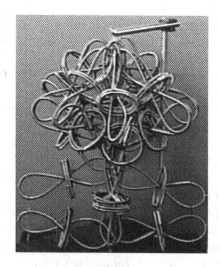

Fig. 8.23

Front view of mechanical mock-up

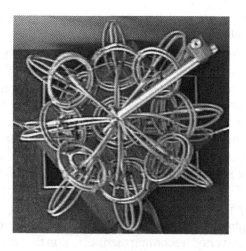

Fig. 8.24

Polar view of mechanical mock-up

8.11 Electron series in the atoms at larger Z number

The electron orbitals of Hydrogen atom appear in a similar way in other atoms, with exception of the Helium. The two electron orbitals of the Helium lie in two well separated planes parallel to the equatorial plane of the nucleus. They may interact magnetically (by the QM spin), but any transition between them are not possible.

The appearance of the Hydrogen-like orbitals in all other atoms exhibits some restriction due to the complicated nuclear configuration. Balmer series is one of the most important series. It appears in the free end club of any valence proton (not involved in a chemical bond). The orbital quasiplane of the Lyman series, however, has to pass through the polar region. Since the valence protons are interconnected in solids, this series could restricted in other aggregate states of the elements: quite much restricted is solids and less restricted in liquids.

While the orbitals in heavier atoms are subordinated by similar quantum orbital rules as in the hydrogen (and deuteron), their energy levels are modified due to the following additional conditions:

- **The Ground State (G.S.) appears shifted and the series range spanned as a result of redistribution of the Super gravitational and electrical field inside the integrated Bohr surface.**

- **The orbital interactions between electronic orbits bound to different protons change the conditions for the electron transitions**

8.12 Spin orbit interaction

The valence electron behaviour of the I-st group alkali atoms is similar as in the hydrogen. The ions, which have lost all the electrons but one, have also the same type of Hydrogen-like spectra (only different energy levels). It is evident, that the electron transition behaviour for all atoms having only one electron in their external shell is similar. In other words, the levels are not split. This means, that the electron from the G.S. for example jumps to upper orbit, when a photon is absorbed, and after the elapsing of the quantum orbit time, falls more probably to the G.S. or less probably on other orbit.

Let us use the terms **"excited"** and **"de-excited"** for characterization of the electron behaviour.

The individual electron behaviour in a neutral atom with more than one electron in its external shell is not identical to the electron behaviour in the hydrogen. Let us consider the electron behaviour in Be atom for the case of spontaneous emission. The configuration of the Be atom with its two valence protons and their electronic orbits is illustrated in Fig. 8.25.

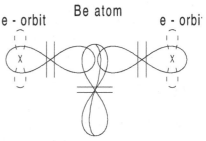

Fig. 8.25

Spin orbit interaction in Be atom. The planes of the two electronic orbits shown by a dashed line are in fact normal to the drawing.

The size of the quantum orbits shown in Fig. 8.25 corresponds to the Balmer type series (but the energy levels are distinctive from the Balmer series level of Hydrogen, because the CL space surrounding the nucleus is shrunk). Let us assume that the two electrons have been in Balmer G.S. orbit, so their QM spins are parallel (matched). In some moment the first electron, for example, is excited and jumps to some upper orbit above the Balmer G.S. After its orbital time is elapsed, it must return to the Balmer G.S. orbit. This process could not be momentary. It should take a finite time, nevertheless, that it may be smaller, than the orbital time. The electron has to pass through a new region of the CL space, not procured like the orbital trace, where it also exhibits different SG forces and inertial mass. The latter parameter is affected by the different value of $\lambda_{SPM'}$, whereas the CL node distance is the same (neglecting the general relativistic CL space shrinkage, that is negligible). The former orbital trace contains CL space energy pumped by the electron quantum motion in a closed loop. This energy could not be contained any more in this space and has to be emitted (as a quantum wave). The pumped energy, however, has to overcome the

Bohr surface barrier. The reaction effect of this barrier, causes, some energy to be reflected back. It is reasonable to consider, that the Bohr surfaces of the paired protons are integrated. In such case, the reflected back energy could be temporally received by the second electron due to magnetic field interaction. Let us suppose, that this energy is smaller in order to excite the second electron to a higher orbit. However, it may be enough to switch its QM spin in anti parallel state. This change will influence automatically the E-filed distribution of the first proton. As a result of this, the quantum condition of the GS orbit will be satisfied for distance from the proton core different, than the initial one (estimated in absolute units of CL node distance). The gradient of the SG field near the proton core is highly dependent on the distance, so small distance change in absolute units means a large potential shift. As a result, the GS level appears shifted. This shift in fact appears symmetrical for the second electron as well. The next excitation for one of both electrons may start from this shifted GS level.

Two additional options exist between the QM spin of every electron and the proton handedness. Their handedness may match or mismatch. Then four combinations are possible (two - from the orbital magnetic field interactions and two - from the electron spin). They define four different GS levels. This is in agreement with the optical atomic spectra. Indeed if we examine the Grotrian diagram for a neutral Be atom, we see, that the lower energy state corresponding to a quantum number 2 has four levels: 2(0) for ns^{1}S; 2(1) for np^{1}P^o; 2(0) for np^{1}S and 2(2) for np^{1}D.

The above described process is a QM spin-orbit interaction in the atoms. One must keep in mind that this is a quantum mechanical spin, attributed to the orbital motion of the electron. It should not be confused with the electron spin momentum described in BSM as a classical feature of the rotating electron.

For atoms possessing a larger number of external shell protons, the QM spin-orbit interactions could provide a large number of level splitting and level shifts. Large shifts are possible especially for the elements from group 14 to 17, because the EB bound external protons may combine positively their QM spins.

8.13 Identification of orbits according to QM notation

The calculations of the atomic and molecular spectra by the Quantum mechanical methods are based on the electronic orbital configuration. They are separated in shells and subshells. Without going into details, we will show that the orbital planes in the BSM atomic models have features that can be attributed to the classification of the QM models. The BSM models, however, are informative about the spatial configuration of all electronic orbits, their freedom and orientation. Presently, the BSM models do not have a developed mathematical technique for a spectral calculations as the QM model, but this is not so important. The main advantage of the BSM atomic models is the possibility to obtain the configuration and the mutual positions of the orbital planes. At the same time they possess the same energetic levels of the excited electronic states and the same ionization potentials, as the QM models.

Figure 8.25.A shows the graphs of the experimentally determined ionization energies of the first 21 elements of the periodic table.

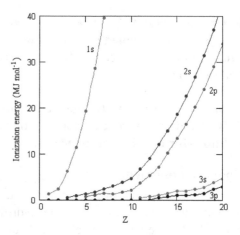

Fig. 8.25.A

Ionization energies of the first 21 elements

The trend between the different states denoted as 1s, 2s, 2p, 3s, 3p are quite apparent. But this trends clearly show some features of the BSM model. They are the following:

The QM orbits denoted **as 1s orbits** correspond to the polar orbitals. The orbital planes are perpendicular to the polar axis (the cosine of the plane vector in respect to the polar axis is equal to 1). They are located in high gradient SG fields. Their quantum levels are strong and grow faster with the Z number, because every new added proton increases the polar SG field. For this reason their QM interaction are also strong and they always have an opposite QM spins. For this reason the quantum level of these two orbitals are exactly the same.

2s and 2p orbits from $_3$Li to $_9$F: corresponds to the **Balmer type orbits connected to the free club of the valence protons (see Balmer series model in Chapter 7).** Having in mind that the potentials are estimated by a photoelectron spectral technique, it is apparent why the trend of *2s* is above the trend of *2p*. It is a result of the Hund's rule orbital pairing. The electron of unpaired orbit is removed easier, than the paired one.

2s and 2p orbits from $_{10}$Ne to $_{18}$Ar: correspond to orbits of the equatorial GBclp bound protons (deuterons). Every orbit contains two electrons with opposite QM spins around two proximity connected proton clubs. For this reason, their quantum energy grows much faster. The spin orbital interaction between these orbits is also much larger, due to their well fixed positions and symmetry in respect to the polar axis. The larger QM spin interactions also contribute to the difference between *2s* and *2p* trends. This is in agreement with the considerations about the strong line pairing in the characteristic spectral lines in X spectral range discussed in §8.4.4.2

3s and 3p orbits from $_{11}$Na to $_{17}$Cl: correspond to Balmer type of orbits connected to the free proton clubs. They are much more peripheral and their removal is easier. Even EB bonds (Hund's rule pairing) are not so strong, so the difference between *3p* and *3s* is much smaller.

Figure 8.25.A shows mostly the bottom range of the Ionization energy in order to emphasize on some features. We may see (a good plotting accuracy is needed) that *2* and *3* (*s* and *p*) trends are not very smooth at some Z numbers.

There is a change of the *2s* trend and especially the *2p* trend for Z = 7. This corresponds to the oxygen, which has two symmetrical GBclp in the equatorial region. The planes of corresponding orbits have cosines in respect to the polar axis approaching 1, as in the case of the *1s* orbits. The QM

spin interactions in such condition are quite strong. Their ionization energy corresponds to the state *2s*. When one electron is removed, the common symmetry is significantly deteriorated and the ionization energy is much lower. This may explain the ionization energy drop of the *2s* trend.

There is a slop change of *2s* and *2p* trend between $_{10}$Ne and $_{11}$Na. At Ne, all protons from the previous external shell becomes GBclp bound in the equatorial region. The number of these protons is even and the normals of the corresponding orbital planes obtain a cosine close to unity. All the electronic orbits of Ar are well aligned, so they may obtain a common QM spin that could be right or left handed in respect to the proton handedness. **For this reason the QM spin has a strong feature apparent in the electronic spectra of argon.** In $_{11}$Na a new proton shell is started. The *2s* and *2p* states for Na lie little bit above the common trend. In such case a possible equatorial shrink of the proximity CL space for the two diametrically situated protons is not excluded (this issue may need a special discussion).

The trends of *3s* and *3p* show a slight change at $_{18}$Ar. For this atom all 16 protons (deuterons) from the external shell are equatorially bound in 8 GBclp type bonds. The difference in the trend is not large because ^{40}Ar has four external neutrons over the four GBclp, while other four GBclp do not have such neutrons. The neutron over GBclp facilitates the electron removal and this contributes to the smoothness of the trend.

Another apparent change of trends *3s* and *3p* is between $_{15}$P and $_{16}$S. This is caused by the GBclp of two proton pairs in the equatorial region in S. In fact the effect is similar as the described above effect for the oxygen atom.

Despite the significant change in the nuclear configuration we see, that the trends of *2s, 2p, 3s, 3p* and so on look pretty smooth. This may lead to the following conclusions:

    - the Bohr surfaces of the internal proton shells are well integrated

    - the smoothness of the ionization energy curves indicates that the SG field is able to redistribute the E-field in the integrated Bohr surfaces to some extent, so the positive charge in the far field looks like a point charge

The above made conclusions are additionally confirmed in the course of BSM.

Discussion: The QM model of electronic configuration of the atoms provides information mainly about the common orbital plane orientation. The BSM model provides the full physical picture of the orbital planes: spatial position and orientation. It is evident that the orbital shells and subshells identified from QM model do not coincide with the physical shells of the proton arrangement and connected to them electronic orbits according to BSM models. This discrepancies increase with the increase of the element Z-number.

It is worth to say that despite the mentioned discrepancies both, the QM and BSM models exhibit the following common features:

- The protons arrangement in shells matches exactly the periodic table of the elements

The BSM model additionally contributes one useful feature that is not apparent in the QM models.

- The orbital plane positions, which are the most important factors for QM spin interactions, are defined by the proton configurations in the nucleus

8.14 Ions

The ions are atoms that have lost or accepted electrons. In the first case, the atom is converted to a positive ion, while in the second case - in a negative ion.

In neutrals, the electrical fields of the system of "protons - electrons" are compensated inside the Bohr surface, so outside of it the atoms appear neutral, despite the different structures of protons and electrons. The neutralization effect is a result of dynamical interaction between the proton static field and the electron's dynamic field. In such case in the absent of disturbing external field we may consider that both - the electrical and the magnetic field of the system are locked inside the Bohr surface. The conclusion made for a system of one proton and electron (H atom) should be propagated for a neutral atom of higher Z-number. Then the concept of Bohr surface defined for Hydrogen could be applied for any neutral atom. Having in mind that the proton clubs in the polar bonding region are very closely spaced, it is reasonable to accept that the in-

dividual Bohr surface of the protons for atomic nuclei with $z > 1$ are integrated in a **common Bohr surface**. This conclusion is in agreement with the concept of the charge appearance in the positive ions, discussed in the next paragraph.

8.14.1 Positive ion

The lost electron might be from a valence proton (external shell proton) or from the internal shell. In the second case, known as Auger effect, the missing in the internal shell electron is replaced by some electron from the valence shell. So we may consider only the case of ions, obtained by losing of valence electrons. From the experimental investigation of the structure of ionic crystals (for example NaCl) and from the anion and cation components in the electrolytic solution, it appears at first glance that the charged ion behaves as a point charge. But at the same time we see that elements possessing a valence shell do not have completely symmetrical nuclei. Then a question arises: how does the positive ion appear as a point-like charge? The physical explanation of this effect is possible if admitting that:

(A) The Bohr surface of the atom is integrated from the individual Bohr surfaces of the protons involved in the nucleus.

According to conclusion (A) the common Bohr surface will sense the missing charge. Inside the integrated Bohr surface some E-filed gradient may exist, due to which any missing Auger electron is replaced by some valence electron. But even one missing electron from the external (valence) shell is able to disturb the integrated Bohr surface. As a result of this, a leak of E-field could emerge trough it. At some finite range from the nuclear centre, the electrical lines get a spatial rearrangement, so the positive ion appears as a point like charge in the far field. The accepted concept leads to the following conclusion:

(B) It is not possible to estimate correctly the atomic nuclear radius from the ionic type of bond.

A positive ion of Na is illustrated in Fig. 8.26

Fig. 8.26

Na$^+$ ion illustrated by a polar section. The
Ne nuclear structure is shown as a dashed oval.

8.14.2 Stable negative ions

Some elements may have stable negative
ions, while others not. The halogens, for example
have stable negative ions. They may participate in
ionic molecular bonds. Other elements, like oxygen, for example, may form a metastable negative
ion.

The explanation of the stability of the negative ion is not so simple as the positive one. Why
the negative ion for some elements is stable, despite the disturbance of the atomic charge neutrality? The stable existence of the negative ion perhaps
involves some mechanism opposing the disturbance of the integrated Bohr surface and assuring a
suitable orbit of the accepted electron.

Let us consider a negative ion of Cl, as a typical representative of halogens. It is not difficult to
guess, that the accepted electron will share the orbit
of the valence electron, but the partial freedom of
the proton may allow it to obtain position similar as
the EB bound protons. This case is different from
the EB bond in a neutral atom where the common
orbit is shared by two electrons. In the latter case
the position of the orbit is kept by two protons. In
the case of negative ion (Cl$^-$) the valence proton,
whose orbit is shared has some angular freedom in
the polar plane. It is reasonable to consider, that
both electrons occupy the Balmer orbital quasiplane. The two electrons will not only neutralise
(dynamically) the positive EQ's around the free
proton club, but will create a negative EQ's in a
close proximity range. Then the free proton club
may approach the equatorial region. **In this position the shared orbit may be oriented in a same
way as the other orbits of the equatorial EB
bonds.** (Notice, that the pairing for Halogens is by
EB bonds). Then the electron configuration of the

external shell may obtain more completed symmetry with the involvement of the accepted electron.
This means, that **the position of the shared orbit
of the valence proton might be kept fixed by the
common magnetic field from all orbits inside the
integrated Bohr surface.** This provides stability
of the negative ion. This is kind of quantum mechanical orbital interaction.

- **The stability of the negative ion of the halogens, might be a result of the quantum
  mechanical orbital interactions inside the
  integrated Bohr surface.**

The possible configuration of the chlorine
negative ion is shown in Fig. 8.27.

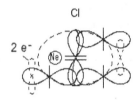

Fig. 8.27

A possible configuration of Cl$^-$ ion illustrated
by a single polar section The Ne nuclear structure,
which is embedded in Cl atom is shown as oval.
The quantum orbit in the left side contains 2 electrons
with opposite QM spin (circling in opposite directions)
while at the same time is magnetically confined to
other 3 EB orbits in the equatorial nuclear region.

8.14.3 Size of the positive and negative ions.

According to the BSM concept, the real nuclear size of the ions does not corresponds to the
size of the ions determined by the extension of the
charge. This is evident from the comparison between Na$^+$ and Cl$^-$ ions. The nuclear structure of
Na$^+$ appears larger than the nuclear structure of Cl$^-$
In the existing so far data about the ionic radii,
however, the radii of negative ions are given larger.
The obtain discrepancy may have the following explanation:

(1) The ionic radius measured experimentally
is a signature of the integrated Bohr surface rather
than the nuclear structure. In the ions, the integrated Bohr surface is disturbed, when compared to the
neutral atoms. This means that the leaked E-filed
lines in ions are rearranged until obtaining some
equivalent radius, which is always larger than the
nuclear structure.

(2) The larger size of the negative ions could be also a result of the asymmetry between the negative and positive prisms that propagates into the CL space features.

From the provided concept, it follows that the nuclear radius could not be estimated from the ionic radius.

8.15 Some aspects of photon emission and absorption

The physical aspects of the quantum wave has been discussed in Chapter 2. The mechanism of photon emission and absorption presented by the Balmer model in Chapter 7 is valid for the other atoms but with additional considerations about the influence of the SG field and the spin-orbit interactions discussed in §8.10.

The most accessible orbits for emission and absorption are those corresponding to the Balmer series in Hydrogen, which are physically connected to the free clubs of the valence protons (deuterons). Due to the SG field, their quantum levels appear higher than in the Hydrogen. The quantum orbits around the equatorial EB and GBclp bonds also are able to emit or absorb photons. Finally the two polar electrons (*s* electrons according to QM) also could generate photons at much shorter wavelength range, due to the larger SG field in the polar region (because of concentration of larger number of protons). But all of this quantum orbits are not able to generate quantum waves (photons) in the X range of the spectrum. Such quantum waves are possible to be generated by the electron-positron oscillations of the orbiting electrons.

One specific feature of the quantum orbits connected with the valence protons is their possibility to be polarized. This means, that they may absorb a polarized photon and emit a polarized one. This is especially valid for the transition between the Balmer GS and the upper quantum level.

One example of such emission is the resonance scattering of Na at 689 nm. If the transition is activated by a laser source with a strong polarization, the emitted photon appears also polarized. The polarized photon exhibits E vector modulated strongly in one direction. This feature is valid for the whole length of the photon's wavepacket. It is quite reasonable to expect that the plane of the orbit

activated by a polarized laser source is tilted from its normal position. This is illustrated by Fig. 8.28.

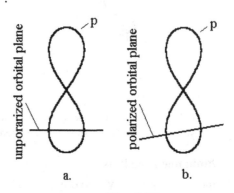

Fig. 8.28
Free valence proton with unpolarized (a) and polarized (b) orbit involved in a resonance scattering process

The process of resonance scattering helps to unveil the physical mechanism of the photon emission and absorption. Let us analyse only the absorption mechanism for the example of Na scattering at 689 nm. The tilting of the orbit points out to two features of the process:

(a) the energy distributed in the wavetrain with transverse diameter of about 689 nm shrinks about 20000 - 30000 times in order to dissipate in the range of magnetic radius swapping the orbit trace.

(b) the process of energy shrink is quite fast; perhaps the SG field interaction process is directly involved

(c) the energy shrink process may be regarded as wavetrain winding: but with a preservation of the spatial information of the polarization

(d) the alignment of the wavetrain shrinkage and the long axis of the valence proton is apparent

(e) in the final moment of the shrunk wavetrain, the atomic nucleus may play a role of a reflector

The feature (b) indicates, that the process of wavetrain shrinkage could not be regarded as a simple winding, but more complicated processes related with the quantum quasishrink effect. This effect appears as static in the case of atomic spectra. It appears as dynamic in the case of molecular spectra, which are discussed in Chapter 9.

The above features could give a bare picture how the absorbed polarised radiation (mainly by laser source) could cause a tilt of the orbital plane.

8.16 Ferromagnetic hypothesis

Let us discuss one known feature related to the explanation of the end product from the nuclear fusion (synthesis) and fission (depletion) reaction. Why the nuclear synthesis and depletion, both lead to the Fe element? Does this feature have some relation to the CL space properties in the vicinity of the atomic nucleus? How the shrunk CL space around the atomic nucleus communicate with the external CL space?

It is known that Iron is one of the very abundant elements in metheorites. The explanation is that both types of nuclear reactions: synthesis and depletion lead to this element. What is else specific for Fe element? Fe is highly feromagnetic. The ferromagnetism appears quite strong also for its neighbours in the Periodic table Ni and Co. In a proper mixture of them the magnetic susceptibility even increases. The opposite effect of the ferromagnetism is the diamagnetism. A proper parameter characterizing the elements for both phenomena is the Magnetic susceptibility of the elements. The magnetic susceptibility of the elements as a function of the Z number is shown in Fig. 8.29.

Note: The magnetic susceptibility parameter is valid when the element is in a solid state.

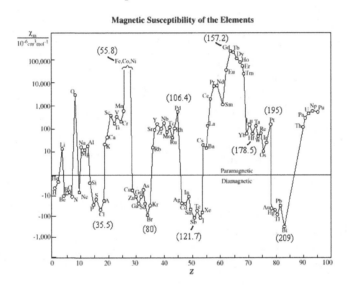

Fig. 8.29. Magnetic susceptibility of the elements

For ferromagnetic materials, the magnetic susceptibility is a positive large number, while for diamagnetic - it is a negative one.

Let us analyse the trend of the magnetic susceptibility as a function of Z number. It has the following important characteristics:

(a) the trend of magnetic susceptibility exhibits a periodical feature but the period is not constant

(b) the positive peaks of the trend are much larger

(c) the trend exhibits a steep change between some neighbouring elements of the periodic table

In Chapter 10 the concept of a local CL space (locally modulated) around the FOHSs of the proton and neutron is discussed. Different elements have different nuclear binding energy, which is expressed by the mass deficiency. According to the BSM concept, even a single atom slightly modulates the surrounding CL space (GR effect in a proximity to the atomic nucleus). In the solid state of the matter one additional feature is added: the atoms are connected in a crystal. This factor is much stronger for the metals where the interatomic distances are smaller. For any macrobody in the Earth gravitational field for which the mass is much lower than the Earth mass the Earth CL space (a modulation of the galactic CL space) penetrates inside the body. Consequently, the solid body and especially a metal solid body could be able to modulate the penetrated CL space. The modulation of the CL space means a slight change of the mean CL node distance for some domains. So the CL nodes of these domains will obtain a different resonance and SPM frequency. It has been shown in Chapter 2, that one of the basic features of the magnetic field is the phase synchronization between the SPM frequency of the involved nodes which is propagated with a speed of light.

Consequently, the domain with a changed SPM frequency will affect the propagation of the magnetic field.

The change of the SPM frequency of such domain obviously should depend on the number of protons and neutrons in the atomic nuclei and the number of free proton clubs. These two parameters are changing simultaneously with the Z-number and contribute to a feature (a) of the trend.

It is obvious that the changed SPM frequency of the affected CL domains will be higher than this of the normal CL space. This is in agreement also with the refractive index change for metals measured by X rays. **Then for some Z-numbers the changed frequency of some particular domains may become a harmonic of the normal SPM frequency. In this case, the magnetic susceptibility could obtain a large value.**

The magnetic susceptibility may not arise only at exact harmonics. The magnetic quasisphere (MQ) of the SPM vector has 6 bumps and dimples between them. The NRM vector spends more time in the bumps, than in the dimples (this is a quantum feature of the CL space). This feature provides a possibility for synchronization between neighbouring domains with a SPM phase difference not only of 2π, but also of $2\pi/6$. (factor 6 is the number of bumps). This of cause is related to a space-time parameter λ_{SPM} and requires a finite length in space. So some stroboscopic effect may appear related to λ_{SPM}.

The above described features may explain the quasi periodical appearance of the magnetic susceptibility trend. The diamagnetism, however, (the

negative value) requires some additional explanation. For this purpose Fig. 8.30 illustrates the 3D shape of the magnetic quasisphere (MQ) viewed from two opposite directions.

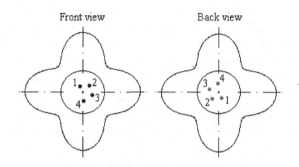

Fig. 8.30

Front and back view of consecutive positions of NRM vector. Positions in the front view are shown as black points, while in the back view by red (or grey) points. The consecutive positions are denoted by the numbers from 1 to 4.

The positions 1,2,3,4 are consecutive positions of NRM (node resonance momentum) vector on two opposite bumps of MQ. So the black and red points are time separated by half of resonance cycle. Every consecutive position is displaced from the previous in an order defined by the rotational direction of the NRM quasiplane. If referencing the position 1 as initial phase position of SPM vector it returns to it after N_{RQ} number of cycles of NRM vector. Then the SPM vector will pass through the same point after a phase of 2π or $n2\pi$, where n is an integer. The rotational direction of NRM determines the two different option for SPM vector related with a clockwise and a counter clockwise circular polarization. In the front view of MQ, the rotational direction is clockwise, while in the back view it is counter clockwise. In the real case, the points are very close because the surface of an equivalent sphere is obtained by $N_{RQ} = 0.8843 \times 10^9$ number of resonance cycle. So there is one important feature:

(A) $(2n - 1)\pi$ **odd phase conditions of the SPM vector**: For a limited number of resonance cycles the direction of rotational motion may not be distinguishable due to some threshold noise. Then for a limited time, the position of SPM vector could appear the same not only for a phase difference of $n2\pi$, but also for a $(2n-1)\pi$.

Let us consider the following two cases:

Case (1): the effect (A) is ignored (due to some threshold noise) and the SPM phase synchronization is at 2π or $n2\pi$. In this case the magnetic susceptibility will exhibit a quasi periodical dependence but only with positive values. When the phase synchronization condition deteriorates for some elements, the magnetic susceptibility declines towards lower values.

Case (2): the condition (A) is valid.

For a limited time, a phase synchronization at $(2n-1)\pi$ may occur for the SPM vectors of some neighbouring domains. For a longer time duration, however, this wrong synchronization could be recognized due to the Zero Point Waves. The wrong temporal synchronization will cause some energy fluctuations tending to remove the **odd phase conditions of the SPM vector** $(2n-1)\pi$. The time duration of such event evidently should be smaller than the time-space constant. The effective force opposing the external magnetic field in this case will be a repulsive one, i. e. the material will appear as diamagnetic.

Note: In case (1), the effective attractive force could be regarded as a result of additive "catch and hold" effect whose strength depends on the degree of SPM frequency change

In case (2), the effective repulsive force could be a result of wrong temporally "catch and hold " effect.

According to the described concept the magnetic susceptibility is directly related to the node resonance frequency of the magnetic domains. The atomic mass is one of the major factor influencing the node resonance frequency of these domains. Then some periodicity between the atomic mass and the peak value of the magnetic susceptibility should exist. In Fig. 8.29 the atomic masses for some of the elements staying in the peak regions are shown as approximate peak values. One may identify:

approximate positive peak values: 55.85, 106.4, 157.2, 195

approximate negative peak values: 35.5, 80, 121.7, 178.5, 209

The trend is pretty disturbed after Gd because of the different trend of the nuclear build. Although some approximate periodicity of about 50 atomic units is apparent. It may correspond to a phase change with 2π. It is also apparent that the phase of negative peak periodicity is shifted at about 25 atomic units. This is in agreement with the presented concept. The variation of the period estimated by the atomic mass might be influenced also by the atomic arrangement in the solids, which may contribute to differences in the specific gravity.

The ferromagnetic hypothesis is additionally discussed in Chapter 10 and 12 in connections to the magnetic fields of the planets and stars.

Chapter 9. Molecules

The purpose of this chapter is to analyse the quantum processes responsible for connection of atoms in molecules by electronic bonds.

Notes:

(1) Some aspects of the BSM concept about the chemical bonds may differ from the existing so far models. For this reason some new interpretations of the experimental data is provided.

(2) The derived vibrational curves in this chapter may differ from the quantum mechanical vibrational curves especially in the limit energy levels of the series. The difference is a result to the the different models models. QM models are built on a current concept of the physical vacuum, while BSM models are built on a new concept. From a point of view of BSM, the Quantum Mechanical (QM) models of the atoms and molecules appear to be not real but mathematical models. For this reason, the Quantum Mechanical models does not provide real information about the geometrical structures of the molecule, since the hardware structure of the elementary particles, atomic nuclei and atoms is missing. The BSM models unveils all these structures, their proximity fields and the behaviour of the electrons in a strong Super Gravitational field of the atomic nucleus. Accurate calculation of the energy levels if using the BSM models is more complicated, these models have other advantages. They gives insight into the physical mechanisms involved in the connection of the atoms in molecules and their oscillations, which signatures become apparent from the analysis of the optical and photoelectron spectra.

9.1 Type of chemical bond

The BSM distinguishes the following types of chemical bonds

     - ionic bond (IB)
     - electronic bond (EB)
     - SG field bond (SGB)
     - Dipole induced bonds

9.1.1 Ionic bond

The BSM concept of the ions has been presented in §8.11. Let us take example with NaCl, whose bond is considered as ionic. The compounds possessing ionic bonds exhibit two major features:

- The ionic compounds are usually dissolved in water. This means that the positive and the negative ions become separated.

- The internuclear distances of the ionic bonds are usually much larger than the covalent bonds. This fact is in agreement with the BSM concept of the ionic bonded molecules.

Fig. 9.1 shows a NaCl molecule in which Na^+ and Cl^- are connected by ionic bond. The elements of the nuclei (protons and neutrons) and the possible quantum orbits are shown in one and a same scale. The zone of the bonded protons is drawn as ellipse. The interatomic distance for this molecule is experimentally determined.

It is apparent from Fig. 9.1 that the two ions could not be connected by a single quantum orbit, because the distance between the atoms is too big. The bonds of such molecules are based on the balance between the Coulomb attraction forces and the properties of the proximity fields of the protons and neutrons (discussed in previous chapters) which was unveiled by the BSM. As a result, the molecules with an ionic bond could not have a vibrational rotational spectra. Such spectra are characteristic features of molecules, in which atoms are bound by common quantum mechanical orbits.

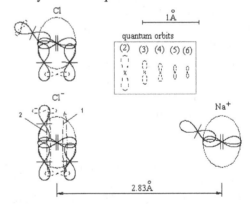

Fig. 9.1. Ionic bond in NaCl; The number in bracket shows the subharminic number of the quantum orbit. The closed loops 1 and 2 are magnetic lines formed by paired eight quantum orbits (each containing 2 electrons with opposite spins) arranged symmetrically around the polar axis of CL^- ion.

In summary, the main distinctive features of simple molecular compound with ionic bonds are the following:

- **The atoms connected by ionic bond does not possesses common electronic orbits. The**

binding energy is a result of attraction between the opposite charges of the ions.
- **Due to the larger bond distance the SG forces are negligible in comparison to the electrical forces.**
- **Ionic bond compound can not exhibit a discrete optical spectrum**

9.1.2 Electronic bond

According to BSM, the electronic bond (EB) between atoms connected in molecules is provided by one or more common quantum orbit. Under this category falls not only covalent bonds (between same elements) but bonds between different elements, as well. Most of the chemical bonds in the organic molecules are of EB type, others are of SG type. From the Atlas of Atomic Nuclear Structures (ANS) we see that the spatial structure of the atomic nucleus is determined by the spatial arrangement of the protons and neutrons. The valence protons posses some angular freedom mostly in the polar plane of the atomic nucleus. The atoms can be connected in molecules by quantum orbits between the valence protons. Such orbits operates by the same physical principles like the quantum orbits of the single atom. Additionally, the SG forces between the atomic nuclei must be considered. It is evident, that the spatial structure of the molecule is determined by the spatial configuration of the atomic nuclei in which the valent protons are interconnected by quantum orbits. In some diatomic or multiatomic molecules, not all valence protons are possible to be connected by EB bonds, due to spatial restrictions imposed by the finite size atomic nuclei.

Electronic bonds are possible, not only between free valent protons, which define the principal valence of the element. In some reactions the weak EB bonds (between protons from the same nucleus) could be broken and involved in the internuclear EB bonds. For instance, the elements with Z number larger than 71 contain underlying proton shell with EB bonds (Lantanide bonds), that are not converted to GBclp. In presence of some strong chemical reagent (or catalyst) they can be broken and involved in internuclear EB bonds. Some compounds of Rn are of such type.

In BSM model, the angular freedom of the valence bonds are defined by the nuclear configuration.

9.1.3 SG bond and dipole induced bond

The SG type of bond is a result of SG attraction. In the dipole induced bond the Wan der Walls forces are involved. Some of the Wan der Walls forces, according to BSM, appears to be SG forces. For this reason dipole induced bond usually could not be separated from SG bond. Compounds like ArHF and Ar2HF contain such type of bonds. Investigating vibrational states of Ar3HF molecule and comparing the results to the theoretical one J. Farrell et al. (1996) found about 11% difference. This difference, according to BSM is due to the SG forces, that are not considered in the QM models.

SG type of bonds exits in the complex molecules especially from heavy elements. The SG type of bonds are usually weaker, than EB type.

In summary:
- **The angles between two atoms in a chemical compound with an EB bond is determined by the nuclear configurations, the angular freedom of the protons involved in the internuclear EB bond and the SG forces between the nuclei**

9.2 Theoretical modelling of chemical compounds

The knowledge of the CL space parameter, the nuclear configuration of the atoms and the quantum orbits provides the opportunity for theoretical structural modelling of chemical compounds. When using the BSM atomic models, the modelling of simple dual or three atomic molecules is a straight forward procedure. An useful verification for the BSM modelling will be a comparison of the obtained molecular structure with the VSEPR model of the molecule if it is known. The VSEPR model provides angles between the atoms in the molecules. This angles are determined experimentally by X-ray crystalography and other methods. While the VSEPR models relies on experimental data, BSM allows prediction and theoretical modelling of possible chemical compounds. While the VSEPR model consider the angular positions of the chemical bonds as a result of a probable wavefunctions, the BSM model permits to find deterministic rules behind the angular positions.

The Atlas of the Atomic Nuclear Structures (ANS) provides the nuclear configuration of the stable elements using symbols for the proton, neutron D and He for drawing simplification. In theoretical synthesis of molecules, however, Hippoped curves should be used for both, the proton and the quantum orbits with their relative dimensions. In this case the twisting properties of the proton and the exact orientation of the quantum orbits could not be shown, but they are apparent. For simple inorganic molecules two or more projection views allows obtaining the needed 3D information. For elements with a higher Z-number, the views are more clear if only the central sections of the atoms are shown. For such elements the central nucleus contains one or more Argon nuclei, which could be drawn as an oval (for a drawing simplification there is no need to show the protons and neutrons in the Argon-like central part of the atomic nucleus). For more complicated organic molecules, however, the proton twisting affects their spatial configuration. In this case, some special drawing program might be necessary.

In the modelling of molecules by using of the atlas of ANS the following considerations are useful:

- The atoms preserve the nuclear configuration and the positions of the quantum orbits connected to bound protons. Only the positions of the valence protons with the connected to them quantum orbits could be changed

- The free valence protons have a limit angular freedom

- The quantum orbits of the atoms connected in a molecule interact magnetically. Their spatial orientation is selfadjusted for a stable energy level, which might be of dynamical type

- the nuclear EB and SG type of bonds may also participate in a chemical bonds. The nuclear EB bonds are stronger than the SG bonds and works in a same principle as the valence bonds.

- The chemical reaction is strongly dependent on the aggregate state, since the latter affect the freedom of the valence bonds

In Chapter 2 the quantum quasishrink effect has been discussed in relation to the photon wavetrain configuration. In Chapter 7, a similar effect was considered for the volume enclosed inside of the Bohr surface. We may refer to this effect as a **quantum quasi-scale change effect**, since it involves a change only of the quantum space parameter the Compton length, while the node distance of the CL space is the same. The degree of the quasiscale change may affect only the size of the bonding quantum orbit, but not the size of the atomic nuclei. Such effect will depend on the SG field of the nucleus, propagated in a close proximity, so it will depend on the nuclear mass of the involved atoms and the distance between them, which will match to a quantum orbit with a proper size. (This conclusion came from the analysis of the atomic spectra presented in later in this chapter).

The equation of the quantum orbit trace length for free CL space was derived in §3.12.3 (Eq. (3.43.i)

$$L_{qo}(n) = \frac{2\pi a_o}{n} = \frac{\lambda_c}{\alpha n} \qquad [(3.43.i)]$$

where: n is the subharmonic number of the quantum orbit

We may use the subharmonic number of the electron motion for identification of the quantum orbit. So a quantum orbit with a subharmonic number of n, is a such orbit for which the electron moves with a velocity corresponding to a quantum motion of subharmonic number of n (see Table 3.1 in Chapter 3 and §7.4 in Chapter 7). Then the ratio between the proton length and the length of the first subharmonic orbit is:

$$L_{pc}/L_{qo}(1) = 2.042685 \approx 2 . \qquad [(7.4)]$$

In this relation, the proton and the quantum orbit, both are assumed to have a shape of a Hippoped curve with a parameter $a = \sqrt{3}$. The shape of the proton, practically could be considered stable. It is reasonable to accept that the atoms connected in molecules by EB quantum orbits may vibrate around an equivalent point, without breaking the quantum orbit.

Initially we may consider that the size of the EB quantum orbit is affected by the distributed E-field inside the Bohr surface. This has been demonstrated by the Balmer series model for the hydrogen in Chapter 7. It is reasonable to accept that the volume enclosed by the Bohr surfaces of the individual atoms connected (by electronic bonds) in a molecule are united (or united Bohr quasispheres). In such way we may explain the lack of irradiative EM field from the molecule, while the electrons are orbiting. **The concept of the united Bohr surface could be explained if accepting that the SG fields of RL(T) structures of the valence protons (deuterons) participating in the electronic bonds are synchronized.** So one question appears: Could a common orbit between atoms connected into a molecule could be quasishrunk as in the case of Balmer model of BSM? (keeping in mind that the quasishrinkage does not mean a change of the CL node distance, but only λ_C. is valid for the quantum CL space). To simplify the answer we may try to find the **condition for which the quasishrink coefficient approaches zero.** It appears that this is valid for the boundary conditions, the signature of which are the limits of the spectral series. It will be shown that nuclear positions for a such case are informative enough in order to obtain the correct molecular structure. Only when necessary, we may use a star notation "*" signifying, that the orbital parameters are quasishrunk.

For drawing purposes, we may use the approximate relation (7.4), connecting the dimensions of the quantum orbit to the dimensions of the proton. In such way we may select combinations of quantum orbits with different subharmonic numbers and serially combined orbits. The possible orbital selections were presented in Table 7.1 (Chapter 7).

It is more connivent to operate with the length and width of the Hippoped curve instead of the orbital trace length. In such case, Eq. (7.4) and Table 7.1 (Chapter 7) could be used.

The proton length as a "length of a Hippoped curve " was determined in Chapter 6 by the use of Eq. 6.57. The same equation is valid for a quantum orbit with a parameter $a = \sqrt{3}$. So we have:

L_p = 0.6277 (A)   - <u>proton length</u>

$L_q(1)$ = 1.366 (A) - <u>length of the first subharmonic quantum orbit (in a free CL space)</u>

where: A (Angstrom) = 1×10^{-10} (m)

If using the approximate ratio of 2 for drawing purposes (see Eq. 7.4), the fractional error is about 2%.

Fig. 9.2 provides drawings of the basic atomic structures and set of available quantum orbits in one and a same scale. The quantum orbits are shown by dashed line and annotated by their subharmonic number. Their lengths have been derived for a free CL space (without electrical field)..

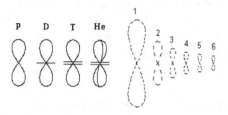

Fig. 9.2
Basic atomic structures and quantum orbits

The basic atomic structure notations are :
p - proton
D - deuteron
T - tritium
He - helium
The horizontal lines in the middle of D , T and He are neutrons.

Figure 9.3 illustrates EB bondings between atoms in the simple molecules H_2, HD and H_3.

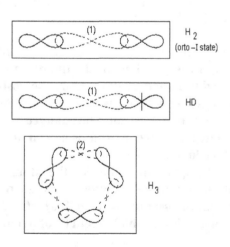

Fig. 9.3
EB bondings in some simple molecules

The H_3 molecule is not very stable at normal conditions. In space environments, however, its ion H_3^+ is observed. The atomic conditions providing stability of this ion are different than those described in §8.11.1 (Chapter 8) about Na^+ ion. The integrity of the H_3^+ molecule is preserved despite the loss of one electron. A possible explanation about such configuration is that the other two electrons circle in a common ring orbit, composed of serially connected quantum orbits (this option is not shown in the drawing). In absence of one electron, the excess positive charge of the H_3^+ may still provide integration of the individual Bohr surface into a common Bohr surface, so the two electron may circle in a common orbit composed of serially connected quantum loops.

The hydrogen molecule appears in two different states - Ortho and Para. Both states are distinctive by their molecular spectra and molar heat capacity. The Para state has a larger molar heat capacity. According to BSM, the larger heat capacity should correspond to a larger SG field, which means that the protons in the para-state should be closer. The Ortho-state may have two substates: Otho-I type and Ortho-II type. The Ortho-II may appear only at very low temperatures. The possible configurations of the Ortho-I and Ortho-II states of H_2 are illustrated in Fig. 9.4 a. and b.

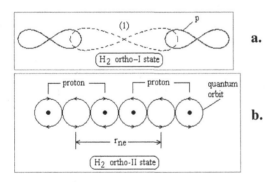

Fig. 9.4. Ortho-I and Ortho-II state of H_2 molecule

Fig. 9.5 shows the possible configuration of the H_2 - para state.

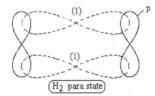

Fig. 9.5
Para state of H_2 molecule

In both ortho states the electrons share one orbit. So they must have opposite QM spins. In Ortho-I, the orbit trace length corresponds to one quantum loop of first harmonic. In Ortho-II the orbit trace length is a sum of two first harmonic quantum loops (serially connected quantum loops). The Ortho-II state may appear only at low temperatures. The optical spectrum usually shows the signature of both - the ortho and the para state. The conversion between the ortho and para state is also more probable at lower temperature, and can be conserved in a room temperature for quite longer time. At normal temperature, the mixture ratio between the ortho-I and the para-states is about 3:1.

9.3 Concept of integrated Bohr surfaces

The integration of the Bohr surfaces between the different protons within atomic nuclei was discussed in Chapter 7. In the chemical EB bonds, a similar conditions are created between the Bohr surfaces of the involved valence protons. Then conditions for quantum orbits are created for the EB bonding electrons. Fig. 9.6 illustrates the shape of the covalent Cl_2 molecule, where the two valence Bohr surfaces are integrated into one.

The Cl_2 molecule is drawn in scale. The covalent bond is a EB type. The equivalent plane of the EB orbit is at angle close to 90 deg in respect to drawing plane. The internuclear distance is known by experimental data. Two of the four GBclp for each atom are also shown. If not considering a quantum quasishrink effect (we will see later that it practically does not affect the internuclear distance), its length corresponds to a third subharmonic quantum orbit (electron energy of 1.51 eV). In the right-down corner of the same figure, the Bohr surface of the valence proton, when not involved in a chemical bond is shown. It is evident, that the two Bohr surfaces of the valence protons from different atoms might integrate into one surface. In this case, the shape of the individual Bohr surface could be modified but the definition condition given by Eq. (7.3.a) is still valid. In such case, the orbital conditions become similar as in the Balmer series model with the exception that the distance between the two cores are not fixed as in the hydrogen.

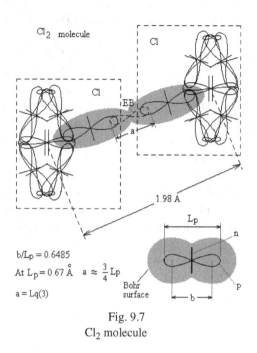

Cl_2 molecule

Cl

Cl

EB

1.98 Å

$b/L_p = 0.6485$
At $L_p = 0.67$ Å $a \approx \frac{3}{4} L_p$

$a = Lq(3)$

L_p

n

Bohr
surface

b

p

Fig. 9.7
Cl_2 molecule

9.4 Molecular spectra as a signature of molecular oscillations

9.4.1 Difference between QM model and BSM model of the molecular spectra

Initial note: In order to distinguish the single electron from multi electron system in the atoms, we will use the following terminology in this chapter:

- electron (an electron system according to BSM, consisting of an electron shell, an internal positron and a central core)

- electron system (multiple electrons for atoms with $z > 1$

The oscillating properties of the physical models of atoms connected in molecules by electronic bonds should provide a molecular spectra with discrete lines. The molecular spectra is the most authentic signature of the molecule. The theory of the molecular spectra is based on the Quantum Mechanical (QM) model of the atom. From the point of view of BSM, the QM is quite good mathematical model providing the possibility for very accurate calculations of atomic and molecular spectra using some experimentally determined constants. Although, as in the case of the atomic structure, we could not make equivalence between the mathematical model and the physical one.

The goal of BSM theory is not to replace the QM model of the molecular spectra, but to find out what are the real oscillating properties of the atoms connected into a molecule.

Under oscillating motion we mean not simple but a complex type of motion. Let us see, what are the parameters, making the QM model different from the physical one.

- The atoms participates in a molecule as a QM atomic model

- The QM orbital arrangement is different, than the BSM orbitals, but the angles between orbital planes in some cases might be similar

- The CL space environment is not considered in the QM models

- The bonding orbital shape is different for QM and BSM models

- The quasishrink factor with its dependence on the internuclear distance is not taken into account in the QM models

- The SG forces controlling the quasishrink space are not apparent in the QM models

The QM model of the molecular spectra is initially based on the concept of a rotated reduced mass whose radius is changeable. While the CL space and the SG forces have not be envisioned, additional empirical rules are involved in the QM model, obtaining in such way the necessary corrections for the missing factors. The big success of the QM model is based on the effect of energy conservation principle, when working in quantum units and orbital plane configurations (by s, p, d .. type of orbits). This approach allows to simulate the real processes by the balance of the apparent energy. The participating SG energy is left hidden. In the QM model, the energy oscillations are described by wave equations in which hamiltonian operators are used. A proper selection of the solutions of these equation known as wavefunctions with number of selection rules and constants provides synthetic molecular spectra matching quite well the observed spectra.

The QM model of the molecular spectra is based on the Bohr type of atomic model. It also deals preliminary with the electron configurations, described by selected wavefunctions, corresponding to *f, p, d, s* type of orbits, spin orbits couplings, and number of additional factors describing the arrangement of the orbits. In fact, the orbits are presented as electron clouds, due to the uncertainty principle used as an basic postulate in QM. **While in the Bohr model all orbits above ground state have a trace length larger than** $2\pi a_o$, **in BSM model, all their possible trace lengths are smaller than** $2\pi a_o$. **In other words, they can be considered as enclosed inside of the Bohr surface.**

The QM model uses the basic parameters h, q, e_o and m_o, which are defined for the external CL space. This approach matches to the Bohr atomic model, where the orbits are larger than $2\pi a_o$. This fact removes the problem of the possible corrections of this parameters in the quantum quasishrink space, inside the Bohr surface. But from the hand : **The quasishrink effect and the SG field are both dismissed in the QM concept of molecular spectra, and replaced by the adopted concept of Born-Operheimer approximation.**

The Born-Operhimer approximation is introduced in QM in order to neglect the nuclear contribution. In fact if considering only the Newtonian masses, their contribution is negligible. In such case the solutions of the equation is simplified significantly. In such approach, however, the quasishrink effect and the SG field parameters become hidden. Later, some additional corrections are involved as a particular rules, in order to correct the spectra, but the above mentioned parameters are still hidden. Therefore, the obtained final solutions in fact describes the energy related to the oscillating motion of the electron systems of connected atoms. **The potential energy between the SG field of the atoms and their influence on the molecular oscillations are not apparent in the QM model.**

According to BSM, there are **two different oscillating systems:**
 - the bonding system
 - the atoms with their electron systems.

The bonding system includes the valent protons (Deuterons) and the bonding electrons. There is a comparatively weak connection between the bonding electrons and the atomic nuclear electrons caused by the magnetic couplings and QM spin.

The real oscillations can be understood only if considering the energy balance in the very basic level of the prisms interaction. This is the balance between the SG (CP) forces and SG(TP) forc-

es. It is more convenient to work with energy balances of SG fields. The system involved in a such balance should include not only the atomic nuclei and bonding system, but the CL space as well. The BSM provides a three dimensional real model of the molecular structure, in which both systems, mentioned above are quite distinctive. The major advantage in this modelling is the knowledge of the dimensions of the proton, neutron, the atomic nuclei and the size and shape of the quantum orbits. Later in this Chapter it will be shown, how both interacting fields SG(CP) and SG(CP) are determined and involved in the molecular oscillation process, leading to emission or absorption of a photon.

The QM model, after applying the Born-Operheimer approximation, deals primary with the adopted configuration of the electron system (s, p, d, ..), described by wavefunctions, in which some orbits are considered as common for both nuclei. The QM model involves a number of constants, whose value are determined from the observed spectra. The adjusted in such way mathematical model provides vibrational rotational spectra that matches quite well the observed spectra of the molecules.

We see that the BSM model has a quite different approach than the QM model. This leads to the following discrepancy between both models:

(a) The "fundamental" frequency or band, according to QM terminology, in fact is a signature of the binding electron system. It does not provide the full picture of the nuclear motion in the molecule.

(b) The real resonance frequency between the two nuclei is much lower, than the "fundamental" frequency.

(c) The oscillation model described by BSM whose signature is the "vibrational-rotational" spectrum, appears different than the mathematical QM model of the molecular vibrational and rotation motion.

(d) While the QM model makes clear separation of the molecular spectra into pure rotational and rotational vibrational, the underlined physical process unveiled by BSM do not show such separation

The adopted QM therm "vibrational-rotational" spectra comes from the accepted initial model of a "rigid rotor".

The proof of the above statements will be presented in the next few paragraphs.

The molecular oscillating motion, according to BSM, is quite different than the QM concept, while providing the same type of spectra. The BSM concept is quite more logical for explanation and does not require the Heisenberg uncertainty principle. The relativistic effects can also be clearly identified and separated. So they could be analysed independently. In such aspect, the BSM model is able to provide analysis by using of classical means.

While the QM model, tuned by number of experimentally determined constants, provides pretty accurate molecular spectra, the BSM model provides an insight about the structural properties of the molecules.

9.5 Molecular oscillating model of BSM

9.5.1 General considerations and features

9.5.1.1 Complexity of oscillations

The complexity of the molecular oscillations increases with the number of valent protons, the atomic Z number and the number of atoms in the molecule. The degree of the oscillation complexity increases in the following order:

(1) homonuclear diatomic molecules with a single valence EB bond

(2) not homonuclear diatomic molecules with a single valence EB bond.

(3) covalent diatomic molecules with an EB valence larger than one

(4) linear three atomic molecules as HCN

(5) diatomic and poliatomic molecules with consumed valences (all valence protons are interconnected)

(6) poliatomic molecules with not consumed valences.

9.5.1.2 Fundamental proper frequencies of the nuclear system and the bonding electron system

All the atoms exhibit high order helicity defined by the twisted proton shape. So their motion in CL space possesses translational and rotational components. The same is valid for the molecules, but simultaneously with rotational motion they vibrate. The dynamics of the molecules could be analysed easier if separating them virtually into a bonding system and nuclear molecular system. This separation is valid only for molecules from atoms with Z number larger than two.

- The bonding system involves those valence protons that are connected by quantum orbits and their electrons.

- the bonding systems are in the middle between the atoms, while the atomic nuclei are in the periphery

- The nuclear molecular system involves the atomic nuclei participating in the molecule together with their electrons. So the connected valence protons with their electrons are not part of this system. The above separation allows also to distinguish the integrated Bohr surfaces of the bonding system from the Bohr surfaces of the individual atoms.

- The SPM vectors of the CL domain enclosed of the bonding system (characterized by integrated Bohr surfaces) and the CL domain of the nuclei may not be synchronized

Differences between the bond system and molecular nuclear system

Now let emphasize some differences between the formulated above two systems of the molecular complex.

No one of the both systems possesses a single proper frequency. There are few reasons for this:

- such type of system could not be considered as a rigid body system

- the quantum interactions with CL space are different for the electrons and protons (and neutrons). The most important parameters involved in the quantum interactions are their magnetic moments, respectively. The magnetic moments of the proton and neutron are quite different than the magnetic moment of the electron.

- a small change of the internuclear distance involves a large SG energy disbalance in comparison to the electromagnetic energy

- the SG field controls the E-field inside the Bohr surface of the bond system (this will be demonstrated by the analysis of the oscillations of the H_2 molecule)

As a result of these considerations, the both systems exhibit not a single proper frequency but set of proper frequencies. For simplification of the analysis, we may consider, that each system posses own **equivalent fundamental frequency.** Such frequency is a centre of mass of the proper frequency set with specific distribution.

The simple molecules, such as H_2 ortho-I and D_2 ortho -I, could be regarded as a single bonding system. It will be shown that these two types of bonding system is a typical for all EB bonds between atoms connected into molecules. The H_2 and D_2 bonding system exhibit own equivalent fundamental frequency. However, due to the quantum interactions it appears as a set of frequency in the far IR range of the spectrum, known as rotational spectrum.

Equivalent molecular fundamental frequency

The equivalent **molecular fundamental frequency,** according to the quantum mechanical considerations for diatomic molecule, is given by the classical equation

$$f_n = \frac{1}{2\pi}\sqrt{\frac{k_n}{M}} \qquad (9.A.1)$$

where: k_n is a force constant expressing the inertial and quantum properties of the atom

M - is a reduced Newtonian mass given by the combination of the two nuclear masses m_1 and m_2.

$$M = \frac{m_1 m_2}{m_1 + m_2} \qquad (9.A.2)$$

It is more convenient to express the spectral features by wavenumbers, $\bar{v}$ in units of cm^{-1}, because this is the most common used parameter in the IR spectroscopy.

$$\bar{v} = \frac{10^{-2}}{2\pi c}\sqrt{\frac{k_n}{M}} \quad [cm^{-1}] \qquad (9.A.3)$$

The Eq. (9.3) is in agreement with the Bohr atomic model, where, the nucleus is very small in comparison to the orbit radius and the orbit is centred around the nucleus.

In the BSM model of Hydrogen, the spatial position of any orbital plane (or quasiplane) is defined by the spatial position of the proton quasi-

plane. Consequently, the large magnetic moment of the electron will influence the inertial properties of the neutral Hydrogen. The same conclusion should be valid for the bond system in any molecule (with electronic bond).

The above considerations require correction of the inertial mass M. participating in the classical Eq. 9.3. For H_2 -ortho-I molecule the equation takes a form

$$\bar{v} = \frac{10^{-2}}{2\pi c}\sqrt{\frac{k_p \mu_p}{m_p \mu_e}} \qquad (9.A.4)$$

where: k_p - is a strength (force constant) of the proton pair bond, m_p is a proton mass, μ_p/μ_e - is a magnetic moment ratio

In most of the chemical bonds, bonding pairs of deuterons instead of protons are involved. Their strength constant is different than the constant k_p for protons bonding pairs and could be denoted as k_d. The Eq. (9.4) could be generalised to any diatomic molecule or group. For a homonuclear molecule it takes a form:

$$\bar{v} = \frac{10^{-2}}{2\pi c}\sqrt{\frac{bk}{M^*}} \qquad M^* = Zm_p\frac{\mu_p}{\mu_e} + Nm_n\frac{\mu_n}{\mu_e} \quad (9.A.5)$$

where: k is a strength of the proton or deuteron bonding pair, b - is the number of the bonding pairs, Z - is an atomic number, N - is the number of neutrons in the atomic nuclei, μ_n is the neutron magnetic moment.

Some aspects about the BSM considerations:

The hadrons (proton, neutron) have a larger inertial mass than the electron, but a smaller magnetic moment. The electron have a smaller inertial mass, but a larger magnetic moment. The nuclear electron system is carried by the nuclear hadron structure, whose SG field simultaneously defines the electrical charge and the E-filed configuration inside the integrated Bohr surface. (Note: the integrated Bohr surface may have a shape of manifold). At the same time, a larger magnetic moment of the electron system means a larger interaction with a CL space. Consequently, the nuclear electron system may influence significantly the oscillation type of motion of the nuclear hadronic system.

The above BSM considerations are not taken into account in the Quantum Mechanical model of the molecules, where the atoms are based on the

Bohr planetary atomic model with a point-like nucleus.

The strength of the molecular bond depends on the internuclear distance. The latter may take different quantum values from the allowed set of quantum orbits. For any single value of this set there are additional quantum conditions due to the quantum quasi-scale change effect that affects the orbital conditions. The kinetic energy is a factor dependable on the quantum orbit parameters. As a result of this dependence, the fundamental molecular oscillations exhibit not one, but **set of molecular proper frequencies**.

The set of the molecular proper frequencies could be considered as a proper resonance frequency modified by the quantum orbit conditions and internuclear distance.

Optical signature:

The molecular proper frequencies are comparatively lower than the bonding system frequencies (discussed in the next paragraph). Then, for the higher energy vibrational levels (identified from the molecular spectra) their contribution is quite small. For this reason, their optical signature may be identified from the lowest vibrational level. The lowest level, according to BSM, corresponds to the internuclear distance at equilibrium. The spectral signature of the molecular proper frequencies at this level appears in the infrared and is known as a "pure rotational" spectrum.

The definitions "vibrational-rotational" and "pure rotational" comes from the Quantum mechanical model. According to the BSM physical model, however, the oscillations are of mixed type. The rotated molecule vibrates simultaneously with the equivalent fundamental molecular frequency. This vibration in fact is involved in the CL space pumping and photon emission, while the rotation energy is constant during the process with an optimal angular velocity dependable on the molecular kinetic energy.

Equivalent proper frequency of the bonding system

The bonding electron system is connected strongly to the bonding protons and weakly to the nuclear electrons by the QM spin and orbital magnetic coupling. These interactions take place in the volume enclosed by the integrated Bohr surfaces. This system is also not a rigid system, so it exhibits multiple proper frequencies. Their optical signature is a set of spectral lines with more complex arrangement in P, R and Q branches (the Q branch is additionally dependable on the bonding type and may be absent in some configurations). The line distribution and the envelop of these branches depend on number of factors, which will be discussed in the following paragraphs. The spectrum of the bonding system is also centred around one average frequency, which could be accepted as an **equivalent proper frequency of the bonding system**. In the Quantum mechanical model of rigid rotor this frequency is known as a "fundamental frequency".

Basic differences between optical spectra generated by the bonding and nuclear systems

The vibrational motions of both systems are mutually dependable, because they have lot of common parameters. The optical spectra, however show some basic differences.

(a). The optical spectra of the molecular proper frequencies ("pure rotational spectra" according to QM model) is in the longer wavelength range (FAR IR) approaching the radio frequency range.

(b) the line width of the optical spectra generated by the binding system frequencies are much narrower, in comparison to those of the molecular proper frequencies (given by the pure rotational spectrum)

About (a): The bonding strength parameter k_n is dependent mostly on the number of connected valences and the possible orbit from the set of quantum orbits. So its variation with Z number is restricted to a finite sets of values. The M parameter in Eq. (9.A.5), however, does not have such restriction. Therefore, the equivalent molecular fundamental frequency tends to increase with the atomic number. This means a tendency of their optical signatures towards longer wavelength range.

About (b):

In the molecular proper frequency oscillations, all electrons of the nuclei are involved in a CL space pumping. The pumping efficiency, however is small, because the nuclear orbits have fixed but different plane orientations. The integrated Bohr surface for any nuclei may have a shape of manifold. In such case, the pumping and radiation efficiency of the system are partly deteriorated. As a result of this, the emitted light is not so monochromatic, i. e. the line width is broader.

For the bonding system, the CL pumping and the radiation conditions are different. A single bond of pair valence protons (deuterons) possesses a common quantum orbit, occupied by two electrons (for a neutral molecule). This provides an optimized conditions for CL space pumping and photon radiation, so the line widths are much narrower.

Summary:

- **A vibrating homonuclear diatomic molecule is characterised by two different equivalent fundamental frequencies: one for the nuclei and a second one - for the bonding electron system**
- **The two equivalent frequencies are dependable on the internuclear distance and the quantum orbit conditions. As a result of this, they appear as sets, denoted as proper frequency sets.**
- **Both frequency sets put own signature on the vibrational motion of the atoms and CL space pumping capability of the system .**
- **The direct optical signature of the molecular frequency set is the "pure rotational" spectrum at the equilibrium distance between the nuclei.**
- **The atoms connected into molecules could be regarded as subsystem connected by H_2 or D_2 bonding system.**
- **The analysis of the simple H_2 and D_2 molecules is quite useful for determination the parameters of the H_2 and D_2 bonding system.**

9.5.2 CL space pumping and radiation (absorption) capability of the bonding electron system

The bonding electron system includes the bonding electrons connected to the bonding valence protons. The vibrational motion between the atoms affects directly the bonding quantum orbits, characterised by their quantum energy in conditions of a quantum quasi-scale length change of the CL space domain. Every bonding pair of protons possesses own quantum orbits with electrons, so it

is able to pumping this CL space domain, which terminates with an emission of photon. The process, however, is dependent and synchronised with other EM bondings and the molecular oscillations.

The main distinctions in the optical spectra related to the proper frequencies of the bonding and the molecular nuclear system are the following:

- the signature of the fundamental frequency of the bonding electron system is a series of lines, spaced much closer, than the frequency set of the fundamental molecular frequency.

- the lines series follows a specific progression

- the lines are much narrower in comparison to the lines generated by the proper molecular frequencies.

Notice: <u>We must point out some difference in the terms "fundamental frequency" used in the BSM model and QM model.</u>

The term "fundamental vibrational frequency" used in the QM model, is relevant to the fundamental frequency of the bonding electron system used by BSM model. In QM model, the Srodinger equation for harmonic oscillator (see D. A. McQuarrie, p.p 162, (1983) does not contain SG forces. Solving this equation for H_2 molecule a frequency called a "Fundamental Vibrational Frequency" is obtained. Then looking at IR spectra of diatomic molecules these frequency are identified and the force constants are calculated for them. The estimated in a such way frequencies for different molecules are shown in Table 9.1. According tot BSM interpretation, however, they correspond to the equivalent fundamental frequencies of the bonding systems of the shown molecules.

"**Fundamental Vibrational Frequency**" according to QM model of harmonic oscillator, coresponding to equivalent proper frequency of the bonding system according to the BSM model   Table 9.1

Molecule	$\bar{v}$ (cm^{-1})	Molecule	$\bar{v}$ (cm^{-1})
		$^{127}I^{127}I$	213
		$^{16}O^{16}O$	1556
$H^{35}Cl$	2886	$^{14}N^{14}N$	2331
$H^{79}Br$	2559	$^{12}C^{16}O$	2143
$H^{127}I$	2230	$^{14}N^{16}O$	1876
$^{35}Cl^{35}Cl$	556	$^{23}Na^{23}Na$	158
$^{79}Br^{79}Br$	321	$^{39}K^{35}Cl$	278

The frequency position of $H^{35}Cl$, for example, is quite accurate, and appears in the middle between P and R branches of the vibrational-rotational spectrum. The synthetic spectrum of $H^{35}Cl$, together with $D^{37}Cl$ is shown in Fig. 9.8. The latter is shown in order to demonstrate the influence of the mass parameter M* involved in Eq. (9.A.6).

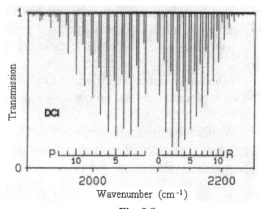

Fig. 9.8
Absorption spectrum of $H^{35}Cl$ and $D^{37}Cl$, distinguished by the position and strength of their P and R branches

The synthetic spectra show the position and intensity of the lines without their linewidths. The shown vibrational rotational spectrum of HCl, according to QM model, is known as a spectrum corresponding to the fundamental frequency of HCL molecule, given in Table 9.1. The two branches P and R are really centred around the value of 2886 cm^{-1}. According to BSM model, this is the equivalent fundamental frequency of the HCl molecule.

The shown spectrum contains only P and R branch. Later we will see, for which type of molecules and bonding systems this type of spectrum is a typical one. The P and R branches exhibit the following general features.

- The line frequency of the R branch are higher, than the P branch.

- The energy distribution of the lines in the branches follows nonlinear orders, which are usually denoted as progressions. They have different span coefficients for P and R branch.

- R branch lines are more intensive, than P branch lines

From the spectrum in Fig. 9.8 we see, that the $D^{37}Cl$ set of lines is slightly shifted from the $H^{35}Cl$

set in the direction of lower frequencies. The equivalent molecular fundamental frequency should have a frequency shift in the same direction according to Eq. (9.1) if only M* is changed, while k_n is not changed. The parameter k_n corresponds to the strength of the bonding system. The strength of the bonding system is determined by the connected pairs of valence protons (or deuterons) and the subharmonic number of the quantum orbit. It varies within a limited range for molecules of different atoms. At the same time, M* may vary in a larger range. This leads to the following conclusion:

Molecules of heavier atoms will exhibit lower equivalent molecular fundamental frequency.

The above made conclusion could be verified by examining rotational spectra of different molecules.

9.5.3 Characteristic features of the molecular oscillations

Relying on the discussed so far considerations we may provide some conclusions about the characteristic features of the molecular oscillations. Their proof will be presented in the following analysis, provided in this chapter.

- **The Bonding electron system, possesses much higher equivalent fundamental frequency, than the molecular fundamental frequency. Its motion generally follows the motion of the nuclei**

- **The SG field of both nuclei will influence their vibrational motion and internuclear distance**

- **The bonding electron system interacts strongly with a CL space and weakly with other electrons connected to the atomic nuclei.**

- **The bonding electron system affects the smoothness of the vibrational motion, due to the stronger quantum interaction with the CL space.**

- **The bonding electron system is sensitive to external factors, such as proximity of photons, electrons, an electrical field, a magnetic field, and a collision of the molecule with another molecule or atom. As a result of** such interactions, the fundamental molecular oscillation may be perturbed.

- **The perturbation of the molecular oscillations provides rich combinational conditions for transitions between different vibrational levels (including additional levels of orbit distortion, discussed later).**

9.5.4 BSM concept about the oscillations of the molecules with EB type of bonds

Balance of forces and its diversity of oscillations

Let us use the example of diatomic molecule consisting of light atoms.

In the Balmer model we have seen, that the electron motion is governed by the balance of the three types of forces: SG forces, Internal Coulomb forces and inertial forces. The Newtonian gravitational force between the electron and proton was neglected. The inertial mass expressed by the centripetal acceleration, is comparatively small, but still plays a role in the force balance.

9.5.4.1 Diversity of the molecular oscillations and their categorization.

In the molecular model, the internuclear distance is in the order of quantum orbit length. A simple test shows, that the Newtonian gravitational energy is much smaller than the bonding and the transition energies. Therefore, its contribution could be ignored.

The problem with the inertial mass is different. The inertial mass participates in the centrifugal forces. For one and a same molecule, these forces are dependent on the molecular velocity (due to the helical interaction with CL space). For different types of molecules, these forces are dependent on the mass distribution in the molecule and the rotational symmetry. This two factors could be defined by the moment of inertia around the axis, passing through the centre of mass point. From this point of view, the molecular oscillations could be divided into two different types:

- I-st type oscillations - for molecules (or fractions of molecules) with a low moment of inertia

- II-nd type oscillations - for molecules (or fractions of molecules) with a higher moment of inertia

It is clear, that the centrifugal forces will obtain a definite value when the angular frequency corresponds to the equivalent fundamental molecular frequency given by Eq. (9.A.5).

In §9.7.5.D it will be shown, that due to the quantum quasishrink effect the change of the internuclear distance estimated by the external CL space length unit is negligible.

The above statement provides one very important conclusion:

(A). The radial component of the vibrating nuclei estimated in CL space unit length is negligible. Consequently, the work for displacement of the CL space node by the FOHSs of the both nuclei should be contributed only for the rotational motion. In fact this is the inertial interaction of the protons and neutrons with the CL space.

The conclusion (A) gives a possibility to estimate the rotational energy by a classical way.

$$E_{rot} = 0.5 I \omega_{rot}^2 = 2\pi^2 I \nu_{rot}^2$$

where I - is the moment of inertia and ω_{rot} is the angular rotational frequency

The moment of inertia for diatomic homonuclear molecule is

$$I = m(0.5 r_n)^2 = 2m_n \frac{r_n^2}{4} = m_p A \frac{r_n^2}{2} \qquad (9.A.6)$$

The rotational frequency ν_{rot} could be estimated by the vibrational frequency ν_{vib} defined as an equivalent fundamental frequency of the molecule. For H_2-ortho-I molecule it is shown in §9.9.2 (Table 9.4) that the ratio between both frequencies is

$$\nu_{vib}/\nu_{rot} = 2:1$$

At such ratio the symmetry of the inertial interactions with CL space is preserved for a complete rotational cycle. For other homonuclear diatomic molecules such condition could be satisfied only for even ratio between both frequencies.

The vibrational frequency, which in fact is the molecular equivalent fundamental frequency, could be estimated by the maximum of the envelope of the "pure rotational" spectrum. The latter is presented as a set of lines whose distribution is given by the QM equation:

$$P_J = (2J+1)\frac{\exp(-(hcBJ(J+1)/kT))}{q_r} \quad (9.A.7)$$

$q_r = \Sigma(2J+1)\exp(-(hcBJ(J+1)/kT))$ - is a partition function

The plot of Eq. (9.A.6) for CO molecule with a constant $B = 1.9225\ cm^{-1}$ at T = 298 K is shown in Fig. 9.9

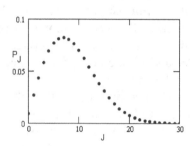

Fig. 9.9

Distribution of population among the rotational states of CO at room temperature

The temperature affects the shape of the curve by a horizontal span factor. The shape of the population curve for different diatomic groups is similar, but with different span coefficients. The shape is also similar to the theoretically obtained curve of the momentary velocity distribution of the oscillating CL space node, discussed in Chapter 2 §2.9.6.A (Fig. 2.29.D). It is shown again in Fig. 9.9.A.

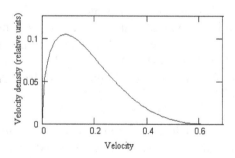

Fig. 9.9.A

Both curves have a shape close to the Maxwel distribution of the velocity, known in the molecular kinetic theory. Having in mind the helicity of the atomic and molecular structures it is evident that the rotational frequency should be correlated with the equivalent molecular fundamental fre-

quency. Consequently, the latter depends on the absolute temperature of the gas substance.

The pure rotational spectrum is usually given in units of cm^{-1}. So it is convenient to estimate the vibrational and rotational frequency in a same units. The rotational frequency is:

$$\omega_{rot} = 2\pi v_{rot} = 2\pi c \bar{v}_{rot}(cm^{-1})10^2 .$$

Then the rotational energy of diatomic homonuclear molecule in (eV) is

$$E_{rot} = \frac{1}{q}\pi^2 m_p c^2 A r_n^2 \bar{v}_{rot}^2 (cm^{-1})10^4 \quad \text{eV} \quad (9.3)$$

where: m_p - is the proton mass, A is the atomic mass, r_n - is the internuclear distance and $\bar{v}_{rot}$ is the rotational frequency in cm^{-1}

In Eq. (9.3) m_p is used instead of the average value between the proton and neutron mass, because the difference between them is quite small.

The rotational energy, calculated by Eq. (9.3) should be compared to the bonding energy for the equilibrium point of the vibration, which from its side is defined by the SG field and the number of involved valence protons and bonding electrons.

If E_{rot} is much smaller than the bonding energy at the equilibrium, the molecular oscillation is of I-st type, otherwise it is of II-nd type.

<u>For molecules with a I-st type of oscillations, the inertial moment is ignored. This simplifies the analysis of some light molecules. For molecules with a II type of oscillations, the inertial moment could not be neglected.</u> Some complex molecules may have both types of oscillations. Some light molecules, also may have large moment of inertia, because it is dependable not only on the Newtonian mass, but on the configuration of the bonding orbit, as well.

The BSM model of vibrational motion for molecule of I-st type is different than the Quantum Mechanical model of harmonic oscillator and its anharmonic corrections.

Type of spatial motion

We may distinguish the following spatial type of motion:

  <u>- linear vibrations (along one axis)</u>
  <u>- quasirotational vibrations</u>

The linear vibrations are typical for molecules with single valence bond. The H_2 ortho states serve as a typical examples. However, linear vibrations may occur also for molecules with larger number of connected valences but with symmetrical positions of the atomic nuclei.

- **Molecules involved in linear vibrations posses only P and R branches in their optical spectrum**

    The quasirotational type of vibrations are possible only for molecules, possessing more than one bonding quantum orbit. The most simple example of such motion is demonstrated by the Oxygen diatomic molecule O_2. The configuration of O_2 molecule in one of its state is shown in Fig. 9.10.

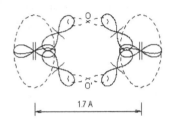

Fig. 9.10
O_2 molecule in one of its states
(2) - quantum orbit of second subharmonic

The Oxygen atoms, has been discussed in Chapter 8. The external proton shell includes two pairs of protons GBclp bonded in the equatorial region and two valence protons. Such configuration provides a large angular freedom of the two valence protons and offers a possibility for different states of the homonuclear oxygen molecules. In any particular state, however, the angular positions are fixed by a proper quantum orbit and internuclear distance. The quantum orbit in the state shown in Fig. 9.10 corresponds to the second subharmonic or energy level of 3.4 eV.

From Fig. 9.10 we see that the points O and O' in the centre of the bonds are completely symmetrical in respect to the polar molecular axis. Let us consider a vibration as a rotational motion around centre O with a limited amplitude. Then the other bond makes motion in a limited arc, that could be approximated by a linear type of motion. For one and a same molecule, the centre of such motion can be alternatively change between O and O'. Such kind of osculations are like rotational motion with limited amplitude and alternatively changed centre of rotation. For this reason it is

called a quasirotational motion (vibration). It is evident, that such type of motion is possible also between molecules with three symmetrical bonds (N_2 molecule for example). Such type of motion is possible in many complex molecules. One specific feature of this motion is their signature in the optical spectrum.

Molecules involved in quasirotational vibrations possess Q branches together with P and R branches in their optical spectra.

Detailed discussion of the above statement will be presented in §9.5.7.4.1.

9.5.4.2 Statistical cycle

In order to analyse the molecular oscillations, from the classical point of view, a concept of **statistical vibrational cycle** containing large number of vibration periods is used. It is called statistical, because one molecule, is barely able to complete such whole cycle. Any absorption, excitation, quenching or emission is able to distort severely the statistical vibrational cycle. The statistical cycle is idealised, because:

- the processes of emission and absorption are ignored (This facilitate the analysis, since it can be regarded as an amplitude modulated vibrational frequency)

- it can be regarded as statistically averaged vibrations from many molecules, showing only the equivalent fundamental molecular frequency (the rotational motion is not shown)

- the vibrations of the statistical cycles are in SG field environment, which is nonlinear

The shape of the statistical cycle is illustrated in the upper section of Fig. 9.11. In the left side of the cycle, the trend of different fields and energies are shown, denoted as:

E_{SG} - is a SG field potential
E_{IP} - is an Ionization potential
E_{DIS} - is a dissociation limit
E - is a momentary system energy

The asymmetrical shape of the cycle is similar as the oscillations of electrical circuit shown at the bottom section of the figure.

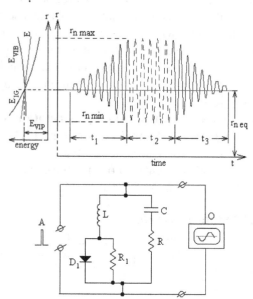

Fig. 9.11. Idealised statistical cycle
and equivalent circuit diagram

If referring the statistical cycle to a single molecule, it is not realistic, because the vibrational-rotational motion of the single molecule could be distorted either from absorption or emission of photons.

Note: Because the molecular vibrational motion is in a nonlinear SG field, it will be shown in §9.7.5.D that the change of the internuclear distance is so small, that it is negligible even in comparison to the proton core thickness. What is changes is the quantum quasishrink space. But in our analysis we may still consider some finite amplitude of vibrations, keeping in mind that they work against the strong SG forces.

There is one specific feature of the vibrational motion, shown in Fig. 9.11 by the statistical cycle. The amplitudes are asymmetrical. At first glance this may look confusing. However it is possible due to the following features, some of which has been already mentioned:

- the quantum quasishrink effect of the E-field inside the Bohr surface

- the quantum conditions of the binding orbits

- the inverse cubic dependence of the SG forces on the distance between the atomic nuclei

The quantum quasishrink effect has been discussed in the Balmer series model in Chapter 7. The case of molecular bonding orbit, however is distinguishable by the following:

(a) In the Balmer series the conditions of orbital length defined by the proton's dimensions are constant.

(b) in the molecular binding orbit, the conditions of the orbital length are defined by the internuclear distance

(c) the internuclear distance for the binding orbit is dependant on the total energy balance and therefore is selfadjusted

(d) we may accept initially that the binding orbit will intercept the proton quasiplane in the locus point (this will be proofed later)

From the provided distinctive features related to the binding orbits of EB type, it is apparent that the quantum quasishrink effect at Balmer series could be identified as a static, while this in the molecular bond, as a dynamic **self adjustable quasi-change effect (for distance)**. The latter exhibits different pumping capability.

In the equivalent circuit diagram, the proper cycle is determined by the inductance L and capacitor C. The Diode D_1, shunted by the resistor R_2, causes the nonlinearity of the amplitude. The resistor R causes the attenuation of the oscillations. If a short pulse A is supplied in the input, an oscillating cycle corresponding to phase t_3 could be observed by the Oscilloscope O.

The presented model of statistical cycle contains three phases:

t_1 - phase of absorption
t_2 - radiation lifetime
t_3 - phase of radiation
They are discussed in the next paragraph.

9.5.4.3 Phases of the molecular oscillations

The phases, indicated as t_1, t_2, and t_3, cover the full statistical cycle of the molecule, including the absorption, radiation lifetime and emission. The normal statistical cycle excludes any possible perturbations, mentioned in the previous paragraph. In analogy with atoms, such cycle corresponds to an optical pumping of the CL space with a spontaneous emission of a photon.

The phase t_1 is an absorption phase, during which the molecule may absorb a photon, whose wavelengths matches the energy level features of the molecule. In this process, mostly the electron system is involved, because it has much richer variety of energy levels, than the nuclear capability of photon absorption. The duration of the phase t_1 is not a constant, because the energy may be pumped by a single line, by few lines and by multiple spectral lines.

The phase t_3 is an emission phase. The main distinctive feature from the phase t_1 is that its time duration is strictly defined by the molecular configuration, and appears as a constant. This is valid for a case of spontaneous emission (but not for a stimulated emission used in lasers, where the lifetime is shortened).

The phase denoted by t_2 is a radiation lifetime of the exited molecule. The duration of this phase may depend not only on the pure vibrational motion, but also on the possibility the bonding electrons to interact with other electrons of the atom. This automatically involves interaction between all electronic orbits and the distributed E-field, controlled by the atomic SG field.

The rotational motion exists, but the system appears fixed to some proper vibrational level. The duration t_2 of this phase is very dependable on the Z number of the bonded atoms and the molecular configuration. In case of H_2 molecule, the duration might be very short. For molecules comprised of atoms with larger number of electrons, the energy exchange capability increases significantly and so the radiation lifetime. In some particular cases, like $O_2(a^1\Delta_g)$ in the night Earth atmosphere above 80 km, the radiation lifetime is about 58 min.

The lack of radiation during the phase t_2 could be explained by the capability of the SG forces to modulate the proximity electrical field around the protons involved in the EB bonding. Additionally, the bonding electron system may interact with the atomic electron shells, around the GBclp, which involves highly spatially oriented orbits in a strong SG field. Having in mind the orbit tilt, the QM spin, and the electron spin momentum of the whole complex, it is possible a temporally stable energy to circulate in a form of running waves in the internal volume of the atoms. All these consid-

erations may provide stable conditions for vibrations in the time interval t_2 without a loss of energy.

9.5.5. Molecules with I-st type oscillations

9.5.5.1. H_2 ortho-I state as a simplest diatomic molecule

For initial proof of the presented concept of vibrational oscillation of a diatomic molecule, we will analyse the simple H_2 molecule - ortho state. The configuration of this molecule with a bonding quantum orbit of first harmonics is illustrated in Fig. 9.12

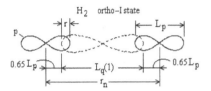

Fig. 9.12
Vibrational scheme of H_2 - ortho-I state

The quantum orbit quasiplane does not coincide with the quasiplanes of the protons. It passes through the locuses of the proton clubs. The following notations are used:

L_p - is a proton length
$L_q(1)$ - is a long side of first harmonics quantum orbit
r_n - is the distance between the Hydrogen atoms
r - distance between the electron and the proton core in the circular section of the orbit

Now let consider one important feature of the vibrational motion of I-st type oscillations: the quantum levels. The quantum levels are caused by the quantized finite dimensions, that the bonding quantum orbit could take in the space of integrated Bohr surface (consisted of the two Bohr surfaces of the individual valence protons (deuterons). It is more convenient to express the quantum levels by the energy instead of internuclear distance. In this case the energy levels inside the integrated Bohr surface could be regarded as an energy levels in external CL space. Similar analytical approach was applied in the Balmer series model of hydrogen (Chapter 7).

Fig. 9.13 shows the vibrational intervals of the H_2 molecule as a typical vibrational curve of a diatomic molecule according to QM model. In the right part of the figure the Photoelectron (PE) spectrum of this molecules is shown. Its reflection by straight line at angle larger than 45° means that the space intervals of the PE spectrum are proportional to the vibrational levels but stretched. This spectrum, discussed later in §9.6.1 is shown in Fig. 9.19. This stretched PE spectrum is likely caused by the lost energy for recoiling of the atomic nuclei during ionization. The largest peak of the PE spectrum corresponds to the largely populated vibrational level.

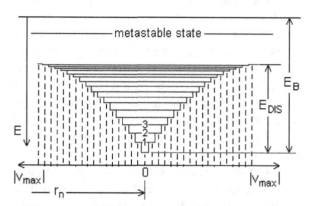

Fig. 9.13.A Vibrational levels as a function of the vibrational quantum numbers

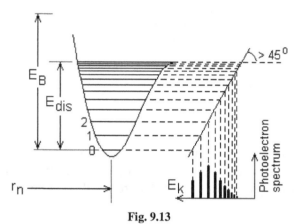

Fig. 9.13
Quantum levels of H_2 ortho-I state as a liner diatomic molecule and its Photoelectron spectrum.

The parameters shown in Fig. 9.13 are: E_B - binding energy (known also as ionization energy), E_{dis} - dissociation energy level, r_n - internuclear distance, E_k - electron spectrum (expressing the PE spectrum); levels 0, 1, 2 - vibrational levels with intervals 1/2, 3/2, 5/2 ...)

The QM vibrational curve is drawn in coordinates of energy level towards radius vector r of anharmonic oscillator. This option is not suitable for BSM analysis, according to which the effect of the quantum quasi-scale change must be taken into account. In order to find the signature of this effect it is more convenient to express the energy levels, shown in Fig. 9.13 as a direct function of the quantum number of the level. The shape of the obtained vibrational curve is shown in Fig. 9.13A.

We may call the step-like curve shown in Fig. 9.13A - a step-like vibrational curve. The E parameter referenced to the bonding (ionization energy E_B) in the further analysis can be regarded as a **momentary energy of the system**. We see that this energy, (connected to the vibrational levels) is quantized. The quantum levels are shown by horizontal blue lines. The steps of the vibrational curve correspond to different quantum orbits, obtained as a result of the quantum quasispace change effect. Referenced to the equilibrium point of level 0, this effect appears as quasishrink for the right side from level 0. While in the Balmer series model of hydrogen the quasishrink is referenced at a fixed distance (because the proton core dimensions defining the quantum orbit are fixed), for the electron bonding orbit of the hydrogen molecule this distance is defined by the internuclear distance between the two atomic nuclei. In the latter case r_n is self adjustable. It is controlled by the total energy balance of the system in which the SG field plays a very important role.

Summary:
- **The BSM model and QM model, both provide the same type of vibrational level transitions, but BSM model shows the real physical process in which the SG field with its hidden (for detection) energy becomes apparent.**

9.5.6 Approximate calculation of the system energy in the equilibrium state

Let us calculate the system energy for H_2 ortho-I state. For this purpose we will take into account the following logical considerations:

The orbiting electrons between two protons are able to neutralize their charges. We may assume that every orbiting electron is able to neutralize one charge, by interconnecting its E-filed lines to the proton E-field inside the Bohr surface. Using the vibrational scheme, shown in Fig. 9.12, <u>let considering the moment, when one of the electrons is in the locus of the left proton and the other one in the locus of the right proton. Their velocity in this case are normal to the direction of the vibration and does not contribute to the momentum energy of the system.</u> This energy, then can be estimated, by considering two unit charges at distance r_n, using the principle of superposition. For this purpose, we may first assume that the left proton and the right electron are both missing. Then the system energy will be $q/4\pi\varepsilon_o r_n$ [eV]. The same results is for an other case (a right proton and a left electron considered as missing). Using the principle of superposition and adding the energies from the two cases we obtain the full system energy.

$$E_{SYS} = \frac{2q}{4\pi\varepsilon_0[L_q(1) + 0.6455L_p)]} = 16.06 \text{ eV} \qquad (9.4)$$

The obtained value is quite close to the parameter Vertical Ionization Potential E_{VIP}, used in the QM model and determined by the Photoelectron spectrum (discussed later and shown in Fig. 9.19).

$$E_{VIP} = 15.967 \text{ eV} \qquad (9.5)$$

The parameter E_{VIP} corresponds to the largest peak of the PE spectrum, so it is easily identifiable. It is a signature of the most populated vibrational energy level. Consequently, we may consider this energy to be equal to the system energy, calculated by Eq. (9.4). From the molecular orbital constants data for H_2 we also have a value of average kinetic energy equal to 15.98 eV. (http://physics.nist.gov/cgi-bin/Ionization/table.pl?ionization=h2)

The match of the theoretical result from Eq. (9.4) with E_{VIP} and kinetic energy value gives us a confidence for using of the theoretical Eq. (9.4) in the further analysis.

Consequently, for practical purposes we may accept, that

$$E_{SYS} \approx E_{VIP} \qquad (9.6)$$

9.5.7 Experimental evidence about the BSM concept of molecular vibrations

9.5.7.1 Cross validation analysis

The presented concept of molecular vibrations in CL space can be proved by analysis of the molecular spectra and the photoelectron spectra for simple diatomic molecules.

Number of methods for obtaining of Photoelectron spectra exist. For our purpose, the most informative method is the He I photoelectron spectroscopy. In this method, the photoelectron spectrum is obtained when molecules of the investigated gas in vacuum conditions are irradiated by He I resonance line at 584 A, possessing an energy of 21.23 eV. The bound electrons are usually with energy below 20 eV, so the photoionization breaks the electronic bonding, providing in such way free photoelectrons. The kinetic energy of the released electrons is measured by an electron energy analyser. The energy spectra is the rate of detected electrons as a function of their kinetic energy.

9.5.7.2 Difference between the ionization mechanism for atoms, and molecules

The photoelectron (PE) spectrum is informative for the internuclear bonds and the vibrational levels of the molecules. If neglecting the motion of the atom of the molecule (referencing to laboratory frame) the energy balance equation is:

$$E_K = h\nu - E_{IP} \qquad (9.7)$$

where: E_K is the kinetic energy of the bound electron; $h\nu$ - is the photon energy of the ionization source; E_{IP} - is an "Ionization Potential" or "Ionization Energy", a term adopted from the QM model.

When applied for the photoelectron spectrum for atoms, Eq. (9.7) is valid for the whole range of the obtained PE spectra. This equation, however, could not be applied directly for a photoelectron spectrum in molecules because of the following differences:

- All quantum orbits in the atom are confined to a fixed distance, defined by the proton dimensions, estimated in absolute scale.

- The quantum orbits of the molecular bonding electrons are confined to a distance that is dependent on the self adjustable internuclear distance.

The above differences affect especially the estimated energy of the bound electron by the measured kinetic energy of the photoelectron.

According to the Photoelectron (PE) spectra theory for molecules, the energy balance is given by Eq. (9.8).

$$E_K = h\nu - E_{IP} - (\Delta E_{vib} - \Delta R_{rot}) \qquad (9.8)$$

where: E_K is the measured energy of the photoelectron.

The terms in the bracket, denoted as vibrational and rotational energy, provides a correction for E_K in order to reflect the internal energy of this electron. In BSM interpretation of PE spectrum, this correction is referred to the SG field between the atomic nuclei and the electronic bonds.

Important feature of the PE spectra is that the range of the energy bands is shrunk. This is apparent from the analysis of the H_2 molecule (provided later) when comparing with the corresponding optical band of same vibrational levels. The reason for such effect is the energy loss of the extracted photoelectron. Losing a electron, the molecule becomes a positive ion. The interaction between this ion and the negative electron causes a partial loss of its energy for recoiling of the ion.

9.5.7.3 Signature of the vibrational bands in the Optical and Photoelectron spectrum

E_{VIP} is easily identified by the PE spectrum of the simple molecule like H_2 ortho-I. Even the corresponding vibrational bands, could be identified from the optical spectrum, and the feature of E_{VIP} parameter as well. (This will be shown later in Fig. 9.19). But while the energy scale of the PE spectrum is shrunk, the energy scale of the optical spectrum is not.

The relations between the vibrational levels of H_2 ortho-I and the optical and PE spectrum are illustrated in Fig. 9.15.B.

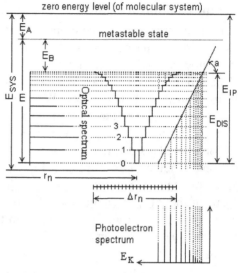

Fig. 9.15.B

Vibrational levels of H_2 ortho-I and their relations to the optical and photoelectron spectra

The following notations are used:

E - is a momentary energy scale

E_{SYS} - system energy

E_{IP} is a ionization potential (known also as a binding energy)

r_n - is an internuclear distance at equilibrium point

Δr_n - is the range of change of the internuclear distance

E_{DIS} - is known as a dissociation energy, according to the accepted QM terminology.

E_A and E_B are energies defining the position of the metastable state

E_K - is a PE spectrum parameter ("electron kinetic energy")

1,2 3 - vibrational bands

It is experimentally found for H_2 molecule that

$$E_{IP} > E_{DIS}$$

The vibrational curve is shown as a function of the vibrational quantum numbers. They are referenced to the change of the internuclear distance r_n. It will be shown later that the range of this change is intrinsically small due to the strong SG field in which the molecular vibrations take place.

The vibrational curve is characterised by increased density of the vibrational quantum numbers in the range of the dissociation limit, but in the following later analysis we will use vibrational levels with a finite quantum numbers.

The optical spectrum corresponding to transitions between the vibrational levels and the metastable state is illustrated in the left side of the step-like vibrational curve in Fig. 9.15. The photoelectron (PE) spectrum is shown below the step-like vibrational curve. The PE levels are reflected by a line at angle larger than 45 deg (multiplication factor < 1) in order to show the shrunk PE spectrum in respect to the optical spectrum, estimated by the energies between the neighbouring lines.

From Fig. 9.15, the functional relation between the Optical and PE spectrum becomes apparent. One and a same shape of the quantum vibrational curve defines two different shapes of the Optical and PE spectrum. Working only by the energy levels, we see that, if applying a proper offset and multiplication factor to the PE spectrum, we will obtains the optical spectrum and vice versa. Consequently, the inverse task - obtaining the vibrational levels from the Optical and PE spectrum is also possible. At the same time, we may get the relation between the vibrational levels and the momentary internuclear distance. This is done in the following later analysis.

The BSM analysis suggests one reasonable explanation of the metastable state of H_2. The quantum orbits for this state may have exactly the same trace length and shape as Lq(1) but it can be formed of the trace length of two quantum loops of second subharmonics connected in serial. Then the distance between the two protons, if not considering any deformation of the orbital shape is unchanged. Keeping in mind that the SG field is able to control the spatial configuration of the proximity electrical field, we may attribute this to a common synchronization of the SG energy of the two protons and the two electrons. Such effect, however, may work up to a limited distance between the protons. In this range the derived later Eq. [(9.18)] is valid. The SG field may be propagated beyond this distance (up to some limit) but the two protons fields are not any more commonly synchronized.

Summary:

- **The Vibrational oscillations of a homonuclear diatomic molecule exhibit unique Optical and Photo Electron (PE) spectra.**
- **The vibrational quantum motion could be analysed by using a proper set of the vibrational transitions and the corresponding PE spectrum.**
- **The PE spectrum exhibits a different energy offset and energy scale factor.**

9.5.7.4 Fine structure of the optical molecular spectra

9.5.7.4.1 Effect of the variations involving a change of the length and tilt of the binding quantum orbit.

Let us consider a diatomic molecule. The system of the electrons (connected to the atom) follows the nuclear vibrational motion, but they are not rigidly connected to the nuclei. Additionally, they have different resonance frequencies, as well. Therefore, their vibrational motion may differ from the vibrational motion of the nuclei. The contributors for this difference are not only their different inertial factors, but also their different magnetic moments (the magnetic moment of electron is quite different than the magnetic moment of the proton and the neutron). The bonding electrons, possessing own spin momentum and QM spin (orbital momentum), interact with the nuclear atomic electrons via magnetic interactions inside the integrated Bohr surface. We may consider, that the bonding electrons are elastically connected to the nuclei. As a result of such connection, the following effects might be possible:

(a) the quantum orbit shape is unchanged but could not follow exactly the vibrational motion of the atomic nuclei

(b) the quantum orbit size is unchanged, but the shape is distorted

(c) the quantum orbit size and shape are the same but phase delay occurs between the dynamical quantum quasishrink effect and the phases of the two electron proper frequencies (electron shell-positron) and (positron- core). (Such difference that is apparent in some experiments is explained so far by the electron spin).

According to the definition of the quantum loop, it is the orbital trace length that defines the quantum orbit. The shape of the orbit could be modified, without disturbing this definition. Then the shape is defined by the E-field configuration (a proximity E-filed of the proton and neutron) and the classical electron spin momentum. The change of the shape, however affects the strength of the quantum effect and the efficiency of the CL space pumping. The strength is maximum, when the spinning electron intercept the magnetic lines of the proton at angle close to 90 deg.

In the analysis (not shown here) it was found that a quantum orbit with a shape of Hippoped curve and parameter $a = \sqrt{3}$ matches quite well to the experimental data. Therefore, **we may consider that the quantum orbit is free of distortion at parameter** $a = \sqrt{3}$

The possible distortions of the bonding orbit are of two types:

- a symmetrical distortion
- an asymmetrical distortion

Both types of distortion are shown respectively in Fig. 9.16, a. and b.

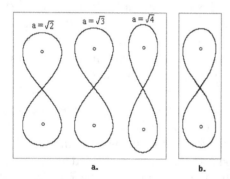

Fig. 9.16. Shape distortions of the bonding quantum orbit, a. - symmetrical, b. - asymmetrical

Considering a vibrational motion of a diatomic molecule, the case of symmetrical distortion appears when the bonding length is changed but the long proton axes are aligned. The asymmetrical distortion appears when the bonding length is not changed but the proton axes get tilting.

The bonding orbits of the molecules with linear type of vibrations exhibit a symmetrical type of distortion. Such type of distortion is valid also for the bonding orbits of the quasirotational vibration, which appears as a quasilinear motion in arc.

The symmetrical orbit distortion may cause a small change of the internuclear distance that could provide some displacement from the defined vibrational level. The asymmetrical effect for displacement from the vibrational level, however, is much smaller.

9.5.7.4.2 Oscillations providing vibrational-rotational spectra with P and R branches only

The equivalent proper frequency of the bonding system is higher than the equivalent proper frequency of the molecule. Therefore, it is the first one that could put its signature on the vibrational cycle. The symmetrical distortion of the bonding orbit causes the appearance of frequency sets on both sides of the vibrational levels (above and below of each level). The signature of these frequency sets are the P and R branches in the vibrational-rotational spectra of the molecule. The signature from the asymmetrical distortions of the quantum orbits appears as Q branches in the vibrational-rotational spectra of the molecule.

Now one question arises:

Why the P and R branches appear at both sides of the vibrational levels (corresponding to the step-like vibrational curve) and with a spacing asymmetry?

For light diatomic molecules, the rotational energy is much smaller, than the bonding energy at equilibrium. Then the inertial interaction of the atomic nuclei could be neglected in the vibrational motion, considering an energy balance only of the quantum quasishrink effect. Such energy balance is between SG(CP) and SG(TP) of CL space (Chapter 2) discussed later in §9.7.5.B, C. The motion of the bonding system is subjected to this balance. The change of SG(TP) defines the proton electrical field inside the Bohr surface, while SG(CP) - the charge unity. It is reasonable to expect, that the motion of the bonding system could not follow exactly the phase of the SG field oscillations. The assumption of a possible phase difference comes from the fact, that the electron has a finite inertial mass and axial spin momentum. **The obtained phase difference may cause overshooting of the vibrational motion of the bonding system in respect to the vibrational level defined by the SG field oscillation. As a result of all these considerations, the**

vibrational motion of the bonding system provides a set of quantum levels around the vibrational level.

The formation of R and P branches from the frequency set of the bonding system is illustrated later by the analysis of H_2 and D_2 molecules. They both could be considered as basic units involved in any EB bonding system between atoms connected in molecules.

The process of P and R branches formation is illustrated by Fig. 9.17, where a section of vibrational curve is shown with three consecutive vibrational levels.

The provided concept of overshooting with the shown phase is in agreement with the analysis of P and R branch formation for H_2 and D_2 molecules (ortho-I state) given later in this chapter (in §9.9.3). The correctness of accepted sign of the phase delay might be a topic of discussion. However it is possible to be verified by study of P and R branches for some linear diatomic homonuclear molecules for which enough optical and PE spectra exist.

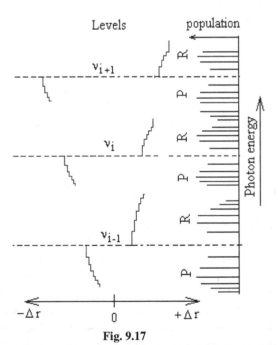

Fig. 9.17

Section of three consecutive levels of the vibrational ladder with fine structure levels from the bonding system frequency set. In the right side the corresponding optical spectrum from transition between these levels and the lowest level is shown

The overshooting process is possible due to the symmetrical distortion of the bonding orbit. We may reference the obtained set of levels as **"orbit distortion levels"** in order to preserve the clearness of the physical concept. They correspond to the J levels of the QM model. For the BSM model, the J numbering does not need to start from 0 and from 1 for *R* and *P* branches, respectively.

In the QM model, these levels are considered rotational due to the accepted initial theoretical concept of "rigid rotor" and they are denoted by J levels.

We will continue to use *J* notation for the orbital distortion levels. They are directly related to the spectral lines contained in the *P* and *R* branches of the molecular spectra.

From the BSM model, the following rules could be formulated:

(1) When the transition is between one vibrational level and the lowest (zero) level, the photon energy of the spectral line is equal to the difference between the instant system energy and the optical boundary energy (see §9.5.7.4 and Fig 9.15). The PE spectrum has a shape corresponding to a set of such type of transitions.

(2) The line strength corresponds to the degree of population of the vibrational levels under consideration

(3) The P and R branches are always separated by a gap without lines because the probability of exact matching the proper frequency set of the bonding system with the vibrational level determined by SG field balance is quite low.

(4) When estimated by external CL space units, R branch corresponds to a symmetrical stretching of the quantum orbit, while P branch to a symmetrical shrinking (assuming the parameter $a = \sqrt{3}$ is preserved). R branch is from the side of larger internuclear distance, while P branch is from the shorter one.

(5) Both P and R branches have nonlinear line spacing counted from lower to higher J numbers. This non linearity is a result of two nonlinear factors: SG field and the reaction of the bonding electron to the orbital distortion.

(6) The P and R branches in the molecule of heavier atoms should be less spread around the middle point of equivalent transitions

(7) For one and a same molecule, the P and R branches, estimated by energy levels, are more compact for $(v_i - v_{i+1})$ transitions, than $(v_i - 0)$ transitions

(8) The optical signature does not show line shape deterioration due to a level jittering, despite the fact that the electron system is elastically connected to the nuclear hadron structure. This may lead to two conclusions:

-The vibrational motion is quite small, while the energy is juggled by the quantum quasishrink effect

-The SG field balance leading to a photon emission (absorption) is a quite fast process.

Some of the lowest J numbers due to the orbit distortion are even apparent in some PE spectra with high resolution. The PE spectrum observed by J. E. Pollard et al., J. Chem. Phys, 77, 34-45, (1982) and given in Fig. 9.19 shows a signature of low J numbers.

From the way the PE spectra is usually obtained, it is clear that the first PE peak is accumulated by electrons from vibrational bands possessing a lower number (see Fig. 9.15). If the band 1 only is scanned with higher resolution, then the obtained PE spectrum will carry the signature of the orbit distortion levels, corresponding to J levels of the optical spectrum. Such experiments is performed by G. K. Cook and M. Ogawa (1965). They obtain absorption spectrum of N_2 in the far UV range. Their method is different than the method which uses He resonance line. They scan a monochromatic line, observing simultaneously the optical spectrum and the ionization current. In such approach, the photoelectron energy range shrink effect does not impose its signature on the measured spectrum. The obtained in such way PE spectrum is shown in Fig. 9.18.

9.5.7.4.3 Molecular oscillations providing vibrational-rotational spectra with P, R and Q branches

Only molecules with two or more EB quantum orbits may have Q branch of their optical spectra. This will become apparent from the following analysis.

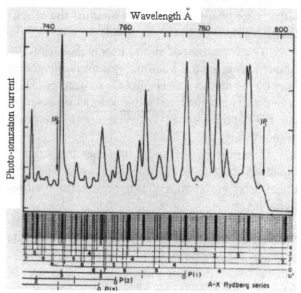

Fig. 9.18.   Photoionization yield-curve and absorption spectrum for N_2 gas (courtesy of G. K. Cook and M. Ogawa (1965).

The concept of the quasirotational motion has been discussed in §9.5.4.1. The main requirement is the number of EB bonds (valences) to be larger than one. At any moment, one of the EB bonds serves as a centre of the motion, while the others vibrate by stretching and shrinking and they interchange their role alternatively. The quantum orbit of the bond with vibrational motion exhibits symmetrical orbital distortions. Therefore, this motion contributes to the P and R branches of the optical spectrum, as described in the previous paragraph. The quantum orbit of the other bond, that is a centre of such motion exhibits an asymmetrical distortion, due to change of the angle between the bound protons. The change of the angle causes some change of the spatial configurational of the E-field around the clubs of the bound protons. This causes a distortion of the Hippoped shape of the quantum orbit. In a first approximation, we may consider that the distance between the bond serving as a centre of rotation and the nuclei of anyone of the atoms is not changed. This approximation is more valid for molecule from heavier atoms. This means that the SG field in this point is one and a same during the quasirotational motion around this centre. Consequently, this type of oscillations will contribute to the Q branch of the optical spectrum. Denoting the bond which is a centre of the quasirotational motion (in a particular moment) as a first bond, we may conclude:

The P and R branches of the optical spectrum is contributed by the quasilinear reversible motion of the second EB bond, while the Q branch - by the quasirotational motion around the first bond. It is apparent that the bonds denoted as a first and second exchange their role alternatively.

The asymmetrical distortion in fact displaces slightly the orbital position from the equilibrium, corresponding to P R branches. Therefore, the Q branch will contain more than one line, additionally spread in a small spectral range. The heavier diatomic molecules with pair bonds exhibit smaller spread of their Q branch, in comparison to the lighter molecules (for example H_2 para molecule). This indicates that the Q branch line spacing is sensitive to the displacement from the equilibrium distance. In heavier molecules, the bonding electrons interact with a larger number of electrons from the internal shells of the atom. They are much more stronger connected to the protons, providing in such way more stable reference point. For this reason, the Q branches of the heavier molecules are less spread.

In many molecules the Q branches appear as more than one set. This is a result of the QM spin orbit interaction between the bonding and other electrons of the atom.

Summary:

- **Molecules with linear vibrations exhibit "vibrational-rotational" spectrum containing P and R branches only**
- **Molecules with a quasirotational vibrational motion exhibit a spectrum containing P, R and Q branches.**
- **The P and R branches of the molecules with quasirotational vibration are from the quasilinear vibrating bond, while the Q branches are from the bond that is a centre of the quasirotational motion.**

9.5.8 A possible mechanism of the Raman scattering and rotational Raman spectra according to BSM

Raman spectra are features only of molecules. The atoms can not posses such kind of spectra. Raman spectra are obtained when the molecule is exited by a strong monochromatic radiation. Such excitation in the optical range is very efficient when the radiation energy is from a laser.

Now let see what happens when the molecule is irradiated by a strong monochromatic radiation. We have seen that the molecular system is very sensitive to a change of the momentary energy balance. The strong monochromatic radiation (with an increased coherency when irradiated by a laser) inputs some amount of energy into the volume of the Bohr surface. The obtained energy disturbs the momentary energy balance of the system, causing a fast transition from one internuclear distance to another. This corresponds to transition from one vibrational level to another directly without following the vibrational curve. This means that the system jumps to another energy level, but the input energy usually is different than the obtained energy between the levels. In such case, the excess energy is released right away as photon. The matched level may be not only vibrational but also some "vibrational - rotational" level. **The process is so fast, that it seems as a scattering. It does not require a finite CL pumping time as in the spontaneous emission.**

The Raman scattering may be referenced not only to the vibrational and vibrational-rotational levels, but also to the quantum levels of the electrons in the nuclei. When the irradiating energy is proper selected and the molecules are in the ground state (lowest vibrational energy) a "rotational" Raman spectra could be obtained.

"Rotational" Raman spectra

The "pure rotational" spectrum is a signature of the proper frequency set. The energy difference between the bottom level and the upper neighbouring level is larger than between neighbouring levels with higher vibrational numbers. Therefore, when the molecule possesses a lowest vibrational energy, the optical signature of pure rotational spectrum might be well separated at the bottom level in order to be identified. Experimental data for such case exist.

Raman scattering

Now let suppose, that molecule in a lowest energy vibrational state is irradiated by a proper quantum energy in order to reach some higher vibrational level pretty accurately. Then the small energy with a proper frequency signature will be translated to a higher vibrational level. If the irradiated quantum energy is properly selected, the condition for photon energy difference will appear exactly in the vibrational level between P and R branches. Therefore, the vibrational levels are possible to be resolved by proper wavelength scan of the irradiating photon energy.

In fact the condition for photon energy difference are fulfilled for two symmetrical levels from both sides of the vibrational curve. Their energy is one and a same, but the internuclear distances are different. It is evident that the emitted spectral lines will have the same signature as the "pure rotational" spectrum, partly modified from the different internuclear distance. The Stoks component corresponds to the longer distance, while the Antistoks - to the shorter one.

The "vibrational-rotational" spectrum is a signature of the proper frequency set of the bonding system. The "rotational" Raman spectrum is a signature of the molecular proper frequencies set. Because they are quite distinct, we may conclude that:

- **The equivalent fundamental frequencies of the molecular system and the bonding system are different.**

The Raman effect is characterized with one specific feature: the photons are emitted right away after the activation (no apparent lifetime of activated state as in the atomic and molecular spectra). For this reason the Raman effect is known as a Raman scattering. The BSM concept is in full agreement with this feature.

9.6 Vibrational bands of H_2 ortho-I state.

9.6.1 Photoelectron spectrum

The H_2 ortho molecule exhibits a typical molecular oscillations of I-st type. This is evident from

the shape of its PE spectra shown in Fig. 9.19 and Fig. 9.29 (both provided by J. E. Pollard et al., 1982).

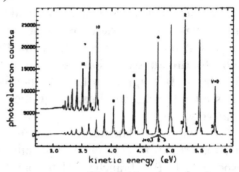

Fig. 9.19

PE spectrum of H_2 molecule (courtesy of J. E. Pollard et al. (1982)

The PE spectrum of H_2 is obtained by using a He I resonance line (21.23 eV). The energy scale is referenced to E_k.

From the PE spectrum in Fig. 9.19 we see, that the largest peak is at 5.26 eV. So

$E_{VIP} = 21.23 - 5.26 = 15.97$ eV. , and according to Eq. (9.6), this is also E_{SYS} (see the energy levels shown in Fig. 9.15).

9.6.2 Identification of the vibrational levels by the Optical spectrum

The optical spectrum for the H_2 ortho-I state, corresponds to the optical spectrum of the system $(B\ ^1\Sigma_u^+ - X^1\Sigma_g^+)$, known also as a Lyman system. The optical bands of this system have only P and R branches, but not Q branches. This is in agreement with the considerations discussed in §9.5.7.4. Good experimental measurements of this system are provided by I. Dabrowsky (1984). Using his data, we may identify the optical transitions between the vibrational levels, by using of the following criteria:

- The photon energy and the line abundance should follow the similar trend as the trend, obtained by the PE spectrum

- The vibrational level corresponding to the E_{VIP} of the PE spectrum should be the most populated one.

Using a cross-validation analysis between the optical and PE spectrum of H_2, we found that there are two sets of strongly populated optical bands, whose levels and trends are very close. The first one is from transitions: (0-1), (0-2), (0-3) and so on, denoted by BSM as (0-v) set. The second one is from transitions: (1-1), (1-2), (1-3) and so on, denoted by BSM as (1-v) set. In both sets, the third consecutive optical bands are the most populated and their population corresponds to the PE energy level trend. We accept, that the difference between the both sets corresponds to the Quantum Mechanical electron spin (the same spin is explainable by BSM model as two electrons circulate in opposite directions, while the proton twisting serves as a reference). The signatures of the two optical sets appear also in the PE spectrum shown in Fig. 91.19, as small peaks from both sides of every large peak. If comparing the energy difference between the two sets with the energy difference of the fine line splitting in atomic Hydrogen, it looks larger. But we do not have to forget the shift effect involving directly the SG energy, that may contribute to the larger energy separation between the two sets of optical bands and PE peaks. From the cross validation between the energy level trends of both types of spectra, we conclude that the PE spectrum of Fig. 9.17 corresponds to both sets.

9.6.3 Identification of the common nonlinear trend between the Optical set and the PE spectrum. Estimation of E_1 parameter.

Firstly we will accept, that the shape of optical set and the PE spectrum are both part of parabolas, but with different coefficients. This assumption will be confirmed later. Then we may apply the rule, that the difference between two parabolic functions is also a parabola. Using the level number as an argument (according to BSM numbering of the vibrational levels), we may identify the common trends of the energy levels by the following procedure.

- digitizing the PE spectrum of Fig. 9.19 into a vector E_{PE} (E_{PE} instead of E_K is used in order to avoid a possible confusion in a later analysis)
- form a vector of (0-v) set, denoted as E_0.
- form a vector of (1-v) set, denoted as E_1.
- make a vector difference $\Delta E_0 = E_0 - E_{PE}$ and $\Delta E_1 = E_1 - E_{PE}$
- fit to a simple nonlinear equation (first try a second order polynomial)
- plot the curves

- find the interception point between E_{PE} and ($E_0 - E_{PE}$) and ($E_1 - E_{PE}$).

From the fine structure model of the molecular spectra presented in §9.5.7.4., it follows that the most accurate quantum energy value of the vibrational bands should be in the middle between the first J lines of the P and R branches.

Table 9.2 provides data for (0-v) set with the related parameters: E_{PE} and ($E_0 - E_{PE}$) vectors; and Table 9.3 - for (1-v) with the similar related parameters.

H_2 ortho-I vibrational quantum levels identified by (0-v) set and PE spectrum **Table 9.2**

Vib. level	transition by QM	Ist R line (eV)	Ist P line (eV)	E_0 (eV)	E_{PE} (eV)	ΔE_0 (eV)
0	(0-1)	10.76	10.65	10.66	5.79	4.87
1	(0-2)	10.186	10.168	10.177	5.52	4.65
2	(0-3)	9. 728	9.71	9.719	5.27	4.45
3	(0-4)	9.298	9.281	9.289	5.025	4.264
4	(0-5)	8.896	8.88	8.888	4.8	4.088
5	(0-6)	8.522	8.507	8.5145	4.58	3.934
6	(0-7)	8.177	8.163	8.17	4.38	3.79
7	(0-8)	7.8626	7.8482	7.855	4.2	3.655

H_2 ortho-I vibrational quantum levels identified by (1-v) set and PE spectrum **Table 9.3**

Vib. level	transition by QM	Ist R line (eV)	Ist P line (eV)	E_1 (eV)	E_{PE} (eV)	ΔE_1 (eV)
1	(1-1)	10.873	10.817	10.826	5.79	5.036
2	(1-2)	10.35	10.33	10.34	5.52	4.82
3	(1-3)	9. 891	9.8739	9.882	5.27	4.612
4	(1-4)	9.461	9.444	9.4525	5.025	4.4275
5	(1-5)	9.057	9.043	9.05	4.8	4.25
6	(1-6)	8.685	8.67	8.677	4.58	4.097
7	(1-7)	8.34	8.326	8.333	4.38	3.953
8	(1-8)	8.0252	8.0117	8.0184	4.2	3.8184
9	(1-9)	7.7409	7.7282	7.7345	4.0375	3.697
10	(1-10)	7.4896	7.4778	7.4834	3.88	3.603

From the Tables 9.2 and 9.3, the following sets of vectors are formed: E_0 and ΔE_0 as a function of the band level; E_1 and ΔE_1 as a function of the band level. Then they and fitted to a polynomial of second order $y = a + bx + cx^2$. The vector E_{PE} is also fitted to the same type of polynomial. The chosen polynomial fits quite well for all these vectors with std of the order of 0.0038. The fitting coefficients are respectively:

for E_0:   a=11.174884   b=-0.52778274   c= 0.01408631
for ΔE_0:   a=5.1015982   b=-0.2386875   c=0.007276785
for E_1:
for ΔE_1:
for E_{PE}:   a=6.0749265   b=-0.29214977   c=0.0072671569

Fig. 9.21 shows the plot of E_0, E_{PE} and ΔE_0 as a function of the vibrational level. The extrapolated curve regions are marked by dashed lines (Note: the curves may not look smooth due to the printing capabilities).

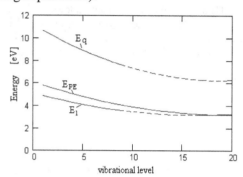

Fig. 9.21

Vibrational levels by the optical band set (0-v) and PE spectrum for H_2 ortho-I. The solid line shows the observed data, the dashed lines show the extrapolated data from the optical spectrum

From the interpolated curves one finds that E_{PE} and ΔE_0 intercepts at 18.147. The first derivative of all three plots at this point are close to zero. At the PE spectrum, the band number 18 (band level by BSM numbering) appears still distinguishable by amplitude and energy difference. If E_0 and E_{PE} were not correlated properly, the interception of E_0 and ΔE_0 would not coincide at the band level 18. So we accept the band level 18 as the last one before the boundary limit convergence. For this band level, from the fitted curve for E_0 we obtain the experimental value of E_B.

$$E_B = 6.239 \text{ eV} \qquad (9.9)$$

$E_B' = 3.163 \text{ ev}$ - shrunk value of E_B estimated by PE spectrum, which has a shrunk energy scale.

Therefore we may accept the last band level to be 18 (this will be confirmed later) and denote it as v_m.

$$v_m = 18 \qquad (9.10)$$

The parameter E_B determined by this approach is approximate. But it helps to identify one important real physical parameter related to SG field (discussed later) and to find its relation to the energy levels shown in Fig. 9.15.

9.7. Theoretical analysis of H_2 ortho-I molecule oscillations

9.7.1 Estimation of rotational energy

The analysis of the H_2 otho-I properties will help to understand the electronic bondings in the molecules. For this reason we will investigate the momentary system energy of H_2. Firstly we will check what type is the molecular oscillation, by comparing E_{rot} and E_1 energy. Using the value of rotational constant, provided by Dieke and Blue (1935) and Dieke, (1958): $B_e = 27.30 \text{ cm}^{-1}$ and applying Eq. (9.A.6), we see that the maximum of the population curve appears for J=2 at temperature 77K and between 2 and 3 , but closer to 2 for 298K. Accepting J=2, corresponding to an equivalent molecular frequency of 54.6 cm^{-1} we may use it for estimation of the rotational energy by Eq. (9.3).

Note: *The experimental value from different experiments shows a variation. The correct choice and some experimental considerations according to BSM are discussed in §9.9.2 and § 9.11.*

Another necessary value is the internuclear distance at the equilibrium point. In the following analysis we will see, that the deviation from the internuclear distance estimated by the free space CL node scale is negligible. So we may use the internuclear distance for the point of dissociation limit. This will correspond to the separation of the two protons of H_2 by a first harmonic quantum orbit $L_q(1)$ as shown in Fig. 9.12. Then the internuclear distance is:

$$r_n = L_q(1) + 0.6455 L_p = 1.795 \text{ A}$$

Using Eq.(9.3) the rotational energy is:

$$E_{rot} = 0.02 \text{ eV}$$

We see, that $E_{rot} \ll E_B$ \qquad (9.11)

The condition (9.11) indicates that the energy of the vibrational motion could not exceed the metastable state. Consequently, this is a molecule with a first type of oscillations, according to the criterion, defined in §9.5.4.1. Additional check of this condition could be made by the shape of its PE spectrum. The bottom edge of the peaks set is not elevated. This is observed even for HD and D_2 molecule, so for H_2 it should be strongly valid. (The PE spectrum of H_2 and HD are pretty close, see J. E. Pollard et al., (1982)),

Consequently:

- **the centrifugal forces in H_2 ortho-I state are negligible in comparison to the bonding forces.**

The above conclusion indicates that H_2 ortho-I state could be considered as a typical case of I-st type molecular vibrations, as defined in §. 9.5.4.1.

9.7.2 Considerations for energy balance involving SG field potentials

The obtained condition (9.11) for H_2 ortho I allows obtaining a simplified expression for the energy balance of the vibrational motion, by neglecting the negligible rotational energy.

The energy balance should be considered for the whole system, comprised of:

     - two protons with their internal RL(T) structures and SG field potentials

     - two electrons of opposite QM spin, sharing a common orbit of first harmonic

     - the CL space enclosed by the integrated Bohr surface from the two protons

     - the internal E-field enclosed in the integrated Bohr surface

All these energetic components are included in a total energy balance between two potential fields: SG field and E-field. The magnetic filed from the orbiting electrons is locked inside the Bohr surface and does not appear in the external CL space. So we may consider that its energy is included in the Bohr surface volume energy, but estimated by the external CL space parameters. <u>This means, that proton's E-field should participate in the balance with its full value, which is equal to the unit charge.</u> From the other side, we know, that the E-field (charge) is contributed and controlled by the SG field. Now we are very close to the possibility to obtain a relation between some characteristic parameter of SG field and the controlled E-field. We may formulate this relation in a following way:

- **The electrical field energy of a helical structure in CL space is adjusted to the energy of unit charge by the balance of SG forces.**
- **The unit charge energy is equal to the Super Gravitational potential of the structure referenced to a free CL space (away from another structure)**

It is difficult to obtain the relation parameters for a single particle like a proton or even for the Hydrogen atom. However, we may obtain the relation by analysis of the motion behaviour of pair of similar charge particle. The two protons involved in the H_2 molecule (together with two electrons) could be regarded as a such pair. The H_2 ortho-I molecule, shown in Fig. 9.12 is suitable for such analysis.

9.7.3 Definition of C_{SG} factor and using it as a characteristic parameter of the SG potentials

The SG potential between two particles can be determined in a classical way, by separating them from their initial distance to infinity. Mathematically, this is expressed by integration of the acting between them SG forces from the initial distance to infinity. Let us determine the SG(CP) potential (CP - denotes the SG vectors of the central part of the prisms) between two hadron structures. Because we accepted that the energy of the electrical charge is a part of the SG energy, we should not take the charge into account. The SG law varies with inverse cubic power of the distance.

Having in mind all this considerations, the SG(CP) potential (Energy) between the two protons or neutrons) can be estimated by the expression.

$$E_{SG}(CP) = -2\int_{r_{ne}}^{\infty} \frac{G_o m_{po}^2}{r^3}dr = \frac{C_{SG}}{\left(L_q(1) + 0.6455L_p\right)^2} \quad (9.13)$$

where: m_{po} is the Intrinsic mass of the proton, G_o is the intrinsic gravitational constant

$$C_{SG} = G_o m_{po}^2 \text{ - SG factor}$$

Note: The factor 2 in front of the integral comes from the two arm branches (along *abcd* axes) of the CL space cell unit. They both are included in the xyz cell unit to which all CL space parameters are referenced. The consistency of all later used equations in which C_{SG} factors is involved confirm the need of the factor 2.

C_{SG} factor for SG forces between two intrinsic masses is similar as Gm_p^2 factor for Newtonian gravitational forces between two Newtonian masses of proton. Despite that the values of G_o and m_{po} involved in the factor C_{SG} are still unknown, this factor is very useful, as will be shown later.

9.7.4 Determination of C_{SG} factor from the SG energy balance of H_2 ortho-I molecule.

We are familiar with the Newtonian gravitational field, electrical field and magnetic field. The SG fields and potentials are quite more strong than any other fields and potentials we are familiar with. Then let formulate the energy balance between the hadron structures by the SG potentials, according to the considerations discussed in §9.7.2. For the whole system including the CL space domain we may expect that the following balance exists:

SG(CP) energy = SG(TP) energy

Applying this balance for the H_2 ortho-I system, we get:

$$E_{SG}(CP) + E_X = 2(E_q) + 2(E_K) \qquad (9.14)$$

where:

$E_{SG}(CP)$ - is the SG(CP) energy

E_q - is the SG energy spent for creating the positive unit charge of the proton

E_K - is the kinetic energy of the electron.

E_X - energy potential contributed by the CL space, which we must identify

The energy E_X may appear involved in the energy diagram shown in Fig. 9.15. Our task is to identify it.

Consideration (a): The Eq. (9.14) characterizes the system without considering the quantum quasishrink effect.

For the electrical charge energy we will use the following well-known expression:

$$E_q = h\nu_c = m_e c^2 \qquad (9.15)$$

The kinetic energy of the electron with first harmonic velocity is

$$E_k = 0.5 m_e \alpha^2 c^2 = h\nu_c \alpha^2 \qquad (9.16)$$

Substituting Equations (9.13), (9.15) and (9.16) into Eq. (9.14) and solving for C_{SG} we obtain:

$$C_{SG} = (2h\nu_c + h\nu_c\alpha^2 + E_X)(L_q(1) + 0.6455L_p)^2 \qquad (9.17)$$

E_X could be some of the energies of H_2 molecule. Expressed in eV it could not exceed 16 eV. For such range the value of the C_{SG} factor (referenced to energy balance in eV) in fact changes insignificantly - only at 5th significant digit.

If $E_X = 0$, then $\qquad C_{SG} = 5.265108 \times 10^{-33}$

If $E_X = 9.719$ eV $\quad C_{SG} = 5.265127 \times 10^{-33}$

It is quite possible E_X energy to be equal to E_B that is the difference between the dissociating limit and the metastable state. It has been determined in §9.6.3 and given by Eq. (9.9). It is more convenient, however, in further analysis to consider the more universal Eq. (9.17) in which $E_X = E_B$. Then the value for C_{SG} factor when used in energy balance in eV is:

$$C_{SG} = 5.26508 \times 10^{-33}$$

9.7.5 Emitted photon energy as an excess energy in the total energy balance involving SG CL space energies

A. Photon emission (absorption)

In the derivation of the factor C_{SG} the quantum quasi-change effect has not been taken into account, but C_{SG} according to Eq. (9.17) is not sensitive to change of E_X in a range of few electronvolts. For determination of the photon energy, however, the E_X energy must be taken into account.

Using again Eq. (9.14) with substituted terms ($E_{SG}(CP)$ from Eq. (9.13) and other terms) and dividing on the unite charge q in order to obtain the balance in electronvolts we get:

$$\frac{C_{SG}}{q(L_q(1) + 06455L_p)^2} = \frac{2E_q}{q} + \frac{2E_K}{q} - E_X \qquad (9.18)$$

The quantum quasishrink effect will affect the orbital length $L_q(1)$. So we have to find the proper function affecting this parameter. In fact there are two ambiguous trends of the system:

Trend (1) The electron tries to fall to lower quantum orbit, whose trace length is distinguished by one Compton wavelength λ_{SPM}.

Trend (2) The system tries to keep the present status by small decrease of the internuclear distance

We will analyse firstly **the effect of Trend (1)** by some approximate method. From Eq. (9.17) we have seen that the SG field energy is much larg-

9-30

er than other energies, so a small change of internuclear distance means a large energy change in the range of system energy E_{SYS}.

Let us analyse the effect starting from the dissociation limit where Lq(1) could be considered unchanged and moving in a direction of shorter r_n. At this point the orbit length is $L_q(1)$ and its trace length contains approximately 137 numbers of Compton wavelengths. Changing the number of the Compton wavelength by one, the next stable quantum orbit may contain 136 numbers, following by 135 and so on. It is convenient to denote the change of these numbers by integer parameter Δ. Let us denote the quasishrink length by a prime index ('). Then for the quasishrink trace length we have:

$$L_{Tq}' = L_{Tq}\left(1 - \frac{\Delta}{137}\right) \approx L_{Tq}(1 - \alpha\Delta)$$

If assuming that the shape of the quantum orbit in a quasishrink space is preserved, then a similar expression should be valid for the size of the quantum orbit, but the term $\alpha\Delta$ should be multiplied by the factor π. This comes from the following considerations: The trace length of the Lq(1) is λ_c/α. We may make from the same length a circles with a diameter $d = \lambda_c/(\pi\alpha)$. If λ_c changes due to the quantum quasi-change effect we have $\Delta d = \Delta\lambda_c/(\pi\alpha)$. We see that α is multiplied by a factor π. Now if considering that the distance is defined by a Hippoped curve with a same trace length, then the condition is the same - α is multiplied by a factor π. Consequently, we arrive to the expression.

$$L'_q(1) = [L_q(1)](1 - \pi\alpha\Delta) \qquad (9.19)$$

The obtained expression is only approximate, because we also need to take into account the ability of the SG field to control the Compton wavelength in a strongly oriented E-field.

The effect of Trend (2)

Eq. (9.19) has only one dimensional spatial dependence whose variable parameter is Δ. But the proximity E-filed controlled by the SG field is a three dimensional. The possible parameters from Eq. (9.19) that could be affected are only α and Δ. Obviously some additional dependence of Δ pa-

rameter should exist. We may try to find this dependence by empirical way. Following the logical considerations about quasi-change space we may define the first unknown parameter as a power degree on α, and the second one as a power degree on Δ.

Then the expected function $L'_q(1)$ will take a form:

$$L'_q(1) = [L_q(1)](1 - \pi\alpha^x\Delta^y) \qquad (9.20)$$

The quantum number Δ (integer) was referenced to the dissociation limit of the step-like vibrational curve. In order to use the vibrational number we substitute

$$\Delta = \upsilon_m - \upsilon \qquad (9.20.a)$$

where: υ - is the vibrational number (also integer) and υ_m - is its maximum value, previously obtained in §9.6.3 and given by (9.10).

If substituting $L_q(1)$ from Eq. (9.18) by $L'_q(1)$ given by Eq. (9.20) and expressing the internuclear change as a function of the vibrational number υ we get:

$$\Delta E = \frac{C_{SG}}{q[[[L_q(1)](1 - \pi\alpha^x(\upsilon_m - \upsilon)^y)] + kL_p]^2} - \frac{2E_q}{q} - \frac{2E_K}{q} + E_X \qquad (9.21)$$

where: $k = 0.6455$ a coefficient defining the locus position of the proton (having a shape of a Hippoped curve with a parameter $a = \sqrt{3}$)

The obtained balance difference in comparison to the first and second terms is quite small. Let us accept that **the system reacts in order to restore the accurate balance by emission (absorption) of a photon.**

When the energy difference of (9.21) is positive, a photon with the same energy will be emitted. In such way, the SG energy balance in the system is restored. Therefore, we may regard the balance change ΔE as a photon energy. Then plotting Eq. (9.21) as a function of υ, we try to obtain a best fit to the (0-v) and (1-v) optical transitions sets, by selecting the unknown parameters x and y.

Fig. 9.24 shows a step-like plot of ΔE as a function of υ with $x = 4$, and $y = 2$, together with the vectors E_0 and E_1. For better visualization, the argument v is shifted at a half step in order the middle of the step to be between the two sets of the observational data given by vectors E_0 and E_1. The vertical scale is drawn reversed in order to reference to energies corresponding to the necessary zero level, which appears in upper part of the step-like curve (the curve is similar as the vibrational curve shown in Fig. 9.15, but only the right-hand section is shown.)

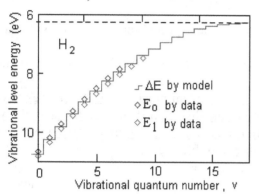

Fig. 9.24

Fig. [9.24]. Calculated, ΔE, and estimated vibrational levels by the optical transitions E_0 and E_1 for H_2 ortho-I

The shown vibrational levels in fact are contributed by the momentary balance disturbance from both sides of the equilibrium point of vibration. The energy difference ΔE works symmetrically for positive and negative quantum displacement, so they appear with same vibrational levels as shown in Fig 9.24. Examining the dependence of the obtained plot from the parameter E_X we find that only its vertical position is influenced (it is a same if E_X is included as part of C_{SG}). The best vertical fit is obtained at $E_X = 6.26$ (eV). In Fig. 9.23 this level is shown by a dashed line. The difference between the theoretical plot and the levels obtained by the optical data is shown in Fig. 9.25.

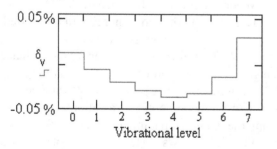

Fig. 9.25. Difference between calculated and experimentally determined vibrational levels

The excellent agreement between the theoretical results and the experimental data permits to ob-

tain the values of the two unknown parameters: $x = 4$ and $y = 2$.

Then the final set of expressions for calculating the vibrational levels of H_2 ortho-I molecule whose configuration is shown in Fig. 9.12 are:

$$E_\upsilon = \frac{C_{SG}}{qr^2} - \frac{2E_q}{q} - \frac{2E_K}{q} + E_X \qquad (9.23)$$

$$r = [[L_q(1)](1 - \pi\alpha^4(\upsilon_m - \upsilon)^2)] + 0.6455L_p \qquad (9.23.a)$$

$$L'_q(1) = [L_q(1)](1 - \pi\alpha^4(\upsilon_m - \upsilon)^2) \qquad (9.23.b)$$

where:

E_υ [eV] is the vibrational level of H_2 ortho-I molecule, Eq and E_K are given respectively by Eqs. (9.15) and (9.16); r - is the internuclear distance as a function of the vibrational number υ

$$E_X = E_B = 6.26 \text{ (eV)} \qquad (9.23.c)$$

The good match between the calculated results and experimental data also allows to identify the energy levels of H_2 molecule in order to define the position of the step-like vibrational curve in respect to other experimentally determined energy levels - the binding and the dissociating energies (see Fig. 9.15).

The energy difference between level 0 and $\upsilon_m = 18$ obtained by Eq. (9.23) is 4.483 eV. This value is pretty close to the dissociation energy obtained experimentally $E_{DIS} = 4.478$ (eV). The energy difference between dissociation limit and the metastable state is $E_B = E_X = 6.26$ (eV). The binding energy of H_2 is an accurately measured parameter: $E_{IP} = 15.426$ (eV) (E. McCormack et al., 1989). Then we obtain $E_A = 4.69$ (eV) that corresponds to the distance between the metastable state and the ionization limit.

It is found that the Eq. (9.23) provides exactly the same results for both cases: (a) the offset value of 6.26 (eV) is a part of C_{SG} expression according to Eq. (9.17); (b) the offset value is included as a separate term in Eq. (9.23). Accepting the second option, allows C_{SG} factor to be defined independently of its involvement in any molecule and to be expressed accurately by the known physical constants. Determined in such way, it could be conveniently used in the further analysis of more complex diatomic molecules.

$$C_{SG} = (2h\nu_c + h\nu_c\alpha^2)(L_q(1) + 0.6455L_p)^2 \qquad (9.24)$$

$$C_{SG} = 5.26508 \times 10^{-33} \left[\frac{m^5}{kgs^3}\right] \qquad (9.25)$$

According J. H. Black and A. Dalgarno, (1976), the excited H_2 by UV radiation (Lyman and Werner system) leads to "fluorecsence to the vibrational continuum of the ground electronic state, thus resulting in two separated atoms". The provided above analysis is in agreement with this concept.

The results obtained by the provided analysis lead to the following conclusions:

(a) The value of C_{SG} factor, expressed by known physical constants, is an accurate enough for the suggested analysis

(b) The equation (9.23) shows the SG field involvement in the total energy balance including also the electrical charges of the system

(c) The H_2 ortho -I state could be regarded as a bonding structure between atoms in the molecules in a case of valence proton without connected to it neutron. (In the case of connected neutron - Deuteron, Eq. (9.23) is slightly modified - see §9.8)

B. Domination of the C_{SG} factor in the total energy balance

The C_{SG} factor could be considered as a basic intrinsic matter parameter. It involves the equivalent intrinsic gravitational constant, G_o, (for the mixture of two type intrinsic matter in CL space and in particles) and the intrinsic mass of the proton or neutron (having in mind that the proton and neutron have one and a the same intrinsic matter quantity, referred in BSM as an intrinsic mass).

The corresponding factor for Newtonian mass is

$$C_N = Gm_p^2 = 1.866772 \times 10^{-64}$$

where: G -is the Newtonian gravitational constant and m_p - is the proton mass.

We must keep in mind that C_{SG} is related to the inverse cubic law, while C_N to an inverse square law, so they have a different dimensions. Then the following expression is valid for length larger than the CL node distance, since the Newtonian gravitation is valid only above this distance.

$$C_{SG}/C_N = 2.8204 \times 10^{31} L \quad \text{(m)} \qquad (9.25.a)$$

Let us get this ratio for a distance L equal to the node distance d_{nb} obtained in §2.11.3 and given by (2.56): $L = d_{nb} = 1.097 \times 10^{-20}$. Then we obtain:

$$C_{SG}/C_N = 3.094 \times 10^{11}$$

The obtained ratio could be considered as valid for the boundary zone of the length scale above which the Newtonian gravitation is possible. We see that at this scale point the SG forces are extremely strong.

If scaling the ratio C_{SG}/C_N for a larger distance we will see that it is still significant for the size of the quantum orbits and even for the internuclear distance of atoms connected in molecules by EB types of bonds. For this reason the C_{SG} factor is important parameter in the energy balance of the vibrational and rotational motion of the molecules. Additionally, one very useful feature of the SG energy balance is apparent from the H_2 analysis:

The bonding system comprised of the H_2 (or D_2) structure has ability to provide a fast and accurate restoration of the balance of the total energy of the molecular system in CL space, by emission or absorption of a photon. It is apparent that the vibrational complex of this system will carry also a signature of the participating atoms.

C. Intrinsic matter parameters involved in the total energy balance and their common relations.

The accurate balance of the total energy according to Eq. (9.23) leads to one question: *What is the driving mechanism that is able to provide so accurate balance?* The possible answer is:

This could be the balance between the energy of SG(CP) and the energy of SG(TP) vectors of the whole system, including the CL space and the local gravitational field.

The system possesses self adjustable parameters such as: the node distance for the local CL space and the E-field distribution inside the Bohr surface. The effect of Bohr surface integration also should be taken into account (involving a synchronization between RL(T) of all FOHSs of the protons and the interaction of this commonly synchronized system with the CL nodes).

In such aspect, it could be considered that the mechanism providing the accurate energy balance of Eq. (9.23) is defined by the balance between the SG(CP) energy and SG(TP) energy of the whole system. In Chapter 2, §2.9.6.B, it was accepted a priory that the energy balance between both types of intrinsic energies embedded in the prisms, forming the CL space, is expressed by the equation:

$$E_{SG}(TP) = 2\alpha E_{SG}(CP)$$

where: α - is the fine structure constant, $E_{SG}(TP)$ - the SG energy associated with the twisting part of the prism, $E_{SG}(CP)$ - the SG energy associated with the central part (untwisted SG component of the prism model)

This acceptance is successfully validated in §9.15.2 for derivation of total energy balance of a simple diatomic molecule.

Let us try to find which of both SG energies the energy terms of Eq. (9.23) are related to.

The term of Eq. (9.23) containing C_{SG} originates from E_{SG}, whose derivation was based on Eq. (9.13) with the presumption that it is related to the SG(CP) field. The latter supplies the embedded energy for the proton electrical charge according to Eq. (9.14). Additionally, it is able to regulate the proximity E-field of the proton (quantum quasishrink effect), according to Eq. (9.23). Following the same logic, we can accept that it is involved in regulation of the E-field inside the Bohr surface of the single not connected proton, defining in such way the unit charge of the proton. In the latter case, however, the SG(TP) field may also be involved. The unit charge according to BSM concept is related to the CL nodes dynamics, so it is defined only for a CL space environment. Then the logical conclusion is that the following components are involved in the unit charge formation:

- SG(CP) field
- SG(TP) field
- CL space parameters

D. Range of vibrational motion of the protons in the H_2 molecule

From Eq. (9.23) for internuclear distance we see that only one term is changing with the vibrational number.

$$\Delta r = \pi\alpha^4(\upsilon_m - \upsilon)^2 L_q(1) \qquad (9.26)$$

The change of the internuclear distance for the full vibrational curve is twice larger or

$2\Delta r = 4.37 \times 10^{-17}$ (m). <u>It is quite small in comparison to the internuclear distance</u> $r_n = 2.224 \times 10^{-10}$ (m) (see Fig. 9.12). The reason for the small scale range of vibrations is the superstrong SG field in which the vibrations take place. Then we may reference the internuclear distance to vibrational number counted by the equilibrium distance, so they will mach to the QM vibration number. This option appears more useful especially for the analysis of homogenous diatomic molecules, other than H_2. Then the convenient expression of the internuclear distance is

$$r = r_n \pm \Delta r \quad \text{- for diatomic molecule} \qquad (9.26.b)$$

$$r_n = L_q(1) + 0.6455 L_p \text{ - for H2 -ortho-I} \qquad (9.26.c)$$

$$\Delta r = \pi\alpha^4 \upsilon^2 L_q(1) \text{ - for diatomic molecule} \qquad (9.26.d)$$

The analysis of other diatomic molecules with discrete spectra (presented elsewhere as a BSM_Appendix 9-1) also shows a quite small vibrational range in a length scale. **This means also that the vibrational motion does not involve a significant inertial interaction with CL space related with displacement and folding of CL nodes.** Then for a practical considerations we may ignore the inertial moment in a such type of vibrational motion. The rotational motion and its anticipated energy however exist. The negligible vibrational range in comparison to the internuclear distance facilitates the theoretical analysis by the following considerations:

(a) The system does not spend much vibrational energy for inertial interactions between the nuclei and the CL space, so they can be ignored from the vibrational energy expression. The transitions between different levels are pure quantum mechanical transitions related to the quantum quasi-change (space) effect.

(b) The rotational motion exhibits inertial interactions in CL space and cannot be ignored.

(c) The small change of the internuclear distance permits to use the internuclear distance for the limiting orbit for energy estimation and size, where the the external CL space parameters are valid.

(d) The above features could be propagated for diatomic molecules with a Ist type of vibrations.

(e) The lack of detectable time delay in the Raman process obtains a physical explanation.

The conclusions (a) and (b) allows to understand, why the rotational and vibrational energies overlaps in the vibrational-rotational optical spectra of the molecules.

The conclusion (c) has been demonstrated by the calculation of E_{SYS} energy by Eq. (9.4) and compared with the value known as Average kinetic energy 15.98 (eV).

The conclusion (e) is confirmation of the discussion analysis about the Raman scattering provided in §9.5.8.

The derived model of H_2 ortho-I could be regarded as a single valence bonding system between two atoms. In most elements, however, the deuterons instead of protons provide their valences. In this case the bonding system is comprised of pair deuterons connected by a common quantum orbits. For this reason a similar analysis of a D_2 molecule is presented in the next section.

Note: The EB bond system between elements could be also of type: proton-deuteron, proton-tritium, deuteron-tritium and proton-tritium.

9.8 D_2 ortho-I molecule as a single valence bonding system in the molecules.

Most of the atoms contain Deuterons in the external valence shell instead of protons. Some heavier atoms contain Tritium.

It is quite logical to expect, that the D_2 will have similar oscillating behaviour, as H_2, but modified by the two neutrons. The neutrons are over the saddle point of each proton so the internuclear distance between the protons and deuterons is one and a same. Logically thinking, both deuterons will affect the degree of the quantum quasishrink space in the integrated Bohr surface. Consequently, we may try to use the Eq. (9.23), derived for H_2 molecule, but searching for a proper correction factor for the term Δ^2. This correction will affect the curvature of the step-like vibrational curve. Another correction, affecting the vertical position of the vibrational levels should be related to E_B for D_2. The C_{SG} factor obtained by H_2 model, however is intrinsic SG field parameter and should be used without change for D_2 case.

The procedure is exactly the same as for H_2 and there is no need to be described in details. It is useful to show the PE spectra of D_2 and H_2 measured at one and a same experimental conditions. Figure 2.29 shows PE spectra of H_2, HD and D_2 obtained at one and a same pressure and temperature, provided by J. E. Pollard et al. (1982).

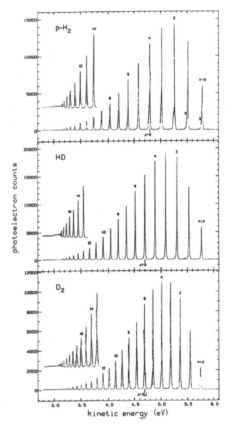

Fig. 9.29

The 584 A photoelectron spectra of H_2, HD and D_2 expanded from 200 Torr at 77K.
(courtesy of J. E. Pollard et al. (1982)

The optical spectrum provided by H. Bredohl and G. Herberg (1973) denoted as Lyman band of Deuterium, matches the spectrum of D_2 ortho-I (according to BSM).

Using the PE spectrum we obtain:

$E_{VIP} = 21.23 - 5.05 = 16.18$ eV. The identification of E_{VIP} only by the most populated bands of the optical spectrum for D_2 is not so obvious as in the case of H_2. But using the simultaneous signature of the vibrational levels in the PE and optical spectra, the identification of E_{VIP} level becomes possible. If assuming that the Energy level E_B is close to the same energy level for H_2, then we have:

16.18 – 6.26 = 9.92 eV . Using the observed data presented by H. Bredohl and G. Herzberg, (1973), the obtained energy corresponds to the mean energy value of the optical (0-4) and (1-4) transitions. The available observed spectral data are from (0-3) to (0-10) for (0-v) set and from (1-3) to (1-12) for (1-v) set.

Following a procedure similar as for H_2, it is found that the optical transitions that match better the PE spectrum are again from (0-v) and (1-v) transitions. The largest vibrational level identified by PE spectrum matches well the identified optical transitions. In order to obtain a best fit between the optical transition sets and the calculated data, the Eq. (9.23) is used but with correction factors for the curvature (multiplying factor for the term $\alpha^4\pi(\upsilon_m-\upsilon)^2$) and for the vertical positions value of E_X. Both correction factors are obtained by tuning of the calculated vibrational curve to the energy levels identified by the optical spectrum of D_2.

The obtained equation for the quantum vibrational levels for D_2 ortho-I state is:

$$E_V(\upsilon) = \frac{C_{SG}}{qr^2} + 6.235 - \frac{2E_q}{q} - \frac{2E_q}{q} \quad (eV) \qquad (9.27)$$

$$r = [[L_q(1)](1 - 0.5\pi\alpha^4(\upsilon_m - \upsilon)^2)] + 0.6455L_p \qquad (9.27.a)$$

where: C_{SG} - is given by Eq. 9.25;

E_q - is a charge energy defined by Eq. (9.15) and E_k is the electron kinetic energy, given by Eq. (9.16), $L_q(1)$ - length of a first harmonic quantum orbit, L_p - proton's length

The energy 6.235 eV is obtained by the best fit of the vertical position, while the factor 0.5 - by the best fit of the curvature of the vibrational curve (calculated by Eq. (9.27) to the energy levels from the spectroscopic data. The theoretical curve and the energy levels from the spectroscopic data are shown in Fig. 9.30.

The correction factor of 0.5 is reasonable if taking into account that every proton has a neutron in its saddle. This neutron may influence the strength and the range of the quantum quasi-change effect (length scale) causing an extended vibrational curve with a different curvature. The energy 6.235 eV is distinguished from the corresponding energy of 6.26 eV (for H_2) only by 0.025 eV.

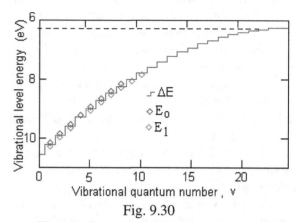

Fig. 9.30

Theoretically calculated vibrational levels, using Eq. (9.27) (step line) and experimentally determined vibrational levels involved in (0-v) and (1-v) transitions (daimon points) for D_2 Lyman bands

For υ_m = 23 - a value identified by the PE and the fitting to the optical spectra and the curve fitting to spectral data, the change of internuclear distance for the full vibrational curve is 2.79×10^{-17} (m) . This is very small in comparison to the internuclear distance of 1.79×10^{-10} (m) (estimated by Eq. (9.27.a)).

Since the valences of most molecules are provided by valence deuterons, the D2 molecule can be considered as a bonding element. Then for diatomic molecules, the internuclear distance is more convenient to be expressed as:

$$r = r_n \pm \Delta r \quad \text{- for diatomic molecule} \qquad (9.27.b)$$

$$r_n = L_q(1) + 0.6455L_p \quad \text{for } D_2 \text{ -ortho-I} \qquad (9.27.c)$$

$$\Delta r = 0.5\pi\alpha^4\upsilon^2 L_q(1) \quad \text{- for diatomic molecule} \qquad (9.26.d)$$

Summary

- **The Intrinsic factor for SG(CP) forces between SG masses of two protons is derived**
- **The derived energy balance equations (9.23) for H_2 and (9.27) for D_2 ,provide a model of a single valence electronic bonding system involving valence protons and deuterons, respectively. The bonding connection is identified by analysis of the optical and PE spectrum of the H_2 and D_2 molecules.**
- **The process of photon emission or absorption is related to the restoration of the**

momentum disbalance of the total energy of the system involving SG filed components.
- **The size of the length scale of the vibrational range is very small, since the inverse cubic SG forces are involved. This feature allows to facilitate the analysis of the vibrational motion by ignoring the inertial interaction between the atomic nuclei.**

9.9 Interactions in quantum quasishrunk space

9.9.1 Affected CL space parameters in the proximity E-field of the proton

In the previous sections we find out that the vibrational levels are related to quantum interactions, which take place in the proximity E-field of the proton, where conditions of quantum quasishrunk space are created. In this section an additional analysis of some spectral features of H_2 and D_2 molecules will allow us to understand the signature of the SG field in the line spacing of the P, R and Q branches of the vibrational rotational spectra of molecules.

In order to visualize the vibrational level for one full statistical cycle of the vibrating diatomic molecule will use a symbolic graphical presentation of the cycle as a plot shown in Fig. 9.31.

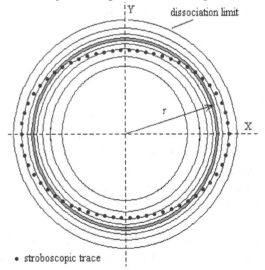

Fig. 9.31
Symbolic plot of the vibrational rotational motion

The full vibrational cycle is presented as discrete energy levels, shown as points, which a radius vector r (with a variable length) makes for a 2π rotation. In fact the internuclear distance is twice the size of the radius vector r. During one cycle, the larger internuclear distance is along a the horizontal axis x, while the shorter one - along the vertical axis y. The density of the vibrational energy levels are shown by concentric circles (closer circles correspond to a larger difference between the energy levels). The closely spaced circles in the central zone correspond to the equilibrium position at the zero energy level of the step-like vibrational shown in Fig. 9.15.A. The points arrange in a shape of ellipse are the allowable vibrational levels. They intercept the equilibrium position twice per one statistical cycle.

For a single molecule, the possible transitions between the vibrational levels in the simultaneous rotational and vibrational motion are such, that radius vector is not necessary to pass through the equilibrium positions. Instead, it may jump from one point of the slop of the vibrational curve to another to a lower energy point from the other slop. In this case, a photon with energy equal to the energy difference between the two levels will be emitted. So the transitions are in fact between both sections of the vibrational curve. This is in agreement with the energy balance discussed in §9.7.4 and Eq. (9.23). Additionally it is possible a transition between one vibrational level from one side of the equilibrium to the same vibrational level on the other side, but at different rotational levels. (The latter case is more probable for molecule of II-nd type of vibrations, at which the inertial interactions are larger). This feature in fact is present in the vibrational-rotational spectra of the molecules.

Let us see what could be the possible relation between the statistical vibrational cycle and the rotational cycle. From the symbolic plot in Fig. 9.31 it is evident that if their periods are equal, the positional molecular momentum is not balanced in respect to the stationary frame. When the rotational period is equal to two vibrational periods, the positional molecular momentum is balanced.

9.9.2 Distribution of the quantum transitions along the vibrational trace

Figure 9.32 illustrates the trace of the vibrational motion as a symbolic curve.

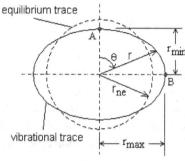

Fig. 9.32

where: $2r_{ne}$ - is the internuclear distance in equilibrium point

Note: The radius vector r in the symbolic plot is not proportional to the internuclear distance in free CL space but to some normalized value in the quasishrunk space. In such aspect r_{min} and r_{max} are referenced to the quasishrunk length unit

Points A and B are symmetrically situated from the both sides of the equilibrium trace. Although they correspond to one and a same energy (of the vibrational level), they are physically separated.

Here we will consider a balance between two types of SG energies: SG(TP) and SG(CP) involved in the following types of interactions:
SG(TP): involved in vibrational energy: for $r > r_{ne}$
SG(CP): involved in attraction energy: for $r < r_{ne}$

It is more convenient to use the ratio between r_{min} and r_{max} $k = r_{min}/r_{max} < 1$

Let us examine the energy distribution in a limited angle of the radius vector around X and Y axes, for example within angle of $\pm 45°$.

Firstly, we may determine the change of radius vector as a function of angle deviation, but normalised to the corresponding value at the positions A and B respectively:

$$\delta r_A = \sqrt{k^{-2}\sin^2(\theta) + \cos^2(\theta)} \text{ near } r_{min}$$

$$\delta r_B = \sqrt{\cos^2(\theta) + k^2\sin^2(\theta)} \text{ near } r_{max}$$

The corresponding SG(CP) energies are inverse square dependent on the distance. The normalization of the radius vector for points A and B in fact is justified from the energy point of view. If

assuming transitions between the corresponding energy levels and the lowest vibrational level, the energy difference will give the photon energy. So the expressions of the energy envelope of the corresponding spectral line are respectively:

$$E_A = [k^{-2}\sin^2(\theta) + \cos^2(\theta)]^{-1} \quad \text{near } r_{min} \quad (9.36)$$

$$E_B = c[\cos^2(\theta) + k^2\sin^2(\theta)]^{-1} \quad \text{near } r_{max} \quad (9.36)$$

where: c - is a normalization factor of the E_B in order to be a unity at point A, since a common vector angle θ is used. It depends on k.

The plotted curves of E_A and E_B (at parameter $k = 0.9$ for which $c = 0.81$) are shown in Fig. 9.33.

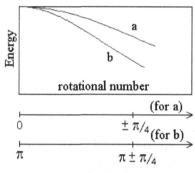

Fig. 9.33

Shape of the energy envelope of the rotational spectral lines for one band of H_2 spectrum
a - near r_{min} (at one side of the vibrational curve)
b - near r_{max} (at other side of the vibrational curve)

The shown plot illustrates what could be the shape of the energy envelope of the spectral lines from P and R branches if plotted towards the rotational quantum number (Note: must not be confused with the P and R branch plot towards wavenumber of wavelength). For linear diatomic molecules the R and P branches are well separate ad the R-branch is with higher energy spectral lines.

E_A - corresponds to R branch from one band

E_B - correspond to P branch from the same band

The P and R branches of the vibrational-rotational spectra of the ortho-I states of H_2 and D_2 molecules exhibit a similar signature. Figure 9.34 shows the energy position of the lines from the P and R spectral branch of H_2 for (0-v) transitions.

According to BSM model, there is not a reason for offsetting the J numbers of P and R branch by one. Therefore, both branches are shown as started from one and a same rotational quantum number. The data are taken from I. Dabrowski, (1984) (Lyman system $(B^1\Sigma_u^+ - X^1\Sigma_g^+)$ $(0 - v'')$

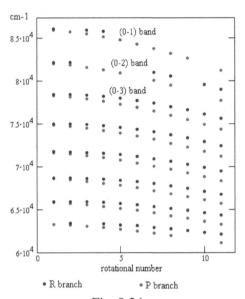

Fig. 9.34
Energy position of the lines from P and R spectral branches of H_2 for (0-v) transitions.

Fig. 9.34 shows a similar plot for D_2 ortho-I state. The spectral data are provided by H. Bredohl and G. Hezberg, (1973).

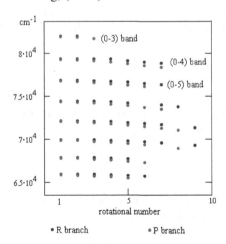

Fig. 9.35
Energy position of the lines from P and R spectral branches of D_2 for (0-v) transitions.

The comparison between both figures indicates, that the eccentricity of the vibrational trace for D_2 molecule is much smaller than the H_2 molecule. This is a result of the influence of the neutron, which is over the proton saddle. This also shows that the confinement between the vibrational cycle and the SPM vector is larger for H_2. When examining the lines from the first few vibrational numbers (directly from the spectra), we find that the difference between them for less energetic bands is quite small. For some bands, the trend for the second and even third vibrational number is reversed. This effect is weaker for H_2 but much stronger for D_2. It appears only for R branch, but never for P branch. The explanation of this effect is the following. For lower vibrational energy, the molecule exhibits a stronger confine interaction with the SPM vector, because of the close proportionality between the vibrational period and the CL time constant. Then a second mode appears in the vibrational trace. This mode is stronger in D_2 molecule because its vibrational trace is with smaller eccentricity. The second mode is illustrated in Fig. (9.36).

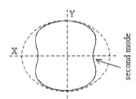

Fig. 9.36
Second mode in the vibrational trace

In order to provide some prove of the quasishrink effect, we can make the following test with H_2 data: displacing the P from R branches with some number ΔJ until obtaining of overlap in most of the points. This is done in Fig. 9.37, where the P branches are displaced by $\Delta J = 3$ The red (grey) spots are over the blue (black spots).

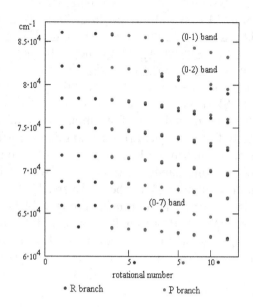

Fig. 9.37

Test of the quasishrink effect by displacement of the P branches in respect to the R branches
The red (grey) spots are over the blue (black) spots

9.9.3 H_2 ortho-II state.

This state of Hydrogen molecule has been shown in Fig. 9.4., which is given again.

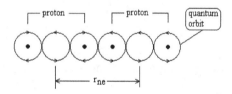

Fig. 9.4
H_2 ortho-II state configuration

The sections of the two proton cores are shown as small black circles. The common quantum orbit is consisted of two serially connected first order quantum loops. The two electrons posses an opposite QM spin. The arrows show the trace direction of one electron and simultaneous positions of the two electrons.

Due to the extended shape of the molecule it could vibrate in two possible modes:

   - stretching (without bending)
   - stretching and bending

The stretching mode is characterized by P and R branches only. The stretching with bending is characterized with P, R and Q branches. The bending causes asymmetrical distortion of the quantum orbit, as discussed in §9.5.7.4.1.

Observations of such spectra is provided by T. Namioka (1964). The data indicated as $(B' - X)$ bands correspond to the first mode, while the data indicated as $(D - X)$ bands - to the second mode. Table 9.4 shows the energy levels of the first J lines from the P and R branches of $(0 - v)$ transitions, from $(B' - X)$ bands.

Optical transitions (0-v) of H_2 ortho-II   ***Table 9.4***

Transition	First J of R (eV)	First J of P (eV)	$h\nu_{av}$
(0-0)	13.7	13.683	13.69
(1-0)	13.936	13.92	13.928
(2-0)	14.149	14. 129	14.139
(3-0)	14.338	14.318	14.328
(4-0)	14.498	14.48	14.489
(5-0)	14.616	14.598	14.607
(6-0)	14.652	14.636	14.644
(7-0)	14.664	14.649	14.656
(8-0)	14.672	14.657	14.664

Without plotting the photon energy, it is evident that the larger energy is at (9-0) transitions.

Now let make a simple analysis of the H_2 ortho-II configuration, shown in Fig. 9.4. If accepting that the orbital shape is made of aligned circles and the sections of the proton core are centred in these circles as shown in the figure, the total trace length is: $3\pi L_p$ = 6.286327 A. Dividing on the trace length of the first harmonic quantum loop we obtained ratio of 1.89067. In order this ratio to be equal to the closer integer 2, we have to correct the length by 2/1.89067=1.0578. With the same factor the internuclear distance should be corrected, or:

$$6 \times \frac{L_p}{4} \times 1.0578 = 1.05833 \text{ A}.$$

Using this value for internuclear distance as for H2 ortho-I, we get:

$$E = \frac{2q}{4\pi\varepsilon_o 1.05833 \times 10^{-10}} = 27.212 \text{ eV} \qquad (9.28)$$

Subtracting from this value the photon energy of the largest optical transition (9-0) we get:

$(27.198 - 14.667) = 12.545$ eV. The half of this value is 6.272 eV, which is very close to E_B for H_2 ortho-I state (6.26 eV), known experimentally and determined theoretically in §9.7.4. This coincidence is one additional confirmation of the BSM concept about the molecular vibrations.

We must emphasize that the stretching and stretching + bending are two different modes, which will have a different vibrational constant referenced to level $\upsilon = 0$. The vibrational constants provided by Namioka, based on the observations are shown in Table 9.4 with identification of the vibrational mode by BSM.

Table 9.5

State	v	observed B_e (cm-1)	mode according to BSM
$B'^1\Sigma^+_u$	0	25.42	stretching only
$B''^1\Sigma^+_u$	0	25.43	stretching only
$D^1\Pi^+_u$	0	31.14	stretching + bending
$D'^1\Pi^+_u$	0	31.20	stretching + bending

The appearance of two different data for every mode according to BSM is a signature of elliptical shape of the vibrational trace in quantum quasishrunk units. The effect discussed in §9.9.3. could be valid also for H_2 ortho-II state.

9.10. Discussion about the bonding energy at equilibrium and the rotational constants of H_2 and D_2 molecules.

In the previous paragraphs we saw that the different states of the H_2 molecule should have different bonding energies at the equilibrium and different rotational constants B_e, referenced to a level zero.

This explains the large variation of the constant B_e, calculated by the observed spectra. BSM is mostly interested of the parameters of the ortho-I states of H_2 and D_2, since they play the role of bonding systems between the atoms in the molecules. For this reason, the evaluated constants are seeded out in order to determine the correct state. To seed out the corresponding value, the following criteria are used:

- the higher energy range of the observed spectrum and its match to the the theoretical calculation of the system energy

- the match between the optical and corresponding photo-electron spectrum (if the latter is available)

- excluding observations, where Q branches appear

The last criterion, for example, excludes number of observations, and they usually provide Be of the order of 30 cm⁻¹.

One of the correct seeded value is $B_e = 27.30$ cm⁻¹, provided by Dieke and Blue, (1935) and Dieke (1958). It is very close to our theoretically calculated value of 27.87 cm⁻¹ (presented elsewhere).

9.11. H_2 para molecule as a most simple example of diatomic homonuclear molecule with a quasirotational vibration

The quasirotational vibration has been introduced in §9.5.4.1 and additionally discussed in §9.5.7.4. The basic requirement for such motion is the molecule to have at least two electronic bonds. For such bonds two spatially displaced quantum orbits are needed. The simple H_2 para molecule satisfies this conditions. Its configuration has been shown in Fig. 9.5. One distinctive feature of the para-state is that the electrons do not share a common orbit. It has been pointed out that the quasirotational motion is identified by the Q branches in the optical spectrum. The P and R branches are contributed by the quasilinear bond oscillation, while the Q branch - by the bond acting as a centre of the quasirotational motion. So the first type bond emits (photons) by stretching (shrinking), while the second one - by bending. These functions exchange alternatively between both bonds. So the equilibrium is in the moment, when the polar axes of the two protons are parallel. At the equilibrium point, the orbital energy could not be smaller than this of the first quantum orbit (13.6 eV). So we arrive to the same definition of the system energy as for the H_2 ortho-I state (factor two in the nominator of Eq. (9.13), but for a nuclear distance defined directly by the length of the quantum orbit. Then the system energy equation of the H_2 para state is:

$$E_{VIP} = \frac{2q}{4\pi\varepsilon_o L_q(1)} = 21.135 \text{ eV} \qquad (2.28)$$

The Eq. (2.28) shows that the photo-electron spectrum from H_2 para state could not be observed by using the He line of 584 A (21.23 eV). The optical spectrum of this state is known as Werner band or $c\,^1\Pi_u \rightarrow X\,^1\Sigma_g^+$ system. Observational data are provided by Dabrowski (1984). The optical transitions of Werner band go to higher energies than the Lyman band. This indicates, that the para state has a higher bonding energy. Without a photo-electron spectrum, however, we can't apply the same method for estimation of this energy, as for H_2 ortho-I state.

9.13. Molecules (or fractions of molecules) with II-nd type of oscillations.

The two types of oscillations introduced in §9.5.4.2 are distinguished by the moment of inertia of the molecule or its structures. For the molecules with I-st type of oscillations, discussed so far, the moment of inertia is quite small and has been neglected.

For molecules with a larger moment of inertia, the centrifugal forces could not be ignored. They contribute to the rotational energy. For homonuclear diatomic molecules this energy is given by Eq. (9.3). If this forces are smaller than the E_B energy, the molecule possesses a I-st type of vibrations, otherwise it may posses a II-nd type of vibrations. **In the II-nd type of vibrations, the inertial forces work against the quantum quasishrink effect. This may cause a deterioration of the quantum conditions.**

The deteriorated quantum conditions may affect the process of photon emission (absorption). The signature of II-nd type of vibrations is not so apparent from the optical spectrum. However, it could be identified by the Photoelectron (PE) spectrum. The distinguishing signatures in the PE spectrum the following:

A. For molecules with a I-st type of oscillations:

(a) The bottom level of the PE spectrum is smooth.

(b) The bonding system of H_2 (or D_2) contributes to a narrowly spaced and well separated peaks

(c) the bottom level of H_2 (D_2) peaks are not shifted

B. For molecules with a II-nd type of oscillations:

(d) appearance of quite wide peaks with larger separation

(e) appearance of H_2 (D_2) signatures with elevated bottom level of the peaks

(f) appearance of H_2 (D_2) signatures overimposed on wide peaks

According to BSM analysis, the wide peaks with larger separation are indicators of deteriorated quantum conditions. If examining the wide peaks with higher resolution PE spectroscopy, it is apparent that some of them have a fine structure, others do not. The fine structure is indication that quantum conditions exists, but they might be from superimposed PE spectrum of some groups containing H_2 (D_2) bonding, possessing a I-st type of vibrations. This conclusion has experimental confirmation, when interpreting the PE spectrum of H_2S and D_2S. Such spectra are provided by Eland, C. J, (1979). They are shown in Fig. 9.41A.

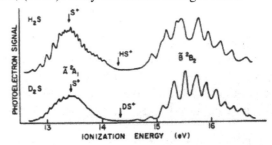

Fig. 9.41A

PE spectrum of H_2S and D_2S (Courtesy by Eland, (1979))

The left part of the spectra is from a neutral molecule, while the right part - from an ion. The H_2S^+ ion is obtained after one of the bonding electrons is lost. The possible molecular structure of this ion is shown in Fig. 9.43.

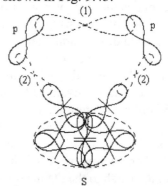

Fig. 9.43

Possible configuration of H_2S^+ ion. The number in bracket shows the subharmonic number of the quantum orbit.

The top quantum orbit could be also of (2) subharmonic
Every quantum orbit contains a single electron.

In the shown possible configuration, the structure of H_2S^+ molecule appears symmetrical in respect to the polar axis of sulphur (S element). The quantum orbits are also symmetrical. The single electron in the top orbit (between both protons), is not paired. The effective positive charge may modulate the surrounding CL space in proximity of the molecule. In comparison to the neutral H_2S, the electrons are not so well paired and their quantum interactions with the CL space should be larger. Then the wide PE peaks in Fig. 9.41.A may correspond to deteriorated quantum conditions, mentioned above. The left part of the PE spectrum contains the familiar PE spectrum of H_2 but with a significantly elevated bottom level. So the left slope of the left wide peak could be contributed by I-st type vibrations of H_2S with well defined quasishrink effect, while the right slope could be a result of deteriorated quantum conditions valid for a II-nd type of vibrations for H_2S^+ molecule. The confirmation of this conclusion comes also from the comparison of the fine structures of the left wide peak for the H_2S and D_2S, which is provided by the bonding electrons. In case of D_2S, this structure is shifted to smaller E_{IP} energies (larger E_K energy). This could be explained by the increased SG forces from the D_2 type of bonding.

Summary:

- **The I-st type of the molecular vibrations could be identified by the PE spectrum**
- **The complex molecules may contain groups with I-st and II type of vibrations**
- **Only the I-st type of vibrations provides a strong discrete optical spectrum**

9.14. Information about the molecular configuration, provided by the photoelectron spectrum

The PE spectrum may provide additional information about the molecular configuration by the following features:

(a) The energy spacing between the peaks in the I-st type of vibrations

(b) Angular distribution of the photoelectrons

(a): The energy spacing between the peaks in I-st type of vibrations is weakly dependent on the interactions between the bonding and nuclear quantum orbits. A stronger interaction corresponds to a larger spacing. This is known from the existing theory of PE spectra. For the BSM model, this feature may provide a valuable information about the atomic internuclear separation. It may allow to identify the possible configuration of different excited states of the O_2 molecule, discussed later in this Chapter.

(b): The angular distribution of the PE, known as an "anisotropy parameter" appears when the molecules are properly oriented in respect to the analyser magnetic field. The anisotropy parameter can be evaluated from the experimental angular distribution of the photoelectrons. Such distribution for H_2 is shown in Fig. 9.44 (From. Carlson, T.a and Jonas A.E (1971).

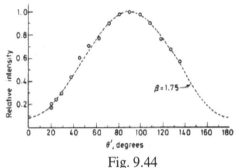

Fig. 9.44

Experimental angular distribution of the photoelectrons from ionization of hydrogen by He I line (From Carlson et al. (1971)

The anisotropy parameter β is evaluated by the least square fit to a theoretical equation:

$$I(\theta) = 1 + \frac{3\beta}{4 - 2\beta}\sin^2(\theta') \tag{9.46.A}$$

where: I - is the relative intensity of PE signal
       θ' - angle of distribution

The anisotropy parameter is measured for number of molecules in different states. It varies from 1.8 (for H_2) to some minus values for some molecules. It also has different value for different states of one and a same molecule. When plotting the function $I(\theta')$ we obtain a sense of the position of the plane of the bonding quantum orbit in respect to the molecular orientation along its vector of velocity. In case of H_2 molecule, the parameter β is a largest one. **Consequently, the parameter β is a signature of the orbital plane orientation in respect to the molecular axis of the rotational mo-**

tion. For two different states of N_2 it obtains value of 0.6 and 0.65.

If the bonding orbits are more than one and with a good rotational symmetry in respect to the molecular axis of rotation, then the parameter β is low. For the noble gases the parameter β is not related with the bonding orbits but with the photoelectrons from the internal completed layer of protons. For this reason the β parameter is contributed by electrons from the nuclear orbits, which are strongly coupled to the completed proton shells. Consequently, the β parameter is informative about the nuclear shape of the noble gases. For example, the shape of the Ar nucleus is closer to a sphere, but the nucleus of Kr or Xe is not. Fig. 9.39 shows plots of β parameter for H_2 and for some noble gases.

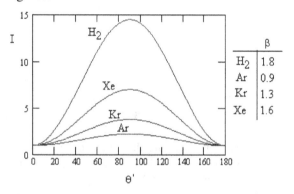

Fig. 9.45

Angular distribution of photoelectrons from PE spectra and corresponding β parameter for different gazes

9.15. SG energy balance for system of diatomic homonuclear molecule

9.15.1 SG Energy of a bonding system

In the presented examples of H_2 and D_2 molecules, the SG energy balance was satisfied for the first harmonic quantum orbit. Let us consider the ortho-I state of H_2, the SG energy balance of which is described by Eq. (9.23). If using $L_q(n)$ instead of $L_q(1)$, where $n > 1$, the Eq. (9.23) is unbalanced quite a bit. The disbalance $E_V(\Delta)$ (eV) for quantum orbits with different subharmonic number n is shown in Table: 9.5

Table 9.5

subharmonic number (n)	$L_q(n)$ [A]	$E_V(\Delta)$ [eV]
1	Lq(1)	6.27
2	Lq(2)	1.6326E6
3	Lq(3)	3.166E6
4	Lq(4)	4.487E6

where: $L_q(n) = L_q(1)/n$

We see that for larger subharmonics, the SG balance of the system is disturbed quite a lot. If the obtained disbalance, however, is compensated by a proper change of the internuclear distance, a new balance could be obtained, valid for the whole system. In such case, the bonding energy would be expressed by the same equation (9.23) but instead of length $L_q(1)$ we will have $L_q(n)$ where n - is the subharmonic number $n > 1$. In this point, however, one question arises: Does a similar bond system of higher subharmonics quantum orbits have a similar vibrational curve? To answer this question we must consider two cases:

Case (A). Assumption that the term $\alpha^4 \pi (\upsilon_m - \upsilon)^2$ is one and a same for quantum orbits with different subharmonic number

In this case, the total momentum energy as a function of the quantum number υ is:

$$B_{H2}(n, \upsilon) = \frac{C_{IG}}{q[r_n(n, \upsilon)]^2} - \frac{2E_q}{q} - \frac{2E_K}{q} + 6.26 \ \text{(eV)} \quad (9.47)$$

$$r_n(n, \upsilon) = [L_q(n)](1 - \alpha^4 \pi (\upsilon_m - \upsilon)^2) + 0.6455 L_p \quad (9.47.a)$$

where; $B_{H2}(n, \Delta)$ is the momentary energy of the bonding system of protons (corresponding to H_2 ortho-I type system), $0.6455 L_p$ - the locus distance from the central point of Hippoped curve, υ - is the vibrational level (referenced to the equilibrium distance),

For bonding system of deuterons, the momentum bonding energy is equivalent to the total energy of D_2 molecule given by Eq. (9.27). Applying for subharmonic numbers of n, we obtain a momentary bonding energy of deuterons, denoted as B_{D2}.

$$B_{D2}(n, \upsilon) = \frac{C_{IG}}{q[r_n(n, \upsilon)]^2} - \frac{2E_q}{q} - \frac{2E_K}{q} + 6.235 \ \text{(eV)} \quad (9.49)$$

$$r_n(n, \upsilon) = [L_q(n)](1 - 0.5\alpha^4\pi(\upsilon_m - \upsilon)^2) + 0.6455 L_p \quad (9.49.a)$$

There are two special points of the vibrational curve defining two special values for the momentary bonding energy:

(a) the equilibrium point defines a **momentary bonding energy at equilibrium** ($\upsilon = 0$).

(b) the dissociation point defines a **momentary bonding energy at dissociation limit** (for some value of υ_m)

Note: QM models accepts an infinite vibrational numbers assuming that, they could not be measured due to a technological limit. However, according to the BSM model, the quantum vibrational numbers have a limited value υ_m. The continuum may contain a large number of quantum orbits, which are not like vibrational levels, but are result of deformation of the orbital shape as discussed in section 9.5.7.4. The continuum is also contributed by emission (or absorption) of photons at deteriorated quantum conditions.

The single valence bonding system between two atoms in the molecule could be regarded as bondings of H_2 or D_2 system. In fact D_2 system bonding is more typical as seen from the Atlas of the Atomic Nuclear Structure (ANS). Therefore, the Eq. (9.47) is valid for a bonding system of protons while the Eq. (9.49) - for a bonding system of Deuterons.

Case (B). Assumption that the term $\alpha^4\pi(\upsilon_m - \upsilon)^2$ is dependent on the quantum number of the orbit

Larger subharmonic numbers means smaller length of the particular quantum orbit. If considering that SG energy of heavier than H_2 and D_2 molecule causes a larger bending of the vibrational curve, then the above term may have some kind of inverse proportional dependence on n. If assuming that this is correct, the same rule should be valid also for the Lq(1). The correct dependence cold be determined only by studying the vibrational properties of diatomic molecules for which photoelectron and optical spectra are available. Some study of this problem by using the photoelectron and optical spectra of O2 molecule is presented separately as BSM_Appendix9_1.

9.15.2 Energy balance in diatomic molecules

The momentary bonding energy could be compensated by a proper SG potential between the atomic nuclei involved in the molecule. For a stable connection of atoms into a molecule, the average SG energy of the total system should be in balance. The average SG energy means that the system may vibrate around the momentary balance equilibrium state in a similar way as the H_2 and D_2 molecules. Then the SG energy may be regarded as comprised of two components: DC (constant component, like "direct current") and AC (alternative component). Such system includes:

- the CL space occupied by the volume of the integrated Bohr surfaces of both atoms and bonding system.

- the elementary particles of both atoms (comprised of helical structures with their internal RL(T) lattices).

- the bonding system comprised of one or more pair of protons(deuterons).

The energy interactions between two atoms in a simple molecule with electronic bonds are schematically illustrated in Fig. (9.46). It is assumed that the bonding system is comprised of deuterons (a more typical case).

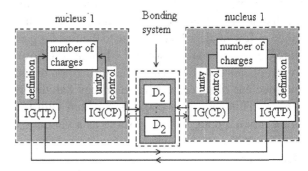

Fig. 9.46

Energy interaction diagram of simple diatomic molecule

The SG(TP) of the atomic nucleus provides a charge definition (by RL(T) of the FOHSs), while the SG(CP) provides a charge unity control for all proton charges. The bonding system, shown in Fig. 9.46 contains two deuteron's pair. In order the molecule to be balanced for particular quantum orbit (from the possible quantum orbit set) of the bonding system, it needs some amount of SG energy of SG(CP) type (SG energy between the two atoms).

The SG(CP) energy should be balanced by SG(TP) energy, in order the whole system (including the CL space) to be brought to a total energy balance. The momentary energy of the molecule may oscillate around the equilibrium point, so the total SG energy should have a DC and AC components. In the AC component the bonding system, similar as H_2 or D_2 is involved. Then the AC energy balance conditions puts the following requirement for the EB bonding system:

EB requirement: At the equilibrium point of vibrations, the momentary disbalance of the total AC type SG energy of the molecule should be equal to zero.

The bonding system disbalance involves SG(CP) energy, so the necessary interaction should be of this type, as shown in Fig. 9.46. It is difficult to estimate directly the SG(TP) energy supplied by both atoms, but it could be estimated by the total balance of the system. This energy is subtracted from the total system energy, so it has to be compensated. The possible compensation could come from SG attractions involving also SG(TP) fields of both nuclei.

The SG energy between two nuclei could be estimated as a work for separation of the nuclei from the internuclear distance to infinity. Practically, the integration converses to a finite value at distance of few nuclear lengths (due to the fast drop of the SG forces with the distance). More useful, however, appears the inverse task: the estimation of the internuclear distance, when the energy balance is known. This parameter could be verified with the observational data, so it may serve as a validation parameter for the unveiled molecular configuration.

It is reasonable to consider that the proportionality between the nuclear atomic mass and the atomic mass unit (valid for the Newtonian mass) is preserved also for the intrinsic masses of the nuclei. For approximate calculations, we may accept the following two simplification:

- neglecting the difference between the masses of neutron and proton and using the proton mass only (in fact they both contain one and a same intrinsic matter)

- neglecting the mass deficiency (nuclear binding energy) especially for atoms with a low Z number.

For neutral molecules, when estimating the SG(TP) forces between the two nuclei at a distance r_n, we should exclude the protons involved in the bonding system, as their forces and energies are directly involved in the bonding system balance. Having in mind that the factor C_{SG} is normalized to the intrinsic proton mass, the SG force between two identical nuclei could be expressed as:

$$F_{SG} = \frac{C_{SG}(A-p)^2}{r_n^3} \qquad (9.51)$$

where: C_{SG} - is the SG factor, defined by Eq. (9.13.a) and expressed by Eq. (9.17), A - is the atomic mass of the participating atom, p - is the number of protons per atom, involved in the bonding system

In Eq. (9.51), only the protons are excluded from the attractive SG forces between the two nucleus. (In fact the distance between neutrons over proton saddle are closer than the internuclear distance but SG field between them could be partly affected by the proximity fields of the protons). This assumption leads to more reasonable final results, tested especially for the Oxygen molecule.

Having in mind that $E_{SG}(TP) = 2\alpha E_{SG}(CP)$ and integrating Eq. (9.32) on a distance r, we get:

$$E_{SG}(TP) = 2\int_{r_n}^{\infty} \frac{2\alpha C_{SG}(A-p)^2}{r^3} dr = \frac{2\alpha C_{SG}(A-p)^2}{r_n^2} \qquad (9.52)$$

The factor 2 in front of the integral takes into account the two branches of aligned prisms along *abcd* axes, involved in one cell unit of CL space.

For a momentary AC type of balance at equilibrium distance, the $E_{SG}(TP)$ energy given by Eq. (9.52) should be equal to the energy $B_{D2}(n, \Delta)$ multiplied by the number of connected protons (per one atom) involved in the binding system. For a particular molecule, the only factor that may change the system disbalance according to Eq. (9.52.) is the deviation from the equilibrium distance. For a bonding system of protons, this deviation is given by Eq. (9.47), while for deuterons - by Eq. (9.49). The total SG(TP) energy necessary for a momentary balance of the bonding system is:

$$E_{SG}(TP) = \frac{2\alpha C_{SG}(A-p)^2}{(r_n \pm [\Delta r(\upsilon)])^2} \qquad (9.53)$$

For internuclear bonding system of deuterons we may use the expression for Δr derived in §9.8.

$$\Delta r = 0.5\pi\alpha^4 \upsilon^2 L_q(1)$$

If equating the energy of $E_{SG}(TP)$ in (eV) with the energy $B_{D2}(n, \upsilon)$ we get AC balance condition only for $\upsilon = 0$. For any $\upsilon \neq 0$ the disbalance energy provides vibrational energy levels. **Therefore, the vibrational energy levels (for D_2) are equal to the difference between the energies given by Eq. (9.49) and Eq. (9.53) for different integer values of υ. We can denote them as** ΔE.

$$\Delta E = \frac{2\alpha C_{SG}(A-p)^2}{(r_n \pm [\Delta r])^2} - B_{D2} \qquad (9.54)$$

The energy balance B_{D2}, defined by Eq. (9.47) is for one valence connection. The diatomic molecules (or groups) usually do not have more than three valence connections. So for more than one valence connection we may express the balance energy by simply multiplying the energy term B_{D2} by the valence factor p. The final expression for the vibrational levels in a diatomic homonuclear molecule obtains the form:

$$\Delta E(p, n, \upsilon) = \frac{2\alpha C_{IG}(A-p)^2}{[r_n \pm [\Delta r(n, \upsilon)]]^2} - pB_{D2}(n, \upsilon) \qquad (9.55)$$

where: A - is the atomic mass in atomic mass units (per one atom), p - is the number of protons involved in the bonding system (per one atom), n - is the subharmonic quantum number of the quantum orbit, r_n - is the internuclear distance at the equilibrium, Δr - is the change of the internuclear distance as a function of vibrational quantum number υ

The participation of $L_q(n)$ parameter in Eq. (9.55) does not exclude a serially connected quantum loops. In the following analysis, however, only single quantum loops are discussed.

The application of (9.55) is limited mainly for diatomic homonuclear molecules with electronic bond system. Such molecules are possible for the elements from 15, 16, and 17 vertical group of the Periodic table (but possible applications for other groups are not excluded).

The disbalance of Eq. (9.55) provides the photon energy. The emitted or absorbed photons from the vibrational-rotational spectra are of the order of a few eV. The other energy terms involved in (9.55) are much larger. Investigating Eq. (9.55) we could see that a balance for $n = 1$ could not be

achieved for molecules with atoms heavier than deuteron, because the internuclear distance become very small, but it is limited by the finite nuclear size. (This corresponds to a lower limit of internuclear bond in QM models). Consequently, **the case of $n = 1$ (first harmonic quantum orbit) is valid for H_2 and D_2 molecules only. Any other molecules are heavier and could not have a EB system with $n = 1$.** This derived rule facilitates the task for determination of the configurations of molecules of heavier atoms by excluding the bonding orbit $L_q(1)$.

The obtained Eq. (9.55) is of great importance for BSM. It provides the opportunity to find the real structure of some simple diatomic molecules, if the optical and PE spectra are available. By careful analysis of such spectra, the parameters n, Δ, and r_n could be determined. The analysis could be simultaneously assisted by finding the proper spatial configuration, working with the known three dimensional shape of the proton, (neutron), and the set of quantum orbits, defined with their shape and dimensions. The Δr parameter is quite small in comparison to r_n, as in the H_2 and D_2 molecule. Therefore, we can use the well defined dimensions of the quantum orbit in free CL space.

Such type of analysis was applied for identification of the possible states of the oxygen molecule.

Note: The vibrational levels calculated by Eq. (9.55) are approximate due to a simplification that the SG mass of the nucleus is located in a point. However, the nucleus of every atom has a finite dimensions according to BSM models, so this is a source for error. Another smaller source for error could be the ignored spin-orbit interactions. For this reason we may not expect an accurate match of calculated vibrational levels.

From Eq. (9.55) we may derive a direct expression for the internuclear distance for a homonuclear diatomic molecule. Having in mind that the change of the internuclear distance for vibrating molecule is intrinsically small, we may use the distance at equilibrium at which $\upsilon = 0$ and the momentary energy is zero $\Delta E(p, n, \upsilon) = 0$ Then solving Eq. (9.55) for r_n we obtain:

$$r_n(n, A, p) = (A-p)\sqrt{\frac{2\alpha C_{IG}}{pB_{D2}(n)}} \qquad (9.56)$$

Eq. (9.56) is very useful, because it allows determination of the internuclear distance for diatomic homonuclear molecule without knowing the any spectral data. It has been used for number of molecules, whose configuration are presented in the next sections. Eq. (9.56) can be easily modified for multiatomic homonuclear molecules or even for heteronuclear molecules.

9.15.3. Electronic bonds with quantum orbits formed by serially connected quantum loops.

We have considered so far that the trace of any quantum orbit is formed by a single quantum loop. The condition of quantum orbit, however, is satisfied for more than one quantum loops, connected serially. This was discussed for Hydrogen in §7.4 and the possible quantum orbits in such case were given in Table 7.1. There are not theoretical restrictions such orbit to exists in the bonding systems between atoms in the molecules. If the quantum orbit, for example, contains two quantum loops of third harmonic, it will still have kinetic energy of $13.6/3^2 = 1.511$ (eV) per electron, but its linear dimension will be twice the linear dimension of third subharmonic orbit. Consequently, quantum orbits of serially connected quantum loops increases the set of possible lengths of quantum orbits. When this set is combined with the limited angular freedom of the polar bound protons, additional options for the total energy balance for atoms connected in molecule are obtained. The derived equations for the bonding length and the vibrational levels are easier modifiable for this option. For the bonding energy expression, the serially connected loops will affect the kinetic energy in Eq. (9.49) and the internuclear distance in Eq. (9.49.a). For the total energy balance of the molecule, the serially connected loops will affect r_n and Δr_n in Eq. (9.53).

The derived equations are used in BSM_Appendix9_1 for analysis of different states of the O2 molecules in order to obtain their configurations.

Summary:

- **The vibrational energy levels for diatomic molecules could be approximately estimated by Eq. (9.55). Using the simplification that the SG mass is located in a point and ignor-**

ing QM spin-orbital interaction, it may provide useful results for estimation of the internuclear distance and the shape of the molecule.

- **Diatomic molecules or groups with an atomic mass larger than 4 could not have a bonding orbit of first harmonic.**
- **By simultaneous cross analysis of the PE and optical spectrum with the help of Eq. (9.55) the real structure of the diatomic molecules and their different states could be obtained.**

9.16. Oxygen molecule in different states.

The analysis of different states of O_2 molecule in order to obtain the corresponding molecular configurations is provided in BSM_Appendix9_1. For this purpose PE and Optical spectra of O_2 molecule are used. Here only some final results are shown.

Fig. (9.48) shows the PE spectrum provided by K. Kimura et al. (1981).

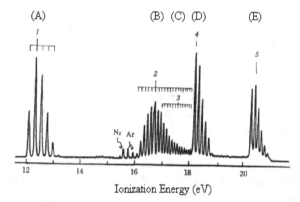

Fig. 9.48. PE spectrum of oxygen molecule excited by He I radiation (Turner et all., courtesy of K. Kimura et al., (1981))

According to BSM analysis, the different bands of PE spectrum correspond to different molecular configurations or states. They are denoted by the alphabets (A), (B), (C), (D), (E) states (BSM annotation), put in the top of the figure. Note that the (C) and (D) states are partly overlapped in the PE spectrum. The possible configurations of the {D} and {A} states are shown in Fig. (9.50).

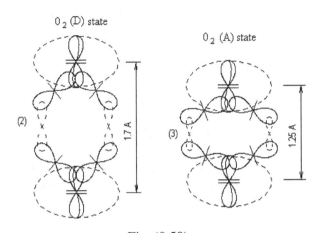

Fig. (9.50)
Possible configurations of O_2 molecule in {D} and {A} states (by BSM)

Fig. 9.52 shows the possible configuration of {E} state of O_2 molecule. Eq. (9.55) is not directly applicable for this state without modifications, because the quantum orbits are not aligned with the common internuclear axis.

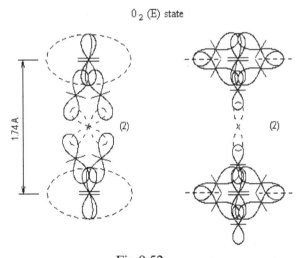

Fig.9.52
A possible configuration of (E) state of O_2 molecule

The calculated values of r_n (rounded) by Eq. (9.56) and those matching the spatial configuration by using a drawing method, are given in Table 9.6.

	Table 9.6

Calculated (rounded) values for r_n

r_n (A)	$L_q(2)$ (1 bond)	$L_q(2)$ (2 bonds)	$L_q(2x)$ (2 bonds)	$L_q(3)$ (2 bonds)
by Eq. (9.56)	2.57 A	1.695 A		1.217 A
by drawing	2 A	1.7 A	1.74 A	1.25 A
state	(B), (C)	(D)	(E)	(A)

The calculated distances could not be considered accurate as some elaborate methods of Quantum Mechanics. They are shown only for identification of the possible molecular configuration.

9.16.5. (OH)⁻ ion.

The possible configuration of (OH)⁻ ion is shown in Fig. 9.54. It is assumed, that the plane of the two bonding orbits are parallel, so they exhibit a maximal QM spin interaction. In such case the polar angular position of the valence protons is influenced by the QM spins of the two orbits even for bonding orbits with different subharmonic number.

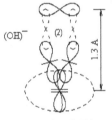

Fig. 9.54
A possible configuration of (OH)⁻ Every bonding orbit contains two electron with opposite QM spin. (the bonding orbits are shown in the drawing plane for simplicity, although they are perpendicular to the drawing).but rotated

The shown configuration is characterized, also, by the following feature: The planes of the three sets of orbits: the He nucleus orbits, the orbits of GBclp protons, and the bonding orbits are mutually orthogonal. The described features might be important factor for the stability of OH⁻ ion.

9.16.6. A special (airglow) state of the oxygen atom

Two optical transitions, known as "forbidden" states of the atomic Oxygen are quite abundant in the Earth atmosphere. They are explained as forbidden because they do not exist in the normal atmospheric pressure or when the free path length between the collisions is not enough long. This state of the oxygen atom is known as an airglow state, but one important fact is that the energy levels of the two strong optical transitions do not match the energy levels of the Grotrian diagram of oxygen.

state	emitting at	radiation lifetime
O (^{1}D)	5577 A	
O(^{1}S)	6300 A.	

The possible configuration of the oxygen atom, according to BSM analysis, providing these two transitions is shown in Fig. 9.60. We may refer it as an airglow state

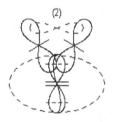
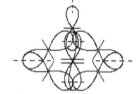

Fig. 9.60

Two views of the possible configuration of oxygen atom in Airglow state responsible for line emissions at 5577 A and 6300 A.

The planes of the orbits that are not parallel to the drawing plane are shown as dashed lines. Each of the common orbits connecting the two valence protons contains two electrons with opposite QM spins. It is likely a quantum orbit of a second subharmonics number. Some of the features related to the long lifetime of the previously discussed state {E} of O2 are valid for the Airglow state of the atomic oxygen.

Some considerations in favour of the described concepts for O_2 {E} state and the Airglow state of the atomic oxygen, come from the atmospheric chemistry. Both states are often involved in commonly dependable reactions, identified by their emission spectra. From the point of view of BSM, the airglow atomic state could be yielded as

a result of O_2 {E} state depletion. The configurations of both states have two similar features by which they are distinguishable from the other O_2 configurations.

- they posses quantum orbits with a same subharmonic number

- the opening angle between the axes of the valence protons (equal to twice the polar angle) has a pretty close value for both states (O2 {E} state and Airglow state of O).

The approximate values of the polar angle for different states of oxygen atom and molecules are given in Table 9.9.

9.16.7 Ozone molecule

The rich optical spectrum of the Ozone molecule O_3 indicates that it possesses electrical bondings. Only one possible configuration of this molecule exists. While the first harmonic orbit is excluded by the same considerations as for O_2 molecule, the most likely possible orbit is a second subharmonic one. The configuration of O_3 is illustrated in Fig. 9.61.

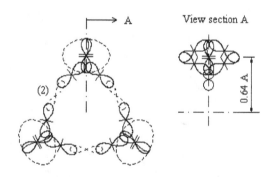

Fig. 9.61
Ozone molecule

The *sectional view A* shows the GBclp bound deuterons of the nucleus, whose quasiplane is at 90 deg in respect to the bonding system quasiplane.

9.16.8 (OH) radical

The (OH) radical, as a neutral molecule, known also as a Hydroxil is quite abundant in the Earth atmosphere. It possesses a rich optical spectrum spread in a large spectral range in the visible and infrared region. The possible configuration of this radical is shown in Fig. 9.62

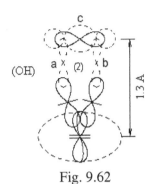

Fig. 9.62

A possible configuration of (OH)
(the planes of bonding orbits are rotated
at 90 deg, but in a real model, they are
normal to the drawing plane)

The two orbits a and b are likely of second subharmonics, each one containing a single electron. The orbit c could be a composite one containing 2 loops of second subharmonic. Additional study for other options could be interesting.

It is assumed, that the plane of the two bonding orbits are parallel, so they exhibit a maximal QM spin interaction. In such case the polar angular position of the valence protons is influenced by the QM spins of the two orbits. The shown configuration is characterized, also, by the following feature: The planes of the following sets of orbits: He nucleus orbits, orbits of GBclp protons, the two bonding orbits (a & b) and the orbit c are mutually orthogonal. The described features might be important for the long radiation lifetime of the (OH) radical observed in the IR spectrum of the Earth atmosphere.

9.16.9 Polar angles of the valence protons in different states of atomic and molecular oxygen.

The polar angle has been defined in Chapter 8, §8.38. as the angle between the nuclear polar axis and the long axis of the polar bound proton.

The limited freedom of the polar angle of the valence protons means a limited flexibility when connected into molecules by EB type of bonding. The change of the polar angle involves a change of the shape of the integrated Bohr surface, especially for the bonding system. This means redistribution of the internal E-filed involving the SG(CP) and SG(TP) fields. It is reasonable to accept that the change of the polar angle of the valence proton will be slower, than the positional change of the connected quantum orbit. This has direct implication about the preferable chemical connections. It will be demonstrated later with the correlation observed between the depletion and creation process between O_2 and (OH) molecules in the Earth atmosphere.

The polar angles could be estimated from the atomic or molecular configurations for different states. Even accuracy of few degree is still informative. Table 9.9 shows the possible value of some diatomic molecules (groups) estimated in this way.

Polar angle of a valence proton
at matched internuclear distance   Table 9.9

molecule, state	polar angle	r_n calculated	r_n (matched)
O_2 {A}	68.5°	1.219	1.25 A
O_2 {D}	51°	1.698	1.7 A
O_2 {E}	17.5°		1.74 A
O airglow	29°		
O_3	68.5°		
(OH)+	16°		

From Table 9.9 we see, that the difference between the polar angles of the oxygen valence protons for O_2 {D} and (OH) is only 1.5°. If the bonding orbits of (OH) is of second subharmonic number as for O_2 {D}, some correlation might be expected between the formation and dissociation of these states. The dissociation of two (OH) groups, for example may lead to formation of one O_2 molecule in a {D} state. Correlation between airglow emission from (OH) and O_2 {D}, corresponding to depletion - creation process is observed in the Earth atmosphere (Arizona Airglow experiment (GLO) from Space Shuttle, September 1995)).

9.17. NH_3 molecule

The Nitrogen atom has 3 free valence protons (deuterons) and two equatorial EB protons in its external shell. Two of the three valence protons are bound to one pole and another one to the second pole. In NH_3 molecule every valence proton of N is connected to one hydrogen atom (proton). The bonding orbits are perhaps of second subharmonic number. The configuration is difficult to be shown by small number of views. The study of the optical spectrum of NH_3 indicates two minima of the vibrational curve, corresponding to lower energy and higher energy states. The two possible states according to BSM have the following configuration:

A. Lower energy state:

This states may correspond to a large polar angles of the three deuterons as in the water molecule in a gas phase. The connected protons are aligned by the valence deuterons.

B. Higher energy state:

This state may correspond to smaller polar angles of the polar bound deuterons of nitrogen atom. The positions of the bonding quantum orbits in this state may play a stabilizing role in the following way. Their long axes become parallel to the polar axis of the nitrogen atom. In such configuration their magnetic fields could be coupled with the magnetic fields of the orbits of the two equatorial EB bound deuterons. In such condition, the molecule obtains four symmetrically aligned quantum orbits in respect to the polar axis. The QM spin interaction in such case should be quite large. This may provide a stable molecular configuration.

9.18 Molecules with folded vibrational-rotational spectra

From the previous paragraph we see, that the valence protons of the nitrogen atom is asymmetrical in respect to the polar axis. When participating in molecular groups it propagates its asymmetrical feature to them. For molecules and groups with large asymmetry, the bending centrifugal forces become so large that the optical spectrum is additionally distorted. The large asymmetry between P and R branches causes a folding of the R branch. A typical example of such molecule is HCN.

The four protons (deuterons) in the Carbon atom posses a polar axial symmetry but rotated at 90 deg. It is apparent that the obtained bonding connection $C \equiv N$ is quite asymmetrical. The single valence bond between the carbon and hydrogen additionally increases the asymmetrical properties of the molecule.

A possible configuration of HCN is shown in Fig. 9.63.

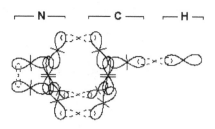

Fig. 9.63. A possible configuration of HCN molecule

Fig. 9.64 shows a stretched P branch and folded R branch of one of the optical transitions of HCN molecule. The folded R branch of HCN molecule evidently is contributed by the $C \equiv N$ bondings and the asymmetrical bending forces around the common axis.

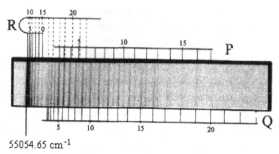

Fig. 9.64. Absorption band of $\tilde{A} \, ^1A'' - \tilde{X} \, ^1\Sigma^+$ transition of HCN molecule (courtesy of P. F. Bernath, (1995)

The first contributing factor is the asymmetry of the $C \equiv N$ bonding system in respect to the molecular axis of rotation and the second one is the asymmetrical position of the proton, bound to the carbon.

For explanation of the folded R branch we may use the following simple example of a three leg chair on a horizontal plane. Let us trip the top of the chair by a horizontal force, in order to cause a flopping rotation around its vertical axis. There will be two different moments of the motion: when two legs are in contact with the horizontal plane and when only one leg is in contact. Now suppose, that the top of the chair is loaded with mass that is displaced from the central axis of rotation. Let us also consider, that the horizontal plane is partly soft. We arrive to a similar conditions as for the bonding connection and SG attractive forces between C and N nuclei.

The folding of the R branch is a result of changed momentary energy balance for one of the electrical bonds of CN group due to a redistribution of the SG forces.

9.19 CO₂ molecule

The carbon atom in CO_2 molecule is in the middle, so this molecule is highly symmetrical. For calculation of the internuclear distance, Eq. (9.56) could be used with small modification.

- considering CO system as a single molecule
- replacing the factor $(A-p)$ with the product $[(A_1-p)(A_2-p)]^{1/2}$, where: A_1 is the carbon atomic mass A_2 is the oxygen atomic mass. The factor 2 in front of p for C atom takes into account the exclusion of 4 protons, as participating in the bonding system.

The obtained internuclear distance for CO group, according to the above consideration is about 1.284 A (angstroms) for a second subharmonic orbit. One view of the obtained configuration is shown in Fig. 9.65.

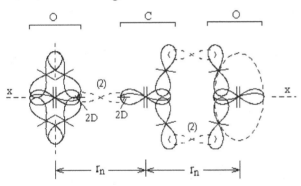

Fig. 9.65 Single view of CO₂ molecule Note: The two proton pairs in the left side of the drawing with two bonding orbits are similar as these in the right side but rotated at 90 deg. So only the projection of one of two bonding orbits is visible in the left side

If the view is rotated at 90 deg around *xx* axis, the left-hand side will be interchanged with the right hand one. The molecule is twisted due to the proton twisting, but this is not shown in the drawing.

CO_2 molecule exhibits both types of above discussed vibrations contributing respectively to the following two types of spectra:

Case (a) Pure R and P branches only, the middle of the gap between them is at 2350 cm^{-1}

Case (b) R and P branch with very strong Q branch at 664 cm^{-1}.

Case (a) is related to a linear type of vibrations, while case (b) - to a quasirotational motion. The latter is a result of molecular bending. Such conditions are possible due to the large length to width ratio of the molecular dimensions (a similar reason for Q branches in H_2 ortho-II molecule, see §9.9.4) The QM model for case (a) is known as antisymmetric stretch (top), and for case (b) as asymmetric stretch (bend). Their spectrum are shown respectively in Fig. 9.66 and Fig. 9.67.

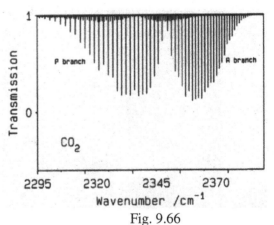

Fig. 9.66
Antisymmetric stretching fundamental band of CO_2

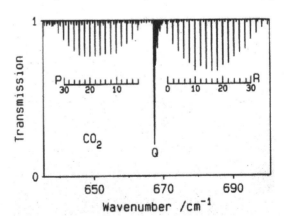

Fig.9.67
Bending fundamental band of CO_2

9.20. Water molecule

Figure 9.68 shows the configuration of the water molecule in one projection. The equatorial GBclp bound deuterons of the oxygen are not shown in this projection.

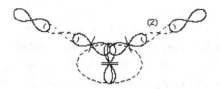

Fig. 9.68
Water molecule

The configuration of H_2O shows a large angular freedom of the oxygen protons (deuterons). In the solid phase this angle is self adjusted in order to match the convenient crystal configuration.

Chapter 10. Time, Inertia and Gravitation

10.1 Origin of time

The prisms are comprised of sub-structural formations organized in hierarchical orders. The level of matter organization below the prisms has been referred as a lower level. The configurations of the lower level structural formations were mentioned briefly in Chapter 2, showing the possible hierarchical orders. Their properties are discussed in more details in Chapter 12. The structural formations of lower hierarchical orders possess higher vibrational frequency. In such aspect one may expect that some structural formation in the lowest order should serve as a primary etalon. It will be shown in Chapter 12 that the consecutive upper order structural formations should possess lower frequency modes obtained by division of the primary frequency etalon. The unique physical mechanism assuring this division is embedded in stable structural formations, where the 3D geometry plays an important role. The analysis in Chapter 12 will put also a light on the most fundamental law, the law of Super Gravitation. Here, some aspects about the time base and time scale will be apriory given.

Let us accept that the theoretically known physical parameters Planck's time (introduced by Planck) is equal to the vibrational period of some primary frequency etalon.

$$t_{pl} = \sqrt{\frac{Gh}{2\pi c^5}} = 5.39 \times 10^{-44} \ (\text{sec}) \qquad (10.1)$$

where: G - is the classical gravitational constant in CL space (related with the Newton's low of gravitation)

We are still not sure is the Planck's frequency a vibrational proper frequency of the fundamental particle or the first geometrical formation - the primary tetrahedron. In Chapter 12 we will show that one of the most fundamental parameter in Physics - the fine structure constant is associated with the frequency of a common vibrational mode in the primary tetrahedron.

It is still a discussion issue at this point which gravitational constant should be involved in Eq. (10.1): G or G_o ? (the latter is a SG constant in a classical void (empty) space). It was envisioned, also, that the two fundamental particles have different intrinsic time constants, which must be involved in the definition of the Planck's time. Then the Planck's time might be a waited value based on both types of fundamental particles.

While the Planck's time is accepted as a primary time base, natural mechanisms embedded in the structural formations of the lower level structure of the prism assures secondary time bases by accurate division of the Planck's frequency on large numbers (discussed in Chapter 12).

The CL space is a formation above the prism level. It characteristic frequencies are lower than the mode frequencies in the prism substructure. For this reason, the CL node proper frequency and the SPM frequency could be considered as upper level frequencies (time bases) above the prism's level.

The levels of matter organisation and their relation to the structure of elementary particles are discussed in Chapter 12. A characteristic frequency could be assigned to every identified level of matter organization. The BSM analysis gives indication that between the Plank's frequency and the CL node frequency should be another level of matter organization with own characteristic frequency. It should be related to the internal structure of the prisms. For this reason, a zero number is assigned for the primary level, which is associated with the Plank's time (frequency).

Table 10.1 shows the guessed level of matter organisation in space based on a criterion of identified characteristic frequencies. These frequencies (or time bases) are obtained by a frequency division of the primary etalon (time base) provided by embedded natural physical mechanism. Since the four time characteristic frequencies cover very large range a natural logarithm of the frequency, $\ln(\nu)$, is also used.

Levels of matter organization				**Table: 10.1**
Level x	Time (sec)	Frequency, ν (Hz)	$\ln(\nu)$	Type of oscillation
0	5.39E-44	1.855E43	99.629	Planck's frequency
1				
2	9.152E-30	1.0926E29	66.86	CL resonance
3	8.093E-21	1.236E20	46.26	SPM, Electron

Fig. 10.1 shows a plot of $\ln(\nu)$ versus the level of matter organization in space, x.

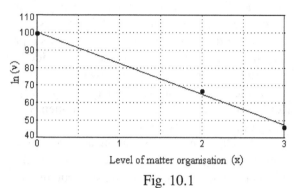

Fig. 10.1

Logarithmic plot of the identified so far characteristic frequencies towards the levels of matter organization in space

A robust line is fitted to the data of three points, corresponding to the identified levels of the matter organization in space. Despite the missing data for level 1, it will be shown in Chapter 12, that this assignment is logically correct. The CL space exists in levels 2 and 3 of the matter organisation in space, but not in level 0 or 1.

The very steep falling trend (having in mind the logarithmic scale) might be explained by the change of the inertial factor of the structures corresponding to the particular level of the matter organization. We see that the relation between the trend and the inertial factor (defined in Chapter 2) follows the rule: a larger inertial factor - a lower characteristic frequency. Then we come to a logical conclusion that the level zero should correspond to a matter organisation with an intrinsically small inertial factor. For a first guess, one may assume that this level corresponds to the bulk primordial matter. Although, in Chapter 12 (Cosmology) we will see that it could be attributed to the simplest material structure possessing oscillation properties.

The time base that we use in our observations and experiments is a **secondary time base**. It is defined by the CL space and more accurately by the **proper CL node resonance frequency**. The latter defines the light velocity and is also involved in the permittivity and permeability of free space (CL space). While the primary time base is a constant, **the secondary time base is subjected to relativistic effects** as they are known by the General and Special Relativity. For the first time, the BSM concept about space permits clear logical understanding of the physical mechanism behind the

relativistic effects. The theoretical models presented in this chapter and the analysis of the astronomical observations in Chapter 12 allows identification and definition of absolute reference frame with a reference point - the centre of any galaxy. This allows clear understanding and solution of many long existing enigma (a twin paradox, for example, related to the formulation of inertial frame in Special Relativity).

10.2 Inertia

The inertia of the matter we are familiar with (from elementary particles to astronomical body) is one of the most controversial and tough issue in contemporary physics. BSM allows a deeper insight into this feature of the matter.

10.2.1 General considerations

The stable elementary particles are formations of helical structures. The external dimensions of the elementary particles are many orders larger than the internode distance of the CL space or the individual prisms. Even the electron, which is the smallest stable elementary particle contains thousands of prisms.

The inertia, we are acquainted with, is a phenomenon related to the motion of the stable particles in CL space environments. Having in mind the finest helical structure formations with their internal lattices, it is apparent that the motion of any elementary particle in CL space invokes unimaginable number of fine interactions between the ordered fine structure of the particle and the CL space. These fine interactions are behind the Newton's inertia. The Newton's first law, known as a law of inertia is a macro effect from these interactions (valid for particles, atoms, molecules and macrobody formations of atomic matter) above the level 2 of matter organization.

Body at rest remains at rest and a body in motion continue to move at a constant velocity unless acted upon by an external force.

In the inertial considerations about the Intrinsic Matter (IM) provided in Chapter 2, it was shown that the definition of the Newton's law of inertia is not able to explain some interactions between bodies of IM in empty space. For this reason the inertial factor was introduced by Eq. (2.7) and different prism-to-prism interactions were discussed. It has been mentioned that the inertial factor between separate prisms is quite small. It arises to some level, when the prisms are combined in nodes and the nodes - in a gravitational lattice. The CL (cosmic lattice) space is a global formation of such lattice filling a huge void space (in a classical meaning) that is the observable Universe. For helical structures moving in CL space, the inertial factor is much larger that the IM inertia in empty space. This is a result of higher orientational order of the twisted structures comprised of FOHSs with their internal RL(T) structures.

The atom is a complex system comprised of oriented helical structures whose electrical field is neutralized in the far field because of the high speed motion of the electrons in quantum orbits. Due to the quantum features of such system, the orbital changes of electrons does not affect the inertial property of the system. The system inertial properties, however, are inseparable from the CL space properties.

It is apparent from BSM concept that the atomic matter (a matter we are familiar with) is quite distinctive from the Intrinsic Matter (IM) by some basic physical attributes, considered so far as fundamental. These are: the mass and the inertia. In order to distinguish them from the similar (but different) properties of the intrinsic matter we will refer to them as mass and inertia (considering Newtonian type), and Intrinsic Matter (IM) mass and inertia. They are quite different.

The inertial mass could be detected only in motion. From a point of view of the twisted prism model, the motion interaction between prisms could be attributed to the SG(TP) filed interactions, while the gravitational interactions - to the SG(CP) interactions. The elementary particles are built of ordered helical structures, in which the same types of prisms are embedded. Then the gravitational interactions between the elementary particles in CL space could be attributed to the SG(CP), while their inertial interactions - to the SG(TP) of the prisms, having in mind that the RL(T) of any helical structure contains the essential fraction of all prisms, from which it is built.

The same considerations could be further propagated to the atoms, molecules and a macro body, keeping in mind the very complex but well organized helical structures assembly.

Consequently, we may consider that the gravitational interaction between any two objects (of atomic matter) in CL space is a result of the SG(CP) field interactions between the helical structures they are built of. In this case, the CL space serves as a mediator for the propagation of the Newtonian gravity.

The propagated SG(CP) field through the _abcd_ axes of the CL space appears as a Newtonian gravitation.

One of the CL space parameter, which is related to the inertia is the equivalent inertial mass of the CL node (m_{node}). In Chapter 2, §2.11.3, this parameter was theoretically estimated by the fundamental physical constants and the derived parameters of the quantum wave. (Eq. 2.48 in Chapter 2). The same expression is shown here.

$$m_{node} = \frac{4h\nu_c k_{hb}^3}{\pi(c)c^2 N_{RQ}^3 k_d} \qquad [(2.48)]$$

The estimated value of the CL node equivalent inertial mass for the local Earth field is about 6.95×10^{-66} (kg). At first glance it seams that Eq. (10.5) involves only CL space parameters, however, the Planck's constant (h) and the Compton frequency (ν_c) are characteristic parameters of the electron, as well. From the other hand, the electron (positron) structure could be regarded as a building element of the stable elementary particles - the proton and neutron (including their internal helical structures). **Consequently, the equivalent inertial mass of the CL node may serve as a relation between the CL space and the elementary particles from which the atomic matter is built.**

The parameter m_{node} is related to the CL node distance. This is in a boundary of the length scale where the IM inertial properties are from the shorter side of the scale and the Newtonian inertial properties are from the longer one. For this reason m_{nod} is denoted as an equivalent parameter, not belonging to other side of the length scale.

This consideration is verified by some theoretical calculations (not provided here) for estimation the number of prisms involved in the electron structure, indicating that the classical inertial moment for example is not applicable for the internode distance length scale and below it.

The equivalence principle, formulated by Einstein, also exhibits a microscale limit when approaching the internode distance of the CL space. This distance, calculated in Chapter 2 is of the order of 1×10^{-20} (m). Most of the analysis in this chapter is relevant for the length scale above this distance.

10.2.2. Equivalence between gravitation and inertia in CL space

The Newtonian gravitational mass, according to BSM, is an attribute of the atomic matter detectable due to the attraction forces between two objects of atomic matter immersed in CL space. Behind these attraction forces are the SG forces propagated in CL space by the *abcd* interconnection axes of the CL nodes. While the time constant of the IM is much shorter than the proper frequency of the CL node (related to the speed of light), the propagation of SG forces for a complex formation (even such as an elementary particle) exhibits the limitation factor of the CL node oscillation. That's why the propagation of the Newtonian gravitation may exhibit the limit of the speed of light. In the case of CL node synchronization as in a closed magnetic line for example, this limit might be exceeded. This means a superluminal propagation of some information (some recent experiments about a phenomenon called "quantum teleportation" are manifestation of such effect according to BSM analysis). In a normal CL space, the SG forces may leak in very close distances of dens atomic objects (some Van der Walls forces between atoms and molecules) or in a case of very well polished surfaces in proximity (Casimir forces).

The equivalence between the gravitational and inertial mass (both of Newtonian type) is established and verified principle in contemporary physics. From the point of view of BSM we may add, that it is valide for particles comprised of helical structures containing a second order of helicity and placed in CL space environment.

The equivalence between the gravitational and inertial mass means that there is a balance between SG(CP) and SG(TP) fields for the system

comprised of the interacting bodies and the CL space involved as mediator.

From the above considerations it follows, that the mass equivalence from a BSM point of view may be formulated as:

- **In CL space environment, the gravitational mass of particle comprised of helical structures is equal to its inertial mass**

10.2.3 Involvement of the fine structure constant in the motion of the electron in CL space

From the Newtonian inertial and gravitational considerations in the previous paragraph it is apparent that the SG(CP) and SG(TP) prisms interactions are involved. Then we may use the relation between these two parameters initially adopted in Chapter 2 §2.9.6.B (Eq. 2.A.17.C) and validated later in Chapter 6 (calculation of the binding energy between the proton and neutron) and in Chapter 9 (the analysis of the molecular vibrations and determination of C_{SG} factor §9.7.5.C).

$$E_{SG}(TP) = 2\alpha E_{SG}(CP) \qquad \text{[(2.A.17.C)]}$$

where: α - is the fine structure constant, SG(TP) and SG(CP) are respectively the SG energies involved in the twisted and central part of the twisting prism model.

The fine structure constant is a characteristic feature of the electron confined motion. In Chapter 3, it was shown that it could be expressed by the CL space parameters and the electron geometrical parameters (equations (3.9), (3.10, (3.11), (3.12). One of the electron structure parameters - the helical step, for example, is completely defined by the CL space parameters, including the fine structure constant.

$$s_e = \frac{\alpha c}{v_c \sqrt{1 - \alpha^2}} \qquad \text{[(3.13.b)]}$$

where: s_e - is the helical step of the electron structure, v_c - is the Compton frequency (primary proper frequency of electron), c - is the light velocity.

Eq. (3.13.b) also shows that the fine structure constant is related to another important parameter of the CL space - the Compton frequency (or wavelength). The Compton frequency appears simultaneously as the SPM frequency (CL node parameter) and the first proper frequency of the oscillating electron. The second proper frequency of the electron (the frequency of the internal positron) appears to be three times the Compton frequency (according to the analysis of the electron behaviour in Chapter 3 and the Quantum Hall experiments in Chapter 4).

The fine structure constant is related also to other important characteristics of the orbiting electron: the orbital length and time (lifetime of excited state). It was shown in Chapter 3 and 7 that the quantum loop is defined by the Compton wavelength, while the time duration (lifetime) by the conditions of two consecutive phase match between the two proper frequencies of the electron), taking into account the relativistic gamma correction.

10.3. Inertial interactions of the atomic matter in CL space

When analysing the inertial interactions of the atomic matter in CL space, the gravitational field must be taken into account. Let us assume that every object of atomic matter is able to cause a slight gradient in the surrounding CL space. We may consider two types of CL space modulation:
(a) relativistic CL space modulation
  - GR effect of space curvature
  - SR effect of mass increase and time dilation
(b) gravitational (Newtonian) modulation

The space curvature of GR effect is obviously caused by a slight change of the internode distance due to the gravitational influence of the body mass. The effect is very weak due to the small influence of the very rarefied atomic matter of the body in comparison to the high density matter of the prisms, from which the CL nodes are formed. The slight change of the internode distance respectively causes a slight change of the CL space parameters including the light velocity.

The SR effect of the mass increase and time dilation is a result of the changed conditions of the confined motion of the electron at very high velocities.

10.3.1. Galactic CL space and gravitational field of material object. Definition of a local CL space and Separation Surface (ESS).

One of the problems of the Special Relativity is the lack of an absolute reference inertial frame.

In such case, any inertial frame could be selected as a reference. This provides an ambiguity, which is demonstrated by the so called "twin paradox". If the first twins is on the Earth, while the second one is travelling with a relativistic velocity and returned back, from the point of view of the first twin the second one will be left younger. From the point of view of the second twin, however, the first one is travelling with a relativistic velocity and he (first twin) must be left younger. This paradox does not exists in the BSM concept about space. Later from this chapter and Chapter 12 it becomes apparent that **there is an absolute reference point in the space - the centre of the local Galaxy**. This conclusion agrees quite well with the new scenario of the Universe and galaxy evolution presented in Chapter 12, which supported by strong arguments from the analyses of accumulated experimental and observational data.

Someone may rase a question that the provided concept contradicts to some postulates in Special relativity: the formulation of inertial frame, relying on the assumption that the propagation light is velocity independent. Here we must claim that the presented so far interpretation of the Michelson-Morley experiment is inconclusive from the point of view of the BSM concept about the physical vacuum. This and other similar experiments admit a methodological error in the interpretation of the experimental results. Presently the methodological error is widely discussed. In fact the first order result seams to support the velocity independence, but this is a result of one effect, which has been firstly recognized by R. R. Hatch: the Doppler shift effect is cancelled by the effect of the clock rate change. R. Hatch is one of the pioneers in the GPS system. The GPS system requires periodical corrections of its clock, which is in conflict with some formulations in Special Relativity.

Presently a number of experiments disprove the Special Relativity formulation of the inertial frame and the velocity independence of the light propagation. Among them are: G. S. Smoot et al. (1977), E. Silvertooth, (1986), D. Duari et al, (1992). Stefan Marinov first measured the Earth motion by laboratory experiments beginning in 1976 (S. Marinov, 1980, 1983). M. Consoli, A. Pagano and L. Pappalardo (2003) re-analysed the data of Michelson-Morley experiment using corrected methodology and extracted the velocity data about the orbital motion around the Sun and the absolute motion around the Milky Way centre. The extracted velocity data match well the data from other experiments. The failure of Michelson-Morley experiment to detect the Earth motion through space is also discussed and explained by R. Hatch (a pioneer in GPS) system and the canadian physicist and researcher Paul Marmet (search Internet for their publications).

In Chapter 12 (Cosmology) quite strong arguments are provided for the stationary Universe. It is shown that the two basic arguments of the concept of the Big Bang and expanding Universe are wrong: The first one is the assumption that the Universe space is homogeneous from which automatically follows the second one that the galactic red shift is of Doppler type. The BSM concept about the physical vacuum, however, leads to the conclusion that the Universe is stationary. This is supported by numerous observational results, when properly interpreted from the point of view of the new concept of the physical vacuum.

All galaxies contain own CL space, connected to the CL space of the neighbouring galaxies, so the CL spaces of the connected galaxies are stationary each other. The galaxy matter is rotated within the own galaxy space around the point of the largest mass identified recently as a supermassive black whole. The galactic CL space is strongly influenced by this mass and consequently it may serve as an absolute reference point. (This is confirmed in the observational analysis provided in Chapter 12).

The solar system with all planets and satellites is immersed in the Milky Way galaxy CL space. It is evident that the rotational velocity of the solar system could be detected if observing extragalactic sources. The average rotational velocity according to Duari (1992) is 238 km/sec. This velocity is quite away from a relativistic velocity, but it means that every object even at the microscale range down to the helical structures of the elementary particles is ablated by flying CL nodes. One may raise a question how this could be possible. When we take into account that the Intrinsic matter possesses an intrinsically small time constant (closer to the Planck's time) which defines the extremely fast propagation of the SG field in empty space,

the presented above explanation is quite reasonable. Now taking into account the influence of the Newtonian gravitation of a heavy object like the Earth on the surrounded CL space we may accept the following approximation:

The CL space in a local gravitational field of material object contributed by its whole atomic matter behaves like a stationary one for this matter.

For atomic matter moving through the galaxy space the CL nodes are disconnecting in the proximity of the RL(T) envelopes of the helical structure (of the elementary particles) following a displacement and after that - a return and reconnection to the CL space. The disconnected nodes are partly folded while the energy of their displacement is preserved as a rotational moment (keeping in mind the intrinsic inertial factor of the prisms). Such envisioned process is in fact behind the inertial properties of the atomic matter in CL space.

In the analysis of the inertial interactions of the atomic matter in CL space, the effects of the General Relativity (GR) and the Special Relativity (SR) could not be ignored. The operation in Minkowski space, however, makes the analysis very complicated. BSM found a simpler way to analyse the inertial interactions without ignoring the GR and SR effects.

First, let explain how the formulated by GR space curvature phenomenon is understood from the point of view of BSM. We may use an imaginary absolute scale (discussed in Chapter 2) whose unit vector is unshrinkable. The estimated (unshrinkable) size of the proton, for example may serve for this purpose. If considering that the CL space around some material object (of atomic matter) is stationary, the space curvature will means a gradually shrinkage of the CL node distance in the vicinity of the surrounding space. This shrinkage is quite small due to the week influence of the object atomic matter on the Intrinsic matter of the prisms, the matter density of which is much superior. The small shrinkage, however, will influence slightly the proper resonance frequency of the CL node which will affect the light velocity and other parameters (permittivity and permeability) of the physical vacuum. **We may refer this space as a local CL space, accepting also that every body**

possessing a Newtonian gravitational field could be regarded as possessing a local CL space.

The above conclusion will allow to analyse the inertial interactions between massive bodies, applying the inertial interactions between elementary particles, which will be derived in this chapter.

In order to simplify the analysis of gravitational and inertial interactions between massive bodies at distances larger than their radius, we will introduce a simplified model of a local CL space with a constant internode distance, but having an equivalent separation surface at which the internode distance changes sharply. Let us consider an ideal case of a heavy spherical material object with own gravitational field immersed in the Galactic CL space but very away from any other gravitational field. We may regard the GR space curvature as influence of the gravitational field of the object on the internode distance of the surrounding CL space. The internode distance of this space will be slightly shrunk with a gradient decreasing with the distance from the object. This of cause will change slightly the proper resonance frequency and the SPM frequency of the CL nodes. For a massive spherical body (with spherical density symmetry) the space curvature can be presented as a surface plot having a bell shape. The section of such plot with a normal plane is a bell shape curve as shown in Fig. 10.2.

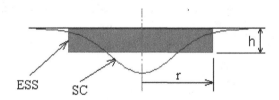

Fig. 10.2. Bell shape curve as a section of the space curvature with a normal plane. The rectangular function with a same area has a height h and radius r. The latter defines the Equivalent Separation Surface (ESS) of a sphere enclosing an imaginary CL space volume with a constant node distance

The bell shape curve could be approximated by a rectangular function possessing the same area as shown in the figure. Then the three dimensional space curvature according to GR could be represented as an **equivalent sphere** characterized by:

- a constant space grid density, corresponding to the parameter h of the rectangular curve

- an equivalent separation surface with a shape of sphere with a radius r

The CL node distance in the equivalent sphere will have a sharp difference at its edge in comparison to the galaxy CL space in which it is immersed. For this purpose we introduce the term **Equivalent Separation Surface (ESS) of the local CL space of the object.** For a spherical body (with a spherical symmetry of its matter density) the size of ESS is defined by the radius r as shown in Fig. 10.2.

The adopted assumption simplifies the analysis of a single body, by associating the properties of the gravitational field to some properties of the surrounded CL space called a local CL space. In a case when a less massive body is in the gravitational field of a more massive one, it appears that the CL space of the less massive one is immersed in the CL space of the more massive one, however, both of them, in fact, are immersed only in the galactic CL space. **For this reason we need to keep in mind that the local CL space is also imaginary, meaning that the surrounding galactic CL space is just locally modulated by the local gravitational field.**

The analysis of number of cosmological phenomena, observations and experiments in Chapter 12 indicates that the CL spaces of the galaxies are stationary. Then the galactic CL space can be considered as an absolute reference frame.

10.3.2 Relation between the gravitational field and the local CL space

A. Atoms and molecules in a gas substance

Single particles, atoms and molecules are attracted by the Earth gravitational field, according to the Newton's gravitational law. They also possess a local gravitational field in close proximity. Then they will cause a slight gradient (space curvature) of the CL space in close proximity. The signature of this is the Lamb shift observed in the atomic spectra of elements. If they are quite away from any massive body, they appear immersed only in the Global (galaxy) CL space. Otherwise we may consider that they are immersed in some "massive body CL space", like Earth, which in fact is a modulation of the "Sun CL space", which is a modulation of the Milky Way absolute CL space.

Note: Any material object in fact is immersed in the absolute CL space of the home galaxy. The theoretical concept of a Sun's local CL space and Earth local CL space is used only for convenience in the analysis.

The elementary particle possesses additionally its own quantum quasishrunk space due to their proximity electrical field (discussed in Chapter 2, 7 and 9). This is valid also for a particles with a proximity locked electrical field (such as the neutron, and the pair pions inside of the proton and neutron).

B. Atoms and molecules included in a massive body

The atoms and molecules included in a massive solid body do not possess freedom as in a gas substance. They may vibrate around a fixed point, but their average positions are fixed. Their individual gravitational fields contribute to the local gravitational field of the massive body. This body has a local CL space contributed by all individual particles. The involved single particle could cause a slight gradient in the local gravitational filed and consequently may possess a local CL space in their proximity, but this effect is much smaller in comparison with the local field. So they could be considered as immersed in the common CL space of the massive body. The particle also possesses own quantum quasishrunk space in proximity, as in the case A.

10.3.3 Inertial interactions of moving FOHS

Let us analyse the inertial interaction of simple FOHS (first order helical structure), taking for example the external positive shell of the proton. The structure is immersed in the CL space of a massive body, for example the Earth. One single turn of this helical structure is exactly the same as the helical structure of the positron. Let us find out what kind of interaction may exist between such simple structure and the CL space. We will consider two cases: FOHS is either in rest or in motion in respect to the external CL space (of the Earth).

A. FOHS in rest: In this case the interaction is only the static pressure that the CL space exercises on the impenetrable volume of the FOHS. This defines its Newtonian mass, according to the derived mass equation in Chapter 2. The mass equation was derived by using of the FOHS volume of

the electron, the dimensions of which are well above the CL node distance. Therefore the mass equation is valid for both: the gravitational and the inertial mass of the elementary particle.

B. FOHS in motion: The displaced from the FOHS volume CL nodes do not preserve their normal shape. The particle motion is accompanied with continuous process of folding and unfolding of the stationary (to galactic space) CL nodes, which ablate the FOHS volume. The folded nodes are not any more connected to the CL space and are displaced from their original position, while spinning. They return to their previous position, unfold and reconnect to the CL space after the FOHS is passed.

The process is illustrated by Fig. 10.3, where second order helical structure (containing FOHS) is shown. The E-field lines around the sections of the FOHS are shown by dashed lines. They define the quantum quasishrunk space of the particle. (Some of the lines in the internal zone are proximity locked as a result of the regulation effect of SG(CP) providing a charge unity).

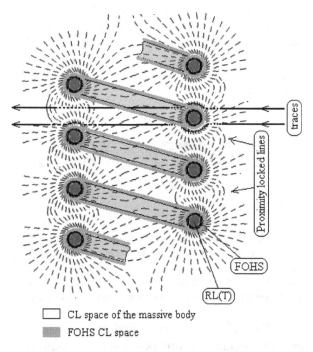

CL space of the massive body

FOHS CL space

Fig. 10.3

Relative trace of two CL nodes in the reference frame of moving second order helical structure. The folded state of the nodes through the quasishrunk quantum space of the structure is shown by dashed lines

The two horizontal dark lines in the figure show two traces of CL nodes, which become disconnected and folded in the curved section of the trace. The folded nodes intercept the denser E-field lines in the local space at angle close to 90°. **In this case the interaction with the local QE quasi-spheres is minimal. In such way the proximity E-field of the structure is able to guide the passing folded nodes.** Even the neutron has a proximity E-field, that is not apparent in the far field, but has its signature - the neutron's magnetic moment. So the proton and neutron proximity E-fields perform a guiding of the passing folded nodes.

The local CL space of single particles depends on the matter quantity in the participating helical structure. For single charge particle, however, the local space is not neutral but populated by EQ type of CL nodes. In Fig. 10.3 this space is illustrated by a light gray colour. The E-field of charge particle, however, is extended much beyond its local field.

The particle may be a part of a large solid body that has a local CL space. Then it is immersed in this local space and simultaneously contributes to it.

We may generalize some of the single particle features to a massive solid body. Then we arrive to the following conclusions:

(a) For elementary particle moving in CL space, the volume of its FOHSs is ablated by folded CL nodes. The folded nodes contain spinning energy of SG(TP) which is equivalent of their SG(CP) energy when connected to the CL space. Therefore, in case of uniform linear motion or motion in equipotential surface, there is not any loss of energy, since the energy spent for folding in the "entrance" is returned back to the system it in the "exit".

(b) The process of folding and guiding is assisted by the proximity field provided by the internal RL(T) of the FOHS.

(c) The folded nodes possess a rotational spin momentum and interact weakly with the CL nodes of the local field. In such way they are able to provide an uniform interaction in the whole volume of the body local space.

(d) The folded shape of the nodes is kept by a dynamical interactions with the normal nodes of the local space.

(e) The number of the folded nodes ablating the FOHS for a non zero velocity is proportional to the velocity of the structure. Larger kinetic energy corresponds to a larger number of ablating folded nodes per unit time.

(f) For a solid body moving in CL space, the folded CL nodes ablate the FOHSs of all elementary particle it is built of. At the same time, these particle are connected in atoms and molecules, held by the strong SG forces. Consequently, their inertial interactions appear as an integral interaction for the whole body.

(g) For a not relativistic motion, the inertial mass appears constant for different velocities. In such case the time for folding and unfolding is much larger than the resonance period of normal CL node. <u>For relativistic velocity the folding/unfolding time becomes comparable to the resonance period of the normal CL node. Then the motion exhibits resistance that contributes to a relativistic mass increase.</u>

10.3.4 Relativistic effect as a physical phenomena

From the above considerations it is apparent, that the folding and unfolding time of the CL node depends on the velocity of the moving elementary particle. It is evident that the following conditions are valid for relativistic and not relativistic motion.

$$t_F \ll t_R \quad \text{- for not relativistic motion} \qquad (10.6)$$
$$t_F < t_R \quad \text{- for relativistic motion} \qquad (10.7)$$

where: t_F - is the folding (unfolding) time, t_R is the resonance period of normal CL node

The effect of mass increase at relativistic velocity, V, according to the Special Relativity is given by the expression

$$m = m_o \gamma \qquad (10.8)$$

where $\gamma = (1 - V^2/c^2)^{-1/2} \qquad (10.9)$

Why the effect appears as continuous function of the velocity instead of a sudden increase, when the condition $t_F = t_R$ is fulfilled? One of the possible explanation is the following: The resonance period determines the light velocity according to light equation (Chapter 2). The trace of the resonance cycle of a single node shown in Fig.

2.26, Chapter 2 is almost a flat curve and may have a finite resonance width. The resonance cycles between the nodes in the local field volume are synchronized by the Zero Point Waves but this is not so strong connection between the individual nodes. As a result of this some domains may exhibit drag of their resonance frequency affected by the interaction with the spinning folded nodes. Such interaction may provides effect of inertial mass continuously increasing with the velocity. Such This phenomenon might be behind the relativistic increase of the mass according to the relativistic equation (10.8).

The gamma factor used in the relativity was derived in BSM by the analysis of the electron confined motion (Chapter 3, section 3.11.A).

From the physical analysis of the relativistic effect, it is evident, that the mass increase is a result of the quantum forces. These forces, however, have some limiting holding range. For a small single particle as the electron, for example, they may succeed to hold the continuity of Eq. (10.8) for large relativistic velocities. If the particle, however, is very fast accelerated by a large pulse energy, the Eq. (10.8) may break. The natural conditions for such event in free CL space, however, are not so common. One example of partial break of Eq. (10.8) is the synchrotron effect observed in the particle accelerators. However, in any conditions of partial break of Eq. (10.8) the principle of the energy conservation is preserved. The energy balance of the system particle - CL space is preserved by releasing of gamma photons from the CL space domains surrounding the accelerated particle.

10.3.5 Body with own local field in rest

Let us analyse the motion of a massive body with its local field (CL space) immersed in the external local field (CL space).

The provided concept of inertia requires a special attention for a body with a local CL space at rest. The problem may be referenced again to the proton in a massive body, when latter possesses an own CL space. Let us consider a body motion with a continuously decreased velocity. We see that consideration (e) of §10.3.3 provides some problem for the zero velocity. According to this conclusion, when approaching zero velocity it will be

some moment, when all folded nodes should be expelled from the volume of the local field. But where they will go? If the expelled folded nodes convert to normal nodes, then the CL static pressure in the surrounding domain will be increased. The CL space could not tolerate also a not uniform distribution of the folded nodes because this would cause large gradients of the background temperature and consequently the ZPE. Such effect is not observed, because the ZPE is kept constant due to the zero point waves. Obviously, the system provides a feedback assuring the energy carried by the folded node in a unit volume of CL space to be close to some preferable value. This will be proved later by the theoretically found optimal ratio between the normal and folded CL node pressures. This effect means that some number of folded nodes always exist in the volume of the local space. But then one problem needs a solution: When the body with a local space is in rest, the folded nodes could not have zero velocity, because this contradicts to the consideration (d) and (e) of §10.3.3. So they must circulate inside the body local field. But if they have a finite velocity they should circulate only inside the local field volume. This means that when the velocity is decreased from some finite value to zero, the state of the motion of the folded nodes will be changed in some point, i. e. the individual velocity vectors will be changed. As a result of this some threshold must exist when accelerating a body from rest, but this is not reasonable. *Consequently the option of circulating folded nodes in a body local field in rest does not provide a satisfactory solution from energetic point of view.*

Practically, the discussed inertial behaviour of a solid body could not be observed in our solar system since it moves with a velocity more than a two hundred km through the Milky Way CL space.

Phenomena related with this motion are analysed later in this chapter.

10.3.6 *Conclusions:*

(a) There is not a body in the Earth and in the Solar system, that is in absolute rest in respect to the Milky Way Global CL space.

(b) The internal space of a body in a rest in respect to the Earth contains passing folded nodes from external CL spaces (Sun's and home

galactic CL space) in which the Earth is immersed.

(c) When operating with local CL spaces (imaginary, but used for convenience) the equivalent velocity vector of the folded nodes is a vector sum of the velocities of every upper level of local spaces until the final one - the home Global space of the Milky Way.

(d) the number density of the folded nodes does not depend on the number of upper local levels, but of the vector sum of their absolute velocities referenced to the motion in respect to the global CL space.

(e) The absolute amount of folded nodes in the local field in rest could be determined if knowing the following parameters: the local field volume, the node density and the resultant velocity from all above levels.

(f) It is more convenient to operate with the energy of the folded nodes instead of their number, because the spin momentum of the folded node is taken into account. In this case the expressions are referenced to the relative velocity in Earth local space.

(g) The inertial interaction at relativistic velocity is increased, since the folding, unfolding time approaches the CL node resonance time (NRM vector frequency).

10.4 Theoretical analysis of the inertia in CL space. Partial CL pressure and force moment

From the conclusions made in §10.3.6 it follows, that we need a reference frame. All the physical constants related to CL space parameters are valid for the Earth local space. For this reason, we must use the imaginary defined local CL space of the Earth, so our analysis and equations about the inertia should be referenced to this frame. In this case the consideration (f) of §10.3.6 is valid.

The basic CL space parameters involved in a definition of the newtonian mass of the particles is the CL static pressure. Relying on the considerations discussed in §10.3.6 we may express, the inertial interaction as a partial CL pressure. This parameter should characterise the interaction between the folded nodes and the particles involved in the moving body. From consideration (e) of §10.3.3 it follows, that this pressure should be pro-

portional to the body velocity referenced to the external CL space.

According to §10.3.3 (g) the motion analysis depends on the velocity range due to the relativistic effect. For simplicity, we may separate the analysis into a not relativistic and relativistic case. For convenience, we may consider the velocity $v = \alpha c$ as a point between relativistic and not relativistic velocity.

10.4.1 Partial CL pressure for a motion with not relativistic velocity and force moment of the folded nodes

The static CL pressure was defined as a real physical parameter of CL space. This parameter was used for definition of the mass equation (Chapter 3). We may define a partial pressure associated with the folded nodes, ablating the helical structures of the elementary particle in motion.

Let us provide a definition of the CL Partial Pressure, the validity of which will be proved in the later analysis (§10.4.2, §10.4.2). We may assign this parameter as an attribute of the amount and the rate of the folding CL nodes, when the elementary particles (comprised of helical structures) moves through CL space. It is apparent that this parameter could depend on the particle velocity, so it should not be a constant like the Static and Dynamic CL pressure. However, it may have an optimal value that will correspond to some kind of optimal interaction between the particle of motion and the CL space. In this aspect we will use again the electron, since its geometrical and dynamical parameters were revealed in Chapter 3. From its motion behaviour in CL space we found that it has a preferable velocities. Then the **Partial Pressure of CL space** could be defines as velocity dependent according to the expression

$$P_P = P_S \frac{v}{c} \alpha \quad \left[\frac{N}{m^2} \right] \qquad (10.10)$$

where: $P_S = \dfrac{h v_c}{V_e}$ - is the static CL pressure   [(3.51)]

where: P_P - is the partial pressure (optima pressure value), v - is a velocity of the structure referenced to the external CL space, V_e is the electron volume, P_S is the static CL pressure defined in

Chapter 3 (the nominator can be expressed also as $hv_c = mc_2$).

While the parameter P_S is defined for the electron, we may expect to use it also for other elementary particles, in a similar way as the Static and Dynamic CL pressures. This consideration is supported by the discovered fact that the electron (positron) structure is implemented in the structure of any elementary particle. One major difference, to be taken into account, is that the electron possesses oscillation freedom, while the proton and the neutron do not possess such. However, every stable elementary particle exhibits a well defined Broglie wavelength, when in motion, so this is an indication of interaction with the CL space.

We see that the right side of Eq. (10.10.a) contains one vector parameter - the velocity (v), while all others are scalars. **Consequently, the CL Partial pressure has a directional component. This is also a distinctive feature from the CL Statics pressure.**

The defined parameter Partial CL pressure expresses the relation between the energy of the folded nodes and the particle velocity referenced to the electron.

10.4.1.A. Inertial force moment of the folded CL nodes (moment of force)

Multiplying Eq. (10.10) by the electron volume we get a parameter with dimensions [Nm]. This is a dimension of energy, but it could be considered also as a **force moment**.

So the inertial force moment for the electron is given by Eq. (10.10.a)

$$E_{IFM} = P_P V_e = hv_c \frac{v}{c}\alpha \quad [Nm] \equiv [J] \qquad (10.10.a)$$

where: E_{IFM} - is a force moment (posessing a same dimensions as energy), V_e - is an electron volume, v - is the relative velocity between the electron and the CL space.

We prefer to use a force moment definition (instead of energy) in order to distinguish its specific nature. It is implicitly valid for the folded nodes only. Physically, it expresses the work for deviation of the folded nodes from their straight trajectory. (Remeber that for a structure with a stable geometry moving with a constant velocity, the

work for folding the CL nodes in the entrance is exactly retutned at the exit - there is not any friction-like losses).

From the Newton's low of inertia it follows that the inertial mass could be estimated only during accelerated or decelerated motion. Only in such case the inertial force appears, according to the Newton's law: $F = ma$. Let us estimate the momentum of this force by using the defined parameter of partial pressure. The dependence of latter from the velocity requires the estimation to be done for a small deviation at some selected velocity. It is convenient to use the first harmonic velocity of the electron, equal to αc. The choice of this velocity matches also the conditions for derivation of the CL static pressure and the mass equation in Chapter 2. In order to eliminate the velocity dependence we will estimate the force moments for two close velocities $(\alpha c + dv)$ and $(\alpha c - dv)$, where dv is a small velocity change. The forces corresponding to these velocities are respectively F_1 and F_2. Substituting the velocity v in Eq. (10.10.a) with these two velocities (in brackets) we get respectively the force moments corresponding to the two cases:

$$E_{F1} = hv_c\alpha^2 + \frac{hv_c\alpha}{c}dv \qquad (10.11)$$

$$E_{F2} = hv_c\alpha^2 - \frac{hv_c\alpha}{c}dv \qquad (10.12)$$

The force moment difference is:

$$E_{F1} - E_{F2} = 2\frac{hv_c\alpha}{c}dv \qquad (10.13)$$

The kinetic energy for the two cases expressed in a classical way are respectively:

$$E_{K1} = \frac{1}{2}m_e(\alpha c + dv)^2 \qquad (10.14)$$

$$E_{K2} = \frac{1}{2}m_e(\alpha c - dv)^2 \qquad (10.15)$$

The kinetic energy difference is
$$E_{K1} - E_{K2} = 2m_e\alpha c dv \qquad (10.15.a)$$

Now comparing (10.13) and (10.15.a) we see that they are equal, because after simplification they lead to an equivalent equation (known as electron-positron "annihilation").

$$hv_c = m_e c^2 \qquad (10.16)$$

Conclusion:

The equality between the force moment difference $(E_{F1} - E_{F2})$ **and the kinetic energy difference** $(E_{K1} - E_{K2})$ **indicates that the concept of the force moment defined by Eq. (10.10.a) is correct.**

The correctness of the force moment definition is confirmed also later in this chapter, where it is used successfully for a motion analysis of astronomical objects, like the planets and satellites in the solar system.

Discussion: The obtained equivalence by Eq. (10.16) matches the "annihilation" energy for the electron rest mass, despite using the velocity value of αc, that approaches the relativistic motion. However, this is the velocity of the optimal confined motion of the electron. The static CL pressure and the mass equation are defined for the same conditions of particle motion. The application of such conditions for derivation of the mass equation in Chapter 3 and its successful application in Chapter 6 confirms the equivalence between the Newtonian mass and the "annihilation energy" While the meaning of the "annihilation" is broadly used in the modern physics it is logically incorrect according to BSM theory.

Knowing the Newtonian mass relation between electron structure and any other particle comprised of FOHSs, we may express the **force momentum of any particle** by the equation:

$$E_F = P_P \Sigma V_{SC} \quad [Nm] \equiv [J] \qquad (10.16.a)$$

where: V_{SC} - is a volume of a single coil FOHS, but multiplied with a proper factor (1 for a negative; 2.25 for a positive external shell of FOHS).

10.4.1.2 Partial pressure for relativistic motion. Relativistic mass increase

From the definition equations of the static and partial CL pressure we see, that the parameter velocity is involved only in the partial pressure. Then the factors influencing the relativistic velocity become separated. Such kind of separation will be of great importance for the physical understanding of some relativistic phenomena. Let us provide a verification test about such important outcome.

It is known from the theory of Special Relativity that the dependence of the mass from the ve-

locity is equal to the rest mass multiplied by the relativistic gamma factor. Using again the electron as a reference particle we have:

$$m_{eR} = m_e \gamma, \text{ where: } y = (1 - \upsilon^2/c^2)^{-0.5}$$

m_{eR} - is the electron mass at relativistic velocities

Using the mass energy equivalence $m_e c^2 = h\nu_c$, we see that partial pressure for relativistic velocity will get multiplication factor γ. Let us obtain the partial CL pressure at the optimum quantum velocity of the electron corresponding to energy 13.6 eV. Its axial velocity is $\upsilon = \alpha c$. Substituting this value for the velocity and in the multiplying factor γ we get the partial pressure expressed by electron geometrical parameters at its optimal confined motion.

$$P_P = P_S \frac{\alpha^2}{\sqrt{1 - \alpha^2}} = \frac{h\nu_c}{V_e} \frac{\alpha^2}{\sqrt{1 - \alpha^2}} \qquad (10.17)$$

It is evident right away that we may obtain the ratio between the Static and Partial pressure at such conditions (of optimal confined motion of electron).

$$P_P/P_S = \alpha^2/\sqrt{1 - \alpha^2} \qquad (10.18)$$

To verify the physical validity of the obtained equation (10.18) let calculate its reciprocal value. We get:

$$P_S/P_P = 18778.4$$

We get very close number to the electron revolutions in one quantum orbit corresponding to the a_o quantum orbit. determined in Chapter 3 by Eq. (3.43.h)

$$\frac{2\pi a_o}{s_e} = \frac{\lambda_c}{\alpha s_e} = 18778.362 \qquad [(3.43.h)]$$

Having in mind that the second proper frequency of the electron is $3\nu_c$ (the frequency between the shell of internal positron and the central negative core) it is evident that the second proper frequency will have a whole number of cycles if the fractional value is 1/3 (condition for standing waves in the short magnetic line conditions see Chapters 3 and 7). Equalizing the equations (10.18) with [(3.43.h)] we arrive to the equation of the helical step of the electron (s_e, that has been derived by different physical approach in Chapter 3.

$$s_e = \frac{\alpha c}{v_c \sqrt{1 - \alpha^2}} \qquad [(3.13.b)]$$

There are two important additional conclusions from the provided analysis:

- the helical step of the electron is completely defined by the CL space parameters

- having in mind the energy mass equivalence equation applied for the electron, ($m_e c^2 = h v_c$) the ratio P_P / P_S defines also the parameters of the first harmonics quantum wave (511 keV).

We may summarize:

- **The Partial pressure is a parameter characterizing the folded nodes. It is directly involved to the definition of the inertial properties of the atomic matter by the vector parameter Inertial Force Moment, E_{IFM}.**

- **The definition concepts of the static and partial CL pressure are in full agreement with the relation between Newtonian mass and inertial properties of the electron. All kind of helical structures included in the proton's (neutron's) structure could be referenced to the electron structure. Consequently the derived concept should be valid for all kind of atomic matter.**

- **The ratio between the partial and static pressure P_P / P_S of the CL space is a constant value determined entirely by the fine constant according to Eq. (10.18). This ratio defines completely the helical step of the electron structure according to Eq. (3.13.b).**

- **The ratio P_P / P_S is a self standing CL space parameter valide for a first harmonic quantum wave (511 keV).**

10.4.3 Specific partial pressure

The CL static pressure expressed by the electron mass density (see §3.13.3 Eq. (3.55) and (3.56), Chapter 3) is given by the expression:

$$P_S = \rho_e c^2 \qquad (10.19)$$

where: $\rho_e = \dfrac{m_e}{V_e} = \dfrac{g_e^2 h v_c^4 (1 - \alpha^2)}{\pi \alpha^2 c^5} \qquad (10.20)$

is the electron's density

For a not relativistic motion, we substitute (10.19) in (10.10) and obtain:

$$P_P = (\alpha c \rho_e) v \qquad (10.21)$$

Both equations (10.10) and (10.21) show that when referenced to the electron FOHS volume the Partial CL pressure is proportional to the velocity. Knowing that the volume of any FOHS structure could be referenced to the electron's volume, the product in the brackets in Eq. (10.21) can be regarded as a specific parameter of the motion of any particle consisted of FOHSs. We may call it a **specific partial pressure, denoted by** ρ_P. Expressed by the physical constants it is

$$\rho_P = \alpha c \rho_e = \frac{g_e^2 h v_c^4 (1 - \alpha^2)}{\pi \alpha c^4} = 3.34348 \times 10^{15} \left[\frac{N \; sec}{m^3} \right] \; (10.22)$$

Having in mind that gyromagnetic factor g_e of the electron is determined by the CL space, it is apparent that the new defined parameter p_p is completely defined by the CL space. From the additional considerations:

(1) both parameters: the static pressure and the specific partial pressure of CL space are referenced to one and same structure - the electron, using the volume of its FOHS.

(2) any FOHS could be referenced to this volume

(3) the volume of any FOHS of a stable particle is a constant in CL space environment

We may conclude:

(a) The specific partial pressures estimated by the electron geometrical parameters is valid for all types of helical structures with second order helicity

(b) The total force moment of any complex particle consisted of helical structures could be estimated if knowing the total volume of its FOHSs.

10.4.4 Force moment of the neutron and proton

From the mass equation we know that the neutron to electron mass ratio is equal to the ratio of their FOHSs volumes with proper correction factors for positive FOHSs and kaons.

$$V_\Sigma / V_e = m_n / m_e$$

For case of not relativistic velocity, applying Eq. (10.10.a) and substituting P_P by its definition in Eq. (10.10) we get:

$$E_{IFM} = h\nu_c \frac{V_\Sigma}{V_e} \alpha \frac{\upsilon}{c} = h\nu_c \frac{m_n}{m_e} \alpha \frac{\upsilon}{c}$$

Substituting $m_e = h\nu_c/c^2$ one obtains:

$E_{IFM} = (m_n c\alpha)\upsilon$	for neutron	(10.23)
$E_{IFM} = (m_p c\alpha)\upsilon$	for proton	(10.24)

Eq. (10.23) and (10.24) are the force moments for neutron and proton, respectively. The parameter in the bracket is a constant. Its values for a neutron and proton is pretty close, so it could be named a **specific force moment of hadron** (proton or neutron).

If using the envelope volume of the proton or neutron (estimated in Chapter 6 as 71.72 times larger than the total volume of all FOHSs of the neutron), the force constant could be expressed by the specific partial pressure, whose relation to the CL space parameters is given by Eq. (10.22). Then the force moment of a moving neutron is given by:

$$E_{IFM} = \rho_P \frac{V_n}{71.72} \upsilon \qquad (10.25)$$

where: the factor 71.72 is the ratio of the volumes V_n/V_Σ estimated in Chapter 6, and V_n is the envelope volume given by:

$$V_n \approx V_p = \pi(R_c + r_p)^2 L_{pc}.$$

where: R_c is a Compton radius, r_p - is the small radius of the positive FOHS, L_{pc} - is the length of the proton (neutron) core.

Both equations (10.23) and (10.25) provides exactly the same value in the case of neutron.

For relativistic motion the force moment is multiplied by the gamma factor.

The force moments given by Eq. (10.23) and (10.24) are more convenient for practical applications, while Eq. (10.25) is useful for a cross validation of the analysis correctness.

10.4.5 Inertial properties of the atoms and molecules

The local field of the FOHS as a part of each elementary particle was shown in Fig. 10.2. From the analysis and derived equations of inertial interactions for electron and neutron (and proton as well) it is evident, that the energy of the node displacement from the FOHSs is only involved. For a constant velocity the energy for the CL nodes folding is equivalent to the energy of unfolding with an opposite sign. In other words, both energies appear as self-compensated <u>reactive energies</u>, localised always at very small distance equal to the small diameter of the FOHS:

$2r_e = 1.768 \times 10^{-14}$ (m)	for negative particle
$2r_p = 1.18 \times 10^{-14}$ (m)	for positive particle

The unveiled reactive energy appears hidden for linear inertial motion, but not for a motion with acceleration. Let us consider now only the inertial motion.

The hardware structure of the stable elementary particles as electron, positron, proton and neutron neutron are not affected from their motion. Then the reactive energy will appear hidden for inertial motion of the atom.

If ignoring, firstly, the general relativistic effect of CL gradient around a mass object, we arrive to the following conclusions:

(A) The inertial force moment of a moving atom is a sum of the force moments of its protons and neutrons.

When the general relativistic effect is taken into account:

(B) Considering the general relativistic effect in atomic nucleus requires a correction of the inertial force moment by the nuclear binding energy. The latter is expressed by the equivalent mass deficiency (a difference between the total mass of neutrons + protons and the atomic mass).

The binding energy expressed by the mass deficiency is: $E_B = m_d c^2$

The atomic force moment could be obtained by applying considerations (A) and (B) with Eqs. (10.23) and (10.24).

$$E_{IFM} = c\alpha\upsilon(Zm_p + Nm_n - E_B/c^2)$$

where: Z - is the number of protons, N - is the number of neutrons, E_B is the nuclear binding energy.

The atomic mass is: $Au = (Zm_p + Nm_n) - E_B/c^2$, so the obtained expression of the atomic force moment is:

$$E_{IFM} = (c\alpha Au)\upsilon \quad \text{for atom} \qquad (10.26)$$

where: A - is the atomic mass, u - is the atomic mass unit.

10.4.6 Inertial properties of macrobody moving with a constant velocity

In order to provide a bridge between the inertial properties of a single particle and material object containing large number of atoms, we need to provide a definition of a macrobody (massive body):

A macrobody is a material object containing a large number of atoms (or molecules) whose integrity is kept by a Newtonian type of gravitation.

The provided definition is quite broad. It may refer to:

    - a self contained gas volume or liquid in a cosmic space.

    - a gas volume enclosed in some volume in the Earth gravitational field (a balloon)

    - a liquid in a vessel

    - a solid body

A special case of interest is the solid body. From the point of view of the inertial properties, two of its characteristics are mostly important: **its integrity** (all atoms are moving together) and its **mass.** The latter parameter is proportional to the quantity of the atomic particles (if neglecting the mass deficiency in the atom expressed by the nuclear binding energy). The expression of the mass by the quantity of atomic particles appears to be quite useful approach for transferring the inertial properties of a single particle to a solid body. At the same time it becomes apparent that the solid body may posses a proper (or local) CL space. The possible existence of such space depends on:

    - is the solid body immersed in some external gravitational field and how strong is it?

    - does the solid body possess enough mass (matter quantity) in order to have the necessary proper gravitational field?

It is also apparent (and will be confirmed later) that a macrobody with a quite large mass may contain extended CL space whose Equivalent Separation Surface is beyond its surface.

A massive macrobody with extended external CL space appears as an astronomical object, defined later in this chapter. (The main distinctive feature of the astronomical object (or body) from a macrobody is that its shape is close to a sphere due to its large mass).

Inertial interactions for a macrobody.

In the general case, the macrobody is consisted of large number of interconnected atoms or molecules, moving as a common volume.

From the conclusions in the previous paragraph, it follows that the force moment of the inertial interaction of macrobody could be regarded as a sum of the inertial interactions of the involved atoms. If ignoring the general relativistic effect, it is evident, that the force moment of the folded nodes inside the body will be proportional to the number of protons and neutrons in one atom and to the number of atoms in a unit volume. <u>Taking into account the General relativistic effect, it is the atomic mass that determines the inertial interactions.</u>

According to the general definition of the macrobody, it is necessary to distinguish the two quite different cases: a volume of gas and a solid macrobody.

10.4.6.1 Inertial properties of a gas enclosed in a finite volume

The number of atoms (molecules) in unit volume (for example 1 m^3) is strongly dependent on the pressure and temperature due to the effect of the Brawnian motion. For conditions of ideal gas this dependence is given by the equation:

$$PV = nRT \qquad (10.27)$$

where: P - is a gas pressure, V - is a volume of the vessel, R is an ideal gas constant, T - is a temperature in [K], n - is a number of *kmoles*.

In case of atmosphere around a planet (or gas volume around an astronomical object) the volume is defined by gravitational considerations.

The conclusions in §10.4.5 are fully consistent with the relation between the mass of 1 kmol of a gas substance and the number of atoms (molecules) given by the Avogadro's number:

$$N_A = 6.022142 \times 10^{26} \text{ [numbers/kmol]} \qquad (10.28)$$

With a good approximation for inertial analysis when using Eq. (10.27), the Earth atmosphere, for example, could be regarded as an ideal gas. It must be mentioned that Eq. (10.27) is not accurate for a temperature closer to the absolute zero, since it takes into account the collision interactions between the molecules (atoms).

10.4.6.2 Solid body

According to the conclusions in §10.4.4, the force moment (of the folded nodes) in the solid body will depend on two factors:
- the atomic mass number
- the number of atoms in unit volume (1 m^3 in SI)

The first factor is one and a same for a gas or a solid body. The second one, however, is dependable on the crystal structure of the solid substance. It also may vary between bodies made of different atoms. This variation is a result of the different spatial configurations of the atomic nuclei. So it is dependable on the number and positions of the valence protons, as they are involved in the connections between the atoms in the crystal structure. There are essential differences also between the crystal structures of metals and insulators.

In order to show how much the second factor influences the force moments, a simple analysis is made by comparing similar physical parameters of two cases of homogeneous matter substances.

Case A. Comparison between silver and gold

Table 10.3

element	Z	N	Z+N	A (u)	$\rho \times 10^3$ (kg/m^3)
Ag	47	61	108	107.87	10.5
Au	79	118	197	196.97	19.32

$\frac{(Z+N)Ag}{(Z+N)Au} = 0.5482$	$\frac{A(Ag)}{A(Au)} = 0.54764$	$\frac{\rho(Ag)}{\rho(Au)} = 0.54347$

Case B. Comparison between Al and Si.

Table 10.4

element	Z	N	Z+N	A (u)	$\rho \times 10^3$ (kg/m^3)
Al	13	14	27	26.98	2.7
Si	14	16	30	28.09	2.33

$\frac{(Z+N)Al}{(Z+N)Si} = 0.9$	$\frac{A(Al)}{A(Si)} = 0.96$	$\frac{\rho(Al)}{\rho(Si)} = 1.158$

In Case A: The two elements are quite distant by their Z number, but they have very similar configurations especially about the valence proton (see Atomic Nuclear Atlas). So it is reasonable to expect that they have a similar metallic crystal structure. This is confirmed by the pretty close ratio value between their atomic mass and density.

In Case B: The two elements are distinguishable only by one proton and two neutrons. Their atomic mass ratio, however is quite different than their density ratio. Such difference may come only from the crystal structure. The hadron density in unit volume for Al is higher, than for Si. This indicates also that the interatomic connections for metals and nonmetals are different. The molecular binding system of Deuterons, described in Chapter 9 should be valid for a solid silicon. But such binding system, evidently, is not involved in all type of interatomic connections between Al atoms in the crystal structure of solid aluminum.

The provided two examples are in good agreement with the consideration, that the force moment for a solid body is proportional to the sum of the volumes of its FOHSs.

Ideal solid body

Definition: A solid body possessing a local CL space for which the volume of its ESS (see Fig. 10.2) is equal to its body volume.

The concept of ideal solid body permits to analyse the inertial interactions by a model, according to which the whole volume of the ideal solid body is filled by its own CL space and the external CL space does not penetrate inside. For convenience we may consider that the **ideal solid body is homogenous i. e. it does not have cavities.**

Practically such body does not exist. All real solid bodies contain microcavities. (The local CL space of real solid body is discussed in the next paragraphs). The definition of ideal solid body, however, allows to transfer the force moment conditions of folded nodes from single particle or atom to a small solid body. The force moment of a homogeneous solid body (comprised of one atomic substance) with mass of 1 kgmol, can be expressed by multiplying the atomic force moment (given by Eq. (9.26)) by the Avogadro number:

$$E_{IFM} = (c\alpha N_A Au)\upsilon \quad \text{[J/kgmol]} \qquad (10.29)$$

where: $N_A Au = m$ - is a mass of 1 kgmol

The force moment of any substance of mass 1 kg is:

$$E_{IFM} = (c\alpha N_A u)\upsilon \quad \text{but} \quad N_A u = 1 \text{ , so:}$$

When working in SI system, the inertial force moment of a body of normal matter, possessing a mass of 1 kg is given by:

$$E_{IFM} = (c\alpha)\upsilon \quad \text{[J/kg] per 1 kg substance (10.30)}$$

Note: The term normal matter means the body is composed only of protons, neutrons and electrons. In latest paragraphs of this chapter it could be discussed that, "crushed" matter in form of kaons, may exist in the womb of the heavy astronomical objects.

While Eq.(10.30) appears referenced to 1 kg Newtonian mass, a solid body of *m* kg (but still complying to ideal solid body definition) will have an inertial force moment of

$$E_{IFM} = c\alpha m\upsilon \quad \text{[J] referenced to the mass (10.31)}$$

$$E_{IFM} = c\alpha\rho V\upsilon \quad \text{[J] referenced to the volume (10.31.a)}$$
$$\text{at constant mass density } \rho$$

where: *V* - is the body volume, ρ - is the density

The final equations (10.31) and (10.31.a) are quite simple and convenient for use for not relativistic motion. In case of relativistic motion, the right side should be multiplied by the relativistic gamma factor.

The Eq. (10.31) shows one important feature of the force moment:

The inertial force moment referenced per one kg of substance is independent from the atomic and molecular composition of the substance.

10.4.7 Relation between the inertial force moment and the first Newton's law of inertia.

Let us demonstrate the relation between the inertial force moment and the Newtons law for inertia for the electron accelerating by a constant force in the range of not relativistic motion. According to the first law of inertia we have $F = m_e a$

where: F - is the force with a constant value, m_e - electron mass and *a* is the obtained constant acceleration.

According to the general equation for the inertial force moment of body with mass m (Eq. (10.31)) we have $E_{IFM} = c\alpha m_e \upsilon$ (10.32) whose dimensional equivalence is:

$$[\text{N m}] \equiv \left[\frac{m}{s}kg\frac{m}{s}\right] \quad (10.33)$$

If Eq. (10.32) is divided by some specific CL space parameter with linear space dimension, we can obtain equation with dimensional equivalence of Eq. (10.32). Such specific CL space parameter is the space-time constant λ_{SPM}. In the Earth local space we have $\lambda_{SPM} = \lambda_c$. So dividing Eq. (10.32) by λ_c we get:

$$\frac{E_{IMF}}{\lambda_c} = \alpha m_e \frac{c}{\lambda_c}\upsilon = \alpha m_e \nu_c \upsilon = \alpha m_e \frac{\upsilon}{t_c} = m_e\left(\frac{\upsilon}{t_c/\alpha}\right) \quad (10.34)$$

Associating (10.34) with the Newtons law: $E_{IFM} = c\alpha m_e\upsilon$, we have: $E_{IMF}/\lambda_c = F$ and $\frac{\upsilon}{t_c/\alpha} = a$.

We see that the obtained Eq. (10.34) has the same dimensional equivalence as the first Newton's law of inertia. This confirms the validation of the introduced inertial force moment as a characteristic parameters of the CL folded nodes. Its validity for the electron as a single coil structure is propagated automatically for the other stable particles : proton, neutron and positron.

10.4.8 CL space inside of real body

Note: It must be kept in mind that the concept of the Local CL space was introduced for convenience. In fact all kind of bodies (comprised of atomic matter) are immersed in the galactic CL space.

Two concepts will be briefly discussed
- frame reference
- CL space continuity in one specific case

10.4.8.1 Frame reference

Case A: A massive body possessing a local gravitational field, which is larger that any external gravitational field in any point of its local CL space.

The extent of the local field around the protons and neutrons depends on the amounts of accumulated atoms. The local space of a massive real body will be a three dimensional manifold, but still continuous. All FOHSs will look as a three dimensional grid immersed in the local CL space of the body. If the body possesses microcavities, they are still part of the body CL space. An example of such massive body is a planet in a solar system. The body is characterized with the following important features:

- **when a massive body is in a relative motion in respect to external gravitational field, the own atoms and molecules are carried by the local field of its proper CL space and do not feel the motion**

Case B. A massive body with a local gravitational field smaller than some external gravitational filed.

In this case the CL space of the external gravitational field is a three dimensional manifold penetrating into the atomic matter of the body down to the level of helical structures. The inertial interactions of atoms and molecules in this case when the body is in motion are different:

- **when the body is in motion in respect to a stronger external gravitational field, the own atoms and molecules feel the motion**

Example: moving objects in earth gravitational field: thrown stone, car, train, aeroplane, rocket, satellite.

10.4.8.2 Continuity of the penetrated CL space

The concept of continuity of penetrating CL space is valid, when the local gravitational field in any point of the body volume is smaller than the gravitational field of external more massive body. The most important feature is:

- **The penetrating CL space inside the solid body is a continuous three dimensional manifold**

Example: any body in Earth gravitational field whose mass may range from single crystal, to large iceberg, artificial satellite and so on. The Moon, however, is not included in this category.

The concept of continuity could be demonstrated by the example of two flywheels, as shown in Fig. 10.4. The first one is solid, while the second one is hollow inside.

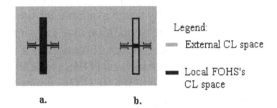

Fig. 10.4
Two flywheels and their local fields

The solid flywheel *a* has a larger moment of inertia than flywheel b. Let us consider that the flywheel b. contains some small parts inside the hollow volume. If this volume is vacuumized and there is not any friction forces between the internal walls and the small parts, they will be always attracted by the Earth gravitational field, independently of the motion conditions of the flywheel. If applying a linear acceleration of the flywheel b. in direction coinciding with the rotational axis, the small parts in the internal inertial hollow volume also feels the acceleration. The inertial properties of the small parts also does not depend of the thickness and matter density of the walls. So they exhibit the same adequate inertial properties as the solid volume. All these considerations lead to a conclusion, that the penetrating external CL space is continuous. <u>This is so because the volume of all FOHSs in the atom is much smaller than the atomic nuclear volume.</u> The interatomic distances are also larger than the atomic nucleus even for most dens metals (for example gold, lead, platinum etc.). From the ANS we see that element Ar possesses one of the most dens atomic nucleus (It is embedded in the nuclei in all heavier elements). Let us calculate the <u>volume ratio</u> between the envelop volume of Ar nucleus and the volume of all of its FOHSs. The envelope volume radius according to Michigan Institute technology data is 0.088 nm. The volume of all FOHSs in a single proton (neutron) is $[\pi(R_c + r_p)^2 L_{pc}]/71.72$. Then the <u>volume ratio</u> for Ar atom is

$$volume\ ratio = 65110$$

The similar volume ratio for any heavy metal could exceed a few hundred thousands. In liquids and solids, however, the interatomic distances are guaranteed by the finite sizes of the valence protons (deuterons) and electronic orbits. Even in the most dens solid the protons and neutrons are surrounded by space available for the CL space structure. In such conditions we may apply for the solid body the concept of motion of a free proton (neutron) as a bunch of FOHS, which was presented in §10.3.3 and Fig. 10.3. This concept is in agreement with the application of Avogadro theory in the case of solid state of the matter.

The presented concept of small body (case B) excludes the possibility of the body to possess own proper CL space.

10.5 Using the concept of the local CL space and ESS for massive objects

For a massive body with a detectable gravitational field it is convenient to use the concept of the local CL space with its presentation as an Equivalent Separation Surface (ESS) defined in §10.3.1.

10.5.1 Relation between the gravitational field and the local CL space

A local (proper) CL space by definition is possible only around a matter. In a normal state this matter is composed of helical structures. The shape of the local field of a single proton could be an envelope around all of its FOHSs without closing the gaps between their second order windings. This is in agreement with the concept of continuous three dimensional manifold presented in §10.4.8.

Accumulation of large number of atoms and molecules may form, for example, a spherical body whose local field is also a sphere but its ESS is extended beyond the volume of the solid body. So we need to define a **massive real solid body** from a point of view of a local CL space concept:

A massive real solid body is a such object for which the ESS of the local CL space is extended beyond its solid surface. The continuous three dimensional manifold in this case could be considered as occupied by the own (equivalent) CL space.

In the process of real body building from atoms and molecules the local gravitational field also arises and in some point it becomes detectable. So we may formulate the following general definition of a local CL space.

(A) If a real body has domains for which the local gravitation field exceeds any external gravitational field, it possesses a local CL space.

It is evident from the above formulation, that the possession of local CL field depends on number of conditions:
- the mass of the body
- the mass of the external (more massive) body providing the external gravitational field
- the distance between the body of consideration and the external more massive one
- the matter density of the body

10.5.2 Local CL space of large astronomical object

In order to analyse and explain some inertial phenomena in astronomical scale, we must define a criterion for considering the object as astronomical. For this purpose the following criterion could be used:

A real body could be considered as an astronomical object, if posessing a detectable gravitational filed, extended beyond its solid surface or self contained gas volume

According to above criterion, large astronomical bodies are: stars, planets, planet's satellites (moons) and asteroids.

10.5.3 Concept of separation surface between local CL spaces

Body with a small gravitational field is usually immersed in another gravitational field of a larger body. Then if the smaller body possesses own CL space, it is immersed in the CL space of the larger body. Even a single body with a local CL space quite distant from other bodies is still immersed in the Global CL space of the home galaxy.

For single astronomical body (as idealistic case) with a spherical shape, the ESS of its local CL space will be also a spherical. In a general case the spin rotation could be considered as a normal state. It is reasonable to accept, that the local CL space will rotate with the same angular velocity. Then for a case of spin rotation, the following boundary conditions are valid:

(a) The radius of the local field becomes limited by the light velocity at the boundary surface. Then the separation surface becomes defined by the ideal separation radius R_s, satisfying the condition:

$$R_s \omega = c \qquad (10.35)$$

where: ω -is the angular velocity and c - is the light velocity

(b) ZPE threshold cut: The radius of the local field becomes limited by the Zero point energy of the free (galaxy) space playing a role of a threshold.

In a real world it is not possible to find a case for a single macrobody, where there is not a second body in the range of the ideal separation radius of the first one. This also means, that the second body affects the size of the CL space of the first one by its gravitational field. Conclusion:

In the real world, the ideal case of ESS of a single body is disturbed by another body with a local gravitational field (possessing a local CL space) in the range of the separation radius defined by the ZPE threshold cut.

10.5.4 Two bodies at rest in respect to the upper level CL space - idealised condition

In this case (possible also in idealised conditions), we consider only two massive bodies in the global (galactic) CL space, both in rest in respect to this CL space. Both bodies are in gravitational interaction, but we may consider an initial moment after they has been hold in rest. In such conditions the inertial force moment could not be defined and only newtonian gravitational forces are considered. If the distance between both bodies is smaller than the radius restricted by the ZPE threshold (considered in the previous paragraph), both bodies will have a common ESS, for every point of which the Newtonian gravitational forces will be equal. If one of the bodies is much less massive, its ESS will have a shape of an egg immersed inside of the ESS of the other (more massive one).

If trying to analyse the folded nodes of the more massive body, penetrated in the CL space of less massive one, a problem appears: the folded nodes are not in motion! This is in conflict with the inertial definition and even the partial pressure could not be defined. This is another reason that make the case idealistic, so the two bodies could not be at rest in respect to the upper CL space - the galactic one.

10.5.5 Two bodies with a constant distance between them but in a common motion referenced to the upper level CL space.

This is a realistic case in which both bodies will rotate around a common centre, whose position depends on their mass ratio. Therefore the definition of local CL spaces and ESS is reasonable and could be used for a motion analysis. In a general case when both bodies are with different masses and close enough so their local CL spaces could not be limited by the ZPE threshold, the ESS has a form of egg and encloses the less massive body. This case could be applied for a planet in a solar system. The separation surface is defined by the condition that the two vectors of the gravitational forces in any point of the system (not taking into account the centripetal forces) have equal magnitudes. For a planet in a solar system the condition defining the separation surface is expressed by Eq. (10.35.a) (based on the Newton's gravitational law).

$$\frac{M_S}{(d - r_s)^2} = \frac{M_P}{r_s^2} \qquad (10.35.a)$$

where: M_S is the solar mass, M_P is the planetary mass, d - is the distance between them, r_s is the distance of the separation surface from the planet.

10.5.6 Two bodies with changeable distance between them.

The two bodies will have elliptical orbits around a common centre of rotation. The CL space of the less massive body is always immersed in the CL space of the more massive one. The ESS enclosing the less massive body will have a shape of egg as in the previously discussed cases, but its size and shape will vary with the distance change between them.

10.6 Total energy balance of moving macrobody

In the previous paragraphs, the force moment of the folded nodes was discussed for a single particle, for an atom or for an ideal solid body. Now we need to examine the validity of Eq. (10.31) for real solid body.

Let us accept initially that: Eq. (10.31) is valid for a large body and even astronomical object. In order to find out is this true or not, we will analyse the relation between the gravitational and inertial interactions of a real body in different gravitational fields. For this purpose some data about the solar system planets (and some of their moons) will be used.

10.6.1 Force moment of real body in a free fall motion

Let us analyse how the energy of a free fall body with a mass m near the surface of an astronomical body (planet or moon) is changing per unit time by comparing its motion energy and its gravitational potential. To simplify the analysis we will consider that the planet has a perfectly spherical shape, so the centre of mass coincides with its geometrical centre. Let the body is initially kept in rest near the planets surface and at moment zero it is released for a free fall motion. By using the definition of inertial force moment (for folded nodes) $E_{IFM} = c\alpha m v$ we may find the work for deviation of the folding nodes for a small time interval. The force moment is proportional to the velocity change for a unit time interval for any moment of the motion. For a motion range much smaller than the planet radius, the gravitational acceleration could be considered as a constant. Then the change of the force moment is a constant for a small time interval Δt in any moment of the motion. At initial time moment of zero, the velocity is zero, so the force moment is also zero. At the end of the time interval Δt the velocity is equal to $g\Delta t$ and the force moment is $\alpha cmg\Delta t$, where g is the gravitational acceleration. Then the change of the force moment of the real body with mass m per unit time is:

$$\Delta E_{IFM} = \alpha cmg \quad [\text{J/s}] \qquad (10.36)$$

Let us assume that the planet is characterised by the following features:

- perfect spherical shape
- constant or linear change of the solid matter density with the radius from the centre
- all the mass is enclosed below the solid surface of the planet (the mass of atmosphere is ignored)
- ignoring the free fall influence from the planet sideral rotation, when the body is closed to the planet surface (practically valid for all planets of the solar system)

The gravitational potential at the same level (considered near the surface of the planet) is given by:

$$U_G = G\frac{Mm}{R} \quad [\text{J}] \qquad (10.37)$$

where: G - is the gravitational constant, M - the mass of the planet, R - a spherical radius of the planet

The gravity at the level defined by R is:

$$g = \frac{F}{m} = G\frac{M}{R^2} \qquad (10.38)$$

Ignoring the centrifugal force, according to the above mentioned consideration and using Eq. (10.36), we obtain the expression for the ratio $U_G/\Delta E_{IMF}$:

$$U_G/\Delta E_{IMF} = \frac{1}{\alpha c}R \qquad (10.39)$$

Let us define the above ratio as a factor K_E.

$$K_E = U_G/\Delta E_{IMF} \qquad (10.39.\text{a})$$

Note: The change of force moment ΔE_{IMF} in Eq. (10.39) is a parameter of a real small body in gravitational field of astronomical object. It should not be confused with a force moment of the massive astronomical object.

Eq. (10.39) leads to some interesting results:

(a) From a pure theoretical point of view

Let us estimate Eq. (10.39) for a time interval Δt equal to the Compton time t_c, which is inverse to Compton frequency, ν_c. We obtain:

$$R = (\alpha c)/\nu_c = 1.7706 \times 10^{-14} \quad (\text{m})$$

Amazingly, this value is quite close to twice the small radius of the electron structure, r_e, which was derived in Chapter 3.

$$2r_e = 1.7685 \times 10^{-14} \quad (\text{m})$$

Conclusion:

- **The obtained coincidence confirms the envisioned concept that CL nodes partly folds and deviates from the moving through space helical structures (possessing impenetrable for CL nodes internal RL lattice).**

The above result is valide for all negative elementary particles since they have a similar FOHS as the electron. It is valid also for the positive particles including the proton's (neutrons's) external shell) using the same correction factor as for the mass equation for positive helical structures (or particles).

(b) From a practical point of view of a system comprised of real astronomical objects:

Eq. (10.39) is very useful theoretical expression that could be tested for the planets and moons from the solar system. The parameters U_g and ΔE_{IMF} can be separately calculated by the data available for them. If the astronomical body (planet or moon) does not have a perfect spherical shape the radius R could be replaced by a volumetric radius R_V. One important feature of the ratio given by Eq. (10.39) is that the planet mass is eliminated. However it participates in the separate parameters U_g and ΔE_{IMF}. The parameter K_E could be ploted against the planetary volumetric volume for convenience, because it is fully defined by the planetary radius. The planets, however, possesses also a sideral rotation and its possible contribution should be taken into account.

For astronomical body with a significant spin rotation, the falling body near the solid surface will get rotational energy that should decrease U_G. If assuming that an unit mass of atmosphere (1 kg) is uniformly distributed as a thin shell at radius R_V, its rotational energy is

$$E_R = \frac{1}{2} I \omega^2 \qquad (10.40)$$

where: $\quad I = \frac{2}{3} m R_V^2 - \frac{1}{2} R_V^2$

and ω - is the angular velocity determined by the sideral period

Then for a real astronomical body (planet or moon) the ratio k_E becomes:

$$k_E = \frac{U_G - E_R}{\Delta E_{IMF}} \qquad (8.41)$$

Using the NASA fact data sheets about the planets and moons of the solar system, the ratio (8.41) in a measurement system unit of SI is estimated for number of planets and moons in the solar system. For this purpose the volumetric radius R_v of the solid surface is used. Since this parameter is not well known for the largest planets Jupiter, Saturn, Uran and Neptun, they are not included in this analysis. the radius at 1 mbar pressure is used for them. All other planets, most of the moons and one asteroid with a spherical shape are included. The data source is:

(http://nssdc.gsfc.nasa.gov/planetary/planetact.html).

For estimation of ΔE_{IMF}, the surface gravity, from the "fact data sheet" is used if it is available. Otherwise, it is calculated by Eq. (10.38). Some of the necessary data and the calculated parameters are given in Table 10.4. The rotational energy (for 1 kg atmosphere) calculated by Eq. (10.40) for the planets of the solar system is quite small, so it is not given in the table. For example: E_R parameter (from the sideral rotation) for Mars and Earth is respectively 1.925E4 and 7.194E4, while U_G is respectively 12.63E6 and 6.256E6. So E_R for this two planets are respectively 0.15% and 1.14% of U_G and could be ignored. For other planets the E_R contributions are much smaller, so they are ignored. Then with a quite good approximation, the Eq. (10.42) could be accepted to be valid for the planets of the solar system:

$$K_E = \frac{U_G}{\Delta E_{IMF}} \qquad (10.42)$$

The plot of K_E versus the volumetric radius R_V is shown in Fig. 10.5.

The planets and moons for which the surface gravity is provided by the fact data sheets are drawn by green (or gray) points. For other objects (mostly moons) the surface gravity is calculated by Eq. (10.38). All planets and moons lie very closely to the theoretical line given by Eq. (10.39).

Planet (moon) data   ***Table 10.4***

No Planet (moon)	M x 10^{23} (kg)	R_v (km)	g (m²/s)	T_{sid} (hr)	k_E
1. Nereid	0.0002	170			0.077
2. Vesta	0.000301	265			0.121
3. Umbriel	0.0117	584.7			0.267
4. Charon	0.019	593			0.27
5. Oberon	0.0301	761.4			0.348
6. Titania	0.0352	788.9			0.36
7. Pluto	0.125	1195		153.3	0.546
8. Triton	0.2147	1352.6		141	0.618
9. Europa	0.4797	1569		85.2	0.706
10. Moon	0.7349	1737.4	1.62	655.7	0.796
11. Io	0.8933	1815		42.48	0.837
12. Calisto	1.076	2403		400.5	1.1
13. Titan	1.345	2575			1.173
14. Ganimede	1.482	2634		171.6	1.21
15. Mercury	3.302	2439.7	3.7	1407	1.11
16. Mars	6.418	3390	3.69	24.62	1.562
17. Venus	48.685	6051.8	8.87	5832	2.766
18. Earth	59.736	6371	9.78	23.934	2.92

The rotational energy E_R (for 1 kg mass) is quite small and not listed in Table 10.4. The position of planet (moon) is notated by the *No* given in Table 10.4.

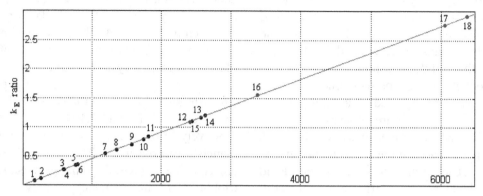

Fig. 10.5

From the analysis of the ploted data, the following conclusions are apparent:

(a) The planets and moons generally align well to the linear theoretical expression given by Eq. (10.39)

(b) A few planets and moons exhibit some small deviation from the theoretical curve. The reason for this could be the difference between their mean density and the theoretical mean density of the small body used as a reference (having a mass of 1 kg).

10.6.2 Anomalous position of Mercury in the plot of the ratio k_E as a function of mean radius

The assigned number for each planet or moon in Tables 10.4 follows the trend of their mass increase. From the data points plotted in Fig. 10.5,

however, we see that Mercury (point number 15) does not follows the trend but takes a reversal direction. In order to study this anomaly we analyse the trend of the matter density of astronomical body, when a large mass is accumulated. For this reason we will analyse the trend of the planetary (moon) mass increase as a function of their volume.

Fig. 10.6 shows the trend in a full volume scale, while Fig. (10.7) shows a portion of the volume scale with increased vertical resolution. The plots in Fig. 10.6 and 10.7 clearly show that the common trend of matter density breaks into two trends. The first trend for planets (moons) with a smaller mass ends up at Mercury. The second trend begins with Calisto and continues for the larger planets.

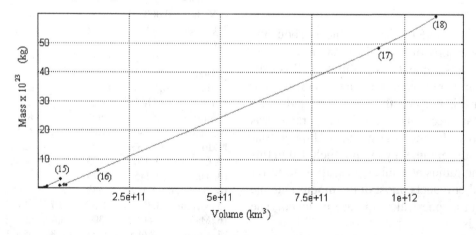

Fig. 10.6

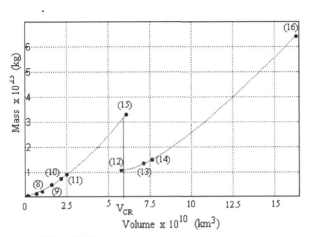

Fig. 10.7. Initial section of the plot shown in Fig. 10.6 with better scale resolution, showing the break zone of the trend

The break of the common trend appears in the region close to $V_{CR} = 5.2 \times 10^{10}$ (km^3). The object 15 (Mercury) is still in the strong gravitational field of the Sun.

The break of the trend has some similarity with the mass efficiency (binding energy) in atomic nuclei as a function of Z number. The possible explanation of the observed effect is the following:

Due to the enormous gravitational pressure the structure of the protons and neutrons in the central zone of a massive astronomical object may brake. The internal pions may convert to straight structures - kaons getting alignment to the central kaon. The obtain bundle of kaons in such enormous pressure might be stable. They may shrink additionally the internal CL space, providing the same effect of mass deficiency as in the atomic nuclei. At the same time the internal RL(T)s of all kaons appear axially aligned. They can modulate strongly the external CL space, providing an excellent conditions for a strong magnetic field.

The above explanation is additionally discussed in the magnetic field hypothesis for the planets proposed in §10.14 and for the stars (discussed in chapter 12).

The suggested explanation may also give an answer, why Mercury overpasses the theoretical break zone. This planet is very close to the Sun and exhibits pulling forces, which continuously vary due to the elliptical orbit. In such conditions, the gravitational pressure in the central zone of the planet is decreased. Additional factor may be the continuously changed conditions in respect to the strong magnetic field of the Sun.

The theoretical threshold separates the common trend into two zones: a zone above, and a zone below the threshold value V_{CR}. The critical mass, M_{CR}, corresponding to V_{CR} could be considered as an average between the masses of numbers 12 and 15.

$$M_{CR} \approx 2.189 \times 10^{23} \quad \text{(kg)} \tag{10.43}$$

The corresponding critical matter density for this case (average for the whole volume) is:

$$\rho_{CR} = \frac{3M_{CR}}{4\pi\alpha^3 c^3} \tag{10.44}$$

10.6.2.A. Theoretical concept for dynamical equilibrium of moving astronomical object in CL space.

The planetary motion in the solar system is quite stable. This fact is based on accurate astronomical observations for many years. Then some dynamical equilibrium should exist. Using the golden rule of energy conservation we may try to express some of the parameters by energy ratio and observe the obtained trend. Such opportunity is provided by the derived below theoretical expression (10.47).

Let a planet with a mass M_P is in a stable circular orbit with radius r around a star with mass M_S. Then the gravitational attraction is equal to the centripetal acceleration: $GM_S M_P/r^2 = M_p \upsilon^2/r$. The tangential velocity is

$$\upsilon = \sqrt{\frac{GM_S}{r}} \tag{10.45}$$

The energy ratio between the inertial force moment (of folded nodes) and the kinetic energy of the planet is

$$\frac{E_{IFM}}{E_K} = \frac{\alpha c M_P \upsilon}{0.5 M_p \upsilon^2} = \frac{2\alpha c}{\sqrt{GM_S}}\sqrt{r}$$

Rising in square we get:

$$\left[\frac{E_{IFM}}{E_K}\right]^2 = \frac{4\alpha^2 c^2}{GM_S}r = C_E r \tag{10.47}$$

where: $C_E = \dfrac{4\alpha^2 c^2}{GM_S} = 1.44238 \times 10^{-7}$ (10.48)

The obtained theoretical Eq. (10.47) is very useful. The term $[E_{IFM}/E_K]^2$ is a linear function of the distance r. This allows determination of the constant C_E by two methods:

- by plotting the Eq. (10.47), as a function of r, by using of the planetary data sheets and fitting to a robust line

- by theoretical expression (10.48) where only the solar mass is necessary. The latter is:

$$M_S = 1.9891 \times 10^{30} \quad (kg)$$

The expression (10.47) could be verified by the planetary data of the solar system. Using the planetary fact sheets data, the calculation of E_{IFM} and E_K is a straightforward. Most of the planet orbits exhibit quite small orbital eccentricity. For planets with larger eccentricity the energy E_K is approximately estimated by using the mean orbital velocity.

The input data from the planetary fact sheets are given in Table 10.5. The mean distance of every planet from the Sun is given as d in astronomical units (1 au $= 1.496 \times 10^{11}$ (m) is the mean distance between Earth and Sun). The distance d corresponds to the radius r in Eq. (10.47).

Planetary data   Table 10.5
===

No	M_P $\times 10^{23}$ [kg]	υ (mean) [km/sec]	d (mean) [au]	ρ (mean) [kg/m^3]
1 Mercury	3.302	47.87	0.387	5427
2 Venus	48.685	35.02	0.723	5243
3 Earth	59.736	29.78	1	5515
4 Mars	6.418	24.13	1.524	3933
5 Jupiter	18986	13.07	5.203	1326
6 Saturn	5684.6	9.69	9.539	687.26
7 Uranus	868.3	6.81	19.18	1270
8 Neptune	1024.3	5.43	30.06	1638
9 Pluto	0.125	4.72	39.53	1750

The calculated parameters E_{IMF} and E_K show quite large variation for different planets (a few orders), although the square values of their ratio exhibits a perfectly linear dependence on the mean distance from the Sun. The plot of this ratio is shown in Fig. 10.8.

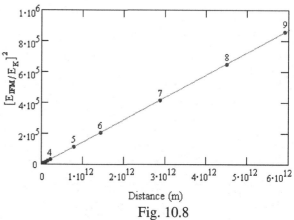

Fig. 10.8

Square value of energy ratio for the planets of the solar system as a function of their mean distance from the Sun

The plot of Fig. 10.8 shows excellent alignment of the planets along the theoretical line. By fitting to a robust straight line the experimental value of the slop is obtained. Its value is 1.44322×10^{-7}. The difference between this value and the theoretical one given by Eq. (10.48) is only 0.06%. This confirm the correctness of the used concept.

Conclusion: The involvement of the fine structure constant in the theoretical equation (10.47), indicates that it plays an important role for the stability of our solar system. This is an important discovery of BSM, not anticipated so far.

10.6.5 Energy involved in the motion of astronomical object in gravitational CL space

10.6.5.1 Balance between the orbital kinetic energy and the inertial force moment (of folded nodes) for astronomical body moving in gravitational CL space

Let us analyse additionally the derived Eq. $[E_{IFM}/E_K]^2 = C_E\,r$, which was verified by the data plot shown in Fig. 10.8. At first glance, it seems that the kinetic energy is not equal to the energy E_{IMF} (inertial force moment of the folded nodes). However **E_{IMF} is defined for a body motion in free space, where the gravitational field from other body is excluded.** So the energy ratio in the above shown expression and the plot in Fig. 10.8 does not take into account the folded node motion in a gravitational field. If writing the above equation in a form $[E_{IFM}/E_K] = \sqrt{K_E}\sqrt{r}$ we see that the ratio between the two energies is proportional to the square root of the distance (orbital radius). In order two energies to be equal, the energy force moment for free space should be multiplied by a common factor, inverse proportional to $\sqrt{r}$. The physical meaning of this could be explained if accepting that:

<u>The folded nodes in closer orbits exhibit larger interaction when passing through the moving body</u>

It is equivalent to;

The folded nodes in a larger gravitational potential exhibit a larger interaction when passing through the moving body.

The above consideration is completely logical if taking into account that a larger gravitational potential corresponds to a slightly smaller node distance of the CL space and a larger CL stiffness. The CL stiffness is directly involved in the definition of the CL node resonance frequency and consequently the light velocity. A very small change of CL stiffness leads to a change of the node resonance frequency. The gravitational effect on the light velocity is quite small because of the self adjustable parameters of the CL space. <u>However, the physical dimensions (length and shape) of the elementary particles proton and neutron are quite stable since they are closed loops structures, so they could not follow a proportional change with the node distance. This may explain why the moving matter in such CL space may exhibit a stronger interactions.</u>

Now the following question arises? What is the physical parameter that could provide the necessary correction factor for E_{IMF} in gravitational field CL space? The answer is: **The possible physical parameter is the spin momentum of the folded nodes.**

Let us consider two logically possible options for the spin momentum dependence of the folded CL nodes:

(a) From the relative velocity between the moving body and the external CL space

(b) From the stiffness of the external CL space, i.e. from the gravitational potential.

From the energy conservation principle it follows that the product of the E_{IMF} and the correction factor for the CL node spin momentum should

be equal to the body kinetic energy when moving in equipotential surface. Such motion is possible if the centripetal acceleration is equal to the gravitational attraction. The circular motion satisfies this condition, so we can use it in the following derived expressions.

Taking a square root of the basic Eq. (10.47), valid for a circular planetary motion in the solar system, we get:

$$\left[\frac{E_{IFM}}{E_K}\right] = 2\alpha c\sqrt{\frac{r}{GM_S}} = \frac{2\alpha c}{\sqrt{U_{Gn}}} \qquad (10.56)$$

$$U_{Gn} = \frac{GM_S M}{r}\frac{1}{M} = \frac{GM_S}{r} = \upsilon_{or}^2 \qquad (10.57)$$

where: U_{Gn} - is the gravitational potential normalized to 1 kg mass (SI system only)

υ - is the orbital velocity for circular orbit

The kinetic energy obtained by Eq. (10.56) is

$$E_K = E_{IFM}\frac{\sqrt{U_{Gn}}}{2\alpha c}$$

According to energy equivalence:

$$E_{IFM}^G = E_K \qquad\qquad\quad \mathbf{(10.58)}$$

where: E_{IFM}^G - is the inertial force moment of the folded nodes in gravitational CL space

$$E_{IFM}^G = E_{IFM}\frac{\sqrt{U_{Gn}}}{2\alpha c} \qquad (10.59)$$

The final Eq. (10.59) is valid not only for a circular motion of a planet around our star, but also for a satellite orbiting a planet. In this case the solar mass M_S in Eq. (10.57) should be replaced by the planet mass, M_P.

Note: The normalized gravitational potential is a theoretical value estimated for the central mass point of the moving object. In such case it does not depend on the size of the moving body.

10.6.5.2 Physical relation between the inertial force moment (of folded nodes) and the gravitational potential.

10.6.5.2.1 Energy equivalence equation.

From the the planetary motion and its theoretical interpretation it is evident, that the folding/ unfolding process involves reactive type of energy borrowed from the external CL space. This energy, however, it is not part of the kinetic energy and does not participate in Eq. (10.59). When using the concept of a local CL space with ESS, this energy appears located in the ESS. For a solid astronomical body with high matter density, the separation surface can be beyond its solid surface.

Eq. (10.59) from the previous paragraph shows that in a gravitational CL space E_{IFM}^G depends on U_{Gn}. It was noticed that the latter parameter is theoretical and is referenced to the central mass point of the moving body. It is theoretical because in that point (and any point inside the body) the estimation of the parameters of the folded nodes (from the external CL space U_{Gn}) is impossible. Therefore, the physical parameters providing U_{Gn} may be attributed to the theoretically introduces Equivalent Separation Surface (ESS). Then one question arises:

How the value of E_{IFM}^G giving the energy of the folded nodes corresponds correctly to the energy of folding/unfolding process at the imaginary ESS?

The above question has the following possible answer:

The reactive energy for the folding/unfolding process is equal to E_{IFM}^G energy. So it is exactly equal to the kinetic energy estimated by the orbital velocity.

$$E_{IFM}^G = E_K = E^R \qquad (10.60)$$

where: E_{IFM}^G - is the inertial force moment of the folded nodes in a gravitational CL space; E_K - is the kinetic energy of the body possessing a local CL space; E^R - is the reactive energy borrowed from the external CL space and located around the ESS (while in the real case it is distributed through the whole mass of the body).

Eq. (10.60) is an energy equivalence equation.

In the theoretical case of motion in orbit with infinite radius the Newton's gravitation is intrinsically small and could be neglected. So this case could be regarded as a linear motion with kinetic energy determined by the body mass and velocity. Then we logically arrive to the following conclusion:

In case of a circular orbital motion, the parameter E_{IFM}^G implicitly includes the energy

of the Newton's gravitation. The linear motion of body in a free space (away from gravitational field) does not include such energy.

The above made conclusion is in complete agreement with Eq. (10.59).

Now another question appears: Is there some upper limit of the radius of a circular motion, when it could be energetically considered as linear motion?

The possible answer is: **The upper limit of the circular orbital motion is determined by the equivalence between the gravitational potential of the large body and the Zero Point Energy of the free space.**

From the provided so far analysis, it becomes evident that the local CL space for any astronomical body (whose ESS is extended beyond its solid surface) exhibits a radial gradient (space curvature, according to GR). Let us use the parameter **CL space stiffness** for describing the extended CL space gradient (space curvature according to GR) of a planet orbiting a star. The CL space stiffness (space curvature) will be larger at its centre and will fall gradually until reaching a zone where it will be equal of the CL space stiffness (Space curvature) of the star. <u>The CL space stiffness at this point, however, is larger than the CL space stiffness in a free space (free CL space is means an ideal case of CL space without any gravitational field).</u>

10.6.5.2.2 Motion with a linear acceleration (deceleration) component

In order to simplify the problem, let consider initially a motion with a constant velocity within a limited time range in order to be considered as a linear type of motion. In such case, the parameter U_{Gn} could be considered as a constant and E_{IFM}^G will not depend on it. If the body obtains an acceleration in a second moment but U_{Gn} is still constant, we arrive to the following relations:

$$\frac{\Delta v}{\Delta t}\sqrt{U_{Gn}} = v_2^2 - v_1^2 \qquad (10.60.a)$$

where: v_1 and v_2 are respectively the initial and final velocity and $\Delta v/\Delta t$ is the velocity change or acceleration if $(\Delta t) \to 0$.

At the same time the energy equivalence according to Eq. (10.60) should be still valid. Then the reactive energy will be changed by the value

$$\Delta E^R = \frac{1}{2}m(v_2^2 - v_1^2) = \frac{1}{2}m\frac{dv}{dt}\sqrt{U_{Gn}} \qquad (10.60.b)$$

Eq. (10.60.b) shows that the change of the reactive energy at the ESS is proportional to the velocity change of the folded nodes, that in fact is the body acceleration.

10.6.5.3 Intrinsic energy balance.

The analysis of the process of molecular vibrations and photon emission/absorption in Chapter 9 has shown that the SG energy balance exists in the system comprised of the molecular structure and the surrounding CL space. Extending the logical consideration for a similar balance in the gravitational/inertial interaction for macrobody we may **formulate the following hypothesis:**

The CL space of a star possessing a planet system with stable long time motion parameters (like our solar system) is characterized by accurate balance between $E_{IG}(TP)$ and $E_{IG}(CP)$ energy of any moving body.

The kinetic energy involved in the interactions between the folded nodes and the local CL space nodes could be regarded as SG(TP) type of interaction.

The energy involved in the folding/unfolding process is involved in the change of the CL node geometry. In this process SG forces from the prism's central part dominate over the peripheral part. The peripheral part interactions between both type of nodes are with opposite helicity and the net balance could be considered as zero. Therefore, the folding/unfolding process could be characterised only by SG(CP) interactions. We may conclude:

• **The energy related to the folding/unfolding processes of the CL nodes is a type of reactive energy borrowed by the CL space. When using the concept of local CL space with ESS, this energy could be considered as allocated in the ESS.**

Using the local CL space concept, the reactive energy related to SG(CP) interactions depends on the size of the ESS area and the velocity. In re-

ality, the folded nodes obtain spin momentum, depending of the relative velocity and the gravitational potential at any point containing a Newtonian mass.

The SG energy balance for the motion of every planet, according to the above defined hypothesis, could be written in a form:

$$E_{SG}(CP) = E_{SG}(TP) \qquad (10.61)$$

In the provided so far analysis the circular motion was only considered, for which Eq. (10.60) is accepted. The same equation should be valid also for any stable elliptical orbit, where the average value of the kinetic energy will appear as a constant. In elliptical orbit, however, the relative velocity, the area of the ESS and the spin momentum of folded nodes will have different values for different orbital points.

The SG energy balance for a planet orbiting around a star is illustrated by the diagram shown in Fig. 10.11. The section of the separation surface for idealistic case (no cavities) is shown.

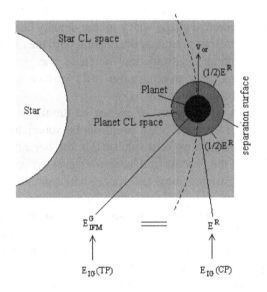

Fig. 10.11
Diagram of SG energy balance of planet in a circular orbit around a star (idealistic case)

The star and planet CL spaces are shown without their stiffness gradient (space curvatures). The section of separation surface is shown by two half rings (in order to show the entrance and exits part of the ESS) properly oriented to the orbital velocity v_{OR}. Using the imaginary concept of Local CL spaces with ESS, we must consider (also imaginary) that the folding process occurs at the lower part of the separation surface, while the unfolding - at the upper part (as shown in Fig. 10.11). The corresponding energy for each part is one half of the total reactive energy E^R. The equivalence between $E_{SG}(TP)$ and $E_{SG}(CP)$ is realised via E_{IFM}^G and E_K. The SG energy equivalence involves the CL space of the planet and the star CL space.

10.9 Mass grow of astronomical body by matter accumulation.

Let us consider an example of a solid body immersed in a gravitational field of a massive astronomical object. If the mass of the real solid body is comparatively small, it does not posses a local CL space. Now let imagine that the body mass increases continuously by accumulation of matter attracted from the surrounding space. Then in some moment it will obtain own CL space beginning from domains with larger matter density and SG field. Let us consider the ideal case that the body has a spherical shape with a matter density larger at the centre and gradually decreasing with the radius. If the matter accumulation increases, the growing CL space may overpass the solid surface of the body at some moment.

In the real case of matter accumulation, the shape of the solid body could not be spherical (no mechanism to make it spherical at the beginning). Then in the process of mass growing, some portion of the growing local CL space, which is also not spherical, may overpass the solid surface, while other not. In such case, assuming the body is **rotating**, the conditions of ESS surface are constantly changing (estimated by the gravitational force equivalence). But this means a change of the reactive energy E^R (defined in &10.6.5.2.2). This energy is quite large because it is in balance with the kinetic energy of the orbital motion of the body. According to the balance diagram of Fig. 10.11 the energy E^R is related to $E_{IG}(CP)$ interactions, which are quite strong. Then the change of E^R will appear as active energy dissipated directly on the solid volume of the body. So it may heat the body volume. If the accumulated mass is quite large but the

shape is not spherical, the change of E^R may dissipate enough energy to melt a part or the whole body. Then the massive body could obtain a shape closer to a sphere, possessing an own CL space.

The asteroids in the solar system and small planetary satellites could serve as proper objects for analysis the growing trend of a massive body. It is more convenient to compare objects with circular orbits arranged in close distances from the main massive object. Otherwise normalization to the distance is required (because it is related to the reactive energy E^R via orbital kinetic energy). The asteroids between Mars and Jupiter orbits are convenient for such kind of analysis. (The planetary satellites require more complicated analysis, taking into account the common motion in the star CL space together with the planetary one). Table 10.7 shows some of the data of 9 asteroids taken from "asteroid fact sheets"

(http:
//nssdc.gsfc.nasa.gov/planetary/factsheets/asteroidfact.html)

Asteroids data **Table 10.7**

No	Name	Mass x 10^{15} [kg]	distance [au]	size [km]	volume [km^3]
1	Eros	6.69	1.458	33x13x13	5577
2	Ida	100	2.861	58x23	262100
3	Mathilde	103.3	2.646	66x48x46	145728
4	Siwa	1500	2.734	103	5.72E5
5	Eugenia	6100	2.721	226	6.04E6
6	Juno	20000	2.669	240	7.238E6
7	Vesta	3E5	2.362	530	7.795E7
8	Palas	3.18E5	2.774	570x525x482	1.44E8
9	Ceres	8.7E5	2.776	960x932	4.43E8

Using the mass and volume data, the average density is calculated (without pretending for high accuracy because the size parameter is used). Fig 10.12 shows the average matter density dependence on the mass in two different scales: **a.** - lower mass range and **b.** - higher mass range. The asteroids are marked by their numbers according to the Table 10.7.

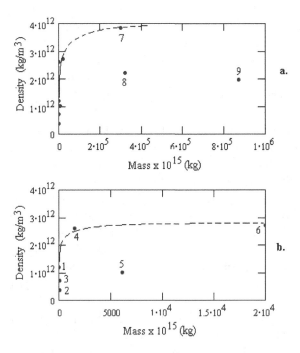

Fig. 10.12

Average matter density of the asteroids as a function of the mass in two different scales: **a.** - lower mass range and **b.** - higher mass range. The asteroids are marked by their numbers according to the Table 10.7.

Figure 10.12 shows that asteroids 1, 2, 3, 4, 5, 6, 7 (according to Table 10.7 numbering) define some trends that is shown by dashed line. Two asteroids 8 and 9 do not follow this trend. The trend is formed by asteroids 2, 3, 1, 4, 6, 7. The asteroid 5 (Eugenia) exhibits lower average density and is aside of this trend. The following conclusions could be made:

- Asteroids residing near the common trend show tendency of a spherical shape with mass increase

- Asteroid deviated from the common trend do not have a spherical shape even if possessing a larger mass

- Asteroid No 5 (Eugenia) is an exception from the observed trend, however, the provided data are not enough for deeper analysis of this case.

The tendency of the astronomical object for obtaining of a spherical shape, when the mass is larger than some limit, could be explained by a partial melting of its internal part. In the case of larger astronomical objects (like planetary satellites), the increased mass leads to more perfect spherical shape.

10.12. Interactions between the folded nodes and the planetary atmosphere.

10.12.1 Atmospheric gravity waves.

Let us regard the atoms and molecules from the atmosphere as simple small microbodies whose protons and neutrons posses own FOHS CL spaces. The protons and neutrons in the nuclei are subjected to a definite order, so we may regard the FOHS CL space as a common **atomic CL space**. Its configuration, however, is quite different than the astronomical object CL space. The molecule contain atoms with their CL spaces but with additional vibrational-rotational motion. In order to analyse the interaction between the folded nodes from the external space and the atomic (molecular) CL space we will consider the following two cases

1. **Idealistic case**: assuming the molecules do not exhibit a Brawnian motion. In such case we may distinguish three types of folded nodes interacting with the molecular local CL spaces:

- folded nodes confined to the Earth CL space
- folded nodes confined to the Sun CL space
- folded nodes confined to the Galactic CL space

Note: Keeping in mind that the introduced local CL space is imaginary, we use this concept for separating the GR effect of space curvature invoked separately by the Earth and Sun gravitational field.

We may apply Eq. (10.59) for a laminar flow of folded nodes. From this equation we see, that if the velocity and the gravitational potential are both constant, the energy balance is constant and the borrowed reactive energy E^R is also constant. Let us assume that the gravitational potential is constant i. e. there is not a vertical motion of the molecules. Then:

(a) <u>if considering the sideral rotation of the Earth, the reactive energy E^R will exhibit diurnal variations (24 hours).</u>

(b) <u>if considering the solar system motion in the Milky Way galaxy, the energy E^R will exhibit semiannual variation due to the Earth orbital motion around the Sun. The latter variations will be additionally asymmetric because the Earth orbit is not perpendicular to the solar system orbital plane</u>

around the Milky Way centre. The magnitude of the semiannual modulation is also a small, because the Earth orbital velocity is about 1/7 of the solar system velocity around the Milky Way centre.

While the constant reactive energy E^R is not detectable, the change of this energy could be detectable by accurate observation of some atmospheric phenomena. Such phenomenon is the observed **waves structure in the Earth atmosphere with a strong semiannual period** (BSM interpretation), shown in §10.12.2.

2. **Realistic case:** The molecules and atoms of the atmosphere posses a Brawnian motion characterised with a velocity direction randomness. This will affect the interaction between the molecules and the folded nodes in two ways:

- **In respect to the gravitational potential** U_G: The vertical component of Brawnian motion will cause slight changes in U_G. Then according to Eq. (10.59) and (10.60) a slight change of E^R could be observed in a global planetary scale.

- **Affecting the relative velocity vectors** between the folded nodes and the momentary velocity of the molecule casing some random velocity component. The features (a) and (b) from idealistic case will be still valide but the randomness will introduce some noise component.

Question: How the change of the reactive energy E^R could appear as an active detectable energy?

In order to provide a physical explanation we will consider a short interval of motion of the oxygen molecule in *nadir* (direction toward Earth centre). The Earth orbital motion velocity could be larger than the molecular one, but the velocity ratio is still finite. Let us consider a very short patch of CL nodes entering inside the protons and neutrons at the time moment 1 and exiting at the time moment 2. The E-field inside the proton and neutron is highly oriented and the folded nodes are strongly guided around FOHSs volumes (occupied by RL(T)). So entering the proton's or neutron's internal volume the folded nodes may lose their common group synchronization while they are inside. In the exit, however, they have to restore their common synchronization matching to the groups that have bypassed the internal proton's or neutron's volume. If assuming that they have not exchange any energy with the molecule the patch of folded

nodes will appear at level with a slightly higher gravitational potential but with a same velocity. Their "running" position will differ from the "running" position of other neighbouring patches that have not passed through the internal proton (neutron) volume. Then a spatial nonuniformity will appear as a result of their spatial positions and different U_G. This nonuniformity has to be removed by the Zero point waves. This means a larger ZPE fluctuation in these zones, which may cause excitations of some electronic states, more probably lower energy transitions in some molecular vibrational-rotational levels. In a nightglow emission layers these excitations could be detected as wavelike structures with large spatial dimensions. This kind of phenomena according to BSM provides **a wavelike phenomena, corresponding to one type of gravity waves** detected in the atmospheric airglow.

Gravity waves are observable phenomena, well known in the atmospheric science. BSM theory provides a different explanation of the physical mechanism for one type of the gravity waves.

10.12.2 Gravity waves in the Earth atmosphere with semiannual period

The phenomena is discovered initially as a semiannual variation of the zonal wind in the equatorial and upper atmosphere ((Reed, 1965). Presently, the gravity waves are modern topic of experimental and theoretical work in atmospheric science. Lot of observational data are accumulated covering a large vertical altitude in the Earth atmosphere. Some of the gravity waves at lower altitudes and near mountains could be influence by vertical tide. In higher altitudes, however, the existence of large scale gravity waves could not been explained by the vertical tides. Then the BSM concept of gravity waves phenomenon might be more relevant for altitudes above 50 km.

Figure 10.13 shows the vertical time-section of the monthly mean zonal wind, in which the semiannual period is clearly apparent (I. Hirota, 1978).

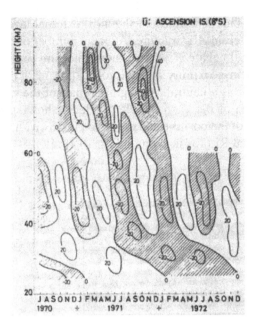

Fig. 10.13.

Vertical time-section of monthly mean zonal wind, in which the semiannual period is clearly apparent (Courtesy of I. Hirota, 1978).

From the above shown plot and from other observations, it is found that the neighbouring semiannual patterns are not exactly equal, while the average ratio between the magnitude of the even and odd patterns is preserved. The possible explanation according to BSM is the following: The Earth orbital plane intercepts the plane of the Milky way Galaxy under angle about 63^0. Then the folded node velocity from the Earth orbital rotation (about 30 km/s) obtains a small cosine in respect to the larger velocity of the solar system rotation around the Milky Way centre (about 320 km/s). According to Eq. (10.60) the energy balance is

$$E_{IFM}^G = E_K = E^R, \text{ where } E_{IFM}^G = E_{IFM}\frac{\sqrt{U_{Gn}}}{2\alpha c}$$

and $E_{IFM} = (m_n c\alpha)v$

For a circular orbit U_{Gn} is a constant. Consequently the change of folded node velocity v affects E_{IFM}^G, but the energy balance (10.60) should be preserved. Then the energy for the wave structure should come from the reactive E^R energy of the space. In fact the process is more complicated, because of the interaction between the solar and Earth magnetic field is additionally involved.

10.12.3 Atmospheric gravity waves at altitudes of 95-100 km.

Fig. 10.14 shows gravity wave structures obtained during ALOHA-90 campaign by wide angle all-sky camera at OI (557.7 nm and Na (589.2 nm) nightglow emissions. (Taylor et al. 1955).

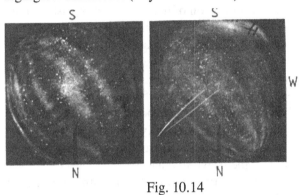

Fig. 10.14

Two examples of all-sky wave structure imaged in the visible wavelength OI (557.7 nm) and Na (589.2) nm nightglow emission during ALOHA-90 campaign. (Courtesy of Taylor et al., 1955)

In the same paper of M. J. Taylor (1997), number of other large scale images obtained from different experiments and authors are also presented. Fig. 10.14.a shows a large scale airglow wave structure from near IR OH emission (G. Swenson et al, 1995).

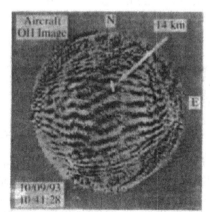

Fig. 10.14.a. All-sky wave pattern of the near IR OH emission (a difference image is obtained by a subtraction of two adjacent images). Measured from an aircraft (ALOHA-90 campaign). Courtesy of G. Swenson, 1995)

So far the solar-terrestrial geomagnetic relation and related phenomena are only considered as factors causing the high altitude gravity waves, while the physical mechanism is not well understood.

10.14 Magnetic field hypothesis (for astronomical objects)

In §10.6.2 and Fig 10.7, the break of the mass trend as a function of the volume of the astronomical body was discussed. To explain the break, a hypothesis was suggested, according to which the protons and neutrons in the central zone of the massive astronomical body break, leading to formation of a superdens nucleus of straight aligned kaons. The concept of such hypothesis can explain two phenomena simultaneously:

(a) a change of inertial property (and consequently the Newton's mass)

(b) creating conditions for a strong magnetic field

The phenomenon (b) could be explained by the strong field alignment generated by the internal RL(T) structures of the **straight shaped kaon**. The axial vector of RL(T) from a straight shaped kaon is much stronger in comparison to the curve shaped R(LT) of the proton (or neutron). Then bunch of straight shaped kaons may form very strong axial component of phased synchronized SPM vectors, much larger than this of the single kaon. Figure 10.17 illustrates a bunch of 6 kaons in a parallel arrangement, while number of such bunches may get a serial arrangement, as well.

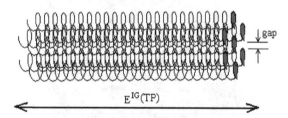

Fig. 10.17. Bunch of six kaons. The drawing shows only the external envelopes of the kaons and their common position. Only portion of the kaon's length is shown, as the length to diameter (of single kaon) ratio is 3451.

A straight kaon nucleus (bundle) can be integrated not only from cut kaons, but also from cut positive and negative pions. They undergo a partial

twisting (discussed in Chapter 6) in which they obtain a charge defined by the most external layer of the internal RL structure. At equal number of positive and negative charged kaons, they are compensated in proximity and the total kaon nuclei will appear as a neutral. A magnified part of the radial section of the kaon bundle (nucleus) showing the order of the positive and negative FOHSs with their internal RL(T)s is given in Fig. 10.18. i

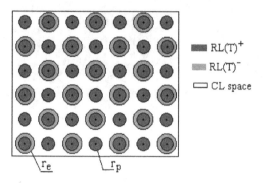

Fig.10.18

Magnified part from the radial section of kaon bundle

The number of positive FOHSs of type $H_m{}^{2-}$:+(-) is equal to the number of negative FOHSs of type $H_m{}^1$:-(+(-). In such case the total charge is equally balanced. The small edge effect charge difference is negligible and cannot affect the total charge integrity. Every positive FOHS with a RL(T) contains a negative core. Every negative FOHS contains with a RL(T), contain inside a positive FOHS with RL(T) with a central negative core (see the detailed description of kaon's and pion's internal structures in Chapter 2). Both types of FOHS are separated by channel-like gaps accessible for the surrounding CL space. The twisted proximity fields of RL(T) provides the inaccessibly gaps between the straight shaped helical structures, satisfying the requirement for a finite gaps, discussed in §6.4.3, Chapter 6. These long gaps are occupied by CL space (internal CL space), which is strongly modulated by the SG field of the RL(T)s of the helical structures. From one hand, this provides conditions for strong phase synchronization between the SPM vectors of the CL nodes along the align gaps between the straight FOHSs (kaons). From the other hand, due to the strong proximity SG field from the RL(T) structures, the resonance frequency of the internal CL space might be a higher harmonics of the free external CL space. It is evident that such arrangement of positive and negative FOHSs is able to provide a super strong magnetic field. The common mode synchronization provides energy flows in closed path in a form of magnetic field (for more details see the magnetic field explanation in Chapter 2). In the ordinary atomic matter built of protons and neutrons, such effect is not possible.

The CL space parameters inside the kaon bundle are strongly influenced by the spatial configuration of the structure. The $E_{SG}(TP)$ vector of the CL nodes will have a strong component aligned with the FOHSs axes. In Fig. 10.15 it is shown as $E_{SG}(TP)$ vector with two arrows, emphasizing its equal behaviour in two opposite directions similar as for a single prism. From these features one important conclusion follows:

(c) The common $E_{SG}(TP)$ vector provides conditions for superstrong magnetic field that may have N-S or S-N direction.

The described kaon structure may influence strongly the surrounding CL space, creating excellent conditions for a magnetic field. The generation of such field then is provoked by the inertial interactions of the moving body in respect to the external CL space. In the case of planetary magnetic field of our solar system, the magnetic filed should be preferentially aligned with the larger velocity vector of the folded nodes. This is the vector of the solar system motion around the Milky Way centre. This condition might be fulfilled in some earlier epoch of the planet evolution, providing the orientation of the kaon nucleus. Once the nucleus is formed it could not be re-oriented in respect to the solid structure body if the rotational axis of the planet or planetary satellite) is changed. So if the spin axis has been changed and the internal region is not melted, the alignment of the previously formed kaon nucleus will be preserved. The large angle between the spin and magnetic axes for the planets Neptune and Uranus confirm this conclusion.

The orientation of the Earth magnetic axis in respect to the velocity vector of the solar system motion around the Milky Way centre could be inferred from Fig. 10.13. Table 10.8 provides some

of the angles for estimation of the angular difference between the axis of the planetary magnetic field and the velocity vector of the solar system motion in respect to the galaxy centre.

Planetary magnetic field orientation   Table 10.7

Planet	magnetic field	tilt to rotational axis (deg)	obliquity to orbit (deg)
Mercury	(*)		0
Mars	(*)		25.19
Venus	No?		177.4
Earth	Yes	11.2	23.45
Jupiter	Yes	~11	3.13
Saturn	Yes	<1	26.73
Neptune	Yes	47	28.32
Uranus	Yes	58.6	97.8
Pluto			122.23

The Mercury magnetic field may not be well defined, for a few reasons:

- Mercury is very close to the Sun and its gravitational field may influence a possible formation of kaon nucleus.

- The eccentric orbit of Mercury involves a large change of the E^R energy, that may also disturb the conditions for formation of kaon nucleus.

Earth, Jupiter and Saturn have well defined magnetic fields. The angle between their axes (taking into account the sideral rotation) and the solar system velocity vector is much smaller than 90 deg. Therefore, their evolution to the present state perhaps did not changed significantly the initial direction of their rotational axes.

The magnetic axes of Neptune and Uranus are in quite different positions. This indicates that once the kaon nucleus is formed its direction in respect to a solid body is preserved even if the axis of their sideral rotation is changed. The rotational axis of Neptune and especially Uranus perhaps have undergone a significant change of their directions as a result of some perturbation with a foreign astronomical body after their kaon nuclei has been forme and after obtaining a spherical shape.

It is a geologically proved fact that the direction of Earth magnetic field has been reversed a few times in the past. This is in agreement with a feature **(c)** of the kaon nucleus.

The presented concept is not able to provide a satisfactory explanation especially for Mars and

Venus. The presented concept, however, may be useful for deeper understanding of the magnetic interactions between our Sun and the individual planet.

The existence of kaon nucleus is not valid only for the planet, but for the stars as well (in some epoch of their evolution). This is discussed in Chapter 12. The Sun also possesses a very strong magnetic field and it is also oriented along the vector of its motion around the Milky Way centre. The kaon nucleus could not be disturbed or destroyed of possible high temperature in the centre of our star (this also becomes apparent from the analysis of some cosmological phenomena in Chapter 12 of BSM)

Further discussions and arguments in favour of the kaon nucleus hypothesis are provided in Chapter 12. The observational data about pulsars indicate that they are huge kaon nuclei released after the death of the star, in which they have been formed.

10.15 Physical explanation of the phenomena of General Relativity:

10.15.1 Gravitational potential and local CL space

The gravitational potential of massive astronomical body for a particle with mass m is given by the equation:

$$U = -G\frac{Mm}{r} \qquad (10.65)$$

The equipotential surfaces could be represented as concentric spheres with radii inverse proportional to the distance from the centre. This dependence has different effect on a body with finite mass and on the photons. The Sun gravitational field, for example, will cause a slight shrink of the local CL space. Then the local CL space parameters might be slightly different than those of a free CL space. The energy exchange however, could be preserved if there is not involvement of mass change in the emission - detection process. In all processes of photon generation and detection (with exception of bremsstrahlung and radioactive decay) the electron particle is involved.

10.15.2 Influence of the gravitational field on the parameters of elementary particles. Explanation of the gravitational red shift.

Only stable particles will be considered: a proton, a neutron (inside the atomic nucleus) and electron (a normal electron containing a positron).

The CL space in the gravitational field may have a slightly changed parameters in comparison to the free space (away from any gravitational field). Such parameters are the static CL pressure, the CL node proper frequency and respectively the SPM frequency and consequently the speed of light. These are basic CL space parameters that will affect all physical constants. The change of the light velocity, for example, could not be detected inside of this local field, because the time base (SPM frequency = Compton frequency) changes accordingly. The same is valide for other parameters and respectively the physical constants. One important thing that must be taken into account is that the first proper frequency of the electron will be also affected because it is always equal to the Compton frequency of the CL node. The electron is an open single coil helical structure, so it is flexible enough in order to accommodate the necessary small change.

The detection of GR effects are possible only remotely from other local gravitational field. For example:

(a) If a photon emitted from a distant star (the source is outside of Sun gravitational field) passes near the Sun disk and is detected on Earth, its initial direction will be changed while its energy should be preserved.

(b) A photon emitted near the Sun and detected on the Earth will exhibit a red shift if the gravitational potential of the point of its emission is stronger than the gravitational potential in the point of its detection.

In case (a), there is not involvement of electron from the Sun gravitational field, so the photon energy is preserved. Only the direction is changed due to CL space geometry. This is a gravitational type of lensing. It is equivalent to passing a light through a lens.

In case (b), the electron involved in the photon emission has a slightly different first proper frequency in comparison to the electron involved in the photon detection. For this reason a red shift is detected.

10.15.3. Advance of perihelion of Mercury

Mercury is orbiting the Sun in an elliptical orbit with eccentricity of 0.2056. The perihelion of the orbit rotates very slowly in respect to the solar system - about 43 arcseconds per century. The phenomenon is known as "advance of the perihelion of Mercury".

While this effect is given as an example of General Relativity explanation for space curvature, its explanation by BSM is more understandable, using the concept of the CL space gradient. The Mercury local space is periodically plunging in the stronger gradient local field of the Sun. This causes changes of the relative velocity between the two objects and consequently the folded node velocity. If applying the concept of local CL spaces, we will see that the area of the separation surface around Mercury will pulsate for every orbital cycle. This will involve a pulsation of the reactive CL space energy E^R around one average value. Physically, the pulsating component is caused from the Mercury immersion in the solar space with a different gradient of the CL space shrinkage (space curvature) between its aphelion and perihelion. The change of E^R appears as an active energy (see §10.6.5 and Eq. (10.59) and (10.60)). This energy causes the slight advance of the perihelion of the Mercury.

The physical explanation if this effect is demonstrated also by the sketch shown in Fig. 10.19. The orbital rotation is highly exaggerated.

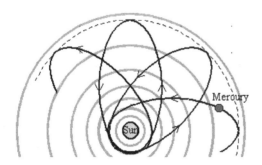

Fig. 10.18

"Advance of perihelion" of Mercury in the Sun gravitational field illustrated by equipotential curves (space curvature). The gravitational potential slightly modifies the CL node distance affecting the CL space stiffness. This affects the CL space parameters including the inertial interactions of the

atomic matter with the CL space. The elliptical orbit of Mercury passes through regions of different CL space stiffness, which is the reason for the observed advance of the perihelion.

10.16 Special relativistic phenomena

10.16.1. Mass increase at relativistic velocity

The mass increase has been discussed in §10.3.4.

10.16.2. Relation between the concept of inertial frame (according to the Special Relativity) and the concept of local CL space (according to BSM).

The folding/unfolding process and folded node interactions with the local space has been already discussed. Here some of the specific features are summarized:

- The deviated folded nodes have a resonance type of interactions with the bumps of the CL nodes quasispheres of the normal CL space.

- The sporadically occurred phase difference between the moving folded nodes is corrected at every node along the path, so the folded nodes obtain a commonly synchronized motion

- The commonly synchronized motion keeps a constant uniformity of the folded nodes flow, providing in such way a constant spatial characteristics of the partial pressure for a motion with a constant velocity. The spatial characteristics allows the initial direction of the folded nodes to be restored after they have been deviated from the FOHS's volumes of the elementary particle.

- The macrobody exhibits uniform inertial interactions as acceleration/deceleration, from every points of its structure, because the spatial characteristics of its FOHS's CL node space participate in both: the Newtonian mass and Newtonian inertia (both valid for the atomic type of matter organization).

We must emphasize the following major difference between the concept of inertial frame, formulated by the Special Relativity and the concept of body local CL space, formulated by the BSM.

(a) The concept of the inertial frame in SR does not have a point of absolute reference, while in the BSM concept such point exist : this is the centre of the home galaxy.

(b) The concept of inertial frame does not take into account the mass of the object and the gravitational potential in which it is immersed. These attributes are taken into account in General relativity, but the relation between SR and GR is not quite apparent.

The BSM concept does not have the above disadvantages. It is able to provide explanations of all relativistic phenomena without contradictions.

10.17. Coriolis force

The oscillation plane for a pendulum located in Earth and moving for a long time precesses about the vertical. This effect was first demonstrated in 1851 by the french physicist Jean Foucault. The forces that cause this rotation are called Coriolis forces. The Coriolis forces are valid also for the satellites motion around the Earth. They are caused by the Earth rotation about the polar axis. The physical effect causing this forces, however, has been not explained so far. According to Einstein theory the Earth should be considered an inertial frame. Then why the Coriolis forces appears for a simple pendulum on the Earth and for the satellites, that are quite distant from the Earth surface?

BSM interpretation of the Coriolis forces is quite logical. These forces are caused by the modulation of the CL space by the Earth gravitational field. In this particular modulation the sideral rotation of the Earth influences slightly the CL space in which the Earth and all material objects are immersed.

Chapter 11. Biomolecules, bioenergy and DNA intercommunication

Abstract One of the amazing features of the biomolecules is their ability to preserve their complex three-dimensional structure in proper environments. This effect cannot obtain satisfactory theoretical explanation from a point of view of the quantum mechanical models of the atoms. Based on an alternative concept of the physical vacuum, a recently developed unified theory titled Basic Structures of Matter (BSM) allows unveiling the real physical structures of the elementary particles and their spacial arrangement in the atomic nuclei. The obtained physical models of the atoms are characterized by the same interaction energies as the quantum mechanical models, while the structure of the elementary particles influence their spatial arrangement in the nuclei. The derived atomic models with fully identifiable parameters and positions of the quantum orbits permit studying the physical conditions behind the structural and bond restrictions of the atoms connected in molecules. The existing data base about structure and atomic composition of the organic and biomolecules provides an excellent opportunity for test and validation of the derived models of the atoms. A new method for theoretical analysis of biomolecules is presented. The analysis of DNA molecule, leads to formulation of hypotheses about the energy storage mechanism in proteins and the DNA involvement in a cell cycle synchronization. The analysis of tRNA molecule leads to formulation of hypothesis about a binary decoding mechanism behind the 20 flavours of the complex aminoacyle-tRNA synthetases - tRNA.

Keywords: *Molecular biology, structure of biomolecules, Structural chemistry, biophysics, biochemistry, VSEPR model, proteins, DNA research, C-value paradox, Levinthal's paradox*

Note: A version of this article was published in Journal of Theoretics (2003) and archive in the National Library of Canada. The numbering of the figures and equations is preserved.

Despite the huge number of possible configurations of the atoms in the proteins, according to the Quantum mechanics (QM), they fold reliably and quickly to their native state. This effect is known in the biochemistry as a Levinthal's paradox. If considering only the Quantum mechanical considerations a protein molecule from 2,000 atoms, for example, should possess an astronomical number of degrees of freedom. In reality this number is drastically reduced by some strong structural restrictions, such as bond lengths, relative bond angles and rotations. The complex secondary and tertiary structures of the biomolecules provide indications about additional restrictions with weaker strength but responsible for their complex shapes.

One of the major benefits of BSM is the revealing of the physical three-dimensional structures of the atoms, which appear to be different than the QM models. The structural features of the atoms are not apparent by the QM models which are based on the planetary concept of the atom and

operate exclusively by energy levels. The revealed BSM atomic model possess the same energy levels but they are secondary features.

The purpose of this chapter is to acquaint the scientists from the filed of molecular biophysics with a new analytical approach for studying the properties of the biomolecules.

11.1. New method for analysis of the biomolecules with identified structure and composition.

11.1. General considerations

The 3D structures and atomic compositions of many biomolecules now are well known. In such structures, the individual atoms are identified as nodes with known coordinates. The angular coordinates of their chemical bonds are also known. This information is sufficient for analysis allowing replacement of the nodes in the 3D structure of any large molecule with the physical models of the atoms according to BSM concept. If BSM models are correct, their spatial configuration and angular bond restrictions should match the 3D structure of the biomolecules. Once the BSM models are validated and corrected if necessary, the complex biomolecules and any macromolecule with known shape could be studied from a new point of view. The application of the BSM models, for example, permits to identify the positions of all electronic orbits. This includes the nuclear and the chemical bonds electronic orbits. Then the conditions for possible interactions, modifications and energy transfer could be analysed at atomic level.

11.2. Ring atomic structures in organic molecules.

Most of the organic molecules contain ring atomic structures. The molecule of benzene could be considered as a simple example of a ring structure. The biomolecules usually possess a large number of ring atomic structures. Figure 11.8. shows the 3D molecular structure of aspirin where the ring structure of 6 carbon atoms is similar as in benzene molecule.

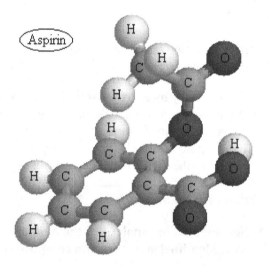

Fig. 11.8
3D structure of the molecule of aspirin (PDB file aspirin visualized by Chime software)

Figure 11.9 shows the same structure of aspirin at atomic level by application of BSM atomic models. The single atoms Deuteron, Oxygen and Carbon and the size of quantum orbit of second subharmonic are shown in the left upper corner. The valence protons (deuterons) of the oxygen atom are in fact in a plane perpendicular to the plane of EB bonded protons (deuterons) but they are shown with reduced dimensions in order to imitate an oblique angle in 3D view. The same is valid also for the valence protons of the carbon atom.

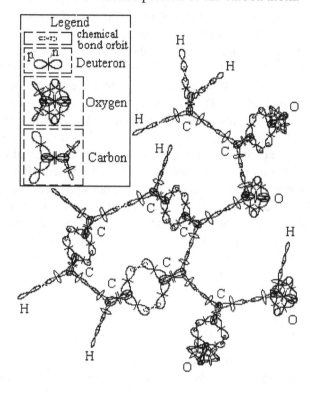

Fig. 11.9 Three-dimensional structure of aspirin using by BSM atomic models

In Fig. 11.9 the electronic orbits providing chemical bonds are only shown. For molecule with known 3D structure and composition the common positions of all electronic orbits with their equivalent orbital planes are identifiable. It is clearly apparent that the 3D structure of the molecule is defined by the following conditions:

(a) a finite size of the involved atomic nuclei

(b) an angular restricted freedom of valence protons

(c) a finite orbital trace length defined by the quantum conditions of the circulating electron

(d) orbital interactions

(e) a QM spin of the electron (the motion direction of the electron in respect to the proton twisting)

The QM mechanical models of atoms are mathematical models in which the features (c), (d), and (e) are directly involved, while the features (a) and (b) are indirectly involved by the selection of proper wavefunctions. In this process, however, some of the spatial and almost all angular restrictions are lost. Let us emphasize now the difference between the suggested BSM models of atoms and molecules, from one side and the QM models, from the other:

- QM model: the electrons participating in chemical bonds are orbiting around both point-like nuclei, i. e. they are not localised

- BSM model: the electrons involved in the chemical bonds do not orbit both nuclei and they are localised

-QM model: the chemical bond lengths are estimated from the electron microscopy assuming the planetary atomic model in which the larger electron concentrations are centred around the poinlike nucleus

- BSM model: the chemical bond length may need re-estimation, because the orbits of the chemical bond electrons do not encircle the bound atomic nuclei.

- QM model: The estimated length of C=C double valence bond is 1.34 A (angstrom), while for a single valence C-C it is 1.54 A. However, these lengths show a small variation when the same type ring groups are included in different biomolecules.

- BSM model: the length of single C-C bond may vary only by the subharmonic number of quantum orbit, while the length in a double C=C bond is additionally dependent on the angular positions of the valence protons.

From a BSM point of view, the adopted and existed so far concept of orbiting delocalised electrons for explanation of the equality between single and double bonds in benzene molecule is not logical. This is quite important for the unveiling of some of specific properties of the ring structures in biomolecules.

Ring structures are very abundant in many biomolecules and are usually arranged in particular order along their chain. DNA and proteins contain a large number of ring structures. Figure 11.10 shows the spatial arrangement of ring atomic structures in a portion of β–type DNA. The positions of some (O+4C) rings from the deoxyribose molecule that is involved in the helical backbone strands of DNA are pointed by arrows.

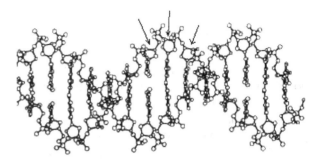

Fig. 11.10 Part of DNA structure with indicated positions of some of (O+4C) atomic rings

Figure 11.11. shows the ring atomic structure (O+4C) from the DNA strand. The bonding deuterons involved in the ring structure practically have some small twisting, but the quantum orbit of single valence bond also could be twisted. This feature gives some freedom for formation of ring structures from different atoms. The rotational freedom of the single valence bonds, however, may be accompanied by some stiffness, which increases with the degree of the orbital twisting.

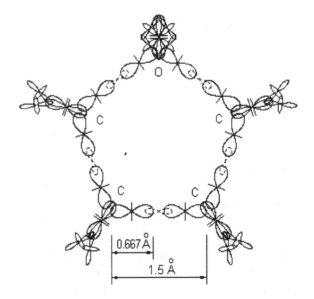

Fig. 11.11 Ring atomic structure from the deoxyribose molecule involved in DNA strand

In DNA molecule, some of the atoms of the ring structure are also connected to other external atoms. All this considerations provide explanation why the ring structure (O+4C) connected to the DNA strand is not flat but curved.

11.3 Weak hydrogen bonds

It is known that a weak hydrogen bond is possible between two atoms, one of which does not possess a free valence. The bond connection is a result of orbital interactions. In such aspects the hydrogen bonds connecting the purines to pyramidines in DNA molecule are of two types: <N-H...O> and <N-H...N>, where the single valence electronic bond is denoted by "-" and the H-bond is denoted by "...". This is illustrated by Fig. 11.12.

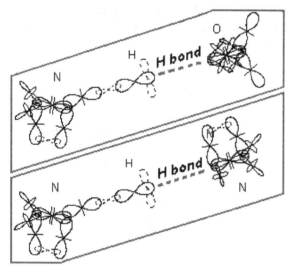

Fig. 11.12. Two types of hydrogen bonds

The BSM concept allows to find out the possible orbital orientation for such type of bond. In a hydrogen bond of type N-H...O the plane of electronic orbit of hydrogen appears almost parallel to the commonly oriented nuclear orbits of oxygen atom in which six electrons are involved (see Table 2 and Fig. 11.6). In a hydrogen bond of type N-H...N the plane of electronic orbit of hydrogen is almost parallel to the equivalent planes of the two polar orbits of N in which two electrons are involved.

It is evident that the hydrogen bond is characterized by the following features:

- the connection is a result of common orbital orientation

- the H-bond requires a critical range of distance

- the H-bond allows rotational freedom in a limited angular range.

These three features permit the DNA molecule to possess excellent folding properties.

11.4. Hypothesis of energy storage mechanism in molecules possessing ring atomic structures.

It is well known from the atomic spectra, that only the alkali metals (Group I) and the positive atomic ions with a single valence electron possess atomic spectrum that could be described by the Bohr atomic model. For elements with more than one valence electron the principal quantum numbers exhibit more than one energy level (degenerate levels), due to the spin orbital interactions. The signature of this feature is apparent from the Grotrian diagrams for atomic spectra. The spin-orbital interaction from a new point view is discussed in Chapter 8 of BSM. The analysis leads to a conclusion that after an orbiting electron is dropped to a lower quantum level, the pumped CL space energy is not emitted in full. Part of it is preserved by the atomic nucleus and redirected to the valence proton, whose quantum orbital plane is parallel to the orbital plane of the consideration. The physics of this effect is explainable if considering the total energy balance including IG field. The latter controls the proton's proximity E-fields distribution, that from his hand defines the orbital conditions of the electron. Therefore, the redirected energy provides some shift of the energy levels, but the effect is stronger for the lower states closer to the ground state of the series (Balmer series has own ground state, according to BSM model). The physical explanation of this effect allows making a conclusion that the released energy prior to formation of a photon is **preferentially guided by the structures of protons (deuterons)** held by the strong IG forces. Applying the same considerations for the ring atomic structures there must therefore be a guiding energy process between the atomic nuclei or protons involved in the ring. Such consideration leads to the following conclusion:

In proper environments, the ring atomic structures in organic molecules may have ability to store energy as an exited state rotating in the ring loop.

The effect of the rotating excited state is possible due to the consecutive re-excitation of the electrons in the separate bonding orbits in the ring. This effect is not apparent by the Quantum mechanical model, where the wavefunctions are complex envelope around the whole nuclei of the involved atoms. However, it is a known fact that the bonding strengths between atoms involved in a ring atomic structure are stronger than between same atoms when not participating in such structures.

Evidently, the condition of rotating excited state in a ring structure could be obtained only for equal energy level differences. Such vision needs to considering excited states not only from same valence bonds but also from single and second valence bonds as well. On the other hand, for a ring structure containing more than two bonds of same valence, excited states may preferentially exist between the same valence bonds. In case of aspirin, for example, such conditions exist for three pairs orbits of second valence and three pairs orbits of single valence bonds. If considering also the fixed nuclear orbits of the atoms in the ring, then twelve polar electrons could be also involved in a ring energy storage effect. Theoretically they may store a much larger energy not only due to their number, but also due to the larger transition energies.

In proper environments, the stored energy in the ring structures of the biomolecules may have the following features:

- a stable cycle of exited state rotation due to a stable finite time of single excited state

- a possibility for interactions with properly oriented neighbouring ring structures in the moment between two consecutive excited states (conditions for synchronization between the rotating states of neighbouring rings)

- a cascade type of energy transfer

Many of the building blocks of the biomolecules or reagents contain rings of type single or attached. For example Adenine (2 attached rings), Guanine (2 attached rings). Vitamin D contains one single and two attached rings. Alpha and Beta tubulins contain groups of: GDP (one single and two attached rings), GTP (one single and two attached rings), TAXOL (4 single rings). The steroids hormones contain usually four attached atomic rings. The ATP, an important energy carrier in the cells

contains one single and two attached rings. It is quite logical to consider that the energy rotating cycles in the attached rings are mutually dependable, so they must be synchronized. Then it is logically to expect that the attached rings may have an increased ability to hold a stored energy in case of environment change.

11.5 Hypothesis of energy flow through the chain structure of the biomolecule.

11.5.1. Energy flow in DNA molecule and its effect on the higher order structural characteristics.

The ring atomic structures in the long chain biomolecules appeared arranged in some spatial order. In the molecule of DNA, for example they are characterised by:

(a) a strong repeating order

(b) a strong orientation in respect to the host strand of DNA

(c) a strong orientational order of the neighbouring rings along the helix

These well known features are easily visualized when rotating the 3D structure of DNA (by programs like: "chime" "Rasmol", "protein explorer" etc.).

The consideration of cascade type of excited state transfer could be applied not only for a ring atomic structure but also for a long chain molecule built of repeatable atomic structures connected by electronic bonds. In this case some more complicated but mutually dependable mechanisms are involved. The following analysis tries to unveil such mechanisms. Let us consider for this purpose one of the backbone strands of DNA. Figure 11.13. illustrates the connection path of the electronic bonds in the strand. Three important features are apparent for every repeatable cascade:

(d) the bond connection path is formed by deuterons or protons connected by electronic orbits

(e) the bond connection path passes through one C-C bond from the (O+4C) ring.

(f) all bonds involved in the bond connection path are single valence

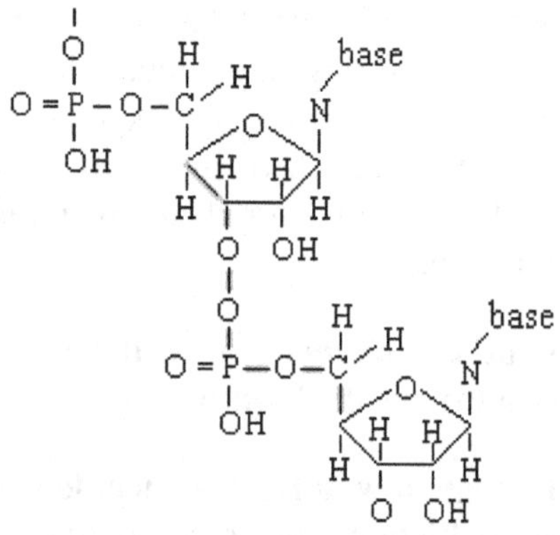

Fig. 11.13. Bond connection path through a DNA strand

> The connection path corresponding to one cascade of the nucleotide is denoted by a thick green line (thick grey line in a black and white graphics)

It is reasonable to expect that the long chain involving single valence bonds may provide conditions for cascade excited state transfer in one direction. The time of every excited state is determined by a quantum mechanical consideration - the lifetime of the spontaneous emission. Keeping in mind the small distance between the neighbouring electronic bonds (of the order of 1-2 angstrom), the transfer time between two consecutive excited states (with a light velocity) is practically almost a zero. Then the time dependence between the two energy process (the cascade energy transfer and the energy rotation period in the ring) is easily obtainable. The bonding path of one cascade contains six bonds total in which one C-C bond from the (O+4C) ring is included. This ring involves five bonds. Then the following condition is valid:

$$0 < (T_R - t_c) < \tau_{av} \tag{9}$$

where: t_c - is a time interval for energy transfer in one cascade of the nucleotide, estimated by the sum of lifetimes of excited states in involved bonds; T_R - is a cycle time of the rotating state in the ring, τ_{av} - is an average lifetime for excited state in a single bond.

The expression (9) means that the cascade transfer and the ring cycle are time dependable, so they should have a proper **phase synchronization.** Additionally, all parameters of Eq. (9) are dependent on temperature, but in a different way. This will impose a temperature range constraint for successful phase synchronization. Consequently:

(g) the whole mechanism will work at optimal temperature within a limited temperature range, defined by the conditions of optimal phase between the cascade energy transfer and the ring energy cycle.

Let us analyse now the conditions that may support the tendency of unidirectional energy transfer. For this reason we will consider a small portion of DNA ignoring it supercoiling. In such case it could be regarded as a linear type of DNA. Let us take into account the following two structural features of Beta type DNA that might be related to the tendency of unidirectional energy transfer:

- It is well known that the nucleotide arrangement in DNA is antiparallel, so the same definition is valid, also, for the bonding paths through the two strands.

- The double helix of DNA is characterised by a minor and major grove. This means that one of the helix is slightly shifted axially in respect to the other.

The concept of energy flow through the single strand of the DNA molecule (with a shape of a helix) could be investigated if associating it with the magnetic field of a solenoid. In such approach, the double helical configuration of DNA could be represented as equivalent circuit of two parallel closely spaced solenoids with a common axis. Now let consider that the cascade energies flow through both strands in opposite (aniparallel) directions. In the equivalent circuit of two solenoids this will correspond to opposite currents. In such case, the magnetic lines in the internal region of the solenoids will have antiparallel direction, while the external magnetic lines will be closed in proximity of both ends of the solenoids.

Let us call this type of field a "**complimentary compensated solenoids type**". The magnetic lines of such field are schematically illustrated by Fig. 11.14. The two solenoids that simulate the two strands are shown by green and red colours. Their magnetic field lines, shown as dashed lines are antiparallel inside the solenoids, while they are connected in proximity at both ends.

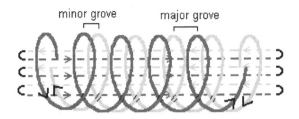

Fig. 11.15. Association of energy flow conditions between DNA strands with complimentary compensated solenoids. The magnetic field lines close in proximity at both ends of the solenoids. The current directions in both solenoids are opposite (shown by black arrows)

If attempting to separate the both solenoids, they will oppose. **In fact the presented solenoid model exhibits some features of bifilar winding.**

The analysis of magnetic field with such spatial configuration reveals the following additional features:

(h) The configuration of the associated magnetic field is independent from the secondary (supercoiling) shape of DNA

(i) Such type of magnetic field will provide an additional strength of the connections between the both strands. It will oppose the separation between the strands because this leads to increase of the close path lengths of the magnetic lines

(j) For a small portion of DNA molecule the complimentary compensated solenoids type of field is axially symmetrical.

The whole mechanism will be characterised by the following features:

(k) the rotated excited states in (O+4C) rings will possess one and a same handedness determined by the direction of the cascade energy flow through the strand to which they are attached

(l) the rotating energy states are phase synchronized along the DNA strand

(m) the rotating energy states sustain the tendency of unidirectional cascade energy flow through the DNA strand

Keeping in mind the features (a), (b), and (c), the rotating energy states in the ring could be commonly dependable. It is apparent, also, that the commonly dependable features (g), (h), (i) and (j) will lead to a self-sustainable mechanism.

The provided considerations may put also a light about the hydrophobic mechanism existing in the space between the two DNA strands. The two bond angles of water molecule are illustrated in Fig. 11.15.A where the positions of the orbits of the two valence electrons are also shown (discussed in Chapter 9).

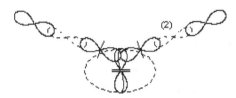

Fig. 11.15.A Water molecule

If a water molecule is placed inside the symmetrical field of the compensated solenoids, the angular positions of the valence electron orbits evidently will be in a conflict with the solenoids field. The interaction of the orbiting electrons with such field may provide expelling forces for such type of molecule. This might explain the hydrophobic environments of the internal region of DNA between the two strands. The hydrophobic environment is quite important for H-bondings between the purines and pyrimidines.

The analysis of compensated double solenoid model for the energy flow through DNA leads to the following conclusion:

(A). The DNA double helix molecule could be easily folded in any shape under influence of external factors.

The external factors could be different kind of proteins.

11.5.2. Magnetic field conditions for proteins.

It is well known, that in proper environments the proteins, which usually possess a complex tertiary structure, preserve their native shape. Although, the linear DNA molecule, when it is free of bending proteins, does not exhibit such feature. The main reason for the different behaviour of the proteins and the DNA of linear type is because of the structural differences between them. The protein is a long single strand molecular chain with diversified sequences of aminoacids. The protein backbone does not contain low order repeatable structures as the DNA nucleotide. It may contain, however, higher order repeats, that may form some helices with small number of turns or other spatial configurations. In such arrangement, the conditions for H-bonding are significantly reduced. All these structural differences indicate that we could not apply the concept of the complimentary compensated solenoids for the proteins, as for the analysis of DNA.

The conclusion, that the diversified sequence of the aminoacids is one of the reason for the different higher order shape of the proteins is supported by the analysis of the shape of the tRNA molecule. Its shape and atomic composition are shown in Fig. 11.16.

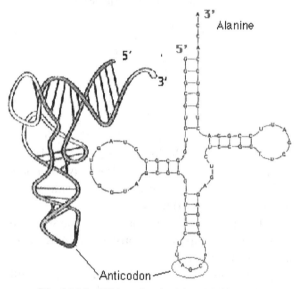

Fig. 11.16. tRNA molecule. (a) - real shape, (b) plane drawing, showing the loops and base pairs

The tRNA molecule possesses a single strand of about 77 base pairs but the repeating order of the nucleotides is similar as in the DNA molecule. Its single strand arrangement obtains a stable shape of "cloverleaf" demonstrating in such way a tendency for arranging parts of the long single strand in proximity.

In such arrangement, the unidirectional magnetic field through the single chain of tRNA becomes of type closer to the complimentary compensated solenoid, so the principle of shortest magnetic lines is satisfied. The magnetic lines through the strand are connected in the proximity at the open ends and this is the area where the tRNA molecule attaches a proper amino acid molecule.

The proteins synthesized according to the DNA code have different sequences of amino acids and exhibit a large diversity in their taxonomy. While the model of the complementary compensated solenoids is not valid for the proteins, they still could be analysed by a single solenoid model. If accepting that a unidirectional cascade state transfer is possible through the bonding path of the chain, then a single solenoid model could be used for the helices of the protein. In such aspect the analysis of secondary and tertiary structures of the proteins leads to the following considerations:

(a) Different types of secondary structures as helices, sheets, beta turns, bulges and so on provide different structures of magnetic fields.

(a) The magnetic field from the unidirectional cascade state transfer may influence the common positions of the secondary structures

(c) The bond angular positions between neighbouring atoms are restricted by the strong conditions of finite nuclear structure and restricted angular freedom of the chemical bonds

(d) Discrete type of rotational freedom with some finite deviation is possible between single valence bonds

(e) The above mentioned considerations and spatial restrictions lead to a conclusion that an overall asymmetry of the magnetic filed from the secondary structures (mainly from the helices) may exist. Then the tertiary structure may tries to restore the asymmetry of the magnetic fields arisen from the diversified secondary structures by obtaining an overall symmetry at some higher structural order. Simultaneously, this will be accompanied with a tendency of shortest magnetic lines.

(f) The cascade energy transfer through the bonding path of the protein chain might be accompanied with some ion current, whose energy could be also involved in the total energy balance.

The taxonomy of 3D protein structures is well described in the book Advances in Protein Chemistry (1981) by C. B. Anfinsen, J. T. Edsall and F. M. Richards.

The stiffness of the secondary order structure is larger than the stiffness of the tertiary one. This matches the strength of the mechanism involved in the restoration of the overall field symmetry. As a result, every protein may have a sustainable native shape when placed in proper environments (temperature, pH, ATP).

The provided analysis helps to explain one of the problems in molecular biology, known like a Levinthal's paradox: "Why the proteins folds reliably and quickly to their native state despite the astronomical number of the possible quantum states according to Quantum mechanics?".

Now let find out how the protein may react if the condition of shortest magnetic lines is temporally disturbed by some external factor, for example, by addition of Adenosine triphosphate (ATP). The ATP carries stored energy in its rings. When the chemical reaction $ATP + H_2O \rightarrow ADP + P$ occurs, the stored energy might be induced in some rings in the protein under consideration. This may change the conditions at which the principle of shortest magnetic lines and compensated field is satisfied. Then the reaction leading to restoration of the energy balance (involving the magnetic energy) could involve some kind of change of shape and motion of the protein. This may eventually explain the protein motility in proper environments (temperature, pH, ATP concentration) that is simultaneously accompanied by a temporary change of its higher order shape.

11.5.3. Environment considerations for preservation of the native shape of the biomolecules.

Many proteins have a complex 3D structure with secondary helix and tertiary structure. They exhibit amazing tendency to preserve their 3D shape in proper environment conditions in which the temperature is one of most important factor.

The electronic bonds and hydrogen bonds allow some freedom of the 3D structure of biomolecules. It is known from the Quantum mechanics that the excitation of particular vibrational-rotational band of the molecular spectra is temperature dependant. Translated to BSM model this dependence means:

- a selection of proper subharmonic number from the available set of the quantum orbit

- a selection of proper quantum number

Consequently, the complex 3D shape of the biomolecules is dependent on two main features:

(a) a proper subharmonic and quantum number of every bonding electron

(b) the same subharmonic and quantum number of the similar interatomic bonds along the molecular chain

The feature (a) provides a requirement for the absolute temperature range for the preservation of the native state of a long chain molecule, while the feature (b) puts a strict requirement for a temperature uniformity along the molecular chain. This, of coarse is not the only factor. Additional factors are the interactions between the highly oriented atomic ring structures discussed in previous paragraphs and the conditions for H-bondings.

Another environmental conditions that are also important, for example, are such as the pH factor and ATP concentration. The first one may assure the proper ion current conditions (according to consideration (f) in section 11.5.2, while the second one - the necessary energy.

11.5.4. Magnetic field involved in the higher order structures of DNA.

The DNA molecules of the single cell organisms are usually of circular type with a supercoiling. The DNA of eukaryotes, however, is of linear type with much more base pairs and complicated secondary and third order structures. According to the principle of the short magnetic lines and compensated magnetic field, the circular DNA should be more resistant, because the magnetic line paths are enclosed inside of the structure. This condition might be partly disturbed only during the transcription. The linear DNA, however, posses some additional properties that may facilitate the process of intermolecular DNA communications, according to a hypotheses presented below.

The length to diameter ratio of DNA is very large. It is extremely large especially for eukariotic DNA. For human chromosomes, for example, the average ratio is of the order of 10^7. In order to be held in a tiny space of cell nucleus such long molecule have higher order structure, known as supercoiling. The mechanisms involved in the formation and sustaining of such a structure has not been completely understood so far. The following analysis put some light about the possible mechanisms.

Let us first present shortly some of the experimental observations about supercoiling features of a long DNA. In the paper "DNA-Inspired Electrostatics" in Physics Today by W. M Gelbart et al. (2000), the authors provide a summary about one important feature of DNA. They say: "Under physical conditions (a 0.1 molar solution of NaCl), a DNA molecule takes on the form of a disorder coil with a radius of gyration of several micrometers; if any lengths of the molecule come within 1 nm of the other, they strongly repel. But under different conditions-in a highly diluted aqueous solution that also contain a small concentration of polyvalent cations - the same DNA molecule condenses into a tightly packed, circumferentially wound torus." Figure 11.17 shows the toroidal shape of DNA from the same paper that has been adapted from O Lamber et al., (2000).

Fig. 11.17 Toroidal DNA condensates, Courtesy of O. Lambert et al. (2000) (Adapted from W. M. Gelbart et al., (2000).

The DNA molecule is usually negative charged, so the repels between the different parts of the long chain molecule in a close distance about one *nm* is understandable. However, why the DNA folds in such packed toroidal structure in a presence of proper polyvalent cations? This effect gets reasonable physical explanation by BSM theory if analysing the magnetic field conditions at CL node level.

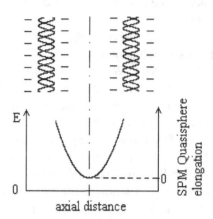

Fig. 11.18. Electrical field and SPM Quasisphere elongation between two parallel parts of DNA

Figure 11.18 shows the electrical field intensity between two parallel parts of DNA. The left vertical scale axis shows the electrical field intensity, while the right vertical axis shows the elongation of the SPM quasispheres in the plane of drawing. One specific feature is apparent from the drawing: the elongation of SPM quasispheres (see Chapter 2 of BSM) becomes zero in the middle between the parallel strands. This is possible only due to the strong spatial orientation of the Electrical Quasispheres (EQ) (contributed to the E-field) along the parallel DNA parts. Consequently, the elongation of all SPM quasispheres in the plane passing through the parallel DNA axes is reduced to zero. Then these SPM quasispheres become of MQ type (Magnetic Quasispheres). In the same time the SPM vector of these MQs might be synchronized, because the interacting EQs from the parallel long chains could be easily synchronized. Then the obtained in such way MQs provides excellent conditions for a permanent magnetic field. In order its direction to be permanent, however, it needs some stabilisation, which can be some interaction with an external current. Such current may be provided by heavy and polyvalent ions, while the masses of Na and Cl are not so different and their charge is not enough large. The current flows of the positive and negative ions are expected to have different paths. This could happen if they have different masses, because they will get different centripetal acceleration in a helical trajectory. The supercoiling shape of the long DNA will provide a necessary condition for such trajectory. In the same time the condition of short magnetic lines will keep the supercoiled DNA in compact configuration, while the negative charge of DNA strands will keep the proper distance between the different parallel parts of the molecule.

The explanation of the DNA supercoiling in eukariotic cells in a natural cell environments requires some additional considerations. The human DNA, for example, is supercoiled around a protein called histon octamer. The shape of the higher order structures of DNA as a nucleosome formation, a Chromatin and a Chromosome are presently well known. They are illustrated by Fig. 11.19.

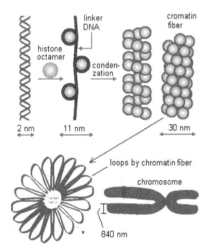

Fig. 11.19. Higher order structures of DNA in Chromatin
and Chromosome

The conclusion (A) in §11.5.1 provides an explanation why the DNA molecule is easily folded as a secondary helix around the histone octamer. The latter, evidently, provides a higher order helical shape conditions. According to the considerations in §11.5.1 and §11.5.2, the histones may also have features of energy flow. If examining the atomic structures of the histones (by using the Protein data Bank) we will find that two type of atomic rings are involved: (6C) and (N+4C). A possible energy flow through the spatially arranged rings may contribute to the bending and coiling of the DNA around the histone octamer.

The Chromatin, known as a chromatin fibber (with diameter of 30 nm) is additionally coiled into a daisy flower shape, called a chromosome mini-

band. The chromosome miniband, shown in the left bottom side of Fig. 11.17 contains 18 loops of daisy flower shape. A stack of daisy flower minibands arranged in a superhelix with a diameter of about $840nm$ forms a chromosome.

The discussed effect of magnetic field creation with sustainable ion current may be part of the complex mechanism sustaining the supercoiled structure of the human chromosome. The balance between the magnetic field, the ion current and the electrochemical potential will depend on the internal cell environments as in the case of proteins. Simultaneously, the principle of the short magnetic lines and compensated magnetic field should also play some role. When one of the both ends of the chromosome is attached to the cell, the ion current and magnetic field will be stabilized, so the supercoiled shape will be stable. If the chromosome, however, is not attached by some reason, a small change in the ion current will cause a reaction of the magnetic field. This may invoke a shape change and a motility of the chromosome.

11.6. Hypothesis about the role of DNA in the cell cycle synchronization mechanism

11.6.1 General considerations

The cell cycle synchronization is an important factor for a normal organ formation in the growing multi-cellular organism. One simple example demonstrating such importance is the eye formation. Some not synchronized cell division in early phase of eye formation may lead to significant defects. Then, how the enormous number of individual cell know the exact moment to enter into mitosis? If only individual clocks mechanisms in the cells are involved, they could be asynchronized after a number of cycles. A synchronization by chemical messengers through the bulk of enormous amount of cells is a not convincing explanation.

It is reasonable to expect that DNA might be involved in the cell cycle synchronization. The analysis of DNA at atomic level by BSM concept led to formulation of hypothesis about electromagnetic intercommunication of DNA molecules located in different cells of same type. The proposed hypothesis suggests a physical mechanism that might be involved in the synchronization process.

Let us use the concept of complimentary compensated solenoids involved in §11.5.1., but regarding the both complimentary parts of the helical system as separate helices. In this case, they could be considered as two solenoids serially connected in circle. Then the direction of cascade energy transfer through both strands of DNA (corresponding to the two separated solenoids) could be clockwise or counter clockwise. This will define two different directions of the magnetic lines. Let us suppose, that the obtained magnetic field is involved also in some interaction processes in the cell, so it could not be reversed spontaneously. This means that the direction of the cascade energy flow in respect to the helical direction will be kept stable. Let us denote the direction of a stable energy cascade through one strand to be +z. This direction regarded as axis +z in fact is not a straight line, but a helix geometrically centred with one of the DNA strands. The stable energy flow through this strand will obviously influence the direction of the rotating energy states in the attached (O+4C) rings. Thus we may accept that they all have a clockwise direction coinciding with the rotating direction of the running energy cascade through the helical DNA strand along +z axis. In other words all stored energies in (O+4C) rings have the same handedness. Note that the definition of handedness is referenced to the direction of the cascade energy flow.

Let us now pay attention about the energy storage capability of Purines and Pyrimidines involved in DNA. The pyrimidins Cytozyn (C) and Thymin (T) have single rings (2N+3C) with electronic bond connection to the DNA strands. The Purines adenine (A) and Guanine (G) both have two attached ring structures (2N+3C) and (2N+4C) that are also connected to the DNA strands. Because, the single rings of C and T are the same as the attached rings of A and G, we may accept that they carry one and a same amount of stored energy.

The Purines and Pyrimidines have stronger bond connections to their own strand, than between themselves as base pairs. The handedness of their stored energies could be also influenced directly by the handedness of the energies in the (O+4C) rings from the strands. Using the same logic, the (2N+3C) rings should get the proper handedness from the strands to which they are strongly con-

nected. The attached to them (2N+4C) rings however will get a complimentary opposite handedness.

Now let consider the normal situation, when the both strands of DNA are closely spaced in a shape of double helix. If aligning the DNA along a new defined axis +Z, the rotated energies in the rings connected to one selected strand, say a first one, will have a clockwise rotation, while from the second one - a counter clockwise. Consequently, when introducing a common direction axis (+Z) for both strands, the energies from (O+4C) connected to the two strands of DNA will look as they have different handedness. (In the further analysis, the common +Z axis will be considered for both strands).

Table 11.1 provides the handedness of the stored energies for all types of rings in DNA molecule, as a result of the above analysis. The stored energy handedness is referenced to a common axis Z+ of the DNA molecule.

Energy and handedness states   **Table 11.0**
==================================

strand A	strand B	Energy states for
E_S/l\	E_S\l/	(O+4C) ring
$(E_1$/l\$)(E_2$\l/$)$ ------- $(E_1$\l/$)$		A---T and G---C
$(E_1$/l\$)$ ------- $(E_2$/l\$)(E_1$\l/$)$		T ---A and C---G

Notations:
  /l\ and \l/ - two states of handedness
  E_S - a rotating state energy in (O+4C) ring
  E_1 - a rotating state energy in (2N+4C) ring
  E_2 - a rotating state energy in (2N+3C) ring
  ----- a connection by a hydrogen bond
  $(E_1$/l\$)(E_2$\l/$)$ - energy states with complimentary handedness valid for attached rings

Let us examine how the stored energies in the rings of different types could be influenced from some change of the cascade sequence. The energies of (O+4C) rings can be stronger affected by a change in the energy cascade flow through the strand, because the strand bonding path passes through C=C bond of every ring. The stored energies in single (2N+3C) rings could be affected by the change of the cascade type energy flow through (O+4C) rings. The stored energies in the attached rings (2N+3C) and (2N+4C), however, is of complimentary type, so it is more resistant to the mentioned above energy flaw changes.

Summarizing the above considerations we may conclude:

(1) The stored energy sequence in (O+4C) rings (with unit value of E_S) is strongly influenced by a change in the cascade energy flow through DNA strands

(2) The complementary energies $(E_1/\backslash)(E_2\backslash/)$ in the attached rings are more stable than the energy E_1 in the single ring.

(3) A change of energy sequence in (O+4C) rings could influence stronger the stored energies in the single (2N+3C) rings than in the (2N +4C) rings.

(4) The energies E_1 and E_2 from connected by H bonds A--T and G--C have always the same handedness.

Now, let consider that the DNA strands are opened in one end only. In such case the paths of the magnetic lines is increased at this end. It is known from physics that if an alternative field components appears in such conditions, some energy will be emitted as EM waves. At this point, however, we must consider some specific features of the double helix structure of DNA:

The (O+4C) rings, arranged in a helix, contains equally spaced gaps. Having in mind the enormous number of these rings, the cascade energy transfer will have a very large unidirectional component but quite a small alternative component for one clock. Then it is reasonable to accept that the generated magnetic field will have the same unidirectional feature like the field from the energy cascade through the strands. This means that every emitted pulse will have energy of $2E_S$.

When considering the energies of single rings (N2+C4), however, the emission conditions are different. They do not form uninterrupted sequence along the chain as the (O+4C) rings, so the sequence of the released E_1 energies cannot possess own permanent component. This might affect the preferred direction of the magnetic field generation. The helicity of DNA and higher order helicity of chromatin and chromosome with some complimentary ion currents may also provide conditions of energy release from (O+4C) rings in one and a same direction for both strands. It is reasona-

ble to accept that the emitting direction could follow the direction of emptying the energies of (O+4C) rings. Let us consider that this direction corresponds to the introduced +Z axes. Figure 11.20 illustrates the sequence of emptying the stored energies in (O+4C) rings (with individual energy E_S) and (2N+4C) rings (with energy E_1). The energy status and the direction of energy emission are shown for two consecutive clocks (i) and (i+1). The energies stored in the two attached rings (2N+4C) and (2N +3C) are not shown in the figure, but they are always complimentary to the (O+4C) rings.

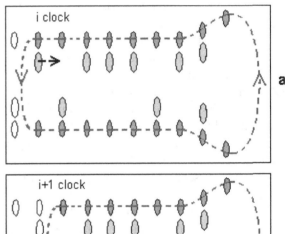

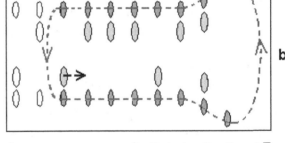

Fig. 11.20 Energy status and direction of energy transfer for two consecutive clocks. The emitting direction is along +Z axis

The handedness of all E_S energies in respect to the direction of the permanent magnetic field (defined by the cascade energy transfer through the bonding path) indicated by arrows in a dashed line is one and a same. When emitted by the common +Z direction, the corresponding E_S energies contribute to a common $2E_S$ pulses. For E_1 energies, however, the emission conditions are different because they do not have a stable unidirectional sequence. Then considering the energy emptying direction +Z the handedness from E_1 belonging to the different strands are different (clockwise or counterclockwise in respect to +Z axis). Let us consider:

(a) The E_1 energies with clockwise handedness in respect to +Z axis provide a photon sequence with a clockwise phase difference between the photons.

(b) The E_1 energies with a counter clockwise handedness in respect to +Z axis provide a photon sequence with a counterclockwise phase difference between the photons.

(c) The emitted photons of E_1 energies from the same strand are entangled with preserved phase difference. The phase advance in one strand (with +Z orientation of its cascade energy) will have a clockwise phase sequence, while from the other strand it will have a counter clockwise.

(d) Anyone of both sequences of entangled photons contains embedded information about the A--T and G--C sequence of DNA referenced to one strand.

(e) The energy of entangled photon sequence is less dispersive in conditions of intercellular medium.

(f) The probability of absorption of the entangled photons is much higher if meeting a similar spatially arranged atomic rings.

(g) If the process of DNA energy release is invoked not by internal factor, but by externally induced synchronization, it might have some phase delay due to a missing part of the initial synchronization sequence.

The photon entanglement is observed in lasers. Firstly, two entangled photons have been observed few years ago. Observation of three entangled photons has been published by Zelinger group (D. Bouwmeester et al, 1999) The authors express opinion for "entanglement between many more particles". The same group later reported observation of four entangled photons. Observations provided with entangled photons show that they fight the diffraction limit valid for a single photon. This may provide explanation of the mentioned above feature (e).

It is reasonable to consider that conditions of uniform temperature and proper spatial arrangement of the ring structures in DNA are necessary requirements for multiple photon entanglements in the described above mechanism. It is known from the single mode narrow line emitting lasers that their coherence time is quite long. Let us assume for example accept that the coherence time in the DNA emitting process is of the order of 1×10^{-10} (s). If an average lifetime for a single bond is about 1×10^{-12} (s), then the time between two E_S clocks is 6×10^{-12} (s) (six bonds). In this case, the entangled sequence will include a binary code corresponding to 600 E_S clocks. Such binary code corresponds to 600 bp (base pairs) and involves 200 codons.

Now let pay attention about one particular problem related to DNA investigation, known as a **C value paradox**.

A substantial fraction of the genomes of many eukaryotes is comprised of repetitive DNA in which short sequences are tandemly repeated in small to huge arrays.

Tandemly repetitive sequences, known as "satellite" DNA's are classified into three major groups;

-satellites - with repetitive lengths from one to several thousands base pairs.

- minisatellites - repetitive arrays of 9 to 100 bp, but usually 15 bp, generally involved in mean arrays lengths of 0.5 to 30 kb.

- microsatellites - repetitive arrays of short 2 to 6 bp found in vertebrate, insect and plant genomes.

The code of repetitive sequence does not encode amino acids. The percentage of this non-informative DNA increases significantly with organism complexity but depends also of other factors. Among them are the living environment conditions of the species.

From a point of view of proposed hypothesis, the C value paradox obtains quite logical explanation:

(h) The repetitive sequences in DNA code provide repetitive synchronization code that may increase reliability of the DNA intercellular communication.

Many diversified features of the DNA code redundancy in different organisms and species could find a logical explanation from the point of view of feature (h).

For example, the shown below repetitive sequence in DNA from Drosophilia corresponds to the following binary codes embedded in two entangled photons, related to both strands:

(AATAT)n, where n - is number of repeats
from DNA strand 1: (00101)n
from DNA strand 2: (11010)n

If the emitted entangled photon from strand 1 contains CW (clock-wise) circular phase sequence of the code, the emitted entangled photon from strand 2 will contain CCW (counter clock-wise) circular phase sequence of the complimentary code.

It is evident that the entangled photons carry embedded information that is dependent on both: the amino acid code and the redundancy code. This dependence may eventually play a role in the immunological response to transplanted tissues or organs. The redundancy code greatly increases the probability of successful synchronization. Then it is reasonable to expect a large abundance of repeatable code near the end side of the linear DNA in eukariotic cells. Such redundancy is really found.

11.6.2. Time sequence in the energy read-out process of DNA and its possible relation to the cell cycle synchronization.

Let us consider some initial state of DNA molecule when all rings are charged with their normal energies E_S, E_1 and E_2. At some particular moment invoked by internal or external triggering, the stored energies begin to clock out with synchronization sequence E_S and two entangled photon sequences. Ones this process is started, it will be self sustained until all E_S energies states are red-out. We may call this a read-out process of DNA. The time duration (t_{total}) for DNA read-out should be

$$t_{total} \approx 6t_{av} \frac{DNA \text{ length}}{0.34 \text{ nm}} \qquad (11.10)$$

where: t_{av} - is the average lifetime of excited states in the bonding path of DNA strand, 0.34 *nm* is the distance between the rings of the neighbouring base pairs.

The total read-out time is very short. For DNA with length of 1 *m* and $t_{av} \sim 1\times10^{-12}$ (s) the read-out time should be only 1.8 ms.

After the DNA is red-out, all E_S energies are emitted. The E_1 energies only of the T and C are emitted, but the complementary state energies E_1 and E_2 of A and G are preserved due to the attached rings. This invokes some type of asymmetry between connected Purines and Pyrimidines by H-bonds. Such conditions may help for separation of the DNA strands for initiating of the replication process.

Now, let analyse the possible involvement of DNA read-out process in the cell cycle synchronization in eukariotes. We will not discuss here the complex processes of cell cycle regulation in which the proteins are involved, but only the conditions of successful triggering and read-out of DNA. We may assume only that every individual cell possesses some kind of triggering mechanism. Such mechanism is necessary but not enough condition in order to initiate the read-out process. The DNA read-out can be successful only if conditions for emission of EM energy exist. Consequently, it will depend on the following conditional states of DNA molecule:

(a) The two strands of one end of DNA must be separated (or unbalanced by promoter) in order the EM quanta (photons) to be emitted

(b) The two strands (not counting the end conditions mentioned in (a)) must be completely symmetrical in order the read-out process to be self-sustainable.

The requirements (a) and (b) exclude the mRNA attachment to DNA or any other regulatory protein, so the transcription process and the regulatory mechanisms should be completed. Then for development of synchronized read-out as an avalanche process the expected optimal conditions should be during the phase S (start of DNA replication).

Now let assume that the internal triggering mechanism provides triggering clocks with period much shorter than the cell cycle. They will not provide successful triggering until the conditions (a) and (b) are satisfied. However, once they are satisfied the DNA read-out will be successful. Then the emitted entangled photons with a large common energy and encoded sequence possess an increased probability for activation of similar read-out processes in the DNA of the neighbouring cells. The extremely fast read-out process could lead to similar read-out process in many DNA molecules. In such case, the emitted entangled photons may additionally interfere and contribute to the avalanche process of synchronization. The avalanche process, however, will be contributed only of DNA molecules that do not have significant differences. In case of tissue and organ transplantation, the avalanche process will not work between the DNA from different species whose synchronization sequences are too different.

11.6.3. Environment considerations for the efficiency of the avalanche process

The phase accuracy of the cell cycle synchronization evidently depends on the individual cell cycle phases and environmental conditions.

11.6.3.1 Phase accuracy dependence of the cell cycle period

If considering the triggering of the synchronization from internal cell mechanism in a proper phase of the cell cycle, the period between two consecutive synchronization bursts will depend on the time lag between the local clock triggering events. Evidently, the period of these events should be much smaller than the cell cycle period.

Let us accept that the individual cell cycle regulation mechanisms provides a Gaussian type distribution of the cell cycle period from many cells. Then the optimum conditions for avalanche read-out should be expected in some moment that is closed but not overpassing the maximum of the Gaussian curve. Once it is initiated, the Gaussian curve will be cut down for all cell contributing to the avalanche. Those who had performed a preliminary read-out and those not activated by the synchronization will not contribute. The cells in which the transcription and regulation processes are not completed in this moment will also be excluded and their mitosis will not be in harmony with the synchronized cells.

Additionally we may not expect that all the synchronized cells will be activated in the very beginning of the read-out of the first activated cells. The repeatable sequences of codons may provide

conditions for larger number of entangled photons. Such photons posses also an increased probability for selective absorption and activation of similar repeats from DNA in other cells. **Consequently, the probability for avalanche from tandemly repetitive sequences of DNA is larger.**

11.6.3.2 Physical factors of cell environment

The most important environmental factor for successful EM synchronization is the temperature.

It is reasonable to assume that the cascade energy transfer through the bonding path of the DNA strand is phase synchronized by the energy rotation cycle in the (O+4C) rings. The duration of this cycle is:

$$T_R = t_O + 5t_C \qquad (11.11)$$

where: t_0 and t_c are respectively the lifetimes of the excited states in oxygen and carbon

The period, T_S, of the synchronization pulses is defined by the sum of the consecutive lifetimes through the bonding path of the strand, shown in Fig. 11.13

$$T_S = t_C + 5t_{av} \qquad (11.12)$$

where: t_{av} is the averaged lifetime value of the bonding electrons involved in the bonding path of the strand.

Then the condition for above mentioned phase synchronization is expressed by the relation:

$$T_S - T_R = t_0 - 4t_C - 5t_{av} = const \qquad (11.13)$$

The expression (13) is a constant for all involved DNA molecules if the same energy levels of the excited states are involved. This is a quantum mechanical condition that will depend only on the temperature. The successful generation of entangled photons, however, is additionally dependent on the correct spatial arrangement of the (2N+4C) rings. If the DNA molecule is sharply bent or it is in proximity to a protein involving magnetic field asymmetry, the condition of multiple photon entanglement will be disturbed. This will affect the efficiency of the synchronization process. Formations of Cruciforms, for example, may affect the emitted sequence. Z-type DNA inclusions and any external influences causing a helical non-uniformity of Beta type DNA may also block the successful read-out process.

The synchronization is possible in limited temperature range and good temperature uni-

formity along the DNA chain. The synchroniza-tion efficiency is dependent of the bending conditions of DNA and the spatial and helical uniformity along its chain. External factors causing any modification of the spatial parame-ters of the DNA double helix may inhibit the re-adout process.

11.7. Established features and experimental results supporting the hypothesis of the DNA involvement in the cell cycle synchronization

The proposed hypothesis is supported by a large number of established features of DNA mol-ecules and experimental observations.

11.7.1 Absorption properties of DNA.

DNA absorbs ultraviolet light in the range of 240 to 280 nm with a maximum at about 260 nm. A temperature increase destabilizes the double he-lix. Experimental observations lead to a conclusion that the thermal stability of DNA depends not only on the hydrogen bonding, but also on the base stacking (Saenger, 1984). This is in agreement with the suggested hypothesis of magnetic field stabili-sation generated by the cascade energy flow through the bonding paths of DNA strands.

The mechanism of energy storage and re-lease, according to suggested hypothesis, is in agreement with the observations, reported by Joseph Lakowicz et al. (2001) in the article "Intrin-sic Fluorescence from DNA can Be Enhanced by Metallic Particles". In their experiments they use silver particles of size about 4 nm put in the surface of two quartz plates with distance between them of 1 - 1.5 um. The DNA placed in this gap has been excited at 287 nm with pulse sequence from dye la-ser with a pulse with of 100 ps. The intrinsic emis-sion following such excitation is in a range from about 330 to 350 nm and it is increased about 80 times in the spots around the metal particles. The detected fluorecsence they believed to be from ad-enine and guanine.

The authors explain the effect by a decreased lifetime in relation to the SERS effect (Surface-en-hanced Ramman spectroscopy). Nevertheless, they also acknowledge the existence of another less un-derstood effect. The obtained results of time de-pendent intensity decay are shown in Fig. 11.21 (corresponding to Fig 5 of their publication).

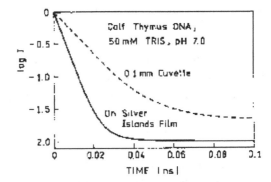

Fig. 11.21 Time-dependent intensity decay of DNA without metal (-) and between silver island film (Courtesy of J. R. Lakowicz et al., 2001).

The explanation of the observed effect from a point of view of BSM theory and suggested hy-pothesis is the following. The laser pulse charges the energy storage rings of DNA. The silver metal islands introduce not uniformity field conditions in the proximity to DNA. In such conditions, the mechanism that may keep the stored energy in the rings is disturbed. Then in the case when DNA is closed to metal islands, a spontaneous read-out process may occur sooner than normal one. This will appear in a shorter time interval between the excited pulse and the fluorescence emission in comparison to the case when DNA is not in prox-imity with such particles. This corresponds to the observed decreased lifetime. It is interesting to note that the fluorescence lifetime is still much longer than the lifetime in a single electronic bond. This confirms the idea that the stored energies in the rings may have a number of rotational cycles.

11.7.2 Increased binding ability of some simple organic molecules

The proteins that bind to DNA are usually very large molecules containing thousands of at-oms. But some comparatively simple organic mol-ecules also have an increased ability for binding. When examining such molecules we see that they posses attached atomic rings. As was mentioned before, the energy in such rings is more stable. This gives them additional ability to interact with DNA molecule based on magnetic field interactions. Such molecules for example are:

Trimethylpsoralen: 3 attached rings
Ethidium Bromide: 3 attached rings + 1 single
Chloroquine: 2 attached rings.

11.7.3 Effects of the salt concentration in helical winding and DNA supercoiling

As the concentration of a monovalent cation (Na^+) or divalent cation (Mg^{+2}) increases to high levels, the DNA double helix becomes wound less tightly.

According to the suggested hypothesis, the ion concentration influences the ion currents involved in the stabilisation of the magnetic field of DNA, responsible for its supercoiling shape.

11.7.4 Role of intermolecular triplexes in genetic recombination

Triplex DNA forming regions have been identified near the sites involved in the genetic recombinations. They usually have a mirror symmetry repeats of base pairs (as AGGAG). Kohwi and Panchenko (1993) found that the formation of intramolecular triplex structures in DNA in vivo can induce genetic recombination between two direct repeats flanking the triplex formed sequence.

According to the proposed hypothesis, the triplex formation in DNA disturbs the symmetry of the compensated field solenoid of DNA. Instead of flowing in the internal region of DNA, some of the magnetic lines are forced to pass through the external region. In conditions of chromosome recombination in vivo, the read-out process may be more often, than in the regulated cell cycle. Then the emitted sequence of entangled photons from one chromosome could be directly induced in the other. This may influence their common position and orientation. In the same time, the magnetic field of the cascade energy transfer through the bonding path of the DNA strand is not compensated in the triplex region. When combining with a similar region from another chromosome, the complex magnetic field may become compensated and stable.

11.7.5. Electronic properties of DNA

This subject is of increased interest of physicists and chemists. Despite the subject is far from

new, it is very controversial. A recent overview paper about this topic is presented by C. Dekker in Physics World, 2001. The experimental results deviate from good insulator to good conductor, while many researchers consider that DNA is able to provide some kind of energy flows.

Currently two possible mechanisms of charge transfer are accepted: a coherent process of single step electron tunnelling and a thermal hoping. It is interesting that a signature of cascade transfer exists in both suggested processes. While this theoretically suggested mechanisms seam reasonable, direct electrical measurements by number of physics groups provide conflicting results. The results reported by Hans-Werner Fink and Christian Schonenberger (1999), for 1 μm long DNA in vacuum indicate a good conductivity. However, this contradicts to the theoretical considerations where DNA is expected to behave as a semiconductor with a large energy gap between the valence and conduction bands. The experiments of Porath and coworkers (2000), for example, on a particular type of DNA - called poly(gG)-poly(dC) DNA show that it behaves as a large gap semiconductor.

One important fact that may influence the experimental results is the different environmental conditions at which the experiments are performed. Additionally, in number of experiments the electronic properties are not directly measured, but derived from the transient absorption spectra.

Let us explain the possible electrical properties of DNA from the point of view of BSM models. In order to have a stable energy cascade transfer, the DNA must be in proper environments (pH, temperature, ATP) and to be enough long. In normal environments, DNA is considered a negatively charged. This charge, however, is uniformly distributed, so it may not influence the energy cascade transfer. In such case, the stored ring energies are synchronized and have a proper handedness. For small voltage along DNA, the electrons do not quit their spatially distributed orbits. For large voltages, however, the proximity proton field defining their orbits are biased, end the valence electron could jump synchronously. This transition, however, is expected to interact with the rotating energy states of the rings. The signature of such interaction might be the effect of "single step electron-tunnelling mechanism". It is evident that such conditions may occur only if a stable energy cascade transfer through the DNA strand exist, but the latter effect is possible only at suitable environments.

11.8. Hypothesis of decoding process in some of the complexes aminoacyle-tRNA synthetases - tRNA.

11.8.1. General considerations and code analysis

The prototype of tRNA molecule is coded in the gene. The current estimate for number of tRNA in animals and plants is up to 50. The typical shape of the tRNA is shown in Fig. 11.16. The tRNA binds to the proper enzyme that charges to its end one of the 20 amino acids corresponding to the anticodon. Aminoacyl-tRNA Synthetase is a family of 20 enzymes whose structures are pretty well known. The enzyme usually recognizes the corresponding tRNA using the anticodon.

Let us analyse the coding of the amino acids based on the anticodons used by tRNA. Instead of anticodons we may use for convenience the most popular RNA codons, given by Table 11.1 In a gen-eral case the anticodon code is directly obtainable from the corresponding codon code. The number of codons is 64. One exclusion from that rule is the anticodon of Alanine, for which the three codes GCU, GCC and GCA are replaced by one codon CGI (I - Ionosine). This reduces the total number of anticodons to 61.

One of the evident features of the amino acid coding is the **code redundancy**. Most of the amino acids are coded by more than one codon. One additional surprise comes from the genome analysis. Some organisms don't have genes for all twenty aminoacyl-tRNA synthetases, but they still use all twenty amino acids to build their proteins.

The synthetase mechanism of the Aminoacyl-tRNA synthetases involves two steps. In the first step they form an aminoacyl-adenylate in which the carboxyl of the amino acid is linked to the alpha-phosphate of ATP by displacing pyrophosphate. In the second step, if only a correct tRNA is bound, the aminoacyl group of the aminoacyle-adenylate is transferred to 2' or 3' terminal OH of the tRNA.

Table 11. 1 Codons involved in RNA

UUU	Phenylalanine	UCU		UAU	Tyrosine	UGU	Cysteine
UUC		UCC	Serine	UAC		UGC	
UUA		UCA		UAA	Stop	UGA	Stop
UUG	Leucine	UCG		UAG	Stop	UGG	Tryptophan
CUU		CCU		CAU	Histidine	CGU	Arginine
CUC	Leucine	CCC	Proline	CAC		CGC	
CUA		CCA		CAA	Glutamine	CGA	
CUG		CCG		CAG		CGG	
AUU	Isoleucine	ACU		AAU	Asparagine	AGU	
AUC		ACC	Threonine	AAC		AGC	Serine
AUA		ACA		AAA		AGA	
AUG	Start; Methionine	ACG		AAG	Lysine	AGG	Arginine
GUU		GCU		GAU	Aspartic	GGU	
GUC	Valine	GCC	Alanine	GAC	acid	GGC	Glycine
GUA		GCA		GAA	Glutamic	GGA	
GUG		GCG		GAG	acid	GGG	

The second step is conditional. This means that if a wrong enzyme is bound to tRNA, the output will be zero (not charging the proper amino acids). If assuming that the process of binding the correct complex of enzyme - tRNA is occasional, then the number of possible combinations for all 20 amino acid codes is $2^{20} = 1048576$. This means such number of produced tRNA and enzymes, so it is unreasonable. Another possible option is the enzymes to recognize or at least to increase the probability to bind to the correct tRNA in which case the above number to be significantly reduced.

One of the currently discussed problems is how these enzymes (aminoacyle-tRNA synthetases) recognize 20 aminoacides. It is estimated that they admit intrinsically small number of errors - about 1 in 10,000. The unmistakeably recognition of 20 different flavours looks as quite intelligent task for a biomolecular structure of only few thousands atoms. This enigmatic feature is known as 20 different flavours of these enzymes.

The above task could not be resolved unless the enzymes have some sensors for preliminary detection before binding and some memory. Having in mind that the enzyme contains only a few thousands atoms it is apparent that such capability should be built very economically using some basic physical properties at atomic and molecular level. Then the sensing should be based on some simplified detection mechanism of energy states, while the memory should be based on basic physical states of binary type. Let us concentrate firstly on the memory feature. The possible binary physical states for a molecule of complexity of the enzymes are following:

(1) Quantum mechanical spin of the electron

(2) direction of magnetic field (S-N or N-S)

(3) handedness of rotating energy states (according to BSM)

(4) electrical charge (+ and -)

The option (4) could be excluded because its realization requires complex structure of semiconductor type. The most reasonable option is (3) with some combination of options (1) and (2).

Let us evaluate the required memory for decoding the codon of any amino acid using the Boolean algebra without minimization. Any codon is a three-digit code. Anyone of these digits needs 4 states in order to present one of the four letters (A,B,C,D). Then the three-digit code of the codon could be presented by 12 bit binary code. This corresponds to a memory map of 4096 bits. Such memory hardly be achieved by any enzyme whose molecule usually includes a few thousands atoms. Additionally, the enzyme must have remote sensors for recognising the tRNA before binding and tools for test of the anticodon.

Obviously, the above general approach for decoding of 20 flavours is not feasible. We must look for some economic natural code hidden in the codons and particularly behind their redundancy. Let us analyse the redundancy using some of the unveiled properties of the involved ring structures

and the possibility for detection of their type. The Purines are distinguishable from the Pyrimidines by the number of rings (able to carry different stored energies), so they could be sensed by some kind of binary test. The possible mechanism of such test will be discussed later. In order to see, how such consecutive tests of the first, second and third letter of the codon will lead to some results, we make the following substitution for the codons in Table 11.1.

Purines (A and G)   -> 2

Pyrimidines (C and U) -> 1

The digits 1 and 2 in this substitution in fact correspond to the number of the atomic rings.

Table 11.2

111	111	121	121
111	111	121	121
112	112	122	122
112	112	122	122
111	111	121	121
111	111	121	121
112	112	122	122
112	112	122	122
211	211	221	221
211	211	221	221
212	212	222	222
212	212	222	222
211	211	221	221
211	211	221	221
212	212	222	222
212	212	222	222

Table 11.2 shows the obtained codes after this substitution. This table does not contain the names of the amino acids, but they are identifiable by the code positions. When examining the Table 11.2 we see that the binary codes (with base states "1" and "2") are in very strict order.

Let us find out how many aminoacids every code from the Table 11.2 is related to. The result is shown in Table 11.3. It is evident that the distribution of the amino acid codes based on the distinguishing features between the Purines and Pyrimidines is pretty uniform. Only the codes for start and stop deviate from the uniformity of the distribution. From this simple analysis, however, one useful conclusion could be made:

The number of possible combinations could be significantly reduced if initial binary detection of Purines or Pyrimidines is performed.

Table 11.3

Code from Table 11.2	No of coded amino acids	Additional coding
111	4	
112	4	
121	4	
122	3	stop
211	4	
212	5	start
221	4	
222	4	

The Purines and Pyrimidines are distinguishable by the number of rings. It will be described later that their possible remote detection have some similarity with the process of DNA readout described in section 11.6.

Let us denote the tRNA anticodon by the triplet *abc*, but using instead the corresponding codons of RNA. The following example shows the reduction of the possible combinations in three consecutive binary tests (Purines or Pyrimidines) applied for the codes of Table 11.2. Let the first test for *a* gives value: *a* = 1. This means that all codons starting with 2 are excluded from the following tests, while those starting with 1 are included. Only 32 from of all 64 combinations are left. This is illustrated in Table 11.4., where the excluded combinations are masked by a grey colour. (Note: the gray colour may look only slightly shaded).

a = 1   **Table 11.4**

111	111	121	121
111	111	121	121
112	112	122	122
112	112	122	122
111	111	121	121
111	111	121	121
112	112	122	122
112	112	122	122
211	211	221	221
211	211	221	221
212	212	222	222
212	212	222	222
211	211	221	221
211	211	221	221
212	212	222	222
212	212	222	222

Let us consider the same test for **b.** If the result, for example is b = 2, the available combinations are additionally reduced in half as shown in Table 11.5.

b = 2   **Table 11.5**

111	111	121	121
111	111	121	121
112	112	122	122
112	112	122	122
111	111	121	121
111	111	121	121
112	112	122	122
112	112	122	122
211	211	221	221
211	211	221	221
212	212	222	222
212	212	222	222
211	211	221	221
211	211	221	221
212	212	222	222
212	212	222	222

In a similar way the third test of c, for example, with a result $c = 2$ will lead to reduction of the available combinations to 8 as shown in Table 11.6.

$c = 2$　　　　　**Table 11.6**

111	111	121	121
111	111	121	121
112	112	122	122
112	112	122	122
111	111	121	121
111	111	121	121
112	112	122	122
112	112	122	122
211	211	221	221
211	211	221	221
212	212	222	222
212	212	222	222
211	211	221	221
211	211	221	221
212	212	222	222
212	212	222	222

We see that by three consecutive tests the available combinations are reduced from 64 to 8, while they address only 4 amino acids. If assuming that this test is performed prior to binding of the enzyme to the tRNA, then the required combinations for a probable correct match are only $2^4 = 16$. Consequently a possible remote sensing will reduce the amount of the necessary number of enzymes and tRNA by a factor of 65536.

The additional tests for identification of the correct amino acids require distinguishing of Adenine from Uracile and Guanine from Cytozine for every letter of the triplet *abc*. The detection for such identification, however, could not be remote and perhaps is performed in the second step of the synthetase mechanism. It is known, for example, that in glutaminyl-tRNA synthetase with its tRNA, the enzyme firmly grips the anticodon, spreading the three bases widely apart. Let us assume that the enzyme keeps in its memory the binary results of the previous remote detection (Purines or Pyrimidines) for every letter. Then the following test after the enzyme has grabbed the anticodon (for every letter of the codon) must be also a binary test of a type: Adenine (A) or Guanine (G) and Cytozine (C) or Uracile (U). How they could be distin-

guished, while they have the same number of rings? When examining their structure we see that the larger ring of Adenine includes three bonds of single valence, while the larger ring of Guanine includes four bonds of single valence. The ring of Cytosine includes two bonds of second valence, while the ring of Uracile includes one bond of second valence. These differences of the ring structures may cause significant differences in their spectral emission-absorption capability. The very narrow temperature range will define a pretty narrow range of the population of the excited states. This may simplify the task for recognizing the differences between (A and G) and (C and U) by their spectral signatures.

After all these binary tests are performed, it seems that the correct amino acid should be decoded. However, we must not forget that the letters *abc* must be read in a correct order (from a most to the less significant digit). If this rule is not observed, the following code pairs could not be distinguished:

UUA	-	AUU
UUC	-	CUU
UUG	-	GUU
UCC	-	CCU
CCA	-	ACC
AAG	-	GAA
UGG	-	GGU

Assuming that the memory map of the enzyme is naturally minimized, the correct code readout obviously must be initially detected prior to the remote sensing described above. In order to find out the possible detection mechanism for the correct code readout direction, we may examine the structure of tRNA that has been already shown in Fig. 11.16. In the left side of the figure, the real shape of the backbone structure is shown. In the right side of the same figure, the tRNA is presented like a flat curve in order to show more clearly some of its structural features such as, the folding of the single strand in cloverleaf shape, the position of loops, the H-bonds, the position of the anticodon and the asymmetrical ends.

The tRNA strand is similar as the DNA strand and contains regularly attached (O+4C) rings. In proper environment the tRNA could be energized, so a cascade energy transfer may occur through its strand in a similar way as in the DNA and proteins.

The asymmetrically terminated end of tRNA may assure the proper direction of the cascade energy transfer. If such process is stable enough for a short time, the rotating energies in (O+4C) rings can be synchronized, obtaining in such way a proper handedness. The same is valid for the H-bonding base pairs and consequently for the codon loop. While the tRNA is not so long (75 - 90 nucleotides) the energy cascade process could not be so stable like in the DNA and the proteins. The disruption of the energy cascade will lead to synchronized readout of the rotating energy states in (O+4C) rings that could be emitted as a coherent sequence of photon pulse whose parameters (phase sequence) will carry the direction of codon readout. This readout of (O+4C) energies from its side might provoke a readout process from the base pairs in a similar way as in the DNA. The conditions for emission from tRNA, however, are different from those of DNA due to the different shape of the tRNA molecule. Let us consider a hypothetical axis (not exactly linear), defined by the twisting shape (known as a secondary structure) of tRNA and passing from the anticodon loop of tRNA to its terminating ends. The two major side loops of tRNA could be regarded as approximately symmetrical pairs in respect to the introduced axis. The supercoiling and tertiary structure may also influence the symmetry of the mentioned pairs, but will not influence the symmetrical features of the anticodon loop. In the same time, all the loops are free of H-bonds. If regarding the emission process as EM filed, it is evident that the two side loops are complimentary and will have a comparatively smaller EM emission capability (because the magnetic lines appear closed in a circle around the introduced tRNA axis). The anticodon loop, however, does not have complimentary symmetry with another loop, so its emission capability could be much larger. Then the emitted radiation from the anticodon loop could be detected remotely by the correct enzyme. The burst of photon sequence from (O+4C) rings readout may serve as a synchronization that prepares the enzyme for detection of the anticodon. In such process the detection could be of synchronized type, so it could allow an increased probability for detection of the anticodon sequence. The tRNA could be re-energize and the readout sequence could be performed a few times. In such

way, the anticodon could be detected by number of corresponding enzymes.

The remotely detected sequence from anticodon loop (carrying information for Purines or Pyrimidines) could be regarded as very fast consecutive tests of the letters in the triplet **abc** prior to binding of the correct enzyme to tRNA. The test results, however, will be additionally needed for the consecutive tests (A or G and C or U) that will be performed after the binding of the enzyme to tRNA. Let us assume that these test results are directly passed to some kind of binary decoder implemented in the enzyme. Then after this remote test the number of codons from 64 is reduced to 8, corresponding to 4 amino acids. This significantly increases the probability of correct binding between tRNA and the corresponding enzyme.

11.8.2 Decoding algorithm

Figure 11.22 presents the decoding algorithm according to the analysis and considerations discussed in the previous paragraph.

The remote sensing includes tests 1, 2, 3, while the other tests are performed after the tRNA is bound to the correct enzyme. The numbers in circle show the enabled codon combinations after each test. Not all the decoding tree is shown, but the algorithm for a possible decoding of all 20 aminoacids is the same. It is evident, that maximum of 6 tests are necessary for decoding of anyone of the aminoacids, but some of them are decoded even at test No 5. (Leucine, for example). Two of the stop codons are also decoded at test No. 5. This provides the opportunity to use the test No. 6 for additional true test, that should increase the decoding reliability. The other stop codon UGA, however is decoded after the test No. 6. Then it could be interesting to study the statistics of UAA and UAG stop codons in comparison to the UGA stop codon (in both the RNA and DNA sequences).

According to the suggested algorithm, the tests from the remote sensing increase the probability for correct binding between tRNA and the proper enzyme, but does not exclude completely wrong bindings. It is logical to expect that only a correct binding will lead to attachment of amino acid after the tests 3, 4 and 5. A wrong binding will provide a zero result.

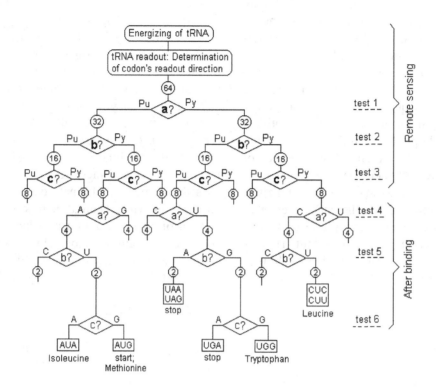

Fig. 11.22. Decoding algorithm according to the hypothesis of decoding process in some of the complexes tRNA - enzyme.

11.9. Discussion.

It has been considered so far that the possible intercommunication between the cells of living organism is based on chemical reactions. With provided analysis and suggested hypothesis, BSM, unveils for a first time that a new type of communication is possible, in which the DNA molecules are directly involved. The DNA has own specific EM signature with embedded DNA code. Together with the revealed energy storage mechanism, this phenomenon puts a light on some unknown so far processes of the living organisms. Further investigation may help to understand:

- How the organism recognises a foreign living cell by their DNA EM signature?

- Why the organism reacts on implemented foreign tissue?

- How the organism recognises and kills the cancer cells in the initial phase?

The brain proteins contain enormous number of ring structures, able to store energy. In proper synchronization of this energy may generate entangled quantum waves carrying coded information.

References:

C. B. Anfinsen, J, T. Edsall and F. M. Richards, Advances in protein chemistry, vol. 34, 1981, Academic Press.

C. Dekker, Physics World, Aug. 2001

H-W Fink and C. Schonenberger, Electrical conduction through DNA molecule, Nature, **398**, 407, (1999)

W. M. Gelbart, R. F. Bruinsma, P. A. Pincus and V. A. Parsegian, DNA-inspired electrostatics, Physics Today, 38-44, (2000).

Y. Kohwi and Y. Panchenko, Transcription-dependent recombination induced by triple-helix formation. Genes Dev. 7, 1766-1778, (1993)

J. R. Lakowicz, B. Shen, Z. Gryczynski, S. D'Auria and I. Gryczynski, Intrinsic fluorescence from DNA can be enhanced by metalic particles, Biochemical and Biophysical Res. Comm., 286, 875-879 (2001)

D. Porath et al., Direct measurement of electrical transport through DNA molecules, Nature, **403**, 635, (2000)

W. Saenger, "Principles of Nucleic Acid Structure", Springer-Verlab, Ney York, (1984)

Chapter 12:

Alternative Cosmology. Stationary Universe and Galactic Cycle.

Hypothesis about hidden cycles of matter evolution.

12.A.1. Introduction

Relying on the idea of a common origin of the material substance in the Universe, we have to build a reliable bridge between the unveiled structure of the physical vacuum and elementary particles, from one side, and the cosmological phenomena, from the other. The analysis in Chapters 2, 3, 4 and 6 revealed the big confusion in the contemporary physics. Based on a wrong concept about the physical vacuum, the famous Einstein equation $E = mc^2$ is wrongly interpreted.

The matter never annihilates, so it never converts to a pure energy. While the Einstein equation is correct, it refers to the mass. The mass according to BSM is not equivalent to matter. The mass is a structure (like a home built of bricks), while the matter is from indestructible basic particles (like bricks). The correct interpretation of the Einstein equation is the following:

$mc^2 \rightarrow E$ - destruction of mass (in particle collider experiments) or hiding of the positron's mass as shown in §3.17.3, Chapter 3).

$E \rightarrow mc^2$ - creation of virtual particles in CL space, not possessing matter (quasiparticle waves)

$E \Leftrightarrow mc^2$ - valid for the binding energy in atomic nuclei, as a result of small change of the CL space node distance in presence of matter.

The virtual particle (single or pair) is a wave in CL space, in which only one type of CL nodes is affected. It moves with a speed of light but does not possesses a matter, and consequently could not be a static elementary particle. They are created either as pairs (Dirac see particles) or from a Beta radioactive decay. In the process of thermalization, the virtual charge particle hits a target. As a result either a low energy electron or positron is extracted, which however is a real particles possessing matter. From the point of view of the correct concept of the physical vacuum, one important thing must be clear: The particles obtained in particle colliders or any high energy experiments are either virtual particles or structural fragments of real particles. No material particles is possible to be created from a pure energy waves. Pure energy interactions may create only virtual particles. Measuring their energy and using the Einstein equation for estimation their of their "mass" and claiming that matter is created is wrong. These "particles" could never become stationary particles with matter, such as proton, neutron, electron or fragments from their disintegration.

From the presented considerations (based on the analysis of all previous chapters), it is evident that real particle can be created only by some cosmological event, the condition of which could not be duplicated in any kind of laboratory. Such event is identified by extensive analysis of different cosmological phenomena and cross validation with the experiments in particle physics, analysed from a new point of view.

In order to provide a realistic scenario about the cosmological process related to formation of elementary particles, we must understand the low level matter interactions, which are simple because only the Super Gravitational law is involved:

- understanding the lowest level structure of the intrinsic matter based on geometrical stable geometrical formations from the two fundamental particles

- understanding the similarity and differences between the two substances of the Intrinsic Matter

- understanding the relation between: the Super Gravitational field, the energy at lowest level of matter organization and the intrinsic time constants for both substances of intrinsic matter.

This chapter is devoted on the analysis of the above problems. The outcomes of such analysis lead to a new concept about the evolution of the galaxies, their connections in supergalaxies and finally to a new vision about the Universe.

The inconsistencies of the Big Bang concept with many observations is discussed in a large number of papers. One of the best summarizing article "Big bang theory under fire", by W. Mitchel, is published in Physics Essay vol. 10., No

2, June 1997. Another large critics from physicists is published in "Open Letter from 36 scientist to scientific community", New Scientist, May 22, 2004). As a result, in 2005 an international conference called "1st Crisis in Cosmology Conference" (CCC-I, June 23-25, 2005, Moncao, Portugal) was provided challenging the Big Bang theory. It is not hard enough to say that the Big Bang must not be called a theory, because it has 101 problems from theoretical and observational aspect. Our goal is to provide a concept understandable and without any contradictions, not only in Cosmology but in all fields of Physics.

12.A.2. Weak points of the Big Bang concept

The weak points of the Big Bang concept are of theoretical and observational aspect. We may denote them as theoretical and experimental problems. They are listed below with assigned number in order to be referenced later.

12.A.2.1. Weak points from a theoretical aspect:

The main theoretical problems are in the following statement:

(1) According to the The Big Bang concept the Universe originates from a single point as a result of gigantic explosion, called a Big Bang. From a physical point of view it inevitably leads to the assumption that the energy could be created from nothing.

(2) The existing theoretical concept about physics of the black holes leads to a singularity problem - disappearing of matter and energy.

(3) The adopted concept about physical vacuum as an empty space but with physical properties required adoption of number unexplainable postulates:

- a constant light velocity, not dependable on velocity

- photon as a carrier of energy propagated without medium, so it may leave the matter forever, carry out the energy

(4) The concept of the relict radiation is unlogical and mind confusing.

(5) Lack of consistency between the Big Bang concept, Quantum mechanics and the Theory of General and Special relativity.

The statements (1), (2) and (3) contradict to the energy conservation principle, the validly of which is verified and well established in all fields of Physics.

Statement (4): The concept of the relict radiation has been invented to support the concept of Big Bang. Its logical explanation is quite controversial. It is not consistent even with the postulate of the constant light velocity. In fact the explanation of the spatial geometry of the relict radiation in consecutive evolutionary phases of the Big Bang is intentionally avoided, because it leads to contradictions.

Statement (5): The problem is known from decades. A number of unified field theories has been proposed but no one of them was able to provide a satisfactory solution of this century old problem. In fact the tendency to fit them to the Big Bang concept leads to not adequate theoretical explanations of the observed phenomena.

12.A.2.2. Weak points from observational aspect:

(1) Discrepancy between the age of some old stars in our galaxy and the estimated age of the Universe

(2) Discrepancy between theoretically estimated mass of the Universe and the observed one

(3) The recently estimated trend of Hubble constant for red shifts larger than 0.8 leads to another paradox - unexplainable acceleration of the galaxies after the Big Bang

(4) Some of the neighbouring galaxies from the local group show blue shift

(5) The Cosmic Microwave Background (CMB), considered as a relict radiation shows anisotropy related to the solar system motion around the Milky way

(6) The galaxy rotational curves are unexplainable enigma requiring an adoption of a "dark matter"

(7) Images from deep space with improved resolution show one and a same picture - same type of galaxies. A recently discovered spiral galaxy only one billion years after the Big Bang contradicts to the Big Bang concept of matter evolution

(8) The red shift periodicity of the galaxies from the local supercluster is not explained

(9) The physics of the globular clusters and the star population of type II are note explained satisfactorily. This refers also for the difference between the classical cepheids and the cepheids of second type.

(10) The pulsation mechanism of the variable stars is not explained satisfactorily

(11) Pulsars ("neutron stars"): steps in rotation rate, spin down rate, micro glitches and timing noise in pulsars - not satisfactorily explained

(12) Pulsars velocities and observed escaping pulsars from Milky way - not explained

(13) Not explained phenomena about the quasars:

- physical mechanism
- broaden lines
- multiple red shifts of absorption lines

(14) Lyman alpha forest - no logical explanation

(15) Not explained phenomena of Supergalaxies formations:

- large non-uniformity in galaxy distribution, cluster formations, voids and "God's effect"
- anisotropy in the galaxy plane orientations

(16) Distortion of CMB in gravitational lensing (Sunaev-Zeldovich effect) - not explainable

(17) Galaxy collisions and related phenomena - not satisfactorily explained

(18) GRB (gamma ray bursts) - not explained

12.A.2.2.A. Main reasons for the inconsistencies in the Big Bang theory

From the BSM point of view, all problems of the Big Bang model are result of the of wrongly adopted concept of the physical vacuum at the beginning of 20th century. Direct consequences of this are the following two assumptions, which are fundamental for the Big Bang model:

(A) considering the Universe space as homogeneous.

(B) assuming that the observed galactic red shift is only of Doppler type

In fact the assumption (B) follows directly the consideration (A).

12.A.3. Introduction into the BSM concept about the Universe:

Our vision about the Universe and its evolution is formed through the prism of the concept about space (physical vacuum). The BSM concept, which has never been investigated before, leads to a completely different vision about the Universe and the formation of the galaxies. The galactic red shift according to BSM is not of Doppler type, but cosmological, based on quite different physical effect, explainable only when a correct concept of the space (physical vacuum) is used. The space concept is so important that it leads to a profound different picture of the Universe and its evolution. Instead of expanding, the Universe appears to be a completely stationary. In such aspect, the cosmological phenomena and their observational characteristics become fully compatible with the concept of the atomic particles (according to BSM). The cross-validation from the particles physics analysis with the analysis of cosmological phenomena, led to understanding the fundamental relations between matter, energy, space, time and fields. In such aspect, BSM analysis allowed to infer the whole process of matter evolution with identification of the most important cosmological cycles - the cycle of the individual galaxy. The average period of this **galactic cycle** appears to be, what is now considered as an age of the Universe - about 12-15 billion years.

The galactic cycle includes the following major phases:

- **particle incubation**
- **active life**
- **recycling**

The BSM analysis of the cosmological phenomena provides strong evidence, that the Universe is stationary, while the observed space is not homogeneous. The latter is cause small difference between the local space of the individual galaxies, as it will be explain later in this chapter. Each galaxy undergoes multiple life cycles with phases of collapse, matter recycling and birth for a new active life in the same place of the Universe. The process of matter recycling goes down to lower levels structures of intrinsic matter, involving even the recycling of the prisms, from which the CL space and the atomic matter of the individual galaxy is built. It takes place in a hidden phase of the galactic cycle, which is completely not observable due to a lack of surrounding CL space (which must serve as a medium for the propagating light). A very low level memory of the intrinsic matter, however, survives the recycling process. It carries the correct information about prisms handedness. This predetermines the new born matter to be of same type (matter) and not of opposite one (antimatter). All the processes from the micro to macro Cosmos are in three dimensional space with a time flowing in a positive direction only. The intrinsic matter never disappear or annihilate. The principle of energy conservation is completely observed and valid for any phase and moment of the galactic cycle.

Before presenting the details of BSM concept about the galaxies and Universe, it is necessary to provide some insight about the lower level structures of intrinsic matter and their interactions based on the most fundamental physical law - the law of Super Gravitation.

12.A.4. Low level structural organization of the intrinsic matter

The more general properties of the intrinsic matter are presented as a "guessed properties of the primordial matter" in §2.3 of Chapter 2. Some of the parameters has been additionally discussed in §6.9.4.2, Chapter 6. The inferred properties are completely consistent with the revealed structures and properties of the CL space and the elementary particles. Now we need to unveil the lower level structural formations, embedded in the prisms. This requires some insight into the very basic structure of the primordial matter and the concept of the Super Gravitation in pure empty space.

12.A.4.1. Inferred basic rules

The following rules are inferred as a logical consequence from a broad aspect analysis using the BSM concept about the space and matter. They will be useful in the further analysis.

Inferred Rules:
P1 The intrinsic matter is composed of two not mixable substances

P2 Spatial geometrical formations of intrinsic matter substances contain internal vibrational energy

P3 A finite quantity of matter is able to handle a finite amount of internal vibrational energy

P4 The Super Gravitation (SG) is a process of external interaction between formations of intrinsic matter. It involves energy that is a complementary to the total internal vibrational energy of the interacting formations. We may refer to this energy as Super Gravitational energy.
SG attraction between spatially separated formations is a manifestation of SG interactions.

P5. A system of formations may handle a finite amount of total energy, equal to the sum of the vibrational and SG energy.

P6 The strength of SG forces are inversely proportional to the cube of distance, while the interaction process is characterized by intrinsically small time constant, not dependable on the distance.

P7 The energy conservation principle is valid from the lowest to highest level of matter organization and from the micro to the macro scale of the Universe.

The initially defined guessed properties and rules are useful for unveiling the structures of the lower levels of matter organization embedded in the prisms. The formulated rules can be indirectly proved by the physical analysis provided by BSM.

12.A.4.2. Structure components of the matter

12.A.4.2.1. Three dimensional structures in the lower level of matter organization.

A. The Primary ball as a most fundamental particle in the Universe

The primordial bulk matter of anyone of the two substance is made of indivisible tiny balls in which the radial dependence of the intrinsic matter density is a nearly bell-shape symmetrical curve.

The balls could not be molted or destroyed in any kind of processes in the Universe. We may call them **primary balls**. Following the volume ratio between right and left handed prisms, determined in §2.3 (Chapter 2 of BSM) as $v_R/v_L = 27/8$, the radius ratio between the primary balls of the two substances is 3/2 (the same as the prism's length ratio). They are characterised by the following features:

- they are indivisible
- all the primary balls of a same substance contain exactly one and a same quantity of matter
- the intrinsic matter density dependence on the radius is a bell shape curve falling to zero at the surface
- the primary balls may congregate into closed packed structures in which the individual balls may vibrate
- the distinguishable matter parameters in the balls of both substances are: the matter density and the radius referred to a common length scale

Inferred conclusion:

The primary balls are the most fundamental particles of the Universe.

The primary balls may congregate in primary structures of tetrahedrons.

The primary structures of tetrahedrons may congregated in additional structures.

The process of congregation is accompanied with a structural change and internal energy change. We may call this process **a growing process belonging to the low level of matter organization, or simply a growing process.**

B. Regular tetrahedron as a basic structure in the Universe.

The SG forces between single primary balls could not be defined until they are collected in congregations. The most compact congregation of balls has a shape of **tetrahedron,** denoted as **TH.** It is shown in Fig. 12.1.

Fig. 12.1. Shape of Primary Tetrahedron
The total number of balls exactly matches the number of balls along the edge

The shape of tetrahedron allows the most compact arrangement of the primary balls. In such configuration the individual balls have a freedom for vibrational motion due to the bell-shape of their radial density. In the tetrahedron configuration, their interaction energy is maximal. Their individual vibrations, however, are mutually dependent. This feature leads to appearance of synchronized **common mode of vibrations**. If considering a sphere instead of tetrahedron a stable common mode could not be obtained, because the spatial momentum is isotropical. The tetrahedron shape, however, assures variations of the spatial momentum if the common mode exhibits a rotational precession. If making analogy with the CL node, the existence of rotational precession of common mode oscillations in the primary tetrahedron seams quite reasonable. Then the common mode of oscillations could be characterised by vectors of resonance and spatial precession momentum similar like NRM and SPM vectors of an individual CL node.

Another important feature of the primary tetrahedron is that **all external shells are completed**. If the primary tetrahedron is in environment of bulk matter containing primary balls a damaged tetrahedron will either get repair or could be destroyed.

C. Vibrational modes of the tetrahedron expressed by vectors.

By analogy to the NRM vector of the CL node, the resonance momentum vector of the tetrahedron will be named **SGRM (Super Gravitation**

Resonance Momentum). Number of SGRM cycles will provide one complete cycle of **SGSPM vector (Super Gravitation Spatial Precession Momentum)**. It has similarity with the SPM vector of CL node. Its behaviour can be described by a quasisphere.

Association:

CL node	Quasipentagon
NRM (vector)	SGRM (vector)
SPM (vector)	SGSPM (vector)
SPM Quasisphere	SGSPM quasisphere

The **SGSPM quasisphere** has a similar meaning as the CL node quasisphere. It is a 3D plot of the spatial density momentum of the SGSPM vector. The SGSPM quasisphere, however, is distinguishable from the quasisphere of the CL node. It possesses four bumps (corresponding to the tetrahedron vertices) and four flats, instead of dimples, (corresponding to the median centre of the tetrahedron planes).

The differences between the CL node SPM quasisphere and the SGSPM quasisphere of the tetrahedron are the following:

CL node SPM vector quasisphere:

- the bumps are aligned to the orthogonal *xyz* axes

- the dimples are aligned to the *abcd* axes (the angle between any of them is 109.5°).

tetrahedron SGSPM quasisphere:

- the bumps are aligned to the *abcd* axes (the angle between any of them is 109.5°.

- the flats are aligned to the orthogonal *xyz* axes.

- the four bumps are connected by riffs corresponding to the tetrahedron edges.

The SGSPM quasisphere shows, that the vector SGSPM spends more time around the bumps and riffs. Then **congregations of same type of tetrahedrons will be preferentially connected by these parts.**

In §12.A.5.3 (Chapter 11 of BSM) it is shown how the number of cycles of SGRM in one cycle of SGSPM vector might be equal to $1/\alpha = 137.036$, where α - is the fine structure constant - a fundamental physical parameter.

D. Quasipentagon

The preferential connection between tetrahedrons leads to the next compact configuration in which five tetrahedrons are congregated into one structure. The new configuration possesses small angular gaps between the tetrahedrons. In fact the tetrahedrons have some limited freedom and the angular gaps may combine into one gap. For simplicity, we may call this structure a **quasipentagon** (QP), because one of its envelope projection is a pentagon. This structure possesses rotational symmetry, so a polar axis could be defined. Fig. 12.2 shows two orthogonal projections of quasipentagon with equally distributed gaps

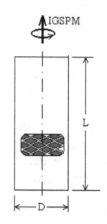

Fig. 11.3

Fig. 12.2. Quasipentagon formed of 5 tetrahedrons

For a single (not integrated) QP, the option of one common gap is more stable. The quasipentagon is still quite compact structure with closed packed primary balls in the embedded tetrahedrons. Therefore, we may accept that the SGSPM of the tetrahedrons are united in a common mode. Then the quasipentagon will posses a own quasisphere of SGSPM. The latter will be distinguishable from this of the tetrahedron mostly by the shape, but the number of precession cycles might be very close to this of the tetrahedron, because the fractional volume of the gaps is small. The SGSPM quasisphere of the quasipentagon exhibits the following features:

- a well defined polar symmetry (using the similarity between the overall shape of the quasipentagon and the oblate spheroid)

- a **spatial anisotropy**: the quasisphere is flatter in the direction of the polar axis

- distributed gaps or single angular gap of 7.355°.

- an azimuthal ununiformity of the quasisphere, additionally affected by the combined angular gap position

The polar symmetry and spatial anisotropy defines the axis of rotating component of SGSPM vector, but the density of the spatial momentum is still stronger around the equatorial plane. The equatorial section of the SGSPM quasisphere (angular density of spatial momentum) is shown in Fig. 12.2.A. for two directions of vector rotation: clockwise and counterclockwize.

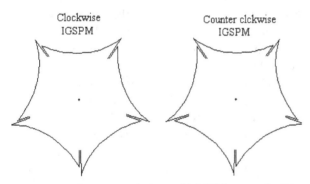

Fig. 12.2. A. Quasisphere of SGSPM vectors in QP

A quasipentagon formed of primary tetrahedron is called a primary quasipentagon.

The angular gap plays important role in the higher order congregations, comprised of aligned pentahedrons. **When number of QPs are integrated into larger structures, the existing gaps provides an unique feature of distortion with ability to preserve the distorted shape.**

E. Column structure of aligned quasipentagons

Now let suppose that a large number of QPs are gathered in a cylindrical column, the length of which is a few time larger than its diameter. The QPs are aligned in a way so their individual polar axes are parallel to the column axis. Such configuration is illustrated in Fig. 12.3.

The angular gaps in the quasipentagons allow the column structure to be additionally right or left hand twisted. In this case, the angular gaps are distributed in a particular order. Once the structure is twisted, this order is held strongly by the SG forces. In such configuration a common mode of SGSPM is obtained with well defined rotational direction. The quasisphere of the common SGSPM, however, is distinguished from the quasisphere of QP, shown in Fig. 12.2.A. It obtains a **stronger gravitational and twisting component along the column axes** and weaker gravitational component in the plane normal to this axis. The azimuthal non-uniformity of the individual QP is smeared in the SGSPM common mode of the structure. It is evident, that **the described structure possesses many of the properties of the prisms introduced in Chapter 2, even when the external shape is not a hextogram prism, but cylindrical.** The most important properties are:

- the **SG spatial anisotropy**

- the **handedness**, defined by the direction of twisting of the embedded lower order QBs.

It is the handedness that provides the left or right handed rotational component of SGSPM vector of this column structure.

In the model of externally twisted prisms, used for simplicity, the twisting properties of SG field has been regarded as contributed by their external envelope. It has been mentioned, however, that the twisting property is a feature cause by the lower level structure of the prism.

F. Quasiball (QB)

The next distinctive congregational structure is a multihedron that could be inscribed in a sphere. We will call such structure a **quasiball (QB)**, for simplicity. There are two options in the growing process for forming a quasiball:

- quasiball option (1) (QB1): comprised of 12 quasipentagons

- quasiball option (2) (QB2): comprised of 20 tetrahedrons

QB1 could be regarded as a regular dodecahedron, in which all flat sides are replaced by QPs.

QB2 could be regarded as a regular polyhedron, but integrated by whole tetrahedrons.

Figure 12.4. shows external and internal sectional views of a mock-up of QB1 type of formation.

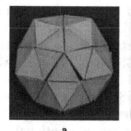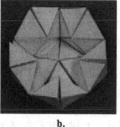

a.   **b.**

Fig. 12.4. Quasiball of type QB1 made of 12 pentahedrals; a. external view; b. internal view of half QB1

(the empty space in the internal view is seen as a pentagon star)

The shown QB1 formation is characterised by the following features:

- it contains exactly 12 pentagons (60 tetrahedrons)

- it contains enclosed empty space

- **it has inevitably a right or left hand twisting**

The twisting property is possible due to the angular gaps of the embedded 12 QPs, which are properly spatially arranged in the QB1 formation. Consequently, **QB1 is a low level 1 bit memory structure of the intrinsic matter**. It is used for memorizing the handedness.

Figure 12.5.shows 3D view of a quasiball of type QB2, formed by 20 THs with equally distributed gaps between the tetrahedrons. The analysis mentioned later, however, indicates that this growing option could be excluded from the possible process of matter self-organization involved in the galactic cycle.

Fig. 12.5. Quasiball of type QB2

G. Congregational order

Quasiballs could congregate in a new tetrahedron, which will belong to next upper order formation from the low level of matter organization. Then the growing process may follow a similar trend. The congregational process from tetrahedron to quasiball could be considered as a common trend, which is repeatable in any upper consecutive order. So we may address the congregational repeatability as congregational order. The primary TH, QP, and QB, for example, are from the first congregational order. In the growing process, the congregational order increases by one for a transition between a QB from lower order and TH from the next upper order.

One specific feature arising from the congregational order is the **handedness preference**. The upper order quasiball will get preferable handedness of the lower order quasiballs. Then for a congregation of p-th order, the handedness is memorised in the quasiballs from all lower orders. From this feature one important conclusion could be made:

• **If the prisms are destructed by number of congregational order, but some quasiballs of lower order are preserved, the memory of the handedness is still preserved.**

The above feature is very important in the process of galaxy recycling, according the scenario provided in this chapter. It gives a reasonable reply of the long time enigma for the theoretical physics: Why the observed galaxies are only of matter?

The meaning of matter and antimatter according to BSM is associated with the two substances of intrinsic matter, so it is some kind of distinctive and specific feature of the real matter. When applied to the proton configuration, it means, that the proton external shell of helical structures, is made of shorter prisms (right handed, if the handedness is correctly assigned) and the electron external shell - from longer prisms (left handed). So if the internal twisting is correctly associated, the meaning of matter and antimatter is defined by the following conditions:

longer prism - lefthanded - negative charge - > matter
longer prism - righthanded - positive charge - > antimatter
where: the sign - > means "leading to".

If our observations identify only a matter and not antimatter, this means that the correct handedness (prisms twisting) is preserved in the recycling phase of the galactic cycle.

H. Propagation of handedness from lower to higher orders

The lowest (I-st) order structure, with a 2 bits memory feature is the primary quasiball (QB). The QB structure is able to obtain a right or left handed twisting due to properly distributed angular gaps in the QPs, from which the QB is integrated. Let us call the axes along which the twisting is made a polar axis of the QB. The QB structure obtains a SGSPM vector aligned with its polar axis. This vector is a common mode of the 12 SGSPM vectors (the SGSPM vectors of the embedded QPs are slightly modified due to the redistributed angular gaps and slidings). The twisted position of the QB is kept stable due to the angular momentum of SGSPM vector.

In the II-nd order TH, the embedded twisted QB's of I-st order have a tendency to align their polar axes according to the quasisphere of SGSPM of the tetrahedron. Then the four apexes of TH obtain the same handedness as the lower order QBs. In the next structure - the QP of II-nd order, the involved five THs provide their handedness to the polar axis of the QP quasisphere. In the next structure - the QB of II-nd order, the handedness of involved 12 QPs causes a twisting of the QB in the same direction. The strength of provided handedness is proportional to the cosine of the angles between polar axes of involved QPs and the polar axis of the QB of II-nd order. The propagation of the handedness in any upper congregational order structure follows a similar way.

The provided concept shows:

(a) The quasiball (QB) is able to be twisted, memorising in such way the handedness. The rotating vibrational mode is described by a SGSPM vector.

(b) The memorised handedness is possible to be propagated from lower to higher congregational orders in the lower level of matter organization.

(c) For a quasiball of any upper order, the strength of the handedness is reinforced by the integrated vibrational modes contributed by the lower level SGSPM vectors.

I. Two growing options in the range of one congregational order

Taking into account the two possible options of QBs structure, we have to investigate which one does fit better to the natural trend growing trend.

    Option (1): TH -> QP -> QB1
    Option (2): TH -> QB2 (see Fig. 12.5)

In option (1), the lower level structures TH and QP are integrated constructively into QB1, while the latter will become embedded in the upper order TH. This option allows any upper order to have the same growing trend.

In option (2) the phase (QP) is excluded, because QB2 could not be integrated by whole number of QPs. Some of the exited QPs have to be disintegrated into THs in order to be integrated in a QB2. Consequently, the option (2) does not have a progressively growing feature as the option (1). The growing trend of option (2) contains a partially reversed process (disintegration of some QPs into THs). From the point of view of the simultaneous process of structure growing, the option (1) will provide much more homogeneity of the growing process, when the structures are segregated in orders, because of lack of reversal growing. Additional considerations about the angular gaps between THs are also in favour of option (1). The angular gaps QPs that are integrated in a QB1 could be kept stable. In QB2 option, some ambiguity exits for the angular gaps, so their position could not be kept so stable. The stability of the angular gaps is important feature for propagation of the handedness in the upper order formations.

From the provided considerations we see, that the growing option (1) is quite more probable.

J. Intrinsic matter quantity of the low level structures

According to defined growing process, the matter quantity of any type of low level structure could be expressed by the number of primary tetrahedrons or primary balls. If assuming that any upper order TH contains N_{PB} number of lower order QBs, then any structure (TH, QP or QB) from any particular order can be expressed by the number of primary balls, N_{PB}, contained in the primary tetrahedron.

Table 12.0 shows the matter quantity of TH, QP and QB from different orders, expressed by N_{PB}.

Intrinsic matter quantity   *Table 12.0*

Congregation order	TH	QP	QB
1-st	N_{PB}	$5N_{PB}$	$60N_{PB}$
2-nd	$N_{PB}(60N_{PB})$	$5N_{PB}(60N_{PB})$	$(60N_{PB})^2$
p-th	$N_{PB}(60N_{PB})^{p-1}$	$5N_{PB}(60N_{PB})^{p-1}$	$(60N_{PB})^{p-1}$

Dozation mechanism:

It is apparent from Table 12.0 that the lower level growing mechanism possesses a unique Dozation properties

Every upper order structure contains an exact number of primary balls. If the growing process contains phases characterized by selection of structures of same type and order, then all of them will contain exactly the same quantity of intrinsic matter. The propagation of the fine structure constant, as characteristic parameter of the PT to any upper order TH and structures is also a signature of the natural dozation mechanism embedded in the growing process. The natural operation of the dozation mechanism is discussed in §12.A.5 to §12.A.8.3 of Chapter 11 of BSM.

Summary:
(a) The three types of low level structures are: tetrahedron, quasipentagon and quasiball (1)
(b) The quasipentagon (QP) provides anisotropy of SGSPM vector
(c) The QB(1) (a Quasiball option 1) possesses a stable twisting that caries the memory of the handedness
(b) Dozation features: Structural formations from one and a same congregational order and type contain precisely equal quantity of intrinsic matter (exact number of primary balls).

12.A.4.3 Super Gravitation and mass-energy balance of the primordial matter

12.A.4.3.1. Stable parameters of the primary tetrahedron

The spectral observations from different galaxies show that the observable Universe is made of common elements. The study of the Universe indicate that the galaxies are made of matter, and not antimatter. It is reasonable to admit that there is a low level structure, which is common for all observed galaxies and caries the necessary 1 bit information about the handedness. Evidently, this memory should be not destroyable in the process of the prism recycling. This process takes place between two consecutive active lives of every galaxy.

The only possible structure capable to carry the memory of handedness is the quasiball QB1. Consequently, some low order quasiballs survive the recycling process. **This automatically means that the primary tetrahedrons and quasipentagons must survive, since they are embedded in the primary QB.**

The surviving of the primary tetrahedron during the galactic recycling, means that it is one and a same for all galaxies and for all of their active lives. Therefore, it is a stable universal structure containing well defined number of primary balls. This conclusion is strongly enforced by the analysis of the vibrational modes of the tetrahedron in §12.A.5.3 indicating that one of the most basic physical parameter is embedded in the tetrahedron - the fine structure constant. <u>In combination with the intrinsic time constant of the primary balls (PB) it may define the condition of a stable PT structure for exact number of PBs</u>

Accepting that the Primary Ball is the most fundamental particle in the Universe, the provided so far conclusion will be valid universally for all galaxies.

$$N_{PB} = const \text{ - number of primary balls} \quad (12.1)$$
$$\text{contained in the primary tetrahedron}$$

The above conclusion provides us with a reliable reference point for further analysis. From a pure geometrical considerations, the number of PBs in the primary QP is $5N_{PB}$ and in the primary QB: $60N_{PB}$. If assuming that a TH of secondary order possesses a same number of QBs as the N_{PB}, then one quasiball of second order should contain $60^2 N_{PB}^2$, and one quasiball from order p should have $(60 N_{PB})^p$ primary quasiballs. Assuming that the same trend is conserved in the upper order congregation, we may reference the parameters of any upper order structure to the parameters of the primary tetrahedron.

12.A.4.3.2 Mean intrinsic matter density of a structure

We may define the parameter intrinsic volume density of the lower level structures as a ratio between the integral volume of embedded PBs and the overall volume of the structure. The integral volume is the volume of the PB multiplied by their number. The three different types of structures within one congregational order posses different intrinsic matter density. For structures like QP and especially QB of higher order, the PBs are not uniformly distributed in the envelope volume, so a mean intrinsic matter volume density is more appropriate parameter. If normalising this parameter to the intrinsic mass of the primary ball we arrive to the following definition:

The mean (intrinsic) matter density of any structure is a ratio between the total intrinsic mass and the external envelope volume of the structure.

In the following expression the mean matter density is referenced to the matter density of the tetrahedron. It could be either the primary one or one from some upper congregational order.

$$\rho_{ST} = \frac{N_T m_T}{V_{env}} = \frac{N_T V_T \rho_T}{V_{env}} \qquad (12.2)$$

where: ρ_{ST} - is mean matter density of the structure; V_T and ρ_T are respectively the tetrahedron volume and the matter density; V_{env} - is the external envelope volume of the structure; N_T - is a total number of tetrahedrons in the structure

It is more convenient to normalize the structure mean matter density to the intrinsic matter density of the tetrahedron. Then we get a dimensionless parameter, called a **structure intrinsic matter number density**. It is given by the equation:

$$\rho_{ST}/\rho_T = N_T \frac{V_T}{V_{env}} \qquad (12.3)$$

If considering that the number of QBs in any upper order QT is equal to N_{PB} (number of primary balls in the primary tetrahedron), then Eq. (12.3) is valid for structure of any order. The volume ratio could be easily calculated if expressing the volume of QP and QB(1) (the envelope volume enclosed by the structure) as normalized to the volume of TH. Using this consideration, the intrinsic matter density is estimated for TH QP and QB from one congregational order as an intrinsic matter number density normalised to the intrinsic matter density of the TH. The results are shown in Table 12.1.

Normalized intrinsic matter number density   *Table 12.1*

structure type	ρ_{ST}/ρ_T
tetrahedron	1
quasipentagon	0.98
quasiball (1)	0.1326

Both parameters: the structure matter density and the number density decrease slightly between the quasiball of lower order and the tetrahedron of the next upper order due to the empty spaces between the quasiballs. If the growing trend is preserved, they increase by a constant factor, which is determined by N_{PB} (number of primary balls in the primary tetrahedron).

Note: It is possible to express the structure matter density of any formation of any congregational order by the matter density of the primary balls if knowing N_{PB} (number of primary balls in the primary tetrahedron).

12.A.4.3.3 Energy well and energy balance

According to the rule P3 (see §12.A.4), the primary tetrahedron could handle a finite amount of energy or it could posses an **energy well.** Let us suppose that the energy of this structure is increased by some external energy, which can be a result of some external pressure. If the obtained total energy exceeds the energy well, the structure has to be freed from this energy. If such structures are quite a lot (as a sea of similar structures) without freedom of motion (as energy interaction) the excess energy might be transfer in new vibrational modes for structures of QPs and QBs. In other words, the excess energy will cause a growing process of type (TH->QP->QB). We see that every type of structure has own energy well. The energy well of the upper type and order structure is larger due to added common mode of vibrations.

For a particular type of structure, the energy well may be filled or not filled with energy. The total energy of the structure is contributed by two types of energy:

    - internal vibrational energy

    - external interactions in form of SG energy

It is apparent that the energy balance for a given structure should fulfil the condition:

$$E_V + E_{SG} \leq E_W \qquad (12.4)$$

where: E_V - is the internal vibrational energy, E_{SG} - is the Super gravitational energy, E_W - is the energy well

While the energy E_V is a constant for one type of structures, the energy E_{SG} depends on mutual structure positions and distances. **Therefore, the condition for existing of SG interactions between same type of structures is the sum of the two types of energies (E_V and E_{SG}) to be smaller than the energy well for that type of structure.**

It is apparent that structures of same type and order occupying a common volume will not only interact by SG forces, but a tendency of equalization of their individual total energies ($E_V + E_{SG}$) will exist. Then the condition (12.4) may be fulfilled for all individual structures, making their common existence more stable.

From the above considerations the following conclusions could be made:

    (a) Every low level structure is characterized by energy well

    (b) The total amount of energy for a stable structure is lower, than the energy well.

    (c) Structures of the same type and order occupying a common volume obtain more stable existence.

    (b) If a sea of low level structures of same type obtains additional energy and the structure energy well is exceeded, a growing process of structural formation of type (TH->QP->QB) is invoked.

    (d) The energy of the SG field is a difference between the energy well and the available energy

The conclusion (d) is quite interesting, since it allows a temporally existence of a negative (repulsive) SG field. Such effect is quite important and will be discussed in §12.A.8.4.

12.A.4.3.4. Mean energy density of the structure of particular type and order

In order to compare the energy wells between structures of different types and orders, it is useful to define a parameter of mean energy density of the structure.

The mean energy density of a structure from a given type and order is equal to the mean energy divided on the volume enclosed by its external envelope surface.

According to the rule P3 and the analysis in the previous paragraph, it is evident that the mean energy density is proportional to the mean matter density.

$$E_\rho \sim \rho_{ST} \qquad (12.5)$$

This relation allows to obtain the energetic parameters using the geometrical parameters of the structure. It is more convenient to estimate the energy of a particular structure as normalized to the energy of some lower order structure, for example, the TH.

12.A.4.3.5. Summary

- **The main characteristic parameters of the low level structures are the structure type and the congregational order**
- **Any upper order structure above the primary tetrahedron is comprised by lower order structures with preserved spatial configuration and properties.**
- **The intrinsic mass distribution is well defined by the type of the structure and its congregational order.**
- **The mean energy density is proportional to the mean intrinsic mass density. Therefore, it could be estimated by pure geometrical considerations. This allows expressing the energy of a given structure by normalized energy units using some lower level structure for talon.**
- **Many parameters of the upper level structures could be normalized to the primary tetrahedron. The latter is a stable structure, capable to survive the recycling process of the prisms.**

- **It is reasonable to look for embedded features of the fine structure constant into the primary tetrahedron** (this is presented in §12.A.5.3.).

12.A.4.4. Relation between the dynamical properties of the lower level structures and some parameters of CL space

Let us consider the three types of formations from one congregational order: a tetrahedron (TH), a quasipentagon (QP) and a quasibal (QB). The mean mass densities, ρ_{ST}, for the first two structures are very close, so the frequency of their SGRM and SGSPM vectors must also be close. Consequently, the SGSPM quasisphere of the QP could be regarded as combined of five tetrahedron quasispheres with vectors possessing almost the same frequencies. The case of the QB(1), which is formed of 12 QPs, however, is different. The parameter ρ_{ST} is reduced. At the same time, the preservation of integrity of the included QPs and THs means a preservation of the frequency of their SGRM vectors. The new feature that will appear in the QB is a common mode of the SGSPM vectors of the involved QPs. <u>The frequency of this common mode could be regarded as obtained by division of the frequency of the SGRM vector by integer (using the analogy with the CL space vectors NRM and SPM).</u> The same feature are propagated for higher congregational order, so we arrive to the following conclusions:

(a) SGRM vector is a feature of the primary tetrahedron and its parameters are preserved in the higher congregational orders

(b) The frequency of SGSPM vector of any structure has well defined dependence on the congregational order and the type of the structure.

The conclusions (a) and (b) might put a light on the physical meaning of one well known theoretical parameter - the Plank's time. In Chapter 2 we have tried to find the relation between the three important physical parameters: the Plank's time (or frequency), the CL node frequency and the Competing frequency, assigning to them three consecutive levels of the matter organization. If putting the Planck constant frequency and the CL node frequency as two consecutive levels of the matter organization, the plot in *ln* scale shows a significant deviation from a robust line. From the provided analysis about the SGRM and SGSPM it becomes apparent that one intermediate level should exists between the Planck's time and the resonance period (frequency) of the CL node. Then the relation between the characteristic frequencies and the assigned level of matter organization are given in Table. 12.1.

Levels of matter organization				Table 12.1.
Level x	Time (sec)	Frequency ν (Hz)	$\ln(\nu)$	Type of oscillation
0	5.39E-44	1.855E43	99.629	(Plank's)
1				SGSPM
2	9.152E-30	1.0926E29	66.86	CL node resonance
3	8.093E-21	1.236E20	46.26	SPM; electron proper frequency

Level "0" is assigned to the Planck's frequency, f_{pl}, (the reciprocal of the Planck's time). The latter is given by the expression:

$$f_{pl} = \sqrt{\frac{2\pi c^5}{Gh}} = 1.855 \times 10^{43} \ (Hz) \qquad [(2.67)]$$

where: G - is the gravitational constant, h - is the Plank constant , c - is the light velocity.

The natural logarithm of frequency versus the assigned level of matter organization is shown in Fig. 12.6. .

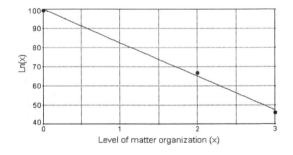

Fig. 12.6.

The three known frequencies now are better aligned in a line. According to the BSM concept, the new identified level 1 may correspond to the SGSPM vector frequency of the prism. Then the following relations are inferred:

    12-13

Level "0": corresponds to the **IGNRM frequency of the primary tetrahedron**

Level "1": corresponds to the **SGSPM frequency of the prism, defined by the highest order quasipentagon embedded in the internal structure of the prism.**

where: **IGSRM** vector (Super Gravitational Structure Resonance Momentum) is similar as NRM vector for CL node; **SGSPM** vector (Super Gravitational Spacial Precession Momentum) is similar as SPM vector for CL space.

Note: Using a fitting by a robust line may allow to calculate the SGSPM frequency for the level 1, but its value might be very approximate (involving a large error).

It is reasonable to accept that all QPs from one prism have synchronized common mode of SGSPM vector. Then if a structure formed of aligned prisms with touching ends exists, the vector SGSPM will propagate between two prisms for one full cycle. This is analogous to the light propagation in CL space (a node distance divided per one resonance cycle). Then the SG field will be propagated with a velocity according to the expression:

$$\frac{\nu_{SGSPM}}{\nu_R} c \qquad (12.5.a)$$

where: ν_{SGSPM} - is the SGSPM frequency of the prism, c - is the velocity of light; $\nu_R \approx 1.09 \times 10^{29}$ (Hz) - is the resonance frequency of a normal CL node (NRM frequency).

Let us take a quite conservative estimate for a minimum value of ν_{SGSPM} using the plot of Fig. 12.6. and test the value of 75 in ln scale, which corresponds to $\nu_{SGSPM} = 3.73 \times 10^{32}$. Then the expression (12.5.a) provides a superluminal velocity of $\sim 1 \times 10^{12}$ (m/s) even for this conservative value. (extensive analyses of cosmological phenomena some of which is presented in §12.B.5.3) indicates that the value of ν_{SGSPM} must be higher).

12.A.5. Formation of upper order congregations in the surface region of the bulk matter.

12.A.5.1. Energy balance between structures of same type but different congregational order

Let us consider the growing option (1), discussed in §12.A.4.2.1.H.

If the primary tetrahedron contains N_{PB} number of primary balls, then the primary QP contains $5N_{PB}$ primary balls and primary quasiball contains $60N_{PB}$. If the same growing trend is preserved in the upper congregational orders, a structure of p order will contain the following number of primary balls:

$$(5N_{PB})^p \quad \text{- for QP of p-th order} \qquad (12.5)$$

$$(60N_{PB})^p \quad \text{- for QB of p-th order}$$

Let us regard the quasiball shape as spherical, for a simplicity. The distance between the QBs is proportional to their radius. The volume of the upper order QBs will grow proportionally with the order number. Then the distance between QBs will grow by cubic root of the volume, while the SG forces are inverse proportional to the cube of the distance. **Consequently, the growing process will not involve consumption of SG energy.** For explanation of this phenomena, illustrations of the relative positions between the bulk matter and two consecutive orders of the same type are shown in Fig12.7.

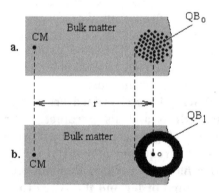

Fig. 12.7. Geometrical comparison between structures of a same type (QB) but of different congregational orders, CM - centre of mass of the bulk intrinsic matter

In case **a.**, lower order quasiballs are shown as QB_0. In case **b.** one quasiball of upper order is shown as QB_1. If the growing trend between the QB_0 and a tetrahedron of upper order is the same as between the primary balls and the primary tetrahedron, then QB_1 will contain $(60N_{PB})$ numbers of QB_0. Therefore, the total intrinsic masses in case a.

and b. are equal. In Fig. 12.7 the position of the geometrical centre of QB1 is indicated by an empty circle. The following features are necessary to be emphasized:

(a) QB_1 contains a large internal empty space

(b) All QB_0 structures are immersed in the bulk matter, near the surface. that could be of lower order congregations. They may contain internal empty spaces from lower order QBs, but the equivalent volume of empty space is enclosed inside of the single QB_1 (see Fig. 12.4).

(c) For QB_1 structures formed at the edge of the bulk matter, the equivalent CM (centre of mass) of QB_1 (shown by a black point) is displaced from the geometrical centre of the structure.

While the feature (c) is not valid for Newton's gravitation, it is very important for the Super Gravitation because the SG forces are inverse proportional to the cub of the distance. (The gravitational potential for the Newton's law is inverse proportional to the distance, while in SG case it is inverse proportional to the square of the distance).

As a result of the above considerations, it may appear that the distance, r, between CM of the bulk matter and the equivalent CMs in cases **a**. and **b**. could be one and a same. **At such conditions, the formation of QB$_1$ structure from QB$_0$ structures will not involve work against SG forces between these formations and the bulk matter.**

The formation of upper from lower order structures, below the prism's level, could be denoted as a **low level growing process**. The therm "low level" is used to distinguish this range of matter organization from the range above the prism's level. More often the therm **growing process** could be only used, knowing that it is related to the low level of matter organization.

The provided example is simplified. In the real case some small energy could be involved. If the concept is true, however, the following phenomena could be formulated:

(A) The formation of upper order structures in the growing process does not involve significant consumption of SG energy.

The above phenomena could be approximately valid due to the simplification of the analysis, but it is of great importance in the process of formation of higher order congregations. It allows formation of shell with higher order QBs without

significant energy expense. At the same time the growing shell will be able to preserve a big portion of the vibrational energy for itself (borrowed from the bulk matter energy).

The energy balance analysis described for the phenomena (A) could be applied not only for QB's of different order, but also for other structures of same types and different order. (THs and QPs).

Note: The same amount of SG energy is referenced to the same number of primary structures.

In the next paragraph it will be shown, that there is a tendency for uniformity of the upper order of the quasiballs.

Conclusions for the growing process within one congregation order.

Considering that the growing option (1) is only valid for the galactic recycling process and accepting the primary tetrahedron as a lowest order structural formation of the growing process, we may summarize:

(a) The primary tetrahedron contains N_{PB} primary balls

(b) The number of quasiballs in any upper order of tetrahedrons is equal also to N_{PB}.

(c) The growing option within the range of one congregational order is: TH -> QP -> QB.

12.A.5.2. Frequency dependence of the SGRM and SGSPM vectors on the type and congregational order of the structure

The inertial properties for structure of intrinsic matter has been discussed in Chapter 2 of BSM. The mass corresponding to such interactions is called intrinsic inertial mass. In the following analysis an assumption is made that the classical equation for the proper frequency is valid also for the intrinsic inertial masses. If this assumption is valid, the vibrational frequency, of the primary quasipentagon, $\nu_{QP}^{(1)}$, can be expressed by the equation:

$$\nu_{QP}^{(1)} = \frac{1}{2\pi}\sqrt{\frac{k}{5N_{PB}m_o}} \qquad (12.6)$$

where: the superscript index (1) means a first congregational order, m_o - is the intrinsic inertial mass of the primary ball, N_{pb} is the number of primary balls in the primary tetrahedron and k - is a force constant between the primary balls.

The parameter k is determined from the SG forces between the primary balls in the primary tetrahedron. For the primary quasipentagon it will be approximately the same, because these two structures have very close intrinsic matter density. Let us assume that the parameter k does not depend on the type and order of the formation. This assumption is based on the consideration that the structure of the primary QP is preserved in all higher order structures. The matter quantity expressed by the intrinsic mass, however, is dependable on the structure type and order.

Let us determine the frequency of the common vibrational mode for a QP of some upper congregational order p. All lower order structures included in this QP preserve their configuration. So the same parameter k between primary balls should be valid (according to the above made assumption), while the mass of the QP of order p could be estimated by the matter quantity given by Table 12.0.

Having in mind Eqs, (12.5) and (12.6) the common mode frequency of the QP of p-th congregational order could be expressed as:

$$\nu_{QP}^{(p)} = \frac{1}{2\pi}\sqrt{\frac{k}{5N_{PB}m_o(60N_{PB})^{p-1}}} \qquad (12.7)$$

We see, that the common mode frequency falls pretty fast with the order number p. Having in mind the quantum features described by the SGSPM quasisphere, the obtained common mode frequency might be considered as a proper frequency of the SGSPM vector.

Making a ratio between $\nu_{QP}^{(1)}$ and $\nu_{QP}^{(p)}$ we get:

$$\frac{\nu_{QP}^{(1)}}{\nu_{QP}^{(p)}} \approx \sqrt{(60N_{PB})^{p-1}} \qquad 12.8)$$

The mass of the primary PQ is five times the mass of primary TH, but the mass to volume ratio (neglecting the small gaps between TH in the QB structure) is approximately the same. Then parameter k for TH and QP is also the same and the expression of the frequency ratio between IGSRM vectors of the primary TH and the SGSPM of highest order QP in the prism will become:

$$\frac{\nu_{TH}^{(1)}}{\nu_{QP}^{(p)}} \approx \sqrt{5(60N_{PB})^{p-1}} \qquad (12.9)$$

Note (1): The factor of 5 refers to the QP's SGSPM, that is a common mode of IGSRM of the included TH's. The real factor might be slightly lower than 5 due to the angular gap in the QP. So SGSPM of TH is approximately equal to 5 times SGSPM of QP (within the same congregational order).

According to the considerations in 12.A.4.4, the Planck's time regarded as a period of IGSRM vector, may correspond to $\nu_{TH}^{(1)}$ or $\nu_{QP}^{(1)}$, while $\nu_{QP}^{(p)}$ could be the SGSPM frequency of the quasipentagons from which the prisms are made.

Let us analyse how the SGRM period changes during the growing process: tetrahedron - quasipentagon - quasiball - upper level tetrahedron.

The SGRM is defined for the primary TH. The period of SGRM may be slightly decreased in the primary quasipentagon, due to the close accumulation of intrinsic mass and obtaining a different shape of SGSPM quasisphere. In a growing process from a QP to a QB within one congregational order the period of SGRM should not be significantly affected, because QPs are connected by small volume sections. In the growing process between a QB and a upper order TH the SGRM period could not be affected significantly because the mean matter density, ρ_{ST}, is approximately the same as of QB. The change of SGRM period in the upper order growing will be smaller and smaller, following a continuously decreasing step function with progressively smaller steps. Consequently:

(a) The change of SGRM of the growing structures is a decreasing steplike function with progressively decreasing steps.

(b) The step change have two periodical progressions:

    - between congregational orders
    - between different type of structures of same order

12.A.5.3. Hypothesis of embedded fine structure constant in the lower level structures of matter organization

Considerations related to the concept of embedded fine structure constant

From the previous chapters it was shown that the fine structure constant is embedded not only in the electron structure but also in its dynamical properties in CL space and many other interactions between the elementary particles and the CL space (for instance: in the quantum motion of the electron (positron); in the quantum orbits conditions for atoms and molecules; in the atomic and molecular spectra). In Chapter 10 of BSM it was shown that α is also involved in the inertial interactions between the elementary particles and the CL space (Eqs (10.36), (10.39), (10.39a). It was even found that the signature of α is involved in the inertial interaction balance of the solar system in our home galaxy - the Milky way (see §10.6.4, Chapter 10 of BSM).

From the analysis of the lower level of matter organization and the concept of the galactic cycle (provided later), it becomes apparent that α is a common fundamental physical parameter for all observable galaxies. Consequently, α is a parameter of very low level structure and its signature is preserved even in the galactic recycling process (discussed later in this chapter). The basic repeatable structure which possesses SGSPM vector is the primary tetrahedron, so it is reasonable to look for a possible signature of α in this structure. One very basic physical parameter of the primary tetrahedron is the number of primary balls. Keeping in mind that all the shells of the tetrahedron should be completed, a simple rule follows that a strong relation must exist between the number of balls along the edge and the total number of balls. For example, if the edge number of balls, N_{edge}, corresponds to the set: 10, 11, 12, 13, 14, 15, 16, 17, 19 and so on, then the total number of balls, N_{tot}, should be respectively: 220, 286, 364, 455, 560, 680, 816, 969, 1140, 1330 and so on.

The experimental value of the fine structure constant is measured with very high accuracy. Finding a theoretical derivation of this fundamental parameter, however, have been one of the most difficult problems in mathematical physics. (see J. G. Gilson)[1]. In fact number of empirical formulae have been suggested, but without understanding what kind of physical mechanism is behind them. One of these formulae (shown as Eq. (F1)) gives a value, which is very close to the measured one recommended by CODATA 98 (if the two involved integer parameters have value: $n_1 = 137$ and $n_2 = 29$):

$$\alpha = \frac{n_2}{\pi}\cos(\pi/n_1)\tan(\pi/(n_1 n_2)) = 7.2973525 \times 10^{-3}$$

$$(12.12)$$

$$\alpha = 7.2973525 \times 10^{-3} \quad \text{(CODATA 98)}$$

Recently another simple expression has been proposed by I. Gorelik[2], as a system of two simple equations:

$$n + k = 1/\alpha$$
$$k(n + k) = \pi^2/2$$

I. Gorelik, mentions the derived method but does not provide any association with a possible physical mechanism. The BSM analysis found an excellent association with the oscillating properties of the lower level structures described by SGSPM vector.

The common mode oscillations of the primary balls embedded in the primary tetrahedron could be described by two vectors: SGRM and SGSPM (see §12.A.4.2.1 C). Making analogy with the CL node dynamics, the primary tetrahedron has the same two set of axes: *abcd* (non-orthogonal) and *xyz* (orthogonal). Then to analyze the dynamics of oscillations of the primary tetrahedron, we may use a concept similar as the CL node dynamics (presented in "Brief intro to BSM..." and Chapter 2 of BSM), but instead of NRM we have SGRM and instead of SPM we have SGSPM. One major difference is that the primary tetrahedron has relatively large intrinsic matter density for the volume it encloses, while the CL node has much smaller intrinsic matter density for the volume it encloses. For this reason we must expect much smaller number of SGRM cycles in one SGSPM cycle in comparison to the CL node dynamics (NRM cycles in one SPM cycles). Due to the two sets of axis, the trace of SGRM will not be circular and will note lie in a plane. We may simplify the analysis if replacing the real trace of SGRM with an equivalent elliptical trace having a dipole moment equal to the real precession moment of SGRM vector for one cycle.

Figure 12.8 illustrates the dynamical behavior of the SGRM vector.

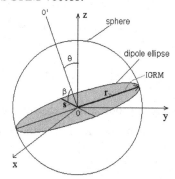

Fig. 12.8. Spatially precessing dipole momentum expressed by SGRM vector.

The origin of the SGRM vector is always fixed at the origin O of a coordinate system *xyz*, while having a freedom to rotate in the spherical space volume. Due to the different stiffness along the two sets of axes xyz and abcd (not shown in this figure), the vector SGRM will perform a helical rotational motion with a very small but constant helical step. This means that after one cycle its tip will not pass through the same point but through a point closer to the previous one, so the distance between them is much smaller than the trace of the vector's tip. We may call this a quasicycle. After many quasicycles, however, the tip of the SPM vector will pass exactly through the same initial point (arbitrary selected). This cycle we may call a full cycle. Then the full cycle will contain many quasicycles, but their number may not be an integer. In this kind of motion, the tip of the SGRM vector will circumscribe a trace, which lies in a spherical surface. It is apparent that for one quasicycle, the trace of the vector SGRM will not lie in a plane, but we may consider an equivalent plane, defined by the condition that the average distances between the points of the vector's tip (through equal time intervals) and this plane is a zero. This will simplify the analysis and will allow us to define the following parameters:

-selection of an initial reference point
-definition of a dipole momentum in a plane
-definition of the step between two neighboring equivalent planes (corresponding to the helical step of the SGRM vector) as an angle between them.

It is apparent that the dipole momentum of SGRM vector could be expressed by an ellipse lying in the equivalent plane. We may call it a "dipole ellipse". The rotational axis OO` will be perpendicular to the major semiaxis r of the dipole ellipse, but not perpendicular to the minor semiaxis. In other words the plane of the dipole ellipse will be rotating with a small pitch angle   of $(\pi/2 - \beta)$ defined by the helical motion of the SGRM vector. Then for one quasicycle, the dipole ellipse will sweep a volume of an oblate spheroid with a major semiaxis r and a minor semiaxis defined by the product: $s\cos\beta$.

In every quasicycle, the dipole ellipse will sweep the same volume, while the initial angle θ (arbitrary selected) will change with one and a same step. This angle is shown for reference only. It could be defined for any one of the orthogonal axes. The rate of θ change will define the number of completed quasicycles within one full cycle. The latter, however, may not contain an exact number of quasicycles but a whole number plus a fraction, so we have:

Full cycle = n + k

Where: n - is the number of completed quasicycles contained in one full cycle, k - is a fraction of a quasicycle

Our goal is to express the fraction parameter (k) as a function of the whole number (n) using the defined model. We will derive expression using the relation between the volume of the circumscribed sphere and the volume of the oblate spheroid.

The volume of the circumscribed sphere is:

$$V_{SP} = (4/3)\pi r^3$$

If the full cycle contains a large number of quasicycles, then: $\cos\beta \ll 1$. We may associate this with the fractional part of $1/\alpha$, so we may write: $\cos\beta = k$. Then, the volume of the oblate spheroid is: $V_{OS} = (4/3)\pi r^3 s\cos\beta = V_{OS} = (4/3)\pi r^3 sk$

The tip of the SPM vector is associated with the point of interception of the dipole ellipse with the major semiaxis. This means that for a full cycle of the SGRM vector, the volume of the oblate spheroid swept by the rotating dipole ellipse will be twice the volume of the circumscribed sphere, or we have $V_{OS} = 2V_{SP}$. The expression corresponding to this is:

$$(4/3)\pi r^3 sk(n+k) = 2(4/3)\pi r^3$$

Multiplying both sides by $1/r$ and using a normalized parameter $s_r = s/r$, we arrive to:

$$0.5 s_r k(n+k) = 1 \qquad (12.13)$$

Now we may look for a possible reasonable value of the product $(s_r k)$, while trying to relate the parameter s_r to π. Knowing that $(n+k)$ is equal to $1/\alpha$, we should have $0.5 s_r k \approx \alpha$. For this purpose we will use the experimental value of alpha given by CODATA 98. Then the normalized minor semiaxis of the oblate spheroid should be close to the value: $s_r = 0.40552$. This value is very close to:

$$\left(\frac{1}{\pi/2}\right)^2 = \frac{4}{\pi^2} = 0.40548$$. The difference between them

is only 0.59%, so we may accept:

$$\sqrt{s_r} = (2\pi) \text{ or } s_r = 4/\pi^2 \qquad (12.13.a)$$

The idea to relate the parameter s_r to π is reasonable if examining the more accurate formula (12.12), where π participates. It complies also with the Feynman's idea[1] that alpha should be somehow connected to the numbers e or π. Substituting (12.13) in (12.13.a) leads to Eq. (12.13.b). It is a quadratic equation. The root leading to a correct expression for alpha is:

$$k = -0.5[(n^2 + 2\pi^2)^{1/2} + n] \qquad (12.13.b)$$

Using the module of the solution (10) and combining with the expression $(n+k) = 1/\alpha$ we get the explicit theoretical expression for the fine structure constant (denoted as α_c)

$$\alpha_c = 2/[(n^2 + 2\pi^2)^{1/2} + n] = 7.29735194 \times 10^{-3} \qquad (12.14)$$

Conclusion: In the obtained equation for theoretical value of the fine structure constant only one number must be selected: n.

Equation (12.14) provides a pretty accurate value for alpha, if the accuracy of its experimental value exceeds some level. This requirement is overly satisfied that is evident from the plot illustrated by Fig. 12.9, according to which we can use

n = 137 with a high level of confidence.

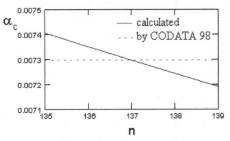

Fig. 12.9. Plot of the fine structure constant, by the theoretical Eq. (11) (blue line) and by CODATA 98 value (red dashed line). The experimental accuracy better than 0.7% allows to use only n =137 for which a quite accurate value for the fine structure constant is obtainable.

Discussion:

The suggested method provides a simplified physical picture of the common oscillating mode in the primary tetrahedron, whose signature is the fine structure constant. Evidently, the fine structure constant is defined by the intrinsic features of the primary ball: an intrinsic time constant and a level of deformation. These two parameter are constant for primary balls of both types of intrinsic matter.

References:
1. J. G. Gilson, Fine structure constant, http://www.btinternet.com/~ugah174
2. I. Gorelik, Formula for fine structure constant, www.geocities.com/Area51/Nebula/3735/fine.html

12.A.6. Super Gravitational Constant

12.A.6.1. Difference between SG constants G_{os} and G_{od}.

The Super Gravitational constant G_0 has been introduced and partially discussed in Chapter 2 of BSM. Now keeping in mind the oscillation properties of the primary balls and TH, QP and QB formations of any congregational order it becomes apparent that SG field could be defined by the interaction energy E_{IG} between two structures of a same type placed in a void space at unit distance. It is reasonable to chose a stable length parameter for a unit distance. For this reason we must be able to scale the chosen unit distance to the dimension of the primary ball. This distance is preserved in all upper order structures. Having in mind the robustness of the formations from intrinsic matter including the prisms it is apparent that the prism length could be also used as unite length. (When analysing phenomena in CL space, however, we consider the internode distance as an unite length keeping in mind that it is only weakly dependable on the mass of a large body (a General Relativity effect).

From the presented in §12.A.4.2 and §12.A.5 scenario of the prisms formation it becomes apparent that the prism is formed of aligned quasipentagons of one and a same order (and from one type of intrinsic matter substance). The analysis made for the prisms should be valid for the structures of the same congregational order. Because we have two types of intrinsic matter substances, we must consider two types of SG constants:

G_{os} - is the SG constant between structures of intrinsic matter substance of same type

G_{od} - is the SG constant between intrinsic matter substances of different type.

The volume ratio between the primary balls from the two substances should be equal to the prisms volume ratio $v_1/v_2 = 27/8$. Then the radius ratio of the primary balls is $r_1/r_2 = 3/2$. Even without knowing the common estimated mass density and the force constant, it is evident that the SGRM vectors of the primary tetrahedrons of both substances will have **different periods** (estimated by a common time base). SGSPM vectors (for TH, QP or QB) of both substances will have a period multiple of SGRM period (for the primary ball).

Let us accept that a complete SG energy exchange between the spatially separated structures (in a void space) is achieved for a finite time, defined by the time constant of the intrinsic matter, t_{IM}. While it is intrinsically small and could be near the range of the Planck's time, it is quite important. As discussed in Chapter 2 of BSM, without such constant the energy conservation could not be defined, while all analysis and observations show that this principle is an iron rule.

Let us associate the intrinsic time constant, t_{IM}, with the cycle of the SGSPM frequency. From the analysis of the SGSPM frequency in §12.A.5.2 we found that it decreases as a step like function in the growing process of lower level of matter organization. Consequently, the secondary time base will be respectively an increasing step-like function.

Let us consider the two types of prisms which play the role of basic elements of CL space. While keeping in mind that the prism has anisotropic SG field, let us focusing on its axial SG field. Since the prisms are formed of aligned QPs, the combined SGSPM of the QPs will define the SGSPM vector of the prism.

Let us consider two cases of spatially separated prisms in a void space.

(1) Case: The prisms are of same intrinsic matter and handedness

(2) Case: The prisms are of different intrinsic matter and handedness

Then the interaction SG energy could be presented as integration of SGRM cycles for the time duration of the SGSPM cycle.

The complex trace of the SGRM vector for a full SGSPM cycle is difficult to be expressed mathematically. We may simplify the problem, by replacing this two vectors with a simplified model of linear interaction between two slightly different frequencies and estimate the product of their interaction. Then the analysed problem could be regarded as the energy transfer between two PLL (phase looked loop) oscillators. The case (1) corresponds to two PLL oscillators with equal proper frequencies ($a = 1$). They need very short time interval in order to get in synchronization and exchange energy. The case (2) corresponds to two PLL oscillators with slightly different proper frequencies ($a \neq 1$). They need a significantly larger time interval and at different moments the biting effect could be constructive or destructive. The associated energy exchange for this simplified model is given by the expression.

$$E_{SG} = \int_0^\theta \sin(2\pi\theta)\sin(a2\pi\theta)d\theta \qquad (12.15)$$

$a = f_1/f_2$ - is the frequency ratio associated to the ratio of SGRM frequencies of the two types of prisms.

For case (1) we have $a = 1$, while for case (2) this factor is $a \neq 1$.

Figure 12.12 shows a plot of E_{IG} for case (1) - the black line ($a = 1$), and for case (2) at three values of a parameter: 0.6, 0.66 and 0.73.

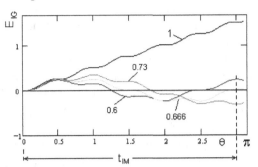

Fig. 12.12. Plot of E_{SG} for different value of a parameter.

While for case (1) the interaction energy is always positive, for the case (2) the interaction energy for elapsed time of t_{IM} (corresponding to SGSPM) could be positive, zero or negative, depending on the value of the parameter a. **When considering that the two prisms are part of a lattice structure from the aspect of the energy conservation principles, the negative interaction energy will cause an increase of the CL node distance until the total energy balance is restored.**

It is interesting to analyse the expression by replacing the factor of 2π with 4π and integrate up to 4π. This may simulate better the SGRM, which in fact rotates in 4π spatial angle. While it is still not equal to the real model, it shows that the plot for $a = 1$ drops to zero for some particular values of θ. This leads to one important conclusion: In some particular arrangement of the prisms, the SG attraction could be lost for a finite time. Evidently such effect could appear only for extremely short time interval, because the SGRM and SGSPM frequencies of the prisms are slightly influenced from their common positions and distances, especially in a lattice configuration.

Despite the simplicity of the model, it allows to make some important **conclusions**.

A. The attraction between the same type of prisms is always stronger, but in some particular structural arrangement it may be decreased significantly

B. In the CL structure where the prisms are arranged alternatively in nodes, the node distance is supported automatically, due to the slight dependence of the SGRM and SGSPM frequency on the CL node distance.

C. The influence of the interaction energy allows deeper understanding the physical meaning of the two SG constants, G_{os} and G_{od}

The conclusions A. and B.put a light on the stability of the LC structure of the physical vacuum and the structural stability of the elementary particles. The conclusion C. puts a light on the non-mixability of the low level formations from the two different substances. This is important feature in the phases of matter evolution during the hidden phases of the galactic cycle.

Note: The evaluation of both SG constants is only possible by the parameters of CL space, so the asymmetrical factor a_{sym} should be taken into account when necessary (see §6.9.4.2 in Chapter 6 of BSM, related to a_{sym}).

12.A.6.2. Intrinsic time constants of the prism.

The provided concept of SG energy exchange between low level matter formations shows, that the SGSPM cycle defines one important feature of the prisms - their **intrinsic time constant**.

When the concept of SG interaction was introduced in Chapter 2 of BSM, it was emphasized, that the intrinsic matter of the prisms should posses intrinsic time constant. Only in such way a finite time could be assign to any interaction process. This is a very important requirement for assuring a finite energy exchange in any kind of interactions, or in other words complying to the energy conservation principle. Having in mind the prism's internal structure of aligned QPs, as shown in fig. 12.3, we arrive to the following conclusion:

The intrinsic time constant of the prism is defined by the SGSPM period of the uppermost quasipentagons, which are embedded in the internal prism's structure.

It has been mentioned in the previous chapters (and used in analysis) that the SG field of the prism exhibits an axial anisotropy. The above defined time constant is valid for the prism's axis direction. In a direction normal to the prism's axis another time constant might be important defining the radial SG field of the prism. This time constant may have a cirality feature, so it could be related to the period of SGSPM vector of the most upper order quasiballs (possessing twisting that defines the handedness) embedded in the QPs from which the prism is built. Having in mind that the QP contains a large number of lower order QBs the radial SG field time constant might be much shorter than the axial SG field one. In such case one important feature of the prisms could be explained: <u>why the prisms of CL nodes do not stack together. They may stack only if the prisms are axial aligned and closer below some critical distance.</u> Such condition may appear only in a very special environment where a crystalization of helical structures from a same type of prisms is possible. The scenario for such process is described in §12.A.11.3.

Summary:

- - both types of prisms have own set of intrinsic time constants: one per axial direction and another one per radial direction
- - the prisms of CL nodes are not stacked due to the different value of their axial and radial intrinsic time constants
- - free prisms may stick along their long side if they are closer below some critical distance. (valid only in suitable environments for crystallization of helical structures from which the elementary particles are built.

12.A.6.3. About the possible equivalence between G and G_o that could allow an estimation of the intrinsic masses of some low level formations.

12.A.6.3.1. Considerations

In CL space the both SG constants could appear as one constant properly corrected by CL space asymmetric factor. The Planck's time (see Eq. [(2.67)]) is defined by the gravitational constant G which is valid for CL space. The analysis in §12.A.5.3 shows that the embedded fine structure constant could be directly related to the Planck's time. Then one may speculate that the Super gravitational constant G_0 (expressed by CL space parameters) might be the same as the universal gravitational constant G. At first glance, this conclusion may seem hypothetical because the SG

forces are quite much stronger than the Newton's gravitation. But when expressed by the equation of the SG law these forces may appear large due to the larger intrinsic masses (involved in the elementary particles) and the inverse cubic dependence of the forces on the distance.

If the above consideration is true, then the intrinsic masses of all structures from the lower level of matter organization including the primary ball could be found (if the parameters N_{tot} and p discussed in §12.A.5.3 are correctly determined).
Note: The intrinsic mass could be estimated only by the units valid for CL space. In such aspect the asymmetric factor should not be discussed here with the presumption that it is valid when distinguishing the right-handed from the left-handed prisms.

12.A.6.3.2. Equivalent intrinsic mass and matter density of the CL node

The two types of prisms are respectively from two different intrinsic matter substances, but the intrinsic mass related to the SG law we may estimate only in CL space using the Newton's mass unit. For this reason we call it an equivalent intrinsic mass.

The factor C_{IG} has been accurately determined in Chapter 9 as:

$$C_{SG} = G_o m_{po}^2 = 5.276867 \times 10^{-33} \qquad [(9.25)]$$

where: m_{po} - is the intrinsic mass involved in the proton (neutron).

The similar factor for a Newtonian gravitation is:

$$C_N = G m_p^2 = 1.866772 \times 10^{-64}$$

where: G -is the Newtonian gravitational constant and m_p - is the proton mass.

We must keep in mind that C_{IG} is related to the inverse cubic law, while C_N - to an inverse square law, so they have a different dimensions. Then, the following analysis could be valid if the assumption that $G = G_o L_{SI}$ is correct, where $L_{SI} = 1 (m)$ is the unit length in the system SI, in which both factors are compared. Then we have:

$$\frac{G m_p^2}{C_{SG}} = \frac{G m_{CL}^2}{G m_{CLo}^2} \qquad (12.15.a)$$

where: m_p is the Newtonian mass of the proton, G_o - is the intrinsic gravitation, m_{po} - is the intrinsic mass of the proton, m_{CL} - is the CL node inertial mass (CL node mass as a Newtonian mass), m_{CLo} - is the intrinsic CL node mass expressed by the Newtonian mass.

The inertial node mass has been determined in Chapter §2.11.3, Eqs (2.48) and (2.57):

$$m_{CL} = 6.94991 \times 10^{-66} \ (kg)$$

Solving Eq. (12.15.a) for the intrinsic mass of the CL node, m_{CLo}, we obtain the intrinsic mass of the CL node:

$$m_{CLo} = m_{CL} \sqrt{\frac{C_{SG}}{m_p^2 G}} = 3.691 \times 10^{-50} \qquad (12.16)$$

Now we may calculate the approximate value of the CL node matter density. In $.211.3, Chapter 2 of BSM the CL internode distance was found to be $d_{nb} = 1.0975 \times 10^{-20}$ (m). Let us accept that the aspect ratio (length to diameter) of this prism is obtained initially by the first crush of the higher order QB into QPs in a way that the QPs become axially aligned and it is preserved in the further process of moulding. Then we may obtain the approximate aspect ratio of the prism using the relative dimensions of the QP shown in Fig. 12.2. One prism will contains 12 QPs, so the obtain aspect ratio is

$(diameter)/(length) \approx 1/5.4$

If assuming a gap of 1/3 of the internode distance we obtain:

prism diameter: 1.04×10^{-21} (m)
prism length: 5.616×10^{-21} (m)
prism volume: 4.45×10^{-63} (m^3)

Keeping in mind that the CL node contains four prisms, we get the CL node matter density.

$$\rho_{node} = 2.07 \times 10^{12} \ (kg/m^3)$$

Having in mind that the two prisms are formed of two intrinsic matter substances with different densities, the obtained value for the prisms density must be considered as an equivalent one.

12.A.6.4. Summary about the gravitation

(A) The Super Gravitation can be regarded as a result of energy interaction process between intrinsic matter objects in empty space

(B) The SG vibrations in QP are characterized by the SGSPM vector. All upper order QPs from one prisms have synchronized common mode of their quasispheres

(C) SG forces of the prisms exhibit anisotropy due to the strong alignment of the higher order quasipentagons

(D) The propagation of SG forces between prisms in empty space is carried out by the SG-SPM vector

(E) SG forces between prisms of the same type (substance) are quite stronger than SG forces between prisms of different types, but in some particular cases they may decrease significantly

(F) The handedness of the SG field of the prism is memorized in its lower level structures. Some of the lowest level memory about the handedness (cirality) is able to survive the prism's recycling process (taking place in the galactic cycle).

(G) The Newtonian gravitation is a propagation of the Super Gravitation in CL space environments for a long range distance. The velocity of its propagation is limited by the CL node resonance frequency, which defines also the light velocity.

The feature **(E)** may explain the refurbishment of the lattices in some particular cases.

12.A.8. Processes of primordial bulk matter of two substances leading to eruption

12.A.8.1. Considerations for the low order structure growing

Let consider a quantity of bulk matter (of primary balls) of a single substance only, possessing the lowest possible energy. In such conditions even a primary tetrahedrons could not be stable, because it posses increased total energy due to the common mode oscillations described by the IGSRM vector. If a second bulk matter of same substance approaches and collide with the first one, the resultant object will obtain energy larger that the sum of both individual energies. Then the energy to mass ratio of the new object of bulk matter may increase to a level when conditions are obtained for creation of PTs and even primary QPs and QBs. The primary QB already have a possibility for 1 bit memory and due to a common mode interactions only one of both states (right or left-hand twisting) will dominate for closely spaced structures of same type and congregational order.

Now let considering a system of common bulk matter containing equal amounts of the two intrinsic matter substances. <u>The formations from these substances are not mixable due to different geometrical dimensions and the difference in their oscillation properties defining the attraction or repulsion.</u> It is known from the classical physics that a system composed of mixture tends to occupy the lowest energy state. Following this principle, both substances should possess, for example, formations of primary QBs with opposite handedness. It is reasonable to accept that the common total energy of such object is properly distributed between both substances. Such formations could be stable if their total energy (based on the mean energy density) does not exceed the energy well of any one of both substances. It is reasonable to accept that a growing process (within one congregational order and more than one order) may take place if the total energy of the existing formations exceeds the energy well of the system. Then part of the substance energy could go into the structural formations.

Conclusion:

Lower level structure growing process is possible only when the total energy of both intrinsic matter objects exceeds the energy well of the system.

12.A.8.2. Formation of homogeneous layer of quasiballs

Let us suppose that the bulk matter has a shape of sphere and contains primary QB's (formed of 60 primary THs) which are not destructible in the process of the prism's recycling. The growing process of the low level structure should be possible due to the feature (a) in §12.A.5.1 and it will start from the surface. In the growing process, the features (a) and (b) will lead to a slow change of the SGRM frequency of the growing layer of THs at the surface of the bulk matter in comparison to the SGRM of the primary THs frequency in the bulk matter. This will cause structures of same type and order to be attracted together and to be segregated from the bulk matter forming in such way a spherical shell at the surface of the bulk matter. It is more reasonable to expect predomination

of quasiballs than tetrahedrons and pentagons, because:

- QB is a complete structure
- tetrahedrons and QP are substructures contributing to building of QB
- the interaction energy between QB's is smaller than between other two types of structure

The growing process will lead to formation of homogeneous layer of same order QBs. The layer will be positioned at the surface of the bulk matter and will have a finite thickness. One important feature of the quasiball discussed in the previous sections is its ability to memorize the handedness. Consequently:

(a) The homogeneous layers of QBs will accept the handedness memorized in the lower level QBs that may exist permanently in the bulk matter.

12.A.8.3. Layers segregation for mixture of two substances

Now let considering a bulk matter containing two substances of intrinsic matter, that are not mixable. The non mixability means, that they could not form structures containing mixed lower order embedded structures from both substances. Having in mind that the ratio of the primary ball radius is 2/3, it is apparent why the substances are not mixable even for the lowest level structure as the primary tetrahedral.

Structures formed from anyone of both substances could be initially spatially mixed but due to their different Super gravitational constants (G_{OS} and G_{OD}), the structures of same substance will tend to increase their spatial homogeneity. This will lead them to be separated in layers. This will allow the bulk matter containing, for example, lowest order QBs to have a perfect spherical shape.

The different values of G_{os} and G_{od} (see §12.A.6.1 and Fig. 10.13) for the two matter substances occupying a common spherical volume will not disturb the growing process. However they will contribute to one additional feature, described below.

Let us accept that one of the two substances has low order structures with higher mean energy density. Then according to the analysis in §12.A.4.3.3, this structure will be preferentially involved in a growing process taking place at the surface. When the homogeneously growing surface layer reaches some critical thickness, it will obtain slightly different SGRM and SGSPM frequencies. As a result of this the attraction forces between this layer and the bulk matter will be decreased. Underneath the surface layer suitable conditions will be created for growing of second homogeneous layer but from the other substance of matter. In a similar way a third layer from the first substance could grow underneath the second layer. Finally a number of alternative homogeneous layers could grow constructively. This is a guessed scenario, but it might be simulated by mathematical models (not developed yet). A simultaneous growing of such formation of alternative layers to higher congregational orders may also be possible. The layers in some moment of the growing process may have a structure as this illustrated in Fig. 12.13, where the following notations are used for the layers:

Fig. 12.13. Layer structure; 1 - external layer of *m*-th order QBs of I-st substance, 2 - layer of *m*-th order QBs of II-nd substance, 3 - layer of (*m-1*) order QBs of I-st substance, 4 - layer of (*m-1*) order of QBs of II-nd substance

The shown formation of layers is characterized by the following features:

- All layers are homogeneous i.e. they contain formations of same type and congregational order.

- The quasiballs from any layer of particular substance accept the handedness memorised in the lower order QBs of the same substance.

- The alternative layer configuration allows to decrease the attraction between the most external layer and the bulk matter (screening of the SG interactions due to the opposite handedness). At the same time it creates better conditions for improvement the homogeneity of the most external layers.

- Every layer carries embedded portion of energy borrowed from the bulk matter in a form of internal vibrational energy (due to a common modes oscillations).

12.A.8.4 Eruption mechanism

The SG attraction between the most external layer and the bulk matter becomes smaller as a result of:

(a) number of alternative layers

(b) decreased interference of the SGRM and SGSPM vectors from the neighbouring layers, which are of different substance and handedness

(c) steplike frequency deviation between the SGSPM vectors of the uppermost layer and the same substance layers underneath.

If some internal layer is of different order (but same type of structures) its SGSPM frequency might be a harmonics of the external layer SGSPM frequency.

Let us suppose, that the internal layer 3 is grown to m-th order QBs like the external layer 1. Then their SGSPM frequencies will be slightly different due to their different positions in respect to the bulk matter and all internal layers. Then the interactions between their SGSPM vectors could be regarded as interactions between two oscillators with very close frequencies. The energy exchange between such systems is characterised by a frequency beating. This may lead to the following possible process:

Eruption scenario:

It is reasonable to consider that the energy well of the structures from the most external layer is not filled, so there is a room for SG field. Despite the screening effect of the alternative layers, this SG field may have some interaction with the lower level layers, and likely with the bulk matter. The quantum features of SGSPM may cause a phase locking effect (described below) between SGRM vectors of the uppermost layer and the same substance layers underneath at phase difference of π for number of SGSPM cycles. So for a finite time duration of few SGSPM cycles the SG interactions may cause an opposite effect - a repulsion instead of attraction.

Phase locking effect:

The conditions for a phase locking at phase difference of π are the following:

- The underneath layer (3) is of same order as layer (1), so their SGSPM frequencies are very close

- The quantum features of the SGSPM vectors provide a "catch and hold" effect on the phase. This removes the possibility for fast phase flipping and increases the stability of the phase locking for a finite time.

The effect of frequency beating and SG reversal is illustrated by Fig 12.14. a, b.   where a. - is a time diagram showing the frequency beating, b. - is a time diagram showing the duration of the phase locking, t_{inv}, at phase difference of π. The catch and hold effect and the phase locking, both provide the necessary conditions for inverting of SG forces. The phenomena has something similar to the diamagnetism, according to the feromagnetic hypothesis (presented in Chapter 8).

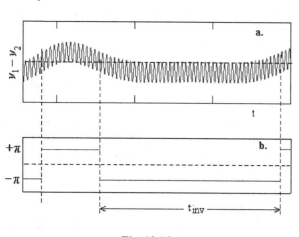

Fig. 12.14

Note: In the classical case of two simple oscillators, a nonlinear media is necessary for a beating effect to occur. In the described above case the nonlinear conditions are created from the quantum interactions of SGSPM vectors. More explicitly it is due to the quasispherical shape of their spatially distributed momentum.

Having in mind the inverse dependence of the SG law on distance, a small separation might be enough for a complete separation of the most external layer. In other words an eruption of the whole layer may occur.

After the first layer eruption the leftover uppermost layer is from the other substance. Assuming that the both substances keep their separate portion of energy, a similar eruption is expected to happens for this layer as well.

As a result of the phase locking effect, the other internal layers lose portion a of their internal energy and further eruptions may not be possible.

12.A.9. Prisms formation

12.A.9.1. Concentric clouds

Only the uppermost shells of the two substances are able to erupt. The shells underneath lose energy during the eruption process and are not possible to erupt.

Let us assume that the erupted shell could not obtain escaping velocity but is diffusively spread in a spherical cloud, due to the predominant SG forces between the same type QBs. It may obtain a common rotating velocity, because the SGSPM vectors of the homogeneous quasispheres now is completely undisturbed from any other formations. The second erupted shell possesses a smaller energy, so it will reach a smaller height. The erupted material from the two layers will be initially concentrated in two concentric clouds around the spherical bulk matter. The whole intrinsic matter of the future galaxy contained in these two clouds. It will be apparent from the following analysis that every well formed galaxy, which is able to have multiple life cycles must contain also an enormous intrinsic matter in its centre (what is now consider as a supermassive black hole in the centre of every galaxy).

12.A.9.2. Column formation and prism molding

The QB's of the layer obtained by the eruption tend to gather together, because their individual SGSPM vectors tend to synchronize. As a result they may obtain a common component which have strong tangential alignment. The inverse cubic law of SG will lead to significant common attraction. From the moment when the QBs gather in a concentric shell, the SG forces will be increased tremendously due to the strong commonly synchronised SGSPM and they will crash into QPs arranged in a shape of rods. The rods are formed by the initial crush of the twelve QPs embedded in every single QB structure. Due to their shape and interacting IGSPM vectors, the twelve PQs become arranged in a column. The common SG pressure, however, might be large enough to cause an additional destruction of the aligned QPs into lower order QBs and QPs. The destruction may happen for a few orders, but the cylindrical formations obtained from the erupted QBs will be arranged in columns. Finally the destruction process will lead to some lower order of QPs for all structures belonging to shell. All QPs will appear aligned in columns as illustrated in Fig. 12.3.

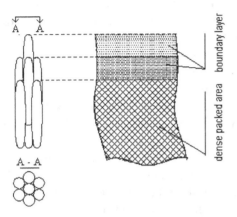

Fig. 12.14.A. Common positions of columns, forming overlapping layers before the formation of the prisms.

It is evident that the increased common strength of SG forces will cause initially formation of cylindrical rods (with simultaneous destruction process of lower order formations) and then will mould them to a shape close to a hexagonal prism. The common positions of the columns obtained after the process of upper orders QB destruction is illustrated in Fig. 12.14.A. The shape of the individual prism has been illustrated in Fig. 2.1. b, Chapter 2.

At the end of the prisms formation phase, their internal structure will be comprised of aligned pentagons of one and a same congregational order. In all next phases of upper level matter organization this internal structure will be preserved, because such strong molding forces will not be available any more.

12.A.9.3. Quantity of the intrinsic matter in a single prism.

The quantity of the intrinsic matter in a single prism is equal to the matter quantity contained in the highest order quasiball that was involved in the eruption process.

During the process of prism molding the highest order QB's may crash a few orders ending

to some lower order QP. **Then it follows that every single prism is moulded from a single high order QB, which is internally crashed by more than one congregational orders.**

It is apparent that the number of crashing orders could not affect the matter quantity of the obtained prisms, but only the degree of their finishing. If the primary tetrahedron, for example, contains N_{tot} = 455 (corresponding to N_{edge} = 13) the grade of finishing for one to three level of destruction are shown in Table 12.2.

Example of prism's finishing grade		**Table 12.2**
Destructed number of QB orders	Number of PQs in a single prism	Finishing grade
1	12	very low
2	$12(60N_{PB}) = 327,600$	good
3	$12^2(60N_{PB})^2 = 1.07\times10^{11}$	extremely high

The destruction order of 2 provides quite good finishing grade. The first option is not acceptable because of very low finishing grade. The third option will require about 5 orders larger molding forces in comparison to the second option. Consequently, the second option is the most probable (for this example). The same option looks more probable even for higher value of N_{PB} as well as for lower values down to 56 (corresponding to N_{edge} = 6, that is a pretty low possible value for the primary TH).

12.A.9.4. A mechanism of same destruction order

The number of crushed orders depends on the common SG forces, which are defined by the SG energy. The latter is involved in the total energy balance normalized to the matter embedded in the structure. We see, that the matter quantity between any two orders of destruction is $60N_{PB}$. This is kind of matter quantity quantization. It allows to remove the effect of possible differences in the molding SG forces. This mechanism is preserved even for eruptions in different galaxies possessing different galactic mass (i. e. different total quantity of erupted matter) and consequently different value of the molding SG forces participating in the prisms formation.

If $N_{tot} = 455$, for example, a variation of the total SG molding forces in range of 27300 will lead to a same order QB's destruction. Therefore, this a kind of Dozation mechanism assuring one and a same matter quantity in every prism. We may call it a **mechanism of the same destruction order**. The importance of this mechanism is not only in the molding quality, but the value of the SGSPM frequency of the upper order QPs. This is the frequency assigned to the level "1" of the matter organization and it defines the node resonance frequency and the SPM frequency of the CL nodes. So the light velocity and the quantum features of CL space are directly dependable on this mechanism.

The same destruction order mechanism assures the prisms from different recycles to posses the same number of QPs from a same type. This is an extremely important feature allowing the interconnections between the CL spaces of the neighbouring galaxies, which are from different formations and have different matter quantity (proportional to their total mass). The same order destruction mechanism, however, could not control completely all the geometrical parameters of the prism. While the variation of the SG forces (for different formations) are not able to affect the order of the crush in the formation of the prisms, they may cause a difference in their molding shape factor. The difference will appear only in the length to diameter ratio of the prisms. This is the basic reason for the appearance of the cosmological red shift of the galaxies from one side and the observed matter segregation after the eruption of some supernovae (which in fact has been a remnant of a previous galactic life, discussed later in §12.B.7.4).

12.A.10. Summary about important structural features and processes at the lower level of matter organization

(A) The fine structure constant is embedded in the lowest level structure - the primary tetrahedron

(B) The acceptance of existence only of matter (and not antimatter) in the observed Universe means that the handedness is memo- rized in some lower order quasiball type of structure, which is not destroyable in the process of the galactic recycling

(C) The initially deposited energy in the bulk matter defines the eruptions of quasiballs to be at predefined order. This assures similar phases of prisms formation and particle crystalization for different galactic formations.

(D) The same destruction order mechanism assures:

- prisms from different recycles (formations) to be formed of one and a same type of quasipentagons

- definition of the upper level frequencies of the matter organization (NRM and SPM frequencies of the CL node)

(E) The variation of the molding SG forces in different galactic cycles causes variation of the length to diameter ratio for prisms from different galactic recycles (relevant to different galaxies and different galactic lives of one and a same galaxy).

(F) The different length to diameter ratio of the prisms from different galactic formations causes a upper level matter segregation between the atomic matter from different galaxies and their CL spaces. One of the effects of this differences is the cosmological redshift.

(G) The upper level matter segregation (valid only for different length to diameter ratio of the prisms) means:

(a) the CL spaces of the different galaxies (and different eruptions) are not mixable

(b) existence of separation surface with infinite thickness between neighbouring galaxies (or matter of different eruptions)

(c) Every observable galaxy possesses own CL space parameters optimized to its total energy.

12.A.11. Protogalactic egg and phases of its internal evolution.

12.A.11.1. Preincubation period

The process of the prisms molding could be quite fast because the time base for the processes between the low level structures is very small in comparison to the processes in CL space.

At the end of the phase of the prisms formation, the expected mass distribution is following:

- a huge amount of intrinsic matter in the centre of the galaxy with a spherical shape
- two concentric shells with finite thickness

The two shells contain all of the prisms from which the elementary particle and CL space of the new galaxy will be made.

The two shells of moulded prisms forms the protogalactic egg. The intrinsic matter in the centre of the protogalactic egg becomes later a galactic nucleus.

The galactic nucleus is energetically isolated and its total energy after the eruption is lower. So it could not provide further eruptions until the next galactic recycling process.

The protogalactic egg, however, undergoes an evolution process. At the end of the prisms molding, the total envelope volume of the material formation is decreased in comparison to the volume it had initially when containing upper order QBs. Although, it contains the same amount of energy. The moulded prisms in the shell are arranged in a partially overlapped layers (along the prisms axis) as shown in Fig. 10.14. As a result of this arrangement, the SG forces are not mutually neutralized and in some final moment of the molding phase the uppermost layer of the internal shell cracks and get momentum toward the external shell. From this moment prism to prism interactions described in Chapter 2 of BSM takes place. The erupted prisms form bundles which propagate and hit the internal surfaces of the external egg-shell. This trigger a similar process from that egg-shell toward the opposite one. Prisms of one type are moving in one radial direction and from the other type in the opposite radial direction. The pair bundles moving in opposite directions may interact between themselves making the dynamics more stable. They are simultaneously rotating and interacting, so these features keep the process self sustainable. In such process, the prisms are released from the compressed shells a layer by layer. The effect is like an "onion peeling", so this name is adopted for a shorter reference to the described process.

The process of "onion peeling", however, should not involve a full destruction of the egg, otherwise, the next phase of particle crystalization

will not be possible. The strength of the peeling effect will be continuously decreasing due to the increased amount of the released prisms in the volume between the two egg-shells. It is reasonable to expect that in the most cases the peeling effect stops completely before the galaxy egg is punched. At this moment most of the prisms are liberated from the shell package, but the galaxy egg still has two solid shell envelopes (external and internal) with finite thickness left from the two concentric egg-shells. The spherical bulk matter is enclosed in the centre of the protogalactic egg.

12.A.11.2. Phase of rectangular lattice

After the peeling effect is terminated, the interior volume of the protogalactic egg (between the two shells) encloses an enormous number of both types of prisms. Due to the high prism's density in this volume, only a rectangular type of gravitational lattice is possible to be formed (see §2.6, Chapter 2). The difference between SG constants G_{OS} and G_{OD}, from one side and the SG anisotropy from the other, allows formation of two types of rectangular lattices formed of RL nodes of the two types of Intrinsic matter substance. The nodes are interconnected in way that the two rectangular lattices are one within the other, occupying a common volume. The neighbouring nodes of the longer prisms are interconnected each other. The nodes of the shorter prisms are between them and they are not interconnected. If the occupied space from both lattices is smaller than the total available space, the new formed lattice formation will be closer to one of the egg-shell. From pure geometrical considerations the lattice of the longer prisms could touch the own egg-shell without a lattice distortion. This is not valid for the lattice of shorter prisms (that is inside of longer prisms lattice). Let us assume that:

The external egg-shell is made of longer prisms

This assumption will be confirmed later by interpretation of some observational data. In Chapter 2 and 3 of BSM it has been shown, that the longer prism is related to the negative charge. For convenience, we may associate the shell type to the electrical charge

- the external egg-shell - negative charge
- the internal egg-shell - positive charge

In the described so far phases the CL space is still missing and the charge appearance is not like the charge we are acquainted with. But prisms to prisms interactions due to a different handedness are still valid (see Chapter 2 of BSM).

The protogalactic egg and the intrinsic matter segregation in the end of this phase is illustrated by Fig. 12.15.

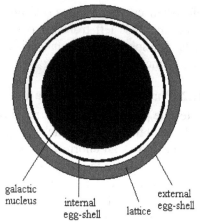

galactic nucleus   internal egg-shell   lattice   external egg-shell

Fig. 12.15. Protogalactic egg

The central galactic nucleus shown in Fig. 12.15 contains formations of low congregational order and only its surface may contain some QBs or QPs. The two types of the rectangular lattices are with a radial symmetry. <u>While the length of prisms are infinitely smaller than the size of the protogalactic egg radius, the formed lattices could be practically considered as perfect cubic type lattices.</u>

12.A.11.3 Phase of crystalization

The handedness is a distinctive feature of both type of prisms, which defines the charge in the CL space. In Chapter 2, the longer prisms was associated to the negative charge. To avoid the problem of incorrect assignment of the handedness (internal structure twisting) we may use the **attribute positive and negative**, keeping in mind that the charge appear only in CL space environment. Using the adopted assignment, the following ratio is valid (derived from the analysis in Chapter 2, 3 and 4 of BSM).

length ratio between a positive and a negative prism = 2/3

Sometimes when the handedness has to be explicitly used for explanation of some process, the following adopted assignment will be used:

negative (longer) prism: right handed internal twisting
positive (shorter) prism: left handed internal twisted

The crystallization process could start from any free prisms, not integrated into the lattice, but after they are consumed the process will be fed by the prisms integrated into the lattice. The prisms nodes from the positive lattice structure are not connected between themselves, so they could be disintegrated much easier from the lattice. Therefore, positive helical structures will begin to crystalize first. It is apparent that the crystallization process will begin near the internal surface of the lattice, because it is not connected to the solid egg shell.

Many features of the crystallization process has been already discussed in Chapter 2 of BSM. In the same Chapter the components of the helical structures and their symbolic notations has been also given.

The crystallization starts with a building of elemental node of 7 prisms, as shown in Fig. 12.16. (More detailed sketches about the geometrical features of the real prisms were shown in Chapter 2 of BSM).

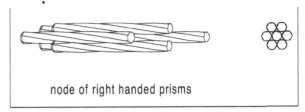

node of right handed prisms

Fig. 12.16. Elemental node of helical structure core (Note: The shown large degree of peripheral twisting is used for the model only. In the real prism the twisting is in its internal structure).

Once started, the elemental nodes will continue to grow in the axial direction forming a zero order structure (core). After it becomes long enough, it will start to bend, so it will continue to grow in a curvature instead of in a straight line. After completion of one full turn, the both ends will not meet each other due the built helicity in the prisms and the possible small external twisting of the prisms. When the number of helical turns is increased suitable conditions will be created for building of quasirectangular type of lattice (with axial helical symmetry) inside the cylindrical space enclosed by the helical structure. (The property of this lattice has been described in details in Chapter 2). This internal lattice structure will have an empty hole along the central axis, where a condition for a trapping hole mechanism discussed in Chapter 2 will be created (due to prism to prism SG interactions). Negative (right handed) prisms will be favoured to penetrate inside the hole, where suitable conditions exists for crystalization of a core. At some point, the external helix will become quite long and will start to bend. It will start to convert to a second order helical structure, but before making a full turn the internal helical core of negative prisms already will determine the direction of the bending (handedness) of the second order helical structure. So the helicity of the second order helical structure of positive prisms, corresponding to the negative (charge) attribute defined by the handedness (right handed accepted) will be determined by the internal core formed of negative prisms. The second order helical structure will be of type $SH_m^2:+(-)$. A portion of such structure is shown in Fig. 2.12., Chapter 2. When the number of turns **m** becomes large enough, the structure will start to bend. This transition is expressed by eqs. (12.18).

Then a full turn from a third order structure of type CH_1^3:+(-) will be formed, which will grow to a multiturn structure according to the Eq. (12.19), where the symbol "->" denotes the crystallization grow process.

$$SH_m^2{:}{+}(-) \to CH_m^2{:}{+}(-) \qquad (12.18)$$

$$SH_1^3{:}{+}(-) \;\to\; SH_m^3{:}{+}(-) \qquad (12.19)$$

It was pointed out in Chapter 2 of BSM, that the higher order structures have a lower stiffness. **The second order structures still have some stiffness that opposes the Super gravitation to attract the neighbouring coils. This allows the structure to be bent.** For the third order structure the stiffness is much lower, so the SG forces may predominate. Therefore, the third order structures are not expected to bent. This also means that formation of structures with order higher than 3 is likely not possible. The analysis of the structure of the elementary particles provided in the previous chapters confirm this, so we may conclude that the highest order structure is of the type SH_m^3:+(-).

According to the provided considerations, the crystallization process is expected to lead to formation of clusters made of parallel neighbouring structures. Figure 12.17 illustrates the mutual positions of such clusters, where SH_m^3:+(-) is the cluster envelope helical structure and SH_m^1:-(+(-) is the internal axial helical structure.

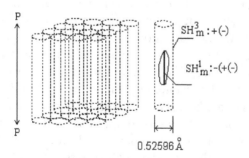

$$SH_m^3{:}{+}(-)$$
$$SH_m^1{:}{-}({+}(-)$$

0.52596 Å

Fig. 12.17. Bunch of clusters

The external diameter of a single cluster is estimated by the proton (neutron) dimensions, calculated in Chapter 6 and verified in Chapter 7 , 8 and 9.

$$DIA = (L_{pc})/\pi + 2(R_c + r_p) = 0.52596{\times}10^{-10} \text{ (m)} \quad (2.20)$$

Figure 12.18 illustrates a single cluster with an exploded views showing the fine structure of the external shell. One may distinguish three different types of spaces enclosed in the cluster's bunch. Spaces "one" and "two" are denotes as internal, because they are inside of the clusters, while the spaces between the clusters are considered as "external".

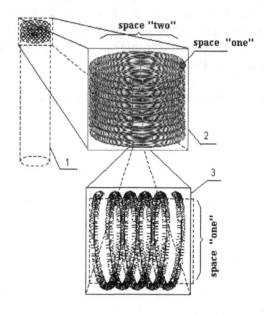

Fig. 12.18
1 - single cluster; 2 - exploded view of the cluster;
3 - exploded view of the helical structure SH_m^3:+(-)

12.A.11.4. Crystallization inside the internal spaces

The external space is not so quiet as the internal one because a motion between the structures could exists. Consequently the internal spaces are much more convenient for crystallization of complex helical structures.

The crystallization in the internal spaces is favoured not only of the quiet conditions but also from the trapping mechanism, discussed in Chapter 2 (with evidence about its existence in some experiments with kaons). This mechanism is quite effective in structures possessing a smaller diameter and a large length to diameter ratio. Consequently, this mechanisms is quite strong in the space "one" and less stronger in the pace "two".

The features of building of internal rectangular lattice in FOHS and the trapping mechanism has been described in Chapter 2 of BSM. Applying these features, it is possible to understand how the initially defined conditions allow a crystallization of well defined helical structures with internal rectangular lattices.

Let us provide some logical scenario for the crystallization process in space "one", which is inside the structure $CH_m^2{:}{+}(-)$. The internal gravitational lattice is of a rectangular type and of the same prism's type as the helical structure. It is able to provide a concentrated SG field (see Chapter 2 of BSM), which in this case is of "positive" type. This field will favour trapping of negative prisms. When enough long core of these prisms is built, it will start converting into a first order helical structure. Then this structure will allow building of its own internal rectangular lattice and a new trapping mechanism will favour inserting of positive prisms leading to building of a positive core inside. When this core becomes long enough it will start to bent and convert into a first order positive helical structure with an internal second order cubic lattice. Now the trapping mechanism inside of the positive structure will favour insertion of negative prisms. Then a negative core will be built. In this case a complete first order compound helical structure will be built in the space "one'. It will be of type $SH_m^1{:}{-}({+}(-)$. This structure grows in the lattice space of $SH_m^3{:}{+}(-)$. Initially it will occupy the central zone where the trapping mechanism exists but in some moment the whole structure $SH_m^1{:}{-}({+}(-)$ will start to bend and convert to a second order helical structure inside the lattice space "1":

$$SH_m^1{:}{-}({+}(-) \; \to \; CH_m^2{:}{-}({+}(-) \qquad (12.22)$$

The structure $CH_m^2{:}{-}({+}(-)$ forming inside of the space "1" is restricted by diameter. It could not obtain the same second order helical step like the external structure $CH_m^2{:}{+}(-)$ because of a minimum bending radius restriction (a critical curvature).

When the process (12.22) occurs, the central zone of the $CH_m^2{:}{+}(-)$ becomes free and then a process of a new second order structure will start. But now the trapping mechanism is inverted because of the opposite handedness SG field of the new built $CH_m^2{:}{-}({+}(-)$ structure. Despite the fact that this structure has less number of turns per unit length in comparison to the structure $CH_m^2{:}{+}(-)$, it

is closer to the central zone. The positive internal radial field of $CH_m^2{:}{+}(-)$ also is not so strong and now the negative field of $CH_m^2{:}{-}({+}(-)$ will predominate (these fields are caused by SG forces propagated through the existed lattice). This means that the next complete built structure will be with a positive external shell. Therefore, the charge sign of every consecutive second order lattice will change. Following this logic, the internal space "one" should be filled with pairs of second order structures with opposite handedness of their external shells until a balance between the opposite SG fields is obtained. Such pairs will later convert to internal pion pairs (negative and positive). (In Chapter 6 of BSM, it has been found that a pair of opposite charged pions with an overall shape of double helix exists inside the proton (neutron)).

The protopion pairs contain even number of completed helical structures, so they could not occupy the central region of space "one'. This region, however, could not be left empty, because conditions for the trapping mechanism still exist. Consequently, the internal crystallization in space "one" will be terminated only by a straight helical structure of first order $SH_m^{1-}{:}{-}({+}(-)$. From this structure the internal kaon of the protoneutron will be formed.

If we review again the structures notified in Table 2.1 and graphically shown in Fig.3.2 we will see that **for structures possessing a central core the higher order helicity is always determined by the core handedness.**

This is outcome from the initial choice that the denser but shorter prism is of a "positive" type (left handed).

The conditions for crystalization in the internal space "2" are not the same as the space "1". The space "2" has much larger diameter and its internal walls are not so smooth. The trapping mechanism for such conditions may not be so effective. But we may accept that when the cluster becomes long enough the space "2" may create conditions for formation of a negative helical structure. Then the evolution of this structure may create conditions for a trapping mechanisms, so a straight structure of type $SH_m^1{:}{-}({+}(-)$ could be formed.

We have seen that the initial core element for starting the crystalization process is quite simple configuration of 7 prisms, but it leads to creation of long clusters. So, it is quite logical to accept that the described crystalization phase involves multiple parallel crystallization processes. They may begin in many places in the internal lattice layer of the protogalactic egg. The speed of such processes might be extremely fast when taking into account how small is the Planck's time which is related to SG processes. In fact the prisms time constant is expected to be between the Plank's time $(5.39 \times 10^{-44}$ s$)$ and the CL note resonance cycle $(9.15 \times 10^{-30}$ s$)$.

It is evident that the cluster growing along the axis P-P shown in Fig. 12.17 will be much faster than the formation of parallel clusters, because the second option requires new crystalization nuclei. This may lead to cluster alignment even between different domains. Then at the end of the crystallization phase, all clusters may get preferential alignment forming aligned superclusters. In such case, the prism's matter in the galaxy egg that had initially a complete central symmetry may obtain some degree of polar symmetry about one axis, named a **galaxy egg polar axis**.

Conclusions:

A. The trapping mechanism of the internal lattice of $CH_m^2:+(-)$ structure is inverted after every built structure of second order. This means that the external shell (expressed by sign associated with the handedness) will be changed according to the order: - + - +

B. The whole space of the internal lattice of $CH_m^2:+(-)$ will be filled with structures with pair charges.

C. The crystalization of the lattice space of $CH_m^2:+(-)$ will be terminated with a final central structure of type $SH_m^1:-(+(-)$.

D. The central cores of all structures are negative as a consequence of the fact that the shorter prism is the positive one.

E. The phase of crystallization involves multiple parallel processes

F. With the advancement of the crystallization process, the clusters may form superclusters

with some degree of rotational symmetry about an axis that could be denoted as a galaxy egg polar axis.

12.A.11.5. Cluster refurbishing

We have seen that in the process of cluster creation and growing, the crystallisation of helical structures is more effective in space "1", quite less effective in space "2" and not effective in the "external space" (the space between the clusters). More effective crystallisation means a larger consumption of prisms. Consequently, the space "3" will be more enriched on prisms. The prisms number density for this space will be above the critical value for a lattice formation (see Chapter 2 of BSM). Hence, the lattice structures will be first created in the space "one", then in the space "two", while a possible lattice creation in the space between clusters will be delayed significantly. This feature gives enough time for the crystallization of the subatomic particles in the spaces "1" and "2".

At some particular moment when the clusters become long enough, the conditions for internal cubic type of lattice in the space "2" will prevail over the CL type of lattice. But there is a geometrical incompatibility between the internal cubic lattice and the helicity of the structure. The effect of this incompatibility for the space "1" could be diminished from the influence of the few second order twisted structures built inside. Although for the space "2", where there is only a one central structure, this effect in some particular moment could become quite strong. The cubic lattice space with a boundary shape of a long cylinder could have strong concentric fluctuations. In some moment the phase of this fluctuations could align across the length. This will cause a strong share stress between the clusters coils. The created shear forces will lead to a cutting of the structure $SH_m^3:+(-)$ in a section parallel to the central core of the cluster. In such process, the internal structures $CH_m^2:+(-)$, $CH_m^2:-(+(-)$ and the central one $SH_m^1:-(+(-)$ will be also cut. Their internal lattices from the first two structures (but not the central one) will get a partial twisting and conversion from a RL(R) to a RL(T) type. The strong SG fields from RL of the cut structures inside the single coil of $SH_1^3:+(-)$ will manage to connect the both ends forming in such way the structures given in the Table 12.4.

Table 12.4

From	Final	Internal RL	description
CH_m^2:+(-) -> (internal π^+)		RL(T)	curled torus
CH_m^2:-(+(-) -> (internal π^-)		RL(T)	curled torus
SH_m^1:-(+(-) -> (internal kaon)		RL(R)	torus
SH_m^3:+(-) -> protoneutron			complex torus

The connection of the free ends is favoured by two features:

- the axial anisotropy of the SGSPM field built in the prisms
- the existence of internal lattice inside space "1".

We may call this phase a **cluster refurbishing**. The external shell expression for such transition is:

$$SH_m^3:+(-) -> n[TH_1^3:+(-)] \qquad (12.23)$$

where: n - is a number of toruses produced from one cluster.

The cluster refurbishing may occur in different domains of clusters, spatially separated. Having in mind the parallel type of the crystallization mechanism discussed in §12.A.11.4, the moments of this transitions in the time scale may have a gaussian type of distribution around some mean moment in the time scale of the matter evolution inside the protogalactic egg.

For any refurbished cluster, the crystallization process is terminated when torus shaped particles are obtained. The internal straight structure SH_m^1:-(+(-) inside the cluster, may still exist, but the new obtained cluster of toruses does not have enough resistance for keeping the new obtained clusters of toruses aligned. Now the alignment of these toruses is kept only by the internal rectangular lattice inside the space "two".

Summary:

The obtained structures after the cluster refurbishment are toruses with external shells of type $[TH_1^3:+(-)]$. These structures are protoneutrons. They are distinguished from the protons and neutrons only by their overall external shape:

- the protoneutron is a torus

- the proton is a twisted protoneutron (as the shape of the number 8)

- the neutron is a folded protoneutron

12.A.11.6. Explosion phase of the protogalactic egg.

12.A.11.6.1. Final phase of the particle crystallization process and explosion

The protoneutrons are kept in the refurbished clusters as long as the rectangular lattice (RL) inside space "2" is not destroyed. After the cluster refurbishments into aligned toruses, the alignment forces are not strong as before. A cluster disalignment may occur, leading to break-up of the RL lattice structure in space "2". As a result, the released RL nodes will migrate to the space between clusters, where the prisms abundance is low. In such environments, the RL nodes might be refurbished into CL nodes. In CL space environment, however, the shape of torus is not stable, so it is converted to twisted (proton) or folded shape (neutron). At the same time the neutron obtains near locked electrical field, while the electrical field of the proton is unlocked (far field). The obtained protons are so close, that the energy from the common repulsive forces is enormous. **This will lead to a huge explosion, which initiates a cosmological event of a galaxy birth.**

We see, that the driving force for the explosion is the creation of CL space from the released prisms of RL space domains, which have been in space "2" during the crystalization phase. The fast creation of CL space appears as an avalanche process. All clusters that has been previously refurbished to columns of toruses are converted to protons and neutrons and contribute to the explosion. At the same time, the central structure SH_m^1:-(+(-), that has been in space "2" brakes into pieces which are initially inside the RL structure environment. In such conditions the central structure may break into number of single coils. These coils become normal electrons once they appear in CL space (their internal RL type lattice converts to RL(T). Therefore, enough quantity of electrons is created just in this violent process. Any new born electron is possible to oscillate and emit X-radiation in CL space (see the electron structure in Chapter 3). The combination of electron and protons provides atomic hydrogen, that is also possible to emit photons.

The described process of explosion initiates the birth of a new galaxy. The further development of the process and the detectable "message" sent in the Universe about this process is discussed in §12.B.5.2.

The described process allows understanding of some aspects about the possible driving mechanism behind the explosion. There is another aspect of this mechanism which is related to the radial direction of the explosion.

12.A.11.6.2. Interaction between the bulk matter nucleus and the internal egg-shell

According to the described concept, the internal egg-shell is positive (see Fig. 12.15). It is comprised of positive prisms, radially aligned in respect to the protogalactic egg centre. Such structure possesses an enormous radial component of SGSPM. This component appears from both sides of the egg-shell towards the bulk nucleus and towards the internal space of the egg (between the two egg-shells). After the eruption of the two most external QB layers from the bulk nucleus, its total energy (of the bulk nucleus) is reduced but some underlying layers of QBs then become external. It is reasonable to expect that the shape of the nucleus is very close to sphere. Then a strong prism-to-prism interaction (see Chapter 2 of BSM) may exist between the QB's layer of the bulk nucleus and the positive protogalactic egg-shell, which is left after the explosion. These interactions are through empty space, but they are still possible due to the high order alignment of the QPs in the prisms that formes the egg-shell. The whole protogalactic egg may posses rotational motion in respect to the bulk nucleus, which is enclosed in the centre of this egg. Then the interaction between the internal egg-shell and the nucleus may be responsible for:

(a) contributing to smother surface of the spherical bulk nucleus

(b) keeping the bulk nucleus in the centre of the internal egg-shell.

The features (a) and (b) are important not only for the correct process of explosion, followed by formation of well developed galaxy. These features additionally define conditions for stable existence of the galaxy during its active life. For this reason **the existence of the internal egg-shell left over the explosion is of crucial importance.**

12.A.11.6.3. Role of the internal egg-shell on the direction of explosion.

The fast created CL space before the explosion could preferentially fill-up the space enclosed between the two shells of the protogalactic egg. The egg-shells may still influence and modulate strongly the CL space with its radial SGSPM component. So the internal egg-shell of aligned positive prisms will behave as a huge positive charge, while the external one as a negative. During the cluster refurbishment, all of not completed toruses (and part of completed) could be destroyed. Their positive and negative FOHS's could lead to a large production of broken helical structures of second order, which will decay to the stable particles electrons and positrons.

All the protons and positrons will be repulsed from the huge positive charge of the internal egg-shell. **This provides initial direction of the explosion in a radial direction. Consequently, the internal egg-shell should be strong enough in order to sustain this initial shock. However, this shock may serve as an initial test for assuring the stable existence of the internal egg-shell during the long active life of the new born galaxy.**

The huge amount of the initially created positrons may contribute to the amount of the positrons still detectable from the space. The large amount of observed cosmic positrons is in agreement with the presented concept.

The other important role of the internal egg-shell is the keeping the bulk nucleus isolated from the new born CL space and matter. It is of key importance about the active life of the new born galaxy. Its behaviour during the active life of the galaxy is further discussed in §12.B.6.1.2.

Summary:
- **The explosion is driven by the repulsive forces between the closely spaced new born protons when they suddenly occur in CL space environment**

- **The birth of a new galaxy is successful if the internal egg-shell of the galaxy egg survives the explosion**
- **The internal egg-shell keeps the bulk matter nucleus isolated from the atomic matter of the new born galaxy. Its existence is of crucial importance during the whole active life of the galaxy.**

12.B. BSM concept of stationary universe

The "Big Bang theory" relies on the assumption that the space we observe is homogeneous. Then the observed red shift is assumed to be only of Doppler origin. However, based on the alternative concept about the physical vacuum, we arrived to the conclusion that space of the observed Universe is not homogeneous. The CL spaces of the different galaxies and different galactic lives differ, because in the moulding process of prisms formation the total detectable galactic mass participates, and it is different for the different galaxies. Therefore, despite the equal matter quantity of the prisms (as shown in 12.A.9.4), their diameter to length ratio may vary between different galactic formations.

The observed galactic red shift, according to the BSM, is not of Doppler type but of cosmological origin (discussed in §12.B.4.2.2). **This is a fundamentally important feature, the understanding of which leads to a profound change of our vision about the Universe.** The presented considerations, supported by a theoretical and experimental analysis inevitably leads to the conclusion that the Universe is a stationary instead of expanding. In such aspect, the cosmological phenomena and their observational characteristics become fully compatible with the BSM concept about the physical vacuum, the rules of the Quantum mechanics and the relativistic phenomena.

In the new vision about the Universe, all observed cosmological phenomena could find logical explanations in a real three dimensional space with unidirectional time scale. Furthermore, the time scale is not reversible. The analysis of observed cosmological phenomena allows also to infer the whole process of the matter evolution with identification of some cosmological cycles. The most important cosmological cycle appears to be the life cycle of the individual galaxy. We may call it a **galactic cycle**.

The galactic cycle includes the following main phases:
   - **active life**
   - **recycling**
   - **particle incubation**

The transition moment between the particle incubation and the active life of the galaxy could be denotes as a **galaxy birth.**

The transition from the active life to the recycling phase could be denoted as a **death of the galaxy**, or a **galaxy collapse**.

Both transition moments posses finite time durations, but they are much shorter, that the time durations of the main phases.

During the recycling phase, the former prisms are destroyed and new prisms are created but with the same handedness and matter quantity. In the ideal case, the recycling must includes all the prisms - those from CL space and those embedded in the structure of the elementary particles. In the real case, however, some islands of matter may brake from the process of galaxy collapse and survive as a remnant from the previous galaxy life. We will see later how important is this exception for assuring of normal process of recycling and multiple active life of any galaxy. In the recycling phase (and any other processes) the intrinsic matter never disappear.

It is evident that we may obtain observational material about the galaxies only during their active life. Only during this phase the galactic CL space exists and appears connected to the CL spaces of the neighbouring galaxies. Only in such conditions we may get photons from the galaxy. In the hidden phases of recycling and particle incubation, any informational exchange with the external world is impossible. Even the gravitational energy exchange could tend to zero, since the SG field falls very fast with the distance and the Newtonian gravitation cannot be propagated in empty space. But we may observe some phenomena preceding the galaxy collapse and immediately after the galaxy birth.

One factor that may influence the future active life of the galaxy is its **proper space** that appears as a pure void space in a classical sense (excluding some small islands that usually escape the collapse). In a typical case, the proper space must be unchanged and a new born galaxy will occupy it again. If the space, however is changed due to some abnormal activity of the neighbouring galaxies, then the active life of the new born galaxy could be affected. So it may not reach a good development like the Seifert type of galaxy. The duration of its active life also could be affected. The new galactic life may also be affected if a significant fraction of the matter have escaped the collapse and consequently the refurbishment. It is evident that after number of cycles, the mentioned phenomena may lead to differentiation of the shape and life duration of the individual galaxies. One informative observable parameter is the shape of the galaxy. The spiral and barred shaped galaxies are well developed, while the elliptical and lenticular galaxies are underdeveloped. The active life of the latter two types could be shorter.

Every normal galaxy contains a huge amount of intrinsic matter in its centre. We may call it a **galactic nucleus**. This is a logical result from the concept about the particle crystallization processes and the eruption mechanism (previously discussed). The collected observational material indicating the existence of massive "black holes" (with enormous masses estimated as equal to billion solar masses) in the centre of Milky Way and other galaxies is in good agreement with the BSM concept about the galactic nucleus.

(see *http://chandra.harvard.edu/xray_sources/
     blackholes_sm.html*)

The matter of this nucleus contains not only primary balls, but also primary quasiballs. These lower level formations, which are never recycled are able to carry the information for the handedness, which is necessary for a correct crystalization process (discussed in §12.A.8).

12.B.1 Recycling and incubation phases

While the physical processes that takes place in the hidden phases of the galactic cycles are without interruption, we may separate them into different subphases for convenience. Starting from the moment of the galaxy collapse, the following subphases are inferred:

- galaxy collapse
- prisms destruction

- process of structural formations in the lower level of matter organization (prior to the eruption of the external layers described in §12.A.8)
- eruption and formation of two concentric clouds
- formation of concentric spherical shells and prisms moulding - a protogalactic egg
- internal process of shell pealing between the two shells of the protogalactic egg leading to release of the newly moulded free prisms
- formation of rectangular (cubic) lattice in the internal space of the protogalactic egg between the two shells
- crystallization of helical structures
- explosion of the protogalactic egg (crack of external core only); formation of a new CL space and expansion of the new born atomic matter (from the crystalized elementary particles)
- a birth of a new galaxy (more accurately a new galactic life)

12.B.2. Galaxy collapse

The galaxy collapse could be regarding as a transition subphase between the phases of active life and recycling

While the active life of the well developed galaxy can be billions of years, it is in continuous energy balance exchange with the neighbouring galaxies, or in other words with the Universe. A disturbed balance will lead to a energy loss (while is not the sole reason for a shorter galaxy life as will be discussed later). The aging of the star population is one factor contributing to the aging of the galaxy. It is known that many stars terminate their life as black holes or pulsars.

The black holes may disturb the radiation balance that will decrease the Zero Point Energy of the home galaxy CL space. The galaxy CL space with less and less moving objects may become poor on ZPE. The zero order waves keep the ZPE to be uniformly distributed over the whole galactic CL space. If the ZPE of a galaxy becomes lower below some critical level in comparison to the neighbouring galaxies, a tension build-up in the separation surface between the CL spaces of this galaxy and its neighbours will appear. The CL space structure of such galaxy could not resist indefinitely to the strong pull-up forces toward the galactic nucleus and at some critical moment the galactic CL space could be separated from the neighbouring CL spaces. This is the last moment when we may get a signal from the collapsing galaxy. The separated CL space shrinks in the following way: The distributed CL space energy is reduced in the break-up zone, due to the energy lost from the break-up. In this zone the prisms of the CL nodes exhibit lower SG energy. This disturbs the SG mechanism supporting the gaps between the CL nodes. Then the most external zone may generate a stream of folding nodes moving to the galactic nucleus. The CL space begins to collapse. The flying folded nodes create inertial forces on any object made of helical structures, so all the galactic matter: from a single particle to a planet and star obtain acceleration towards the galactic nucleus. The process may develop very fast. In the ideal case the bulk nucleus will collect the prisms of the whole galactic CL space and the whole matter (built by elementary particles). **If the galactic CL space, however, gets some internal break during this violent process, some islands of star formations may escape the collapse.** They will likely loose part of their CL space, which will cause a significant disturbance of the optimal ratio between the Static and Partial pressure given by Eq. (10.18), (Chapter 10 of BSM). While this ratio is an intrinsic parameter of the CL space with atomic matter it may affect the escaped island that may contain a large number of stars now completely isolated from the whole Universe. Such star formation may undergo some peculiar evolution tending to restore the optimal value of the above mentioned ratio. This, however, could be done for the expense of reduction of the previously occupied volume. Some destruction of older stars in the central part of the region will contribute for the partial restoration of the optimal ratio between the Static and Partial CL pressure. **Finally, the escaped from the galaxy collapse islands will be converted to compact Globular Clusters, capable to survive the phase of recycling and particle incubation of the former host galaxy. In the new born host galaxy they will be remnants from the previous galactic live.** At the same time, the stars in the **Globular Clusters** may exhibit a strange behaviour of their gravitational-inertial interactions in the environment of the new born CL space. The signatures of the strange be-

haviour of the stars from the Globular Clusters are later discussed in this Chapter.

It is very important that the galactic nucleus is intact during the whole active life of the home galaxy. A potential danger for the preliminary ending of the active life comes from the speeding pulsars. BSM analysis of pulsars (§12.B.6.4) unveils that they possess active jets (one or two in an opposite direction). One jet pulsar behaves like a rocket with an active engine during its lifetime. Fortunately both, the galactic nucleus and the pulsar possess superstrong magnetic fields which combined with the specifics of the pulsar jet deflect the speeding pulsars away from the galactic nucleus.

12.B.3. Preservation of the information about the prism's matter quantity and handedness

The processes of matter evolution in the hidden phases of galaxy recycling have been described from §12.A.8 to §12.A.11. Among the important phenomena assuring the same matter quantity in the prisms from different formations (recycling phases) is the **matter dozation mechanism (or mechanism of the same destruction order of QBs).** It is valid for different galactic lives and for different galaxies, as well.

Another important feature for the prisms recycling is the **lower level memory**. The necessary 1 bit memory for preservation of the correct handedness of the prism is provided by the handedness of the preserved QBs. During the recycling process the prisms are destructed but their internal structure may not be destructed to the lowest level of the matter organization. If the QBs structures from a primary order or any upper order are able to survive the destruction, they will carry the necessary (1 bit) memory about handedness. The information of such memory could be effectively propagated from lower to higher order QBs, which are involved in the process of the prisms formation. The memory of the handedness assures creation of matter and not antimatter in the different galactic cycles. All astronomical and cosmological observation indicate that the Universe is made of matter and not of antimatter.

12.B.4 Variation of diameter to length ratio parameter for prisms from different recycles

12.B.4.1 Condition for interconnection between CL spaces from different galaxies and galactic recycles

While the matter dozation mechanism assures equal matter quantity for prisms from different galactic cycles, the total mass differences between the different galaxies means a different strength of the SG forces involved in the prisms moulding. As a result, the diameter to length ratio (usually less than unity) for prisms from different recycles could be different, as shown in Fig. 12.19. The difference of diameter to length ratio is self-compensating by the self-adjusted parameter *node distance*. If estimating the physical constants of any galaxy in units defined by the own CL space, they will look exactly the same. But if estimating them by the CL space units of only one galaxy, they will exhibit differences. These differences will be caused only by the diameter to length ratio difference of the prisms moulded in different recycles (formations). This difference, of cause, could not be estimated from a phenomenon taking place in the host galaxy (because the CL space here is homogeneous). The signature of the difference can appear only if the investigated phenomenon is created in the CL space of one galaxy and detected in the CL space of another galaxy. For example we have to investigate by optical methods a specific phenomena occurred outside of our home Milky way galaxy in order to find a difference. The study of such phenomena are discussed later in this chapter.

Now let consider that the prisms of two neighbouring galaxies have a small difference in their diameter to length ratio. This will affect the interconnection of their CL spaces. In order to avoid the node dislocations in the connection zone, the internode space of the first CL structures will be slightly shrunk, while for the second one - slightly stretched from their optimal node spacing. Consequently, the connected galaxy CL spaces will contain a finite thickness zone. We may call this zone a Galactic Separation Surface (GSS). **The GSS will allow the propagation of the photons (and any EM radiation) but with some small energy** loss due to the wavetrain refurbishment. This energy loss, according to BSM, appears as a cosmological redshift.

The GSS zones are infinitely small in relation to a single galaxy CL space. Evidence of such zones are discussed in §12.B.4.2.5 and §12.B.12.

12.B.4.2 Origin of the Cosmological red shift

12.B.4.2.1 How the diameter to length respect ratio of prisms from different formations may affect the fundamental CL space parameters

Let us analyse the conditions for interconnection between the CL space of two neighbouring galaxy whose prisms have different diameter to length ratio, as shown respectively in Fig. 12.19a. and b.

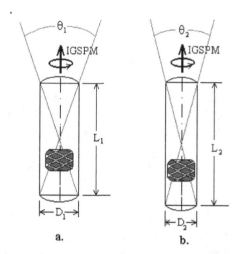

Fig. 12.19. Difference between two prisms of same substance and handedness but from different formations (recycles)

In the analysis of the process of the prisms moulding we concluded that the internal QPs of the prisms are aligned along the prism's axis. For prisms with different diameter to length ratio the subtended angles θ_1 and θ_2 are also different. Consequently, the SGSPM in both cases will have one and same integral strength but the spatial phase diagrams for both cases will be different. The dynamical properties of the CL nodes built of such prisms will also exhibit some differences. One possible difference is illustrated by Fig. 12.13. where the SPM quasispheres for prisms of two different formations (illustrated in Fig. 12.19) are shown.

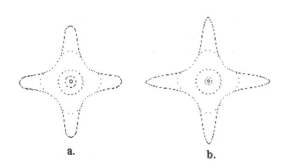

Fig. 12.20. SPM quasispheres of CL nodes from prisms of different formations with a different diameter to length ratio

The quasisphere of SPM vector for case b. exhibits sharper bump regions. The different steepness around the bump regions may cause a difference in the N_{RQ} cycles (number of NRM cycles for one cycle of SPM). Different N_{RQ} parameter means a different permeability of the physical vacuum (see §2.11.3 of Chapter 2 of BSM), which is involved in the definition of the light velocity according to the well known expression $c = (\varepsilon_0\mu_0)^{1/2}$. It is evident that the difference between prisms will lead to a difference in the dynamical properties of the CL nodes formed from them. Then the corresponding two CL spaces will have different CL space parameters. The difference in the CL space parameters would appear as a difference in the following physical constants:

- space time constant λ_{SPM} and consequently the light velocity
- Planck's constant

The difference between the N_{RQ} parameters of the galactic spaces is identified by the analysis of some observed phenomena discussed in §12.B.4.2.5.

Now let considering the conditions for interconnection between neighbouring galactic CL spaces based on the following considerations:

(a) Prisms of different formations contain one and a same number of QPs and one and a same quantity of intrinsic matter (due to the intrinsic matter dozation mechanism). This means that the internode distances for different prisms formations should be the same. Consequently, the interconnection requirements for the CL spaces of the neighbouring galaxies are satisfied.

(b) The same number of aligned QPs in prisms from different formations will provide the same frequency of the common mode SGSPM of the prisms for both formations

The conclusion in (a) about the same node distance is in agreement with the propagation features of the photons through CL space. They do not exhibit a scattering from GSS (otherwise GSS will have a visible signature), but only an energy loss.

The same CL node matter quantity and internode distance for CL spaces of different prism's formations allow correct node to node interconnections between both spaces. At the same time the both parameters determine the CL node resonance frequency. Consequently

(c) both CL spaces should have the same resonance frequency ν_R (NRM vector frequency)

(d) the difference between both CL spaces will be in their N_{RQ} parameters

(e) both CL spaces will have different space-time constants $\lambda_{SPM}(1)$ and $\lambda_{SPM}(2)$ respectively.

(f) both CL spaces will have different light velocities (because ν_R and node distance is the same but λ_{SPM} is different)

The space time constant is a main characteristic parameter of quantum waves (see Chapter 2 about the equations of the light velocity). From conclusion (e) and (f) it follows that the next relation is valid

$$\frac{c(1)}{\lambda_{SPM(1)}} = \frac{c(2)}{\lambda_{SPM(2)}} \qquad (12.24)$$

where; $c(1)$ and $c(2)$ are the light velocities for both spaces referenced to the common parameter - the node distance

(g) Eq. (12.24) is valid for any photon whose frequency is a same subharmonic number of ν_{SPM}

12.B.4.2.2 Energy of photons from other galaxies

A. Energy loss of photon passing through GSS

Figure 12.21 illustrates part of the CL spaces of two neighbouring galaxies G1 and G2 and their common Galactic Separation Surface (GSS).

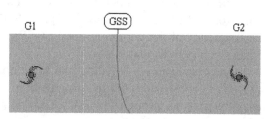

Fig. 12.21. GSS between CL spaces of two galaxies

The geometrical and interaction parameters of the elementary particles composed of helical structures are complied to the parameters of their home galaxy CL space (built of prisms from the same formation). Consequently, **the quantum features and the interaction energies are optimized for particles interactions taking place in own galaxy CL space.** This is valid for EM quantum waves (photons) emitted by own particles and propagated in own CL space and for generated electrical and magnetic fields, as well. Then if a photon emitted in a CL space "1" is detected in the same CL space, its energy should be one and a same (excluding the Doppler shift). The same is valid for photons in the CL space "2". If a photon emitted in a CL space "1", however, is detected in CL space "2" (or vice versa), its parameter are not optimized for a propagation through CL space "2", because of the difference between the N_{RQ} parameters of the both spaces. It will exhibit a loss of energy that will appear as a wavelength change towards the longer wavelength. This is so called redshift, which from the BSM concept about space appears to be not of Doppler type but cosmological. Having in mind the consideration (g) of §12.B.4.2.1, the energy loss corresponding to the cosmological red shift could be expressed by the relation

$$E = \left| \frac{h_o c_o}{\lambda_0} - \frac{h_1 c_1}{\lambda_1} \right| > 0 \qquad (12.25)$$

where: index "0" denotes the parameter in the space where the photon is emitted and "1" - where the photon is detected.

In the light of the presented consideration, Eq. Eq.(12.25) means that the constants for both CL spaces (h_0 and h_1) are different. Their common relation, however, is difficult to be estimated, because they are defined for their own space parame-

ters only. For this reason the module of the difference is used.

It is apparent, that the conditions of the propagated photon in the new CL space are always less optimal than in the own CL space, where they are generated. Consequently, the process will always be accompanied with some energy loss, but never with energy gain. Such loss may occur only at the GSS, because the CL spaces are homogeneous. The concept of the energy loss in the GSS is confirmed by some observational data discussed in following sections.

Consequently:

(a) A photon passing through a GSS exhibits a small energy loss

(b) The small energy loss is detected as a red shift, the passed photon emerges with a slightly decreased wavelength. We will refer this effect as a Cosmological red shift.

(c) the observed large Z-shift for the distant galaxies is contributed from the cosmological red shifts.

Note: The GSS is a real spatial domain. It should not be confused with the imaginary separation surface introduces in Chapter 10, for analysis of inertial phenomena of massive astronomical bodies in the CL space of the home galaxy.

B. About the proper energy of the photon

The duration of the active life of the galaxy depends mostly on its own factors. In very rare cases it could depend on the neighbouring galaxies. One of this factors is the energy lost by emitted radiation if not equivalent absorbed radiation from the Universe is obtained. The energy stored in the galactic nucleus supply energy to the host galaxy during its active life, but the power of this energy may decrees with the time. (Now energy emitted from a supermassive black hole in the galactic nucleus is experimentally confirmed). Then it is reasonable to accept that the CL space of a new born galaxy has slightly higher static Zero Point Energy (ZPE-S) energy (see Chapter 5 of BSM about ZPE-s and ZPE-D) in comparison to the CL spaces of its neighbours. It follows from this that a photon emitted from this galaxy may be detected in the neighbouring galaxy as a blue shifted, despite that it will undergo some red shift when passing the GSS. If the same photon, however, is detected in a more distant galaxy, it will appear as red shifted, because it will undergo multiple redshifts at each of GSSs through which it passes. From these considerations we arrive to the following conclusions:

(1) The cosmological Z-shift is comprised of two components, which are not of Doppler origin: a blues shift and a red shift.

(2) The red shift cosmological component is related to the energy losses of EM radiation when passing through the GSSs between the galaxies.

(3) The blues shift cosmological component is a signature of the hidden ZPE-S of the CL space from where the EM radiation originates

(4) The proper energy of the photon is its energy if it is measured in the same galactic CL space where it is emitted.

The large redshift observed for distant galaxies corresponds to the case (1). The small blue shifts observed from a limited number of small galaxies close to our home galaxy corresponds to the case (2). In the presently adopted concept about space and the Big Bang theory, all cosmological shift is assumed as a Doppler shift. We see that one and a same observational data leads to quite different picture, when examined from a different concept of the physical vacuum.

12.B.4.2.3. Hypothesis of accumulated energy at the Galactic Separation Surface (GSS)

Every photon passing through the GSS loses a fraction of its energy. The amount of the lost energy according to Eq. (12.25) depends on the difference of their constants, which are dependent on the different diameter to length ratio of the prisms from different formations. Evidently, some energy could be accumulated on the GSS region. From logical considerations, we may assume that this energy in fact occupies a region from both sides of GSS with a thickness equal to $\lambda_{SPM(1)}$ and $\lambda_{SPM(2)}$, respectively. These regions could be regarded as **two energy regions of GSS**. In these regions the passing wavetrain of the photon will be refurbished. Photons with different wavelength will occupy different refurbished volume, proportional to the quadrature of the transverse wavetrain radius multiplied by $\pi/4$.

12.B.4.2.4. Cosmological Z shift with dominating redshift component

All observed galaxies (with exception of two closer ones) exhibit red shifted lines. According to the presently adopted concept about space and the Big Bang theory the space between all galaxies is considered as homogeneous, so the observed red shift is accepted to be of Doppler type. The observed red shift is denoted as a Z - shift, a dimensionless parameter defined by the formula:

$$Z = \frac{\lambda_{obs} - \lambda_o}{\lambda_0} \qquad (12.25.a)$$

where: λ_o and λ_{obs} are respectively the wavelength of the emitted and the observed photon

The same formula can be written in a form:

$$\lambda_{obs}/\lambda_o = Z + 1 \qquad (12.26)$$

According to the BSM concept, the space homogeneity is valid only for the home space of a single galaxy, but not for the Universe. From this aspect the observed photons from other galaxies will contain a wavelength shift component that is not of Doppler origin. Then the observed Z shift must include two major components: a cosmological one and a Doppler shift. The cosmological component contains a blue shift and a red shift as discussed earlier, but for a distant galaxies the redshift is predominate significantly. It predominates also the Doppler shift which is usually inside the home galaxy. For this reason in the following analysis we may address the cosmological redshift only as a cosmological shift, which term is largely adopted in the observational data.

The cosmological red shift component is strongly dependent on the number of crossed GSS. While Eq. (12.25) provides the photon's energy loss for one GSS, the same photon may pass through number of GSSs, exhibiting in such way multiple energy losses. This means that the redshift cosmological component of Z shift will increase with the distance (photons pass through many GSSs). At the same time the contributions from a possible small blue shift or a Doppler shift component becomes comparatively very small and insignificant. For this reason all distant galaxies appear as red shifted.

Figure 12.23 illustrates a photon propagated through a number of galaxies. The GSSs between the galaxies are only shown.

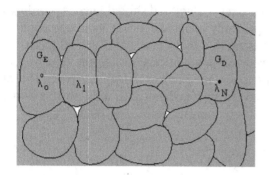

Fig. 12.23. A photon trace (green line) passing through number of galaxies. The galactic CL spaces are shown as gray areas; the GSSs between them are shown as black lines.

The galactic CL spaces are shown as gray areas, while some empty spaces between them (which are not excluded from a theoretical point of view) are shown white. GSS are shown as black lines between the galactic CL spaces.

A single photon with wavelength λ_o is emitted in the CL space of galaxy G_E and detected in the CL space of Galaxy G_D. It may pass only through CL spaces, but not through empty spaces. Due to energy losses in each GSS its wavelength will undergo consecutive red shifts from λ_1 to λ_N, where N is the number of crossed GSSs. We may apply an analogy of a photon emitted in a medium with one value of refractive index and detected in a medium with another value of refractive index. Applying this analogy for two neighbouring CL spaces we may introduce the parameter **GSS quasirefractive index**. It is reasonable to assume that the diameter to length ratio for the prisms from different formations (galaxies) varies around one mean value. Then the GSS quasirefractive index should also vary around one mean value that can be expressed by Eq. (12.27).

$$\bar{n} = \frac{1}{N} \sum_i^N \left(\frac{\lambda_i}{\lambda_{(i-1)}} \right) \qquad (12.27)$$

where: $\bar{n}$ is a **mean GSS quasirefractive index,** N - is the number of crossed GSSs, i - is an index number of the crossed GSSs.

The mean GSS quasirefractive index could be regarded as a valid for a single GSS between CL spaces distinguished by the average value of the prism's diameter to length ratio.

The ratio between the observed and the emitted wavelengths is:

$$\lambda_i / \lambda_o = (\bar{n})^N \qquad (12.28)$$

Combining Eq. (12.28) with Eq. (12.26) we get the **equation of the cosmological Z -shift (redshift only).**

$$(\bar{n})^N = Z + 1 \qquad (12.29)$$

where: N - is the number of crossed GSSs by a single photon, $\bar{n}$ - is the mean quasirefractive index of GSS

Eq. (12.29) shows the relation between the total cosmological Z shift and the number of GSS.

Number of observations exists for which the Z parameter is estimated. The parameter N also could be determined. According to BSM, the parameter N is directly related to the number of absorption L_α lines in the Lyman alpha forest observations (see the physical explanation of the phenomena in §12.B.12). Then the parameter $\bar{n}$ is calculable by Eq. (12.29).

12.B.4.2.5 Signature of GSS existence as a result of CL space parameters variations estimated by the red shift periodicity

A significant observational material exists about the red shift periodicity. It could be classified into two groups:
- observations from quasars
- optical observations

The red shift periodicity investigation is pioneered by G. Burbidge (1967, 1968). Studying samples of QSO (quasistellar objects = quasars) and radio and Seyfert galaxies he concluded in 1968 that the distribution of red shift has a peak at $\Delta Z = 0.061$ and integral multiples of it. Later Duary et al. (1992) found that the observed periodicity is described by the expression $Z_n = 0.0035 + 0.0565 n$, where n is integer. For $n = 1$, the strongest peak at 0.06 is obtained.

B. Guthrie and W. Napier (1996) write in the abstract: "**Persistent claims have been made over the last 15 years that extragalactic red shift, when corrected for Sun's motion around the Galactic centre occurs in multiples of 24 km or 36 km**". Investigating 40 spiral galaxies in a range up to 1000 km/s (distance scale used in the presently adopted concept for expanding Universe) and applying the above mentioned correction they found a strong evidence for z-shift periodicity of 37.6 km. Some of their results are illustrated in Fig. 12.23.A showing the relative distribution of red shift differences corresponding to 103 redshifts from spiral galaxies.

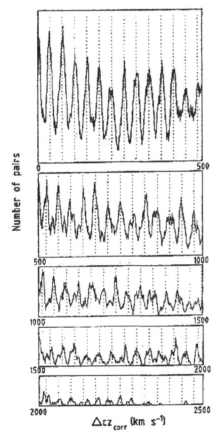

Fig. 12.23.A. Relative distribution of redshift differences. Vertical dotted lines represent the best-fitting periodicity (Courtesy of B.N.G. Guthrie and W. N. Napier (1996)

The red shift periodicity can not be explained by the concept of the expanding Universe. Burbidge in 1968 expressed the idea that the red shift has a cosmological (not a Doppler type) origin.

From the BSM point of view, the explanation of the redshift periodicity is the following:

GSS exists between the CL spaces of any neighbouring galaxies whose prisms (as a rule) are from different formations and exhibit differences in their diameter to length ratio. This difference may lead to a slight variation of the space-time constants t_{CL} for the CL space of any galaxy. Then the photon's wavetrain undergoes a process of refurbishing when crossing the GSS. Photons from a distant galaxy cross many galactic GSSs. Let us assume that the variation of the time-space constants of the crossed CL spaces leads to variation of the time delay of the passing photon due to a wavetrain refurbishing process in the GSS. Then for a set of observations where the photon emitters (emitted at known spectral lines) are distributed over a large distance, we may expect some kind of stroboscopic effect with embedded signature of some of the fundamental CL space parameters. In other words we may expect that the accumulated time delay may appear as a whole number multiplier by a term, which is strongly correlated to the space-time constant of our galaxy. If investigating a large number of such sources, the signature of this low number periodicity may become apparent.

The observations show that two types of periodicity exists:

- One type of periodicity appears as related with the variation of the time-space constant of the different galaxies

- Another type of periodicity appears as a signature of the fine structure constant

Table 12.5 provides the results of the analysis of the first type periodicity: 24 km 36.2 km and 72.46 km. The estimation of this periodicity is provided by Guthrie and Napier (1996).

Redshift periodicity of first type   **Table 12.5**
(related to the signature of the space-time constant)

1	2	3	4	5
72.46	2.382E-4	2.6936E-13	9.0065	1
36.2	1.2067E-4	1.365E-13	17.77	1.966
24	8E-5	9.046E-14	26.81	2.978

where:

Column 1: (Δz c) - periodicity in [km/s]
Column 2: Δz - red shift periodicity

Column 3: $\frac{\Delta z}{N_{RQ}}$ - redshift normalized to one resonance cycle referenced to SPM period

Column 4: $\frac{t_{CL}}{\Delta z/N_{RQ}}$ $\left[\frac{\#\ cycles}{Hz}\right]$ - time delay per normalized redshift

Column 5: value of column 4, normalized to the smallest periodicity value of the column.

The following CL space parameters (for our galaxy) are used:

$$t_{SPM} = t_c = 8.0933 \times 10^{-21} \quad \text{(s) SPM (Compton) frequency}$$

$$N_{RQ} = 0.8843 \times 10^9 \quad \text{- number of resonance cycles per one}$$
SPM cycle (see the calculations of
CL space parameters BSM, Chapter 2)

The close values of the term of column 5 to integer indicate a strong correlation between this term and the parameters t_{CL} and N_{RQ}. The process is similar to a stroboscopic effect between two closely spaced frequencies. When the difference between the frequencies is smaller, the periodicity is larger. In our case we have more than two frequencies: these are the N_{RQ} frequencies of the different CL spaces. **Measured by the period of SPM vectors in own spaces they are constants. Every CL space has own SPM frequency (and time-space constant) defined by own CL space and prism's parameters. But if their parameters are estimated by the CL space parameters of our Milky Way galaxy they will exhibit variation. So their N_{RQ} parameter will appear different when referenced to the N_{RQ} parameter of our home galaxy.** The strongest factor contributed to the stroboscopic effect is the ratio between N_{RQ} frequency of our galaxy and some closer neighbouring galaxy.

Another redshift periodicity found initially by Burbidge and analysed latter by Karlsson (1977) exhibits a different set: 0.3, 0.6, 0.96, 1.41, 1.96. In order to remove the error from the solar system motion around the Milky way center we must apply a galactocentric correction (see Duari, 1992):

$$z_{GC} = \frac{1 + z_{obs}}{1 - v\vec{u}/c} \tag{12.30}$$

v = 238 (km/s) - is the mean velocity of the solar system motion used by Duari (1992).

$\vec{u}$ - is unit vector from Earth to QSO

z_{GC} and z_{obs} are the galactocentric and observational redshift, respectively

Red shift periodicity of the second type (related to the signature of α)			**Table 12.6**	
1	2	3	4	5
0.3	0.301	3.404E-10	140.3	1.024
0.6	0.601	6.799E-10	280.2	2.045
0.96	0.962	1.087E-9	448.15	3.27
1.41	1.412	1.597E-9	658.05	4.802
1.96	1.962	2.219E-9	914.59	6.677

Columns:

1: z_{obs} - observed redshift periodicity from the Earth

2: z_{GC} - galactocetric corrected redshift according to Eq. (12.30)

3: z_{GC}/N_{RQ} - redshift per one resonance cycle estimated by SPM period time base

4: $\frac{z_{GC}}{N_{RQ}t_{CL}}$ $\left[\frac{\#\ cycles}{Hz}\right]$ - column 3 normalized to time-space constant

5: $\frac{z_{GC}}{N_{RQ}t_{CL}}\alpha$ $\left[\frac{\#\ cycles}{Hz}\right]$ - column 4 normalized to $1/\alpha$.

The plot of column 5 is shown in Fig. 12.23.B.

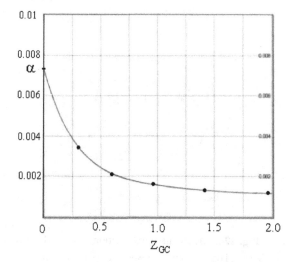

Fig. 12.23.B. Plot of the terms of column 5 (of Table 12.6) vs the observed red shift periodicity. The robust straight line shows a strong correlation between the involved CL space parameters.

The analysis of the data and the plot in Fig. 12.23.B shows that:

(1) the low red shift periodicity is closely proportional to the fine structure constant

(2) the proportionality for higher redshift periodicity decreases, but the slop of the linear dependence is constant

The possible explanation of feature (2) is a following: The fine structure constant (α) is a basic low level structure parameter. For quantum waves generated and detected in own CL space α is always apparent. If considering a single GSS between two galaxy, the cosmological red shift from it is dependent on the ratio between the total masses of the galaxies. The fine structure constant, however, defined by the low level intrinsic matter structures is not related to the total masses of the galaxies. From the other hand, the fine structure constant defines the ratio $E_{IG}(TP)/E_{IG}(CP)$, that is one and a same for any galactic CL space (due to the self adjusted CL node distance) and is related with the quantum energy transfer. Consequently the cosmological redshift (also related to the quantum energy transfer) will contribute to the decrease of the α periodical signature from signals emitted in more distant galaxies. This trend evidently is systematic and the alignment of the data points in a straight line as shown in Fig. 12.23.B is a strong argument for the correctness of the provided concept.

12.B.4.3 Summary

- **Every new born galaxy creates an own new CL space, which assures an optimal environment for the new crystalized particles**
- **The galactic CL spaces are separated by Galactic Separation Surfaces (GSS)**
- **The photon crossing through a GSS exhibits a small energy loss as an increase of its wavelength due to an wavetrain refurbishing. The passed photon appears redshifted. The photons crossing a number of GSSs exhibit multiple energy losses, which appear as a wavelength redshift, an observed cosmological phenomenon which is not of Doppler origin.**
- **The accumulated energy in the GSS from the photons energy losses is reemitted by GSS as a diffused X-ray radiation**
- **In the stationary Universe the mutual motion of the galaxies is an exclusion and a**

wide spread Doppler shift component from a galaxy motion is not expected. The real Doppler shift should be usually from the relative motion of the astronomical objects referenced to the host galaxy CL space.

- **For larger distances in the observed Universe the Z-shift of any object is completely dominated by the cosmological redshift.**

Additional observational data confirming the concept of the cosmological **z**-shift and the role of GSS are presented in the next sections.

12.B.5. Phenomena indicating a death (collapse) or a birth of a galaxy in the Universe

The concept of a stationary Universe and galaxy recycling, means that the death or birth of a galaxy are common phenomena in the Universe. The average frequency of such phenomena depends on the mean life of the galaxies and could be estimated if we know the radius of our observational perimeter. The possible signature of such events should be randomly distributed in the observational sphere of 4π srad.

Signatures of death and birth of a galaxy are really observed, but they have not been correctly recognized so far. **The detected phenomena are known as Gamma Ray Bursts (GRB).**

According to some observational features, we may divide the observed GRBs into two categories:

(a) GRB with an optical counterpart

(b) GRB without optical counterpart

According to BSM:

GRB of type (a) corresponds to the phenomena of galaxy birth (denoted by BSM as GRB(B)

GRB of type (b) corresponds to the phenomena of galaxy collapse (denoted by BSM as (GRB(C)

Both types of GRB have also a number of distinctive features. Their decoding may provide a knowledge of exclusive importance. Once these specific features are understood, the identification of the GRB type can be made without detection of the optical counterpart.

BSM identifies one common feature about both type of GRB. They carry a **signature of gravitational shock wave.** The latter is caused by a large pulse of SG(CP) energy moving with gravitation velocity that may exceed the speed of light. Both types of GRB exhibit such features, but their signatures are different.

From the previous analysis and the considerations in §12.B.4.1 it becomes apparent that the front of such shock wave may preserve its integrity moving through number of CL spaces.

12.B.5.1 GRB without optical counterpart

A gamma ray burst (GRB) with a specific signature and without optical counterpart is indication of a galaxy collapse

This phenomenon occurs in the moment of separation of the collapsing galaxy CL space from the CL spaces of its neighbours. It causes interactions between central part of the prisms carrying SG(CP) forces. The CL space separation is in enormous large area, so the generated SG(CP) pulse in the neighbouring galaxy CL space is also enormous. Such strong SG(CP) pulse is propagated through the CL spaces of connected neighbouring galaxies with the velocity of the gravitation. This is a **gravitational kind of shock wave.** Its main feature is the ability to generate gamma and X-ray directly by the CL space (even without atomic matter). **For this reason the gamma ray burst does not carry a signature of thermal radiation.**

The missing of optical counterpart is reasonable (the matter of the collapsing galaxy is already separated from the external world).

Specific features of GRB from a collapsing galaxy: The gamma ray burst is composed of a large amplitude initial pulse followed immediately by a package of closed spaced pulses with slow decreasing amplitude. One example of detected GRB by BATSE (The Burst and Transient Source Experiment) with such features is shown in Fig. 12.23.C. The shape of GRB, shown in the figure, is composed of multiple X rays covering a spectral range up to a few hundreds KeV. It could be considered as a signature of specific gravitational shock wave, the formation of which is described below.

The most important features of the gravitational shock wave are the following:

- isotropical propagation with a light velocity
- finite lifetime of the front formation

- spectral dependence of the generated gamma and X-rays from the region

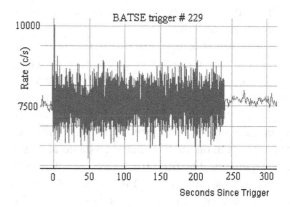

Fig. 12.23.C. GRB from a collapsing galaxy by BATSE (the pulse is truncated at 240 s due to a lack of memory of the recording device)

The Newtonian gravitation, we are familiar with, is a propagation of the Super Gravitation in CL space environment. Between the most important feature of CL space that we must keep in mind is the oscillation properties of the CL node. It is characterised by the NRM and SPM vectors and their frequencies: the node proper resonance frequency and the SPM (Compton) frequency. It is reasonable to expect that these two frequencies may affect the propagated SG field if the gravitational change is too fast. In such aspect we may consider two cases: (A) slow transient state related with a weak gravitational disturbance and (B) fast transient state related with a strong gravitational disturbance.

Gravitational disturbances might be of different cosmological disturbances in CL space and their propagation in CL space may invoke a slow or a fast transient state in the domain where the detector is placed.

In case (A) the propagation of the weak gravitational disturbance could be affected by the CL node NRM and SPM in the following way depending on its strength

(a) Not affecting by NRM and SPM frequencies

(b) Affected only by the SPM frequency

In case (a) there is not generation of X or gamma rays from the event, but it may contribute slightly to the ZPE in the GSS that could be later emitted sporadically.

In case (b) a generation of X or low energy gamma rays from the event may occur. Such rays, for example, are generated in the Earth atmosphere and they are related to some gravitational changes in our Sun.

In case (B) the propagation of the strong gravitational disturbance could not be fully attenuated by the oscillation properties of the CL node (NRM and SPM vectors). In such case, part of the strong disturbance could be propagated with a very high hyperlight velocity defined by the SGSPM vector of the prism, the frequency of which is much higher than the NRM frequency of the CL node. **This case is valid for the GRB from the event of galaxy birth or galaxy collapse.**

Let us analyse the beginning of the galaxy collapse event and what kind of signal could be detected in a distance much larger that the size of the collapsing galaxy CL space. In the moment of the local CL space separation from the neighbouring galaxy a large energy could be dissipated in the broken part of the GSS which is connected to the neighbouring galaxy. This should be a very strong gravitational disturbance. According to the above provided considerations, it will propagate isotropically, but within a solid angle defined by the initial conditions (the propagation is possible only through the connected galactic CL spaces). This signal will arrive to the distant detector very fast (with a hyperlight velocity) but attenuated, in a way that its strength may fall with the inverse square of the travelled distance. We may call this kind of energy propagation a **primary gravitational shock wave**. **(PGSW)**. When crossing the GSSs closer to the collapsing galaxy the energy of the gravitational shock wave is still quite strong, so it may invoke strong transient states in these GSSs. Then they may become sources of **secondary gravitational shock waves (SGSW)** propagating also with a hyperlight velocity. Obviously their amplitudes will be smaller because the process of their formation is different and the involved energy is smaller. One **distinctive feature** of the SGSW is that they are generated by a large number of GSSs (while the PGSW is generated only from one GSS). This feature, namely, provides a package of multiple pulses at the detector. If assuming that a very small time

interval is necessary between the moment when the PGSW strikes the particular GSS and the generation of the SGSW, the SGSW will arrive at a distant detector as a package of closely spaced gamma or X rays pulses. The package will start with the strong pulse contributed directly by the PGSW. One may say that the amplitudes of the pulses in the package should be much smaller than the amplitude of the first pulse (pulses). However, we must take into account that the collapsing galaxy is usually quite distant (estimated by the Z shift). Then for the detector, the subtended angle for the source of the PGSW is much smaller than for the sources of the SGSW. Additionally the strength of the sources of SGSW will also fall with the distance from the collapsing galaxy. Then the detector will still detect the signals from SGSW within a finite angle of view.

The presented analysis provides one additionally possibility for studying the GRB events apart of their spectrum. The study of the spatial distribution of the GRB packet with simultaneously measurement of the amplitudes might be also informative. It such way a possible differentiation of the PGSW from the SGSW could be achieved.

In the provided analysis it was assumed that the PGSW and SGSW are both propagated with a hyperlight velocity to the detector. If the detector, however, is very distant their energy may fall below some critical value so they could not be propagated with a hyperlight velocity. Then their temporal and spatial characteristics could be preserved but their energy could be much lower and they will propagate with a light velocity. This may contribute for a weak but very long tail, which follows the GRB event.

12.B.5.2 GRB with an optical counterpart

A gamma ray burst of type (B) with optical counterpart with specific spectral and time characteristics is a signature of a new born galaxy.

The galaxy birth begins with the explosion of the galaxy egg, but the external world could not detect this moment until the new CL space is built and connected to the Universe. The event consequence scenario after the break up of the external egg-shell of the galaxy egg is a following:

(1) building (expanding) of a new CL space

(2) Interconnection of the new CL space to the CL spaces of its neighbours (all or few of them), causing a gravitational shock wave, which generates a GRB(B)

(3) X-ray generation from the new obtained electrons

(4) UV and VIS radiation from the first formed atoms, which are highly excited by the ZPE fluctuations of the newly formed CL space.

(5) Broad band radiation from normally excited H and D atoms and the first simple molecules: H_2 and D_2.

The mechanism of GRB(B) phenomenon from the gravitational shock wave is similar as in the case of GRB(C) event from a collapsing galaxy.

GRB from the new born galaxy is generated in the moment when the new built CL space is connected to its neighbours. Therefore, the events of galaxy collapse and galaxy birth provide two distinctive GRB phenomena:

- GRB(C) from a galaxy collapse is generated by the GSS of its neighbours with their neighbours

- GRB(B) from a galaxy birth is generated by the new formed GSS between a new born CL space and its neighbours

- GRB(B) signal from a galaxy birth may be followed by a strong radiation around 511 KeV, generated from the excited single electrons (electron-positron system), while this component should be missing in the GRB(C) from a galaxy collapse

Specific features of GRB(B) from a new born galaxy:

(a) GRB(B) may have an optical counterpart,

(b) Two large pulses in the X-ray region are very common feature

(c) The pulse packet shape of GRB(B) is different than the pulse packet shape of GRB(C).

Optical counterpart:

In first we have to point out that a missing optical counterpart from some GRB(B) could be a detection problem (very distant galaxy or obscured).

The optical counterpart may contain two time distinctive sources of radiation with distinguishable spacial and time features:

(1) The initial phase of GRB(B) with an exponentially decreased intensity is followed by a detectable strong "point like" optical source with a larger pulse time duration.

(2) A detectable extended light source with a relatively small but almost constant brightness

The optical source in case (1) is formed by highly excited new born atoms (mainly H and small portion of D) during the process of the matter expansion after breaking of the galaxy egg. The detection of the rising edge of the pulse indicates, that the new CL space is just created and connected to its neighbours (in other words - connected to the Universe).

The detectable extended source in case (2) is from the newly born and spread atomic matter of the galaxy away from the galaxy bulge. Firstly, it may obtain a large momentum from the explosion. Secondly, the new CL space initially is without node gaps and does not possess a normal static and partial CL pressure. This phase called a **phase of Transition CL Space** is discussed later. It delays the process of inertial interactions between the newly born atoms and folded nodes, but At the same time it allows the huge potential energy stored in the protogalactic egg to be transferred and distributed in the vast volume of the newly created CL space. The newly born atomic matter obtains a significant portion of this energy.

The simultaneous observation of GRB with an optical counterpart is of great importance for unveiling the process of the galaxy birth. GRBs has been observed for more than 30 years. It is reasonable that they appear completely random in the observable Universe. The observation of optical counterparts of some GRB became possible after the development of BATSE Rapid Burst Response system (http://www.batse.msfc.nasa.gov) and ROTSE project. Fig. 12.24 shows GRB(B) in few spectral ranges.

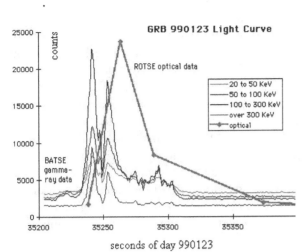

Fig. 12.24. GRB 9901123 light curves
(Gamma and optical range). Sources:
(http://laastro.lanl.gov/rotse/grb/990123)
(from http://laastro.lanl.gov/rotse/grb/990123/)

The shown light curves possesses a feature of signature from a new born galaxy.

The expected extended source of radiation is more difficult for observation, because of its lower intensity in comparison to the "point source". Fortunately, the observation of one GRB with counterpart provided such possibility in 1998. This is the GRB980613 with an observed optical counterpart. The extended source has been identified by the Hubble space telescope. Images from HST 26 and 39 days after the gamma burst are published in Nature v. 387 (29 may 1997) by Kailash et all. In the abstract of the paper the authors write: "The optical counterpart appears to be embedded in an extended source..." Figure 12.25 shows image accumulated from two HST observations after 26 and 39 days from the detected GRB. The size of the image is 11.5 x 11.5 arcsec. The observational data from two other gamma ray bursts: GRB980613 and GRB990123 also show typical characteristics of a new born galaxy. The estimated Z-shifts (cosmological) are respectively 1.096 and 1.6. So the new born galaxy in the first case is closer and the identification of the "extended" source is possible.

More details about the observed phenomenon GRB970228, are provided also in:
http://antwrp.gsfc.nasa.gov/apod/ap970407.html
Number of observations of GRBs with optical counterparts confirm the BSM hypothesis that

GRB(B) is a signature of a birth of a new galaxy in the observable Universe. S. Holland et all. (2001) have investigated the optical light curve of the host galaxy of GRB 980703 by data from HST taken 17, 551, 710 and 716 after the GRB event. The authors identified very intensive star formation with rate of 8 - 13 M_O/yr in the host galaxy (which must be the newly born galaxy according to BSM). They also write: "This suggests that the host galaxy is undergoing a phase of active star formation similar to what has been seen in other GRB host galaxies". According to BSM concept, the new born stars are the first stars of the new born galaxy.

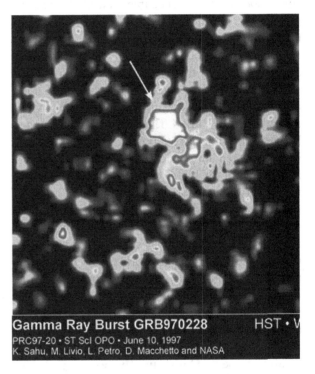

Fig. 12.25. Extended source from a new born galaxy
(according to BSM concept)
source: http://www.seds.org/hst/gb970228.html

12.B.5.3 Phase of CL space formation. Hypothesis of transition type of CL space.

From the moment of the eruption of the protogalactic egg, the kept inside large amount of SG energy begins to transfer to the newly created CL structure and supply the interaction energy between the new elementary particles

It is logically assumed that in the initial phase of the CL space formation the prisms of the neighbouring CL nodes touches each other (no gaps between the CL nodes as in the case of a normal CL space). This is a temporally state of CL structure that we may call a **Transition Cosmic Lattice (TCL).** TCL structure could not posses the same node oscillation capability as CL one. It will not have also the same type of energy well as the normal CL space.

The first domain with TCL could be formed near the internal egg-shell which is not destroyed after the protogalactic egg explosion. This is the first empty space to be occupied during the egg explosion and the first domain where the TCL structure will be converted to CL one.

The conversion of the TCL to a normal CL space means that the CL nodes become properly separated, providing conditions for a normal node oscillations and equalization of the Zero Point Energy by the ZPE waves. The necessary gap distance between the (alternative type) CL nodes will become adjusted automatically as a result of the self adjustment of the high frequency modes of the SG forces between the two types of prisms.

The released energy in fact will provide the major boost for the expansion of the new born matter into the new created CL space. After the external egg-shell is broken, the process of TCL buildup will continue in a free empty space environment. The TCL, however, is characterised with strong dynamics of expansion, so its existence is temporal. It is spherically expanding, while leaving a newly created CL space in the volume it enclose in any moment. **Consequently, we may expect that the TCL is like a hollow sphere whose thickness continuously decreases with the expansion, while the new born CL space forms inside this hollow sphere.** Then at the time when the TCL reaches the CL spaces of the neighbouring galaxies, its thickness could be quite small.

Let us analyse some features of the TCL. It is reasonable to expect that the growing TCL in an empty space exhibits a spherical shape until reaching some boundaries. Let us admit that in the time of explosion the protogalactic egg has been near the centre of its proper (void) space. Figure 12.25.A illustrates the moment when the expanding TCL reaches the boundaries of this void space.

In this moment the spherical shape of TCL and the created behind it CL space begin to convert to the shape of the void space, which has been preserved for this new born galaxy. In that time the following expected phenomena will take place:

(a) the spherical expansion of the TCL structure, which at this moment must be comparatively thin (in comparison to the galactic proper space) will be stopped by the boundaries of the CL spaces of the neighbouring galaxies.

(b) the energy momentum from the stopped expansion of TLC will cause an interconnection of the new born CL space to the neighbouring galactic CL spaces and a creation of a shock gravitational waves.

The feature (b) means that the interconnection with the neighbouring galaxy CL space requires some energy pulse. The TCL is incompressible and should be able to provide such energy pulse. As a result a strong gravitational shock wave is created. It is a quite fast transitional event for the CL spaces of the interconnected galaxies, so it may be propagated with a hyperlight velocity.

From the provided considerations and analysis it follows that in the moment of generation of the gravitational shock wave (the beginning of CL space interconnection) the TCL space may still exists for a limited short time.

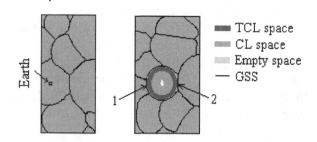

Fig.12.25.A. Illustration an expanded TCL domain during the phase of CL space formation

We may look for a proof of the suggested concept by studying a large set of GRB(C) data and looking for the following features:

- a large first pulse with a finite duration (the first pulse of the GRB), assuming that it corresponds to the moment when the TCL reaches the end 1 (see Fig. 12.25.A).

- a smaller pulse with a finite duration displaced from the first pulse by a time interval t_1.

The described signature of two pulses is clearly evident in the GRB curves shown in Fig. 12.24. The energy components of the two pulses could be also reasonably explained. The reflected pulse does not exhibit so strong accumulated shock wave for generation of hard gamma rays because it does not have a single straight path.

BATSE project provides excellent GRB data web site with curve plotting capabilities:
http://www.batse.msfc.nasa.gov/batse/batse-home-meantime
(**Note:** The truncated pulse packet of the GRBs shown in the BATSE data base is an artifact due to a lack of recording memory).

One very interesting phenomenon may happen between the moment, when the new built TCL reaches the neighbouring galaxy CL space. If assuming that the galaxy nucleus is approximately in the middle of the void space the new TCL space will reach almost simultaneously the two opposite ends 1 and 2 of the previously void space. The obtained gravitational shock wave from end 1 (closer to the Earth) will generate the first pulse of the GRB that will be propagated with a hyperlight velocity. At the same moment, the gravitational shock wave from end 2 will be partially reflected back, but it will propagate through the temporally existed TCL space. It will arrive to the Earth also with a hyperlight velocity. The both strong pulses will have their signature in the created GRB(B). **Then the time duration between the generated two shock waves will provide us information about the value of the hyperlight velocity through the TCL if we know size of the CL space of the new born galaxy.**

For very approximate estimate of the hyperlight velocity, denoted as C_{HL}, we may assume a spherical shape of the TCL in the moment of interconnection to neighbouring spaces. Then we have:

$$C_{HL} = \pi D_G / t_1 \qquad (12.31)$$

where: D_G is the average size of the galactic proper space (the void space before interconnection to neighbouring CL spaces), t 1 - is the time difference between two strong pulses from the GRB(B).

One good example of well separated two pulses of GRB(B) is shown in Fig. 12.25.B.

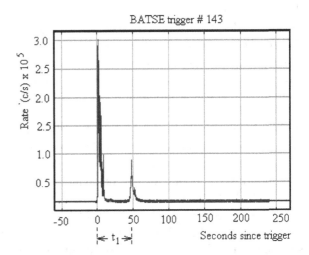

Fig. 12.25.B. GRB from a new born galaxy

The first pulse in Fig. 12.25.B (starting with the gamma ray) indicates the moment of connection from the side closer to the Earth, while the second pulse is a back reflection from the opposite side. It passes the spherical shell of the TCL space for time t_1.

The GRB(B) from a new born galaxy could be distinguished from a GRB(C) from a collapsing galaxy even without optical counterpart. Examining the curves from large number of GRBs we may see, that some of GRB(C) (from collapsing galaxy) have prepulses before the main gamma burst, referenced at zero time (the moment of triggering). This feature is explainable if considering that the separation of the collapsing galaxy CL space could be preceded by consecutive pull-offs and indicates at the same time **that the process of galaxy collapse is triggered by some major event: perhaps the break-off of the galactic nuclear shell (leftover from the internal egg-shell of the protogalactic egg)**.

We may detect a GRB(B) with clearly identified two peaks only at the described above ideal conditions. If the galactic nucleus appears not in the central domain of the void space, we may not detect a clearly identified two peaks of GRB(B). The same reason may provide also an answer why the number of identified GRB(B) by the two peaks criteria is much smaller than the number of the total registered GRBs.

Let us provide some theoretical insight into the hyperlight velocity propagation in the temporally existed TCL structure. In a normal CL space, the gaps between the CL nodes allows their oscillation properties, which are important for the known quantum features of the physical vacuum. Among them are the constant propagation of the light velocity, defined by the internode distance and the period of the proper resonance cycle of the CL node. (The light velocity in Chapter 2 of BSM was estimated as a energy momentum transfer between two neighbouring CL nodes for one resonance cycle). Behind the EM waves in fact is the SG field, but the velocity of its propagation in a normal CL space is restricted by the CL node resonance cycle. Such restriction is missing in the TCL structure and we may accept that the SG waves is propagated between two neighbouring nodes for one SGSPM cycle of the prism. Then we may solve the following problem: **approximate estimate of the hyperlight velocity through the TLC from a properly selected GRB data corresponding to a birth of a new galaxy, or GRB(B).**

In the case of approximate estimate we may accept that the internode distances for a normal CL space and TCL are equal. Then the ratio between the hyperlight velocity (denoted here by C_{HL}) and the light velocity will be approximately equal to the ratio between the SGSPM frequency v_{IGSPM} and NRM frequency v_R.

$$(C_{HL}/c) = (v_{SGSPM}/v_R) \qquad (12.32)$$

where: $v_R = 1.0926 \times 10^{29}$ (Hz) (derived in Chapter 2 of BSM), c - is the light velocity

Combining Eq. (12.31) and (12.32), we may obtain the approximate value of the prism's v_{SGSPM} frequency

$$v_{SGSPM} = (\pi D_G v_R)/(ct_1) \qquad (12.33)$$

The average diameter of the proper void space D_G in fact is equal to the average distance between the galaxies. It must be estimated statistically for a large number of galaxies including small and large galaxies. In order to avoid a large error from estimation of the intergalactic distance for a stationary Universe, we will use data about the local clusters galaxies without taking into account the home galaxy, which size is pretty large. The aver-

age intergalactic distance in such case is about 100,000 LY.

The average time t_1 can be estimated from a large number of identified GRB(B). Figure 12.25.C shows a histogram of 37 identified GRB(B) from the "light curves" provided by BATSE.

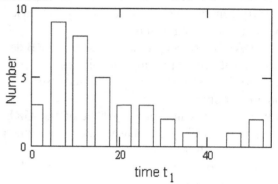

Fig. 12.25.C. Histogram of 37 identified GRB(B) from the "light curves" provided by BATSE. The time t_1 is an indicator for the size of the galactic CL space.

The mean time duration of the set shown in Fig. 12.25.c is 19 sec. Then using Eq. (12.33) we get:

$$v_{SGSPM} = 5.697 \times 10^{40} \text{ (Hz)}$$

Now we may put the missing value of the frequency for the level 1 of matter organization in Table 12.1 as

$$\ln(5.697 \times 10^{40}) = 93.84$$

The new plot of the frequency vs the level of matter organization is shown in Fig. 12.25.D. The plotted curve is a Gaussian fit.

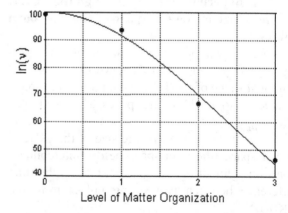

Fig. 12.25.D Frequency vs the level of matter organization with approximate estimate of the frequency for level 1

The shown estimate for the frequency of level 1 is very approximate and may contain error of few orders, but being between the levels 0 and 2 it partly confirms the suggested hypothesis about the formation of the galactic CL space.

Now let see what happens with the enormous energy that has been in a form of potential energy in the protogalactic egg. After the explosion of the protogalactic egg part of the energy is transferred to the TCL. When TCL expands it creates a CL space structure inside of the volume it encloses, while its thickness is decreases. Part of the TCL energy obviously might be transferred to the created CL space.

In Chapter 10 of BSM we found out that for a normal CL space with atomic matter (particles, atoms, molecules, solids) the ratio between the Partial and Static CL pressure tends to satisfy the relation:

$$P_P/P_S = \alpha/\sqrt{1-\alpha^2} \qquad [(10.18)]$$

Consequently, the tendency to approach the above ratio will cause an involvement of the newly created atomic matter into a motion in a newly created CL space. Therefore, the process of the energy transfer is continuous during the time period from the protogalactic egg explosion to the interconnection of the new CL space to its neighbours. After it is interconnected and some transitional fluctuation process of the new CL space is over, this space appears in an absolute rest in respect to the interconnected galactic spaces - or the CL space of the Universe. Then the transferred energy momentum to the new born atomic matter will provide a rotational momentum of this matter in respect to the galactic nucleus. The new born (visible) atomic matter begins its own evolution leading to creation of stars, solar planetary systems and so on.

12.B.5.4 The observable Universe as a conglomeration of galaxies with interconnected CL spaces

The possibility to get information from huge number of galaxies means that their CL spaces are interconnected. The void spaces (empty space in a classical sense) are rear but the option of their existence is theoretically feasible. Such spac-

es may belong to a galaxies that are currently in its hidden phases of evolution or some of them could be a leftover between some galaxies. Logically the second case could be rare but still possible, because the galaxies are of different size. Additionally the larger galaxies (possessing also a larger matter quantity) may have a longer active life and consequently a longer cycle period. **Then it is reasonable to expect that in a long time of many galaxy cycles the larger galaxies will be interconnected for a longer average time than the smaller galaxies. Then the larger galaxies will form the supergalactic skeleton of the Universe, while smaller galaxies will fill the gaps between them.** Two identified features from GRB data are in favour of this conclusion:

(1) the time variation between the two peaks of GRB as seen from the histogram given in Fig. 12.25.B indicates a different sizes of the galactic CL spaces.

(2) When studying the angular distribution of the observed GRBs in the whole observable sky we see that they are not completely random. Some spatial correlation is observed. This is demonstrated by Fig. 12.25.D showing the angular distribution in galactic coordinates of 1635 BATSE bursts (Meegan et al. 1997)

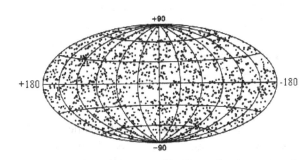

Fig. 12.25.D. The angular distribution in galactic coordinates of 1635 BATSE bursts (Meegan et al. 1997)

12.B.5.5 Galactic nucleus during the phase of recycling and incubation

12.B.5.5.1 The void space obtained after the collapse of the galaxy

From the properties of the boundary zone of GSS it becomes clear, that the connection of the new born CL space with the CL space of its neighbours will involve some **connecting energy**. This energy is evidently supplied by the new born galaxy, more accurately, by its TCL structure. The signature of this energy is the dilation of the first pulse of GRB(B) from the hard gamma ray pulse at zero triggering.

12.B.5.5.2 The galactic nucleus during the phase of recycling and incubation is kept inside the void space

The galactic nucleus has to stay inside the void space during the phases of matter recycling and incubation. It could be a disaster if the nucleus is drifted in the neighbouring galaxy space. How it is kept insight the own empty space?

One probable explanation is that it is kept inside by the Globular Clusters (GC). They are discussed in §12.B.7. The Globular Clusters are formed from islands that have escaped from the matter collection process during the galaxy collapse. This is possible for large star clusters if the CL space around them gets broken during the galaxy collapse. Left disconnected, they may undergo specific evolution converting them to globular clusters with their own CL space, partial of which is generated after the galactic collapse. Globular clusters are observed mostly in the hallo of almost all spiral galaxies. In our galaxy their number is about 150. Massive and tightly packed GCs typically contain from 100,000 to million and more stars. Their age, estimated by the existing methods appears to be in order of 15 billions years. This age is not in agreement with the estimated age of the Universe (~ 12 Billion years) that has been unsolved paradox so far.

The galactic nucleus is much smaller than the size of the void space. It is also very distant from the large masses of the neighbouring galaxies. Having in mind the inverse cubic law of SG and its specific properties for the low level structures (discussed in previous paragraphs) it is evident that any interactions with the neighbouring galaxies are practically eliminated. In such case the SG interactions of the nucleus with the globular clusters could be enough to influence the position of the nucleus. Globular clusters (GC) are discussed in §12.B.7 where it is shown that they possesses lower total energy. Consequently, they could not be able to connect themselves to the CL spaces of the neigh-

bouring galaxies, because a connection energy is required.

From the provided analysis it becomes apparent, that the GCs are able to keep the protogalactic nucleus inside the empty space during the incubation period. GCs may surround the protogalactic nucleus (during incubation period) but not all of them could be swallowed by it. Once it enters the phase of the matter segregation, the peripheral SG energy becomes concentrated in the QB structures and the approaching GC will be not swallowed, but could be kept close. At the same time the nucleus have spin velocity in respect to the Universe and the surrounding GCs may gradually obtain a similar spin. In such way the globular clusters may serve as a buffer zone of the protogalactic nucleus (in the phase of atomic matter incubation). The whole assembly may drift, but if it reaches the boundary zone of neighbouring galaxy CL space it will be reflected back.

The provided concept is in agreement with some GRB(B) data after proper interpretation. If the galactic nucleus (and respectively the protogalactic egg) is well centred in the empty space, the following signature of a GRB(B) should be observed: strong forward pulse and weak reflected pulse separated by the time t_1. BATSE GRB with trigger number 143, shown in Fig. 12.25.A is a such a case. Examining the BATSE lightcurves from other bursts we may find number of cases when the second (reflected) pulse is larger than the forward one. This is an indication, that the galactic nucleus has been biased from the empty space centre. The lower intensity first pulse shows, that the expanding TCL has reached the neighbouring CL space but a complete CL interconnection is not occurred (only a fraction) and it has been reflected back (together with the whole new galaxy including the nucleus) and then a larger CL connection occurred after the expanding TCL has reached the opposite end of the own space. For this reason the second pulse may appear stronger, despite the fact, that it is a reflected pulse for the observer.

Note: We should not be surprised that a superhigh acceleration and velocities are possible in a void space. The Newton's mechanics and especially the inertia we are familiar with are not valid for this type of space and the intrinsic matter formations.

Another confirmation of the role of GCs during the phase of recycling and incubation comes from the observational data from the Milky way galaxy centre. K. I. Uchida and R. Gusten, (1995) discuss the observed linear filaments aligned with the strong magnetic field within the inner 1° radius of the Galactic center. Large negative velocities are observed from the filaments. Inspection of Bell Labs (Bally et al., 1987) survey data show, there is no other strong negative-velocity emission within immediate region. According to BSM, the emitting low temperature gas (0.4 - 1.6K) in the filaments is a remnant from a GC, which has been very close to the protogalactic egg. The logical conclusion is that the CL space boundary of this GC has been broken during the galactic explosion, however, the CL structure of this remnant is not mixable with the new CL structure of the host galaxy. The trapped matter in the filaments operates in own CL space and is much brighter. The emission is excited by the X-rays emitted from the strong electron oscillations. The red shift is of cosmological origin. The filament structures and their characteristic features are similar as in the Crab nebula. They are typical for the remnants from the previous galactic life.

12.B.5.6 Summary:

- **GRB(C) type is a signal from a collapsing galaxy**
- **GRB(B) type is a signal from a new born galaxy**
- **GRB(B) carry a signature of hyperlight velocity propagated through the new born TCL (the transition phase of the new CL space)**
- **Some correlation between the time recycling sequences of the neighbouring galaxies exists**
- **Globular clusters may play a role of keeping the galactic nucleus inside the own former space (a void space) during the phase of recycling and particle incubation**

12.B.6 Active galactic life

The active galactic life is the apparent phase of the galactic cycle. This is the phase in which we are able to exist and observe. The galaxy during this phase also have evolutional process and the observed phenomena are quite rich. For this reason only the most important phases of the evolution and the related phenomena will be briefly discussed.

12.B.6.1 Some features of evolution after the explosion of the protogalactic egg

We could not try to divide the active galactic life into phases. Some of the phenomena following the eruption of the protogalactic egg has been described in §12.B.5. They include the CL space formation and interconnection, (characterised with a transitional phase, denoted as TCL) and the observed "point" and "extended source" related to the GRB phenomena. The latter is of larger interest as it is related to the expanding new matter that will develop into a new galaxy with one of the following shapes: disk, lenticular, spiral or bared spiral type. The shape that the new galaxy will obtain depends also on the volume and the shape of the void space that has been inherited and preserved from the former galactic life.

The protogalactic egg does not have strong interaction with the neighbouring galaxies and their CL spaces, so the rotational momentum that he eventually get during the galaxy collapse could be preserved. After the explosion the new born matter get the orientation of its spin axis. The spin energy, however, will get interaction with own and neighbouring CL spaces, after the interconnection and transition of the TCL structure into a CL space. Then the matter thrown at larger distance from the galactic nucleus will appear with larger momentum. Once the CL space is interconnected it could not rotate any more. At this moment the energy from the eruption is converted into interaction energy between the FOHS's of the new born atoms and the new born CL space. In this interaction process, the CL node are forced to fold and deviate by the envelope volume of the FOHS, which contain denser Rectangular Lattices. These is a type of inertial interaction, which may provide both: momentum and excitation of the electrons of the atoms, so they could emit a broad band radiation.

The spiral type of the new born galaxy is obtained as a result of unfolding the transformed clusters which has been obtained in the crystalization phase inside the protogalactic egg.

It is evident, that the final shape of the galaxy depends on some factors , such as:

- a proper balance between the quantity of the crystalised particles and the quantity of the free prisms which will built the CL space

- a proper balance between the galactic matter and the preserved void space in which a new CL space will be created

Optimal balance conditions will lead to a development of galaxy of a spiral type.

Not optimal balance conditions may lead to a disk or lenticular galaxy.

12.B.6.1.1. Interaction of the galactic nucleus with the visible matter of the host galaxy during the active life of the galaxy

The formation of the galactic nucleus, comprised of the internal egg-shell and the enclosed inside intrinsic matter, were explained in §12.A.11.6. The galactic nucleus exists through the whole active life of the home galaxy and the importance of such existence will become evident in the discussion provided later. In §12.A.11.6 it was explained how the egg-shell is able to keep the bulk matter nucleus inside. A new question that needs an answer is: What is the interaction of the galactic nucleus with the visible matter of the host galaxy during the its active life?

The protogalactic egg together with the enclosed bulk intrinsic matter possesses some spin momentum obtained during the collapse of the former galaxy. After the eruption the spin axis of the new galactic nucleus gets feedback initially from the CL space interconnection. Later it gets feedback from the inertial interaction between the spiral arms and the stationary (already interconnected) CL space. In this process one major factor is **the magnetic field created by the rotating internal egg-shell.** The egg-shell is comprised of highly oriented positive prisms (see §12.A.11.6), so in the established CL space it behaves as a huge rotating electrical charge (dynamo). Such charge is able to provide a huge magnetic field through the galaxy. At the same time, the positive egg-shell is surrounded by a huge amount of electrons obtained from the negative FOHS's that has not been involved in the final product of the matter crystalization - the protoneutrons. The positive FOHSs

which has not be involved in the protoneutrons are also thrown during the explosion of the protogalactic egg. They could not be closer to the positively charged external shell of the new born galactic nucleus, so they contribute to the flux of the cosmic positrons.

The electron structure and its oscillation capability has been extensively discussed in Chapter 3 of BSM (referenced as a electron system due to its complexity). Figure 12.25.E shows again the shape of the electron structure, comprised of an external FOHS with a negative RL(T), an internal FOHS with a positive RL(T) and a negative central core. The internal FOHS with a positive RL(T) and a negative central core is a positron.

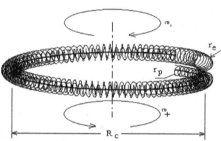

Fig. 12.25.E. Oscillating electron structure
R_c - compton radius, r_e = 8.84E-15 (m),
r_p = 5.89E-15 (m). (the internal negative and positive RL(T)'s structures are not shown)

It has been shown in Chapter 3 of BSM that if the electron is strongly excited it performs multiple oscillations with exponentially dumped amplitudes. When the excitation of such type is with a maximum amplitude, it leads to a process of CL space pumping that is followed by a release of two gamma quants 511 KeV each. Strong oscillations with smaller amplitude usually lead to generation of lower energy gamma photons, usually by 3-gamma photon process. Oscillations between one free positron and one electron also lead to emission of two photons of 511 KeV and the obtained particle is a neutral one with a mass of 1.022 MeV (not detectable so far due to its neutrality). (In the modern physics the process of 511KeV gamma radiation is known as "annihilation", but according to BSM, there is not any type of matter annihilation in such process).

The CL space surrounding the galactic nucleus could not rotate as it is connected to the neighbouring galaxies CL spaces. A large cloud of electrons are attracted by the positive galactic nuclear shell (the internal shell of the protogalactic egg) while they are in the CL space. The rotating positive external shell of the galactic nucleus will attract the electrons and drag in rotation while they will have inertial interactions with the CL space. The rotating electron cloud will cause formation of a large magnetic field that will comply to the rotational axis of the galactic nucleus (the internal positive shell of the protogalactic egg). At the same time the strongly excited electrons (as mentioned above) will emit mostly a 511 keV radiation. Due to the inertial interaction with the CL space the rotating electrons will provide a feedback to the positive galactic nucleus, so the direction of its rotational axis will be kept stable.

The rotating positive egg-shell creates conditions for uniformly distributed electron clouds in angular range centred about the galactic plane, but not near the poles. The rotating and wobbling electrons near the two polar regions of the rotating galactic nucleus will get different magnetic force component: one will be attractive, while the other will be repulsive. For this reason the electron clouds around both poles will have different number density. This difference will appear as a signature of radiation at 511 KeV.

The observations of 511 KeV emission from the central region of Milky way is in complete agreement with the above presented concept. Uchida and Gustein (1995) estimate a large scale magnetic field in the galactic centre by measurements of Zeeman splitting in 1665 and 1667 MHz. Recent results from OSSE (Oriented Scintillation Spectrometer Experiment) provides a map of the galactic emission at 511 KeV (reported by Purcell et al., (1997). Figure 12.25.F shows the map of the radiation. By using data from number of experiments and applying a maximum entropy method L.X. Cheng et al. (2001) suggest that the 511 KeV radiation consists of two components: bulge centroid with $FWHM \approx 5°$ and galactic plane component. The bulge centroid, according to BSM, is from the cloud surrounding the positive galactic nucleus.

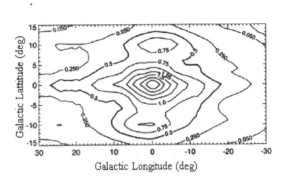

Fig. 12.25. Galactic emission at 511 keV (after Purcell et al. (1997)

Additional observational data indicating a rotating positive galactic nucleus are evident from the measurements of molecular clouds velocities near the Milky way centre. Figure 12.25.G shows the cloud velocity versus the galactic longitude (T. M. Bania, 1977). The measured cloud velocity is from a $^{12}C^{16}O$ molecular cloud in the galactic longitude range of $0 \pm 10°$ . .

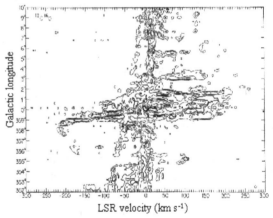

Fig. 12.25.G Longitude-velocity contour diagram of the $^{12}C^{16}O$ emission observed at $b = 0°$ for the region $10° \geq l \geq 352°$. Contour levels correspond to T = 1.4, 3.0, 5.0, 10, 15 K (After T.M. Bania, 1977)

We see that the rotational velocity in the range $0 \pm 2°$ changes between 0 and 200 km/s. The explanation of BSM is the following: The rotating positive galactic nucleus shell induces rotational field component in the stationary CL space through the bearing gap (empty space). The rotational field directly or through the existing electron clouds provides momentum to the molecular clouds. The inertial interactions of the dragged in this way molecules excite them and they emit a not thermal radiation (at very low temperatures). This mechanism obviously has a limited radial range and is influenced by the strong magnetic field from the spinning galactic nucleus positive shell (rotating together with the bulk nucleus of the intrinsic matter enclosed inside).

The assembly of the positive galactic nucleus shell and the internal bulk matter is further referenced as a **galactic nucleus**.

12.B.6.1.2 Kinetic energy storage mechanism of the galactic nucleus

The provided concept shows that the positive galactic nucleus shell together with the internal bulk matter are kept in the centre of the active galaxy by a complex mechanism in which the following processes are involved:

(a) SG interactions of (CP) and (TP) type between the positive galactic nucleus (the internal shell from the protogalactic egg) from one side and:

- the proximity CL space (strong electrical and magnetic field generation)

- the free electrons in the CL space closer to the egg-shell

(b) radiation of high energy photons from the excited free electrons

(c) radiation from molecular clouds in the range of ±2° galactic longitude (data valid for Milky way)

(d) long range interaction between the generated strong magnetic field and the galactic matter in the bulge and the disk of the galaxy.

The energy of these interactions comes from one source: the rotational energy of the galactic nucleus in respect to the stationary CL space. The latter is referenced to the stationary Universe. The rotational kinetic energy is like the kinetic energy of a classical flywheel. Evidently this energy is enormous (having in mind the huge matter density of the galactic nucleus). This kind of stored energy may compensate the energy lost from the continuous radiation from the stars, an energy that escapes the home galaxy. When considering an intergalactic energy balance, the stored energy may compensate the inevitably energy imbalance. In such way

it provides conditions for a long active life of the host galaxy.

Figure 12.25.H illustrates the model of the interacting galaxy nucleus according to the presented concept. The following notations are used:

1 - bulk nucleus with low level structures from two intrinsic matter substances
2 - external growing layers of QB's with alternative handedness
3 - empty space gap
4 - internal positive shell from the protogalactic egg
5 - empty space gap
6 - boundary zone of the CL space
7 - electron cloud

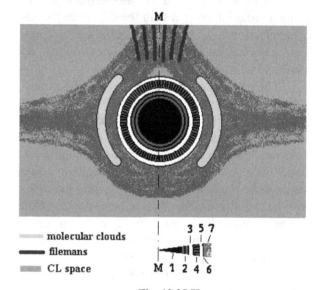

Fig. 12.25.H
Model of Milky way galaxy nucleus illustrating formations involved in various type of interactions

The bulk nucleus contains the very low level structures from two type of intrinsic matter, that are still able to carry the memory of the correct handedness.

The alternative layers 2 are preserved from the phase of the intrinsic matter segregation before the eruption (a phase preceding the formation of the protogalactic egg). The most external of the alternative layers 2 is a negative. (It is not possible to infer only: are the structures of the most external layer still of QB type or destructed to QP type?).

The interactions between the positive shell of the galactic nucleus and the internal intrinsic matter (with a spherical shape) could allow the both formations to have almost zero velocity. In such way the positive galactic nucleus shell will probably tend to reach the rotational velocity of the nucleus, obtained during the galactic collapse.

The gap 5 serves as an ideal bearing between the stationary CL space and rotating positive shell of the galactic nucleus. At the same time it allows the kinetic energy of the nucleus to be transferred to the surrounding electron and molecular clouds and finally to the galactic matter. The boundary zone is of similar type as the TCL space, but it is permanently stable. It protects the gap 5 from penetration of migrating electrons or any elementary particles including folding CL nodes.

The galactic matter could surround the nucleus but in the normal CL space, i. e. outside of the boundary zone. The molecular clouds (green area) showing a rotational velocity in a range of -200 km/s to +200 km/s are in the galaxy longitudinal range about ±1.5°. The free electrons occupy the closer area (red coulor) with asymmetrical tails for the two magnetic poles. The filaments (blue area) are remnants of these globular clusters, which has been destroyed during the explosion of the protogalactic egg. The CL space of this matter is not mixable with the new galactic matter.

While the stored energy in the galactic nucleus is perhaps exhausting slowly during the active galactic life, the magnetic field that it supports will be also decreasing with the time. The magnetic field may play one important role: deviating the pulsars that occasionally move inside the bulge of the galaxy from falling on the galactic nucleus. The moving pulsars posses a jet propulsion mechanism and superstrong magnetic field (see §12.B.6.4) and if some of them are not properly deviated they may hit the positive shell of the galactic nucleus. If the latter breaks as a result of a such bombardment, the galaxy collapse will occur prematurely. For galaxy nucleus with large kinetic momentum the pulsars could be deviated not only by the galactic magnetic field but also by the radiational pressure from the X-rays emitted by the electron clouds (see §3.2 in Chapter 3 of BSM for X-ray emission from oscillating electron).

The provided BSM concept about the stored energy in the galactic nucleus and the mechanism of its radiation is in agreement with the recently observed phenomenon in the galaxy MCG-6-30-15. Jörn Wilms of Tuebingen University, Germany, and an international team of astronomers identified a phenomenon of "power trapping" by observing a supermassive black hole in the core of the galaxy named MCG-6-30-15 with the European Space Agency's X-ray Multi-Mirror Mission (XMM-Newton) satellite. In a paper posted by NASA at 22 October 2001, titled: "New Study Shows Black Hole Belching Energy" it is written:

Scientists for the first time have seen energy being extracted from a black hole. Like an electric dynamo, this black hole spins and pumps energy out through cable-like magnetic field lines into the chaotic gas whipping around it, making the gas - already infernally hot.
(http://www.space.com/scienceastronomy/astronomy/blackhole_energy_011022.html)

12.B.6.1.3. Summary:

- **The galactic nucleus possesses enormous kinetic energy. This energy may determine the duration of the active life of the galaxy.**
- **The galactic centre is a region of complex phenomena, whose signatures, however, are observable parameters leading to a reasonable logical interpretation from a point of view of BSM concept about the space and the Universe**
- **The superstrong galactic magnetic filed may play a role about the galactic matter evolution and its integrity during the active life of the galaxy**

12.B.6.2. Galaxy rotational problem

The galaxy rotation refers to the phenomenon of the discrepancy between the observed rotational motion of the matter in the galactic disk from the predictions of the Newtonian dynamics. This is illustrated by Fig. 12.25.I.

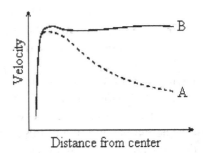

Fig. 12.25.I. Galaxy rotation problem. A - expected rotational curve, B - observed rotational curve

Curve A shows the expected rotational curve based on the Newtonian mechanics. Curve B shows the typical shape of a rotational curve for a spiral galaxy. It appears from the galaxy rotation curve that part of the galactic disk behaves as a solid body. It is in a sharp contrast from the rotation of the planets around the Sun. This has been a big confusion so far. Attempts to resolve this problem led to a suggesting of hypothesis that this is caused by a huge amount of invisible matter in a form of multiple black holes, so called "dark matter" Another hypothesis suggested a "modified Newtonian dynamics". Avoiding the speculations of the mentioned two hypothesis, BSM analysis leads to quite logical solution of the problem.

The initial spread of the matter occurs after the break-up of the external shell of the protogalactic egg. At the same moment a building process of expanding TCL begins with expanded CL space enclosed inside. The spreading new atomic matter can not be involved in a large inertial interactions with the new built CL space until it is not interconnected to the neighbouring galactic CL spaces. In the phase of CL space interconnection, the new CL space becomes spatially fixed in result of which, the spread new matter appears in motion in respect to the own CL space. But the protogalactic egg generally has been in a relative rotational motion in respect to the neighbouring galaxies. In result of this, all material objects of the new galaxy (particles, atoms, molecules, solids) may obtain a large momentum with two components: a rotational and a radial one. The Newtonian type of gravitation will compensate the radial component of expansion, while the rotational component will be preserved and even increased. This is kind of transition process, which may have extremely long transition period.

The presented scenario leads to a **conclusion that the observed rotational curves of the spiral galaxies is not a stationary but a transitional process.**

In fact all of our observation are taken during an intrinsically small period of the galactic active life. The transitional process could be even comparable with the active life duration of the galaxy (or may be it does not reach a fully stationary rotation). The finite lifetime of the stars ending with explo-

sion and birth of pulsar also may influence the rotational motion.

Figure 12.25.J demonstrates a rotational curve for NGC4138 as a position-velocity map, according to Broelis. A. H. (1995).

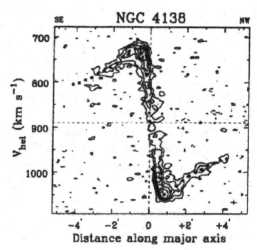

Fig. 12.25.J. Position-velocity map of NGC according to Broelis. A. H. (1995).

It is apparent, that after the CL interconnection the new born galaxy gets feedback in respect to the stationary CL space and stationary Universe. The feedback energy depends mainly of two factors: the amount of the rotational energy momentum and the matter quantity in the new galaxy. One particular case deserves attention. Part of former space of the new born galaxy could be annexed by its neighbours due to some cosmological events in the galaxy neighbourhood. In such case the new galaxy may not have enough room to develop as a spiral type. Rather it will obtain an elliptical or lenticular shape. Then it is quite possible that a part of the newly formed prisms (with some particles) may penetrate in the foreign CL space of its neighbours, where they may disturb the local CL space. The thrown matter will obviously occupy the most peripheral part of the galaxy lying on the galaxy plane. The penetrated matter could not operate effectively in a foreign CL space so it could stay much colder. The possible signature of such phenomena could be a sharp dark features in the galactic disk periphery. The best condition for observation of this hypothetical effect is when the galactic plane is aligned with the line of sight. Number of observed disk type galaxies show sharp dark features in their disk, but they are interpreted so far as strong dust absorptions.

12.B.6.3 Some features in the processes of star formations and their evolution

The current observations from our galaxy may not contain information from its evolution immediately after its birth. Galaxies at such phase of evolution might be quite distant, so we may lack of good observational data from this phase. Then we may extrapolate our vision using some secondary observed features, for example, the abundance of hydrogen and deuterium in the deep space. Due to the extremely large dynamics in the immediate phase after the galaxy birth, more complex atoms or molecule are less likely to be formed. The galaxy birth is accompanied with a huge initial burst of gamma and X-ray. After the release of this energy, one may expect formation of mostly Hydrogen, less likely deuterium and much less likely helium. The only possible reliction of this is the abundance of these atoms in the deep space.

The process of star formation has been studied from many decades and it is not a main issue of the BSM theory. Only some particular moments of the star evolution will be presented for which some new mechanisms are identified. They are related mostly to the evolution phase described as a Main Sequence, which in fact is very time distant from the galaxy birth. Before the Main Sequence a Pre-Main Sequence exists, in which a protostar is formed. (The Pre-Main and Main Sequences a features of the star evolution).

A protostar could be born from condensed gas clouds and its typical evolution is described by the Pre-Main Sequence. The probable scenario is a following:

Cold hydrogen molecules concentrate into giant clouds with 200,000 to 1,000,000 times the mass of our Sun. The matter density of such formation increases slowly until the newtonian gravitation start to play a role causing formations of gas fragments. The cores of this fragment (with a mass much smaller than the gas cloud mass) converts into smaller denser bodies. When the internal mass density reaches some critical level, the increased internal pressure may invoke a fusion reaction. A

ptotostar is born. The most probable nuclear reactions are:

$$(p \rightarrow n) + p = D + energy \qquad (12.34)$$
$$D + D = {}^4He + energy \qquad (12.35)$$

About 25% of the protostars are born from gas clouds in a star formations, known as open clusters. Most of the born stars escape from the cluster. It is quite probable some energetic balance between the cluster matter and the CL space environment to be involved in this process. The evolution process of the protostar may take from 100,000 years to 1 million years in the Pre-Main Sequence. It depends mostly on the mass of the protostar. A smaller mass protostar spends much more time than a larger mass until reaching the Main Sequence.

12.B.6.3.1 Main sequence and particular points.

The Main Sequence is a typical process of the star evolution in which the relation between the surface temperature and the luminosity follows particular curves described by the Hertzsprung-Russel diagram (H-R diagram). The observation material about this phase of the star evolution is richer than the Pre-Main sequence. H-R diagram is shown in Fig. 12.26.A.

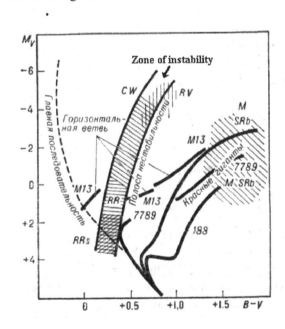

Fig. 12.26.A. Hertzsprung-Russell diagram

According to the classical concept, the evolution of the stars in the Main sequence of H-R diagram depends mostly on its mass. In such aspect, three main cases are distinguishable"

  (a) case: the stellar masses is between 0.8 and 11 solar masses

  (b) case: the stellar masses is between 11 and 50 solar masses

  (c) case: the stellar masses is greater than 50 solar masses

  If the born star corresponds to case (a) it spends about 100,000 to 10,000,000 year (according to contemporary astronomy) until reaching the Main Sequence. But stars more massive than 11 Suns are essentially born on the Main Sequence.

  The time evolution in the Main Sequence follows the direction from the upper left end of the curve to the down right end. Stars within different mass range, spend different time on the main sequence. Low or intermediate-mass star spends between 80 to 90 % of its life in the Main Sequence phase.

  BSM provides analysis only for some particular points or sectors of the Main Sequence. One such sector is the range of instability. It is intercepted by a zone of instability as shown in H-R diagram of Fig. 12.26.A. A star from a Main Sequence in the sector of instability behaves as a variable star. It may pass this critical sector continuing on the Main Sequence as a stable star or it may deviate in a zone of instability. Cepheids are typical representative of the variable stars in the range of instability. They exhibit well defined Period-Luminosity relationship. The cosmological science has not provided so far clear convincing explanation about the mechanism of this instability. The observations shows that the star may have a loop in this region with two exiting options:

  - following the main sequence of H-R diagram.

  - following a horizontal path of H-R diagram and converting to red giant

  The BSM analysis of the star behaviour in the sector and zone of instability, unveils that a hidden so far physical process is behind the instability and the stars known as variable stars.

12.B.6.3.2 Physical process related to the zone of instability in H-R diagram, according to BSM.

  The instability sector of the Main sequence of H-R diagram is well after the beginning of the curve, as shown in Fig. 12.26.A. The provided below analysis is focused on this sector.

  From a point of view of BSM, the balance between the gravitational energy of the star and the CL space environment is important factor for the star evolution in the main sequence of the H-R diagram. The gravitational pressure in the central region of the star is quite big. The conversion of Hydrogen into He in these region leads to increasing the density. Then the SG forces between the hadrons (proportional to inverse cube of the distance) begin to contribute significantly to the effective gravitational pressure in the central region. At some critical value of this pressure it may affect the structure of the elementary particles proton and neutron from which the atomic nuclei are built. The external positive shell (envelope) of the proton (neutron) is a third order helical structure. Its stiffness is lower than the stiffness of the second and first order helical structures. Therefore, it is reasonable to accept that this structure (of the proton and neutron) may break first. In this process its internal RL(T) structure is obviously destroyed into partly folded LR nodes, which are of the size of the CL nodes and easily escape out. The internal pions and kaons, however, are more resistant on the increased gravitational pressure. They are only cut (in one or two pieces). The most of pions loos their second order helicity and are converted to straight FOHSs structures like the central kaon structure. All they form a bundle of kaon-like structures (FOHSs). In such case **the star obtains a kaon nucleus**. This concept of kaon nucleus in stars is in full agreement in the analysis of the pulsars, presented in §12.B.6.4. (Chapter 12 of BSM), while a small kaon nucleus in planets is discussed in §10.6.2 (Chapter 10 of BSM). The size of the nucleus increases with every pulsation, but it is always surrounded by a large quantity of atomic matter. The latter protects the ends of the kaon nucleus, not allowing destruction of the internal RL(T) structure (as in the decay of a single kaon in CL space - see Chapter 6 of BSM). The axially aligned RL(T) from the aligned FOHSs provide a conditions for a **super strong magnetic field of the star** (this conclusion is apparent from the pulsar analysis in §12.B.6.4). According to BSM, the magnetic field of the star may have important role about the nuclear reactions in the star residing on the main sequence of H-R diagram after the region of instability:

(A). The star's magnetic field probably creates conditions for a larger variety of fusion reactions leading to building of atomic nuclei with higher Z numbers. It helps the protons, neutrons and deuterons to get a suitable alignment for such type of nuclear reactions.

One may argue: Why the critical mass of the planet for obtaining a kaon nucleus (see Chapter 10 of BSM) is much smaller than any stellar mass? The possible answer is: The conditions for formation of such nucleus depends not only of the mass but of the atomic mass density of the astronomical object, because SG forces are also involved. The planets are formed of higher Z number atoms than the stars (especially stars from this part of the main sequence), so the planetary density in the central region is much higher. Consequently in the case of star, much larger mass is necessary for crushing the protons and neutrons in the central region.

The process of crushing the protons (and neutrons) appears in steps, so the kaon nucleus increases in steps. During such event the nodes of the destroyed RL(T) structures fly folded through the star, but they are distinguished from the folded CL nodes (RL nodes are formed of 6 prisms, while CL nodes are formed of 4 prisms). Flying folded RL nodes however may cause excitation of the star matter. It may perhaps cause an eruption of matter from some internal regions and pulsation of the most external gas layer. These phenomena provides possibility for detection of:
- luminosity in function of periodical phase
- estimation of upper layer pulsation by measuring the Doppler shift of some spectral lines

The flying folded nodes (passing also through the molecules, atoms, proton and neutron) may simulate moving of emitting or absorbing atoms with high velocities, while in fact they do not have such. This is a specific effect, theoretically inferred by BSM analysis, that may simulate a rela-

tivistic Doppler shift. Then the shifted spectral lines could be mistakenly attributed to a high velocity gas motion. **According to the BSM conceptual analysis, the large radial velocities measured for the cepheids are contributed by such effect.**

Now it is apparent that the emitted radiation is proportional to the quantity of the released RL nodes from the crushed RL(T) structures. The experimental evidence about this is provided later.

The described features are valid for variable stars known as **cepheids of I-st type**. The described process is in agreement with the inferred postulated rules P3 and P5 about energy balance for structures of Intrinsic Matter. The Newtonian mass we are familiar with, is proportional to the Intrinsic Matter. From this and from presented concept the following conclusion follows for these variable stars:

(A) Larger stellar mass -> larger quantity of crushed protons and neutrons -> larger quantity of destructed RL(T) -> larger emitted radiation (larger luminosity) (A)

(B) Larger stellar mass -> larger period of pulsation

The above relations (A) and (B) lead to a logical explanation why the period of the fluctuation is directly related to the luminosity of the cepheid star.

The kaon nucleus is much denser and heavier than the atomic matter. Then it may play a role for triggering a nuclear reaction in the star. Let us assume that a critical volume quantity of kaon nucleus (a critical mass) is necessary for this purpose. The formation of kaon nucleus is accompanied with an eruption and lost of some star material. This, however, causes some decrease of the pressure on the central region. So the obtaining of the necessary critical mass of the kaon nucleus for effective nuclear reactions and the loss of matter are two ambiguous processes. For this reason the variable star could make an open loop in this region. If the first process prevails, the star continues on the main sequence path of H-R diagram. If the second process prevails, it continues in the horizontal branch and ends up its evolution as a red giant. In this case it may not have enough kaon nucleus mass and strong magnetic field, necessary for more effective nuclear reactions.

Our Sun follows the main sequence after the region of instability. It possesses a strong and stable magnetic field, so it perhaps has a well formed kaon nucleus.

12.B.6.3.3 Evolution of stars with masses between 11 and 50 Suns.

Evolution process

The evolution process in this case follows the phases:

- red or blues supergiant with a helium core
- or red supergiant with iron core
- supernova
- pulsar ("neutron star")

Note: BSM concept uses pulsar instead of a "neutron star", putting the latter term in quotation marks for reasons explained below.

For a star belonging to the Main sequence after the range of instability, the primary elements as H and D are converted to elements with higher Z numbers. It is well-known fact that the mass deficiency (binding energy) increases with the Z number. This is related to the Newton's gravitational mass. So the inertial mass of the star changes in the same direction as it is changed by the increase of its kaon nuclei. This tendency could be expressed also by the change of the matter density. The increased matter density involves increasing of the intermolecular SG forces and consequently the internal gravitational pressure. At the same time the star have lost a lot of energy due to radiation so the average excited state of the atoms continuously decreases. This provides a compacting effect on the atomic matter. The inverse cubic law SG forces between the atoms then get additional increase. At some point the kaon nucleus may increase by crushing additional matter of the internal surrounding layer. If the amount of the crushed matter (protons and neutrons to kaons) is significant, it may blow out almost all of the atomic matter layers. This phenomenon is a **collapsing star** (according to BSM concept). The process may appear as a huge explosion. Since this phenomenon is strongly influenced by the kaon nucleus, which always have and extended shape, the shape of the explosion is not spherical. Consequently:

The phenomenon of collapsing star shows specific features indicating existence of kaon nucleus:

(a) two jets in a common jet axis

(b) atomic matter thrown around the plane normal to the common jet axis

(c) hardware neutrino particles

The phenomena (a) is from the destructed RL(T) structures of crushed external shell of the protons and neutrons. The released RL nodes (known as neutral current in Electroweak theory) obtain preferential direction along the superstrong magnetic field created by the kaon nucleus.

The phenomena (b) is the released atomic material by the explosion.

The phenomena (c) is mainly from the crushed FOHS envelop of the external proton's (neutron's) shell.

The Hubble Space telescope has provided an excellent opportunity for observation of such phenomena. Pictures corresponding to the described events are shown in Fig. 12.27.A, Fig. 12.27.B and Fig. 12.27.C

Egg Nebula
Hubble Space Telescope • WFPC2 • NICMOS

Fig. 12.27.A dying star
(www.seds.org/hst/97-11.html)

The process of star dying may take finite time. It could be interrupted and then starting again. Finally it will lead to simultaneous crush and release of atomic matter until one or two ends of the kaon nucleus become completely uncovered from the possible atomic matter. In the latter case the explosion could be even more powerful. The phenomena of supernova is one option of star dying process. Stars whose initial mass is within the mentioned range inevitably provide a pulsar or object of "black hole". The physical mechanisms of the jets end the eruption become apparent in the BSM analysis of the pulsars provided in the next section.

2. Additional analysis of observations from a point of view of BSM is presented in the next sections of Chapter 12 of BSM (enclosed in CD ROM).

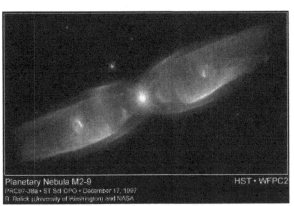

Fig. 12.27.B. Picture of Dying star
(http://oposite.stsci.edu/pubinfo/pr/97/38.html)

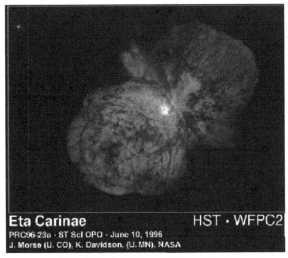

Fig. 12.27.C.
(http://oposite.stsci.edu/pubinfo/pr/97/38.html)

The terms "egg nebula" and "planetary nebula" are adopted only due to geometrical similarities, but does not involve any physical similarities with the galactic formations.

Notes:

1. The references used in this chapter refer to the references of the BSM theory, which are provided in the CD ROM included to this book.

12.B.6.4 Pulsars

Extensive BSM analysis of accumulated observation material about pulsars leads to the conclusion that the pulsar is an extended object formed of dens kaons-like structures, so it is denoted as a kaon nucleus. The conditions leading to formation of kaon nuclei were discussed in Chapter 10. The formation of kaon nucleus begins in the womb of the star in a particular moment of its evolution, which corresponds to the phase of instability of the R-H diagram. It is a result of crushed protons and neutrons cased by the increased gravitational pressure in the central region of the star. The kaon nucleus, formed of aligned straight FOHSs is much denser than the atomic matter. In the process of dying star, the atomic matter covering the kaon nucleus is thrown. The free kaon nucleus becomes a pulsar possessing unique properties, which will be discussed in the following sections. The new born pulsar from a dying star may still carry some atomic material around its external surface.

12.B.6.4.A. Kaon nucleus

The kaon nucleus is a large bunch of aligned FOHSs. During a normal star existence the both ends of kaons bundle are covered by atomic matter. The structure of the kaon nucleus inside of the star was discussed in §10.14, Chapter 10, in connection to a hypothesis about the magnetic field of a star. A straight kaon nucleus (bundle) can be integrated not only from cut kaons, but also from cut positive and negative pions. Their helical structures undergo a partial twisting (discussed in Chapter 6) in which they obtain a charge defined by the most external layer of the embedded RL structure. At equal number of positive and negative FOHSs with a same integrated length, their charges are compensated in proximity and the total kaon nuclei will appear as a neutral.

Fig. 12.28 illustrates a magnified portion of a sectional view of the kaon nucleus.

i

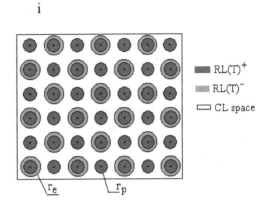

Fig.12.28. A magnified portion of a sectional view of kaon nucleus

The number of positive FOHSs of type $H_m{}^{2-}$:+(-) is equal to the number of negative FOHSs of type $H_m{}^{1}$:-(+(-). In such case the total charge is equally balanced. The small edge effect charge difference is negligible and cannot affect the total charge integrity. Every positive FOHS with a RL(T) contains a negative core. Every negative FOHS contains with a RL(T), contain inside a positive FOHS with RL(T) with a central negative core (see the detailed description of kaon's and pion's internal structures in Chapter 2). Both types of FOHS are separated by channel-like gaps accessible for the surrounding CL space. The twisted proximity fields of RL(T) provides the inaccessibly gaps between the straight shaped helical structures, satisfying the requirement for a finite gaps, discussed in §6.4.3, Chapter 6. These long gaps are occupied by CL space (internal CL space), which is strongly modulated by the SG field of the RL(T)s of the helical structures. From one hand, this provides conditions for strong phase synchronization between the SPM vectors of the CL nodes along the align gaps between the straight FOHSs (kaons). From the other hand, due to the strong proximity SG field from the RL(T) structures, the resonance frequency of the internal CL space might be a higher harmonics of the free external CL space. It is evident that such arrangement of positive and negative FOHSs is able to provide a super strong magnetic field. The common mode synchronization provides energy flows in closed path in a form of magnetic field (for more details see the magnetic field explanation in Chapter 2). In the ordinary atomic matter built of protons and neutrons, such effect is not possible.

The portion of the kaon nucleus radial section, illustrated by Fig. 12.28 is without any structural defect. In this case the supperstrong magnetic field inside of the structure should be completely homogeneous. For such structure it is obvious that:

(A). The total energy of the magnetic field of the kaon nucleus is proportional to the quantity of FOHS's.

Note: If the kaon nucleus is not burning and not involved in any interactions with matter, it still possesses a total magnetic field energy in form of reactive energy of CL space. The same is valid for any single permanent magnet.

While the kaon nucleus of the star begins in the zone of instability of H-R diagram, it may continue to grow slowly after that, since the nuclear reaction leading to formation of heavier elements increases the mean specific gravity of the star.

Fig. 12.28 illustrate also another important structural feature of the kaon nucleus: It is formed of equal quantity of positive FOHSs and negative compound FOHSs. The latter contain internal positive FOHSs, but they are hidden inside and could not contribute a positive charge in the surrounding CL space. While the kaon nucleus is intact (inside of the star) its appears as neutral. However, when a process of destruction begins, the internal RL(T) structures begins to be released and the charge neutrality is disturbed. The process of destruction is similar like the destruction process of the kaons, pions and muons, discussed in Chapter 6. For the kaon nucleus, however, the destruction process, which we may call "burning" will be influenced by the its super strong magnetic field. Mixed RL nodes (positive and negative) will provide a flow, that according to the electroweak theory corresponds to a "neutral current". The both types of neutral current may drag and excite the atomic matter around the kaon nucleus envelope. The latter could be ionized, providing a broad band emission spectrum of lines. These are the observed jets from the dying star, in which the process of the kaon nucleus gets a final grow leading to eruption of the atomic matter.

12.B.6.4.B. Environment conditions in the process of kaon nucleus destruction

Let us consider the destruction of a charged balanced kaon nucleus. The intrinsic mass quantity of the positive straight FOHS is included mainly in its internal RL(T) structure. The negative FOHS, however, contain both types of internal lattices RL(T)- and RL(T)+. Their energy ratio is equal to the energy ratio of the same structures in the electron external shell and the positron (see §6.9.1 and 6.9.3.3 in Chapter 6):

$$(1.44 \text{ GeV})/(1.7778 \text{ GeV}) = 0.0234 \ (E_D^+/E_D^-)$$

The destruction of any negative FOHS will provide a neutralized flow plus 2.34% charged component, according to the above relation. (The neutralised component is equivalent to the "electroweak neutral current" according to the therminology of electroweak theory, known also as V-A theory, while the charged component is equivalent to a "charged electroweak current"). The destruction of any positive FOHS will provide only a positive component (the matter quantity in the negative central core is negligible). **Consequently, the total flow of electroweak current from the kaon nucleus destruction will be dominated by a large positive component**.

The flow of the positive component ("electroweak charge current") will be affected by the strong magnetic field and will be located in the jet region. At the same time, the flow of the balanced component ("electroweak neutral current") will be not affected by the axial magnetic field and could be released in a much larger spherical region. The kaon nucleus of a young pulsar may contain some atomic envelope of matter, which could be excited differently from both types of electroweak currents.

12.B.6.4.1 Birth of pulsar

During a normal star existence, both ends of kaons bundle are covered by atomic matter. After the final explosion, however, at least one of both ends becomes uncovered. Then a process of continuous destruction of RL(T) structure begins, forming a jet from the free end. **A pulsar is born**. The destruction process has some similarity with the process of single kaon decay in atomic coliders where the released energies, according to BSM, are embedded SG energies corresponding to the energy equivalence of the masses of "bosons" (Chapter 6). From one side, the released RL(T) structures of kaon nucleus interacts with the CL space in a similar way as the energy of bosons in the particle coliders. From the other side, the process is influenced by the super strong magnetic field. As a consequence, the pulsar (burning kaon nucleus) possesses a continuous operating propulsion mechanism. It is active until all kaon nucleus is exhausted. We may call the process a **kaon nucleus burning**. It has the following distinct characteristics in comparison to the characteristics of the dying star:

(a) most of the atomic matter is erupted, but small portion may still exist in the earlier age of the pulsar

(b) some inclusions of atomic matter could be trapped in the kaon nucleus

(c) the released energy from the kaon nucleus burning, in comparison to the energy of atomic matter is enormous.

The feature (c) could be easily proved by the unveiled structure and properties of the electron (Chapter 3). The maximum optical radiation from the oscillating electron is 2 x 511 KeV, while the embedded SG energy, estimated by the released energy from destruction of both RL(T)s is $1.444 + 1.7778 = 3.22 GeV$. So the ratio between both types of energies is 3150 times in favour of RL(T) energy.

In the active stars like our Sun, the magnetic polar axis coincides with the spin axis due to the inertial interaction with the Milky way CL space (see Chapter 10). In the dying star, however, the spin axis may get a change due to the reactive momentum from the erupted material. Then the spin axis of the new born pulsar could not coincide any more with the magnetic field axis and the kaon nucleus with its magnetic axis will get wobbling. In such way the reactive force from the nuclear burning will provide two components of thrust momentum - a rotational and an axial one. The pulsar motion in this case will be similar as a coning motion of a rocket without stabilizers. This kind of rotational motion creates radiowaves, which are detected in packets with a period equal to the rotational period of the kaon nucleus. The BSM analysis identified a mechanism of spin stabilization

accompanied with emission of periodical radio frequencies. This issue is discussed the next few sections.

12.B6.4.2. Idealized model

The idealized model of pulsar is a kaon nucleus with a shape of cylinder formed of aligned FOHSs.

In the idealized model, the axis of kaon nucleus with a shape of cylinder coincides exactly with the axis of the magnetic field. In the real model (discussed later in §12.B.6.4.7), this condition is usually not fulfilled. Figure 12.29 illustrate the overall shape of the real (a) and idealized (b) pulsar model. The latter is convenient for analysis of the pulsar interaction with the CL space.

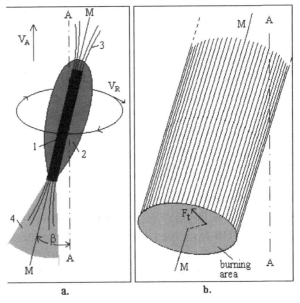

Fig. 12.29

a. Pulsar model; b. Magnified version of kaon nucleus (an idealized version)

In section **a**. of the figure the idealised pulsar model is shown, where:

1 - is the kaon nucleus (idealised shape)

2 - is atomic matter

3 - is a magnetic line

4 - is a jet from destructed RL(T) structures

A-A - is the spin axis

M-M is the magnetic axis of kaon nucleus

V_A and V_R are respectively the axial and the rotational velocity

β - angle between the magnetic and rotational axes

In section b. of the figure the burning end surface is shown.

12.B.6.4.3 Pulsar's features

The pulsar has the following characteristics:

General features:

(a) a superstrong magnetic field

(b) a proper motion

(c) a very stable secular period (estimated by integrated pulse profile)

(d) a very slow period increase with the time

(f) a high degree of polarization of emitted frequencies

(g) glitches in the pulse sequence

    - a subpulse phase drift in respect to the integrated pulse profile

    - a sudden small period increase with a slow restoration of the general trend of increasing period.

Specific features:

(h) double pulse profile

(i) short time pulse fluctuation and discontinuities

Features (c), (d), (e), (f), (g), (h) are observable, while the features (a) and (b) are revealed from the BSM analysis.

Kaon nucleus burning

If considering the disintegration of a straight FOHS from the kaon nucleus, the process is similar as the disintegration of the kaon structure, described in Chapter 6. From the analysis of the particle physics experiment, it was found that the FOHS disintegrates with a constant rate. Firstly, the RL(T) is released as a continuous structure without a helical boundary, but still exhibiting a charge. The disintegration of the individual RL nodes (containing 6 prisms) into CL nodes (containing 4 prisms) may be a long time and distance process, keeping in mind the large penetrating features of the partly folded RL nodes. However such refurbishment affects directly the Static CL pressure, which is a constant CL space parameter for the whole galaxy. Then it is reasonable to formulate the following rule:

(A). The refurbishing rate of RL nodes (Λ)into folded CL nodes in unit CL space volume (V) is a constant value.

$$\Lambda / V = const \qquad (12.36)$$

where: $\Lambda = N_{rf}/t$ - number of refurbished nodes per unit time

The released RL(T) structure from the high velocity jet has a highly ordered twisted shape, but without boundary core. Such structure exhibits a strong charge and consequently interacts with the strong magnetic field. This interaction leads to CL space pumping and generation of quantum waves which are in the radiowaves frequency range. In this process it first releases a portion of its energy (due to its twisting) before undergoing the process of node refurbishment, which might be spread in a vast space region. Similar processes, according to BSM, exists in the particle coliders (see destruction energies of RL structures in Chapter 6). Consequently the released by the jet RL(T) structure may posses a limited life. This life will depend on the strength of the magnetic field, because it has axial alignment to the FOHS's axes in the burning area. So the released RL(T) structures are well guided by the magnetic field. Then the strong magnetic field will extend the finite life of the released RL(T) structure. At the same time, the density of the released RL(T) structures will decrease with the time, because of the large velocity and the divergence of the jet (due to the magnetic field). Then for a stronger magnetic field the beginning of the refurbishing process will start at a larger volume (V). In such case according to conclusion (A) expressed by Eq. (12.36) the rate of the node refurbishment Λ will be larger. But in the described process it is evident, that the debit flow of the jet is controlled by the refurbishing rate.

Consequently we arrive to the conclusion:

(B). The flow debit (D_F) from the jet is proportional to the total energy of the magnetic field (E_M).

$$D_F = \frac{\Phi_{RL}}{A} = \frac{N_{RL}}{At} \sim E_M \qquad (12.37)$$

where: Φ_{RL} - is a flow rate, A - is the burning area, N_{RL} - number of released RL nodes, t - is unit time.

The total energy of the magnetic field is proportional to the quantity of the FOHS's in the kaon nucleus, according to the conclusion (A) in §12.B.6.4.A. The amount of the released matter from the kaon burning for one rotational period is intrinsically small in comparison to its total matter. Consequently:

(C). For a time interval much smaller than the pulsar life the magnetic field could be considered as a constant. According to Eq. (12.37) the flow debit in this case is a constant.

Inertial interactions

The aligned FOHSs' with their RL(T) in the kaon nucleus modulate strongly the aligned gaps of the CL space between them. For this reason the kaon nucleus exhibits highly anisotropical inertial factor. This means that the motion in a direction aligned to its axis exhibits much less inertial interaction, in comparison to a motion in any other direction. For the idealised nucleus, the kaon axis coincides exactly with its magnetic axes. So the anisotropy of the inertial factor could be conveniently referenced to the magnetic axis.

$I_F = f(\beta)$ - kaon nucleus inertial factor
where: β - is the wobbling angle shown in Fig. 12.29.a.

The kaon nuclear inertial factor has anisotropy along the kaon nucleus axis. More specifically the axial inertial component is smaller than the radial one. Then the inertial interactions with CL space could be easier analysed if regarding the translation-rotational motion of the nucleus as two separate motions: an axial motion and a rotational motion around the axis A-A. We see that the inertial interactions with the CL space increases with the angle β. This involves some energy of CL space, which is not directly observable, so it is some type of reactive energy. It is analogous to the reactive energy at the imaginary CL space separation surface, for astronomical body with a local gravitational field (see Chapter 10). Using the same analogy, this reactive energy is equal to the kinetic rotational energy and could be detected at the moment when the latter is changed.

(D). The rotational motion of the kaon nucleus involves a reactive energy of the CL space, equal to the kinetic energy of the rotation.

12.B.6.4.4 Energy radiation process and period stabilization

The mechanism of EM pulse generation by the pulsar is more specific and unique. Optical counterparts are observed only for a new born pulsars indicating that initially it may contain some atomic matter covering the kaon nucleus. With the aging of pulsar, most of such atomic matter is lost and if some thin layer is left it should be highly ionised. The CL space in the proximity to the kaon nucleus is heavily biased and the ionised atoms could not emit (if all electrons are lost). So emission in the optical range is missing. This concept is in agreement with the observations.

12.B.6.4.4.1 Concept of energy radiation

The observations of pulsars show short and long term variations of the integrated pulse profile and period. In order to unveil the process of the energy radiation we have to provide analysis in a proper selected time interval. Let us accept a time interval much larger than the rotational period but quite smaller than the pulsar life. In such case **the total magnetic field can be considered as a constant.**

The emission of radiopulses is due to a mechanism quite different than the existing mechanisms of the atomic matter. It is a result of the relative velocity obtained between the rotating magnetic field and the expanding charged component of released RL nodes. If neglecting the rotating magnetic field, the jet of RL nodes should have a similarity with the process in the particle colider experiments related with Z and Higs bosons.

The spin axis of the pulsar is tilted in respect to the magnetic axes. The released RL nodes stream is expanded in a cone. Such stream exhibits a strong electrical charge (as in the case of bosons experiments). The refurbishment of RL to folded CL nodes (6 prism to 4 prisms) requires a finite time, so the flowing charge needs a finite time for its dispersion. Then the released RL stream should form a conical spiral, the shape of which is illustrated in Fig. 12.30.

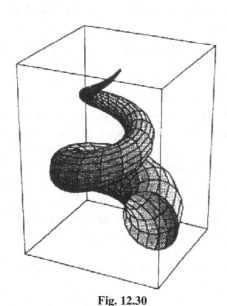

Fig. 12.30

Shape of released RL node stream from a pulsar jet. The jet is oriented downward

The stream of RL nodes from the burning kaon nucleus of the pulsar is different than from the dying star. The kaon nucleus contains equal number of positive and negative FOHS's but the positive FOHS has only a positive internal RL(T) while the negative FOHS contains both: a negative and a positive RL(T) (inside the negative one, see Fig. 12.28). Then the released flow will be comprised of dominated positive RL nodes. Using again the destruction energies values from RL(T)'s of normal electron we get:

$$\frac{1.7778 + 1.7778}{1.44} = 2.469 \qquad (12.38)$$

- disintegration energy ratio between positive and negative RL nodes.

Consequently, the energy flow from the released positive nodes dominates. Using the terminology of Electroweak theory, the released stream of RL nodes will contain components of "neutral electroweak current" and "charged electroweak current".

The observations show that the pulsars away from the galaxy disk are still able to emit radiopulses. So we may analyse the process in free CL space away of any atomic matter. In a free CL space, there is not any disturbing factor for the pulsar magnetic field and it will rotate synchronously with

the kaon nucleus (assuming the peripheral velocity of its extent is much lower than the light velocity). The same rule is not applicable for the stream of RL nodes. Even ignoring the magnetic field, which is induced by the charged component, the stream additionally exhibits inertial interactions with the CL nodes of the galactic CL space. Consequently the spiral cone shown in Fig. 12.30 will rotate but with some constant phase delay. Then the rotating magnetic field will provide a continuous push on the rotating spiral cone. The relative motion between the rotating magnetic field and the phase delayed spiral cone will have a component moving towards the spiral origin. This process will cause two effects:

(a) pumping of the CL space in the large volume region of the spiral cone followed by a pulse type of radiation in the radio spectral range

(b) feedback from the emitted radiation to the rotational speed of the pulsar

In comparison to the CL space pumping from electrons in quantum orbits, the pumping mechanism of the pulsar is active in a very large CL space volume. Then despite the smaller photon energy (corresponding to radio spectrum) the total energy of radiation is enormous. This energy appears as a feedback in the process of kaon nucleus burning and spin frequency (period) stabilization. It may affect the angle between the rotational and magnetic axes. From the analysis of all described features we arrive to the conclusion:

The stable rotational period of the pulsar is a result of active stabilization mechanism.

The process of spin frequency stabilization could not be explained if ignoring the interaction with the CL space. It relies on the conclusion of the constant debit, given by Eq. (12.37).

12.B.6.4.4.2 Coherence

The coherence is important feature allowing observations using interferometric methods.

By definition, the specific intensity (I_v) from a coherent emission source containing N particles must satisfy the condition $I_v > N I_{v,i}$, where $I_{v,i}$ is the intensity from a single particle. The coherence mechanism requires a presence of particle bunches of dimensions less than a wavelength, separated by distances greater than a wavelength.

The RL nodes have intrinsically small inertial factors in comparison to the electron and positron. So they are able to obtain relativistic velocities. Having in mind, the high degree of their spatial ordering at the burning surface, their grouping tendency (because of twisting) and their high velocities, it is evident that the coherence conditions could be obtained in a large spatial volume.

12.B.6.4.5 Factors involved in the pulsar period stabilization

From the previous paragraphs it is evident, that the pulsar motion in a free CL space involves the following interactions:

- reactive force from the pulsar's jet

- specific inertial interaction with the CL space, taking into account the anisotropical inertial factor of kaon's nucleus

- interaction between the positive charge dominating component of (RL^+) nodes and the pulsar magnetic field

The reactive force from the jet, called also a thrust force has two components in respect to A-A axis (see Fig. 12.29.a):

- Axial component contributing to a pulsar proper motion

- Radial component tending to increase the angle β. It is opposed by the inertial reaction from the CL space.

The interaction between the RL^+ component and the magnetic field provides a rotational momentum in respect to the axis M-M. This momentum interacts with part of the radial component leading to a rotational momentum around the axis A-A.

The rotational momentum in respect to the axis M-M is a result of the initial twisting of RL(T) that predefines the motion of the released RL nodes. While expanding in a free CL space, their spatial conformity is preserved for a finite time duration.

Another important feature is the overall charge of the kaon nucleus. When it is inside of the star (without active jet) it is a neutral. Once the kaon nucleus is free and the jet starts, this neutrality is disturbed. The released quantity of the positive RL(T) structure is larger than the quantity of negative RL(T) structure. Then the interaction compo-

nent with the strong magnetic field of the pulsar will be predominated by the positive (RL(T). The interaction will take place in an intercepted conic volume, adjacent to the burning area with some finite conic height. Let us accept that the interaction energies are proportional to the destruction energies. Then according to Eq. (12.38) the ratio 2.469 is valid for $RL(T)^+$ to $RL(T)^-$ components. If the dominated positive component corresponds to a left hand twisting, then it will generate a counterclockwize rotating force. All such forces from individual FOHSs averaged for one rotational period could be represented by one tangential force, F_t, normal to the magnetic field and lying in the burning surface at arm a from the central axis (see Fig. 12.29.b). This force combined with a small fraction of the radial force from the trust provides the rotational force F_{rot}, which is responsible for the kaon rotation about the axis $A - A$ and generation of radiowaves.

From the kaon nuclear section shown in Fig. 12.28, it is evident that the charged component of RL nodes is well mixed with the neutralised component. Consequently, the magnetic field will influence the shape of the whole stream. More specifically, it will influence directly the jet cone angle. The energy of the total magnetic field is proportional to the quantity of the kaon nuclear matter. Then for the idealised cone shape, the strength of the magnetic field appears proportional to its length. This parameter is changing during the pulsar aging.

Stabilized spin frequency with short term variations

The proper motion of the pulsar along axis A-A is provided by the axial component $F_{ax} = F_{tr}\cos(\beta)$, where F_{tr} is the total trust component of the jet along the magnetic axis M-M.

The radial component is $F_{tr}\sin(\beta)$, but part of it is compensated by the anisotropical type of inertial interaction. Then the residual radial component is $(1 - k_1)F_{tr}\sin(\beta)$, where k - is anisotropic inertial coefficient.

Let us consider that the described process provides stabilization around a mean frequency $\bar{\omega}$ with some variations $\Delta\dot{\omega}$. , so the momentary frequency is $\omega = \bar{\omega} \pm \Delta\omega$ Let us assume also that the variational time constant is larger than the rotational period. Without going into details about the constants, the residual radial component will provide an angular momentum

$$(1 - k_1)F_{tr}\sin(\beta) \quad \rightarrow \quad L_1 = m\omega r^2 \qquad (12.38)$$

where : m - is the equivalent mass at radius r

Dividing L_1 on some time base we get the torque, which has dimensions of energy potential. Let us take a stable time base by using the secular period of integrated pulses, estimated for observational period much larger than the time constant of the variations. Then we have a torque (energy potential):

$$\tau_1 = \frac{m(\omega \pm \Delta\omega)r^2}{\overline{T}} \qquad (12.39)$$

The torque could be regarded as a rotational kinetic energy of the nuclear motion in which the anisotropy inertial factor (I_F) of the nucleus is implicitly involved.

$$E_K = \tau_1 = f(I_F) \qquad (12.40)$$

The torque from the interaction between the released twisted RL(T) and the magnetic field could be expressed by

$$\tau_2 = k_B aF_tB \qquad (12.41)$$

where: B - is the magnetic field and k_B is a coefficient of proportionality with dimensions of [1/T]. The latter may have similar behaviour as the kaon inertial factor I_F.

The difference between the two torques, τ_1 and τ_2 provides the energy of radiated pulses. Its observational estimate is the integrated pulse profile (E_p). Therefore, we obtain:

$$E_P = \frac{m(\omega \pm \Delta\omega)r^2}{\overline{T}} - k_B aF_tB \qquad (12.42)$$

The obtained expression is the **apparent energy balance equation of the pulsar** (or energy balance only). In this form it provides the total energy of emitted individual pulses. The therm apparent is intentionally used because there is another hidden balance. This is the balance between the kinetic energy (E_K) and the reactive energy (E^R) of the CL space:

$$E_K = E^R \qquad (12.43)$$

The expression (12.43) is the **hidden energy balance**. It is similar as the hidden energy bal-

ance between the kinetic energy of an astronomical body and the reactive CL space energy around its imaginary separation surface. Applying the same approach, the reactive energy (first derivative) will appear as active energy, the signature of which can be identified.

Equations (12.42) and (12.43) general expressions, permitting to understand the most basic features of the pulsars.

In a general case, the condition $\Delta\omega > 0$ is valid but within some limit and the integrated profile fluctuates around its mean value. In this case subpulse drifting is observed. The phenomenon of subpulses is explained in §12.B.6.4.6. If the drift is systematic, the condition $\Delta\omega = 0$ occurs periodically. The emitted energy in this case is

$$E_P = 2\pi mr^2 - k_B aF_tB \qquad (12.44)$$

The average time interval at which this condition occurs gives the time constant of the stabilization mechanism. The observations show, that the difference between the two terms of Eqs (12.42) and (12.44) is always positive. The first term involves inertial interactions related to the motion of the nucleus as a whole. So it always has some **reserve of accumulated kinetic energy**. The second term does not have a such.

What could happen if the magnetic field B is decreased suddenly with a large value? Then the energy reserve of the first term might be exhausted and the pulse emission will be stopped for a finite time. The stabilization mechanism, however, still work. It is supported by the inertial type of interaction between the nucleus and the CL space. In this case Eq. 12.43 is still valid. The reaction energy E^R, however, could not be changed suddenly. It requires some finite time. The signature of this effect can be detected from the trend of the consecutive pulses during the time of this phenomenon.

The described phenomenon may have a signature of **absent single pulses**. There is another phenomenon in which absent pulses are observed for a time interval of number of pulses. The possible explanation of this phenomenon is the following: The angle β may not be a constant but fluctuating, due to the nuclear inhomogenity described later (see §12.B.6.4.7). If it becomes zero for some finite time, the described emission mechanism will be not valid any more. In such case missing pulses will occur. During this time the equality between E_K and E_R might be disturbed by oscillation about the exact value (due to the reactive energy properties). So after a finite time, the steady state conditions of Eq. (12.43) could be restored, and the process of pulse emission will continue. The observable phenomenon with such signature is known as a **pulse nulling**.

Detailed analysis of the pulse nulling problem is reported by Ritchings, R.T. (1976). He investigated this phenomenon in 32 pulsars and found that the radiation from many of the observed pulsars is generated in bursts of typically 50 pulses and pulse nulls corresponds to the time intervals between these bursts. The burst occurs less frequently as a pulsar grows older. According to BSM concept, this behaviour is a result of the nuclear inhomogenity and the nuclear composition as a bunch of subnuclei (see §12.B.6.4.7 Real kaon nucleus). Ritching used autocorrelation method and built separate histograms for the missing and detected pulses (after correction of scintillation effect - a pulse distortion from the propagation through the galaxy). The shape and offsets of the histograms are very informative about the real mechanism of pulse generation, according to BSM. The histograms of pulsars with single and multiple (in most cases two) pulse profile are also clearly distinguishable. The double jet pulsar described in §12.B.6.4.9. corresponds to the case of a multiple pulse profile. The BSM analysis fits very well to the results provided by Richtings (1976).

Factors influencing the secular period change with the pulsar age

The total magnetic field is proportional to the nuclear matter quantity. So it decreases continuously with the aging of the pulsar. The flow debit according to Eq. (12.37) is also decreasing. It is evident, that the change of the secular period will change continuously with the pulsar age. The exact dependence is not derived here, but some of the factors influencing the stabilization process are following:

- decrease of the magnetic field strength and increasing of the line divergence outside the kaon nucleus

- change of contribution of the twisting force to the pulsar rotational momentum
- change of the rotational kinetic and reactive energy
- possible change of angle β
- change of the trust force components due to the increased divergency of the magnetic field.

12.B.6.4.6 Integrated pulse profile and drifting subpulses

It is apparent that the emitted energy in form of radio pulses is directly involved in the stabilization process. Having in mind that it covers some spectral range the most accurate estimation of the energy pulse is to measure simultaneously all emitted frequencies. This parameter in the field of pulsars observation is called **integrated pulse profile.** Despite the practical difficulties for observing a large spectral range, integrated profiles are obtained for many observed pulsars. Many of them show one typical feature: **subpulses drifted systematically** across the profile.

Figure 12.31.a illustrates the integrated pulse profile and its components for the pulsar PSR0809+74, while Fig. 12.31.b shows the subpulses drift (Kuzmin, 1992 (an English translation)).

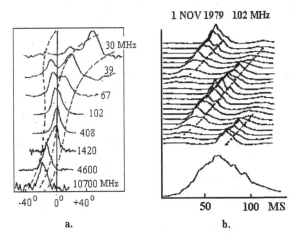

Fig. 12.31
Change in the spatial position and shape of the integrate profiles with frequency for pulsar PSR 0809+74
Courtesy of Kuzmin, 1992.

The BSM explanation of the integrated profile shape and the subpulses is the following: The radio frequency emission is a result of the relative motion between the rotating magnetic field and the phase delayed rotating spiral cone, illustrated in Fig. 12.30. <u>It can be regarded as a result of the energy pumping of the CL space from the released RL(T) structure.</u> The phase delay (should be not confused with the stabilization time constant) is a result of the inertial interaction of the expanding RL nodes in the jet with the CL space. If examining the mutual cross section between them in a plane normal to the proper motion, the phase delay will move in the direction towards the smaller conical section. This causes a radio frequency emission from the moving charge component of RL nodes in respect to the magnetic field. If separating this region into sectors as shown in Fig. 12.30, we will understand that the relative motion has a different mean tangential velocity for each sector. From the other hand, the sectors includes large spatial volumes with a different rate of CL space pumping. Consequently, every sector will emit separate radiopulse with a spectral width centred around the optimum mean value for this sector. The uniformly distributed energy of the RL nodes in the sectors and the simultaneous emission from a large volume could provide increased coherence of the radio signal. At the same time, the close interactions between the neighbouring sectors and the rotational motion of the pulsar might be the cause for the strong polarization of the subpulses.

Figure 12.31.a shows the signature of such pulses with different mean wavelength but the number of observational frequency is not equal to the number of the sectors. For this reason some pulses appear shorter while other doubled.

The rotation of the cone spiral and the partly coherent radiation from the volumes of the sectors cause a not isotropic type of radiation. In such way, the swapping beam and the source of the emitted signal (the CL space domain) appear rotating with the nucleus around the axis A-A. There is not triggering mechanism, however, for the moment of the quantum wave generations from the column sectors. But this components are exactly involved in the stabilization mechanism discussed in §12.B.6.4.5 They provide the drifting pulses. Consequently:

The drifting pulses are direct signature of the spin rate stabilization mechanism of the pulsar.

The drifting pulses could be considered also as harmonics of the main pulse carrying the necessary small portions of the energy that provides the energy balance according to Eq. 12.39 and keeps the frequency deviation $\Delta\omega$ within some limit. If the kaon nucleus is smooth as in the idealised model, a systematic drift will be observed. The effect of the frequency drift is very similar as the drifting of single frequency dye laser in a free run mode without locking.

Figure 12.33 illustrates a systematic subpulses drift for two pulsars. Subpulses in these pulsars first appear at the trailing edge of the pulse profile and then they drift toward the leading edge, disappearing several periods later. The effect is more apparent for PSR0031-07, where an interruption region (missing pulses) also is apparent between 37 and 39 pulse number. The effect of missing pulses has been mentioned in §12.B.6.4.5, while the probable mechanism is discussed in §12.B.6.4.10.

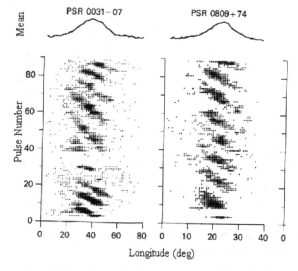

Fig. 12.33

Longitude-time diagram for two pulsars with drifting subpulses. (After Taylor et al., 1975)

12.B.6.4.7 Real kaon nucleus

A. Internal structure

The process of kaon formation and growing is possible only in steps. The analysis of the star evolution using the features of the Hertzsprung-Russell diagram (Fig. 12.26.A) indicates that this process begins and is quite intensive in the phase of instability of the main sequence. Later its is quiet until getting a final boost in the dying star. Due to the step-like formation, we can not expect the internal structure of kaon nucleus to be so perfect as shown in the Fig. 12.28. Some atomic matter could be trapped inside the kaon nucleus, causing structural defects. In the process of radial nuclear growing, the most internal atomic inclusions may crash and defects could be partly repaired. This however is possible if the kaon is not a continuous structure along the rotational axis but bunch of subnuclei. Then the radial section will show less defects in the central region than in the periphery. The possible radial section of a kaon nucleus with defects is illustrated in Fig. 12.34.a, while Fig. 12.34.b shows a magnified part from its axial section.

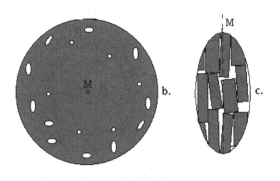

Fig. 12.34. Defects in a real kaon nucleus
a. Radial section with defects;
b. Magnified part from the axial section illustrating the spatial positions of the kaon subnuclei

B. Pulse energy variation

The defects in the real kaon nuclear structure appearing as a bunch of subnuclei will cause the position of the magnetic axis of the pulsar to be dependant on the positions of the individual subnucleus. Then in the process of nucleus burning two types of **sudden changes of the magnetic axis are possible:**

     - small angular shift: - change of the parameter β

     - small radial shift: - change of the parameter: k_B

Both types of change will affect the rotational component of the thrust force, causing a period change, but the mechanism of spin rate stabilization will correct it by adjusting the energy

of the emitted radiopulses. This is one of the reason for the observed **pulse to pulse energy variation.**

C. Sudden small period increase with a slow restoration of the general trend

Some large atomic inclusions near the circumference of the nucleus section may disturb the magnetic field homogeneity in this area. Part of the magnetic lines may exit outside of the nuclear surface (the effect is similar as in the magnetic bar with defects caused by air inclusions). Then this part of the field will be also involved in a quantum wave generation but with much smaller energy. The energy of this radiation is part of the total radiated energy. When the burning process overpasses this region, the kaon nucleus become suddenly more homogeneous and the polar oriented magnetic field becomes more concentrated. This will affect the jet parameters by slightly decreasing the apex cone angle. Then the period of the spin rotation and pulse emission will be slightly increased. From 1968 to 1969 three sudden small drops of the period are observed for Vela pulsar, PSR0833-45. The observational data showed that the deviation from the general trend of the increasing period decayed slowly away after about a year. This may indicate that the pulsar magnetic filed is influenced stronger by a nuclear defect when it is closer to the burning end.

D. Real nuclear shape

So far all the features of the pulsar has been explained by the idealised model, assuming a cylindrical shape of the kaon nucleus. The massive astronomical object, however, are spherical and obtaining of kaon nucleus with a cylindrical shape is less probable. The most likely shape could be a prolate spheroid composed of multiple kaon subnuclei, closely aligned to the common magnetic axis. Fig. 12.35 illustrates a possible shape of such nucleus with magnetic axis aligned to the major axis of the spheroid. Having in mind the feature (c) of §12.B.6.4.2 and (Eq. 12.36), the analysis provided for the idealized case of cylindrical nucleus should be valid for the real shape nucleus, as well.

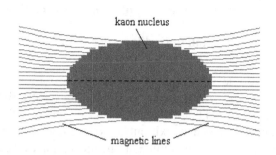

Fig. 12.35

A probable real shape of kaon nucleus
with magnetic lines

There is one problem that should be correctly solved: Some kaon subnucley will be involved in the burning process not in the beginning but in proper time later. How the ends of the FOHS's of the peripheral subnucley will be preserved?

Some possible explanations are the following:

(1) The external subnucley are covered by thin layer of atomic material. The layers is formed of highly ionised atoms that have lost all their electrons. Such ions could not emit radiation. The repulsion between these ions is significantly reduced by the strongly modulated CL space in the proximity of the superstrong magnetic field and the increased IG forces in that region. During the burning process the positive ions may contribute to the positive component from the RL nodes. Their charge component may become active at some critical distance from the kaon nucleus.

(2) The external subnuclei are not covered by atomic matter, but RL(T) destruction in the open ends of FOHS's does not start due to a strong biased CL space from the kaon magnetic field. The destruction is possible in the area of common burning where the fine structure of the magnetic field is disturbed by the turbulence from the destruction, i. e. this might be a self sustainable effect.

12.B.6.4.8 Proper motion of the pulsar

The proper motion of the pulsar is driven by the axial force component of the jet that is much stronger than the radial one.

Map of pulsars with periods from 0.1 msec to 10 sec is shown in http://pulsar.princenton.edu/pulsar/map/PulsarMap.html.

It is evident from the provided concept, that the propulsion system of the pulsars is active for a long time. Then their travelling distance can be significant. This allows their motion to be traced, but mainly for a velocity direction. One method for estimation of their velocity, called **pulsar scintillation**, relies on the interstellar scattering and Faraday rotation effect. The method is not quite accurate and may provide some higher velocities, but number of correction procedures exist. Interferometric measurement are also provided. Most of the measured velocities are in the range 100 - 200 km.

Fig. 12.36 illustrates the calculated traces for large number of pulsars in our galaxy whose velocities are measured by the scintillation method (see Cornell news: Tracking pulsars; *http://www.news.cornell.edu/releases/June98/scintillation.deb.html*

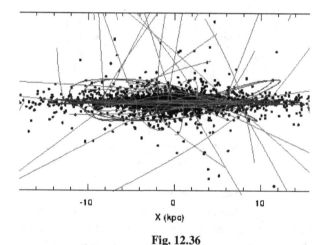

X (kpa)

Fig. 12.36

Pulsars in motion: The red lines indicate the orbits of fast-moving pulsars, some of which will escape from the galaxy. The blue lines show the orbits of slower moving pulsars. The black dots are ordinary stars that delineate the Milky way galaxy
www.news.cornell.edu/releases/June98/scintillaion.deb.html

One important consequence of the pulsar proper motion is **that they may pass through GSS and continue their motion in a foreign CL space**. In such conditions, however, their dynamics is different. The balance between the nucleus rotational kinetic energy and the reactive CL space energy might be disturbed. In the foreign CL space the released RL nodes after their conversion to CL nodes may form a long pipe of CL space not mixable to host CL space. Then a stable pipe can be formed in which the atomic matter may get a specific type of interactions. Observational evidence of such phenomena is discussed in §12.B.11.

12.B.6.4.9 Pulsars with two opposite jets

While the uncovering of one end of the kaon nucleus is the most probable case, the another option when the two ends are uncovered is not excluded. In this case two jets at $180°$ become active. Figure 12.37 illustrates the burning nuclear of such type of pulsar.

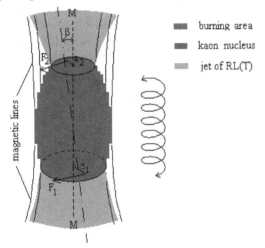

	burning area
	kaon nucleus
	jet of RL(T)

Fig. 12.37
Pulsar with two jets

Figure 12.37 shows a general case, when the two burning surfaces have different areas. Only the peripheral magnetic lines of the nucleus and of the burning areas are shown. Such pulsar will have some specific features:

(a) no proper motion

(b) a small angle β between the magnetic and spin axes

(c) separate energy pulses from the two jets

(d) a comparatively smaller period

(e) a faster aging

According to the conditions given by Eqs. (12.36) and (12.37), the thrust forces of the opposite jets will be equal. Consequently, the pulsar will not have a proper motion. (It may have some small

proper motion if the internal nuclear homogeneity in both sides is different). For equal internal homogeneity along the pulsar axis, the magnetic field from both sides is equal, but the burning area of upper jet is within smaller area of the magnetic field. One specific difference between them is the torque from twisting forces (obtained by the interaction between the positively dominated twisted RL(T) and the magnetic field). The angular momentum from both twisting forces are with opposite direction, but the lower one is larger than the upper one. Then the frequency stabilization effect will be of differential type. The rotational period will be stabilized by the pulse emission generated from both jets. For a long term the degree of the period stabilization will be proportional to the difference between emitted energies from both jets. It is clear that the burning rate of the double jet pulsar is twice higher than for a single jet. So it will age faster. The secular period will show a faster increase.

Another important feature is that for nucleus possessing a good rotational symmetry and low internal defects, the angle β tends to approach zero, because the twisting forces across the burning areas will have a good rotational symmetry. In such case, the pulsar's dynamics could appear differently. If accepting some rotational asymmetry, however, the same analytical approach applied for the single jet pulsar could be used. In any case, the angle β for the double jet pulsar should be much smaller than for the single get. Then for a same reactive energy E^R, the rotational frequency will be larger. The conditions for $\beta \approx 0$, however will exhibit a longer time constant. It could be identified by investigating the null states and the pulsar behaviour after a sudden frequency jump of detected pulses (due to a burning of some inclusion or disaligned subnuclei). This problem is discussed in the next paragraph.

The analysis of R. Ritchings (1976) about the pulse energy around the pulse nulling phases confirms the BSM concept of a double jet pulsar and allows to infer the mechanism of the pulse generation. It supports the concept of the reactive energy of the surrounding CL space. The signature of the differences between a single and a double jet pulsar is apparent from their pulse energy histograms. Fig. 12.38. **a** and **b**. show respectively the histograms for a single pulse (single jet) and a double pulse (referenced from Ritchings as multipulse due to a separation of one of the pulses in subpulses components). The BSM explanation of subpulses has been provided in 12.B.6.4.6).

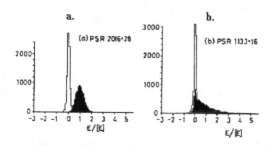

Fig. 12.38. Pulse energy histogram:
  a. from a single pulse pulsar
  b. from a multiple pulse pulsar (double pulse
  with a constant subpulse, according to BSM)
  The abscissae are in units of mean pulse energy
  [Courtesy of Ritchings (1976)]

The black area histogram is from the detected pulses while the white area - from the missing pulses.

A typical example of a double jet pulsar is the **Crab pulsar in the Crab nebular.** Its secular period is 33 ms. The Vela pulsar in the Vela nebular exhibits similar characteristics.

The Crab and Vela pulsars exhibit also a number of additional features, which however are caused by some specific features of their host nebula. The Crab pulsar and nebula are discussed in §12.B.10.

12.B.6.4.10. Long relaxation time constant of the pulsar and its signature

The inertial interaction between the moving and rotating kaon nucleus and the galaxy CL space is a mechanism in which the anisotropical inertial factor is involved. It is reasonable to expect, that it may put a signature on the stabilizing process by some time constant, larger than the rotational period. In §12.B.6.4.7 B. it has been explained how the position of the magnetic axis could be suddenly changed. This means a change of the angle β that will lead to a change of the rotational kinetic energy (E_K). At the same time, the balance according to Eq. (12.43) should be satisfied, so the reactive energy should be also changed. The reactive energy (E^R), however, could not be changed quite fast, because it is located in a large space volume (the vol-

ume occupied by the nucleus, the released RL structures and the magnetic field). Therefore, the time duration of (E^R) change could be much longer than the rotational period, so its signature could be detectable as a relaxation time constant.

A signature of the relaxation time constant could be identified from the analysis of the pulse nulling effect and from the period glitches of the observed pulses.

The analyses of R. Ritching (1976) about the pulse nulling effect provides clearly indication about the existence of the relaxation constant and its behaviour with the pulsar aging.

Relaxation time is observed and reported by Lohsen (1975), after speed-up of the Crab pulsar by a frequency jump of $\Delta \nu = 0.097$ periods on February 4, 1975. A sharp increase in the pulse frequency is observed, resulting in an initial phase shift of 0.1 periods. The frequency shift slowly decayed, approximately following an exponential low. Figure 12.39 shows the time residuals trend before and after the glitch, the time of which is referenced as a "0" day.

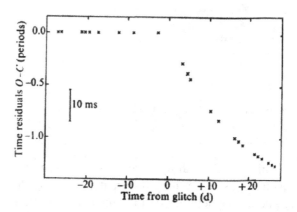

Fig. 12.39
Observed pulse arrival times minus arrival times computed from cubic fit to the data before the glitch (Lohsen (1975)

From the same data Lohsen estimated the natural logarithm of the frequency differences $\Delta \nu (\nu_o - \nu_c)$ and found a relaxation time of 17 days for the frequency jump.

The observed data indicates presence of two relaxation time constants: a longer one, corresponding to the exponential shape of the frequency change (longer than 20 days) and a shorter one corresponding to small oscillations around the frequency decrease trend (~17 days).

According to BSM, the frequency jump of the pulses does not mean that the real rotational frequency is jumped. During the relaxation time, the pulse energy and its moment of emission are both regulated by the reactive energy E^R. In such way, the event will start with an initial phase shift, followed by a slow phase drift during the relaxation time. An initial phase shift is really observed and reported by Lohsen (1975).

16.B.6.4.11 Summary:

- **The pulsar is not a neutron stars, but a kaon nucleus possessing one or two jets caused by the disintegration of the FOHSs with their RL(T) structures.**
- **The kaon nucleus creates a superstrong magnetic field**
- **The rocket-like spinning motion of the pulsar through the galactic space is driven by the thrust forces from its jet**
- **The pulsar jet contains both: a neutralised and a positive charged component corresponding respectively to "a neutral and a charged electroweak current" (according to the Electroweak Theory, known also as a V-A theory).**
- **The generation of quantum waves (detected radiopulses) is a result of the angular speed difference between the rotating magnetic field and the charge component mixed in the jet volume with a shape of a conical spiral.**
- **The spin frequency of the pulsar is stabilized by a self adjusting mechanism.**

12.B.6.5 Star evolution termination options: a black hole, a binary pulsar, or a supernovae

From the concept of the pulsar as a burning kaon nucleus we may distinguish the following options of the dying star:

(a) a kaon nucleus with uncovered one or both ends

(b) a kaon nucleus with covered both ends

(c) a kaon nucleus of both types ((a) and (b)) in a binary system

(d) an explosion due to a radial crash of the FOHSs of the kaon nucleus

The option (a) corresponds to the active pulsar described in the previous sections.

The option (b) corresponds to a different astronomical object with observational features of accretion disk (considered so far as a "black hole").

The option (d) corresponds to a phenomenon known as a supernovae.

12.B.6.5.1 A passive kaon nucleus in a role of a "Black hole"

If the mass of the star is below some critical limit it could not be able to form a large kaon nucleus. From the other hand, the quantity of the erupted material in the process of kaon growing is proportional to the kaon increase. Consequently, the star with a mass closer but below some critical limit may die more quietly. If the star mass is just above this limit, but closer, it is possible the kaon nucleus to be freed in the final eruption phase while still covered by thin atomic layer. Such object will behave quite differently than the pulsar. We may call it a **passive kaon nucleus** (not possessing a jet).

The passive kaon nucleus will have the same strong magnetic filed as the pulsar. Its initial spin momentum will be exhausted from the interaction between the rotating magnetic field and the surrounding matter. The atoms of the thin atomic layer could not operate normally, because of the large biasing of the surrounding CL space from the strong magnetic field. Such object will not have the radiation characteristics of the pulsars.

The passive kaon nucleus will exhibit one specific feature, which is not so apparent in the pulsars. For any radial section of the kaon nucleus the total electrical charge is balanced. Outside the kaon nucleus, however, in a direction of the magnetic lines the positive charge will dominate. This is a result of the hidden positive FOHS inside the negative FOHS (see Fig. 12.28). The strong axial alignment of the positive component may pass through the thin atomic matter, creating a huge dipole of positive charge in the external CL space. Consequently, the passive kaon nucleus will have a strong positive electrical field aligned with the strong magnetic axis. The latter will allow the electrical dipole to be stretched out at significant distance from the nucleus. But for enough large distance it will behave again as a point positive charge in a similar way as the proton behaviour. This is an unique kind of object.

The described object could be discovered only by the following feature: If some atomic matter passes nearby the object the atoms could be heavily polarised and ionised. The free electrons then could be attracted by the strong positive charge. Approaching the kaon nucleus, the trapped electrons will be guided by both axially aligned fields - magnetic and electric. So the electrons will form a narrow beam with a conical helical trace. Simultaneously they will be continuously accelerated and the beam will become narrower. In such conditions, the positron-electron system of the normal electron could be activated (see Chapter 3) and an emission of X-rays may occur. Such phenomenon is observed and it is known as an **accretion disk**.

It has been shown that the inactive kaon nucleus (not possessing a jet) could provide conditions for the superstrong magnetic field with either N-S or S-N orientation referenced to the kaon nuclei. The stable appearance of one or another orientation of this field is supported by external interactions in which the magnetic field is involved. Let us assume that the magnetic field during the phase of a stable direction of the magnetic field provides conditions for an accretion disk from one side with a shape of right handed conical spiral. Then the shape of the other accretion disk (from the other side) will be a left handed conical spiral (because of the opposite direction of the magnetic field). In such case, both accretion disks will provide rotational momentum of the nucleus in one and same direction. If the pulsar is in a close gravitational interaction with a giant star (they has been a binary stars), the kaon nucleus will ionize the external layer providing continuous source of electrons for the accretion disks. At the same time it will posses spin rotation and common rotational motion. Therefore, it will carry a significant reactive energy E^R as in the case of the active pulsar (possessing a jet). While the energy E^R of the pulsar in free CL space is kept constant, now the proximity of the nucleus to the star will ionize and drag atomic matter from the external layers of the star, so the latter will be spinning in the equatorial re-

gion around the nucleus. At the same time the rotational period, depending on the accretion disk, could not be stable as in the case of an active pulsar. Consequently, fluctuation may exists. <u>The described binary system will exhibit strong X-ray radiation from accretion disk and some optical radiation from the spinning ionized atomic matter.</u>

Double objects with features, described above, have been detected but explained differently, so far. <u>The observed Cygnus X-1 pair object of supergiant star HDE226868 and X-ray companion has features matching to the described case of passive kaon nucleus.</u> So far, the invisible partner of the system has been categorized as a "black hole" (see the image in the left side of Fig. 12.40 and the presented so far explanation at the right side).

i

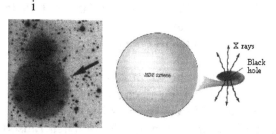

Fig. 12.40

Similarity between the features of the black hole concept and the undeveloped pulsar (BSM concept) with a companion of giant star
(Adopted from Serway College physics)

The observed radiation detected by Rossi X-ray Timing Explorer has a very common signature to the described above effect. It 1997 the observed phenomena is announced as a "frame dragging effect" and is reported in the Sky § Telescope magazine, December, 1997 under the title "Seeing black hole spin".

From the presented concept it is evident that the passive kaon nucleus may become a pulsar in some particular moment if one of its ends is uncovered. This may happen due to the continuous bombardment from the accelerated electrons. The electrons (possessing RL(T) structures) may crush when hitting the polar region. The released RL(T) structures may oblate the atomic matter that cover the end of the kaon nucleus. The uncovering of the end enables the jet process, so a pulsar will be born. Before the pulsar is born, a transition phase may take place in which the object may emit an optical radiation from the gas clouds in the vicinity. Such clouds can't be left from the explosion, but formed later from the leftover atomic matter that cover the kaon nucleus. A possible example of such object is the supernova SN1986J in the galaxy NGC891 30 million light years away. The supernova explosion has occurred in 1983. (for more info, search for "SN1986J supernova").

Summary:
- **The considered so far Black hole with accretion disk appears to be a passive kaon nucleus**
- **The passive kaon nucleus is a predecessor of a pulsar**

12.B.6.5.2 Binary pulsar

Binary pulsars are not rare in our galaxy. This is the possible evolutional end of one of the binary stars. For such system the probability for quiet end of one of the stars against a birth of a supernova is larger due to the gravitational pull-up from the second star.

The normal binary pulsar is formed of kaon nucleus with one burning end. According to the presently adopted concept about the pulsars, the binary pulsars have been considered as "neutron stars", while their motion has been considered as a result of initial velocity obtained in the process of their formations. Such concept could not explain the long track motion of many pulsars in our galaxy, some of which may escape from the our galactic CL space. BSM unveils that the pulsar possesses a continuous propulsion energy due to its jet. For this reason BSM does not use the mass estimation of the pulsar based on an initial energy momentum.

One characteristic feature of the binary pulsar is the x-ray radiation. The explanation according to BSM is a following:

Binary stars are quite abundant in the Universe. Let us consider a binary star system with dissimilar masses, while both have a mass above the critical one (for evolving into a pulsar). The heavier one will evolve earlier, so its jet of released RL(T) nodes will irradiate periodically the other star. This radiation may continuously evaporate and ionize matter from the star companion. The positive component of RL(T) (equivalent to electroweak charge current) in the far field of the pulsar will attract the electrons from the ionized matter so they may form a shell, which will be kept between the star and the pulsar. Such pulsar will have two types of rotational motions: one around its axes (with a short period of the order of ms) and a second one with the star companion around a common axis. Then the electron shell may be periodically irradiated by the pulsar jet. The electrons could be strongly excited by the positive RL(T) component of the jet. This excitation causes emission of X-rays from the electrons (electron shell-internal positron oscillations, see Chapter 3). The radiation is in X-ray range with a burst period equal to the proper pulsar period the proper pulsar period. The common rotational period of the two bodies around their barionic centre is usually much longer.

From the presented concept it is evident that the described mechanism does not involve accretion disk that is a feature for continuous X-ray radiation. It is also evident that the X-ray pulses and the possible optical pulses are not generated by the same mechanism as the radio pulses from the pulsar. Therefore, such radiation is not involved in the pulsar stabilization mechanism and if such pulses are eventually detected they will be phase shifted. The observation of X ray, optical, and radio pulses from Vela pulsar PSR0833-45 are in full agreement with this concept.

One additional phenomena may also take place. The refurbished CL nodes may carry part of the momentum of the RL nodes. <u>Then a flow of folded nodes with a large group velocity may pass through the star's gravitational field. This will cause a relativistic effect exhibiting a large Doppler shift.</u> If some atomic layers are in that zone, they may show Doppler shifted absorption lines even if they are not moving. Some observations shows large (relativistic) Doppler shifts in a quite similar situation. **The assignment of the observed Doppler shift to object velocity in such case should be reconsidered.**

12.B.6.5.3 Supernova

In the normal kaon nucleus, all FOHS's are separated by CL space gaps as shown in the radial nuclear section given in Fig. 12.28. There is a critical minimum value for the gap (see Chapter 6). Below this value the radial stripes of RL(T) between neighbouring FOHSs will start to interfere by their IG forces (inverse proportional to the cube of the distance). Such interference will inevitably invoke a triggering of a destruction of the whole FOHS, accompanied with a release of the internal RL(T)s. Now it is apparent that this may happen in the final process of the dying star when the kaon nucleus obtains the last growing portion. <u>Because the kaon nucleus growing is directly related to the gravitational pressure, it is obvious that this option is dependable on the initial mass of the star.</u> But we saw that larger disalignments of the kaon subnucley may also lead to a radial destruction even for smaller nucleus. Consequently the expected initial critical mass for such option may have some variance around a mean value. How large could be this variance is a question, because the occurring of such phenomenon does not mean that all helical structures will be destroyed. If the mass is not large enough and the process begins in zones of subnuclear disalignments, the whole nucleus may break into separate nuclei, so they will undergo their evolution as pulsars.

If the mass of the star is well above the critical one, the kaon nucleus may obtain a significant grow before the final eruption. Then in some moment of the star's evolution, the minimal critical distance value between FOHS's could be overpassed and the radial destruction might be initiated. If the quantity of the atomic matter is still large, the process of destruction may involve also a large portion of the kaon nucleus. The process will be developed as a huge explosion that will throw all the atomic matter away. Most of the matter will be evaporated due to the enormous flux of RL nodes released in all directions. **This phenomena is supernova.** In any case, some small portions of the kaon nucleus may survive, but both ends of such nucleus will be likely uncovered. <u>Such type of kaon nucleus will behave as a pulsar with two opposite jets and very small proper (linear) motion. It will show faster period increase, because its nucleus burns twice faster</u>.

In some supernova explosions, before the pulsar is born, a transition phase may take place in which the kaon nuclei might be enveloped with some atomic matter, so a jet could not be developed. Such object may emit an optical radiation from the gas clouds in the vicinity. The gas clouds

might be obtained after the explosion: for example they may originate from the leftover atomic matter that covers the kaon nucleus. A possible example of such object is the supernova SN1986J in the galaxy NGC891 30 million light years away. The supernova explosion has occurred in 1983. Its picture is shown in Fig. 12.40.A. (for more info, search for "SN1986J supernova").

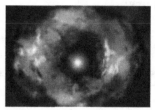

Fig. 12.40.A Supernova NGC891 (from public domain)

One example of the described scenario of supernovae with a born pulsar is the **Crab nebular and the Crab pulsar**. The former star (or stars), evidently has been of II-nd population, so it has been a remnant from the former galactic life. All globular clusters, for example, are such remnants and their features are discussed in §12.B.7. Most of the features of the Crab pulsar are explainable by the presented general concept. Some specifics, however, are influenced by the features of the II-type star population and their CL spaces. The same is specifically valid for the matter in the Crab nebular and the interactions between the pulsar and the atomic matter from the former star. The specific features of the Crab nebular and pulsar are discussed in §12.B.10.

12.B.6.5.4 Summary about the evolution of stars with masses between 11 and 50 solar masses

(a) The matter of the Globular Clusters (stars, dust, CL space) is a remnant from the previous galaxy life, which have escaped the phases of the galaxy collapse and recycling. The stars from this matter does not follow the main sequence of H-R diagram.
(b) The stellar evolution in the main sequence of H-R diagram (known as I-st type galactic population) is valid for the matter of the present galactic life.

(c) The physical mechanism behind the zone of instability in the Hertzsprung-Russell diagram is the formation of kaon nucleus inside the star.
(d) The kaon nucleus assures the strong magnetic field of the star. The latter provides conditions for synthesis of elements with higher Z-numbers
(e) In the phenomena of star dying the kaon nucleus gets additional final increase.
(f) In the final eruption of a dying star with a large mass, the most amount of the atomic matter is lost and at least one end of the kaon nucleus becomes uncovered. The observable phenomenon is a supernova that may give a birth of pulsar.
(g) The pulsar is characterized by a process of continuous kaon nucleus burning (disintegration of the FOHSs). It obtains a rotational motion with an actively stabilized period and an axial thrust force. It emits enormous energy by generating of radio frequencies from a large CL space volume.
(h) The single jet pulsar possesses a linear (proper) motion, while the double jet pulsar - not.
(i) The single jet pulsar may pass a significant distance in the CL space of the home galaxy and may even migrate in the CL space of a neighbouring galaxy.

12.B.7. Remnants from the previous galactic life

12.B.7.1 Theoretical concept

In §12.B2 it was mentioned that during the phase of the galactic collapse the galactic CL space may get some internal break. In such case some islands containing a large number of stars (as the open clusters) may escape from the collapse. Left isolated in a pure void space (in a classical meaning), they will undergo a fast evolutional process. The space surrounding the isolated star cluster will lose a big amount of its CL structure. Once separated, the self restoring feature of the broken CL space will invoke shrinking of the volume occupied by the leftover CL space. This will move the stars much closer than before. This is a **process of cluster collapse, leading to formation of Globular Cluster (GC)**. (We must keep in mind that the Newtonian gravitation occurs only in CL space).

The process of GC formation, however, may have some transitional phase. In Chapter 10 it was found that the intrinsic condition for a stable complex of CL space and atomic matter (stars, planets dust and so on) is when the ratio between the Partial CL pressure (related to the inertia in CL space) and the Static CL pressure (defining the Newtonian mass) approaches the relation:

$$P_P/P_S = \alpha^2/\sqrt{1-\alpha^2} \qquad (10.18)$$

In the process of the galaxy collapse, the folded nodes surrounding the FOHS of the elementary particles could escape easier because they are not strongly connected to the CL structure. At the same time, the amount of the folded nodes, could not be below some critical level (as discussed in Chapter 10), but when it is much lower than the optimal ratio provided by Eq. (10.18), this will means a reduced dynamic energy of the whole system. Consequently:

- **The CL space of the globular cluster will exhibit a decreased Maxwellian energy due to the reduced P_P/P_S ratio (a reduced amount of the necessary folded nodes).**

At the end of the evolution process, the escaped from the collapse open cluster will be converted to a globular cluster (GC) where the stars will be very closely spaced.

The process of the GC formation from an escaped star cluster may be tremendously shorter than the process of the star formation during the active life of the galaxy. However, the globular clusters may survive all hidden phases of the galactic recycling. At the same time, they play a very important role during the hidden phases of the galactic cycle: surround the protogalactic nucleus and keep it inside the own void space previously occupied by the CL space of the former galaxy.

When the new galaxy is born, some of the globular clusters, which has been closer to the protogalactic nucleus, could be broken and will form filaments (because the old CL space structure usually is not mixable with the new one). GCs which are more distant from the protogalactic nucleus, however, may survive. The CL space of the survived GCs will be interconnected to the new CL space in a similar way as the neighbouring galaxies. The local CL space of GCs, however, is preserved, because the diameter to length ratio of the prisms in the GC will not be exactly the same as for the prisms from the new galaxy formation. Consequently, a separation surface similar as a GSS surface will exist between the CL space of the GC and the CL space of the new host galaxy. This means that the photons emitted from the matter in the GC will appear red shifted when detected outside. Then the observable red shift will be caused by two physical phenomena:

- a lower Maxwellian energy of the GC

- an energy loss from trespassing the photon through the GSS between the GC and the CL space of the new host galaxy.

According to the existing so far categorization, the GC has been considered as similar as the small size lenticular or disc galaxies. According to the BSM, however, they are result of quite different evolution. One of major distinctive features is even observable: The Globular Cluster does not possesses a galactic nucleus, which corresponds to the supermassive black hole now unveiled to exist in every well developed galaxy. These feature is very important for the behaviour of the GC in the new galactic CL space. It means that the lost energy of the GC could not be compensated by the same way as in the normal galaxy, which obtains a constant energy supply by the galactic nucleus. The reduced total energy of the GC means that the CL space parameters including the Planck's constant are affected. The latter is involved in the definition of the Newton's mass of the matter.

Another specific feature of the Globular clusters is their dynamics in the new CL space. GCs have a large number of very closely packed stars from thousands to millions. They possess own CL space. Then this CL space will behave like a CL space of a neighbouring galaxy: it will be stationary in respect to own galactic centre and neighbouring galaxies. This means that the GC will have:

- the internal dynamics of the stars will be somewhat similar as in the star dynamic in a galaxy: they will rotate around a common centre, but within its own CL space

- the GC system velocity in fact will be zero, while its red shift will be not from Doppler but from a cosmological origin as discussed above.

Since the GC cluster does not have an energy supply like the ordinary galaxy, it will continuously loose energy by radiation. The compensation of the required kinetic energy of the stars will come either from dying of some very old stars or from escaping stars. In the second case part of the kinetic energy of the escaped star will be preserved for the GC total energy. In order to keep the compatibility with the CL space of the new host galaxy, the GC matter may undergo a particular evolutional process. The matter density and the energy of the GC could be segregated in concentric layers of stars. Such effect is a typical for the GCs and is known as a mass segregation. Some stars from the periphery could be lost, living part of their rotational energy to the total energy of GC.

The matter of GC may also undergo reversed evolutional process in respect to the evolution of the stars from the main sequence. This means, that the metalicity may go partly in direction of decreasing instead of increasing. In other words, elements with larger Z numbers may convert to lower Z number (by nuclear reactions) if the reaction leads to energy release. For the same reason, the star released by GC and occurred in a foreign CL space may undergo again a phase of instability. **According to BSM, such star behaves as a cepheid of II-nd type.**

Massive and tightly packed GCs typically contain from 100,000 to millon and more stars. Approximately 150 of these huge clusters populate the halo of the Milky way galaxy. Their age, estimated by the existing methods appear to be of the order of 15 billions years. This parameter by itself has been one of the big puzzles of the Big Bang theory (because the estimated Universe age appears smaller). The estimated age of some GCs in Andromeda galaxy appear to be in the range between 10 and 20 billion years. For more information about this issue see http://www2.astronomy.com/astro/News/News/020400Hubble.html.

GCs exist in all large and well developed galaxies. In some of them their number reaches thousands. In the next section, the provided above considerations are confirmed by analysis of observational data about the GCs and the Cepheids type II.

12.B.7.2 Observational data analysis

12.B.7.2.1 Reduced energy to matter ratio and CL space energy of the GC in comparison to the host galaxy

King's study (1996) have shown that a lowered Maxwellian energy dependence appears to be a good approximation to the solution of the Fokker-Plank equation describing the phase-space diffusion and evaporation of stellar systems like GCs. This model fits the density profile of GC quite well, but it has not been accepted at that time because a "lack of physical meaning". Later, the model has been modified, considering that the GC system is consisted of two separate subsystems:

- a slowly rotating halo subsystem
- a rapidly rotating disk subsystem

According to BSM, the first subsystem might be in the own GC's CL space, while the second one - in the foreign CL space (the CL space of the host galaxy).

Figure 12.41 shows the comparison of the surface brightness in NGC 3379 with a theoretical curve, according to the early King's model (King, 1996).

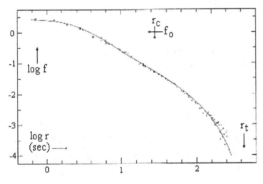

Fig. 12.41

Comparison of the surface brightness in NGC 3379 with a theoretical curve according to King's model (Courtesy of King, 1996)

Another indication about a lower GC's total energy comes from the comparison between the period-luminosity relations obtained for classical cepheids and for cepheids of II-nd type. The latter type of cepheids is considered as a typical for cep-

heids from GCs. Figure 12.42 shows the relations for both type of cepheids.

A similar difference between both types of cepheids is valid for other galaxies as well. Baade W. and Swope, H.H. (1965) have shown this for Andromeda galaxy.

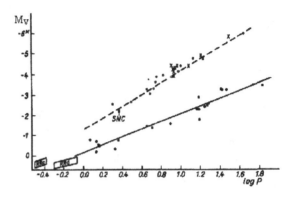

Fig. 12.42

Period-Luminosity relationship for a classical cepheids

(above) and for W Virgins stars (below). o - variables in galactic clusters, x - variables in the Large Magelanic Cloud.

(Courtesy of Dickens and Carey (1967))

Another observed phenomena, characterizing the cepheids of II type (from GC) is the **spectral line doubling**. It usually occurs during the phase of eruption. The two lines are often referenced as a blue and red component, respectively. The line doubling is usually combined with **another effect:** the radial velocity curves estimated by different spectral lines show a phase displacement. The existing so far theories for explanation of this phenomena accept a concept of a shock wave moving upward and intercepting the falling inward external layer. This concept, however, is not quite convincing in the explanation of the following effects:

- if the shift is of Doppler type only, why the lines are not smeared?

- the blue component is always stronger: this means, that one layer is always above the other.

The line splitting and the phase displacement for a II-nd type cepheid is firstly reported by R. Sanford (1952). Fig. 12.43 shows the calculated radial velocity assuming a Doppler shift for a W Virginis cepheids (type II) from GC M15 (NGc7078), (by R. Sanford, 1952).

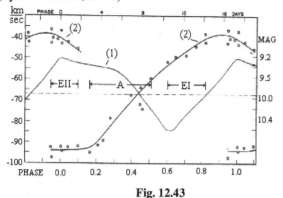

Fig. 12.43

Circles are absorption-line velocities of W Vir. One complete velocity curve is show, together with the end of the preceding and beginning of the following velocity cycle. The Gordon and Kron's light curve is shown by a thin line, the systematic velocity is shown by a broken line.

Figure 12.44 shows absorption line doubling for W Virginis (II type of cepheids) by Wallerstein (1959).

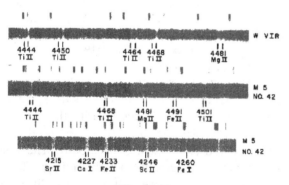

Fig. 12.44

Selected region of the spectra of W Vir and M5 No. 42 near maximum light

Abt (1954, pp. 86-87) shows that the excitation temperature derived from the violet component is always higher than the temperature derived from the red component. Wallerstein (1959) suggested that the violet component must be formed at a greater depth. He also mentions that H_γ emission component detected in W Virginis stars is missing in the classical cepheids.

12.B.7.2.2 BSM concept of the processes in cepheids of II population:

If a star from a GB quits the cluster it must carry its own surrounding space, but the radius of this CL space is much smaller than the stars of I-st type population. Its total system energy is also smaller. It begins a reverse evolution leading to a phase of instability, so it converts to a cepheid of II-nd type. The kaon nucleus increases in steps, accompanied by crashing the external shells of the protons and neutrons that contributes to the release of RL nodes (consisted of 6 prisms). These nodes fly through the external atomic matter as folded nodes. They cause a broad band emission from the internal gas layers of the star. We may detect a signal not only from this layer, but also some absorption signal from the external gas layers. Some of the external layers are dragged by the flying folded nodes. So they may have a Doppler component but, we can not ignore the contribution also from the component, caused by the high velocity RL nodes (explained below). Due to the multiple periods of this phenomena, some of the upper layer material could be dragged beyond the GSS (between CL spaces of GC and the host galaxy). Let us use the data shown in Fig. 12.43. The curve (1) could be from the layer inside of GSS, while the broken curve (2) - from the layer thrown outside GSS. It is quite reasonable to assume that the gas layers, whose absorption is observed, are optically thick. From phase 0.1 to 0.95 the flying nodes are still inside GSS and the absorptive features of the internal layer are only observed. From a phase 0.95 to 0.1 the flying nodes (passed through GSS) activate emission and absorption in the external layers, which are outside of the GSS. The spatial distribution of the flow of flying RL nodes could not have a spherical symmetry, because of the extended shape of the kaon nucleus. Then in a phase range of 0.95 to 0.1 we may observe absorptive lines simultaneously from layers inside and outside GSS. This means a simultaneously observation of two lines or line doubling. The line doubling is caused by the energy difference of the photon generating and photon propagating mechanism described below.

The photon emission and absorption process that takes place inside the GSS is optimized by the own CL space parameters of GC. So the observed photons from the layer inside of the GSS cross once this surface and exhibit a cosmological z-shift. The absorption layers outside of the GSS are formed of atoms, possessing own FOHSs but operating in a foreign CL space. The proton and neutron are closed helical structures and could not exhibit a geometrical change in the foreign CL space. The electron is an open structure, adjustable to CL space, but the electrons originated in GC are from the previous galactic life with different length to diameter ratio of the prisms they are built of.

It becomes evident that the process of photon emission and absorption in the layer outside of the GSS will be less effective than the same processes in the host galaxy. **Consequently, it will exhibit a specific cosmological z-shift**. This shift could be even larger than the z-shift of a photon generated inside GSS (in own CL space), and passing through GSS.

According to the above considerations the detected absorption signal should be comprised of:

(a) photons generated inside GSS with missing strong absorption lines (both processes in own CL space), exhibiting a Z-shift

(b) photons generated inside and outside GSS with missing absorption lines from a photon absorption outside of GSS (absorption in foreign CL space)

The physical process causing a z-shift from the GSS of GC is identical to the z-shift process between the galaxies. So with some approximation we may accept that the quasirefractive index of GSS of GC is equal to the mean quasirefractive index of GSS, $\tilde{n}$, (discussed in §12.B.13.2.). The relation between $\tilde{n}$ and z-shift was given by Eq. (12.29)

$$(\bar{n})^N = z + 1 \qquad [(12.29)]$$

Let us assume that the z-shift from GSS of GC is approximately equal to the normalized GSS quasirefractive index (normalized to one pass of GSS) discussed in §12.B.13.2.

Then we may find the quasirefractive index of GSS of GC as an energy ratio between the red and the violet component of the observed lines from the spectrum shown in Fig. 12.44. Using Ti II lines of M5 spectrum we get

$$\frac{E_1}{E_2} = \frac{\lambda_2}{\lambda_1} = 1.0002 = 0.02\,\% \qquad (12.45)$$

The obtained value of the quasirefractive index is quite approximative, because a possible Doppler shift could be indistinguishable from the cosmological z-shift in this simple interpretation. However, it still may indicate that the quantum efficiency of the process of emission/absorption of GC matter in a foreign CL space is different than in own CL space. Some useful insight into the process could be obtained also by the analysis of the hydrogen emission component H_γ.

The cepheids of type II exhibit much shorter period than the classical cepheids. Their evolution, however, may also lead to a pulsar as the case of a dying star from a I-st population.

12.B.7.3 Summary:

(a) The globular clusters are remnants from the previous life of the host galaxy

(b) Every GC is separated from the host galaxy by a GSS

(c) The similarity between the GC and a small galaxy is that both have a GSS, but the GC lacks of galactic nucleus as a source of energy and its CL space volume is much smaller.

(d) The star evolution in the GC may go partly in a reverse direction (from the point of view of the feromagnetic hypothesis)

(e) If some star is released from a GC it may occur in a foreign CL space. In a such environment it may undergo again through a phase of instability and become a cepheid of type II.

(f) The main distinguished features between the classical cepheids and those of type II are the differences in their metalicity, their period range and the slope of luminosity- period relation.

12.B.7.4 Conglomeration of remnants from the previous galactic life

In the previous paragraph we see, that individual GCs are formed of islands, which have escaped the former galactic collapse. Some of this islands could be quite large with not uniform matter distribution. Having such shape they may not convert to a compact individual GCs during the recycling phases. As a result, they may form an irregular looking galaxy in which some individual GC's could be identified. Congregation of isolated islands into one common island is also possible during the phases of the galaxy recycling. Irregular galaxy with such signature is the Sagittarius Duarf galaxy (SagDEG) discovered in 1994 by R. Ibata, G. Gilmore and M. Irwin. Its shape is shown in Fig. 12.45.

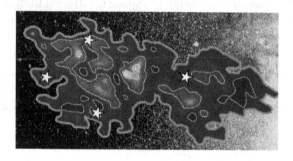

Fig. 12.45
Sagittarius Duarf galaxy

SagDEG is inside the Milky way CL space about 80,000 LY from the solar system. The M54 coincides with one of the galactic two bright regions and exhibits the same z-shift. Another three GC's from the galaxy are identified (Arp2, Terzan 7 and Terzan 8).

The researchers who studied the SagDEG found a large systemic velocity corresponding to galaxy circling inside the Milky way in an almost a "polar" orbit (if accepting that the Milky way galaxy disk defines the equator). However, they estimated the systemic velocity as a Doppler red shift, which is wrong, according to BSM, since the Sag-DEG has an own CL space. As a result of this mistake, they calculated that the SagDEG has made many orbits around the dens galactic bulge. Then how this galaxy has not been disintegrated. They arrived to unsolvable enigma.

Since the discovery of SagDEG, researchers have noticed that some of its stars are strikingly similar to the stars from the Large Magelanic Cloud (LMC), which is another nearby satellite galaxy located slightly further out from the Milky Way. A team led by Patrick Cseresnjes of the Paris Observatory found strong similarities in a certain class of highly evolved, old stars seen in both of these satellite galaxies. An opinion is expressed that both satellites may have a common galactic ancestor. The bulk of the SagDEG stars are relatively

metal-rich and their age has been estimated in the range between 10 to 14 billion years (Fahlman et al., 1996). Such age contradicts to the currently estimated age of the Universe (about 12 billion years). Number of theories exist trying to solve this paradox but their arguments are not enough strong (see *http://map.gsfc.nasa.gov/m_uni/uni_101age.html*).

According to BSM, the SagDEG is not a separate galaxy, but a formation from the former life of the Milky way galaxy, which has escaped the collapse. The systemic velocity of the SagDEG is zero, while the individual stars have a proper motion. The similarity between the SagDEG and the very close irregular galaxy of Magelanic Clouds leads to the hypothesis that they both are remnants from the former life of a galaxy, which occupied the space of the current Milky way and Magelanic clouds galaxies.

12.B.7.5 Effect of GC detectability in distant galaxies

Why the GCs from distant galaxies are detectable?

The line emissions from the GCs after passing through the GC GSS appear slightly red shifted in respect to the lines that could be emitted or absorbed in their host galaxy. In such way the photons with shifted wavelengths escape from the strong absorption. The absorption is mainly due to the Doppler broadened line profile so it is not eliminated but reduced. This is valid also for the broad band emission, because it is consisted of numerous individual lines. Passing through consecutive GSS the emitted lines are additionally red shifted. This effect, however, allows to observe the GCs from very distant galaxies. This provides excellent opportunity for determination of Hubble constant for distant galaxies by distinguishing the light from the I-st and II-nd type population and applying the knowledge of the difference in the luminosity functions, determined by the Milky way and neighbouring galaxies. The method is proposed by William Harris (1988) and successfully used by Tonry et all (1990).

In §12.B.13 a new BSM method of data interpretation is presented for determination of distances to very distant galaxies in the stationary Universe. In this method the Hubble constant is implicitly involved, so it appears useful even for the concept of a stationary universe.

12.B.7.6 Energy loss effect of GC

The mismatch of the emission and absorption lines between GC and the host galaxy leads to a constant energy imbalance between them. But in own CL space of the GC, the energy balance is preserved. The GC, however, may loose faster energy in comparison to the host galaxy, because the lack of own galactic-like nucleus. The continuous starvation for energy might be partially compensated by two mechanisms:

(a) dropping some peripheral stars whose energy is sucked by the system

(b) decrease of the binding energies of the atoms (decay nuclear reactions)

(c) obtaining of energy stored in the atomic particles by crashing some atomic material

The case (a) is observable phenomena

The case (b) means disintegration of heavy elements

The case (c) is related to the growing of kaon nucleus in the star womb and releasing of RL(T) structures from crashed FOHSs.

12.B.8 Interacting galaxies

According to BSM, the case of interacting galaxies is considered as deviation from the normal galactic cycle. Their percentage is relatively small. The possible conditions leading to interaction between the galaxies are the following:

(a) uneven collected matter during the galaxy collapse

(b) some defects of the galactic nucleus.

The case (a) means that the spherical symmetry of the collapsing events is so severe disturbed that it may lead to form of GCs and escaped islands in a small spatial angle from one side of the galactic nucleus than in the full 4π angle - valid in the normal case. This may happen if the CL space is separated unevenly from the neighbouring galaxies during the collapse. This may effect both: the position of the protogalactic egg (with included inside nucleus of bulk matter) and the directions of the expanding new born matter after the explosion of the protogalactic egg. Then a portion of the released new matter, together with the new CL space

may trespass in the CL space of one of its neighbouring galaxies. **This concept provides an answer why most of the observed colliding galaxies are of odd shape.** One example of colliding galaxies with irregular shape is the Antennae galaxies.

In a normal case the prisms from different galaxies have different length to diameter ratio. So their CL spaces and matter are incompatible. Any matter (atomic structure) will operate in a foreign CL space (this has been discussed for the cepheids from the GCs). Then the congregation of atomic matter is not so effective and it may not lead to formation of stars as in the case of main sequence of H-R diagram. From this point of view, the CL spaces and matter from different prism's formations will exhibit a feature of not mixability. This means that two colliding galaxies may form a manifold of CL spaces separated by GSS. The image from such galaxies will look as they have multiple separation surfaces. Such features a observed in the colliding galaxies. Fig. 12.46 shows the image of Antennae colliding galaxies obtained by the Hubble Space telescope. The enclosed part of the lower resolution left image is shown as a high resolution image in the right side of the figure

(from: http://astrowww.phys.uvic.ca/~patton/openhouse/
antennaehst.jpg).

Note: Download the original image for better resolution.

The low resolution image shows radiation from regions quite away from the large mass concentration. This is a signature of CL space interactions. In fact it is from a gas substance around the regions adjacent to the GSS. The high resolution image unveils nonmixability of the two CL spaces. Multiple stripes with sharp contrasts are quite distinguishable. They correspond to penetration of matter from one galaxy into the CL space of another one. Note that the stripes are quite extended and look connected. The interpretation of the dark stripe as a dust is not reasonable. A possible dust component could not exhibit such spatial features with a sharp contrast.

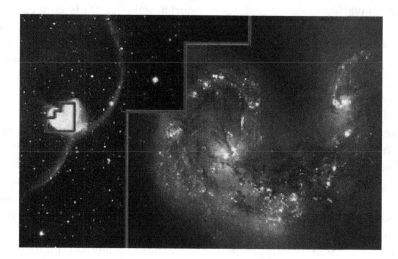

Fig. 12.46

12.B.9 Cosmological anisotropy.

The pancake shape of the galaxy is predetermined by the axial rotational symmetry of the clusters in the galaxy egg. It defines the preferable direction of expansion of the new formed CL space. The shape of the new CL space is restricted by the existing empty space, but the island escaped the collapse, such as the SagDEG may modify the shape of the available space. Additionally some small empty may exist between some neighbouring galaxies. Then for a extended time duration of multiple galactic cycles, the neighbouring galaxies may get preferential orientation of their minor axes. In such case, larger formation of galaxies may exhibit a spatial orientation effect. Such effect, called an anisotropy in the galactic clusters is really observed.

The spatial anisotropy provides one specific feature when observing radiation from distant quasars. The photons from any quasar with high z-shift passes through multiple GSSs, exhibiting small energy loss from any one. The multiple refurbished wavetrain of the photon may carry accumulated features from the incidence angle on every GSS. The alignment between galaxies is valid up to some range, but once the passing quantum wave is polarized it could not be depolarised if a scattering is not involved. Only the vector of polarization will have different angle for different distances from the source. In such case the detected signal will get polarization that will show increase with the z-shift. A polarization dependence on the z-shift is really observed in the radiation from distant quasars (Antonucci R, Hurt T. and Kinney A., 1994).

12.B.10 Crab nebular and pulsar

The bright pillar structure of the Crab Nebular was one of the big mysteries, until the explanation provided by BSM.

The Crab nebula is a remnant of supernova explosion observed by the Chinese as a "guest star" in 1054 a. d. Figure 12.47 shows clearly the emitting pillars in a red light photography (Courtesy of Lick Observatory).

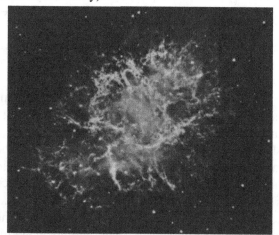

Fig. 12.47 The Crab nebula photographed in red light [Courtesy of the Lick Observatory]

The Crab nebular and its pulsar are result of explosion of supernova of second population star (or more than one star). The Crab pulsar exhibits all features of a double jet kaon nucleus described in §12.B6.4.9: a lack of proper motion , a very short period, a comparatively fast aging (estimated by the increasing rate of the secular period).

In many scientific articles the strange behaviour of the Grab Nebula with its pulsar is identified as a natural synchrotron.

From the point of view of BSM, many features of the Crab Nebula indicate that it is a remnant from one of the previous lives of our galaxy. The exploded star probably is from a globular cluster, containing own CL space. The prisms of this CL space are from the old formation (having slightly different diameter/length ratio). Such CL space structure is not mixable with the Milky galaxy CL space. For this reason it is spread in form of pillars. Many of these pillars are likely interconnected, forming a spatial manifold. Some quantity of the remnant atomic matter could be settled inside the pillars, other - outside of them. The atomic matter outside of the pillars has to operate in foreign CL space, with significantly decreased quantum efficiency. The matter inside the pillars operates in own CL space with normal efficiency. Many observational properties of the Crab pulsar indicate that it is a double jet kaon nucleus. Some observers accept that the pulsar rotates around some massive object. Then, according to BSM, it will have an wobbling motion and its two jets will irradiate the spread atomic matter. The jets containing high velocity RL nodes not only irradiate the pillars, but supply them with a constant flow of flying RL nodes. If accepting that a fraction of them is refurbished into folded CL nodes, a hallo of own CL space will be surrounding the pulsar. In such way the remnant matter that is in this hallo operates in own CL space and could be observable as a diffuse radiating region.

Figure 12.47.A shows an X-rays image from the Crab nebula obtained by Chandra X-ray Observatory in 1999. The analysis of the observed image is in agreement with the above presented concept.

The pillars of Crab nebula are more distant from the central hallo region but the intensity of their radiation is higher. Their radiation mechanism is different. Let us suppose that the radiating pillars are connected in manifolds. The manifold sections nearer to the pulsar will be stronger irradiated by the jets than the more distant sections. Consequently, the wobbling rotation of the pulsar will supply different energy to the connected pillars maintaining in such way a constant ZPE potential difference. This will cause a continuous motion of atomic matter inside the connected pillars. In such process it could be highly ionized and will emit a broad band spectrum. **Then the system of the manifold structure in the Crab Nebula pillars and the Crab pulsar may provide a synchrotron-like radiation.**

Fig. 12.47.A. The Crab nebula in X-rays
Credit: Chandra X-ray Observatory, NASA
(http://antwrp.gsfc.nasa.gov/apod/ap990929.html)

The ZPE potential between the closer and the distant pillars in respect to the pulsar will contribute also to the gradient of the cosmological shift between inner and outer regions of the nebular. This is the reason for identification of larger expansion velocity in the more distant pillars, when considering that the observed red shift is only of Doppler effect.

12.B.11 Quasars

Fig. 12.36 shows that the number of pulsars detected in the Milky way galaxy is quite large. All they are in motion due to own propulsion mechanism suggested in §12.B6.4.2. and illustrated by Fig. 12.29. Cepheids of type II are typical for the GCs. From the concept presented in §12.B.7.2.2 it is evident that their evolution also leads to formation of pulsars. The released nodes of kaon nucleus RL(T) structures from the moving pulsar will be refurbished to CL nodes, but they could not mix with the CL space of the host galaxy. Consequently, the moving pulsars will build a pipe of a local CL space inside the home galaxy CL space. It is possible one pulsar born from a cepheid from one GC to arrive to another GC. In such case the CL spaces of both GCs could be connected by a pipe of CL space from their own type of prisms. The pipe could be quite long of the order of kpc.

Let us analyse the case of two GCs, which have been isolated for a long time and in some moment they get a connection through a pipe of own CL space, provided by a moving pulsar. Since they carry an older CL space, they will have a lower CL space energy. Their CL space energy also can be different, since they are isolated (the ZPE waves equalize the energy only in own CL space). Then in the moment of their interconnection by a pipe, an energy potential will occur between them. This potential will cause an energy transfer that will be carried by accelerated atomic matter. The transferred atomic matter may rotate inside the pipe and the atoms and ions might be excited. Since the electrons have much larger magnetic moment, the process will generate a strong magnetic field.

The cosmological length of the pipe and the rotational motion of the transferred atomic matter may cause appearance of longitudinal modes with long wavelength coinciding with the emission of the hydrogen or some other atoms in the radio spectral range. In such case the emitted radiation obtains some kind of coherence. It is reasonable to expect that the trace of the flowing and excited atomic gas form spiral shape with density closer to the pipe wall. This is illustrated in Fig. 12.48.

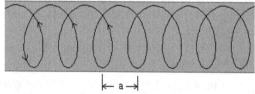

12.48 Atomic motion in the pipe

It is obvious that for the distant observer, such type of motion will exhibit a large Doppler broadening of the emitted lines. If accepting a laminar flow we may expect that the emitted radiation in the radiowaves frequency range may appear synchronized. The longitudinal mode defined by the helical trace step *a* may become a whole number of some wavelength of some of the transitions of the atoms (or molecules) involved in the gas flow. In such case the helical trace is stabilized and may provide a coherence radiation. The emitted radiation will be characterised by:

- cosmological z-shift in respect to the home galaxy (comparatively low in respect to more distant quasars)
- coherence from a large volume
- a large Doppler line broadening
- a polarisation from the strong magnetic field

The coherence provides the opportunity the radiation to be effectively observed by technique known as a Very Long Base Interferometry.

The described phenomenon is a quasar. The emitted lines from all quasars are broaden. Compared with the variable stars the quasars are comparatively rare phenomena. For these reason they has been observed preliminary in large distances (large cosmological z-shift). For this reason they have been mistakenly categorised as "remnants from the early Universe" according to Big Bang concept. In 1994, however, a similar phenomena has been observed in the radiogalaxy Cygnus A only at 600 million light years from us. The emitted spectral lines from this quasar are also broadened, but not so strongly polarized, as from the more distant quasars. This difference, according to BSM, is a result of smaller number of intercepted GSSs, which also affects the polarization (discussed in §12.B.9). Figure 12.49 shows the image of Cygnus A quasar.

Fig. 12.49. Cignus A quasar as a connecting pipe between two cosmological formations (BSM interpretation). Pub. domain: http://home.achilles.net/~jtalbot/news/3C405.html
Note: Download the original image for better view of the pipe

12.B.12 Lyman Alpha Forest

It was shown in §12.B.4.2.3 that the observed large cosmological z-shift (in a stationary Universe) is a result of accumulated red shifts obtained from the consecutive passing of the photon (or any type of quantum wave) through a large number of GSSs. It will be very useful if we are able to identify signatures from these multiple red shifts. Such opportunity, fortunately, is provided by the distant quasars. Now BSM is able to provide a correct interpretation of the phenomenon know as a **Lyman alpha forest**. The BSM explanation of this phenomenon is the following:

The Lyman alpha forest appears when observing the Lyman alpha line from a distant quasar. It appears as multiple line structure as shown in Fig. 12.51.A. The quasar phenomenon was explained in §12.B.11. The Hydrogen as one of the most abundant and light element, in the galaxies flows through the pipe of the quasar, driven by the ZPE potential difference between the CL spaces, which are connected by the pipe. The strongest line of the excited atomic Hydrogen is L_α (Lyman alpha) at 121.6 nm. The line emitted from the flowing Hydrogen in the pipe is significantly Doppler broaden, due to the helical motion, illustrated in Fig. 12.48. (lines from other elements are also Doppler broaden). The emitted photons after passing through the first GSS (between the pipe and host galaxy) appear red shifted in the host galaxy. Then the normal absorption of the narrower L_α line appears superimposed on the spectral profile of the broaden and red shifted line from the quasar. This feature is red shifted constructively when the photons pass through the GSS of the closer to us galaxies. Additionally in every consecutive galaxy the signature of the absorbed L_α is superimposed. The broaden quasar line appears red shifted in every consecutive GSS, serving as an illuminating source for the narrower absorption L_α line in the galaxy through which the photons pass. The absorption lines does not have the broadening mechanism of the quasar emission line. So they are distinguishable. The degree of their separation depends only on the degree of difference between the prisms of neighbouring galaxies (different diameter/length ratio of the prisms, see §12.B.4). Fig. 12.51 illustrates the formation of L_α forest in the detection side.

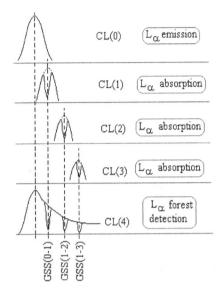

Fig. 12.51 Formation of L_α forest from a distant quasar

The following notations are used:
CL(0) - space of quasar pipe
CL(1) - space of the host galaxy of the quasar
CL(2), CL(3) consecutive galaxy spaces
CL(4) galaxy of detection (Milky Way)

The following features are apparent for a distant quasar, where the possible Doppler shift component is negligible in comparison to the total cosmological z-shift

(a) the number of absorption components are equal to the number of intercepted CL spaces, but the quasar pipe CL space is not included

(b) the redshift from a single GSS corresponds to the difference between diameter/length ratio of the prisms for the neighbouring galaxies

(c) the envelope of the detected signal is affected by: the number of crossed GSSs, the attenuation from dust in every galaxy and the absorptions in every passed galaxy.

(d) if neglecting the attenuation from the dust, the amplitude of any absorption component is determined by the absorption through the current galaxy CL space.

(e) the first absorption component is from the host galaxy, where the quasar pipe is located.

Fig. 12.51.A illustrates Lyman alpha forest observations reported by A. Songalla, E. M. Hu & L.L. Cowie (1995).

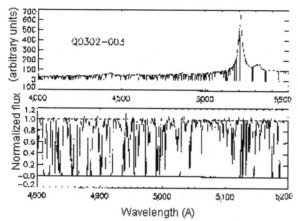

Fig. 12.51.A. Top, high resolution spectrum of the quasar QO302-003 (emission redshift Zem = 3.286), which is the sum of eight 2,400-s exposures made with the HIRES spectrograph at the Mauna Kea, Hawaii on the nights of 1994 September 16 and October 14 UT. Courtesy of A. Songalla et al..

12.B.13 Estimation of cosmological distances

12.B.13.1 General considerations

Number of methods are used so far for measuring of cosmological distances, but according to the concept of Big Bang the distance could be considered only as a momentary distance at fixed moment. For this reason the distance has been often referenced as a "distance of expanding Universe". This is a result of not correct interpretation of the cosmological red shift, assuming that it is of Doppler type.

In the BSM concept of a Stationary Universe the distance is a stable real parameter in a classical sense. The cosmological component of the Z-shift was discussed theoretically in §12.B.4.2. It was shown how this component obtains significant increase for large cosmological distances. If the z-shift is assumed as a Doppler shift it provides completely wrong picture of the Universe. BSM clearly shows, that the Universe is a stationary with galaxies undergoing a quasiperiodical recycling process. Fortunately, the Z-shift parameter still could be used for aproximative estimation of large cosmological distances, if a proper interpretation is used.

In §12.B.4.2.3 and Fig. 12.23 the consecutive red shifts from GSS has been presented. The energy loss of a photon crossing GSS was discussed in §12.B.4.2.2. The signature of the cosmological red-shift is apparent from many observational effects:

(a) from the analysis of cepheids of II population

(b) from the motion analysis of stars in the globular clusters

(c) from the large red shift in the filament structures of the Crab nebular

(d) from the large red shift in the filament structures in the galactic Centre

(e) from the z-shift periodicity

(f) from the interpretation of the forest Lyman alpha

The large red shifts for cases a, b, c, d are results of emissions from remnants of previous galaxy life (in respect to the host galaxy).

12.B.13.2 Lyman alpha forest method

This is the simplest aproximative method for estimation of large cosmological distance. For a distant quasar (with a large z-shift) we have a large number of absorption L_α lines. In such conditions we may accept that the average intergalactic distance is equal to the average length of the galaxy CL space. We may call this parameter a **mean size of galaxy space**. Then the number of absorption lines should correspond to the number of GSS minus one in the line of sight. So if we know the mean intergalactic distance, $\tilde{L}$ (equal to a mean size of galaxy space) we can determine the approximate distance (r) to the quasar simply by the product:

$$r = (N-1)\tilde{L} \text{ - for quasar} \qquad (12.46)$$

$$r = N\tilde{L} \text{ - for emitting objects of I-st population} \quad (12.47)$$

where: - $\tilde{L}$ is a mean intergalactic distance, N - is a total number of the absorbed L_α lines in the forest for a range of $0 < z < z_Q$, z_Q is the quasar z-shift

The relations (12.46) and (12.47) are approximative because the the galaxy sizes varies more than one order, but the accuracy is improved for a large statistics, i. e. if we estimate data from objects with a large z-shift.

The total number (N) of absorbed L_α lines can be determined from the equation of the cosmological z-shift (see §12.B.4.2.4):

$$(\tilde{n})^N = z + 1 \qquad [(12.48)]$$

Solving for N we get:

$$N = \frac{\ln(z+1)}{\ln(\tilde{n})} \qquad (12.49)$$

where: $\tilde{n}$ is a **mean GSS quasirefractive index** (defined in §12.B.4.2.4 by Eq. (12.27) for N >> 1).

Substituting N given by Eq. (12.49) in Eq. (12.47) we obtain **the equation of the cosmological distance as a function of Z-shift.**

$$r = \tilde{L}\frac{\ln(z+1)}{\ln(\tilde{n})} \qquad (12.50)$$

Layman alpha forest observations usually does not cover the whole spectral range of redshifted absorption lines. The more easier directly observed parameter is the number density of absorbed L_α lines. The theoretical value of this parameter is obtained by differentiating the Eq. (12.49) on Z.

$$\frac{dN}{dZ} = \frac{1}{(z+1)\ln(\tilde{n})} \qquad (12.51)$$

The estimation of the line density, according to BSM, is dependable on the proper selection of the observed L_α set. For this reason, different investigators provide a different empirical estimate for the number density as a function of z-shift. W. L. W. Sergent et al. (1980) found that the line density is pretty independent from the z-shift and corresponds approximately to 60 lines per unit redshift at $z = 2.45$ Murdock et all.(1986) investigating the line density in a number of sets found that the individual density trend in a single set differs from the common trend of the sets. The reason for such difference according to BSM is that the parameters $\tilde{n}$ between neighbouring galaxies are strongly correlated.

The BSM concept shows, that the better method for estimation of line density and the parameter $(\frac{dN}{dz})_{.0}$ is:

- finding the fitting equation from every separate forest set measured by own quasar

- obtaining a common fit equation by assembling the separate fitting equations

The more accurate obtaining of $\tilde{n}$ is dependable also on the spatial position of the line of sight, due to the spatial anisotropy effect and the inhomogeniety of the observable Universe (according to BSM concept) as a transparent medium. If the above considerations are taken into account, more realistic value of $\tilde{n}$ could be obtained by fitting the theoretical expression (12.50) to the data.

The line density vs z-shift has been studied by number of scientists, but the above mentioned considerations has not been evident because the z-shift parameter has been considered so far as a Doppler shift. Then it is not a surprise that different authors obtain quite different empirical dependence of the line density vs z-shift. Eq. (12.50) shows that the standard variation of this parameter for low z-shifts (below 1 and approaching zero) will be much larger, than for a larger z.

For estimation of large cosmological distances, when z-shift is known, (by Eq. (12.50)) the knowledge of mean galaxy length and mean quasirefractive index are both necessary. The latter parameter is directly obtainable from Eq. (12.51), while its physical meaning is discussed in the next paragraph.

12.B.13.3 Theoretical concept of the Universe optical inhomogeniety

12.B.13.3.1 Light propagation from energetic point of view

Energy transfer as EM waves between the matter from different connected galaxies is possible only through CL space. The connected CL spaces of the separate galaxies form the global CL space of the Universe through which we get information about the connected galaxies. Therefore the CL space of the observable Universe is a conglomerate of CL spaces of the connected galaxies. The prisms of the individual galaxies contain equal amount of intrinsic matter but they are formed in separate processes under different formation forces (defined by the total mass of the galaxy), so they may have different diameter/length ratio. Consequently, the space of the visible Universe is inhomogenious. The degree of inhomogeniety depends on the difference of the diameter/length ratio of the prisms in the interconnected galaxies.

The **mean GSS quasirefractive index, $\tilde{n}$,** (defined in §12.B.4.2.4 by Eq. (12.27)) could serve as a parameter of the Universe inhomogeniety.

$$\tilde{n} \approx \frac{\lambda_1}{\lambda_0} \approx \frac{\lambda_2}{\lambda_1} \approx .... \frac{\lambda_i}{\lambda_{(i-1)}} > 1 \qquad [(12.27)]$$

In the previous paragraph we found that this parameter could be estimated by the absorption line density. Solving Eq. (12.51) for $\tilde{n}$ we obtain

$$\tilde{n} = \exp\left(\frac{1}{(dN/dz)(z+1)}\right) \qquad (12.52)$$

Taking into account the considerations (in the previous paragraph) about the estimation of the line L_α density, we may calculate $\tilde{n}$ only for some limited points of z-shift from carefully selected observational data. W. L. W. Sargent et al, (1980) have investigated L_α forests from six QSOs in the quasar z-range $1.7 < z < 3.3$ and found that the mean line density per $\Delta z = 1$ at $z = 2.45$ is about 60. Then the mean quasirefractive index ($\tilde{n}$) is:

$$\tilde{n} = \exp\left(\frac{1}{60(2.45+1)}\right) = 1.0048 \qquad (12.53)$$

H.S. Murdoch et al. (1986) provide different empirical equation for estimated line density.

$$\frac{dN}{dz} = 4.06(1+z)^\gamma,$$

where $\gamma = 2.17 \pm 0.36$. Then for their sample at z = 3.5

$dN/dz = 106$ (lines per $\Delta z = 1$). The corresponding mean quasirefractive index from this data is:

$\tilde{n} = 1.0021$ (by Murdoch et al.,1986)   (12.54)

The estimated value for dN/dz varies significantly from the observational samples. It seams to be dependable on the spatial direction as well. Then the parameter $\tilde{n}$ will also exhibit variations.

Comparing the obtained above two values of $\tilde{n}$ with the value of 1.0002 for quasirefractive index of GC M5 (obtained by the spectrum - see §12.B.7.2.2), we see that the latter is much smaller. This is reasonable when considering that the GC is a remnant of the previous life of the home galaxy and the prisms diameter/length ratio should be much closer in this case.

From the above mentioned observational data we may accept that the probable range of the mean GSS quasirefractive index is

$1.0021 < \tilde{n} < 1.0048$ \qquad (12.54.a)

12.B.13.4 Existing methods for determination of cosmological distances independently from the concept of the expanding Universe.

The cited below methods are used for calibration of the Hubble "constant", which according to the BSM concept appears as a constant only in a limited cosmological distance. Different methods cover different range of cosmological distances that are offset by different distances from the observer. As a result, the real cosmological distances could be estimated only by correct intercalibration between these methods. These approach is often referred as a **cosmic distance ladder**. The ranges that the separate methods cover according to this approach are illustrated by Fig. 12.52.

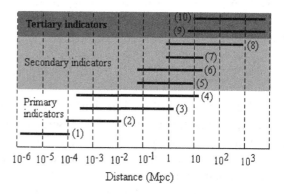

Fig. 12.52.
Methods for estimation of cosmological distances and their positions in the cosmic distance ladder scale

Some of the used so far methods should be reconsidered due to the following phenomena discovered by the BSM theory:

(1) The cosmological origin of the red-shift

(2) Globular Clusters (GC) in distant galaxies looks brighter (effect discussed in §12.B.7.2.2)

(3) The exciting of the external layers of the cepheids is caused by the released RL(T) structures (effect discussed in §12.B.7.2.2).

(4) The rotational motion of the galactic matter may not be in a stationary phase

These phenomena will affect the following methods of distance estimation:

- Cluster H-R diagram fitting (affected by the phenomenon (1))

- Cepheids (affected by the phenomenon (3)

- Brightest stars in galaxies (affected by the phenomenon (3))
- Globular clusters (affected by the phenomenon (2)
- Tully-Fisher relation (affected by the phenomenon (4) and a-parameter dependence on the observed wavelength (see L. Bottinelli et al (1983)).
- Supernovae (affected by the phenomenon (3)).

The distances estimated so far according to the Big Bang concept are based on the assumption that the Hubble distance-velocity relation of the Hubble law is a constant (due to the Doppler interpretation of the observed z-shift). The distance-velocity relation is empirically obtained by measuring the distances by the methods show in Fig. 12.52.

12.B.13.5 Controversial interpretation of the Hubble law. Recent experimental results.

The correct value of the Hubble constant has been and continues to be a debated topic for about a 100 years. Enormous observational material is obtained, but the Hubble "constant" continues to be one enigma. For small z-shift observations, the Hubble law shows a deviation from linearity, but this has been attributed to a statistical error due to small number of observations. However, recent observations for z-shifts up to 0.8 show an additional deviation from linearity in a completely unexpected direction.

The classical plot of the Hubble law is a velocity (km/s/Mpc) vs Distance (Mpc). In the scientific papers, the Hubble plot is given more often as an apparent magnitude (of the observed galaxies) vs z-shift. It preserves the "linear" appearance of the classical Hubble plot if both axes are linear or logarithmic. The apparent magnitude has a logarithmic relation to the distance according to the expression

$$m - M = 5\log(d) - 5 \qquad (12.61)$$

where: m - is the apparent magnitude, M is the absolute magnitude, d - is the distance.

Figure 12.53 shows a Hubble plot for z-shifts up to 0.8 provided by Perlmutter et al., Astrophys. J. 517: 555-586, 1999; available also in http://laml.arxiv.org: astro-ph/9812133. The data are ob-

tained from a ground based telescopes. The parameter m_B, called an effective magnitude is a restframe magnitude, corresponding to the observed brightness, corrected for width luminosity relation. The four theoretical curves shown in Fig. 12.53 are characterized by two parameters Ω_m and Ω_Λ, denoting respectively the mean mass and the dark-energy density of the Cosmos, normalized in a way that their sum is precisely one.

The observations lead to the conclusion that

$$\Lambda = (\rho_{vac}/\rho_{cr}) > 1 \qquad (12.62)$$

where: ρ_{vac} and ρ_{cr} are respectively the vacuum energy density and a critical vacuum energy density.

The parameter Δ is related to the cosmological constant introduced initially by Einstein in General Relativity. However, later he dropped it from his equations, referring to it as his greatest blunder. The cosmological constant and the Δ parameter, however, are still used in the Big Bang concept, so the obtain result $((\Lambda > 1)$ means that the Universe should be expanded forever, or in other words, it will be disintegrated. Such conclusion, is completely illogical and brought a new problem for the Big Bang concept. In order to save this highly contradictable theory, a new concept of a "vacuum energy of unknown character" and even a "negative energy" have been suggested.

The enigma of expanding forever Universe is based on interpretation of observations provided by the Hubble space telescope and published recently by A. G. Riess et al. (2004),Astrophys. J. (in press), also in: (http://arXiv.org/abs/astro-ph/0402512).

Figure 12.55 provided by A. G. Riess et al. (2004) provides a Hubble plot of data for z-shifts up to 1.75 together with some theoretical plots. The data interpretation leads to a conclusion that it has been deceleration in the initial phase of the Big Bang, while at $z = 0.46 \pm 0.13$ a cosmic "jerk" changes the deceleration to acceleration. Such cosmological behaviour, however, is completely illogical. Attempts are made to explain this cosmic paradox by evolution of the "dark energy". For this reason an equation of state with parameters "ratio of negative pressure to energy density" is used but the explanation is away from the common sense logic.

From a point of view of the BSM, however, the obtained experimental data appear to fit quite well to the the analytical results from the BSM concept about the stationary Universe. The derivation of BSM results are described in the next paragraph.

12.B.13.6 New interpretation of the Hubble law.

12.B.13.6.1 General considerations:

According to the concept of the Big Bang theory the z-shift is of a Doppler type, so the distance of the "expanding" Universe is determined by the simple relation:

$$r = \frac{zc}{H_0} \quad [m] \tag{12.63}$$

where: z - is the observed redshift, c - is a light velocity and H_0 is a Hubble constant (velocity/distance).

For a not relativistic velocity ($z < 1$) the product zc provides the expanding velocity using the classical Doppler shift formula. For observations with $z > 1$ the classical Doppler shift formula provides velocity larger than the speed of light, but quasars with z-shift up to 4 has been observed. In order to save the concept of the Big Bang a relativistic formula for Doppler shift is used in which case the velocities could not exceed the speed of light in any value of $z > 1$. More distant galaxies shows a larger z-shift, however, numerous exceptions from this rule exist, known as peculiar velocities (or z-shift) for which no convincing explanation has been suggested. According to Sten Odenwald and Rick Fienberg, (1993), "the adoption of the special relativistic Doppler formula by many educators has led to a peculiar 'hybrid' cosmology which attempts to describe Big Bang cosmology using general relativity ".

The Big Bang concept of expanding Universe relies also on the assumption that the parameter H_0 is close to a constant. The parameter H_0 is formulated as a velocity-distance relation by the great astronomer Hubble. **Hubble did not like the idea of expanding Universe**, but this interpretation has been adopted by Modern physics because it fits to the suggested in that time concept of the expanding Universe.

12.B.13.6.2 The Hubble constant from the BSM point of view.

A. general considerations

In the Big Bang concept the relation product $((zc)/H_0)$ is regarded as a **momentary distance**, while $H_0 = zc/r$ is denoted as a Hubble constant. According to the BSM concept, however, it is not a constant in a perfect sense.

B. Derivation of the Hubble law from a BSM point of view

The distance in the expanding Universe is given by Eq. (12.63) while for a stationary Universe we derived Eq. (12.50). For an infinite small range of distance (or z-shift), the instant distance according to the concept of the expanding Universe will be equal to the real distance according to the BSM concept of stationary Universe. Then equating (12.63) and (12.50) we get:

$$r = \frac{zc}{H_o} = \tilde{L}\frac{\ln(z+1)}{\ln(\tilde{n})} \tag{12.64}$$

From where:

$$H_o = \frac{zc\ln(\tilde{n})}{\tilde{L}\ln(z+1)} \tag{12.65}$$

$$\text{Let: } A = \frac{c\ln(\tilde{n})}{\tilde{L}} = const \tag{12.66}$$

For an infinite small range of distance (differentiating Eq. (12.64)) we have

$$dr = \frac{c}{H_o}dz \tag{12.67}$$

Multiplying Eq. (12.65) by Eq. (12.67) and dividing the result on H_o we get:

$$dr = \frac{Ac}{H_o^2}\frac{z}{\ln(z+1)}dz \tag{12.68}$$

Integrating Eq. (12.68) in a range from 0 to z we obtain

$$r = \frac{Ac}{H_o^2}\int_0^z\frac{x}{\ln(x+1)}dx = \frac{c^2\ln(\tilde{n})}{\tilde{L}H_o^2}\int_0^z\frac{x}{\ln(x+1)}dx \tag{12.68}$$

Let $r_n = r(Ac/H^2_o)$ is a normalized distance, a dimensionless parameter, defined by the following constants: c - a light velocity, $\tilde{n}$ - an average quasirefractive index of GSS, $\tilde{L}$ - an average distance between the galaxies, H_o - the Hubble constant. It is evident from Eq. (12.66) that r_n appears normalized on the average distance between the galaxies $\tilde{L}$. Then we obtain:

$$r_n = \int_0^z \frac{x}{\ln(x+1)} dx \qquad (12.69)$$

The expression (12.69) provides the relation between the normalized distance to a distant galaxy and the measured z-shift. The equation (12.69) can be regarded as a theoretical expression of the Hubble plot.

Equation (12.69) is calculated numerically and plotted in Fig. (12.54) in a log-log scale and in Fig. (12.56) in a log-lin scale. Figure (12.54) is plotted in a same scale like the Fig. (12.53), for comparison with the shown experimental Hubble plot. For the same reason Fig. (12.56) is plotted in a same scale like the Fig. (12.55) for comparison. The derived theoretical equation of the Hubble law, according to BSM concept, is in excellent agreement with the experimental data. (For a fitting test the theoretical plots can be copied and past over the experimental data).

12.B.13.7 Observational evidence about the inhomogeniety of the Universe

Some of the observational evidences about the inhomogeniety of the Universe has been discussed:

- The lower Maxwellian energy in the Globular Clusters discussed in §12.B.7.2.1
- The red shift periodicity discussed in §12.B.4.2.5
- The quasar phenomenon as a pipe between two CL spaces with different ZPE immersed in a host galaxy CL space
- The Lyman alpha forest phenomenon discussed in §12.B.12.

One additional phenomenon which might be explained by inhomogeniety is discussed in the next paragraph.

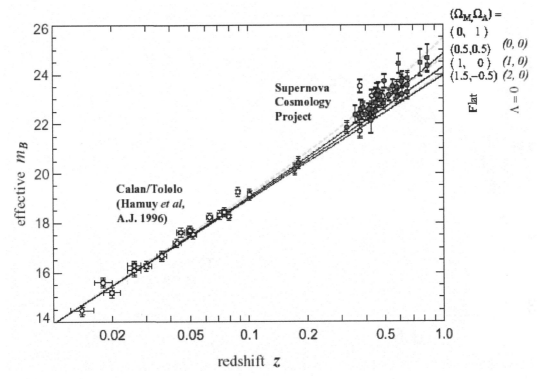

Fig. 12.54.   Hubble plot for 42 high red-shifted Type Ia supernovae. Courtesy of Perlmutter et al., Astrophys. J. 517: 555-586, 1999; available also in http://laml.arxiv.org: astro-ph/9812133

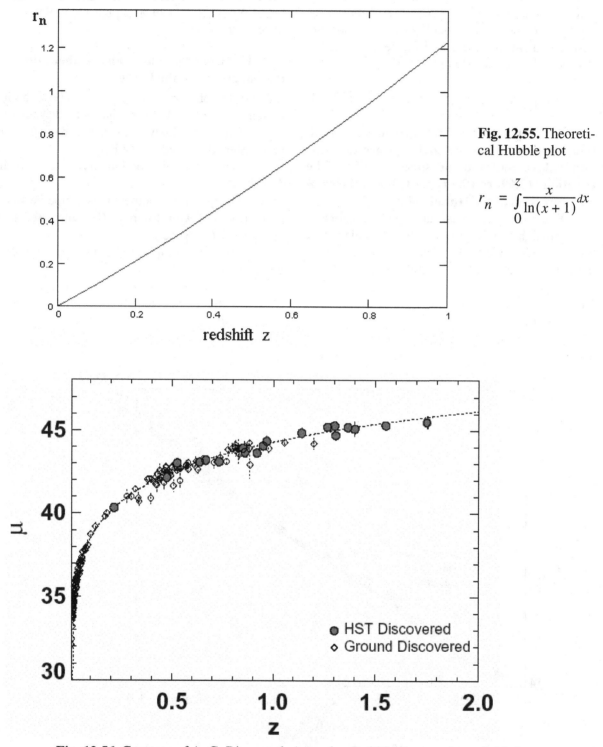

Fig. 12.55. Theoretical Hubble plot

$$r_n = \int_0^z \frac{x}{\ln(x+1)} dx$$

Fig. 12.56. Courtesy of A. G. Riess at al, Astrophy. J., 2004 (in press), available in http://arXiv.org/abs/astro-ph/0402512

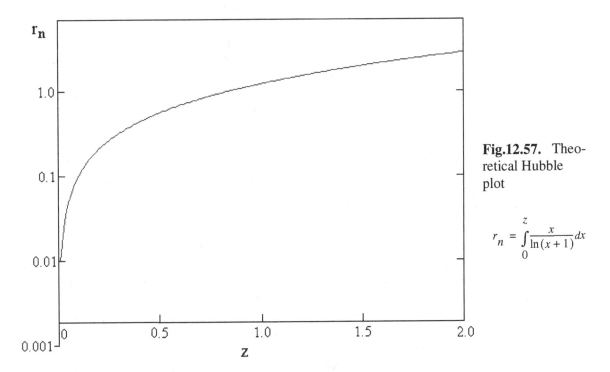

Fig.12.57. Theoretical Hubble plot

$$r_n = \int_0^z \frac{x}{\ln(x+1)} dx$$

12.B.14. Stability of the Cosmic Lattice structure

One of the amazing feature of CL structure is its stability. It means, that any temporal damage of this structure is repaired (however, this process may take a finite time, depending on the size of damage). This effect is subordinated by the SG forces and the intrinsically small time constant of the SG types of interactions. These are kind of transition processes for CL space, which may be invoked only by very fast and strong interactions. Such interactions take place, for example, in the atomic coliders when FOHSs are destroyed. In this case a large energy appears providing infinities in the Feynman diagrams. But where this energy does come from? The possible answer is: It is extracted from the CL space. The destructed FOHS causes a temporal local change of the static CL pressure, which is equalized by the CL space in an extremely fast way. The speed of the process is controlled by the high frequency parameters of CL space. The time-space constant, used also as a CL space relaxation time is obviously involved in such transition

process. The nuclear weapon causing a massive fission or fusion reactions is one of the most violent destruction of CL structure, but it is still repaired (fortunately). The huge energy effect comes again from the CL space. One interesting fact deserves attention: we get energy from both type of reaction: fission (nuclear depletion) and fusion (nuclear synthesis). This means that CL structure reacts on both action: local abnormal increase or decrease of the static CL pressure. The invoked SG waves in such reaction cause a large production of gamma waves from the CL space. Consequently the CL structure is in very strict balance. But any energy balance requires at least two separate components. What are these components of CL space? The possible answer is: **The two energy components are the energies of the left and the right handed node systems. This means that despite the alternative arrangement of the lefthanded and right handed nodes in the CL space they are connected pretty strongly.**

Having in mind the difference between G_{OS} and G_{OD} we may consider the total energies of left handed nodes and right handed nodes as separate energy components. So we may regard the CL sys-

12-105

tem as comprised of two subsystem with their individual total energies. Then the mean energy of any individual node of normal CL space may posses energy proportional to its fraction in the subsystem it belongs. If considering a free space (without particles) then we may expect that a sudden change of the static CL pressure in some spatial domain is distributed as interaction energy faster in the same handedness node system than between them. Then the final effect of the static pressure change could appear as restoration of the energy balance between the both subsystem. The secondary effect of such restoration should be an emission of neutral gamma waves. The predominant energy released in the case of a nuclear explosion is perhaps in a form of such type of radiation. So we may accept that the right and the left handed parts of the CL structure posses **own individual subsystem energy level**, the sum of which provides the total energy of the system. In a normal CL space both subsystems have equal total energies.

According to the concept developed in §12.A.4.3.3. the total energy of the susbsystem is the internal vibrational energy in all vibrational modes of the low level structures included in the prisms. It should be in balanced conditions with the saturation energy level and the SG interaction energy. According to this balance and the energy limiting conditions of the saturation level it is reasonable to expect that any change of the balance could involve a slight change of the IGSPM frequency or at least its phase. But then a condition for SG repulsion may occur in a similar way described in §12.A.8.2. CL space structure, however, is very well spatially distributed, due to conditions created by the magnetic protodomains and the zero point waves. Consequently the time-space parameter (or relaxation constant) t_{CL} should be involved in a mechanism of proper IGSPM frequency or phase change. So we may accept that a suitable frequency or phase change of IGSPM vectors of both subsystems might keep a zero gravity between the neighbouring CL nodes of different type, **but only for well defined node distance**. Then we arrive to the following conclusion:

The distance between neighbouring nodes in CL space is self adjustable in order to keep the proper phase or frequency of IGSPM vector between the opposite type nodes.

The self adjusting mechanism is a finite process in which the time-space constant t_{CL} is involved

The self adjusting node distance mechanism does not disturb the NRM and SPM vectors whose periods are much shorter than the time-space constant t_{CL} (CL space relaxation time).

The group dynamical behaviour of the CL nodes and the Zero Point waves assure the constancy of the two important parameters of the physical vacuum: the permittivity and permeability of the free space.

12.B.15. A new life for the Ether concept of the space in which we live and observe

The vision of some great physicists about the space (physical vacuum) is different from the currently adopted concept. Michael Faraday and James Maxwell, on which equations the modern electrodynamics relies, have been rigorous supporters of the Ether concept. In A Treatise on Electricity and Magnetism vol. II James Maxwell concludes on the last page in favor of the Ether:

".....whenever energy is transmitted from one body to another in time, there must be a medium or substance in which the energy exists after it leaves one body and before it reaches the other"

Attacking the opponents of his concept Maxwell say: **My further researches lead me to find that these 'eminent men' who take upon themselves the task of ignoring anything that contradicts their cherished beliefs, follow what is called Scientism. And Scientism is well known by some people as a corruption of Science that is really a 'pseudo religion.' With so many 'eminent men' following their religion of Scientism and pretending it to be Science, it is little wonder that the world is in a very 'sorry state' of affairs.**

In most of the standard modern physics textbook it is written that Einstein disproved the Ether (aether) when it talks about the Michelson Morley experiment. However, if you look at the book: Sidelights on Relativity - Einstein says he did not disprove the ether, just showed that one version of it was wrong [2]. Furthermore, in a publicly documented motion (the movie clip is now available)

Albert Einstein claims: **To deny Ether is ultimately to assume the empty space is not (with) physical quality. The fundamental facts of (Quantum) Mechanics do not harmonizes with this view. According to the General Relativity, space is embodied with physical quality. In this sense, therefore, there exists Ether. According to General Relativity, space without Ether is unthinkable.**

Really the "ether" concept evolved in General relativity and became called "space-time." But that interpretation gets lost in confusion as people try to think from the formulated postulates in Quantum Mechanics, and they reinterpret General relativity in a wrong way. In fact Einstein did not agree with the 1925 theory of Quantum Mechanics [1,2,3]. Today physicists are taught that the 'ether' concept is nonsense, but in fact the natural media or "ether" in Quantum Mechanics is replaced by some of its attributes, such as: quantum fluctuations of the physical vacuum, vacuum polarization, zero-point energy, space-time metrics and other names [1].

The adopted formulation about the space-time created a paradox that we have to study the properties of something real while the existence of the carrier of these properties is denied. Presently, it is assumed that the vector forms of the Maxwell's equations describe entirely the EM field properties of the physical vacuum. Maxwell, however, suggested his equations in quaternion form that according to some recent analysis allows prediction of scalar (or longitudinal) waves. The existence of such waves is in a favor of the Ether concept. Nikola Tesla, shearing the Maxwell's vision about the space and talking about it as a natural medium, provided experiments unexplainable from a point of view of the Modern physics. Today many of his experiments became replicated and the effects described by him are confirmed.

Apart of the discussed in this chapter cosmic paradoxes that are enigma for the Big Bang theory, experiments exist confirming the absoluteness of the space in favor of the stationary Universe. Prof. Stefan Marinov, for example, proposed two laboratory experiments for measuring the velocity of the Earth through the absolute space. These experiments overperform methodically and technologically the classical Michelson Morley experiment

made 100 years ago. In the second experiment published in the article "Measurements of the Laboratory's Absolute Velocity (General Relativity and Gravitation, Vol. 12, No. 1, 1980) Stefan Marinov successfully measured the velocity of the Earth rotation around the Sun and the solar system rotation around the Milky way galaxy centre with unprecedented accuracy. His first experiment performed in 1976 is published as "The interrupted "rotating disc" experiment" in J. Phys. Math. Gen, 16 , p. 1885-1888), (1983). This experiment has been repeated and the results are confirmed by E. W. Silvertooth, sponsored in part by Air Force Systems Command, Rome Air Development Center, Griffiss AFB and the Defence Advance Research Projects Agency. E. W. Silvertooth, Experimental detection of the ether, Spec. in science and Tech. Vol. 10, No 1, p. 3.(1986).

The Earth motion is detected also by the anisotropy in the blackbody radiation (G. F. Smoot, M. V. Gorenstein, R. A. Muller, Phys. Rev. Lett. **39**, 898 (1977)).

The Earth motion is also detecting by the VVLBA technique. The following extract is from the website of the National Radio Astronomy Observatory
(www.nrao.edu/pr/1999/sagastar/)
"Whit this precision, the astronomers were able to detect the slightest apparent shift in position of Sagitarius A* compared to the positions of much more -distant quasars behind it. This apparent shift was caused by the motion of the Solar System around the Galaxy's center. "From these measurements we estimate that we are moving ar about 135 miles per second in our orbit around the center of the Milky Way", Reid said. "Even though it take more that 200 million years for us to complete an orbit of the Galaxy's center, we can detect this motion in ten days, observing with the VLBA!".

Chapter 13. Potential and special applications. Hidden space energy and prediction of a new propulsion mechanism.

13.1 Brief summary of advantages provided by the BSM - a Supergravitation Unified Theory.

The treatise Basic Structures of Matter (BSM) is based on an original idea about physical vacuum, which has never been investigated before. It follows the recommendation of James Clerk Maxwell expressed in his "A Treatise on Electricity and Magnetism" vol. II, section "A medium necessary":

In fact, whenever energy is transmitted from one body to another in time, there must be a medium or substance in which the energy exist after it leaves one body and before it reaches the other... Hence, all these theories lead to the conception of a medium in which the propagation takes place

The Basic Structures of Matter (BSM), a Super Gravitation Unified Theory, unveils the relation between the forces in Nature by adopting the following framework:

- Empty Euclidian space without any physical properties and restrictions

- Two super dens fundamental particles able to vibrate and congregate

- A Fundamental law of Super Gravitation (SG) - an inverse cubic law valid in pure empty space.

An enormous abundance of these two particles, with energy beyond some critical level, are able to congregate into self-organized hierarchical levels of geometrical formations, based on the fundamental SG law. This leads deterministically to creation of space with quantum properties (known as physical vacuum) and a galaxy as observable matter. All known laws of Physics are embedded in the underlying structure of the physical vacuum and the structure of the elementary particles. The fundamental SG law is behind the gravitational, electric and magnetic fields and governs all kinds of interactions between the elementary particles in the space of physical vacuum.

The structure of the physical vacuum, called a Cosmic Lattice (CL) distinguishes from the old ether-like concept by number of features, such as: a high stiffness and pressure, quantum mechanical and space-time features and folding properties. As a result, its complex but well-defined behavior permits explanation of the enigmatic phenomena in

Particle Physics, Quantum mechanics, Relativity and Cosmology. The space-time relativistic features of the physical vacuum are result of modulation effects caused by the immersed material objects (GR effects) and by their motion (SR effects). They become apparent when analyzing the behavior of the single elements of the CL space - the CL node and the motional behavior of an elementary particle, for example the electron. Such analysis leads to definition of the basic physical parameters of the CL space: a Static CL pressure, a Dynamic CL pressure and a Partial CL pressure. The first one defines the Newtonian mass of the elementary particle (a mass equation is derived in BSM). The second one defines the Zero Point Energy (ZPE) related to the Electrical and Magnetic fields. The third one is related to the inertial properties of the elementary particles. The unveiled features allow making analysis beyond the Newton's laws about gravity and inertia and beyond the theory of Special and General Relativity. Additionally, the existence of two types of Zero Point Energy (ZPE) is revealed: a static one (ZPE-S) and a dynamic one (ZPE-D). The first one is related to the Newtonian mass and the effects of GR, while the second one - to the Electrical and Magnetic fields. One important feature of the CL nodes is their ability of self-synchronization with identified experimental signature - the Compton wavelength. This phenomenon is involved in the definition of the permeability and permittivity of the physical vacuum and it is responsible for the constancy of the velocity of light.

The Russian academician G. V. Nikolaev arrived to similar conclusions about the physical vacuum (Nikolaev, G. V, *Electrodynamics of the physical vacuum* (in Russian), Tomsk, 2004)

13.2. Theoretical achievements in support of emerging practical applications

The unveiled Law of Super Gravitation (SG) is the most fundamental law in Nature. Its functionality is based on the ability of the fundamental particles and their formations to possess vibrational energy (see Chapter 12). It is given by the expression

$$F_{SG} = G_0 \frac{m_{01} m_{02}}{r^3} \qquad [(2.1)]$$

where: F_{SG} - SG force, G_0 - intrinsic SG constant, m_{01} and m_{02} - SG (intrinsic) masses, r - distance

The Basic Structures of Matter - Super Gravitation Unified theory permits:

(a) Understanding the fundamental relation between matter and energy

(b) Understanding the structure of the physical vacuum - Cosmic Lattice (CL) and its static and dynamic behaviour permitting the definition of space-time concept

(c) Understanding the physical relation between the Gravitational, Electric and Magnetic fields

(d) Solving the boundary condition problem for the photon as a quantum wave in a structured space of the physical vacuum

(e) Explaining the rules and effects of Quantum mechanics and General and Special relativity by a classical approach (solving a long time existing problem for the missing relation between Quantum mechanics and General and Special Relativity)

(f) Unveiling the physical structure of the electron, its oscillation properties and quantum features.

(g) Unveiling the physical structure of the elementary particles, the atomic nuclei and the real quantum orbits. Understanding the cause of radioactivity.

(h) Unveiling the correct interpretation of the Einstein formulae $E = mc^2$.

(i) Recognizing the existence of hidden space energy of non-electromagnetic type as a primary source of the nuclear energy

(k) Building of alternative Cosmology without contradictions based on the new space concept and reinterpretation of the observations

(l) Unveiling the levels of matter organization in the Universe based on geometrical formations in hierarchical orders.

(m) Deriving expressions showing the relation between the known physical constants and unveiled structural parameters of the physical vacuum and the elementary particles including their mutual interactions

(n) Showing that the Newton's laws (about the universal gravitation and inertia) and the Einstein's Special and General Relativity appear as special cases of the Super Gravitation Unified theory in CL space environment.

(o) Unveiling the existence of longitudinal waves and their properties

(p) The suggested model of two fundamental particles and one fundamental law allows excellent opportunity for computer modelling of the unveiled geometrical formations and their rich vibrational properties.

Let us discus briefly some of the most important results:

(h): Revealing the correct interpretation and use of the Einstein equation $E = mc^2$.

Firstly, the Newtonian mass is not equivalent to matter. It is an attribute of formations from fundamental particles. The fundamental particles are kind of intrinsic matter that could never disappear. The mass is an attribute of a particular formation matter that might be modified or disintegrated, while the energy is embedded in such a formation. Therefore, the Einstein formula is correct above some particular level of matter organisation. Consequently, the interpretation of this formula as annihilation (disappearance) of matter is incorrect. As a result, the theoretical derivations based on such interpretation lead to a significant departure from the reality. The discrepancy between the reality and the present concepts appears quite serious in the fields of Particle Physics and Cosmology, where this formula is used as a creation or disappearance of matter. In Particle Physics, such use leads to an enormous number of particles, definition of rules and violation of rules. In Cosmology, it leads to creation of an elusive picture of the Universe, based on a Big Bang model. From the considerations and analysis presented in BSM it becomes evident that:

The matter never annihilates, so it never converts to a pure energy. The Einstein equation is correct when applied for the mass, keeping in mind that the mass is not equivalent to matter. The correct interpretation of the Einstein equation is the following:

$mc^2 \rightarrow E$ - destruction of mass (in particle collider experiments) or hiding of the positron's mass inside of the electron's structure (see §3.17.1, Chapter 3)

$E \rightarrow mc^2$ - creation of virtual particles in CL space, not possessing matter (corresponding to the Dirac see idea)

$E \Leftrightarrow mc^2$ - valid for the nuclear binding energy as a result of small change of the CL space node distance in presence of matter - General Relativistic effect in a microscale

The above interpretations have also a bottom length scale limit, which is of the order of the CL node distance ($\sim 1 \times 10^{-20}$(m)).

The virtual particle (single or pair) is only a wave of charge propagating in CL space. While the photon is a quantum wave in which the two types of CL nodes are dynamically affected, the virtual particle affects only one type - positive or negative (the left-handed and right-handed formations are related to the sign of the electrical charge). It moves with a speed of light but does not possess matter, and consequently could not be a static elementary particle. Virtual particles are created either as pairs (Dirac see particles) or from a Beta radioactive decay. In the process of thermalization of such a particle from Beta decay, the latter hits a target. As a result, a low energy real particle possessing a matter is extracted (electron or positron). From the point of view of the correct concept of the physical vacuum, one important conclusion emerges: Using the Einstein equation for estimation of the "mass" from measured energy interactions and claiming that matter is created, is wrong. The particles obtained in particle colliders are either virtual particles or structural fragments of real particles. No material particles are possible to be created from a pure energy in a form of quantum wave (even from high energy gamma rays). Such "particles" are only waves, so they could never become material particles, such as proton, neutron, electron or fragments from their disintegration.

From the presented considerations (based on the analysis of all previous chapters), it is evident that real particles can be created only by some cosmological event, the condition of which could not be duplicated in any kind of laboratory. Such an event was identified by extensive analysis of different cosmological phenomena in Chapter 12, while the conclusions are in excellent agreement with the results from the Particle Physics experiments and the cosmological observations.

(o) Longitudinal waves.

The unveiled structure of the photon wavetrain (a neutral quantum wave) was described in Chapter 2, while solving also the problem of the boundary conditions, which are important for preservation of the photon energy during its travel. It was found also that the momentum of pointing vector is a result of the wavetrain helicity . In a normally generated EM wave, the E and H vectors have a small longitudinal component in the direction of propagation, which is in accordance with the contemporary mathematical treatment. However, in some special cases of EM generation, this longitudinal component can be increased significantly.

Accepting the existence of CL space structure means that longitudinal waves (LWs) are possible as compression-like waves, different from the ordinary EM waves. From a point of view of the Classical Electrodynamics, the existence of LWs is explainable only if using the original forms of Maxwell equations (quaternions). This is now theoretically proved by number of theoreticians (Van Vlaenderen K. J. and Waser A. (2001)).

Longitudinal waves, firstly observed by Nikola Tesla 100 years ago, are now theoretically envisioned and experimentally confirmed. There are number of theoretical treatments (related to LWs phenomenon) discussing the Ampere's law. , P. T. Papas, (1983), (H, Aspden, (1985). A good source of reported experiments involving LWs is the New Energy Technology Magazine, edited by A. Frolov, www.faraday.ru).

Understanding the wavetrain structure of the photon (and EM waves) permits to guess what might be the structure of the **longitudinal waves (LWs).** They should possess a stronger longitudinal component. One way to understand these waves is to imagine that they contain counter rotating E and H vectors. While the EM waves for example could be generated by a solenoid, one may guess what might be the configuration of the solenoid or antennae for generation of LWs.

Additional technical considerations exist for generation and reception of LWs, which will not be discussed here. LWs are naturally generated from a lightning. They cause an EM noise known as transients which has a large penetration capability. These transients contain LWs, which have a broad frequency spectrum. They may pass through EM filters and could destroying sensitive equipment.

BSM theory envisions three types of LWs depending of the way they are generated and the conditions of their propagation:

(a) Isotropic LWs

(b) LWs embedded in EM waves

(c) LWs in closed magnetic lines

The isotropic LWs are propagated in close distances only.

The LWs embedded in EM waves may appear hidden for ordinary EM receiver. However, they are able to carry independent information. They are propagated with a light velocity.

The LWs in closed magnetic lines may propagate with hyperlight (superluminal) velocity.

The three types of LWs are particular useful for the special applications discussed later in this chapter.

13.3. Potential applications in different fields of Natural Science

The derived BSM models might have potential applications in the following fields:

(1). Using the BSM atomic models in Structural Chemistry and biomolecules for further study of their properties (Chapters 8, 9, 11).

(2). Unveiling the mechanism of stored energy in biomolecules and predicting the possibility for intercommunication between the DNA molecules (Chapter 11).

(3) Using the BSM atomic models in nanotechology (in book "Beyond the Visible Universe", Chapter 6 p.6-11 to 6-16).

(4) Nuclear reactions: achieving fusion and transmutation of elements by proper manipulation of the structure of the physical vacuum.

(5) Understanding the physical mechanism of the tornado and the hurricane (A large area biasing of CL space structure is likely involved. The SG law plays an important role in these phenomena).

(6) Understanding the real danger from the nuclear explosion.

Let us consider briefly (4), (5) and (6).

Nuclear reactions as transmutation of elements in normal temperature and cold fusion experiments are reported in international conferences. The nuclear reactions in both cases are consequences of properly disturbed CL space. Since the methods are quite diversified, we may briefly mention only one of them which is based on generation of a shock event, for example a spark discharge in liquid. Such type of experiment has

been described firstly by Nikola Tesla 100 years ago, while experimental evidence of nuclear reactions in such conditions are confirmed in the last decade. The SG law in a normal CL space (physical vacuum) is behind both: the nuclear binding energy and the charge of the elementary particles. Consequently, if the CL space is properly disturbed, a nuclear reaction becomes possible without emission of neutrons.

The above conclusion becomes evident if examining the process of alpha emission from heavier atomic nuclei, as described by the author's article: "New vision about controllable fusion reaction" ISBN 0973051523, electronic archive AMICUS, National Library of Canada).

(5) Understanding the physical mechanism of the tornado and the hurricane cannot be achieved if ignoring one very essential factor: biasing of some CL space parameters in which the hidden SG energy is involved. This factor creates the necessary forces, which compensate the centrifugal forces of the rotating air mass. The air and water molecules are only a part of the system. The yearly losses from tornados and hurricanes cost billions of dollars. A significant spending for study of these events is done without a progress. If using the BSM concept about space, a fraction of this spending may lead to significant advance of our understanding about these devastating phenomena.

(6). **Understanding the real danger from the nuclear explosion is vitally important for preserving the living conditions in our planet.** Presently, the predictions of the possible consequences from a nuclear war do not provide a real picture. The predicted nuclear winter is only one of the possible outcomes. Even a single explosion in atmosphere is a potential danger for an unforeseen disaster. Let us find out what it could be.

The nuclear explosion is the most powerful pulse disturbance of the CL space (physical vacuum). The consequence is not only radiation and contamination of the atmosphere. The nuclear explosion is accompanied by a huge and strong tornado-like formation. From the adopted so far concept that the space surrounding the Earth atmosphere is void, it follows that a tornado-like formation could not be extended into space. However, such option is possible if taking into account the hidden material structure of the physical vacuum (presently, the

the missing barionic type of matter is envisioned only in the distant galaxies). Then if an atomic explosion eventually causes an extended tornado-like formation, part of the Earth atmosphere could be sucked into space. In such a fatal event, the Earth may become like Mars, which means a global extinction of the life. Extended tornado-like formations pointing outwards are observed on our Sun and they are big puzzle for the astrophysicists.

Another less dangerous, but not envisioned so far effect from an atmospheric nuclear explosion is a temporal damage of the CL space. Despite the ability of the CL space for restoration after strong disturbance, some micro-holes might exist for a long time, while drifting and decaying. (The possible existence of such holes, without explanation, has been envisioned by Wilbert Smith, Canadian researcher, 50 years ago). When a solid body passes through such holes, firstly, the chemical bonds are easily destroyed and secondly, the metallic lattice in metals gets damage. If an aircraft passes through such drifting holes, an instant structural damage will occur. This might cause an unexplainable crash of an aeroplane. The underground nuclear explosion is another danger, especially for the Earth magnetic filed. Presently, its origin is not well understood. BSM puts a new light by offering a concept which is in excellent agreement with the observations (Chapter 10 and 12).

13.4. Special applications

The special applications are related to the most important predictions of the BSM theory.

(1) The physical vacuum contains a hidden energy of non-EM type, which in fact is the primary source of the nuclear energy. Alternative ways for accessing this energy are possible.

(2) The gravitational mass and inertial mass are not unchangeable properties of the atomic matter. They are possible to be changed by proper modulation of some parameters of the physical vacuum.

(3) Two forms of supercommunication are possible with features different from the presently known type of communication: (a) With LWs propagated with a hiperlight velocity in special magnetic conditions; (b) with LWs embedded in EM waves

(4) The biomolecules possess the ability to store energy, while the DNA is likely involved in inter-communication between the cells of the living organism. (Chapter 11 is devoted on this issue).

13.4.1. Hidden Space Energy

The energy of the physical vacuum has been discussed in Chapters 4, 5, 12 and elsewhere in BSM theory. It has been shown that the CL space contains two types of Zero Point Energy:

ZPE-D - a dynamic Zero Point energy
ZPE-S - a Static Zero Point Energy

The first type, ZPE-D, is related to the dynamical properties of the CL node. It corresponds to the parameter Dynamical CL pressure, derived in Chapter 3. Its measurable signature is the temperature 2.72K estimated by the observed Cosmic Microwave Background (CMB). The theoretical derivation of this parameter from a BSM point of view is presented in Chapter 5. ZPE-D is directly responsible for the permeability and permittivity of free space. It corresponds to the ZPE envisioned by Quantum Mechanics.

The second type, ZPE-S, is embedded in the connections between the CL nodes. The alternative CL nodes are connected by their *abcd* axes, in which the SG law is directly involved. In a normal non-disturbed CL space the SG forces are well balanced, so they are hidden for EM interactions. When an elementary particle is immersed, the CL space exercises strong SG forces on its impenetrable volume of the First Order Helical Structures (FOHS). The static energy from this pressure is related to the Newtonian mass by the Einstein equation $E = mc^2$. This pressure called a Static CL pressure is estimated in Chapter 3 by analysis in which the unveiled structure of the electron is used. Its estimated value is:

$$P_S = 1.3736 \times 10^{26} \ (N/m^2) \ \text{- Static CL pressure}$$

While the obtained value of the Static CL Pressure is very large, one must take into account that it could be exercised only on the volume of the FOHS, since it contains a more dens internal lattice. For the electron, this volume, V_e, is calculated by its identified physical dimensions as a cut toroid with a large radius R_c - (Compton Radius) and a small radius $r_e = 8.8428 \times 10^{-15} (m)$.

$$V_e = 2\pi^2 R_c r_e^2 = 5.96 \times 10^{-40} \ (m^3)$$

According to the mass equation (3.48) derived in Chapter 3, the mass of electron is:

$$m = (P_S V_e)/c^2 = 9.109 \times 10^{-31} \ (kg)$$

Using Einstein equation $E = mc^2$ we have:

$$E = P_S V_e = 8.187 \times 10^{-14}(J) \equiv 511(KeV)$$

Scaling this energy to 1 cubic meter we obtain the value of ZPE-S energy in system SI:

$$E_S = 1.3736 \times 10^{26} \ (J)$$

How such enormous energy is hidden? In fact, ZPE-S is composed of two potential energies related respectively to the SG forces between the left and right-handed CL nodes, which are behind the positive and negative charge. In a non-disturbed CL space, the opposite forces are in accurate balance. It is evident from Einstein equation, that ZPE-S is accessible if the mass is changed. This in fact is the binding nuclear energy. The energy from the nuclear power stations is a result of changing the binding energy.

13.4.2. How the ZPE-S energy is accessible by the nuclear reaction.

The elementary particles are formed of helical structures, which are built of prisms. The SG mass of the prisms participate in the SG law, defined by Eq. [(2.10)] in analogical way as the Newtonian masses in the Newton's law of gravitation. By analysis of the vibrational properties of the molecules in Chapter 9, the following therm as a part of the SG law was derived:

$$C_{SG} = G_0 m_0^2 = (2h\nu_c + h\nu_c\alpha^2)(L_q(1) + 0.6455L_p)^2 \quad [(9.24)]$$

$$C_{SG} = 5.26508 \times 10^{-33}$$

where: m_0 - intrinsic (SG) mass of the proton (valid also for neutron), h - Planck constant, α - fine structure constant, $L_q(1)$ - size of first harmonic quantum orbit, 0.6455 - form shape parameter of proton, L_p - proton size (the last three parameters are derived in Chapter 3 and 7 and cross-validated elsewhere in BSM)

The product C_{SG} is verified by calculation of the nuclear binding energy of the deuteron in Chapter using only BSM models. It agrees with the experimental value within 3.6% accuracy.

The ratio between the intrinsic SG mass of the proton and its Newtonian mass is

$$C_{SG}/(Gm_p^2) = 2.82 \times 10^{31}$$

The strong SG field of the larger concentration of the prisms in the internal impenetrable structure of the elementary particles modulates the CL space dynamically and statically. The dynamical modulation of the CL space from a single elementary particle creates a charge. The statical modulation affects slightly the distance between the CL nodes. A larger accumulation of closely spaced elementary particles, for example protons and neutrons in the atomic nucleus, provides a measurable effect of shrunk CL space. This effect discovered by BSM is fully consistent with the General Relativity phenomenon of Space Curvature, so it is called a **space microcurvature of the atomic nucleus.** The measurable signature of this effect is the nuclear binding energy. It is known that the atomic mass of every element (isotope) is smaller than the sum of embedded protons and neutrons. The nuclear binding energy is just the difference between them. (Another spectroscopic signature of this effect, discussed in Chapter 7 is the Lamb shift, which increases with the number of protons). Consequently, the **space microcurvature** in the microscale range depends on the number of the accumulated stable particles proton and neutron. Because the inverse cubic Super Gravitational law is directly involved, a very small change of the CL node distance causes a measurable change of the CL space parameters. One of most important affected parameters is the CL node resonance frequency, which together with the CL node distance defines the velocity of light. The other affected parameters are the permittivity and permeability of free space. A proof of presented concept of the nuclear binding energy was given in §6.14.1, Chapter 6, where the obtained theoretically calculated binding energy between the proton and neutron agrees with the experimentally measured one within accuracy of 3.6%.

For the nuclear reactions and radioactivity, the process of nuclear modification is quite fast. Therefore, **a fast change of the space microcurvature** occurs in both cases. As a result, the CL structure of the surrounding space is shaken. In this case, the Super Gravitational forces are directly involved and a fraction of the hidden ZPE-S is released as strong longitudinal waves. They are converted to gamma rays, which evaporate molecules and atoms and ionize and excite them so they emit a broadband optical radiation. In the case of nuclear reactor, the radiation is converted to heat and the latter rotate the electric generators. This is what we get as a nuclear energy. Consequently:

- **The primary source of the nuclear energy is the hidden ZPE-S of the CL space (the physical vacuum).**
- **The fast nuclear reaction is a kind of physical mechanism for accessing the hidden space energy**

13.4.3. Alternative methods for accessing the hidden space energy

The BSM theory predicts alternative methods for accessing of the hidden ZPE-S, different from the methods of nuclear reaction. In these methods, however, usually the ZPE-D could be directly accessed, while the primary energy source ZPE-S could be accessed only indirectly as it will be explained later. ZPE-D energy is directly responsible for the ε_0 and μ_0 of the physical vacuum, so it could be accessed by some kind of EM interactions. The nominal value of the ZPE-D energy is supported by the ZPE waves, permanently existed in CL space. However, there is a volume and time limit on the energy debit extracted from the ZPE-D source. If a fraction of ZPE-D is taken out, a finite time will be necessary for its equalisation to a normal value (this is confirmed by the BSM analysis of the experiments of A. Chernetski and other experiments). The restoration of the normal ZPE-D will be provided by the ZPE waves, while the energy will be taken from the primary source ZPE-S.

(A). One way of tapping the ZPE-D is to access the oscillation frequencies of the CL nodes. BSM analysis unveiled the following characteristics frequencies of the CL node (Chapter 2):

$\nu_R = 1.093 \times 10^{29}$ (Hz) CL node resonance (NRM) frequency

$\nu_{SPM} = \nu_c = 1.2356 \times 10^{20}$ (Hz) CL node SPM frequency

In Chapter 2 it was shown that the frequency ν_R and the node distance define the velocity of light, while the frequency ν_{SPM} is related to its constancy. Both frequencies are quite high for direct access, but the BSM theory was helpful to find a possible way. It has been shown in Chapters 2 & 3 that the SPM frequency of the CL node is equal to the Compton frequency, measured in the Earth

gravitational field. At the same time, the first proper frequency of the oscillating electron is also equal to the Compton frequency. Consequently, we may access the ZPE-D by using electrons. Based on the discovered frequency match and the unveiled dynamical interactions between the CL nodes and the oscillating electron, a mechanism called a **heterodyne method for accessing the space energy** is suggested.

According to this method, the SPM frequency of the CL space can be accessed by invoking comparatively low frequency oscillations of electron gas with a proper spatial and geometrical arrangement in vacuum conditions. The expected interaction of the electrons with the CL space will be optimal at proper quantum velocities (discussed in Chapter 3 and the author's article about the electron, published in Physics Essays (S. Sarg (2003). We may access the ZPE-D energy by invoking controllable quantum interactions in plasma. This method is in the category of plasma experiments. Details about experimental arrangement of this method are given in a separate monograph.

B. One additional method of accessing the ZPE involves a fast rotation of magnetic fields with properly intercepted magnetic lines. While the technical realization may have many diversified options, there is a common physical mechanism, identified by BSM analysis. How the method of rotating magnetic fields may work? In Chapter 3 and 7 the conditions for the quantum orbit has been analysed. It was found that the magnetic environment should be stable during the lifetime (excited state) of the orbiting electron. Otherwise, the electron will fall prematurely to a lower quantum level. To assure such disturbance, properly oriented magnetic fields must be rotated with a high tangential velocity. While this is still not enough, a mode of higher frequency rotating magnetic field must be obtained by assuring a stroboscopic effect between the primary magnetic fields. The system must include also a rotating inertial mass and energy phase delay. There are additional considerations for obtaining a conflict between the magnetic fields, which is understandable from the spatial features of the CL node dynamics defining the Magnetic Quasisphere (MQ) of SPM vector (Chapter 2).

Technically, the access to the ZPE-D requires a closed system comprising of: a rotating magnetic fields, a rotating inertial mass, an electrical generator, an energy buffer and an energy phase delay. A significant amount of energy must be rotated in the closed system, in order to extract a small amount of space energy. One very important feature, not well recognised by the researchers is the energy phase delay. It should be of the order of the average lifetime of the excited states. That's why the transients play an important role. The energy buffer could be a rechargeable battery. One must keep in mind that in case of battery charge by voltage spikes a complex interaction might occur in the electrolyte, since the ions and electrons have different inertial and magnetic moments. In a proper mode of operation, the extracted power must be larger than the consumption from the battery. Such mode of operation is known by the ZPE researchers as overunity. For stable overunity operation, the energy debit must be below the limit mentioned at the beginning of this section. One of the monitoring effects for stable operation is the device temperature. The functional module should stay cool and even below the ambient temperature. This is explainable in the following way: If the excited electron falls prematurely to a lower state due to the fast changed magnetic field, the pumped CL space energy could not be emitted as a photon, but will be channelled to the electromagnetic feedback. In overunity operation, only part of this energy is spent for the battery recharge, while other is consumed by the load. Increasing the energy consumption, however, has a restriction up to some limit, because the speed of energy transfer from ZPE-S to ZPE-D has a volume and time limit as mentioned in the beginning of this section. As a result, the temperature of the magnets and coils might go below the ambient one. Obviously, a proper temperature gradient and energy debit should be observed for stable operation. Once the limit is reached, some other side effects might appear, for example a change of the gravitational mass. Near or above such a limit the energy extraction mechanism might not be stable.

The above considerations and conclusions has been reached after the analysis of a large number of reported experiments.

Some researchers relying on "search and try" experience, eventually succeeded to access the ZPE-D, but they often are not able to repeat their own result because of the lack of physical understanding and aware about the existed restrictions. For the plasma experiments, some additional restrictions and technical problems exist. In these experiments, the eventually obtained energy is usually as a short burst. If the integral energy of the burst is above some limit, part of the electrons may lose their internal positron (see Chapter 3). Such electrons lose their normal oscillation properties, which means that they are excluded from the quantum mechanical interactions with the ZPE-D. The restoration of their normal oscillating structure may need a long time, or in some cases, they may not be restored. As a result, the device may work sporadically or may stop to work for a long time or permanently. The mentioned effect may appear also in electromagnetic systems without rotating mass, where the energy is in a form of a burst. The unwanted effect is more easier avoided in the rotating magnetic systems mentioned in (B) with a moderate energy extraction.

The analysis of the existing experiments shows a large variety of their arrangement. In plasma experiments involving sharp discharges, a transmutation of elements (nuclear reactions) occurs on the cathode. First discovered by Thomas H. Moray, more than 50 years ago, now this is confirmed by many researchers. This agrees with the BSM concept that the protons and neutrons are held in the atomic nucleus by SG field, which is synchronised by a frequency higher than the CL node resonance frequency. Then a fast shock wave may disturb the synchronization and consequently the nuclear binding forces.

Using BSM analysis, one of the physical mechanisms discussed in (A) or (B) is identifiable in most of the successful experiments. Other mechanisms, however, are also possible. One common feature is the involvement of the SG field. The CL space is characterised by a high resonance and a Compton frequency and a short relaxation time. The ZPE could be accessed either by very fast changes or by invoking of some conflict effects in the CL node dynamics. In such aspect the following conflict situations are envisioned:

- conflict on CL node dynamics by intercepted and fast changing electrical or magnetic fields

- conflict on CL node dynamics from fast accelerated charge particles

- conflict on CL node dynamics by accelerating opposite charge particles with different masses through obstacles

In number of successful ZPE experiments or other unconventional experiments, a phenomenon of "cold electricity" is observed. This is quite interesting phenomena, firstly demonstrated by T. H. Moray, while some features have been envisioned by Nikola Tesla, when mentioning an ethereal current. According to BSM analysis the carrier of "cold electricity" are not real charge practices, but virtual ones. They are only guided by a conductor or a dielectric guide and they convert to real charge particles in the load, where the energy is dissipated. The virtual charge particles are only waves in CL space, the charge of which is not balanced like the ordinary EM waves and photons. They are similar to the charges from the radioactive Beta decay, before their thermalization (in the process of "thermalization", according to BSM, the virtual particle converts to a real one by expelling real particles from a target).

Serious researchers may save a great amount of time and funds if understanding correctly how the energy is stored in space and the possible physical mechanisms for accessing this energy.

Another alternative way for accessing ZPE-S via ZPE-D is the cold fusion reaction. In this category fall also the methods known as nuclear transmutations. A characteristic feature of all these methods is that there is not a strong disturbance of the CL space. This means that emission of neutrons and gamma radiation should be missing. A discussion about the physics of alpha decay and a possibility for a controllable fusion reaction $D + D \rightarrow He$ is presented in article included in Appendix D.

At the present time, the existence of ZPE-S as a primary source of the nuclear energy is not officially recognised. For this reason, the search of alternative energy from space is often met by scepticism. This leaves the researchers without a guiding theory and many of them try to build own theory. As a result, the number of such "theories" is much larger than the number of successful experiments and they of cause are not helpful. Suggesting a speculative theory for explanation of the origin of ZPE usually jeopardize the experimental results. Presently, more valuable for the scientific advance is the description of the experiment. One

of the useful publications overviewing a large number of experiments (apart of suggested theoretical explanation which are not always useful) is the book "Quest for Zero Point Energy" by Moray King. Another useful book is, "Energy from the vacuum" by Tom Bearden. Quite useful are the following periodical journals: "New Energy Technology", edited by Alexander Frolov, "Infinite Energy Magazine" founded by the late Eugene Mallove and "Journal of New Energy", edited by Harold Fox. Useful seminars and conferences are organised by Thomas Valone, a director of Integrity Research Institute. Another coordinator of alternative energy material and projects is PACE, Canada, directed by Andrew Michrowski.

Presently, a number of experimentally oriented laboratories are dedicated on investigation of non-conventional phenomena related to ZPE and gravity. Among them are: The Laboratory of George Hathaway in Canada; the Faraday Laboratory in St. Petersburg, Russia, directed by Alexander Frolov; the Laboratory of Jean - Louis Naudin in France. There are many other laboratories or individual researchers in different countries. Presently, extensive information about ZPE research is available in Internet, but the reliable sources must be carefully selected.

Summarizing the presented methods for accessing the hidden space energy we may claim:

- In nuclear reaction, the ZPE-S is directly accessed. This is a strong disturbance of the CL space, in which a radioactivity is present.

- In alternative methods for accessing the space energy, the ZPE-S is accessed indirectly by accessing the ZPE-D. In this case, a radioactivity may not be present. In this category are also the cold fusion experiments and the effect known as a nuclear transmutation.

The presented above methods do not exhaust the whole variety of accessing the hidden space energy. Accessing the hidden space is among the most difficult tasks. It requiring extensive knowledge about the way this energy is stored and complete understanding of the physical mechanism for successful access.

13.5. Hypothesis for control of the gravitational and inertial mass of a solid object

Presently, the possibility for control of the gravitational and inertial mass of a solid object is out of vision, so such issue is not discussed in the mainstream journals and media. The attempt to access this issue, while relying on the space concept adopted 100 years ago, usually leads to speculative ideas accompanied by highly abstractive mathematics but without any useful practical recommendations. Advance in this field could not be achieved unless the problem is accessed from a new concept of the physical vacuum.

BSM theory is able to provide understandable relation between the gravity and inertia and between the gravitational, electrical and magnetic fields, using the derived static and dynamical properties of the CL space. This permits to envision what parameters of the physical vacuum must be manipulated in order to affect the gravitational and inertial mass, and what kind of technical methods could be used.

In Chapter 2 it was described that the self oscillating CL nodes can be regarded as Phase Look Loop (PLL) oscillators. Such oscillators possessing a proper frequency are easily synchronised by phase. BSM analysis envisioned the existence of ZPE waves as CL nodes synchronized by a phase propagating with a velocity of light. The average length of the ZPE waves is equal or multiple to the Compton length $\lambda_c = 2.426 \times 10^{-12}$ (m). This is the distance that the phase of SPM vector passes with a light velocity per one SPM cycle of the CL node, the period of which is equal to the Compton time $t_c = 0.809 \times 10^{-20}$ (s). For a stationary frame, the ZPE waves appear as continuously recombining, so they are responsible for the equalization of the ZPE-D and for the space-time properties of the physical vacuum. They are also involved in the definition of the permittivity and permeability of free space, which are responsible for the constancy of the velocity of light.

How the Newtonian gravitation of a heavy astronomical body like the Earth attracts a material object? The SG forces between the Earth and the object are propagated through the CL space structure. More specifically, the Super Gravitational field is propagated through the *abcd* axes of the CL

nodes (Chapter 2), which are always aligned and separated by automatically supported small gaps (the latter phenomenon is defined by the specific properties of the SG field, which is discussed in Chapter 12). At the same time, every CL node vibrates with its proper resonance frequency v_R. The SG field of the prism is characterised by the propagation of SGSPM vector, the frequency of which is obtained by division of the primary Planck's frequency (Chapter 12).

$$f_{PL} = \sqrt{\frac{2\pi c^5}{Gh}} = 1.855 \times 10^{43} \text{ (Hz)}$$

While the CL node resonance frequency is $v_R = 1.09 \times 10^{29}$ (Hz), the frequency of SGSPM vector is higher (the mechanism of frequency division is based on stable frequency modes defined by stable geometrical structures, see Chapter 12). Consequently, the CL node frequency makes an attenuation effect for the long-range propagation of the SG field in CL space. This conclusion of BSM is in agreement with the theoretical derivation of H. E. Puthoff in his article "Gravity as a zero-point-fluctuation force, Phys. Rev. A, v. 39, 2333-2342, (1989). Starting from the Planck's frequency and using one hypothesis of Saharov, he derives the Newtonian gravitation when attenuating the higher frequencies.

The above considerations lead to the following conclusions:

(A). The long-range propagation of the inverse cubic SG field in the CL space appears as a Newtonian gravitational field, which is inverse square dependent on distance)

(B). When analysing the SG propagation through the CL space, the oscillating CL nodes could be regarded as static due to their intrinsically small inertial factor (see Chapter 2 and 12)

(C). The resonance frequency of the CL node imposes some attenuation effect on the propagation of the SG field through the CL space

The feature (B) is very important for understanding the properties of the inertial frame formulated in Special Relativity. It obtains sense when referring to a local mass object. From the other hand, the discovery of this feature permitted a successful analysis and unveiling of the wavetrain shape of the photon and the structure of the magnetic lines (Chapter 2).

From features (A) and (C) it is evident that if the CL nodes between the two material objects are synchronised, the propagation of the SG field will be facilitated. In other words, the propagated SG forces will be stronger, in a case of synchronised CL nodes, in comparison to a case of not synchronised ones. In space environment with a normal ZPE-D, the ZPE waves always keep the synchronization of the CL space microdomains, which keeps the uniformity of the ZPE-D energy. The effect from this is the constancy of the light velocity. Consequently, the Newtonian gravitation regarded as a propagation of the SG field will depend on the permanent existence of the ZPE waves. Then what will happen if the synchronization of the CL nodes is disturbed? Obviously, the propagated strength of the SG field will decrease, which means a decrease of the Newtonian gravitation between the Earth and the object. This exactly is what is necessary for manipulating the gravitation.

Now let us consider how the synchronisation of the CL nodes, or in other words the ZPE waves, could be disturbed in the space surrounding the solid body, in order to decrease its gravitational (and probably inertial) mass. Without entering into details, we may envision the following methods:

- disturbing of the CL node synchronization by emission of longitudinal (scalar) waves (LWs)

- disturbing of the CL node synchronization by gamma rays

- disturbing of the CL node synchronization by invoking a conflict of magnetic line directions, based on the properties of the CL node oscillations included in the magnetic line (MQ SPM is discussed in Chapter 2).

- disturbing of the electrical field of accumulated charge by invoking a conflict of the electrical lines formed by EQ CL nodes (EQ SPM is discussed in Chapter 2).

- disturbing of the CL node synchronization by using the oscillating properties of the electrons

Considering the case of using LWs, we must keep in mind that they contain a longitudinal component resulted from a compressible effect of the CL space, in which the strong SG forces are directly involved. From one hand, the LWs interact directly with the strong hidden ZPE-S energy, so they may carry much stronger energy than the ordinary EM waves and for this reason they are very

penetrative. From the other hand, they may effectively destroy the CL node synchronization for a finite time interval, during which the following effects will occur:

(1) a decreased gravitation between the object and the Earth

(2) disturbed EM waves in the space surrounding the object

(3) a blurred appearance of the object in the visible image

The effect (1) is what we need in order to manipulate the gravitational mass. At the same time, the disturbed synchronization affects the permittivity and permeability of the surrounding space, so the EM field and the propagation of the light in this zone will be also disturbed. This causes the side effects (2) and (3).

Now let us see what might happen with the inertial mass. When an elementary particle, such as proton, neutron or electron moves through the CL space, its impenetrable FOHS structures are ablated by the CL nodes (see Chapter 10). This physical phenomenon defines the inertia of the particle. All elementary particles contain FOHSs of the same type (Chapter 2, 3 and 6). Therefore, the inertial phenomenon is valid for all real particles, atoms, molecules, gas, liquids and solids. If the velocity approaches the speed of light, the elementary particle or the solid object experiences an increasing resistance. The reason for this is that the rate of separating CL nodes (converted temporally to folded) approaches the CL node resonance frequency. This effect is behind the relativistic increase of the inertial mass according to Einstein Special Relativity (the relativistic gamma factor for the electron was derived in Chapter 3, based on its structure).

From the analysis of the astronomical observations in Chapter 10 and 12, it becomes evident that the space of the Milky Way (and other galaxies) could be considered as an absolute reference frame. This is confirmed by a large number of properly arranged experiments. Among them is the laboratory experiments of Stefan Marinov (1975, 1980), who was able to detect our motion around Milky Way centre and measure the velocity. One of his experiments is repeated by E. W. Silvertooth, (1986). Even the re-estimation of the original data from the Michelson-Morley experiment by M. Consoli and E Costanzo (2003) proves this . This is in full agreement with the BSM scenario of Alternative Cosmology presented in Chapter 12, as a consequence from the new concept of the physical vacuum. It demonstrates that the galactic redshift is not of Doppler type, while offering an explanation, which is in excellent agreement with many observed cosmological phenomena. To open a bracket, Edwin Hubble, the discoverer of the galactic redshift, did not believe in expanding Universe until the end of his life.

Understanding the existence of absolute frame of reference is important issue for further analysis of our motion through space, which helps to unveil the possible velocity restrictions.

Now let us analyse the inertial mass from a point of view of the acceleration. In Chapter 10 it was proved that the inertia of the solid object is directly related to the number of folded CL nodes and their relative momentum (the latter parameter is presented by a force moment vector). In a uniform translational motion, these two parameters are constant - there is no acceleration. In a uniform rotational motion, only the direction of force moment vector is changing, so a centrifugal acceleration is felt. In the case of linear motion with acceleration, the magnitude of the force moment vector is continuously changing, so a continuous force is felt. Normally all CL nodes which are in the path way of the translating object oblate the elementary particle to the level of their impenetrable FOHSs. Now, we must emphasize one important feature of the folded CL nodes. They do not have strong connections between themselves such as the normal nodes connected into CL structure. Then we may conclude:

(D). A fraction of folded nodes could be deviated and guided by a strong magnetic field with a proper configuration

(E). The properly deviated and guided folded nodes will cause a displacement of the object without feeling a force as in a normal acceleration (or at least - feeling a reduced force). We may call this predicted effect a **manipulated displacement.**

(F). If the maximal velocity in a manipulated displacement is in the range of our rotational velocity around the Milky Way centre (including the Earth orbital motion), we are sure that the equivalence between the gravitational and inertial mass will be preserved.

The feature (F) means that both - the inertial and the gravitational mass will appear equal but smaller during a manipulated displacement, which leads to the following important conclusions:

(J). In manipulative displacement the object will be able to make sharp turn or reversal of the direction without feeling an excessive acceleration

(H). The acceleration of the object in the case of manipulative displacement will require less force and energy

13.6. Hypothesis about spacecraft based on a new propulsion mechanism

Below is a briefly presented hypothetical version of a new propulsion mechanism, while focusing only on the physical principle and some secondary effects. From the considerations discussed in the previous section, it is apparent that the geometrical shape of the spacecraft is important. Figure 1 shows two views of such spacecraft. .

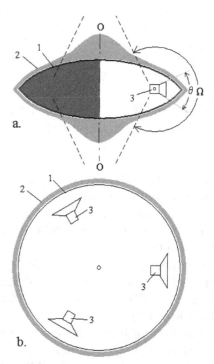

Fig. 1. Spacecraft with a new propulsion mechanism.
1 - spacecraft envelope, 2 - super-strong magnetic field, 3 - radiators of longitudinal waves, θ - instant angle of radiation, Ω - coverage angle (by rotation of the radiator)

The spacecraft enclosure 1, made of proper material, must transmit the LWs from the radiators

3. At the same time, it must be part of superconductor able to generate a super-strong magnetic field 2.

The restoration of the normal CL space parameters after the disturbance takes some finite time. This is theoretically envisioned by the revealed relaxation time constant in Chapter 2, which existence is apparent from the analysis of some experiments and some observed phenomena. This means that the radiators 3 may operate in short burst mode. The duration and the repetition rate of such mode must correspond to the spacecraft manoeuvre characteristics. The radiators may also rotate to cover the field angle Ω or alternatively, a larger field angle could be achieved by proper refractive features of the spacecraft enclosure. The radiators should have also a phase and intensity control of the emitted high frequency. The super-strong magnetic field also could be guided inside of the spacecraft enclosure near the surface. Since the magnetic field and LW have some contradicting features, the researchers and designers of such spacecraft must be well acquainted with the BSM theory.

A spacecraft with properly design shape, material and propulsion system, will be capable to achieve a fast displacement (acceleration), while the crew inside will not feel the acceleration.

The trip to a planet or a distant solar system will contain three phases: an acceleration, a travel with a constant velocity and a deceleration. The maximum velocity and acceleration may have a limit, defined by the intrinsic features of the CL space (physical vacuum).

Let us consider the most conservative option, relying on our motion through the CL space of the Milky Way with a velocity about 220 km/s. Since the Earth orbital velocity is about 30 km/s we are sure that we could not have some unknown biological effect, for a velocity $v \leq 30 (km/s)$, referenced to Earth. Obtaining of such a velocity with acceleration of 9.81 m/s will take about 41 min, so it is insignificant. At one of the closest positions to the Earth (for example in November 2005), the distance to Mars is about $70 \times 10^6 (km)$. Then one way trip should take about 27 days.

In the above conservative option, we excluded the possibility of much faster acceleration and larger velocities because of the uncertainty about some unknown biological effects. Such restriction

may not exist, but this could be verified only by future experiments. If so, from a point of view of BSM, we may consider only one restriction, which is related to the electron velocities in quantum orbits. If taking into account that the normal human temperature is below 40 C, the electrons will be in lower level orbits. We may take the 3rd subharmonic orbit as an upper limit for which the electron energy is 1.51 eV, corresponding to a velocity of 729.7 (km/s). (Note: BSM atomic models show that the orbital electron velocities in the heavier elements are not much different from those of the hydrogen. The higher energy levels known from the atomic spectra comes not from the kinetic energy of the electron but from the SG field potential energy). Then we must consider two cases:

(a) the folded CL node velocity inside the spaceship is equal to the spaceship velocity (the crew feels the acceleration)

(b) the folded CL node velocity inside the spaceship is managed to be smaller within acceptable limit and not dependable on the spaceship velocity (the crew does not feel the extensive acceleration)

In case (a), even for an average distance to Mars about 238×10^6 km, the acceleration (or deceleration) phase with 9.81 m/s is less than a day, while the phase with the constant velocity of 729.7 (km/s) is less than 4 days.

In case (b), the keeping of the folded node velocity within some acceptable limit depends on the spaceship design. This option is suitable for distant space travels. While the biological species might be vulnerable, robotic spacecraft may operate in conditions of higher acceleration and velocity in order to shorten the time of space travel.

Now let us discus some of the features of the spacecraft illustrated in Fig. 1. The zone around the axis OO is a zone of interference of the emitted LWs, while the round zone between the large circular section and the axis OO is suitable for sensors monitoring the motion. The effectiveness of the manipulated displacement (achieving of larger acceleration) will be larger in a direction with a small cosine between the velocity vector and the plane of the larger sectional area of the spacecraft, since the folded nodes must be deviated at smaller angle. When moving horizontally, the spacecraft will be able to move in zigzag with a large sudden change of direction and acceleration. For a fast acceleration, at angle in respect to the horizon, the spacecraft must be initially properly tilted and then accelerated. It is evident that a spacecraft with such a shape is suitable for distant space travels with large acceleration and velocity. For near Earth motion, a spacecraft with other shape may also work.

13.7. Physical effects accompanying the described propulsion mechanism

When moving in the Earth atmosphere, the spacecraft will appear blurred, especially when accelerating. The air molecules surrounding the spacecraft will be ionised and will emit a broadband radiation, the major contribution of which is from the oxygen and nitrogen). If the surrounding CL space is uniformly disturbed, a gradual lens effect will be obtained, in which the whole or part of the spacecraft may look as semitransparent. In such cases, some radars may not be able to detect the spacecraft. Radar bursts, in a way they are generated, contain some LWs embedded in the EM pulse. Radar pulses with a stronger LW component will be able to detect such spacecraft more effectively.

The disturbed synchronization of the CL nodes destroys temporally the magnetic protodomains, which in fact are embedded in the magnetic lines. Consequently, the week magnetic field of the Earth will be locally affected. The affected space region might be significantly extended beyond the spacecraft body and it will contain fragmented domains. When the spacecraft passes away, they will get restoration for a finite time. The Earth magnetic lines get different paths trying to avoid the disturbed domain. The measuring effect will be such that a magnetic compass will be unstable and will show different directions. When the spacecraft is very close to some electronic instruments, their operation might be temporally disturbed, due to affected EM properties of the surrounding zone. Staying near outside of operating spacecraft should be avoided. The LWs may affect the energy storage mechanism of the biomolecules (discussed in Chapter 11).

A number of other not conventional microeffects may take place near the spacecraft, but they will be discussed elsewhere.

The described spacecraft does not need any atmosphere or to be near a massive astronomical object. Its propulsion mechanism might be even more effective in deep space.

13.8. Comparison between BSM predictions and observed physical phenomena

A large number of UFO related publications, containing description of the observed physical features has been analysed from the BSM point of view. While this issue has not been officially in the focus of the mainstream science, the author was careful in the selection of the published material. Among the reliable sources of the selected material are the proceedings of the Workshop "Physical Evidence Related to UFO Reports" held in Pocantico Conference Center, Tarrytown, N.Y., Sep 29 - Oct 4, 1997. and the book *The UFO Enigma, a new review of the physical evidence*, by Peter A. Sturrock, a distinguished astrophysicist and Emeritus Professor. The results and predictions of the BSM theory are in excellent agreement with the observed physical phenomena discussed in this book and other reliable sources. One observed case is given below.

Peter Sturrock describes one well-documented observation made by a government aeroplane mapping the cost in Costa Rica on September 4, 1971. In one of the consecutive frames, an object appears as shown in Fig. 13.2. The object is pictured from the above against an uniform background of water. The picture is accompanied by instrumental record of orientation, coordinates and local time.

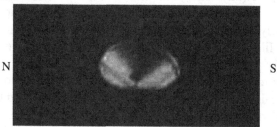

Fig. 13.2. Picture of observed object (Adapted from P. A. Sturrock, The UFO enigma, a new review of the physical evidence, p. 202, Fig. 25-17)

Extracted text from the Sturrock's book.

First, the disk image appears to possess light/dark shading that is typical of a three-dimensional object that is illuminated by sunlight.

Second, the generally triangular dark region on the right-hand side of the disk cannot be a solar shadow cast by the (assumed) opaque disk from the right-hand side. If the disk is an opaque, flat conical section of revolution (the dark spot being the tip of the cone) and if the right side is tipped upward, then the entire surface of the disk should be dark. It is more likely that the light and dark regions are surface markings...

Forth, while the right-hand edge of the disk image is in very sharp focus, the left-hand edge is diffuse and appears to be an irregular boundary which almost transits the light of the background in a transparent manner. It is of interest to note that the general orientation of this left-hand boundary of the image runs north and south rather than being parallel with the visible longitudinal axis of the disk.

Fifth, the entire image is in sharp focus suggesting that (a) the shutter speed was fast, (b) the disk was not moving relative to the Earth background, or both. It is known that the exposure lasted 1/500 seconds, which would "stop" a slowly moving object but not necessary a fast-moving one.

... The 4.2 mm length of the image is equivalent to an object 210 m in length, or 683 feet."

The above-cited observation is only one among the documented material gathered during the past 50 years, but the described effects are very common. From the point of view of presently adopted concept about space, they appear mysterious. As a result, many speculative "explanations" has been suggested, such as: "materialization" and "dematerialization", other dimensions, wormholes, human hallucination etc. Such explanations are completely wrong. Reliable physical records exist indicating that the UFO are real objects. What was missing so far is the physical explanation of the observed phenomena.

The BSM predictions for control of the gravitational mass are in agreement also with the observed, but explained so far Hutchison effect.

13.9. Supercommunication

Currently, the BSM theory predicts two new methods for a distant communication and one for micro-communication.

(A). Distant communication in a closed magnetic lines with a hiperlight velocity

(B). Distant communication with longitudinal waves embedded in EM waves

(C) Micro-communication between DNA molecules in the living organism

In case (A), the receiver and transmitter, while separated by some distance, must be connected by closed DC magnetic lines. It has been

shown in Chapter 2 that the magnetic lines are formed by linear (or curvilinear) arrangement of MQs of the CL node SPM vectors, which are phase synchronized by the speed of light. In such arrangement, all included CL nodes oscillate in phase. This removes the speed of light restriction for the propagation of the SG field. Then a properly modulated SG field may propagate with a velocity thousands times faster than the speed of light. The transmitter and receiver must be of type different from the known EM type. The connected magnetic field must not be of AC but of DC type, while the information carriers are longitudinal waves, modulated by the information that should be transmitted.

In Case (B), the DC magnetic lines are missing. The information is carried by LWs, which are embedded in ordinary EM waves. Such combination, however, might have much larger penetrative capability.

The case (C) has been extensively discussed in Chapter 11, where hypotheses are presented for energy storage mechanism in the biomolecules and the participation of the DNA in a communication process between the cells of the living organism. The micro supercommunication involves directly the genetic code embedded in DNA. It may play an important role in the immune system.

A supercommunication of type (B) using LWs is what is suitable for the distant interplanetary communications and travels. The first SETI (Search for Extraterrestrial Intelligence) project is initiated by Frank Drake in 1960. SETI program has been established also in Arecibo Observatory, PR, but then it was abandoned after a few years of unsuccessful search. (The author was acquainted with this program in 1990, when he was a visiting scientist in Arecibo Observatory, but the SETI program was not operational at this time). From a few years, SETI program has been revived in other places, but convincing results are not reported so far. The researchers expect to find eventual intelligent signals as EM type of communication. No SETI programs, so far has been looking for interstellar communications by LWs. From one hand, the SETI researchers do not envision other alternative way of communication than by EM signals. From the other hand, efforts for development of such technology are not officially made, due to misunderstanding of the physical principle. Presently, the advanced technology of communication is designed to filter out the transients, in which a potential information carried by LWs might be buried. The LW detectors must be build on a quite different principle.

13.10. Opportunity for interplanetary and interstellar travels.

The advantages of the suggested propulsion mechanism over the currently existed ones are overwhelming. The achieving of energy extraction directly from space might allow not only distant travels but also building of colonies in planets and satellites without atmosphere or with an atmosphere different from that of the Earth. Planets like Mars and some satellites of Jupiter and Saturn are such potential options.

The new space travel technology could be developed in the very near future, if adequate funding is allocated. Presently, many developed countries have the potential for development of such new technology. In parallel a serious research has to be conducted on the possible harmful effect of the LW radiation on the living organisms.

13.11. Endnote and disclaimer

The suggested Unified Theory permits to use classical methods in a real 3D space, which allows logical understanding and application of simplified mathematical modelling. At the same time, BSM does not undermine the achievement of the established theories in Modern Physics, such as, Quantum Mechanics, Special and General Relativity. From the suggested framework, they appear as special cases based on mathematical modelling, so they provide useful quantitative calculations. The methods used in BSM theory are closer to the Newtonian mechanics, because the use of a classical approach, while at the same time going beyond the Newton's law of gravity and motion. Free of unnecessary abstractness, the BSM theory has a reliable connection to different fields of natural science and technology. This book provides the fundamental base of this new approach with the necessary supporting arguments. It is written in a way to be understood by theoretically and experimentally oriented physicists, engineers and researchers. The researchers must be well acquainted

with the new BSM concept. A small but important fraction of the theory was published in the peer-reviewed journal Physics Essays, dedicated to fundamental questions in Physics [6]. Extracts from BSM theory (earlier version) have been published in the physics archive [12], in on-line Journal of Theoretics [13,14,15] and in the book "Beyond the Visible Universe", as a more popular version [16].

Disclaimer:

The special applications are in completely new field. The longitudinal waves have been out of vision of the scholar physics. While firstly discovered by Nikola Tesla, they are only partly investigated by him and some other researchers. Their may cause a harmful effects on the human body. Experiments related to special applications must be provided, by qualified researchers on their own risk. The author of this book is not responsible for any injury and loss or damage of property.

References to this chapter

[1]. J. C. Maxwell "*A Treatise on Electricity and Magnetism*", Vol. II. (1893) Dover Publications, N.Y.

[2] V. Koen J. and A. Waser, Generalization of classical electrodynamics to admit a scalar field and longitudinal waves, Hadronic Journal, **24**, 609-628, (2001).

[3] P. T. Papas, Nuovo Cimento **B76**, 189 (1983)

[4] H. Aspden, Physics Letters, **111A**, 22-24, (1985)

[5]. P. Graneau & P. N. Graneau, Phys. Lett. **97A**, 253 (1983)

[6] S. Sarg, A Physical Model of the Electron according to the Basic Structures of Matter Hypothesis, Physics Essays, **16** No. 2, 180-195, (2003)

[7]. S. Marinov, Phys. Lett., **54A**, No 1, 19-20, (1975)

[8]. S. Marinov, J. Phys. A: Math.Gen. **16**, 1885, (1983)

[9] E. Silvertooth, Spec. Csi. Tech., **10**, No 1, 3-7, (1986)

[10] M. Consoli, E. Costanzo,arXiv:astro-ph/0311576 (2003)

[11]. P. A. Sturrock, *The UFO Enigma, a new review of the physical evidence*, ISBN:0-446-5265-0, Published by Time Warner Company, 1999).

[12] S. Sarg, New approach for building of unified theory, http://lanl.arxiv.org/abs/physics/0205052 (May 2002)

[13]. S. Sarg, Brief introduction to Basic Structures of Matter theory and derived atomic models, Journal of Theoretics, January, 2003

[14]. S. Sarg, Atlas of Atomic Nuclear Structures according to the Basic Structures of Matter Theory, Journal of Theoretics March, 2003)

[15]. S Sarg, Application of BSM atomic models for theoretical analysis of biomolecules, Journal of Theoretics, 2003

[16] S. Sarg, *Beyond the Visible Universe*, 2004

APPENDIX: A

ATLAS OF ATOMIC NUCLEAR STRUCTURES

1. S. Sarg © 2001, "Atlas of Atomic Nuclear Structures", monograph

Archived in the National Library of Canada (April 2002)
http://www.nlc-bnc.ca/amicus/index-e.html
(AMICUS No. 27106037)
     Canadiana: 2002007655X
     ISBN: 0973051515
Classification:
     LC Class no.: QC794.6*
     Dewey: 530.14/2 21

2. S. Sarg, "Atlas of Atomic Nuclear Structures According to the Basic Structures of Matter Theory, Journal of Theoretics, Extensive papers, 2003.
http: www.journaloftheoretics.com

Atlas of Atomic Nuclear Structures

Abstract The Atlas of Atomic Nuclear Structures (ANS) is one of the major output results of the Basic Structures of Matter (BSM) theory, based on an alternative concept of the physical vacuum. The atlas of ANS contains drawings illustrating the structure of the elementary particles and the atomic nuclei. While the unveiled physical structures of the elementary particles exhibit the same interaction energies as the Quantum Mechanical models, they permit revealing the spatial configurations of the atomic nuclei, the atoms and the molecules. The unveiled structural features allows to understand the cause of radioactivity. The proposed physical models could find applications in different fields, such as chemistry, nuclear reaction, nanotechnology and biomolecules.

Tables of contents

Table 1: Page location of the elements

Z	Page		Z	Page
1 - 2	II-1		53 - 56	II-11
3 - 7	II-2		57 - 61	II-12
8 - 13	II-3		62 - 67	II-13
14 - 18	II-4		68 - 72	II-14
19 - 24	II-5		73 -78	II-15
24 - 29	II-6		79 - 84	II-16
30 - 35	II-7		85 - 89	II-17
36 - 40	II-8		90 - 95	II-18
41 - 46	II-9		96 - 101	II-19
47 - 52	II-10		102-103	II-20

Notes: The symbols used for notation of the protons and neutrons and their connections in the atomic nucleus are given in Page II-).

Notations:
  Z- number of protons in the nucleus
  N - number of neutrons in the nucleus

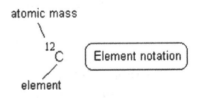

References:

S. Sarg, *Basic Structures of Matter*, monograph, http://www.helical-structures.org,
  also in http://collection.nlc-bnc.ca/amicus/index-e.html (AMICUS No. 27105955), (first edition, 2002); second edition, 2005)
S. Sarg, New approach for building of unified theory about the Universe and some results,
  http://lanl.arxiv.org/abs/physics/0205052 (2002)
S. Sarg, Brief introduction to BSM theory and derived atomic models, Journal of Theoretics,
  http://www.journaloftheoretics.com/Links/Papers/Sarg.pdf
S. Sarg, A Physical Model of the Electron According to the Basic Structures of Matter Hypothesis, Physics Essays (An international journal dedicated to fundamental questions in Physics), v. 16, No. 2, 180-195, (2003)

Introduction

The Atlas of Atomic Nuclear Structures (ANS) is one of the major output results of the Basic Structures of Matter (BSM) theory, based on an alternative concept of the physical vacuum. While the physical structures of the elementary particles obtained by analysis according to the BSM theory exhibit the same interaction energies as the Quantum Mechanical models, they allow unveiling the spatial configurations of the atomic nuclei, atoms and molecules. The unveiled structural features of the atomic nuclei provide explanation about the particular angular positions of the chemical bonds. Such features are in good agreement with the VSEPR model used in the chemistry. Number of other intrinsic features defined by the structural composition of the nuclei provides strong evidence that the proposed models are real physical atomic structures. The arguments for this claim are presented in the BSM theory and more particularly in Chapter 8. The proposed physical models allows understanding the radioactivity. They could be useful in different fields, such as chemistry, nuclear reactions, nanotechnology and biomolecules.

The atlas of ANS contains two parts. Part I illustrates the geometry and the internal structure of the basic atomic particles, built of helical structures. (The helical structures have common geometrical features. Their type and classification are shown in §2.7, Chapter 2 of BSM). Part II illustrates the three dimensional atomic nuclear structures of the elements in a range of $1 < Z < 103$, where Z is the number of protons in the nucleus. Only the stable isotopes given in the Periodic table are shown. In order to simplify the complex views of the nuclei they are shown as plane projections of symbols. For this purpose two types of symbols are used: symbols for hadron particles (proton, neutron and He nucleus) and symbols for the type of the nuclear bonding of the hadrons. The symbolic views contain the necessary information for presenting the real three-dimensional structures of the atomic nuclei by different sectional views. This is demonstrated in page 21 of the atlas, where nuclear sectional views of some selected elements are shown.

The rules according to which the protons and neutrons are arranged in shells in the nuclei are discussed in Chapter 8 of BSM. The trend of consecutive nuclear building by Z-number follows a shell structure that complies strictly with the row-column pattern of the Periodic table. The periodic law of Mendeleev appears to reflect not only the Z-number, but also the shell structure of the atomic nuclei. The latter becomes apparent in the BSM analysis. The protons (deuterons) shells get stable completion at column 18 (noble gases). The separate rows of the Lanthanides and the Actinides are characterised by a consecutive grow and completion of different shells. The nuclear structures of all stable elements (isotopes) possess a clearly identifiable polar axis of rotational symmetry. One or more He nuclear structures are always positioned along this axis. The most abundant sub-nuclear compositions are deuterons, tritium and protons. The strong SG forces hold them together, while the proximity E-fields play a role for their orientations. The identified different types of bonds are shown in the atlas by symbolic notations. For more details, see Chapter 8 of BSM. In the same Chapter, the conditions for instability of the short-lived isotopes are also discussed. They are partly apparent from the Atlas drawings - especially for the alpha decay. The growing limit for stable high Z-number elements is apparent from the shelf completion and the obtained nuclear shape.

The electronic orbits are not shown in the nuclear drawings, but their positions are defined by the spatial positions of the protons (or deuterons). The Hund's rules and the Pauli exclusion principle are both identifiable features related to the available positions and mutual orientations of the quantum orbits. The quantum velocity of the orbiting and oscillating electron, defines the length trace of any quantum orbit (see §3.12, Chapter 3; §7.4, Chapter 7 of BSM)

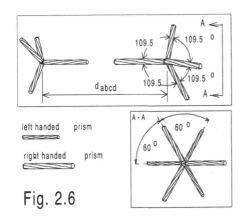

Fig. 2.6

left handed   prism

right handed   prism

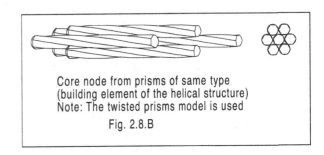

Core node from prisms of same type
(building element of the helical structure)
Note: The twisted prisms model is used

Fig. 2.8.B

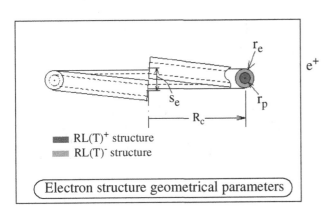

Electron structure geometrical parameters

RL(T)$^+$ structure
RL(T)$^-$ structure

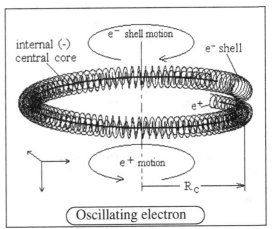

Oscillating electron

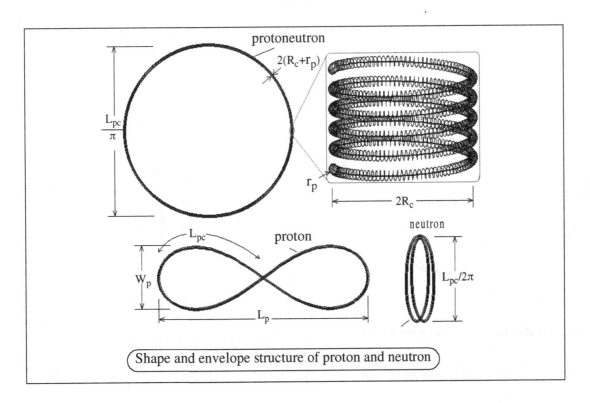

Shape and envelope structure of proton and neutron

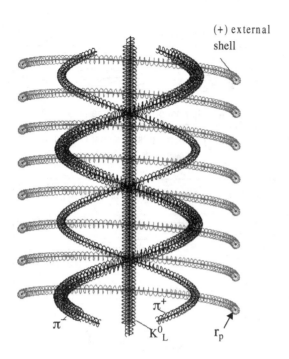

Fig. 2.15.B. Axial sectional view of proton (neutron) showing the external positive shell (envelope) and the internal elementary particles - pions and kaon. All of them are formed by helical structures possessing internal RL structures (not shown).

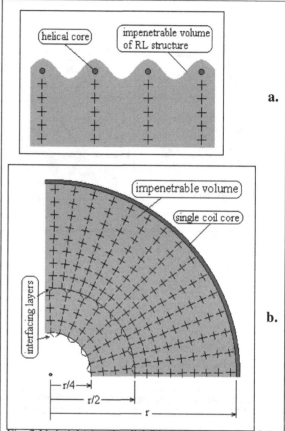

a.

b.

Fig. 2.16 Axial (a) and radial (b) section geometry of the internal RL structure of FOHS (not twisted). The real number of radial layers is large since the prism's length is much smaller than the boundary radius r.

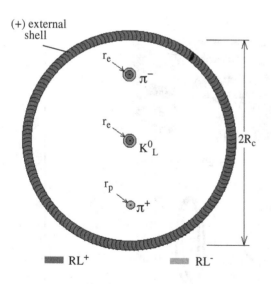

Fig. 2.15.A. Radial sectional view of a proton (neutron) core with internal elementary particles and their internal RL structures. The RL structures are not twisted for the kaon, partly twisted for the pions and fully twisted for the external shell.

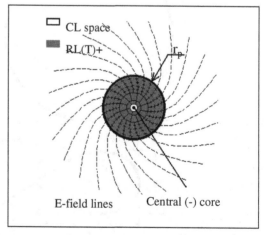

Fig. 2.29.E. Radial section of positive FOHS with twisted internal RL(T)$^+$ structure generating E-field in CL space. The radial section of the FOHS envelope core and the central core is formed of 7 prisms. r_p - is a radius of the FOHS envelope.

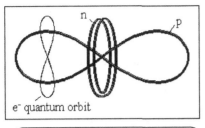

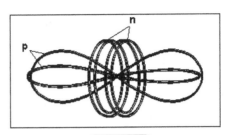

Deuteron with Balmer series orbital — He nucleus

p - proton; n - neutron; e⁻ - electron

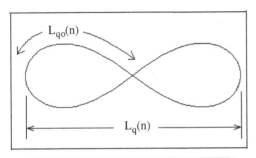

Simple quantum orbit (n - is the subharmonic number)

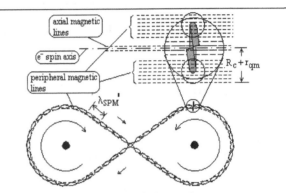

Electron orbit in Balmer series

Idealized shape of stable (quantum) orbit defined by the quantum magnetic line conditions. The peripheral and axial magnetic lines are generated by the screw-like confined motion of the electron in CL space (see §7.7 in Chapter 7 of BSM)

The equation of the quantum orbit trace length, L_{qo} is derived in §3.12.3 (Chapter 3 of BSM).

$$L_{qo}(n) = \frac{2\pi a_o}{n} = \frac{\lambda_c}{\alpha n} \qquad (3.43.i)$$

where: n is the subharmonic number of the quantum orbit; λ_c - is the Compton wavelength; α - is the fine structure constant; $2\pi a_o$ - is the length of the boundary orbit (a_o - is the Bohr model radius)

The shape of the orbit is defined by the proximity E-field of the proton. The most abundant quantum orbit has a shape of Hippoped curve with a parameter $a = \sqrt{3}$. Orbits of such shapes serve also as electronic bonds connecting the atoms in molecules (see Chapter 9 of BSM).

The trace length L_{qo} and the long axis length L_q of the possible simple quantum orbits (formed by single quantum loops) are given in Table 1.

The estimated distance between the CL nodes in abcd axis is: $d_{abcd} \approx 0.549 \times 10^{-20}$ (m).

Table 1:

n	L_{qo} [A]	L_q [A]	e⁻ energy [eV]
1	3.3249	1.3626	13.6
2	1.6625	0.6813	3.4
3	1.1083	0.4542	1.51
4	0.8312	0.3406	0.85
5	0.665	0.2725	0.544
6	0.5541	0.2271	0.3779

The calculated geometrical parameters of the stable atomic particles: proton, neutron, electron and positron are given in **Table 2**. The last reference column points to the BSM chapters, related to the calculations and cross validations of these parameters.

Table 2:

Parame-ter	Value		Description	Calculations and cross validations in:
L_{PC}	1.6277	(A)	proton (neutron) core length	Chapters 5 and 6
L_P	0.667	(A)	proton length	Chapters 6, 7, 8, 9
W_P	0.19253	(A)	proton (neutron) width	Chapters 6, 7 ,8 ,9
r_e	8.8428E-15	(m)	small radius of electron	Chapters 3, 4, 6
s_e	1.7706E-14	(m)	electron(positron) step	Chapter 3
r_p	5.8952E-15	(m)	small radius of positron	Chapters 3, 4, 6
$2(R_c + r_p)$	7.8411E-13	(m)	thickness of proton (neutron)	Chapters 6, 7, 8, 9

Notes:

(1) $R_c = 3.86159 \times 10^{-13}$ (m)   - is the Compton radius of the electron.

(2) $1A = 10 \times 10^{-10}$ (m)   - is the Amstrong unit for length

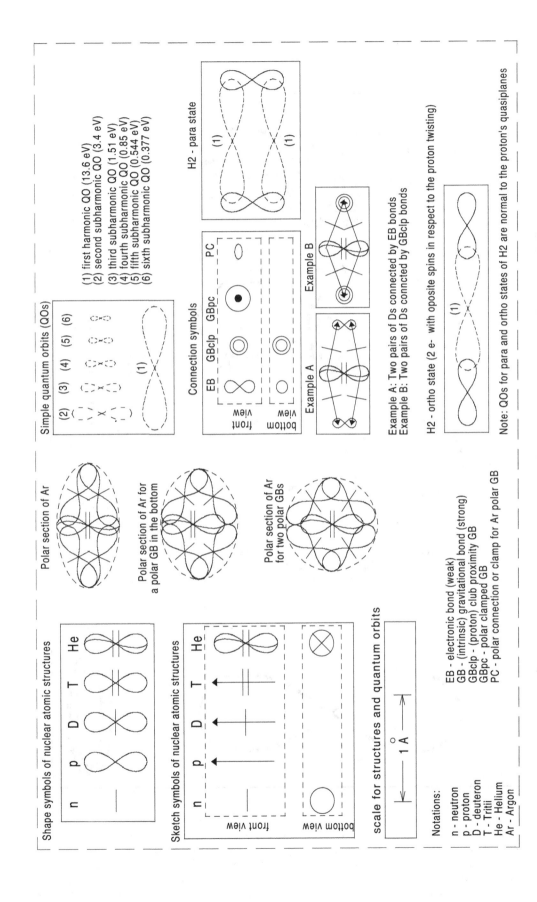

Shape symbols of nuclear atomic structures

n   p   D   T   He

Sketch symbols of nuclear atomic structures

n   p   D   T   He

front view

bottom view

scale for structures and quantum orbits

1 Å

Notations:

n - neutron
p - proton
D - deuteron
T - Tritii
He - Helium
Ar - Argon

EB - electronic bond (weak)
GB - (intrinsic) gravitational bond (strong)
GBclp - (proton) club proximity GB
GBpc - polar clamped GB
PC - polar connection or clamp for Ar polar GB

Polar section of Ar

Polar section of Ar for
a polar GB in the bottom

Polar section of Ar
for two polar GBs

Simple quantum orbits (QOs)

(2) (3) (4) (5) (6)

(1)

(1) first harmonic QO (13.6 eV)
(2) second subharmonic QO (3.4 eV)
(3) third subharmonic QO (1.51 eV)
(4) fourth subharmonic QO (0.85 eV)
(5) fifth subharmonic QO (0.544 eV)
(6) sixth subharmonic QO (0.377 eV)

Connection symbols

EB   GBclp   GBpc   PC

front view

bottom view

Example A   Example B

Example A: Two pairs of Ds connected by EB bonds
Example B: Two pairs of Ds conncted by GBclp bonds

H2 - ortho state (2 e- with oposite spins in respect to the proton twisting)

(1)

H2 - para state

(1)

(1)

Note: QOs for para and ortho states of H2 are normal to the proton's quasiplanes

Atlas of Atomic Nuclear Structures   Part II

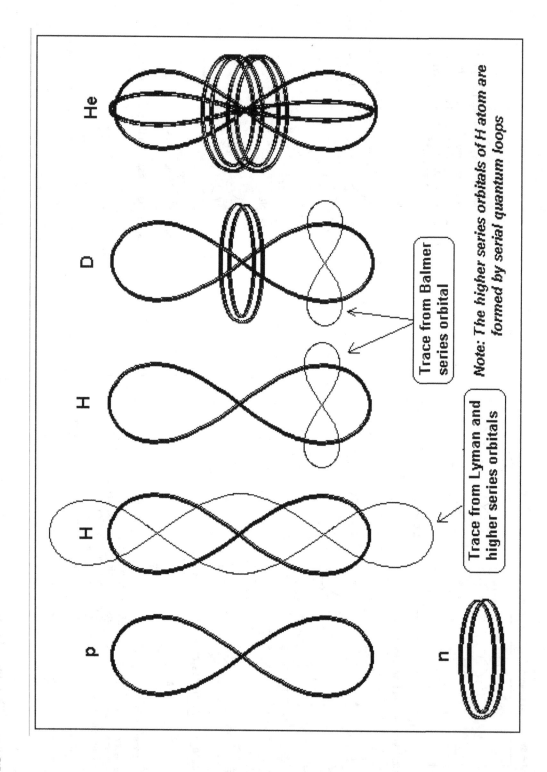

p

H

Trace from Lyman and
higher series orbitals

H

D

Trace from Balmer
series orbital

*Note: The higher series orbitals of H atom are
formed by serial quantum loops*

He

n

^{6}Li	^{7}Li	^{9}Be	^{11}B	^{12}C	^{14}N
Z=3 N=3	Z=3 N=4	Z=4 N=6	Z=5 N=6	Z=6 N=6	Z=7 N=7

Atlas of Atomic Nuclear Structures   Part II

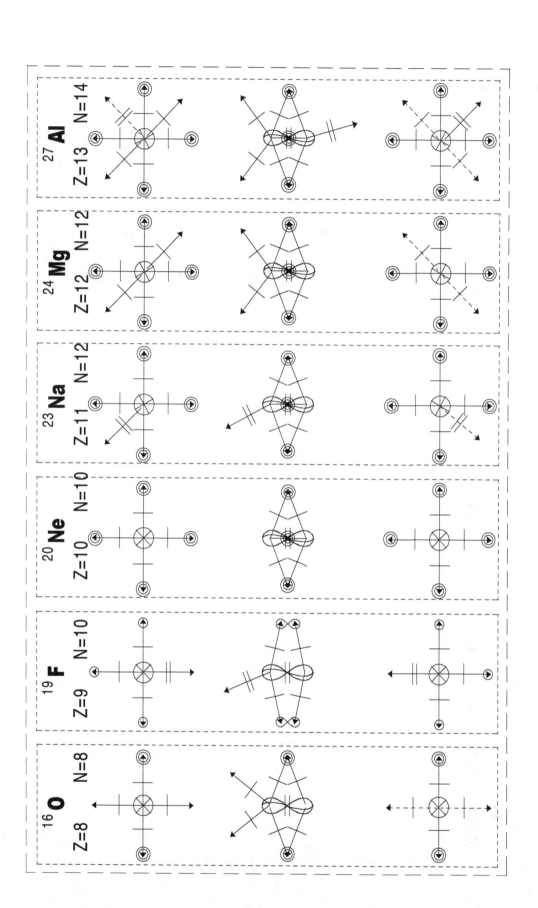

Atlas of Atomic Nuclear Structures   Part II

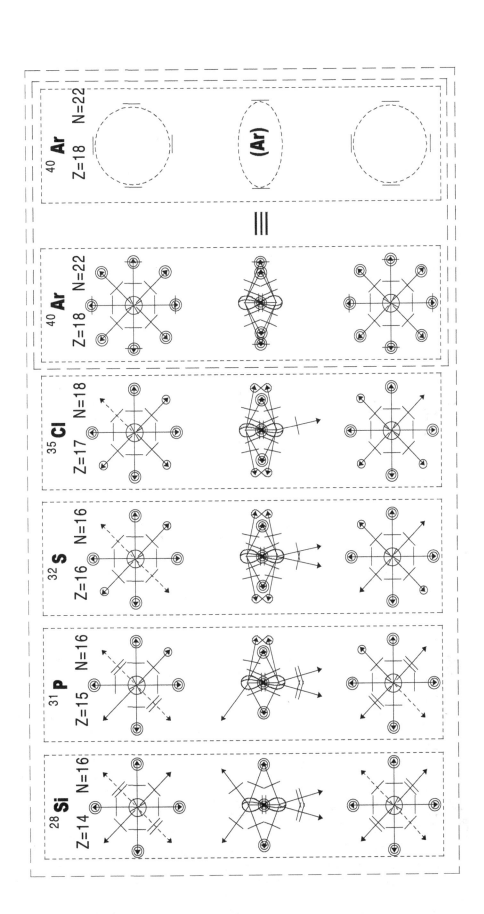

Atlas of Atomic Nuclear Structures   Part II

BSM

Atlas of Atomic Nuclear Structures   Part II

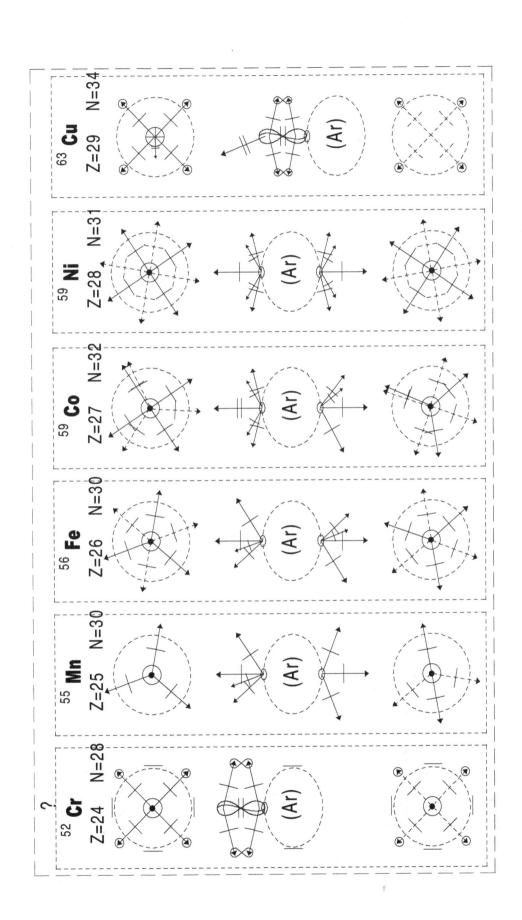

Atlas of Atomic Nuclear Structures   Part II

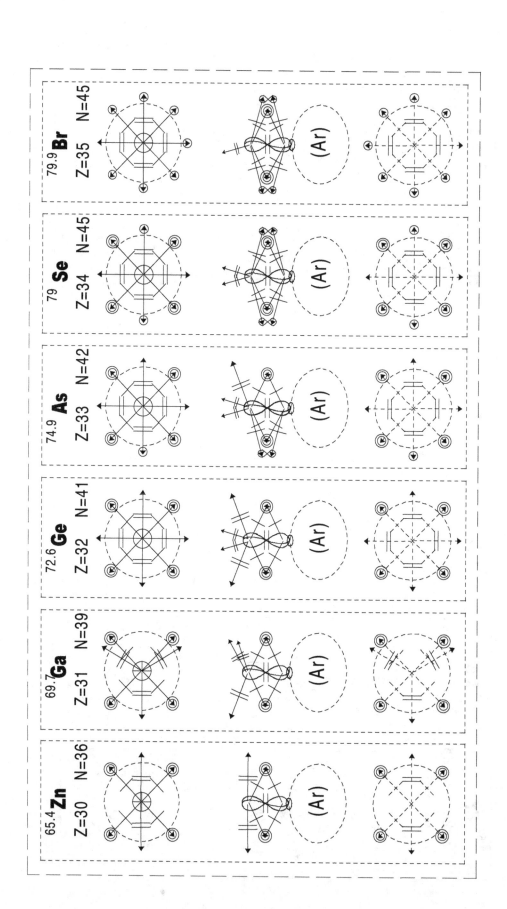

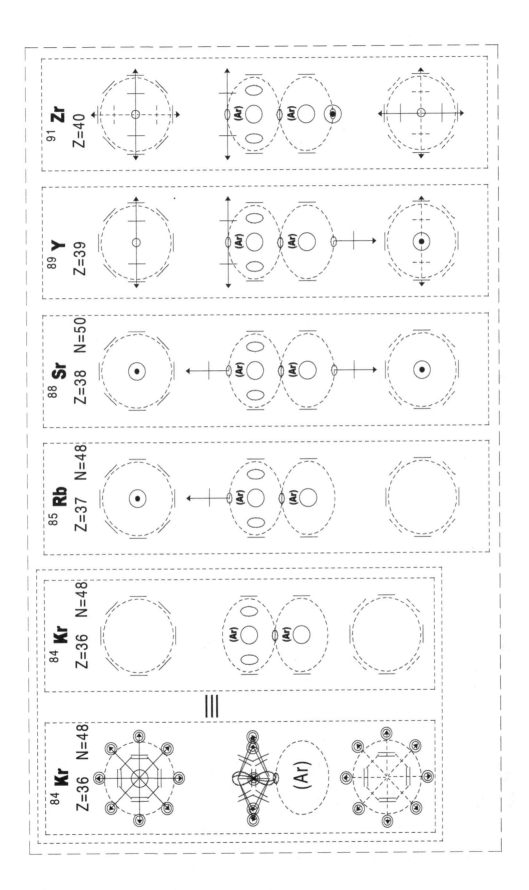

Atlas of Atomic Nuclear Structures   Part II

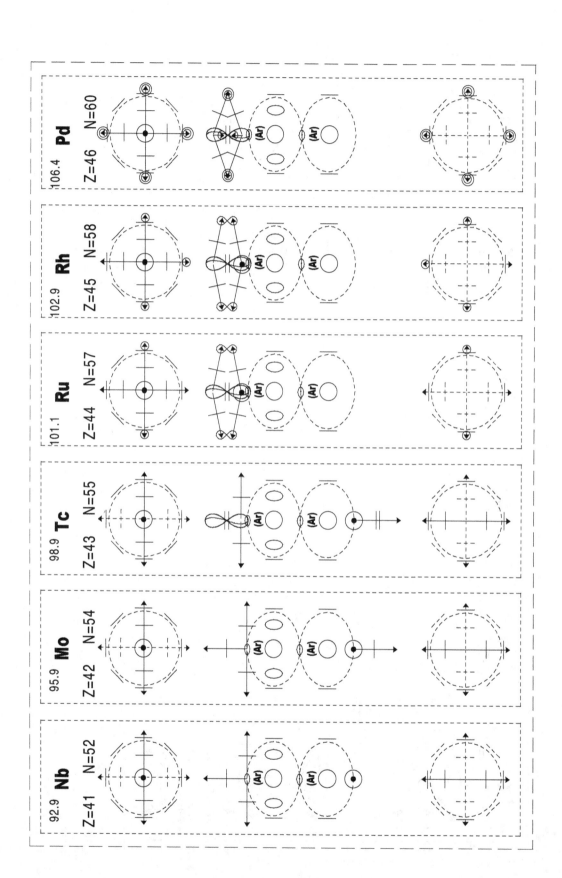

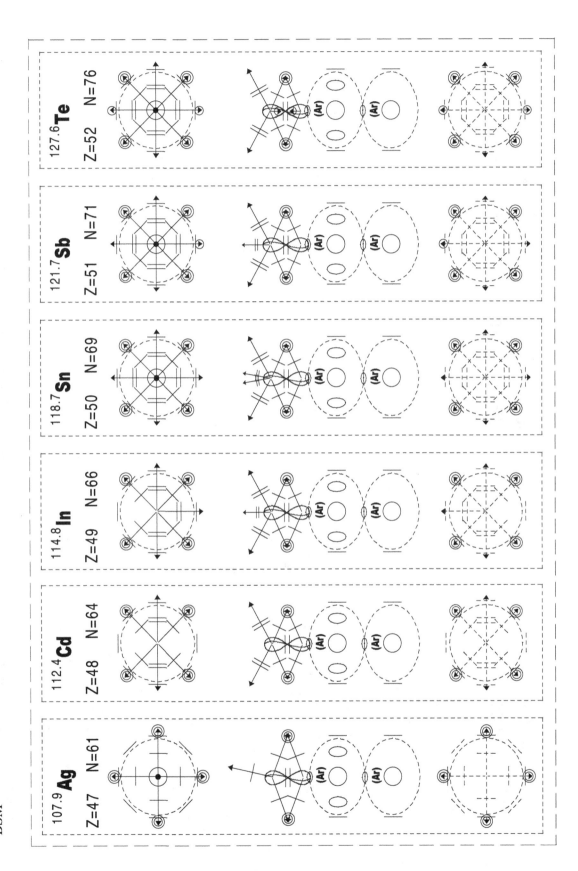

Atlas of Atomic Nuclear Structures   Part II

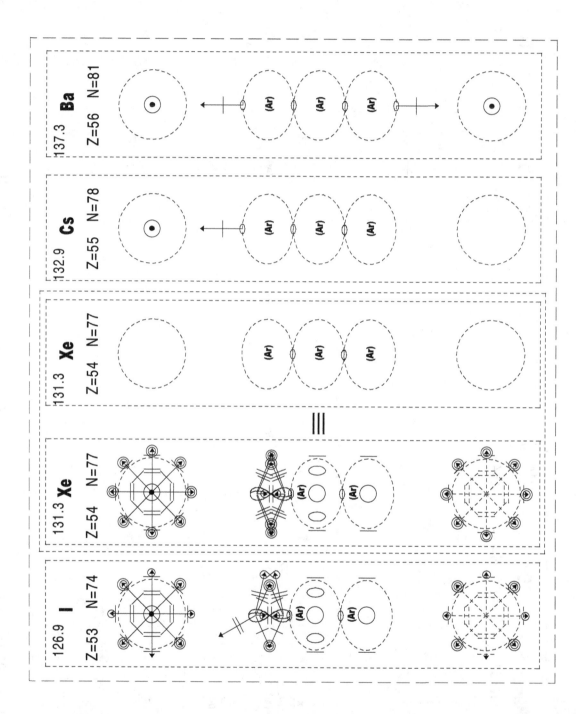

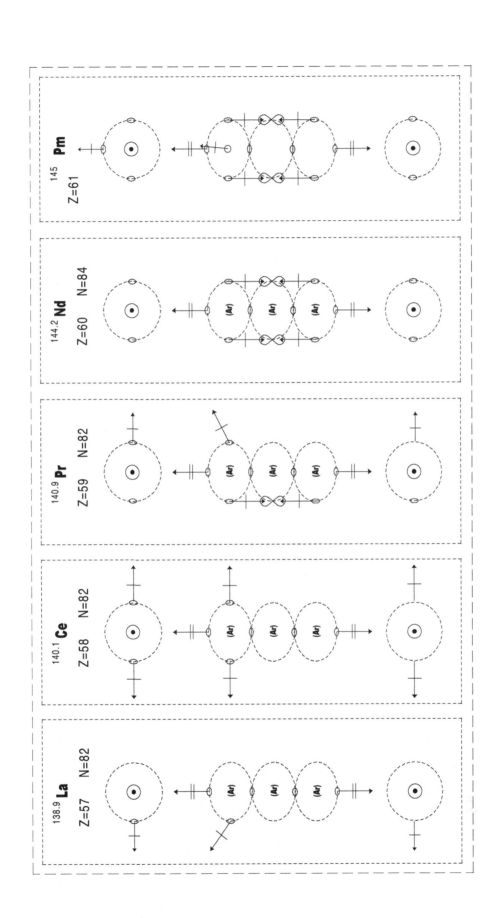

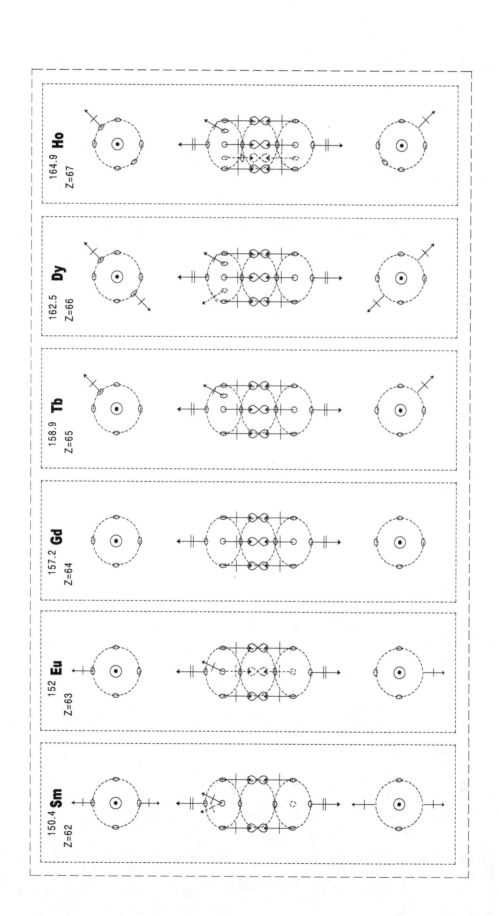

Atlas of Atomic Nuclear Structures Part II

BSM

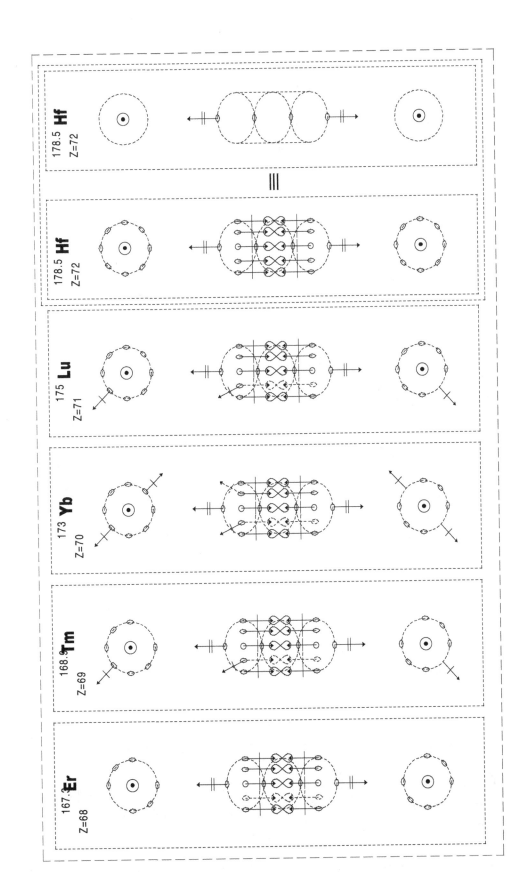

178.5 **Hf**
Z=72

178.5 **Hf**
Z=72

175 **Lu**
Z=71

173 **Yb**
Z=70

168.9 **Tm**
Z=69

167.2 **Er**
Z=68

Atlas of Atomic Nuclear Structures   Part II

BSM

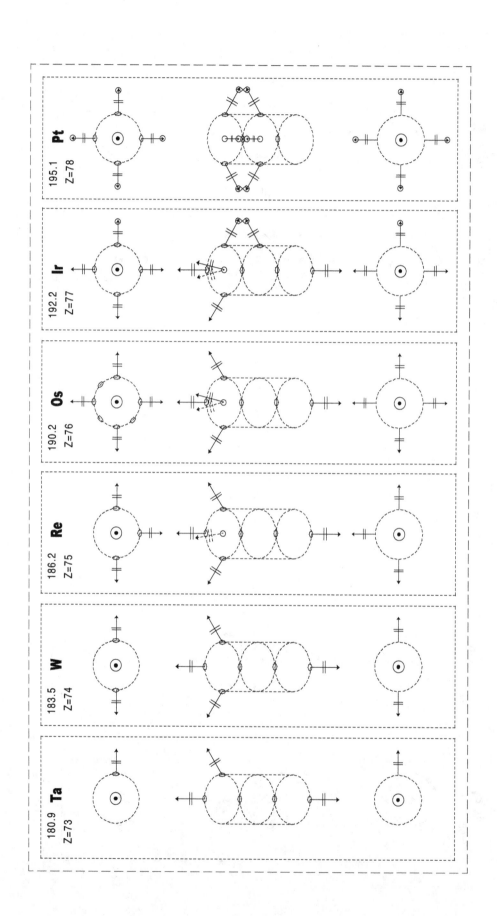

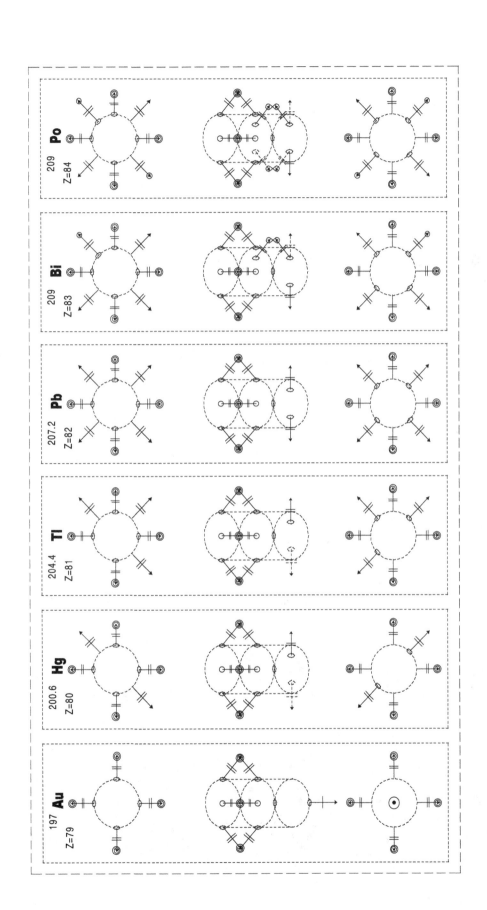

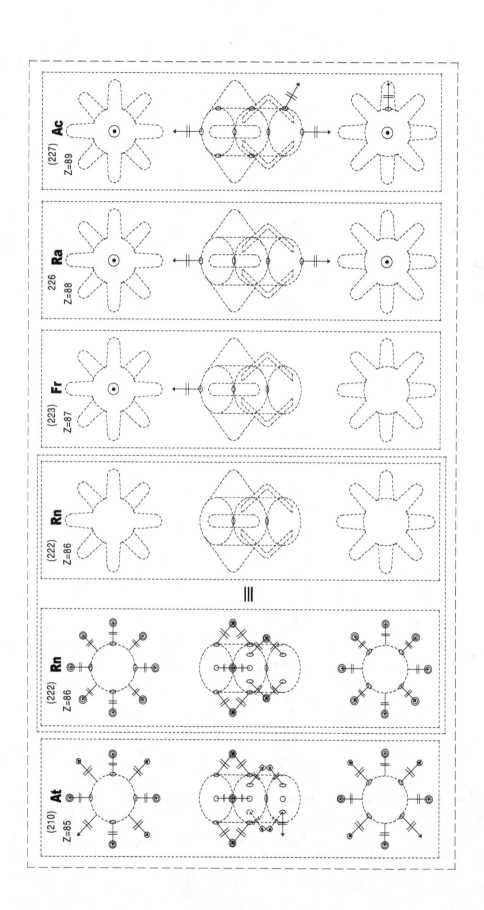

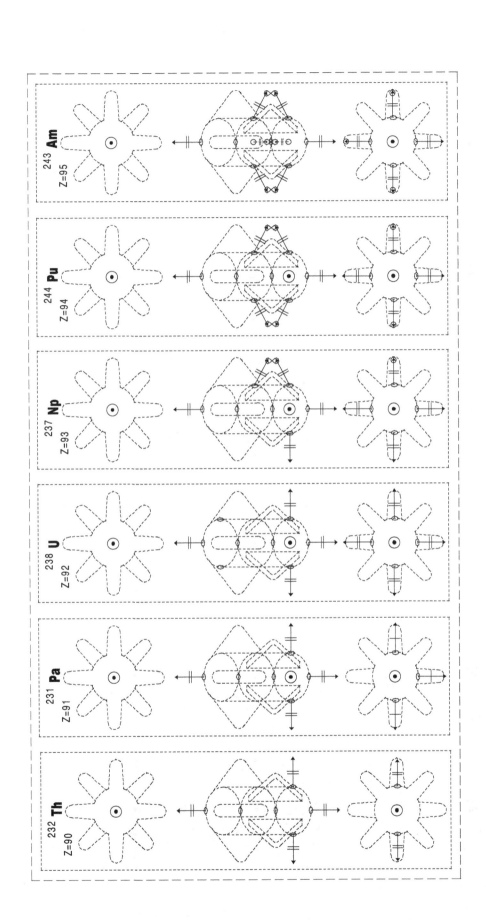

Atlas of Atomic Nuclear Structures   Part II

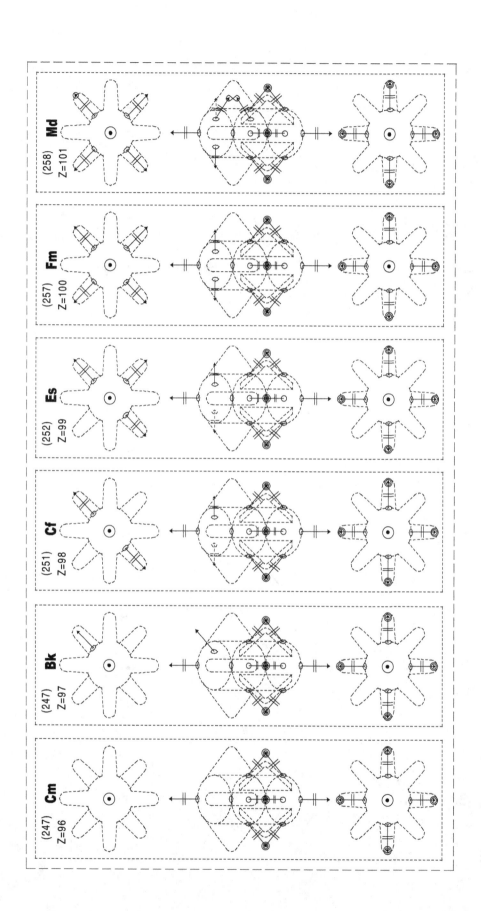

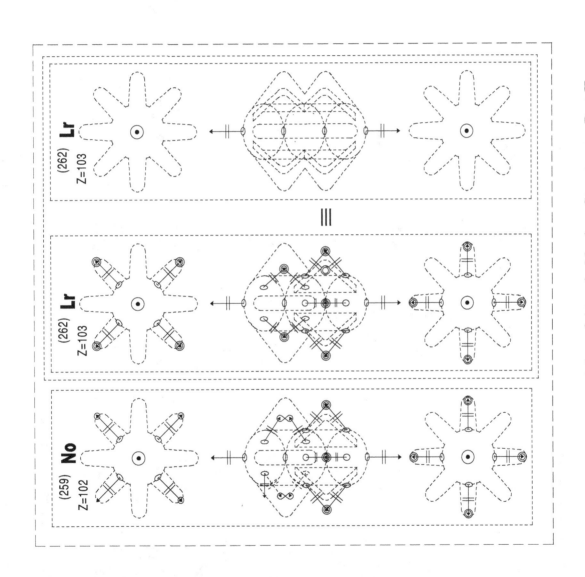

Atlas of Atomic Nuclear Structures   Part II

BSM   Atlas of Atomic Nuclear Structures   Pojection views of selected elements

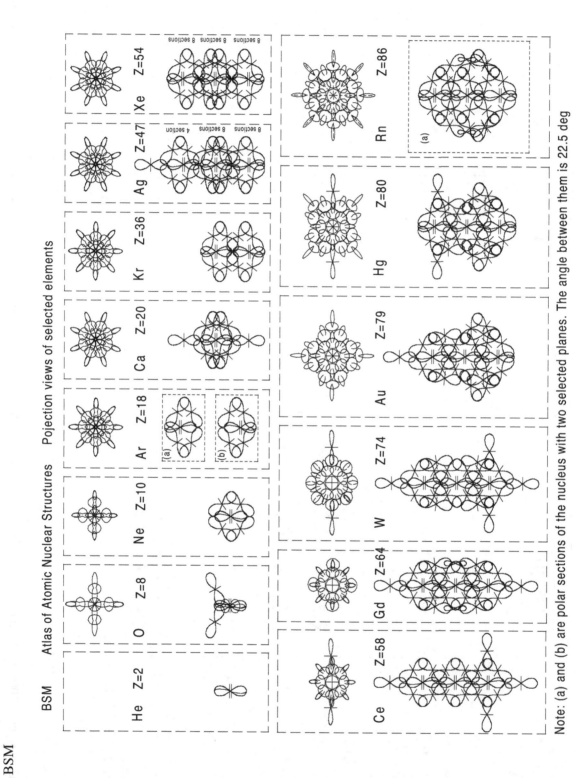

Note: (a) and (b) are polar sections of the nucleus with two selected planes. The angle between them is 22.5 deg

Atlas of Atomic Nuclear Structures   Part II

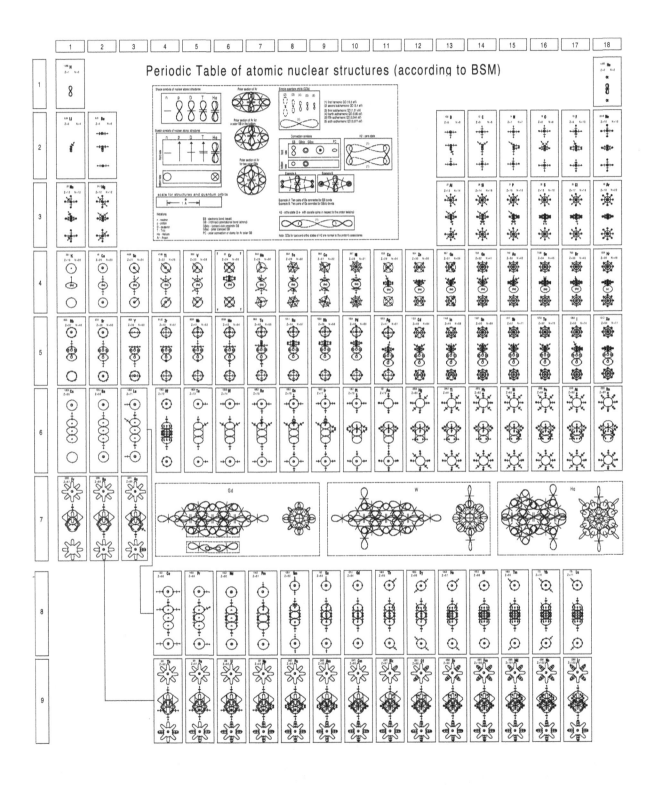

Periodic Table of atomic nuclear structures (according to BSM)

Appendix C (to Chapter 2)

CL node return forces and oscillations

Note: Figures and equations shown in Chapter 2 are shown here by the same number put in square bracket [...].

1. Node configuration of CL structure

Fig. [2.20] illustrates a geometry of a single node in a position of geometrical equilibrium and the axes of symmetry.

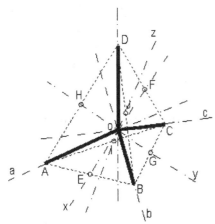

Fig. [2.20] CL node in geometrical equilibrium position The two sets of axes of symmetry are: *abcd* and *xyz*

The four prisms of the node are shown by thick lines. The point where they are connected (by attracting SG forces) is the CL node origin. The prisms are aligned along *a, b, c, d* axes intercepted at point O. The free ends ABCD of the prisms form a tetrahedron ABCD if the angle between each one of *abcd* axes is 109.5°. This state of the CL node is called a geometrical equilibrium. Then *abcd* axes are regarded as axes of symmetry. However, the CL node has also another three axes of symmetry **x, y, z**, which passes through the middle of tetrahedron edges. This axes *xyz* are orthogonal each other. In a geometrical equilibrium the **xyz** axes intercept the axes of **abcd** at angle of 54.75° (half of 109.5°). In CL structure every single node is connected to 4 neighbouring nodes of another intrinsic matter, which prisms have opposite handedness. In geometrical equilibrium, both set of axes (abcd and xyz) between neighbouring nodes coin-

cide. the abcd axes coincide even in non equilibrium position. In not geometrical equilibrium

2. Node displacement along anyone of *abcd* axes

The return forces acting on displacement in two opposite directions along anyone of *abcd* axes are not symmetrical. For this reason expressions for the return forces are derived separately for left and right displacements (denoted also as (-) and (+) displacements in respect to the point of geometrical equilibrium.

2.2 Displacement in a negative direction (left side displacement)

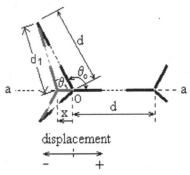

Fig. 1. Displacement along one of abcd axes. Displacement in minus x direction is shown as a left side displacement. The black thick lines are the position of the prisms prisms at geometrical equilibrium of the CL node, while the gray thick lines show the prisms at displaced positions

Notations:
O - geometrical equilibrium point
$\theta_0 = 109.5^0$ - angle between prisms in point O
θ - angle between prisms at displaced position
d - is a node distance along anyone of *abcd* axes at equilibrium position
x - displacement from point O along anyone of *abcd* axes.
d_1 - changed node distance at displaced position

Applying the law of cosines we get:

$$d_1^2 = d^2 + x^2 - 2dx\cos(\pi - \theta_0) = d^2 + x^2 + 2dx\cos\theta_0 \quad (1)$$

$$d^2 = d_1^2 + x^2 - 2d_1 x\cos\theta \quad (2)$$

Using Eqs (1) and (2) and solving for $\cos\theta$:

$$\cos\theta = \frac{d_1^2 + x^2 - d^2}{2d_1 x} = \frac{x + d\cos\theta}{\sqrt{d^2 + x^2 + 2dx\cos\theta}} \qquad (3)$$

The inverse cubic Super Gravitational force is

$$F = G_0 \frac{m_0^2}{r^3} \qquad (4)$$

where: G_0 - an intrinsic Gravitational Constant between the nodes of different type. It is unknown, however, it will be eliminated in the final expression; m - intrinsic mass of the node, r - distance

The hypothetical origin of SG field is discussed in Chapter 12. It leads to the consideration that the sign of SG field between the opposite types of CL nodes in CL structure may change the sign. Let analyse the return forces only at attraction SG forces between neighbouring nodes (of opposite intrinsic matter). For attracting SG field, the force between the node and its right side neighbour is

$$F_R = \frac{G_0 m_0^2}{(d+x)^3} \qquad (5)$$

The resultant force from other three attracting forces from the other neighbouring nodes along the axes of *abcd* set pull in opposite direction. Anyone of these three forces acts upon an angle of $(\pi - \theta)$ in respect to the axis *a-a*. Therefore the three contributions are obtainable by multiplying the attractive forces by a cosine of $(\pi - \theta)$. Then the resultant force F_L from the three neighbouring nodes is expressed by:

$$F_L = 3G_0 \frac{m_0^2}{d_1^3}\cos(\pi - \theta) = -3G_0 m_0^2 \frac{\cos\theta}{d_1^3} \qquad (6)$$

The return force for displacement in (-) direction F(-) is a difference between F_R and F_L forces:

$$F(-) = F_R - F_L = G_0 m^2\left(\frac{1}{(d+x)^3} + 3\frac{\cos\theta}{d_1^3}\right) \qquad (7)$$

Substituting d_1 from Eq. (1) and $\cos\theta$ (given by Eq. (3)) in Eq. (7) and normalizing to the product $G_0 m^2$ one obtains an expression of the normalized return force acting on the node for displacement in (-) direction in respect to geometrical equilibrium.

$$F(-) = \frac{1}{(d+x)^3} + \frac{3(x + d\cos\theta_0)}{(d^2 + x^2 + 2dx\cos\theta_0)^2} \qquad (8)$$

Having in mind that the analysed displacement could be along anyone of axes *abcd*, the Eq. (8) appears to be valid for anyone of these axes. Because the node distance parameter is unknown, it is more convenient to normalize the deviation to *d*. In such case we may substitute $d = 1$ and consider that x is in fact *x/d* parameter. Then theoretically $x < 1$ but practically its upper limit is lower. Eq.Eq. (8) simplifies to Eq. (9) possessing only one argument.

$$F(-) = \frac{1}{(1+x)^3} + \frac{3(x + \cos\theta_0)}{(1 + x^2 + 2x\cos\theta_0)^2} \qquad (9)$$

2.3 Displacement in positive direction (right side displacement)

Figure 2 shows displacement in right hand (+) direction along anyone of *abcd* axes.

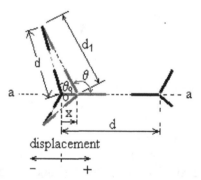

Fig. 2. Displacement in right hand (+) direction. The black thick lines show the prisms of the CL node at geometrical equilibrium. The thick grey lines shows the prisms of the CL node in displaced position

Applying the law of cosines we get:

$$d_1^2 = d^2 + x^2 - 2dx\cos\theta_0 \qquad (10)$$

$$d^2 = d_1^2 + x^2 - 2d_1 x\cos(\pi - \theta) = d_1^2 + x^2 + 2d_1 x\cos\theta \qquad (11)$$

Solving Eq. (9) and (10) for $\cos\theta$ we get

$$\cos\theta = \frac{d^2 - d_1^2 - x^2}{2d_1 x} = \frac{d\cos\theta - x}{\sqrt{d^2 + x^2 - 2dx\cos\theta}} \qquad (12)$$

The SG force pulling in right direction along axis *a* is

$$F_R = \frac{G_0 m_0^2}{(d-x)^3} \qquad (13)$$

The resultant force from the three attracting forces of other neighbouring nodes along the axes of *abcd* pulls in an opposite direction. Anyone of these three forces acts upon an angle of $(\pi - \theta)$ in respect to the axis *a-a*. Therefore the three contributions are obtainable by multiplying the attractive forces by a cosine of $(\pi - \theta)$. Then the resultant force F_L from the three neighbouring nodes is expressed by the same Eqs (6).

Substituting d_1 from Eq. (10) and $\cos\theta$ from Eq. (12) in Eq. (6) we obtain an expression of the return force acting on the node for displacement in (+) direction in respect to geometrical equilibrium.

$$F_L = G_0 m_0^2 \frac{3(x - d\cos\theta_0)}{(d^2 + x^2 - 2dx\cos\theta_0)^2} \qquad (14)$$

The return force for displacement in (+) direction F(+) is a difference between F_R and F_L forces. Dividing by the unknown factor $G_0 m_0^2$ we obtain the normalized value of this force. Additionally substituting $d = 1$, we may consider x as normalized parameter on d. Then the return force for displacement in a positive direction is:

$$F(+) = \frac{3(x - \cos\theta_0)}{(1 + x^2 - 2x\cos\theta_0)^2} - \frac{1}{(1 - x)^3} \qquad (14)$$

2.4. Plotting the return forces for negative and positive displacement

Note: We must keep in mind that in both cases (positive and negative displacement) we considered x as a positive parameter theoretically restricted in a range $0 < x < 1$. (In a real case the deviations are in much smaller range (this is evident in the dynamic oscillation analysis of CL node in Chapter 2 and 4)). Therefore Eq. (8) and (14) are defined only for positive values of x. So, when plotting the forces as function of x we must consider that:

- for the deviations in a positive direction, x increases from left to right

- for deviations in a negative direction, x increases from right to left.

Figure [(2.23)] shows the plot of the return force in a relative scale along anyone of *abcd* axes

as a function of displacement x, normalized to the internode distance.

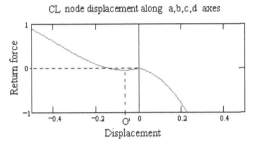

Fig. [2.23] Return force of normalized force in function of displacement normalised to the internode distance

Conclusion:

- **Under inverse cubic law of gravitation, the return force for a positive and negative deviation along anyone of *abcd* axes is not symmetrical**

3. Node displacement along anyone of *xyz* axes.

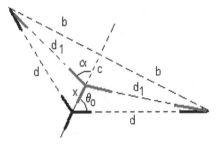

Fig. 3. Displacement of CL node along one of *xyz* axes (*x* axis is shown) The projections of the two prisms in the left down corner of the figure coincide. If the structure is rotated at 90 deg around the x axis the projection of the upper two prisms will coincide.

Following a similar approach and applying the Pitagor theorem and cosine laws leads to derivation of the expression along anyone of *xyz* axes.

$$\text{[(2.14)]}$$

$$F = 2\left[\frac{x + d\cos\left(\frac{\theta}{2}\right)}{\left[x^2 + d^2 + 2xd\cos\left(\frac{\theta}{2}\right)\right]^2} - \frac{d\sqrt{0.5(1 + \cos(\theta))} - x}{\left[x^2 + d^2 - 2xd\cos\left(\frac{\theta}{2}\right)\right]^2} \right]$$

The plot of Eq. [(2.14)] for positive and negative displacement is shown in Fig. [(2.21)].

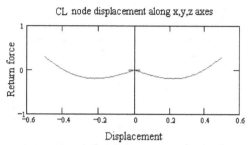

Fig. [2.21] Return force for displacement along anyone of xyz axes

The return forces plot is symmetrical and have two valleys along anyone of xyz axes, positioned symmetrically in respect to the point of geometrical equilibrium 0.

4. Complex oscillations due to different return forces along both sets of axes: *abcd* and *xyz*.

It is not difficult to imagine what kind of oscillations the CL node will have. The symmetrical return forces along *xyz* will contribute to a close to a planar type motion cycle with four bumps and four dimples. The trajectory of such cycle, however, will be not a closed, since the influence of the return forces along *abcd* axes, so we may call it a quasicycle. The asymmetrical return forces along *abcd* axes will cause continuous rotation of the quasicycle mentioned above until a closed trajectory is obtained. The multiple trajectories of the quasicycle will circumscribe a 3-D surface with 6 bumps, aligned with the xys axes and 4 dimples aligned with abcd axes. This is further discussed in Chapter 2, §2.9.2.

The complex oscillation could be regarded as consecutive displacements in any angle in 4π in which the displacement along *abcd* and *xyz* axes are only particular cases. If for one particular displacement we integrate the expression of the return force on displacement in a range from the lowest point to the point of geometrical equilibrium we will obtain expression of energy well valid for displacement along the chosen axis. Following this approach the total energy well could be estimated. as a average integral from all possible directions in 4π range.

Note: This type of oscillations provide AC type of Zero Point Energy of CL space that is relat-ed to Electrical and Magnetic fields. (AC is as alternative current in electrodynamics). The CL grid contains a DC type of energy well, that is much larger (DC is as direct current in electrodynamics). It could appear only if we imagine that we force to separate CL nodes. For analogy the surface waves of the ocean could be regarded as AC type energy, while the gravitation from the water column as DC type of energy.

New vision about a controllable fusion reaction with efficient energy yield

ISBN: 0973051523

1. Short overview

We live in time of energy deficiency and ecological problems. At the same time we know that unlimited energy is locked by Nature. Each litter of water contains about 2 gram of deuterium that is equivalent to the energy of 3000 litters of gasoline if a controlled fusion reaction of type $D + D \rightarrow He$ is achieved. Such process with a positive energy yield has not been achieved so far, despite some 40 years of worldwide efforts. The current attempts are mostly concentrated on achieving of high temperature plasma in a range of hundred millions of degrees in order to create conditions close to those in the stars.

According to the existing scientific concept this type of reaction is not possible at lower temperature. But an alternative physical study indicates that such reaction is possible.

A new theoretical study shows that the vacuum space possesses an underlying material structure. This automatically makes some of the adopted basic physical laws and postulates not universal. It broadens the vision about the space-time and matter-energy relations. Then the quantum features of the space, charge and magnetic features become explainable, while the constancy of the light velocity is derivable parameter. The illogical phenomena from the Quantum mechanics obtain a logical sense. A mass equation is also derived and successfully used for estimation of the geometry and dimensions of the elementary particles. The scattering experiments has been reconsidered for correction of the missing initial condition. Finally the new analysis led to the conclusion that the most effective fusion reaction $D + D \rightarrow He$ is possible at room temperature.

2. The new point of view of BSM theory

BSM theory provides a different view about the vacuum, and the matter. It presents a scientific proof about existence of vacuum structure with unique features, referenced in BSM as a CL (Cosmic Lattice) structure. Applying a new physical approach and using the discovered CL space

structure as a frame of reference BSM succeeded to provide logical explanation of the rules in Quantum mechanics and the Relativistic effects. At the same time it uncovered the real physical structures of the atomic and subatomic particles. The Bohr atomic model appears to be only a mathematical model providing correct energy levels, but it is not identical to the physical one. When taking into account the vacuum structure and the structure of elementary particles, the physical model of the Hydrogen and all stable elements looks quite differently. Having a frame of reference and using a new approach, the BSM theory was able to separate the space from time parameters at low level of matter organization and to perform physical analysis in a real three dimensional space. From BSM point of view, the interpretation of the scattering experiments does not provide correct real dimensions, because both: the vacuum structure and particle structures are not taken into account. BSM theory found that the stable particles like proton, neutron and electron (and positron) posses well defined spatial geometry. They are built of complex but understandable 3 dimensional helical structures whose building blocks are the same as those forming the vacuum structure. Analysing the interaction between the vacuum and matter structures, the BSM theory was able to derive number of useful equations, like the light velocity, Newtonian mass (the mass we are familiar with), vacuum energy (zero point energy) and so on. It succeeded also to explain physically what is the electrical charge and magnetic field energy. One of the most useful result with practical importance is the Atlas of Atomic Nuclear Structures (ANS). It shows that the protons and neutrons follow a strict spatial order in the nuclei with well defined nuclear building tendency related to the Z number. The signature of this tendency matches quite well the row-column arrangement of the Periodic table, the Hund's rules and the Pauli exclusion principle. Figure 1 shows the spatial geometry of the Deuteron, where p -s is the proton and n - is the neutron. The neutron is centred over the proton saddle due to the Intrinsic Gravitation (IG) field and the proximity electrical fields.

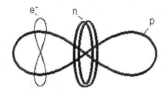

Fig. 1. Deuteron with electron in Balmer series orbital (according to BSM)

The nucleus of the Helium is shown in Fig. 2

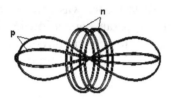

Fig. 2. Helium nucleus (according to BSM)

3. Physical mechanism of alpha decay.

BSM theory provides understanding about processes as radioactivity and nuclear reactions. The possibility about such processes becomes apparent even from the Atlas of ANS. For a drawing simplification of the nuclear structure, the protons and neutrons are presented by symbolic patterns as they are shown in Fig. 3. Two types of symbolic notation is adopted in the Atlas of Atomic Nuclear Structures. They could not show, however, the nuclear helicity (twisting) around the polar axis.

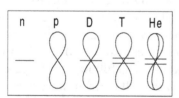

Fig. 3 Symbolic notations used in the Atlas of ANS

Figures 8.6 and 8.7 (from Chapter 8 of BSM) show respectively the axial sections and polar views of the nucleus of Gd and Rn. In the polar section views the protons and neutrons lying in proximity to the section plane are only shown (by their symbolic notations).

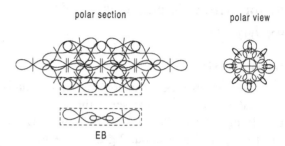

Fig. 8.6 of BSM. Nucleus of Gd

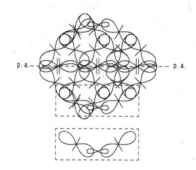

Fig. 8.7 of BSM. Sectional view of Rn nucleus. For a 45° rotation around pp axis the sectional view is a mirror image

In both figures two axially aligned deuterons are shown in the extracted box below the axial views of the nuclei. These deuteron pairs occupy the equatorial region of the nuclei for the elements with Z > 59. If investigating the building tendency related to the Z number of the elements, we see that this equatorial structures are built in the Lantanides. After Hf they are still in the same positions but partly overlapped by a new external shell of deuterons, as shown in the Atlas of ANS. These pairs of axially aligned deuterons in fact provide emission of alpha particles in the radioactive decay. The possible sequence of formation of alpha particles (He nuclei) from deuterons is the following:

The two deuterons are polar bonded, while their second clubs are connected by electronic bonds (a quantum orbit containing a pair of electrons with opposite QM spins). The polar bonding means that the Intrinsic Gravitational (IG) energy modes of the internal RL(T) structures of FOHS (see Chapters 2, 6 and 8) of the protons in polar range are in synchronization. For the whole atomic nuclear system, however, the intrinsic energy balance (the balance between the intrinsic gravitational field and the EM fields) is not stable for a long term (statistically related to the half time decay). In some particular moment, the polar bond fails. The released IG energy is quite big in comparison to the EM fields energy. It disturbs the vacuum structure parameters for a short moment and the repulsive electrical forces between the two oriented protons are also disturbed. Then the IG forces that are inverse proportional to the cube of the distance (in a classical empty space) attract the axially alined deuterons and they merge into a He nucleus. The bonding electrons are lost due to the deteriorated vacuum space conditions mentioned above. So, the obtained He nucleus is a positive ion. Part of the released IG energy goes for emission of gamma radiation, while another part is spent for momentum contribution to the He nucleus and the released electrons.

4. Experimental evidence about the positions of the aligned deuterons

The structures of all atomic nuclei posses well defined polar axis of rotational symmetry and twisting. The twisting is defined by the twisting of the single proton and this feature is propagated in all stable nuclei. The polar symmetry is quite evident from the Atlas of ANS, while the twisting is not shown due to the drawing difficulties. The both features however play important role in the confined motion of the atom and especially for the motion of accelerated atoms in a magnetic field. In the confined motion of the atomic nucleus its polar axis is well aligned to the tangent of its trajectory and it possesses a spin momentum. In such condition the Broglie wavelength is a characteristic parameter.

The ternary fission effect is well known in the fission experiments with heavy atoms. Number of such experiments clearly show that the released alpha particles are predominantly emitted in a spatial angle centred (with some shift) around the equatorial plane with respect to the fission axis. Figure 4. presents a typical angular distribution of the emitted alpha particles in respect to the fission axis.

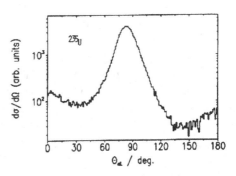

Fig. 4 Ternary alpha particle yields as a function of the emission angle with respect to the direction of light fission fragments. (From Theobald, J., Report IKDA 85/22 Technische, Darmstadt. FRG, 1985)

The observed angular distribution, from the point of view of BSM model, indicates that the emitted alpha particles originate from the axially aligned deuterons existing for all elements with Z>59. Consequently a fusion process have taken place contributing to the emission of alpha particles with such angular distribution. In the fission experiments such process is actively provoked, but the concept of fusion from aligned deuterons is the same as in the alpha radioactive decay. Then the alpha production in both cases: the natural and the provoked one indicates that the fusion reaction of deuterons could be achieved at room temperature.

The envelope shape, the structure and dimensions of the stable particles (proton, neutron, electron and positron) are obtained and cross-validated by the analysis presented in BSM. The proton envelope shown in Fig. 1 is a twisted torus with a shape close to a figure 8, while the neutron is a same structure but in shape of double folded torus. More accurately the plane projection of the proton is quite close to a Hippoped curve with parameter $a = \sqrt{3}$. The interactions between the proton, neutron and electron as physical entities with the vacuum structure are investigated and their geometrical parameters are expressed by the known physical constants. The dimensions are given in the Atlas of ANS, part I. (for the proton: length 0.667 A; width 0.1925 A, thickness 0.0078 A). The electrical field of the neutron is locked by IG field in the proximity range to its envelope and is not detectable in the far field. This is a result of the overall shape symmetry. The locking mechanism, however, does not work well when it is in confined motion in the structured vacuum and the neutron exhibits a magnetic moment. The electrical field of the proton is unlocked. So in the far range it appears as emerging from a point charge, but in the near field it is distributed over the proton envelop. When taking into account the two features of the proton: a finite geometrical size and the distributed proximity electrical field it is evident that the Coulomb law is not valid in the range of near field. Then in proper axial alignment and spin the integration energy for fusion of deuterons into He should not by calculated by the classical way with a classical Coulomb law down to a range of 10E-15. This makes a huge difference in the theoretical estimation of the necessary energy for successful fusion reaction.

5. Conditions and requirements for successful fusion reaction at room temperature

The binding energy of proton and neutron into a deuteron is calculated by approximative method in Chapter 9, &9.12.1 of BSM theory. The calculated energy, according to BSM concept is 2.145 MeV. This is pretty close to the experimental binding energy of 2.2245 MeV. The calculation method applies a disintegration approach of the neutron along the proton polar (long) axis. The binding energy has been obtained by using of IG energy balance in which one of the needed IG parameter has been obtained by the analysis of molecular vibrations in Chapter 9 of BSM. The applied method also indicates that the energy of the interacting proximity electrical field of the neu-

tron (locked) and the proton (unlocked) is part of the IG energy.

One approximate method (of BSM) for calculation of the necessary energy for the fusion is based on the use of the classical Coulomb law but the distributed charges of the two protons in the He nucleus configuration are preliminary converted to two point charges with a proper distance between them. Then the necessary energy is obtainable by integrating from the initial finite distance to infinity. The obtained total energy is below 1 MeV. This result however is valid if the following requirements are fulfilled:

(a) The deuterons are axially aligned and posses opposite linear momentums

(b) The difference between the spin momentum of both deuterons is close to zero:

$$\omega_1 - \omega_2 \to 0$$

The conditions for opposite linear momentums could be better achieved if positive D^+ and negative D^- deuteron ions are accelerated and collided as counter propagating beams.

The requirements (a) and (b) are demonstrated by Fig. 5 where two counter propagated D^+ and D^- are shown (p - proton, n - neutron).

Attempts for fusion of deuterons projectiles has been performed perhaps quite extensively, but the requirement (b) has not been obvious, before the BSM theory. If this requirement is not fulfilled a process of scattering instead of fusion will take place. Additional consideration related with the relaxation constant of the vacuum structure are also important, but they are not presented in this document.

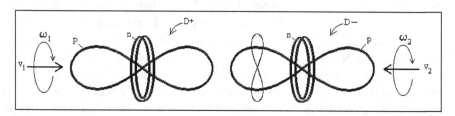

Fig. 5. Axial orientation and spin momentum conditions for counter propagated deuteron ions.

References:
1. S. Sarg, BSM theory, 2001
2. S. Sarg, Atlas of Atomic Nuclear Structures, 2001
www.helical-structures.org (BSM theory and Atlas of ANS)
www.heliconstruct.com   (educational site)
(A larger list of references is included in the BSM theory)

Appendix: Derived equations and calculated physical parameters

1. Derived equations:

Notes: 1. *The equation numbering is the same as in BSM chapters. The first digit corresponds to chapter's number.*

2. Equations (3.13.a), (3.21.a) and (3.42.F), known from the modern physics, are also derivable from the theoretical models of BSM.

$$R_c = \frac{c}{2\pi\nu_c} \quad \text{Compton radius (BSM derivation)} \quad (3.13.a)$$

$$s_e = \frac{\alpha c}{\nu_c\sqrt{1-\alpha^2}} \qquad \text{Helical step} \qquad (3.13.b)$$

$$r_e = s_e/g_e \qquad \text{small electron radius} \qquad (3.13.c)$$

$$r_p = \frac{2}{3}r_e \qquad \text{small positron radius} \qquad (3.13)$$

$$c = \frac{\nu_R d_{nb}}{k_{hb}} \quad [\text{m/s}] \quad \text{light velocity by CL space} \quad (2.75)$$
$$\text{parameters}$$

$$\text{where:} \quad k_{hb} = \sqrt{1 + 4\pi^2(0.6164^2)} = 4 \quad (2.20.a)$$

$$\mu_o = \frac{4\pi m_{CL}k_{rd}^2 c\nu_c^3}{N_{RQ}} \quad \left[\frac{N}{A^2}\right] \qquad (2.52)$$

$$\varepsilon_o = \frac{N_{RQ}}{4\pi m_{CL}\nu_c^3 c^3 k_{rd}^2} \quad \left[\frac{C^2}{Nm^2}\right] \qquad (2.53)$$

$$h = \frac{\pi(c)c^2 m_{CL}N_{RQ}^3 k_d}{4\nu_c k_{hb}^3} \quad [\text{N m s}] \quad (2.58)$$
$$\text{where:} \quad k_d = \frac{\tau_{511KeV}}{\tau_{CL}}$$

$$q = \frac{N_{RQ}^2}{2\nu_c^2}\sqrt{\frac{c\alpha k_d}{2k_{rd}^2 k_{hb}^3}} \quad [\text{C}] \qquad (2.58.a)$$

$$P_S = \frac{g_e^2 h\nu_c^4(1-\alpha^2)}{\pi\alpha^2 c^3} \quad \left[\frac{N}{m^2}\right] \qquad (3.53)$$

$$\rho_e = \frac{m_e}{V_e} = \frac{g_e^2 h\nu_c^4(1-\alpha^2)}{\pi\alpha^2 c^5} \quad \left[\frac{kg}{m^3}\right] \qquad (3.55)$$

$$P_D = \frac{g_e h\nu_c^3\sqrt{1-\alpha^2}}{2\pi\alpha c^3} \quad \left[\frac{N}{m^2}\right] \qquad (3.62)$$

$$m = \frac{P_S}{c^2}V_H \quad [\text{kg}] \quad \text{- mass equation of FOHS} \qquad (3.48)$$

$$m = \frac{h\nu_c}{c^2}K_V \quad [\text{kg}] \quad \text{- mass equation of FOHS} \qquad (3.58)$$

$$\sigma = \frac{\nu_c\alpha^2}{2c} = 1.09737315\times10^7 \quad [(\text{m}^{-1}] \qquad (3.21.a)$$

$$\gamma = (1 - V^2/c^2)^{-1/2} \qquad (3.42.F)$$

$$T = \frac{N_A^2 h\nu_c(R_c + r_p)^3 L_{pc}^2}{S_w \quad 2cR_c r_e R_{ig}}\left(\frac{\mu_e}{\mu_n}\right) \quad [\text{K}] \qquad (5.6)$$

$$T = \frac{N_A^2 hc^2\left(3g_e\sqrt{1-\alpha^2} + 4\pi\alpha\right)^3}{864\alpha^3\nu_c^2\pi^2 g_e^2(1-\alpha^2)R_{ig}}\frac{\mu_e}{\mu_n} \quad [\text{K}] \qquad (5.12)$$

$$r^2 = b^2(1 - a^2(\sin(\theta))^2) \qquad (6.54)$$

$$L_{pc} = \left[\left(\frac{\mu_e}{\mu_\mu}\right)^2(4\pi^2 R_c^2 + s_e^2) - 4n^2\pi^2 R_\pi^2\right]^{1/2} \qquad (6.61)$$

$$E = \frac{2q}{4\pi\varepsilon_0[L_q(1) + 0.6455L_p)]} = 16.01 \text{ eV for } H_2 \text{ ortho} \quad (9.4)$$

$$C_{SG} = (2h\nu_c + h\nu_c\alpha^2)(L_q(1) + 0.6455)L_p) \qquad (9.17)$$

$$\text{where:} \quad C_{SG} = G_0 m_{n0}^2$$

$$E_V = \frac{C_{SG}}{q[[[L_q(1)](1-\alpha^4\pi\Delta^2)] + 0.6455L_p]^2} - \frac{2E_q}{q} - \frac{2E_K}{q} \quad (9.23)$$

$$\Delta r = L_q(1)\alpha^4\pi\Delta^2 \quad [\text{m}] \qquad (9.26)$$

$$\Delta E(p, n, \Delta) = \frac{2\alpha C_{SG}(A-p)^2}{[r_n \pm [\Delta r(n, \Delta)]]^2} - pB_{D2}(n, \Delta) \qquad (9.55)$$

$$P_P/P_S = \alpha^2/\sqrt{1-\alpha^2} \qquad (10.18)$$

$$p_p = \alpha c\rho_e = \frac{g_e^2 h\nu_c^4(1-\alpha^2)}{\pi\alpha c^4} = 2.19\times10^{-25} \left[\frac{N \sec}{m^3}\right] \quad (10.22)$$

$$E_{IFM} = P_P V_e = h\nu_c\frac{\nu}{c}\alpha \quad [Nm] \equiv [J] \qquad (10.11)$$

$$E_{IFM}^G = E_{IFM}\frac{\sqrt{U_{Gn}}}{2\alpha c} \quad [J] \qquad (10.59)$$

$$r = \tilde{L}\frac{\ln(z+1)}{\ln(\tilde{n})} \quad [\text{m}] \qquad (12.50)$$

$$\tilde{n} = \exp\left(\frac{1}{(dN/dz)(z+1)}\right) \qquad (12.52)$$

1. Table of derived equations   Table 1.

Eq. No	Parameter	Name
(3.13.a)	R_c	Compton radius of electron
(3.13.b)	s_e	helical step of the electron
(3.13.c)	r_e	small electron's radius
(3.13)	r_p	small positron's radius
(2.75)	c	light velocity by resonance CL node param
(2.52)	μ_o	permeability of free space
(2.53)	ε_o	permittivity of free space
(2.58)	h	Planck's constant by CL space parameters
(2.58.a)	q	unit charge by CL space parameters
(3.48)	m	(Newtonian) mass of FOHS expressed by its volume V_H
(3.53)	P_S	Static CL pressure
(3.55)	ρ_e	Intrinsic electron density estimated by the Newtonian mass
(3.58)	m	mass of FOHS expressed by volume ratio K_V normalized to the electron envelope volume
(3.62)	P_D	Dynamical CL pressure
(3.57)	m	Newtonian mass of helical structure particle
(3.21.a)	σ	Rydberg constant by CL space parameters
(3.42.F)	γ	Relativistic gamma factor (derived from the electron quantum motion
(5.6)	T	CL background temperature (by e^- param)
(5.12)	T	CL background temperature (by CL space parameters)

Note: The calculated parameter T is for the Earth local field

(6.54)		Hippoped curve in polar coordinates
(6.61)	L_{pc}	Length of proton (neutron) core
(9.4)	E	Calculated energy potential for H_2 ortho-I, corresponding to experimental value of EVIP from PE spectrum
(9.17)	C_{SG}	Product of SG constant and the square of intrinsic proton mass
(9.55)	$\Delta E(p, n, \Delta)$	Equation for vibrational levels for diatomic homonuclear molecules, where: A - is the atomic mass in atomic mass units (for one atom); p - is the number of protons involved in the bonding system (per one atom);

n - is a subharmonic quantum number of the quantum orbit; r_n - is the internuclear distance at the equilibrium; Δr - is a deviation from the equilibrium point; Δ - is a vibrational quantum number, referenced to equilibrium

(9.26)	Δr	Range of vibrational motion of protons in H_2 ortho-1 molecule

(10.11)	E_{IFM}	Inertial force moment of folded CL nodes
(10.18)	P_P/P_S	Ratio between Partial and Static CL pressure
(10.22)	P_p	Specific partial pressure of CL space
(9.23)	E_V	Energy levels of H_2 ortho-I molecule
(10.59)	E_{IFM}^G	Inertial force moment in gravitational field
(12.50)	r	cosmological distance between distant galaxies where: z - redshift; $\tilde{L}$ - mean intergalactic distance;
(12.52)	$\tilde{n}$	mean quasirefractive index of GSS; where: (dN/dz) - is a line density measurable from Lyman alpha forests

2. Used physical constants   Table 2

Constant		Name
$\alpha = 7.29735308 \times 10^{-3}$		fine structure constant
$c = 2.9979245 \times 10^8$	(m/s)	light velocity
$\nu_c = 1.2355898 \times 10^{20}$	(Hz)	Compton's frequency
$h = 6.6260755 \times 10^{-34}$	(J.s)	Planck's constant
$q = 1.60217733 \times 10^{-19}$	(C)	elementary charge
$\mu_0 = 4\pi \times 10^7$	(N/A^2)	permeability of free space
$\varepsilon_0 = 8.8541878 \times 10^{-12}$	(C^2/N m^2)	permittivity of free space
$g_e = 2.0023193$		electron's gyromagnetic factor
$\mu_e = 9.2847701 \times 10^{-24}$	(A m^2)	electron's magnetic moment
$\mu_n = 9.6623707 \times 10^{-27}$	(A m^2)	neutron's magnetic moment

Notes: (1) Additionally to the above constants, the rest masses of elementary particles as: proton, neutron, electron, pions, muon, kaon are used. Two parameters from Electroweak theory also are used - the Fermi coupling constants, G_F and the effective mixing parameter θ_{eff}^{lept}. The measured mass equivalent energies of the bosons and tau are also used.

(2) Large observational data material used by BSM is not presented in the abstract paper.

3. Calculated parameters   Table 3

```
==========================================
Parameter                    Name
------------------------------------------
```

$R_c = 3.8615932 \times 10^{-13}$ (m) Compton radius of electron

$s_e = 1.77061164 \times 10^{-14}$ (m) helical step of the electron

$r_e = 8.842805 \times 10^{-15}$ (m) small radius of the electron

$r_p = 5.895203 \times 10^{-15}$ (m) small radius of the positron

$L_{pc} = 1.6277 \times 10^{-10}$ (m) - proton's (neutron's) core length

$L_P = 0.667 \times 10^{-10}$ (m) - proton's length

$W_P = 0.19253 \times 10^{-10}$ (m) - proton's width

$2(R_c + r_p) = 7.841 \times 10^{-13}$ (m) - proton's core thickness

$N_{RQ} = 0.88431155 \times 10^9$ number of resonance cycles contained in one SPM cycle

$\tau_{CL} = 0.0242631 \times 10^{-10}$ (s) space-time constant of CL space

$k_d = 51.518$ ratio between the CL pumping time for 511 KeV photon and the space-time const

$k_{hb} = \sqrt{1 + 4\pi^2(0.6164^2)} = 4$ - derived from the concept of wavetrain width and Airy disk in diffraction limited optics (Eq. 1.20.a)

$d_{nb} = 1.0975 \times 10^{-20}$ (m) *xyz* node distance of CL space

$k_{rd} = 0.15$ equivalent trace radius of vibrating MQ type of node normalized to a node distance

$m_{CL} = 6.94991 \times 10^{-66}$ (kg) inertial mass of the CL node estimated from Eq. (2.58)

$T = 2.6758$ (K) CL space background temperature for the Earth local field

$P_S = 1.3735811 \times 10^{26}$ (N/m^2) - Static CL pressure

$\rho_e = 1.52831 \times 10^9$ (kg/m^3) - intrinsic electron density

$P_D = 2.0257865 \times 10^3$ (N/m^{2}Hz) - Dynamical CL pressure

$E = 16.06$ (eV) - theoretically derived system energy, corresponding to $E_{VIP} = 15.967$ eV from PE spectrum

$C_{SG} = 5.26511 \times 10^{-33}$ - (J m^2) SG product

$\Delta r = 4 \times 10^{-16}$ (m) - range of nuclear vibrational motion for H_2 ortho-I molecule

$p_p = 3.343482 \times 10^{15}$ (N sec/m^3) - specific partial CL pressure

4. Used abbreviations

BSM	Basic Structures of Matter (theory)
CL	Cosmic Lattice
CL space	Cosmic Lattice space
EQ	Electrical Quasisphere
FOHS	First Order Helical Structure
FQHE	Fractional Quantum Hall Effect
SG	Super Gravitation
MQ	Magnetic Quasisphere
NRM	Node Resonance Momentum (vector)
RL	Rectangular Lattice
RL(R)	Rectangular Lattice (Radial)
RL(T)	Rectangular Lattice (Twisted)
SPM	Spatial Precession Momentum (vector)
SOHS	Second order helical structure
ZPE	Zero Point Energy

REFERENCES

Abrams G. S. et al., Physical Review Letters, v. **33**, No 23, 1406-1410 (1974)

Agnello M. et al., Physical Review Letters v. **74**, 371-374 (1995)

Antonucci R, Hurt T, and Kinney A., Nature 371, 313-314 (2002)

Aubert J. J. et al., Physical Review Letters, v. **33**, No 23, 1404-1406 (1974)

Baade W. and Swope H. H., Astronomical Journal **70**, 212 (1965)

Bailey et al., Nuclear Physics **B150**, 1 (1979)

Bally J. et al., Astrophysical Journal Suplement Series, **65**, 13 (1987)

Bania T. M., Astrophysical Journal, **216**: 381-403 (1977)

Bender D. et al., Physical Review D, V. **30**, 5151-527 (1984)

Benvenuti A. et al., Physical Review Letters, **32**, 800-803 (1974)

Binggeli B., Astronomy & Astrophysics, **107**, 338-349 (1982)

Bortoletto D. et al., Physical Review Letters, **71**, No 12, 1791-1795 (1993)

Bottinelli L. et al., Astronomy & Astrophysics, **118**, 4-20 (1983)

Black J. H. and Dalgarno A., Astrophysical Journal, **203**, 132-142 (1976)

Bredohl H and Herzberg G, Cananian Journal of Physics, **61**, No 9, 867-887 (1973)

Broelis, A. H. , in Dark Matter, AJP Conference Proceed. #336, eds S. S. Holt and C. L. Bennet, p 125 (1995)

Bromley et al., Physical Review **105**, 957 (1957)

Burbidge, G., Astrophysical Journal, **147**, 851 (1967)

Burbidge, G., Astrophysical Journal, **155**, L41 (1968)

Carlson T. A. and Jonas A. E., Journal of Chemical Physics, **55**, 4913 (1971)

CERN-EP-2000-055, Apr. **25**, 2000

Cheng L. X. et al., .Astrophysical Journal Letters, submitted (2001)

Coffin T. et al., Physical Review. **109**, 973-979 (1958)

Compton A., Physical Review, **VII**, Second series, No 6, 646 (1916)

Consoli M and Constanzo E., Old and new ether-drift experiments: a sharp test for a preffered frame, Nuovo
        Cimento, B119, 393-410 (2004); also in http://arxiv.org/abs/gr-qc/0406065

Consoli, A. Pagano,L. Pappalardo, Vacuum condensates and 'ether-drift' experiments, Physics Letters A, **318**,
        292-299, (2003)

Cowan C. L., Jr., Reines, F. et al., Nature, **124**, 103-104 (1956)

Dabrowski I., Canadian Journal of Physics, **62**, 1639 (1984) Cook G. K. and Ogawa M., Can J. Phys., **43**, 256 (1965)

Deaver B. S., Jr, and Fairbank. W., Physical Review Letters, **7**, 43, (1961)

Degrassi G. et al., Physics Letters, **B394**, 188-194 (1997)

De-Piccioto R. et al., Nature, **389**, 162 (1997)

Dickens, R. J., and Carrey J. V., R. Obs. Bull. Greenwich No 129, E340 (1967)

Dieke, G. H., J. Mol. Spectroscopy, **2**, 494 (1958)

Duari D. et al., Astrophysical Journal, **384**, 35-42 (1992)

Edwards C. et al., Physical Review Letters, **49** No 4, 259-262 (1982)

Einstein A., Podolsky B. and Rosenet N., Physical Review, **47**, 777-780 (1935)

Eisenstein J. P., and Stormer H. L., The Fractional Quantum Hall Effect, Science, **248**, p. 1510

Eland J. H. D., International Journal of Mass Spectrometry and Ion Physics, **31**, 161 (1979)

Essenwanger P., Gush H. P., Canadian Journal of Physics, **62**, 1680-1685 (1984)

Farrel J. T., Jr. and Nesbitt D., Journal of Chemical Physics **105** (21), 9421- (1996)

Faynman R. and Gell-Man M., Physical Review, **109**, 193-195 (1958)

Ford C. J. B. et al., Physical Review, **B38**, 85515 (1988)

Forward R. L., Physical Review B, **30**, No 4, 1700-1702 (1984)

Franklin J. D. F. et al, The Aharonov-Bohm effect in the fractional quantum Hall regime, EP2DS- Xi, Nottingham, Aug. 1955

Ghez A. M. et al., Nature, **407**, 349-351 (2000)

Gilman F. J. and Rhie Sun Hong, Physical Review D., **31**, No 5, 1066-1073 (1985)

Gloersen P. and Dieke G. H., J. Mol. Spectroscopy, **16**, 191-204 (1965)

Gregor M., *The enigmatic electron*, Kluwer Academic Publisher, ISBN 0-7923-1982-6, (1992)

Guthrie B. N. G. and Napier W. M., Astronomy & Astrophysics, **310**, 353-370 (1996)

Harris W. E., Globular Clusterssystems as distance indicator, In Proccedings of the 1988 ASP Meeting on the Extragalactic Distance Scale, editted by S. van den Bergh and C. Pritchet (PASP S. Francisco, 231-254

Hirota Isamu, Journal of Atmospheric Science, **35**, 714-722, (1978)

Hoffmeister C., Richter G., Wenzel W., *Variable stars*, Spring-Verlag Berlin Heidelberg Ney York Tokio, 1985

Holland S. et al., Astronomy & Astrophysics, 3 Mar 2001, The host galaxy and the optical light curve of the gamma –ray burst GRB 980703

Hubbelll J. H., Seltzer S. M., *Tables of X-rays mass attenuation coefficients*, NIST

Israelashvili J. N. and Ttabor D., Proc. R. Soc. Lond., **A331**, 19-38 (1972)

Kadyshevski, V. G., *Fundamental length hypothesis and new concept of Gauge vector field*, FERMILAB-Pub 78/22– THY (1978)

Kallash C. Sahu et al., Nature, **387**, 476-481 (1997)

Kawasaki T. et al., Applied Physics Letters **76**, No 10, 1342-1344 (2000)

Kennedy T. A. et al., Solid State Communication. **22**, 459 (1977)

Kimura K. et al., Handbook oh He I PE Spectra of Fundamental organic molecules, Japan Scientific Societies Press (1981)

King Ivan R., Astronomical Journal, **71**, No 1, 64-75 (1996)

Lifshitz E. M., Soviet Physics JETP **2**, 73-83 (1956)

Lngbein D., Van Der Walls Attraction, v. **72** of Springer Tracts in modern Physiscs, Springer, N.Y. (1974)

Lohsen E., Nature, **258**, 688-689 (1975)

London F., Z Phys. **63**, 245 (1930)

Loram J. W. et al., Physica C 282-287, p. 1405-1406 (1997)

Marciano W. J. & Sirlin A., Physical Review Lett, **61**, 1815-1818, (1988)

Marinov S., How to measure the Earth's velocity with respect to absolute space, Physics Letters, **41A**, No 5, 433-434, 1972

Marinov S., A reliable experiment for the proof of the space-time absolutness, Physics Letters, **54A,** 19-20, (1975)

Marinov S., Measurement of the Laboratory's Absolute Velocity, General Relativity and Gravitation, **12**, No. 1, 57-65, (1980)

Marinov, S., The interrupted 'rotating disc' experiment, Journal of Physics A: Mathematical and General, **16**, 1885-1888, (1983)

Marinov S., Experimental disturbance of the principle of the relativity (in Russian), Physical thoughts, No. 2, 52-57, (1995)

Maxwell, J. C., *A treatise on Electricity & Magnetism*, (1983) Diver Publications, Ney York ISBN 0486606368 (Vol. 1) & 0486606376 (Vol. 2)

Maxwell, J. C., *A Dynamical Theory of the Electromagnetic Field*, Scottish Aacademic Press, Edinburgh

McKellar A. R. et al., Canadian Journal of Physics, **62**, 1673-1679 (1984)

McQuarrie D. A., *Quantum Chemistry*, University Science Book, Mill Valey, California (1983)

Mills A. P., Jr. et al., Physical Review Letters v. **34**, No 25, 1541-1544 (1975)

Mills, A. P., Jr, Physical Review Letters, v. **50**, No 9, 671-674 (1983)

Mills, A. P., Jr., Berko S. and Canter K. F., Physical Review Letters, **34**, 1541, (1975)

Murdoch H. S. et al., Astrophysical Journal, **309**, 19-32 (1986)

Murdoch H. S. et al., Astrophysical Journal, **309**, 19-32 (1986)

Namioka T., Journal of Chemical Physics, **41**, 2141-2152 (1964)

Pollard J. E. et al., Journal of Chemical Physics, **77**, 34-46 (1982)

Procario M. et al., Physical Review Letters, **70**, No 9, 1207-1211, (1993)

Purcell W. R. et al., Astrophysical Journal, **491**, 725-748 (1997)

Reed R. J., Journal of the Atmospheric Sciences **22**, 331-333, (1965)

Saminadayar L. et al., Physical Review Letters, **79**,2526 (1997)

Sarg. S., *Basic Structures of Matter*, monograph, (ISBN 0973051507 - Ist edition; ISBN 0973051558 - IInd edition)

Sarg S., *Atlas of Atomic Nuclear Structures*, monograph, ISBN 0973051515

S. Sarg, New approach for building of unified theory about the Universe and some results, http://lanl.arxiv.org/abs/ physics/0205052 (May 2002)

S. Sarg, A Physical Model of the Electron According to the Basic Structures of Matter Hypothesis, Physics Essays, **16**, No. 2, 180-195, (2003)

Sargent W. L. W. et al., Astrophysical Journal Suplement SeriesS, **42**, 41-81 (1980)

Schultz P. J., Lynn K. G., Review of Modern Physics, **60**, No 3, (1988)

Shafroth S., Austin J., Accelerator –Based Atomic Physics Techniques and Appl., ISBN 1-56396-484-8, (1997)

Songallia, A., Hu E. M. Cowie L. L., Nature, **375**, 124-126 (1955)

Stormer et al., Bulletin of Atmospheric Physical Society, **38**, 235 (1993)

Suchard S. N. and Melzer J. E., Spectroscopic Data, vol. 2, Publ. By Plenum Press, London (1976)

Tinbao Chang, Tang Hsiaowei and Li Yaoqing, ICPA **85**, p. 212 (1985)

Titchings R. T., Monthly Notes of the Royal Astronomical Society, **176**, 249-263 (1976)

Tonry J. L. et al., Astronomical Journal, **100**, 1416-1423 (1990)

Trombka J. I. and Fitchel C. E., Physics Reports, **97**, No 4, 173-218 (1983)

Tsui D. C. et al., Physical Review Leters, **38**, 1559 (1982)

Tsui D. C. et al., Physical Review Letters, **48**, 1559 (1982)

Turner D. W. and May D. P. , Journal of Chemical Physics, **45**, 471-476 (1966)

Uchida K. L. and Gusten R., The large scale magnetic field in the Galactic center, A & A, **298**, 473-481 (1995)

Vallance Jones A. and Gattinger R. L., Planetary and Space Science, **11**, 961 (1963)

Wadlunt, Physical Review, **53**, 843 (1938)

Wallerstein G., Astrophysical Journal, **130**, 560-569 (1959)

Wallerstein, G., Astrophysical Journal, **127**, 588- (1958)

Wang L. J. et al., Nature, **406** No 6793, 277-279 (2000)

Willett R. L. et al., Physical Review Letters, **65**, No 1, 112 (1990)

Williams R. W. and Williams D. L., Physical Review D, **6**, 737-740 (1972)

Wilson B. A. et al., Physical Review Letters, **44**, 479 (1980)

Yao W. M., FERMILAB – Conf-99/100-E, CDF and DO

About the author

Stoyan – Sarg Sargoytchev is a Canadian citizen emigrated from Bulgaria in 1991. He obtained a degree of electrical engineering at the Technical University - Sofia in 1971 and worked initially in the manufacturing industry as an electrical engineer. In 1974 he began to collaborate in a new established space research program in Bulgarian Academy of Sciences (BAS) and in 1976, he began a full time work in the Central Laboratory for Space Research (CLSR) in BAS where he was enrolled in a PhD program. He defended a Ph.D. in Physics on a thesis "Satellite and Rocket apparatuses for investigation of the Earth atmosphere". From 1976 to 1990, he held scientific positions in Bulgarian Academy of Sciences and was actively involved in different international Space Research programs (Program Intercosmos, Bulgarian–Indian program, Bulgarian-France and Bulgaria-USA collaboration). He has been a designer or co-designer of six scientific instruments (two for geophysical rockets, two for satellites, one for orbital station and one for interplanetary mission). Over 40 scientific papers about his work in this period have been published in English and Russian journals. For his pioneering work in space research programs Dr. Sargoytchev has been rewarded with medals from the government and from the international organization Intercosmos.

In 1990, Dr. Sargoytchev was invited as a visiting scientist by Cornel University and was involved in building of Doppler wind lidar at Arecibo Observatory, Puerto Rico, a project funded by NSF (USA). In September 1991, he was employed by the Canadian Network for Space Research and was involved in the building of Rayleigh and Sodium scattering Doppler lidar at University of Western Ontario, Canada. After one-year work, his family (wife and son) obtained a status of landed immigrants and after 3 years – Canadian citizens. In 1994, he began to work in space projects as a project scientist in the Institute for Space and Terrestrial Science (now CRESTech) and later in the Centre for Research in Earth and Space Science, York University, Toronto, Ontario, Canada. He was also involved in development, manufacturing and global positioning of automatic ground based instruments for remote measurement of temperature and gravity wave in the atmosphere (80-90 km) with a purpose of long period climate study. From 2002, he is working in the Space Instrumentation Laboratory, York University, where he is involved in space projects, coordinated by the Canadian Space Agency.

Working in diversified projects and fields, Dr. Sargoytchev was able to obtain valuable experience and knowledge in different fields of Physics. In parallel with his research, he was highly interested about the secrets of Nature and particularly some fundamental questions and problems in Physics. During his long work in academic institutions, he was able to participate in many conferences and seminars, while understanding, that many unsolved theoretical problems wait for solution. At some point, he studied extensively the history of physics in the past three centuries. He paid particular attention about unsolved mysteries – physical phenomena, for which the contemporary physics did not offer explanation. The accessibility to extensive scientific information by the university libraries and Internet allowed him to understand that the main problem is in the concept about space, adopted 100 years ago. He developed a new original concept independently from existed dogmatic postulates. For a few years of intensive but morally awarded work he was able to develop and publish in 2001 his original unified theory, called Basic Structures of Matter (Supergravitation Unified Theory). Since the work was based on a challenging revolutionary new idea, it did not match to the currently established and funded programs, so he published it as a monograph under name Stoyan Sarg. Despite his research work with over 70 scientific publications (including eight patents), the author considered his theoretical work as a major achievement in his life. The reason is that he was able to solve fundamental problems in Physics in a way that has been considered impossible in the past 100 years: building of successful and understandable unified theory in a real 3-dimensional space with revealed relations between the gravitational, electric and magnetic fields. The monograph "Basic Structures of Matter" has two electronic editions archived in the National Library in Canada and a few publications in scientific journals. This book with the same title is a third corrected edition, including a new Chapter 13. The author published also a short popular version of some extracts from the theory as a book "Beyond the Visible Universe" (2004). The BSM theory and all related papers and monographs the author published under name Stoyan Sarg. His original works are broadly referenced in Internet.

Without numbering the potential applications emerging from the BSM theory for different fields of natural science, just a few of most important predictions deserve to be mentioned:

(1)   Unveiling of hidden space energy of not EM type – a primary source of the nuclear energy;

(2)   A possibility to control the gravitation and inertia of material object;

(3)   A possibility for supercomunication by waves different than the EM waves.

The first two predictions envision a development of completely new mechanism for distant space travels with an independent energy resource. The third one predicts a new type of communications. The author foresees the beginning of a new era for the human civilization. Firstly, some planets and moons from the solar system will be extensively explored and colonized. If adequate funding is allocated, the new space travel technology will be developed in the very near future. Presently many developed countries have the potential for development of such new technology.

For more information: Search the Internet by the following keywords: S. Sarg; Stoyan Sarg; Basic Structures of Matter; Super Gravitation Unified Theory; helical structures; structure of electron; structure of elementary particles; Atlas of atomic nuclear structures; oscillating electron; ZPE-S; ZPE-D; hidden space energy; energy storage mechanism of biomolecules; Cosmic Lattice; Static CL pressure; Dynamic CL pressure; Partial CL pressure; level of matter organization; heterodyne method of ZPE access.

Author's publications and reports related to the Basic Structures of Matter theory

I. In peer reviewed journals

[1]. S. Sarg, A Physical Model of the Electron according to the Basic Structures of Matter Hypothesis, Physics Essays, International Journal Dedicated to Fundamental Questions in Physics, vol. 16 No. 2, 180-195, (2003); http://www.physicsessays.com

[2]. S. Sarg, Brief introduction to Basic Structures of Matter theory and derived atomic models, Journal of Theoretics (Extensive papers), January, 2003; http://www.journaloftheoretics.com

[3]. S. Sarg, Atlas of Atomic Nuclear Structures according to the Basic Structures of Matter Theory, Journal of Theoretics (Extensive papers, March, 2003); http://www.journaloftheoretics.com

[4]. S Sarg, Application of BSM atomic models for theoretical analysis of biomolecules (with three formulated hypotheses), Journal of Theoretics, May 2003; http://www.journaloftheoretics.com

II. In electronic archives

[5] S. Sarg ©2001, *Basic Structures of Matter*, monograph, http://www.helical-structures.org
also in: http://www.nlc-bnc.ca/amicus/index-e.html (First edition, ISBN 0973051507, 2002; Second edition, ISBN 0973051558, 2005), (AMICUS No. 27105955), Canadiana: 20020076533; LC Class no.: QC794.6*; Dewey: 530.14/2 21

[6] S. Sarg © 2001, *Atlas of Atomic Nuclear Structures*, monograph ISBN 0973051515, http://www.helical-structures.org
also in: http://www.nlc-bnc.ca/amicus/index-e.html (April, 2002), (AMICUS No. 27106037); Canadiana: 2002007655X, LC Class: QC794.6*; Dewey: 530.14/2 21

[7] S. Sarg, *New vision about a controllable fusion reaction*, monograph, www.nlc-bnc.ca/amicus/index-e.html (ISBN 0973051523, April, 2002), (AMICUS No. 27276360); Canadiana: 20020075960

[8] S. Sarg, New approach for building of unified theory, http://lanl.arxiv.org/abs/physics/0205052 (May 2002)

[9] S. Sarg, *Theoretical analysis of biomolecules using BSM models*, ISBN 097305154X, (AMICUS 27749841, LC Class OH506, Dewey 577.8 21, (2002).

III. International conferences

[10] S. Sarg, How an alternative concept of the physical vacuum leads to a different vision about the Universe, "Universe, Nature and Humankind", 28 – 30 Nov 2003, Sofia, Bulgaria

[11] S. Sarg, Basic Structures of Matter Hypotheses based on an Alternative Concept of the Physical Vacuum, World year of Physics,- 2005 Physics for the Third Millennium: II, Conference organized by NASA, 5-7 Apr 2005, Huntsville, Alabama, USA

[12] S. Sarg, Basic Structures of Matter Hypothesis Based on an Alternative Concept of the Physical Vacuum, 12th Annual conference of Natural Philosophy Alliances, 23-27 May 2005, Storrs, CT, USA.

[13] S. Sarg, Alternative concept about space (physical vacuum) leading to a different vision about the Universe, Thirteenth Biennial Meeting of the International Society for the Study of Human Ideas on Ultimate Reality and Meaning, 3-6 Aug 2005, Toronto, Canada (accepted report).

IV. Books

[15] S. Sarg, *Beyond the Visible Universe*, (a popular presentation of BSM concept and some potential applications) Helical Structures Press, 2004 (legal deposit in the National Library of Canada)

Picture of the brochure of the conference "Physics for the Third Millennium: II" with the abstracts. The abstract of the poster report "Basic Structures of Matter Hypothesis based on an Alternative Concept of the Physical Vacuum is shown on page 25. The conference, organized by NASA, was held in the Von Braun Center,

Office de la propriété   Canadian
Intellectuelle   Intellectual Property
du Canada   Office

Un organisme   An Agency of
d'Industrie Canada   Industry Canada

Droit d'auteur   *Copyright*

Certificat d'enregistrement   *Certificate of registration*

The copyright in the work described below has been registered as follows :
Le droit d'auteur sur l'oeuvre décrite ci-dessous a été enregistré comme suit :

Registration No. - Numéro d'enregistrement :   493479

Date of Registration - Date d'enregistrement :   August 27, 2001

First Published - Première publication :   unpublished

Title - Titre :   "Basic Structures of Matter (Thesis about matter, space
                       and time)"

Nature - Nature :   - Literary -

                       (BSM is a physical theory of unification type. It is based on a new
                       approach for study of the Universe in 3 + 1 dim.)

Owner(s) - Propriétaire(s):

                       Stoyan Sargoytchev (Sarg)
                       Toronto, Ontario

Author(s) - Auteur(s):

                       Stoyan Sargoytchev (Sarg) - Toronto, Ontario

Bureau du droit d'auteur Copyright Office

Copyright certificate from the Canadian Intellectual Property Office

Rewards of Stoyan (Sarg) Sargoytchev for his peoneering work in the Space Research instrumentation operating on board of geophysical rockets and spacecrafts

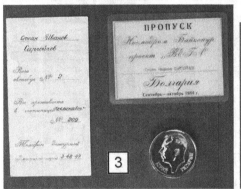

1. Medals from the international organization Intercosmos for contributions to space reserach
2. Medals from Bulgarian gevernment for active participation in space reserach programs
3. Admittance documents to the Baikonur space launch facility during the launch of the Vega-Halley mission 1984-86 and a medal
4. Medal of excelency from the Bulgarian Aacademy of Sciences
5. Certificate of acknowledged contributions to the scientific program from the board of the space station Salut-6
6. A working space research group during a visit of the Space City, near Moscow (1986). The spaceman is Pyotr Klimuk, with triple space flights (1973, 1975 and 1978).

Picture material about the author and his participation in space research programs

Ten countries—Bulgaria, Hungary, the German Democratic Republic, Poland, The Soviet Union, Czechoslovakia, France, Austria, West Germany, and the United States—are taking part in the U.S.S.R.'s two-probe Vega project. In addition to the Vega spacecraft which will study Comet Halley in 1986, two probes from Japan and one probe by the European Space Agency are designed to fly by the celestial object. Shown in photo (left to right), standing in front of a Vega spacecraft: France's Jean-Pierre Lepage, the Soviet Union's Alexander Krysko, and Bulgaria's Kunio Polazov and Stoyan Sargoïchev—all specialists for the Vega mission. (Photo: A. Pushkaryov/O. Kuzmin, TASS)

1. In front of the sattelite Intercosmos-Bulgaria-1300, launched in 1981. From left to right: I. Georgiev, S. (Sarg) Sargoytchev G. Gdalevich (USSR), A. Atanasov, C. Gogosheva, N. Petkov.

2. In front of the payload of the Vega-Halley mission 1984-86. From left to right: J. P. Lepage (Fr), A. Krysko (USSR), K. Palazov (BG), S. (Sarg) Sargoytchev (BG). (Adapted from Space World, p. 32, March 1985)

3. K. Palazov and S. (Sarg) Sargoytchev in fromt of a launching pad in Baykonur, Kazakhstan, USSR (1984).

4. Stoyan (Sarg) Sargoytchev (1981).

5. From left to right: J. P. Lepage (FR), S. (Sarg) Sagoytchev (BG) and K. Palazov (BG) in front of the Tree-channel Spectrometer included in Vega-Halley mission, 1984-86.

6. S. (Sarg) Sargoytchev as a visiting scientist of Cornell University (1990-1991) in front of the radiotelescope in Arecibo Observatory, Puerto Rico.

7. The author's participation in the Physics for the IIIrd Millennium conference, organized by NASA, Huntsville, AL, USA, 5-7 Apr, 2005, where the BSM treatise was firstly presented as a poster report (after a published paper in Physics Essays)

8. Invited talk about BSM, given by the author in the university Asen Zlatarov, Burgas, Bulgaria, 25 March 2004

9. The author in front of the ground-based SATI instrument (imaging interferometer) for temperature measurement of the middle atmosphere, (1987).

10. The author, his wife and son, (2003)